D1770006

TK7867.5
I573
1983

NOISE IN PHYSICAL SYSTEMS AND 1/f NOISE

Proceedings of the 7th International Conference on
'Noise in Physical Systems' and the
3rd International Conference on '1/f Noise',
Montpellier, May 17-20, 1983

Edited by

MICHEL SAVELLI

GILLES LECOY

and

JEAN-PIERRE NOUGIER

Centre d'Electronique de Montpellier
Université des Sciences et Techniques du Languedoc
Place Eugène Bataillon
34060 Montpellier Cedex
France

NO LONGER THE PROPERTY
OF THE
UNIVERSITY OF R.I. LIBRARY

N·H
P C

1983

NORTH-HOLLAND
AMSTERDAM • OXFORD • NEW YORK • TOKYO

100 72658

© Elsevier Science Publishers B.V., 1983

All rights reserved. No part of this publication may be reproduced, stored in a retrieval system, or transmitted, in any form or by any means, electronic, mechanical, photocopying, recording or otherwise, without the prior permission of the copyright owner.

ISBN: 0 444 86800 3

Published by:

North-Holland Physics Publishing

a division of

Elsevier Science Publishers B.V.

P.O. Box 103
1000 AC Amsterdam
The Netherlands

Sole distributors for the U.S.A. and Canada:

Elsevier Science Publishing Company, Inc.

52 Vanderbilt Avenue
New York, N.Y. 10017
U.S.A.

Printed in The Netherlands

PREFACE

Although scheduled for inclusion in these proceedings, a number of contributions are not included due to the non-attendance of the authors at the conference. Others are not included as they have already been published elsewhere. Finally, some of the sessions, which due to pressure of time were split up during the conference, have been regrouped in these proceedings.

A listing of all titles of papers accepted for oral presentation is therefore included in the 'Detailed Program of the Conferences' which can be found at the end of this volume.

The members of the Organizing Committee (Editors of these proceedings) would like to thank very much the Chairmen and the Referees for the important work they carried out during the conference.

M. SAVELLI
G. LECOY
J.P. NOUGIER

ORGANIZING COMMITTEE

Chairman, M. SAVELLI
Vice Chairman, G. LECOY
Vice Chairman, J.P. NOUGIER

PROGRAM COMMITTEE

A. AMBROZY (Hungary)
J. CLARKE (U.S.A.)
E. CONSTANT (France)
L. DE FELICE (Italy)
P.H. HANDEL (U.S.A.)
C. HEIDEN (F.R.G.)
F.N. HOOGE (The Netherlands)
B.K. JONES (U.K.)
P. MAZZETTI (Italy)
T. MUSHA (Japan)
M. SAVELLI (France)
H. SUTCLIFFE (U.K.)
H. THOMAS (Switzerland)
A. VAN DER ZIEL (U.S.A.)
C.M. VAN VLIET (Canada)
D. WOLF (F.R.G.)
R.J. ZIJLSTRA (The Netherlands)

SPONSORING ORGANIZATIONS

European Physical Society
Société Française de Physique

SUPPORTING ORGANIZATIONS

Office of Naval Research
European Research Office
Centre National de la Recherche Scientifique
Centre National d'Etudes des Télécommunications
Ministère Français de l'Industrie et de la Recherche
Université des Sciences et Techniques du Languedoc
Town of Montpellier
Conseil Général de l'Hérault
I.B.M. France
EFCIS
Caisse d'Epargne de Montpellier

TABLE OF CONTENTS

THEORY 1/f NOISE

CHAOTIC SYSTEMS

QUANTUM NOISE

HOT CARRIERS AND SUBMICRON DEVICES

1/f NOISE IN RESISTORS AND THIN FILMS

NOISE IN DIODES AND TRANSISTORS

1/f NOISE IN DIODES

OSCILLATORS

OTHER PHYSICAL SYSTEMS

NOISE IN PHYSICAL CHEMICAL AND BIOLOGICAL SYSTEMS

METROLOGY

CLOSING ADDRESS

OPENING SESSION

NOISE IN PHYSICAL SYSTEMS AND 1/f NOISE
M. Savelli, G. Lecoy and J-P. Nougier (eds.)
© Elsevier Science Publishers B.V., 1983

OPENING ADDRESS OF THE JOINT NOISE CONFERENCES

Michel SAVELLI

Université des Sciences et Techniques du Languedoc
Centre d'Electronique de Montpellier
34060 Montpellier Cedex
FRANCE

Ladies and Gentlemen :

It is a great pleasure for my colleagues, Professor LECOY and Professor NOUGIER, and I to welcome you to Montpellier.

I have been requested by the President of the University, in his absence, to wish you a good meeting and a pleasant stay in our city this week.

For the first time these two conferences - the International Conference on Noise in Physical Systems and the International Conference on 1/f Noise - are being held simultaneously. I do not know if this decision, which was taken in Washington, will also be taken in years to come, but I hope that this meeting in Montpellier will provide an opportunity for contact between the two participating groups.

As regards the scientific organization of the conference the organizing committee decided the following :

- invited papers must be of interest to all the participants;
- the congress must be held within four days;
- the conference excursion will be held on Thursday afternoon to provide a break in the working sessions.

We have received 136 proposed papers. Despite the fact that two parallel sessions were organized, we have selected only 91 contributed and 11 invited papers, time being the deciding factor. This decision should ensure ample time for the presentation and discussion of each paper.

This selection was done by the organizing committee using a notation system accepted by the program committee. Fortunately, without any manipulation, the number of contributed papers was about the same for each conference.

I am sure that, in the future, it will be necessary to set-up a poster session to extend the possibility of scientific communications.

A broad range of topics will be examined from both a theoretical and experimental point of view.

As far as the noise theory is concerned I would like to mention :

- non equilibrium state
- transition and relaxation phenomena
- chaotic systems
- nature fractal
- various models to represent 1/f noise or charge carriers mechanisms.

A wide variety of devices will be presented, such as :

- supraconductors, interferometers, Josephson diodes and SQUIDS for quantum noise
- resistors and thin films
- diodes, photodiodes and lasers
- bipolar and field effect transistors
- oscillators
- hot carriers and submicron devices with works on new electronic materials (GnAs, InP...)

I would also like to mention the presentation on 1/f noise in various physical and chemical systems :

- α particule radiative decay
- electrochemicals solutions and systems
- organic devices
- 1/f noise and reliability

At the proposal of the program committee, we selected 11 invited papers from the most important topics on noise.

I would like to thank the invited speakers for the work and the time they have put into the preparation and presentation of their papers.

Another important point: to ensure the scientific level of the conference proceedings, the accepted and presented papers will be reviewed during the conference. The Chairmen of the sessions will mediate between the referees and the authors to preserve their anonymity. The reviewing work should be achieved before the end of the meeting.

I recall that, for the presentations, we have scheduled 45 minutes for invited talks and 20 minutes for contributed talks; this also includes the time necessary for questions, answers and comments.

This morning 150 participants registered from 17 different countries. Unfortunately, our collea colleagues from the Soviet Union and the Eastern European countries were not able to attend the conferences.

I would like to thank, in particular, the following supporting organizations :

- Office of Naval Research (USA)
- European Research Office, United States Army
- Centre National de la Recherche Scientifique
- Centre National d'Etudes des Télécommunications
- Ministère Français de l'Industrie et de la Recherche
- Université des Sciences et Techniques du Languedoc
- the town of Montpellier, which offers the Conference Banquet on Wednesday evening
- the Conseil Général de l'Hérault
- I.B.M. France
- EFCIS
- la Caisse d'Epargne de Montpellier

Before opening these two Conferences, I would like to thank the two Vice Chairmen who actively worked to achieve the success (I hope so!) of these Conferences - Professor Gilles LECOY as financial manager, and Professor Jean-Pierre NOUGIER as organizing manager.

I now declare the two Conferences open!

Thank you very much for your attention.

THEORY

NOISE IN PHYSICAL SYSTEMS AND 1/f NOISE
M. Savelli, G. Lecoy and J-P. Nougier (eds.)
© Elsevier Science Publishers B.V., 1983

NOISE OUT OF EQUILIBRIUM
(Invited Paper)

Carolyn M. Van Vliet

Department of Electrical Engineering
University of Florida, Gainesville, FL 32611, USA

and

Centre de Recherches Mathématiques
Université de Montréal, Montréal, Qué H3C3J7, Canada

We discuss six properties of stochastic processes which are generally valid only in thermal equilibrium systems. In part A of this paper we discuss the first and most fundamental property, viz. the fluctuation-dissipation theorem. Kubo's and Van Vliet's quantum statistical mechanical derivations are reviewed. Next we discuss Gupta's thermodynamic theorem, which may hold for a class of nonequilibrium systems. In part B we discuss the other properties which mainly center around the principles of mesoscopic reversibility and detailed balance, and the symmetrization property of the two terms in the generalized second moment relaxation theorem (also called generalized g-r theorem, generalized Einstein relation or Λ-theorem). We then give two examples for which this symmetrization property does not hold. First we review Van Vliet's model of nonequilibrium optical pumping in photoconductors with traps. Finally, we review Tremblay et al.'s derivation of Brillouin light scattering as an example of a non-equilibrium hydrodynamic steady state.

1. THERMAL EQUILIBRIUM PROPERTIES

There are a number of properties which render the description of noise in thermal equilibrium considerably simpler than in a nonequilibrium, or driven equilibrium, steady state. The main features for fluctuation processes are as follows:

1) Validity of the fluctuation-dissipation theorem;
2) The principle of mesoscopic reversibility;
3) The principle of detailed balance;
4) The symmetrization property of the "generalized second-moment relaxation theorem";
5) The equilibrium Einstein relation between generalized mobility and diffusivity;
6) The Onsager relations for reciprocal flow problems.

The derivation of (1) will be briefly given below. The fluctuation-dissipation theorem is a genuine consequence of the existence of the microcanonical or canonical ensemble to describe equilibrium properties. It can be derived from a purely thermodynamic point of view (Nyquist 1928, Callen and Greene 1951) or from a quantum statistical mechanical approach (Kubo, 1957).

The other five properties are somewhat more subtile and have a less general form in that we must distinguish between a-variables, which are intrinsically time reversible such as the position, and b-variables which are odd, i.e. change sign under time reversal, such as the velocity (Casimir 1945). The details of this necessary distinction are often overlooked, and a number of errors occur even in recent literature. A much more careful consideration than hitherto given is necessary, and errors occur both in Lax (1960) and the author's papers (1965) in this respect. Since the space of this conference paper is limited, we shall make no attempt to rectify these errors here, by restricting ourselves to a-variables. Some remarks on the other cases will be made in some instances. The five properties 2) to 6) all hang together in a unique way. One can show that the validity of any one of the five implies the validity of the others. Again, the writing of a comprehensive survey paper on this topic would be extremely useful, but space prevents us here since we have been asked to write on nonequilibrium.

This paper will therefore be divided as follows. In section 2 we review the standard derivation due to Kubo, and its recent modification by Van Vliet (1978, 1979) of the fluctuation-dissipation theorem. We also discuss how a true general extension to nonequilibrium systems should be obtained. Lacking such a general extension at this moment, we review in section 3 Gupta's thermodynamic attempt to obtain a meaningful extension for electrical nonlinear circuits under a set of restricted but useful conditions (Gupta 1978, 1982).

In the rest of the paper we discuss the extensions for nonequilibrium when any of the properties 2) to 6) does not hold. This is relatively easier, since full solutions for Markovian nonequilibrium systems exist since the initial work on optically excited generation-recombination statistics in semiconductors by Van Vliet and Blok in 1956 (a & b). The first true nonequilibrium example involving the "generalized nonequilibrium second order moment relaxation theorem" (also called generalized Einstein relation by Lax 1960, generalized g-r theorem by Van Vliet and Fassett 1965, and Λ-theorem by Van Vliet 1971) was given by Van Vliet in 1964. This is discussed in sections 4 and 5.

In recent years renewed interest in the nonequilibrium problem arose from light scattering in the presence of a small temperature gradient, causing nonsymmetric Brillouin peaks. This problem was treated by a number of authors (Procaccia, Ronis and Oppenheim 1979, Kirkpatrick, Cohen and Dorfmann 1979, Tremblay, Siggia and Arai 1980). We will here present the treatment of the latter paper (section 6). It will be shown that this example is another straightforward nonequilibrium application of the "generalized nonequilibrium second order moment relaxation theorem, as originally set forth by the author and coworkers (1954, 1965, 1971).

PART A. THE FLUCTUATION-DISSIPATION THEOREM AND EXTENSIONS

2. QUANTUM STATISTICAL DERIVATIONS

Kubo considers systems with a Hamiltonian

$$H_{total} = H - AF(t). \tag{2.1}$$

Here H is the Hamiltonian of the system proper and $-AF(t)$ the coupling to an external field. For electrical conductors, $F(t) \to q\underset{\sim}{E}(t)$ and $A = \sum_i \underset{\sim}{r}_i \cdot$ where $\underset{\sim}{r}_i$ are the position operators of all of the particles. The Von Neumann equation for the density operator now reads

$$\frac{\partial\rho}{\partial t} + \left(\frac{i}{\hbar}\right)[H,\rho] = \left(\frac{i}{\hbar}\right) u(t)F(t)[A,\rho] \tag{2.2}$$

where $u(t)$ indicates that the field was switched on at $t = 0$. If we are now near a thermal equilibrium state, we can substitute $\rho \to \rho_{eq}$ in the r.h.s. of (2.2) where ρ_{eq} is the canonical density operator $\rho_{eq} = e^{-\beta H}/\mathrm{Tr}\, e^{-\beta H}$, $\beta = 1/kT$. The equation (2.2) can easily be solved for the average response of a flow quantity $\dot{B}$ where B is an operator in the system. One finds in a straightforward way (Van Vliet 1978)

$$\langle \Delta\dot{B}(t)\rangle = \mathrm{Tr}\,\rho(t)\dot{B} - \mathrm{Tr}\,\rho_{eq}\dot{B}$$
$$= \int_0^t d\tau\, \phi_{\dot{B}A}(t-\tau)F(\tau)\ . \tag{2.3}$$

Here ϕ_{BA} is the response function given by the commutator

$$\phi_{\dot{B}A}(t) = \frac{1}{\hbar i}\mathrm{Tr}\,\{[A,\dot{B}(t)]\rho_{eq}\} \tag{2.4}$$

where the time dependence of ρ(Schrodinger picture) was transferred to the time dependence of $\dot{B}$ (Heisenberg picture):

$$\dot{B}(t) = e^{i\mathcal{L}t}\,\dot{B}^S = e^{iHt/\hbar}\,\dot{B}^S\,e^{-iHt/\hbar} \tag{2.5}$$

where $\dot{B}^S$ is the Schrödinger operator (the superscript S will henceforth be omitted); $\mathcal{L}$ is the Liouville superoperator acting in the Liouville space that contains all operators A, $\dot{B}$, ... In particular, let $\dot{B}$ be the current $\hat{\underset{\sim}{J}} = \dot{B} = \frac{q}{\Omega}\sum_i \dot{\underset{\sim}{r}}_i = \frac{q}{\Omega}\dot{\underset{\sim}{A}}$, where Ω is the sample volume. If we take the Fourier-Laplace transform of (2.3), then we have, denoting by ^ the transform,

$$\hat{\underset{\sim}{J}}(i\omega) = q\hat{\underline{\Phi}}_{(q/\Omega)\dot{A},A}(i\omega)\cdot\hat{\underset{\sim}{E}}(i\omega) \tag{2.6}$$

so that the conductivity becomes

$$\underline{\underline{\sigma}}(i\omega) = \frac{q^2}{\Omega}\frac{1}{\hbar i}\int_0^\infty dt\, e^{-i\omega t}\,\mathrm{Tr}\,\{[\underset{\sim}{A},\dot{\underset{\sim}{A}}(t)]\rho_{eq}\}$$
$$= \Omega\int_0^\infty dt\,\frac{e^{-i\omega t}-1}{\hbar\omega}\,\mathrm{Tr}\,\{[\underset{\sim}{J},\underset{\sim}{J}(t)]\rho_{eq}\}\ . \tag{2.7}$$

We have here the beginning of Nyquist's theorem: the conductivity is related to the correlation expression,

$$\langle [J,J(t)]\rangle = \langle JJ(t)\rangle - \langle J(t)J\rangle \tag{2.8}$$

where $J \equiv J^S = J(0)$.

Now to obtain the correlation function proper, it must be defined as an anticommutator, since the product $JJ(t)$ must be symmetrized:

$$\Phi(t) = \langle [J,J(t)]_+\rangle\ . \tag{2.9}$$

It is now possible to relate the commutator to the anticommutator with some transformations involving contour integration for the Fourier transforms. It is then possible to express the full (two-sided) Fourier transform of the correlation anticommutator into the single-sided Fourier transform given by (2.7). The result is the spectral expression

$$S_{J_\nu J_\mu}(\omega) = 4\,\mathcal{E}(\omega,T)\,\{[\sigma'_{\nu\mu}(\omega)]^s + i[\sigma''_{\nu\mu}(\omega)]^a\} \tag{2.10}$$

where ν and μ denote either two different (but correlated) currents or cartesian current components, $\sigma = \sigma' + i\sigma''$, and $\sigma^s_{\nu\mu} = \frac{1}{2}[\sigma_{\nu\mu} + \sigma_{\mu\nu}]$, $\sigma^a_{\nu\mu} = \frac{1}{2}[\sigma_{\nu\mu} - \sigma_{\mu\nu}]$; further, $\mathcal{E}(\omega,T) = (\hbar\omega/2)\coth(\beta\hbar\omega/2)$ is the mean energy of an harmonic oscillator of frequency mode ω. Eq (2.10) is the fluctuation-dissipation theorem in all of its glory.

We have criticized Kubo's derivation in our 1978 paper, since nowhere was a form for the Hamiltonian H, commensurate with dissipation, introduced. In particular, the Heisenberg operators do not represent the required approach to equilibrium. As we indicated in our paper, a partitioning of the Hamiltonian is essential; i.e., instead of (2.1) we write

$$H_{total} = H^{o} + \lambda V - AF(t). \qquad (2.11)$$

Here H^o is the largest Hamiltonian that can be diagonalized for the many-body system, e.g., an electron-phonon system; V represents the interactions which randomize the energy over the states of H^o. In the Van Hove limit ($\lambda \to 0$, t large, $\lambda^2 t$ finite) and large system limit, such randomization leads to irreversibility (the Poincaré cycle becomes "off limits" for observation in such a system). In our second paper on linear response theory, we indicated that (2.2) now is carried over in an irreversible master equation, which for the reduced ρ^R (i.e., after the Van Hove limit) now reads

$$\frac{\partial \rho^R}{\partial t} + (\Lambda_d + i\mathcal{L}^o)\rho^R(t) = F(t) \text{ [times a functional of } \rho(t) \text{ and A].} \qquad (2.12)$$

Here Λ_d is the master operator

$$\Lambda_d K = -\sum_{\gamma\gamma'} |\gamma\rangle \langle\gamma|W_{\gamma\gamma'}[\langle\gamma'|K|\gamma'\rangle - \langle\gamma|K|\gamma\rangle] \qquad (2.13)$$

where $\{|\gamma\rangle\}$ is the set of many-body states of H^o, $\mathcal{L}^o$ is the interaction Liouville operator, $\mathcal{L}^o K = \hbar^{-1}[H^o,K]$, and $W_{\gamma\gamma'}$ is the transition probability according to the go1 ` n rule, $W_{\gamma\gamma'} = (2\pi\lambda^2/\hbar)|\langle\gamma|V|\gamma'\rangle|^2\delta(\mathcal{E}_\gamma - \mathcal{E}_{\gamma'})$. Equation (2.12) can be used similarly to (2.2) to obtain the transport coefficients σ, χ, and others, to be expressed in correlation functions involving the reduced operators $J^R(t)J(0)$, where now $J^R(t) = \exp[-(\Lambda_d + i\mathcal{L}^o)t]J(0)$. The main problem with (2.12) is that the r.h.s. has so far only been evaluated for a near-thermal equilibrium state. For a driven equilibrium, the Hamiltonian (2.11) should be further partitioned:

$$H_{total} = H^{o} + \lambda V - AF - Af(t) - Fa(t). \qquad (2.14)$$

Here AF would be the steady driven equilibrium coupling to the sustaining field F , and f(t) would be a small time-dependent field perturbation. Could equation (2.12) be reformulated for this Hamiltonian with only the part -Af(t) occurring in the r.h.s., then we could obtain a steady-state fluctuation-dissipation theorem. Such a program has not yet been carried out. One statement can be easily deduced from such a theory, however. We note that in the above the two Hamiltonians λV and AF are on an equal footing. The noise due to both is a function of the small signal transport coefficient caused by f(t). Since traditionally λV causes thermal dissipative noise and AF causes current or shot noise, it is clear, then, that in a generalized nonequilibrium fluctuation-dissipation theorem, we will no longer be able to separate "thermal" and "shot noise"; both will be contained in the new fluctuation-dissipation theorem. This is born out by the theory due to Gupta, reviewed below.

3. GUPTA'S THERMODYNAMIC RESULT

Since F is the driving parameter which is fixed, typically a large d.c. field, the entropy will contain intensive variables, as foreseen by Gibbs (1902)--a fact overlooked by Callen in his well-known monograph (1962). Thus, properly the entropy is a Massieu-type Legendre transform of Callen's entropy. From (2.14) we arrive at a Gibbs entropy production

$$T\,d\dot{S} = d\dot{\mathcal{E}} - \dot{A}\,dF \qquad (3.1)$$

where $T\,d\dot{S}$ represents the dissipation caused by λV. This form satisfies Gupta's thermodynamic theory. His basic assumptions are listed in his 1982 paper. The most important are:

(a) The entropy depends, besides on $\mathcal{E}$, on another variable F, which can pass through the port of the system and which is fixed.

(b) The system is purely resistive, so no energy is stored in $\dot{A}\,dF$. We should therefore be able to write $\dot{A}$ as a function of the instantaneous values of F: $\dot{A} = C_1\langle F\rangle \quad C_2\langle F^2\rangle$, etc.

(c) The system has no quantum correction. This is, in our case, born out by the use of the Λ operator.

For the present case of electrical conduction, $\dot{A}$ can again be interpreted as $(\Omega/q)J = IL/q$ where I is the current and $F = qE = Vq/L$. Thus $\dot{A}\,dF = IV$. Gupta's theorem for the noise now reads

$$S_{FF}(\omega) = 4kT\,P_{ex}/\overline{\dot{a}^2} \qquad (3.2)$$

or

$$S_{VV}(\omega) = 4kT\,P_{ex}/\overline{i^2} \qquad (3.3)$$

where P_{ex} is the excess power dissipation caused by the presence of a periodic zero average small signal a (or i) superimposed on the driven steady state. For a nonlinear resistor with $V = V(I)$

$$P_{ex} = \overline{\left(V_o + \frac{dV}{dI}\,i + \frac{1}{2}\frac{d^2V}{dI^2}\,i^2\right)(I_o + i) - V_o I_o}$$

$$= \left[\frac{dV}{dI}\,\overline{i^2} + \frac{1}{2}\frac{d^2V}{dI^2}\,I\,\overline{i^2}\right]_{I=I_o} \qquad (3.4)$$

where the overhead bar means time averaging. Clearly, from (3.3) and (3.4)

$$S_{VV}(\omega) = 4kT\left[\frac{dV}{dI} + \frac{1}{2}\,I\,\frac{d^2V}{dI^2}\right]_{I_o}. \qquad (3.5)$$

Similarly, one finds

$$S_{II}(\omega) = S_{VV}(\omega)(dI/dV)^2$$
$$= 4kT\left[\frac{dI}{dV} - \frac{1}{2} I \frac{d^2I}{dV^2} \Big/ \frac{dI}{dV}\right]_{I_o} . \quad (3.6)$$

Define now

$$\beta = \frac{1}{2}\left[\frac{d^2I}{dV^2} \Big/ \frac{dI}{dV}\right]_{I_o} . \quad (3.7)$$

Then (3.6) reads

$$S_{II}(\omega) = 4kT\ [g(I_T) - \beta I_T] \quad (3.8)$$

where we replaced I_o by the more useful symbol I_T (for total average current).

Application

Schottky barrier diodes or p-n junctions at low frequencies (no energy storage in extra term). Then,

$$I = I_1(e^{qV/kT} - 1), \quad (3.9)$$

$$g(I_T) = (dI/dV)_{I_T} = (q/kT)(I_T + I_1)$$
$$\beta I_T = (q/2kT)I_T . \quad (3.10)$$

Thus (3.8) reads,

$$S_{II}(\omega) = 2q\ [\ (I_T + I_1) + I_1], \quad (3.11)$$

indicating shot noise of forward and reverse current. Clearly, this comprises the thermal noise, for at zero bias

$$S_{II}(\omega)\Big|_{V=0} = 4qI_1 = 4kT\,g\Big|_{V=0} . \quad (3.12)$$

As we noted before, no distinction can be made between thermal noise "proper" and shot noise.

Whereas Gupta's theory is highly specialized, it has opened the way to show that steady-state dissipation-fluctuation theorems do exist, contrary to previous pessimistic claims (for a survey see Van Kampen 1965).

Gupta's result (3.6) or (3.8) is not a trivial result. Others have tried to generalize the fluctuation-dissipation theorem on more heuristic grounds. Van der Ziel (1973) considered the following possibilities for Nyquist's theorem in nonlinear devices:

a) $\langle v_n^2(f)\rangle = 4kT\,B\ (V/I)$;
$\langle i_n^2(f)\rangle = 4kT\,B\ (V/I)\,|y(f)|^2$.

b) $\langle i_n^2(f)\rangle = 4kT\,B\ (I/V)$;
$\langle v_n^2(f)\rangle = 4kT\,B\ (I/V)/|y(f)|^2$.

c) $\langle v_n^2(f)\rangle = 4kT\,B\ (dV/dI)$;
$\langle i_n^2(f)\rangle = 4kT\,B\ (dV/dI)\,|y(f)|^2$.

d) $\langle i_n^2(f)\rangle = 4kT\,B\ (dI/dV)$;
$\langle v_n^2(f)\rangle = 4kT\,B\ (dI/dV)/|y(f)|^2$.

e) None of the above.

In view of Gupta's result, possibility e) is the answer to this multiple-choice question, even though it can be shown that some devices satisfy some of the other possibilities (which then must be equivalent to (3.6) or (3.8)).

PART B. THE GENERALIZED NONEQUILIBRIUM SECOND MOMENT RELAXATION THEOREM (GENERALIZED G-R THEOREM, GENERALIZED EINSTEIN RELATION, OR Λ-THEOREM)

4. THE MESOSCOPIC MARKOV PROCESS

In statistical mechanics the variables are the microscopic variables p_i and q_i or their quantum operators. In stochastic processes we deal, however, with much more coarse-grained variables, which are averages over large ranges of quantum states, even though the fluctuations in these variables cannot be seen until after suitable amplification. Brownian motion and fluctuations in carrier populations (conduction band, traps) are typical examples. We call these variables mesoscopic (term of Van Kampen 1962). They can be pictured in the "a-space" (De Groot and Mazur 1962).

From the microscopic master equation, see (2.12), one can derive a mesoscopic master equation. The equation reads

$$\frac{\partial P(a,t|a')}{\partial t} = \int d^s a''[P(a'',t|a')Q_{a''a} - P(a,t|a')Q_{aa''}] \quad (4.1)$$

where $Q_{a''a}$ is the transition probability per unit time for a change $a'' \to a$. Assuming that $\underset{\sim}{a}$ is a set of variables on $(0,\infty)$, one can Laplace transform the master equation, and one easily finds for the characteristic function (Van Vliet 1983)

$$\frac{\partial}{\partial t}\langle e^{-\underset{\sim}{s}\cdot\underset{\sim}{a}(t)}\rangle_{a'} = \sum_{n=1}^{\infty}\langle e^{-\underset{\sim}{s}\cdot\underset{\sim}{a}(t)}\ \frac{(-1)^n}{n!}\ s^n : F_n(a(t))\rangle_{a'} \quad (4.2)$$

where F_n is the nth order Fokker-Planck tensor

$$F_n(a'') = \int (\underset{\sim}{a} - \underset{\sim}{a}'')^n\, Q_{\underset{\sim}{a}''\underset{\sim}{a}}\, d^s a$$

$$= \langle \underset{\sim}{a}(\Delta t) - \underset{\sim}{a}''\rangle^n_{\underset{\sim}{a}''} / \Delta t \ . \qquad (4.3)$$

Differentiating now repeatedly with respect to s, and setting $s = 0$, one finds the moment equations. The first moment equation is the phenomenological equation,

$$\frac{\partial}{\partial t} \langle \Delta\underset{\sim}{a}(t)\rangle_{\underset{\sim}{a}'} = -\underset{=}{M} \langle \Delta\underset{\sim}{a}(t)\rangle_{\underset{\sim}{a}'} \qquad (4.4)$$

(sub a' means a conditional average in an ensemble with $\underset{\sim}{a}(0) = a'$ fixed); M is the phenomenological relaxation matrix; it is defined by

$$M_{ij} = -\left[\frac{\partial}{\partial a''_j} \sum_{\underset{\sim}{a}} (a_i - a''_i)\, Q_{\underset{\sim}{a}''\underset{\sim}{a}}\right]_{\underset{\sim}{a}'' = \underset{\sim}{a}^o} \qquad (4.5)$$

where $\underset{\sim}{a}_o \equiv \langle \underset{\sim}{a}\rangle$, the unconditional stationary average. By differentiating twice to s, setting $s = 0$ and letting $t \to \infty$, one finds from (4.2) the generalized nonequilibrium second moment relaxation theorem:

$$\langle \Delta\underset{\sim}{a}\Delta\tilde{\underset{\sim}{a}}\rangle \tilde{\underset{=}{M}} + \underset{=}{M} \langle \Delta\underset{\sim}{a}\Delta\tilde{\underset{\sim}{a}}\rangle = \underset{=}{B}(\underset{\sim}{a}^o) \qquad (4.6)$$

where $\underset{=}{B}$ is the second order Fokker-Planck moment; $\tilde{M}$ is the transpose, $\Delta\underset{\sim}{a}$ is a column matrix and $\Delta\tilde{\underset{\sim}{a}}$ a row matrix.

The theorem was first derived for generation-recombination noise in multilevel semiconductors or photoconductors (note that the theorem holds out of equilibrium as in a photoconductor!); in this case $\underset{\sim}{a}$ represents the various populations of carriers, $\underset{\sim}{a} = \{n_1 n_2 \ldots n_s\}$. The theorem was called the generalized g-r theorem (Van Vliet-Blok 1956a). For three-dimensional Brownian motion the theorem takes a special form (M.C. Wang and Uhlenbeck 1943). The phenomenological equation for this type is $d\underset{\sim}{v}/dt = -\beta\underset{\sim}{v}$. Thus $M \to \beta$. The second order Fokker-Planck moment is easily found to be $\underset{=}{B} = 2D\beta^2\underset{=}{I}$, where $\underset{=}{I}$ is the unit tensor, D is the diffusivity. Further, $\langle \Delta\underset{\sim}{v}\Delta\tilde{\underset{\sim}{v}}\rangle = (kT/m)\underset{=}{I}$. Thus (4.6) leads to $mD = kT/\beta$, which is the Einstein relation for Brownian motion. (For carrier motion in a simple semiconductor model $\beta = 1/\tau$, $\mu = q\tau/m$, so the above leads to $qD = \mu kT$.) For this reason some authors (Lax 1960) have called (4.6) the generalized Einstein relation. When dealing with transport processes, $\langle \Delta a \Delta\tilde{a}\rangle$ is to be replaced by the covariance kernel $\langle \Delta a(r)\Delta\tilde{a}(r')\rangle$ and the corresponding theorem has been called the Λ-theorem (Van Vliet 1971). To us, at present, the name "generalized nonequilibrium second moment relaxation theorem" seems most appropriate.

We now state the equilibrium properties 2) to 6) mentioned in the introduction.

a) Microscopic reversibility $W_{\gamma\gamma'} = W_{\gamma'\gamma}$ follows from the golden rule in quantum mechanics. The mesoscopic transition probabilities will be denoted by $Q_{aa'}$. Let now $\chi(a)\, da$ be the number of quantum states $|\gamma\rangle$ when $\underset{\sim}{a}$ varies between $\underset{\sim}{a}$ and $\underset{\sim}{a} + d\underset{\sim}{a}$. We then clearly have, since $\sum_\gamma \to \int \chi(\underset{\sim}{a})\, da$,

$$\chi(\underset{\sim}{a}'')\, Q_{\underset{\sim}{a}''\underset{\sim}{a}} = \chi(\underset{\sim}{a})\, Q_{\underset{\sim}{a}\underset{\sim}{a}''} . \qquad (4.7)$$

This is the principle of mesoscopic reversibility. In the microcanonical or canonical equilibrium ensemble the probability $W(\underset{\sim}{a})d\underset{\sim}{a}$ is proportional to the number of accessible quantum states in $d\underset{\sim}{a}$, i.e., $\chi(\underset{\sim}{a})d\underset{\sim}{a}$. Thus (4.7) leads to the property of "mesoscopic reversibility"

$$W(\underset{\sim}{a}'')\, Q_{\underset{\sim}{a}''\underset{\sim}{a}} = W(\underset{\sim}{a})\, Q_{\underset{\sim}{a}\underset{\sim}{a}''} . \qquad (4.8)$$

b) The above leads to detailed balance (Van Vliet 1964). Consider that $\underset{\sim}{a}$ labels the occupancy of a set of quantum levels, i.e., let $\underset{\sim}{a} = \underset{\sim}{m}$, $\underset{\sim}{n}$, or $\underset{\sim}{k}$, which represent discrete population vectors. In particular we consider transitions $\underset{\sim}{k} \leftrightarrow \underset{\sim}{m}$ such that

$$\{k_1 \ldots k_i, k_j \ldots k_s\} \leftrightarrow \{m_1 \ldots m_i, m_j \ldots m_s\}$$

$$= \{k_1 \ldots k_i + 1, k_j - 1, \ldots k_s\}. \qquad (4.9)$$

Let p_{ij} be the transition rate from levels $\mathcal{E}_i$ to levels $\mathcal{E}_j$ (governed by mass-action laws or similar rules). Then

$$Q_{\underset{\sim}{k}\underset{\sim}{m}} = p_{ji}(\underset{\sim}{k});\ Q_{\underset{\sim}{m}\underset{\sim}{k}} = p_{ij}(\underset{\sim}{m}) . \qquad (4.10)$$

Eq (4.8) reads in this notation

$$W(\underset{\sim}{k})\, Q_{\underset{\sim}{k}\underset{\sim}{m}} = W(\underset{\sim}{m})\, Q_{\underset{\sim}{m}\underset{\sim}{k}} . \qquad (4.11)$$

We sum this result over all $\underset{\sim}{m}$. Then from (4.9) and (4.10)

$$\sum_{k_1=0}^{\infty} \cdots \sum_{k_i=0}^{\infty} \sum_{k_j=1}^{\infty} \cdots \sum_{k_s=0}^{\infty} W(\underset{\sim}{k})\, p_{ji}(\underset{\sim}{k})$$

$$= \sum_{m_1=0}^{\infty} \cdots \sum_{m_i=1}^{\infty} \sum_{m_j=0}^{\infty} \cdots \sum_{m_s=0}^{\infty} W(\underset{\sim}{m})\, p_{ij}(\underset{\sim}{m}) \qquad (4.12)$$

or also

$$\langle p_{ji}(\underset{\sim}{k})\rangle = \langle p_{ij}(\underset{\sim}{m})\rangle . \qquad (4.13)$$

The variables $\underset{\sim}{k}$ and $\underset{\sim}{m}$ are dummy variables; thus, (4.13) simply says $\langle p_{ji}\rangle = \langle p_{ij}\rangle$. This is detailed balance for the rates between any two sets of quantum states in the system.

c) From Bayes' theorem one easily shows that (4.8) leads to the symmetry of the pair correlation function for small intervals Δt

$$W_2(\underset{\sim}{a}, \Delta t;\ \underset{\sim}{a}'', 0) = W_2(\underset{\sim}{a}'', \Delta t;\ \underset{\sim}{a}, 0) . \qquad (4.14)$$

With the linearized phenomenological equations this leads to

$$\langle \underset{\sim}{a}(\Delta t)\tilde{\underset{\sim}{a}}(0)\rangle = \langle \underset{\sim}{a}\tilde{\underset{\sim}{a}}\rangle - \underline{\underline{M}}\langle \Delta\underset{\sim}{a}\Delta\tilde{\underset{\sim}{a}}\rangle \Delta t \qquad (4.15)$$

and

$$\langle \underset{\sim}{a}(0)\tilde{a}(\Delta t)\rangle = \langle \underset{\sim}{a}\tilde{\underset{\sim}{a}}\rangle - \langle \Delta\underset{\sim}{a}\Delta\tilde{\underset{\sim}{a}}\rangle \tilde{M}\Delta t . \qquad (4.16)$$

When in (4.14) we multiply by $\underset{\sim}{a}\tilde{\underset{\sim}{a}}$" and sum over both variables, we easily see that the ℓ.h. sides of Eqs (4.15) and (4.16) are equal. Thus, in thermal equilibrium the r.h. sides of (4.15) and (4.16) must also be equal, which is the symmetrization postulate of the "generalized second moment relaxation theorem": in (4.6) both sides are now equal

$$\langle \Delta\underset{\sim}{a}\Delta\tilde{\underset{\sim}{a}}\rangle \underline{\underline{\tilde{M}}} = \underline{\underline{M}}\langle \Delta\underset{\sim}{a}\Delta\tilde{\underset{\sim}{a}}\rangle . \qquad (4.17)$$

This property gives a tremendous simplification for thermal equilibrium processes; for now (4.6) can at once be solved

$$\underline{\underline{M}}\langle \Delta\underset{\sim}{a}\Delta\tilde{\underset{\sim}{a}}\rangle = \frac{1}{2}\underline{\underline{B}}(\underset{\sim}{a}^{o}) \qquad (4.18)$$

or

$$\langle \Delta\underset{\sim}{a}\Delta\tilde{\underset{\sim}{a}}\rangle = \frac{1}{2}\underline{\underline{M}}^{-1}\underline{\underline{B}}(\underset{\sim}{a}^{o}). \qquad (4.19)$$

Since (4.18) is for Brownian motion nothing but the Einstein relation between mobility and diffusivity, it is clear that some authors (Lax) refer to the full theorem (4.6) as the <u>nonequilibrium</u> generalized Einstein relation. Well, whatever the name of Eq (4.6), we do note that (4.6) is a genuine nonequilibrium result, while (4.18) or (4.19) is an equilibrium result. Thus, the properties

$$\langle \Delta\underset{\sim}{a}\Delta\tilde{\underset{\sim}{a}}\rangle \underline{\underline{\tilde{M}}} = \underline{\underline{M}}\langle \Delta\underset{\sim}{a}\Delta\tilde{\underset{\sim}{a}}\rangle \qquad (4.20a)$$

$$\langle \Delta\underset{\sim}{a}\Delta\tilde{\underset{\sim}{a}}\rangle \underline{\underline{\tilde{M}}} \neq \underline{\underline{M}}\langle \Delta\underset{\sim}{a}\Delta\underset{\sim}{a}\rangle \qquad (4.20b)$$

delineate clearly between equilibrium and nonequilibrium behavior.

We still mention the analog for transport processes (Van Vliet 1971). Suppose we have a stochastic variable depending on a (vector) parameter y, which is continuous in $D(-\infty,\infty)^{s}$. The stochastic process $a(y,t)$ is then infinite dimensional, though we assume it to be still Markovian. The Langevin equation is of the matrix operator form

$$\frac{\partial \underset{\sim}{a}(y,t)}{\partial t} + \underline{\underline{\Lambda}}_{y}\, \underset{\sim}{a}(y,t) = \underset{\sim}{\xi}(y,t) \qquad (4.21)$$

where $\underline{\underline{\Lambda}}$ is some (matrix) integral or differential operator describing the transport process. Let the covariance matrix kernel be

$$\underline{\underline{\Gamma}}(y,y') = \langle \Delta\underset{\sim}{a}(y)\Delta\tilde{\underset{\sim}{a}}(y')\rangle . \qquad (4.22)$$

The analog of the matrix theorem (4.6) is now the Λ-theorem

$$\underline{\underline{\Lambda}}_{y}\underline{\underline{\Gamma}}(y,y') + \underline{\underline{\Gamma}}(y,y')\underline{\underline{\Lambda}}_{y'}^{tr} = \frac{1}{2}\underline{\underline{S}}_{\xi}(y,y') \qquad (4.23)$$

where $\underline{\underline{\Lambda}}^{tr}$ acts from the right on $\underline{\underline{\Gamma}}$. This theorem was proven using Hilbert-space methods by the author. Notice that $\underline{\underline{S}}_{\xi}$ is the white spectral density of the Langevin sources. Again, when both terms on the ℓ.h.s. of the Λ-theorem (4.23) are unequal, we deal with a true nonequilibrium state.

In the next two sections we give both a matrix and a transport example of a true nonequilibrium situation, which, moreover, has been experimentally verified.

5. OPTICAL PUMPING IN A NONEQUILIBRIUM STEADY STATE

In 1964 we considered a model of a photoconductor in which free electron hole pairs were created by optical absorption, whereas all of the recombination was occurring via intermediate states, see Fig. 1.

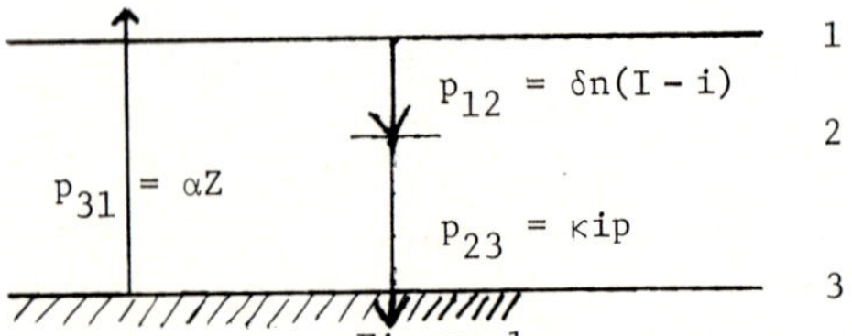

Figure 1.

δ and κ are capture constants. Clearly, there is no detailed balance, so we have a genuine nonequilibrium state. The M matrix follows from the linearized mass action type kinetic equations for $d\Delta n/dt$ and $d\Delta i/dt$ ($\Delta p = \Delta i + \Delta n$ is a dependent variable). One obtains

$$M_{11} = \delta(I-i_o), \quad M_{12} = -\delta n_o$$
$$M_{21} = -\delta I + \delta i_o + \kappa i_o, \quad M_{22} = \delta n_o + \kappa n_o + 2\kappa i_o . \qquad (5.1)$$

For the B-matrix one finds likewise

$$B_{11} = 2\delta n_o(I-i_o)$$
$$B_{12} = B_{21} = -\delta n_o(I-i_o) \qquad (5.2)$$
$$B_{22} = 2\delta n_o(I-i_o).$$

Moreover, in the steady state one has

$$\alpha Z = \delta n_o(I-i_o) = \kappa i_o(n_o+i_o) . \qquad (5.3)$$

One can now solve the nonequilibrium moment relation (4.6), expressing i_o into n_o via (5.3). Since (4.6) is homogeneous in δ/κ, the relative variance $\langle \Delta n^2\rangle/n_o$ can be computed as a function of $\alpha Z (=\mathcal{L})$ for various values of the ratio $y = \kappa/\delta$. The results are given in Figure 2. The peculiarity of this model is that $\langle \Delta n^2\rangle/n_o$ may obtain values >>1, as are indeed experimentally observed in cadmium sulfide. For a

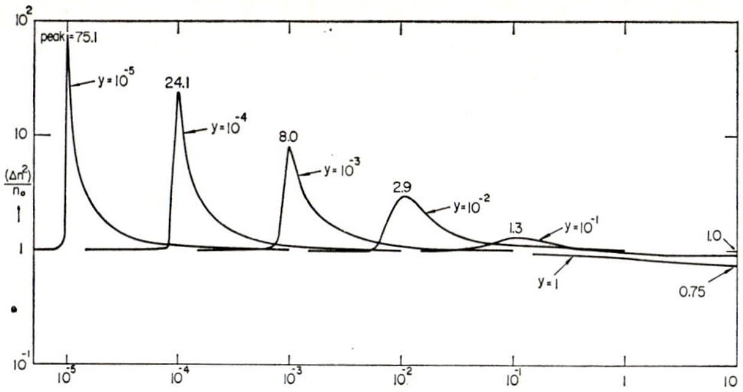

Fig. 2. The relative variance $\langle \Delta n^2 \rangle / n_o$ vs. $\mathcal{L}$.

thermal equilibrium situation, on the contrary, we always have $\langle \Delta n^2 \rangle / n_o \leq 1$, according to Fermi-Dirac statistics. Thus, the model of this section represents a true nonequilibrium situation, in which familiar concepts like Fermi-Dirac statistics are entirely eclipsed.

6. BRILLOUIN SCATTERING AS AN EXAMPLE OF FLUCTUATIONS ABOUT A HYDRODYNAMIC NONEQUILIBRIUM STATE

Tremblay et al. (1980) consider fluctuations in hydrodynamic modes $\underset{\sim}{A}(\underset{\sim}{k})$ governed by a Langevin matrix equation:

$$\frac{\partial \underset{\sim}{A}(\underset{\sim}{k},t)}{\partial t} + \underline{\underline{M}}(\underset{\sim}{k}) \cdot \underset{\sim}{A}(\underset{\sim}{k}) = \underset{\sim}{\xi}(\underset{\sim}{k},t). \qquad (6.1)$$

Notice that $\underline{\underline{M}}(\underset{\sim}{k})$ is the $\underline{\underline{\Lambda}}$ operator of the last part of section 4. This is an example of our transport formalism for infinite dimensional stochastic processes ($y \to \underset{\sim}{k}$) as considered in detail in our 1971 paper. For the Langevin forces correlation we write as usual

$$\langle \underset{\sim}{\xi}(\underset{\sim}{k},t)\underset{\sim}{\xi}(\underset{\sim}{k}',t') \rangle = \frac{1}{2} \underline{\underline{S}}_{\underset{\sim}{\xi}}(\underset{\sim}{k},\underset{\sim}{k}')\delta(t-t')$$

$$\equiv \underline{\underline{D}}(\underset{\sim}{k},\underset{\sim}{k}')\delta(t-t'). \qquad (6.1)$$

Solving from the Langevin equation for the spectrum one finds

$$\underline{\underline{S}}_A(\underset{\sim}{k},\underset{\sim}{k}',\omega) = 2[\underline{\underline{M}}(k) + i\omega \underline{\underline{I}}]^{-1} \underline{\underline{\Gamma}}(\underset{\sim}{k},\underset{\sim}{k}')$$

$$+ 2\underline{\underline{\Gamma}}(\underset{\sim}{k},\underset{\sim}{k}')[\underline{\underline{\tilde{M}}}(\underset{\sim}{k}) - i\omega \underline{\underline{I}}]^{-1}. \qquad (6.2)$$

In a nonequilibrium state $\underline{\underline{\Gamma}}$ can be eliminated by pre- and post-multiplying with the resolvent; hence

$$[\underline{\underline{M}}(\underset{\sim}{k}) + i\omega \underline{\underline{I}}]\underline{\underline{S}}_A(\underset{\sim}{k},\underset{\sim}{k}',\omega)[\underline{\underline{\tilde{M}}}(\underset{\sim}{k}) - i\omega \underline{\underline{I}}]$$

$$= 2\underline{\underline{\Gamma}}(\underset{\sim}{k},\underset{\sim}{k}')\underline{\underline{\tilde{M}}}(\underset{\sim}{k}) + 2\underline{\underline{M}}(\underset{\sim}{k})\Gamma(\underset{\sim}{k},\underset{\sim}{k}') = 2\underline{\underline{D}}(\underset{\sim}{k},\underset{\sim}{k}') \qquad (6.3)$$

where we used the full nonequilibrium second moment relaxation theorem (4.23). Inverting (6.3) we have

$$\underline{\underline{S}}_A(\underset{\sim}{k},\underset{\sim}{k}',\omega) = 2[\underline{\underline{M}}(\underset{\sim}{k}) + i\omega \underline{\underline{I}}]^{-1}$$

$$\underline{\underline{D}}(\underset{\sim}{k},\underset{\sim}{k}')[\underline{\underline{\tilde{M}}}(\underset{\sim}{k}) - i\omega \underline{\underline{I}}]^{-1}. \qquad (6.4)$$

(For comparison with Tremblay et al., notice our $\underline{\underline{S}}_A(\underset{\sim}{k},\underset{\sim}{k}',\omega)$ is their $2\underline{\underline{\chi}}^{\omega}(\underset{\sim}{k},\underset{\sim}{k}')$ with $\underline{\underline{\chi}} = \| \underline{\underline{\chi}}_{\alpha\beta} \|$; our Eq (6.4) is their Eq (7).)

We now come to their specific model. They consider a fluid with a fluctuating stress tensor, in a small temperature gradient. The linearized hydrodynamic equations for such a fluid are

$$\frac{\partial p(\underset{\sim}{k},t)}{\partial t} = -i\rho c \underset{\sim}{k} \cdot \underset{\sim}{v} \qquad (6.5)$$

$$\frac{\partial \underset{\sim}{k} \cdot \underset{\sim}{v}}{\partial t} = -i(k^2/\rho)p - [(\zeta + 4\eta/3)/\rho]k^2(\underset{\sim}{k} \cdot \underset{\sim}{v})$$

$$+ (i/\rho)\underset{\sim}{k} \cdot \underline{\underline{S}} \cdot \underset{\sim}{k} \qquad (6.6)$$

where p is the pressure, ρ is the density, c is the sound velocity, ζ the bulk and η the shear viscosity. There is no Langevin force associated with (6.5), but there is with (6.6), see their Eq (9). Equations (6.5) and (6.6), taken together as a column matrix equation, are of the form (6.1), though the details are quite complex. One can thus solve for $S_{pp}(\underset{\sim}{k}',\underset{\sim}{k}'',\omega)$. They obtain

$$S_{pp}(\underset{\sim}{k}',\underset{\sim}{k}'',\omega)$$

$$= \frac{c^4 2T(k'-k'')[2\eta(\underset{\sim}{k}' \cdot \underset{\sim}{k}'')^2 + (\zeta - 2\eta/3)k'^2 k''^2]}{(\omega^2 - c^2 k'^2 + i\omega D_\ell k'^2)(\omega^2 - c^2 k''^2 - i\omega D_\ell k''^2)} \qquad (6.7)$$

where $D_\ell \equiv (\zeta + 4\eta/3)/\rho$. Due to the temperature gradient contained in $T = T(\underset{\sim}{r})$, the two Brillouin light-scattering peaks obtain an asymmetrical height. One finds a difference spectrum $\delta S_{pp}(\omega)$ for the peaks located at $|\omega| = ck$, see Tremblay et al. Eq (14). The experimental values observed are in agreement with the theory, thus once more confirming the basic nonequilibrium result (4.6) or (4.23).

ACKNOWLEDGEMENTS

This review paper was written and its production supported by AFOSR contract #82-0226 and by grant A9522 from NSERC, Ottawa.

REFERENCES

1. H.B. Callen (1960), Thermodynamics (John Wiley, New York).

2. H.B.G. Casimir (1945), Revs. Mod. Phys. 17, 343.

3. S.R. de Groot and P. Mazur (1962), Non-equilibrium thermodynamics (North Holland Publishing Co.).

4. J.W. Gibbs (1902), Elementary principles in statistical mechanics (reprinted Dover).

5. R.F. Greene and H.B. Callen (1951), Phys. Rev. 83, 1231.

6. M.S. Gupta (1978), Phys. Rev. A18, 2725.

7. M.S. Gupta (1982), Proc. IEEE 70, 788.

8. T. Kirkpatrick, E.G.D. Cohen and J.R. Dorfman (1979), Phys. Rev. Lett. 42, 862.

9. R. Kubo (1957), J. Phys. Soc. Japan 12, 570.

10. M. Lax (1960), Revs. Mod. Phys. 32, 25.

11. H. Nyquist (1928), Phys. Rev. 32, 110.

12. I. Procaccia, D. Ronis and I. Oppenheim (1979), Phys. Rev. Lett. 42, 287.

13. A.-M.S. Tremblay, E.D. Siggia and M.R. Arai (1980), Phys. Lett. 76A, 57.

14. A. van der Ziel (1973), Solid State Electr. 16, 751.

15. N.G. van Kampen (1962), in Fundamental problems in statistical mechanics, E.G.D. Cohen, ed. (North Holland Publishing Co.), p. 173.

16. N.G. van Kampen (1965), in Fluctuation phenomena in solids, R.E. Burgess, ed. (Academic Press), p. 139.

17. K.M. van Vliet and J. Blok (1956a), Physica 22, 231.

18. K.M. van Vliet and J. Blok (1956b), Physica 22, 525.

19. K.M. van Vliet (1964), Phys. Rev. 133, A1182 (corrected in Phys. Rev. 138, 3AB (1965).

20. K.M. van Vliet and J.R. Fassett (1965), in Fluctuation phenomena in solids, R.E. Burgess, ed. (Academic Press), p. 267.

21. K.M. van Vliet (1971), J. Math. Phys. 12, 1981.

22. K.M. van Vliet (1978), J. Math. Phys. 19, 1345.

23. K.M. van Vliet (1979), J. Math. Phys. 20, 2573.

24. C.M. Van Vliet (1983), "Fluctuation phenomena," lecture notes, U. of Florida.

25. M.C. Wang and G.E. Uhlenbeck (1945), Revs. Mod. Phys. 17, 323.

NOISE IN PHYSICAL SYSTEMS AND 1/f NOISE
M. Savelli, G. Lecoy and J-P. Nougier (eds.)
© Elsevier Science Publishers B.V., 1983

MICROSCOPIC SPATIAL CORRELATIONS AT THERMAL EQUILIBRIUM IN NON POLAR SEMICONDUCTORS

J.P. Nougier, Ch. Gontrand, J.C. Vaissière

Centre d'Electronique de Montpellier
Université des Sciences et Techniques du Languedoc
Pl. E. Bataillon 34060 Montpellier - France

The local noise sources at two different points were till now supposed to be uncorrelated. The authors show that correlations do exist over short distances. The expression of the correlation is given at thermal equilibrium when randomizing scattering occur. The expression of the correlation function is given, and the spectral density (noise source term) is computed as a function of distance and frequency, for p-type silicon at thermal equilibrium at 300K. The usual expression of the diffusion noise source is a particular case, giving an approximation valid for long devices. For submicron devices, two points correlations may not be negligible

1. INTRODUCTION

The noise of a device can be calculated using the impedance field and the local noise sources [1]. These noise sources at two different points are always supposed to be uncorrelated. The purpose of this paper is to show that this is not exact : we shall give the expression of the two points correlation function, in thermal equilibrium, of the diffusion noise source of carriers in homogeneous non polar semiconductors. We shall suppose that :

a) the noise is due to velocity fluctuations,

b) the velocities of two different carriers are uncorrelated,

c) the collisions are instantaneous,

d) the scattering mechanisms are randomizing, which means that the state of a carrier after a collision is uncorrelated with its state before the collision. Therefore impurity scattering and polar optical scattering will not be considered here.

In thermal equilibrium, the local current density is $\vec{j}(\vec{r},t)$, and its two points correlation function:

$$C_{j\alpha\beta}(\vec{r},\vec{r}',\theta) = \overline{j_\alpha(\vec{r},t)\, j_\beta(\vec{r}',t+\theta)}^{\,t} \tag{1}$$

where the subscripts α and β mean projections along the directions α and β , the bar means time averaging, or ensemble averaging, using ergodicity. The noise source term is :

$$S_{j\alpha\beta}(\vec{r},\vec{r}',\nu) = 4\int_0^\infty C_{j\alpha\beta}(\vec{r},\vec{r}',\theta)\cos 2i\pi\nu\theta . d\theta \tag{2}$$

Usually, one supposes :

$$C_{j\alpha\beta}(\vec{r},\vec{r}',\theta) = C_{j\alpha\beta}(\vec{r},\theta)\,\delta(\vec{r}'-\vec{r}) \tag{3}$$

so that eq.(2) gives :

$$S_{j\alpha\beta}(\vec{r},\vec{r}',\nu) = K_{\alpha\beta}(\vec{r},\nu)\,\delta(\vec{r}'-\vec{r}) \tag{4}$$

$K_{\alpha\beta}(\vec{r},\nu)$ is the local noise source [1].

2. EXPRESSION OF THE TWO POINTS CORRELATION FUNCTION

Let $\vec{v}_i(\vec{r},t)$ be the velocity of the ith carrier inside the volume d^3r around $\vec{r}$:

$$\vec{j}(\vec{r},t) = q\sum_i \vec{v}_i(\vec{r},t)/d^3r \tag{5}$$

where the sum extends over all the carriers in the volume d^3r substituting eq.(5) into eq.(1), since two carriers are uncorrelated, one obtains.

$$C_{j\alpha\beta}(\vec{r},\vec{r}',\theta) = \frac{q^2}{d^3r d^3r'}\sum_i \overline{v_{i\alpha}(\vec{r},t) v_{i\beta}(\vec{r}',t+\theta)}^{\,t} = \sum_i \overline{\Gamma_i}^{\,t} \tag{6}$$

The individual correlation, for a given carrier, is not null if and only if this carrier, which was at point $\vec{r}$ at time t, has reached the point $\vec{r}'$ at time t+θ , without being scattered during that flight, which means that $\vec{r}$ and $\vec{r}'$ should be linked by the relation :

$$\vec{r}'-\vec{r} = \theta\,\vec{v}_i(t) = \theta\left[\vec{\nabla}_k \varepsilon(\vec{k})\right]_{\vec{k}_i(t)}/\hbar \tag{7}$$

where $\vec{k}_i(t)$ is the state of the carrier. Γ_i is then:

$$\begin{cases} \Gamma_i = \dfrac{q^2}{d^3r d^3r'} v_{i\alpha}(\vec{r},t) v_{i\beta}(\vec{r}',t+\theta), & \text{probability } p_i p'_i \\ \Gamma_i = 0, & \text{probability } 1-p_i p'_i \end{cases} \tag{8}$$

$$\begin{cases} p_i = \text{probability}\{\text{no collision occurs during } [t,t+\theta]\} \\ p'_i = \text{probability}\{\vec{r}' \text{ given by (7) within } d^3r'\} \end{cases} \tag{9}$$

Eq.(6) gives then, taking the average of eq.(8).

$$C_{j\alpha\beta}(\vec{r},\vec{r}',\theta) = \frac{q^2}{d^3r d^3r'}\sum_i p_i p'_i\, v_{i\alpha}(\vec{r},t)\, v_{i\beta}(\vec{r}',t+\theta) \tag{10}$$

Let $P(\vec{k}_i,\vec{k'}_i)$ be the collision rate between states $\vec{k}$ and $\vec{k'}$ of the ith carrier. The probability that a collision occurs in the time interval $d\theta'$ is:

$$d\theta' \int P(\vec{k}_i,\vec{k'}_i)d^3k'_i = d\theta' / \tau(\vec{k}_i)$$

Hence the probability that no collision occurs during the time interval θ is given by :

$$p_i = \exp[-\theta/\tau(\vec{k}_i)] \tag{11}$$

Now, from eq.(9), $p'_i = 1$ if (7) is satisfied within the volume d^3r', and $p'_i = 0$ otherwise. This gives :

$$p'_i = \delta(\vec{r'}-\vec{r}-\theta\vec{v}_i)d^3r' \tag{12}$$

Eqs.(11) and (12), carried into (10) give :

$$C_{j\alpha\beta}(\vec{r},\vec{r'},\theta) = \frac{q^2}{d^3r}\sum_i v_{i\alpha}v_{i\beta}\delta(\vec{r'}-\vec{r}-\theta\vec{v}_i)\exp[-\theta/\tau(\vec{k}_i)] \tag{13}$$

this shows that $C_{j\alpha\beta}$ depends on $\vec{R}=\vec{r'}-\vec{r}$, and not separately on $\vec{r}$ and $\vec{r'}$.

Let $f_0(\vec{k})$ be the thermal equilibrium distribution function, normalized to unity : the average value of a given function $\Phi(\vec{k})$ is $\bar{\Phi}$, and one has :

$$\sum_i \Phi(\vec{k}_i) = nd^3r\bar{\Phi} = nd^3r\int\Phi(\vec{k})\, f_0(\vec{k})d^3k \tag{14}$$

eq.(13) writes then, taking into account eq.(14):

$$\begin{cases} C_{j\alpha\beta}(\vec{R},\theta) = q^2n\int d^3k\ f_0(\vec{k})\ v_\alpha v_\beta\,\delta(\vec{R}-\theta\vec{v})\ \exp[-\theta/\tau(\vec{k})] \\ \vec{v} = \frac{1}{\hbar}\vec{\nabla}_k\varepsilon(\vec{k}) \end{cases} \tag{15}$$

Eq.(15), carried into eq.(2), gives then :

$$S_{j\alpha\beta}(\vec{r},\vec{r'},\nu) = 4q^2n\int d^3k\int_0^\infty d\theta f_0(\vec{k})v_\alpha v_\beta \cos 2\pi\nu\theta \times \delta(\vec{R}-\theta\vec{v})\exp[-\theta/\tau(\vec{k})] \tag{16}$$

Eqs.(15) and (16) clearly show that C_j and S_j are not identical with eqs.(3) and (4), but instead are a superposition of Dirac functions, thus having the shape of a bump spreading in a small region of $\{\vec{r}\}$ space.

Because of the Dirac function, eqs.(15) and (16) can be easily integrated, using the variable change $\vec{k}\rightarrow\vec{v}$. The Jacobian matrix $[J(\vec{k})]$ of the transformation has elements $J_{\alpha\beta}(\vec{k})$:

$$J_{\alpha\beta}(\vec{k}) = \frac{1}{\hbar}\partial^2\varepsilon/\partial k_\alpha\partial k_\beta \tag{17}$$

Defining then $\vec{k}_0$ as $\vec{v}(\vec{k}_0)=\vec{R}/\theta$, that is :

$$\frac{1}{\hbar}[\vec{\nabla}_k\varepsilon(\vec{k})]_{\vec{k}_0} = \vec{R}/\theta \tag{18}$$

Eqs.(15) and (16) can be integrated in $\{\vec{k}\}$ space, and give :

$$C_{j\alpha\beta}(\vec{R},\theta) = \frac{q^2nR_\alpha R_\beta}{\theta^5\det[J(\vec{k}_0)]} f_0(\vec{k}_0)\exp[-\theta/\tau(\vec{k}_0)] \tag{19}$$

$$S_{j\alpha\beta}(\vec{R},\nu) = 4q^2nR_\alpha R_\beta\int_0^\infty \frac{f_0(\vec{k}_0)\exp[-\theta/\tau(\vec{k}_0)]}{\theta^5\det[J(\vec{k}_0)]} \times \cos 2\pi\nu\theta\, d\theta \tag{20}$$

3. APPLICATION : OHMIC DIFFUSION NOISE SOURCE

In fact $f_0(\vec{k})$ is negligible for $v \gg v_{th}$ where v_{th} is the thermal velocity $v_{th}=(3k_BT/m)^{1/2} \lesssim 10^7$ cm/s, and the exponential factor in eq.(15) vanishes for $\theta>\tau_m$ where τ_m is the mean free flight between collisions, $\tau_m \simeq 10^{-12}$ s in silicon. Therefore C_j and S_j are negligible when $|\vec{R}| \gg v_{th}\ \tau_m = 100$Å. Hence, for usual devices, of dimensions larger than 1µm, $C_{j\alpha\beta}$ and $S_{j\alpha\beta}$ are not null over negligible distances, so that $\delta(\vec{R}-\theta\vec{v}) \simeq \delta(\vec{R})$, and eqs.(15) and (16) give :

$$C_{j\alpha\beta}(\vec{r},\vec{r'},\theta) = q^2n\,\delta(\vec{r}-\vec{r'})\int f_0(\vec{k})\ \exp[-\theta/\tau(\vec{k})]d^3k \tag{21}$$

$$S_{j\alpha\beta}(\vec{r},\vec{r'},\nu) = K_{\alpha\beta}(\vec{r},\nu)\delta(\vec{r}-\vec{r'}) = 4q^2n\ D_{\alpha\beta}\delta(\vec{r}-\vec{r'}) \tag{22}$$

$$D_{\alpha\beta} = \left\langle \frac{\tau(\vec{k})v_\alpha v_\beta}{1+\omega^2\tau^2(\vec{k})} \right\rangle \tag{23}$$

which is the usual diffusion coefficient.

4. EXAMPLES : ONE SPHERICAL PARABOLIC BAND SEMICONDUCTORS

If the dispersion law is $\varepsilon=\hbar^2k^2/2m$, then $\vec{v}=\hbar\vec{k}/m$ and according to eqs.(17) and (18) : $\vec{k}_0=m\vec{R}/\hbar\theta$, $\det[J]=(\hbar/m)^3$, and eq.(19) writes, with $f_0(\vec{k}) = (2\pi mk_BT/\hbar^2)^{-3/2}\exp-(\hbar^2k^2/2mk_BT)$:

$$C_{j\alpha\beta}(\vec{R},\theta) = \frac{q^2n}{\theta^5}\left(\frac{m}{2\pi k_BT}\right)^{3/2} R_\alpha R_\beta \exp-\left[\frac{mR^2}{2k_BT\theta^2} + \frac{\theta}{\tau(mR/\hbar\theta)}\right] \tag{24}$$

4.1. Acoustical phonon scattering

When only acoustical phonon scattering is involved, $\tau(\vec{k})=mL/\hbar k$, where L is the mean free path. Then, eq.(24) writes, taking for example $\alpha // \beta // \vec{R}$: $C_{j\alpha\beta}(\vec{R},\theta)=C_{j//}(\vec{R},\theta)$ with :

$$C_{j//}(\vec{R},\theta) = \frac{q^2n}{\theta^5}\left(\frac{m}{2\pi k_BT}\right)^{3/2}R^2\ \exp-\left[\frac{mR^2}{2k_BT\theta^2} + \frac{R}{L}\right] \tag{25}$$

For a given value of R, $C_{//}(R,\theta)$ is null for: $\theta=0, \theta\rightarrow\infty$, and is maximum for $\theta = R\sqrt{5m/(k_BT)}$. Because of the factor exp-R/L, $C_{j//}(R,\theta)$ vanishes for L larger than a few mean free paths.

For a given value of θ , $C_{//}(R,\theta)$ is null for R=0 and $R\rightarrow\infty$, and maximum for $R=R_{max}$ given by:

$$R_{max} = \left[\left(\frac{1}{L^2} + \frac{8m}{k_BT\theta^2}\right)^{1/2} - \frac{1}{L}\right]\left(\frac{k_BT\theta^2}{2m}\right) \tag{26}$$

For low values of θ, $1/L^2 \ll 8m/(k_B T\theta^2)$, then $R_{max} \simeq \theta(2k_B T/m)^{1/2}$ is only related to the thermal velocity.

For large values of θ, $1/L^2 \gg 8m/(k_B T\theta^2)$, then $R_{max} \simeq L$ depends only on the scattering parameters.

Even in this simplified case, $S_{//}(\vec{R},\nu)$ cannot be obtained analytically, but should be computed numerically.

4.2. p-type silicon

In its simplest approximation, the valence band of silicon is considered to be a single spherical parabolic band, with scattering involving one acoustic and one non polar optical phonon. The transport coefficients obtained using this model are realistic and are given in ref.[2], and the parameters used are given in ref.[3]. One is then able to compute numerically $C_{//}(R,\theta)$ given by eq.(20) when $\alpha // \beta // \vec{R}$ so that $R_\alpha R_\beta = R^2$.

Figure 1 shows $C_{//}(R,\theta)$ versus θ for various values of R. In p-type silicon, the mean free path L is about 100A. It can be seen that, over lengths of a few hundred Angströms, the maximum of $C_{//}(\theta)$ lies around a few tenth of a pico second. $C_{//}(\theta)$ is negligible, whatever be θ, when R≫L.

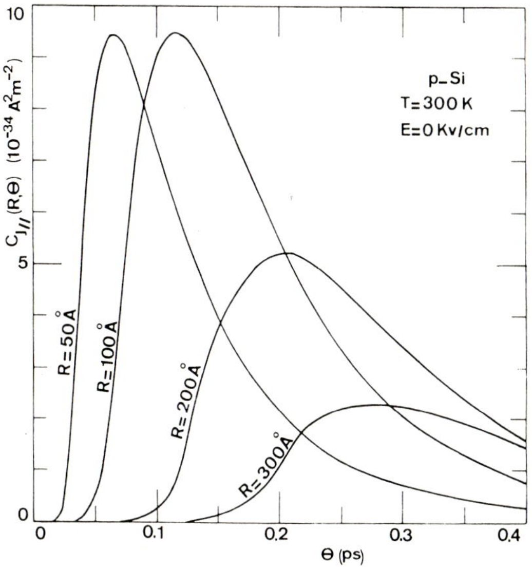

Fig.1. Longitudinal two points correlation function $C_{//}(R,\theta)$ versus time θ for various distances R, for p-type Si at thermal equilibrium at 300K.

The Fourier trentform (see eq.2) of $C_{j//}(R,\theta)$ can be easily computed in order to give $S_{j//}(R,\nu)$. The results are given on figures 2 and 3.

Figure 2 shows $S_{j//}(R,\nu)$ versus R at various frequencies. All the curves obtained at $\nu \lesssim 10^{11}$ Hz are identical, corresponding to a white diffusion noise. Obviously they are not Dirac functions of R, their maximum lies around L ≃ 100Å, however for R ≳ 10L = 1000Å, $S_j(\vec{R},\nu)$ can be approximated by a Dirac function of $\vec{R}$. As a consequence, the usual expression of the diffusion noise source term is expected to be a good approximation for long devices. For short devices (≲ 1000Å in p-type Si), two points correlation should be taken into account. At frequencies higher than 10^{11} Hz, fig.2 exhibits a decrease of $S_{j//}(R,\nu)$ due to the cut-off frequency of the diffusion noise source (see eq.23)).

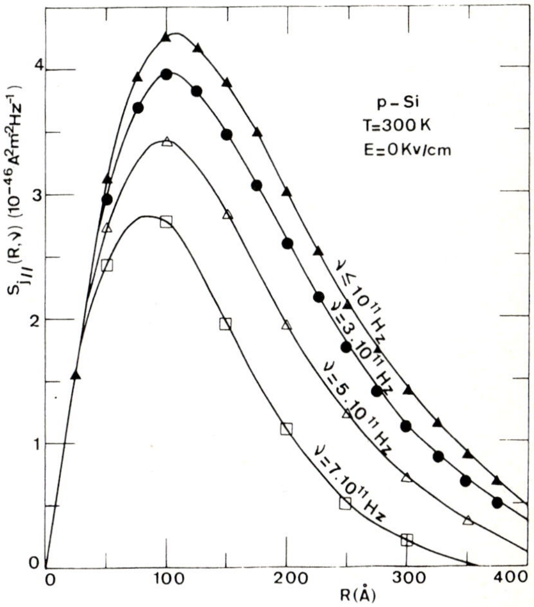

Fig.2. Spectral density $S_{j//}(R,\nu)$ versus R at various frequencies, for p-type silicon in thermal equilibrium at 300K.

Figure 3 shows the set of curves $S_{j//}(R,\nu)$, plotted now versus ν for various values of R. Below 7×10^{11} Hz, one finds of course the results of fig.2. Beyond 7×10^{11} Hz, figure 3 exhibits an unexpected behaviour : $S_{j//}(R,\nu)$ becomes negative and oscillates, versus ν, while damping. This is due to the fact that the correlation function actually differs from its standard expression as $\exp[-\theta/\tau(\vec{k})]$, since it involves θ also in f_0 through $\vec{R}_0$ (see eq.19)), which clearly appears on eq.(25) for example.

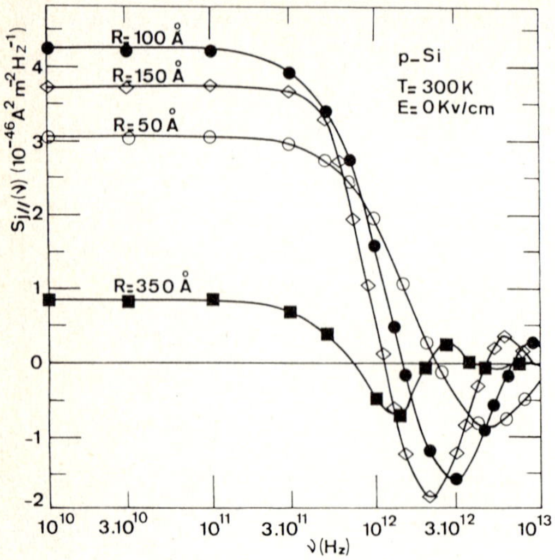

Fig.3. Spectral density $S_{j//}(R,\nu)$ versus ν at various distances R, for p-type silicon in thermal equilibrium at 300K.

Finally figure 4 shows that the cut-off frequency ν_c, defined as $S_{j//}(R,\nu_c) = S_{j//}(R,\nu=0)/\sqrt{2}$, lies in the range 300-900GHz for $R \geqslant 50$Å : in usual operating condition ($\nu < 100$GHz) the diffusion noise is white (see fig.3).

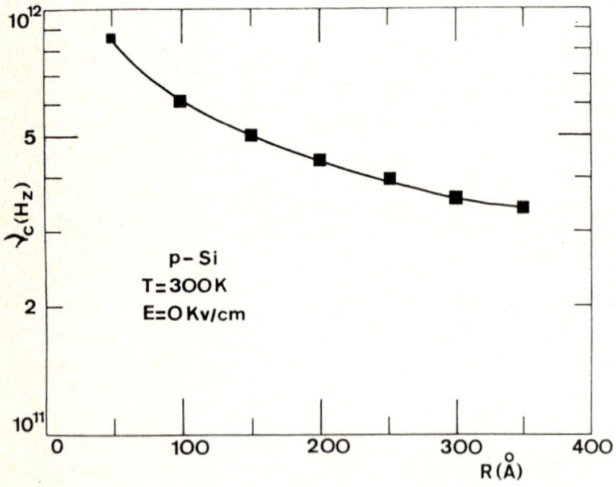

Fig.4. Cut-off frequency ν_c versus distance R in p-type silicon, in thermal equilibrium at 300K.

5. CONCLUSION

This paper clearly shows that correlations of noise sources between two different points do exist, over short distances. The expression of these correlations and of the spectral density could be given, (eqs.(15) and (16)) at thermal equilibrium, for diffusion noise when randomizing scatterings are involved.

The usual expression of the diffusion noise source term (eqs.(21) to (23)) appears to be a particular case of the previous expression, to be used for long devices. However, for short devices, this expression is not correct and the two points correlation should be taken into account.

The calculation performed on p-type silicon showed that the correlation length is about a few mean free paths, that is a few hundred Angströms in p-type Si. In GaAs, since the mean free path is much larger, one may expect that the correlation extends over $R > 0.1\,\mu$m. However, the theory presented here is not valid for GaAs, since then polar optical scattering is preeminent : in such a scattering, the velocity of an electron after a collision is correlated to its velocity before the collision, which increases the correlation length with respect to the case studied in this paper. It may then be expected that the correlation length in GaAs is much larger than 0.1 m, and thus may have non negligible effect on submicron devices.

REFERENCES

[1] J.P. Nougier, Proc. 6th Int. Conf. Noise in Physical Systems, Washington, 1981, National Bureau of Standards, Special publication n° 614 (1981).

[2] L. Regianni, J.C. Vaissière, J.P. Nougier, D. Gasquet, Proc. 3rd. Int. Conf. Hot carriers in semiconductors, Montpellier (1981). J. Phys. (Paris) C7, 357 (1981).

[3] G. Ottaviani, L. Reggiani, C. Camali, F. Nava, A. Alberigi-Quaranta, Phys. Rev. B12, 3318 (1975).

NOISE IN PHYSICAL SYSTEMS AND 1/f NOISE
M. Savelli, G. Lecoy and J-P. Nougier (eds.)
© Elsevier Science Publishers B.V., 1983

SIMULATION OF DIFFUSION NOISE IN A DEVICE

B. BOITTIAUX, E. CONSTANT, A. GHIS

Centre Hyperfréquences et Semiconducteurs
LA CNRS N° 287, UNIVERSITE DE LILLE 1.
Bât P3, 59655 VILLENEUVE D'ASCQ CEDEX, France

This paper deals with the use of Monte Carlo simulations for investigating noise phenomena in submicron devices.
In the first part, Monte Carlo procedure and assumptions are presented and discussed, then the studied structures are described.
In the second part, the possible influence of the space mesh and time interval used in the simulation is pointed out. Then, the current noise properties for various bias voltages and operating temperatures are studied.

INTRODUCTION

It is shown in this paper that Monte Carlo techniques are suitable to study problems regarding transport and noise phenomena in submicron dimensioned devices. By studying the detailed transport history of each simulated carrier in the structure, we obtain the I(V) characteristic of the device as usual, and we can also easily get the current fluctuations to study the diffusion noise as well. Two models of submicron devices are investigated : a GaAs n^+ i n^+ structure with an active region length of 0.48 μm and a n^+ GaAlAs/i n^+ GaAs heterojunction. The effects of the applied voltage and the device temperature on the current noise spectrum are discussed.

I. STUDY OF NOISE BY MONTE CARLO SIMULATION

The advantage of the Monte Carlo technique is twofold : it is based on microscopic model of carrier dynamic and it gives macroscopic informations with suitable averages of the various parameters.

The individual motion of each simulated electron depends at any time on local electric field and on various possible interactions that the carrier may undergo with the lattice and the impurities. Thirteen interactions such as intervalley polar and non polar optical and acoustic phonons are considered in accordance with the energy and the wave vector of the particule. Instantaneous energy, velocity and position in real space are computed at regularly spaced time $t_i = n_i \, \delta t$ thanks to the knowledge of the conduction band. If $\delta t < \tau_1$ where τ_1 is the shortest free path duration, it is obvious that this method is well suited to take into account strong electric field spatial non uniformities or fast dynamic space charge reactions.

The number of carriers required to study a component is determined with the following assumptions : on the one hand we assume the impurities are totally ionized, on the other hand, we assume that the carriers crossing one of the ends of the component are automatically injected again at the other end. In these conditions, the number of simulated carriers n_e in a component with a given cross section A is given as :

$$n_e = A \int_0^W N_D(x) \, dx$$

where W is the length of the component and n_e remains constant and equal to the number of ionized centers in the device. In the following we can consider a one dimensional treatment for space charge reaction if n_e is such as : $\sqrt{A} > W$. The local electric field is calculated at each time step by solving Poisson's equation as we know the instantaneous distribution of the electron in the device. Keeping the applied voltage at the terminals constant in time, the current in the structure is time dependent and has been shown [1] to be :

$$I(t) = \frac{q}{W} \, u_T(t)$$

where $u_T(t)$ is the sum of component velocities of the n_e individual carriers at time t in the current direction. From the evolution of I(t) during a period of time T, the temporal current autocorrelation function is estimated and its cosine Fourier Transform is the current noise spectrum.

The first simulated structure is as follows: a thin GaAs undoped active region of thickness L = 0.48 μm is sandwiched between two n^+ layers. In the second one, the first n^+ layer is a GaAlAs material such as the energy barrier height $\Delta E_c \simeq 0.3$ eV at the interface. In the two cases we assume abrupt homo or heterojunction at the interface and the number of particules employed is 1040. It has been shown [2], that for suitable bias voltages the first kind of component could present overshoot electron transport and in the second kind a ballistic motion of carrier can be obtained. Here, we intend to study the diffusion noise at ambient and nitrogen temperature.

II. RESULTS

All the results presented in the following are obtained when the voltage applied on the structure is constant and the transient regime has vanished.

On Fig. 1, the J(V) characteristics for three structures are presented, the electric field strength achieved in the active region corresponds to the warm electron range and overshoot or ballistic phenomena occur with suitable bias voltage [2] (V_o = 0.5 V)

On Fig. II, typical current autocorrelation functions obtained for several values of time step δt are shown. It can be seen for $\delta t < 0.02$ ps that the initial downswing and the first part into negative region is almost independent of the chosen δt. In this connection, it is interesting to note that the total noise power generated in the whole spectrum : $<\delta J^2> = C_J(0)$ is obtained with less than 10% of error when δt is chosen between 0.2 ps and 0.0025 ps.

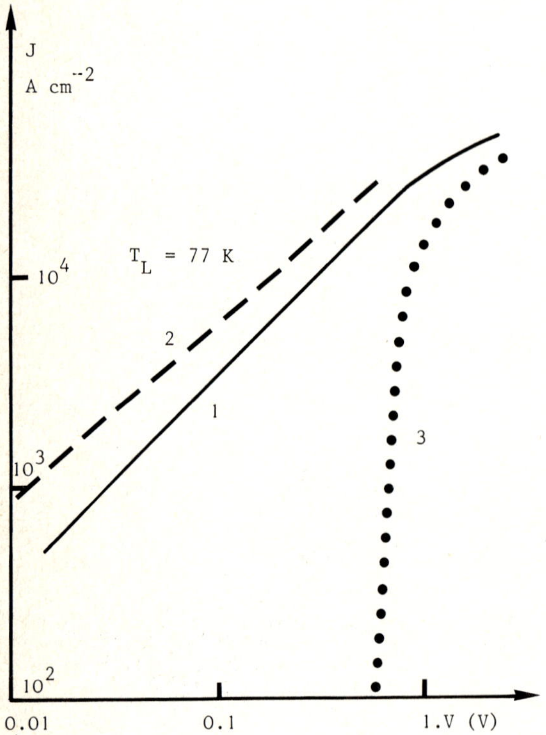

Fig. I J(V) characteristics for three structures

(1) n^+ i n^+ GaAs

(2) n^+ GaAlAs/i n^+ GaAs

(3) n^+ GaAlAs/i GaAs/n^+ GaAlAs

device length : 1 μm, non doped zone : 0.48 μm

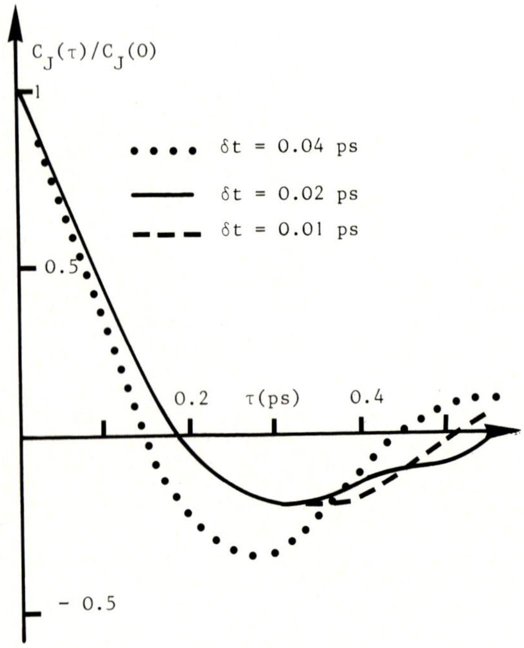

Figure II : Current autocorrelation functions for a n^+ i n^+ GaAs structure under constant bias voltage 0.5 volt at 293 K. δz = 0.01 μm

The same remarks can be made on the curves shown on Fig. III, where the same autocorrelation function is presented for several values of space mesh δz. From these results, we can deduce the upper limit of the mesh size ($\delta z, \delta t$) and also note the special shape of the autocorrelation function with negative values. This fact has been explained in a previous work [1] where the effects of the mean momentum relaxation time of carrier and the dielectric relaxation time of material on the noise properties in bulk silicon have been pointed out.

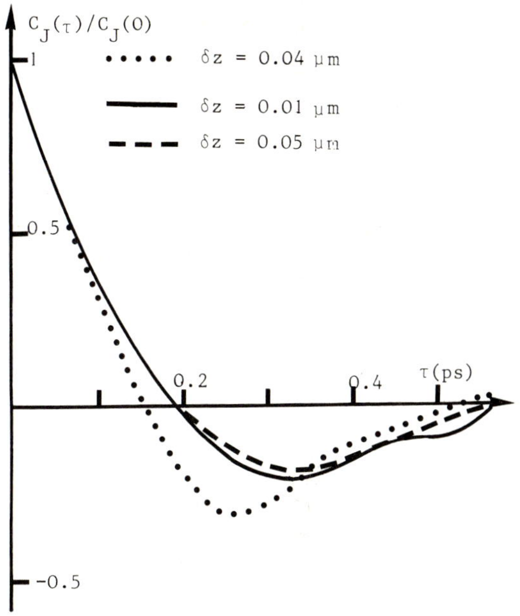

Figure III : Current autocorrelation functions for a n^+ i n^+ GaAs structure under constant bias voltage : 0.5 V at 293 K. δt = 0.01 ps.

In Fig. IV are presented the current fluctuation noise spectra $S_J(F)$ deduced from the previous autocorrelation functions at two lattice temperatures by taking the cosine Fourier Transform. The most important parameter for usuers is $S_J(0)$: the low frequency noise power which can be compared with Nyquist's thermal noise : $4\ k_B\ T_L\ G(0)$ where G(0) is the real part of the small signal differential admittance of the device at low frequencies and k_B is the Boltzmann constant.

Unfortunately, the formula :

$$S_J(0) = 4 \int_0^\infty C_J(\tau)\, d\tau$$

shows that a realistic calculation of this parameter require the knowledge of $C_J(\tau)$ over a very large time. This a computer time consuming method which can lead to an overestimated figure for $S_J(0)$, the lower the noise power is, the bigger the error. This can be seen when comparing the $S_J(0)$ as given by Monte Carlo with the Nyquist formula on Fig IV, at ambient and low temperatures. Nevertheless, the excess of noise power generated when velocity overshoot occurs in the device can be easily extracted.

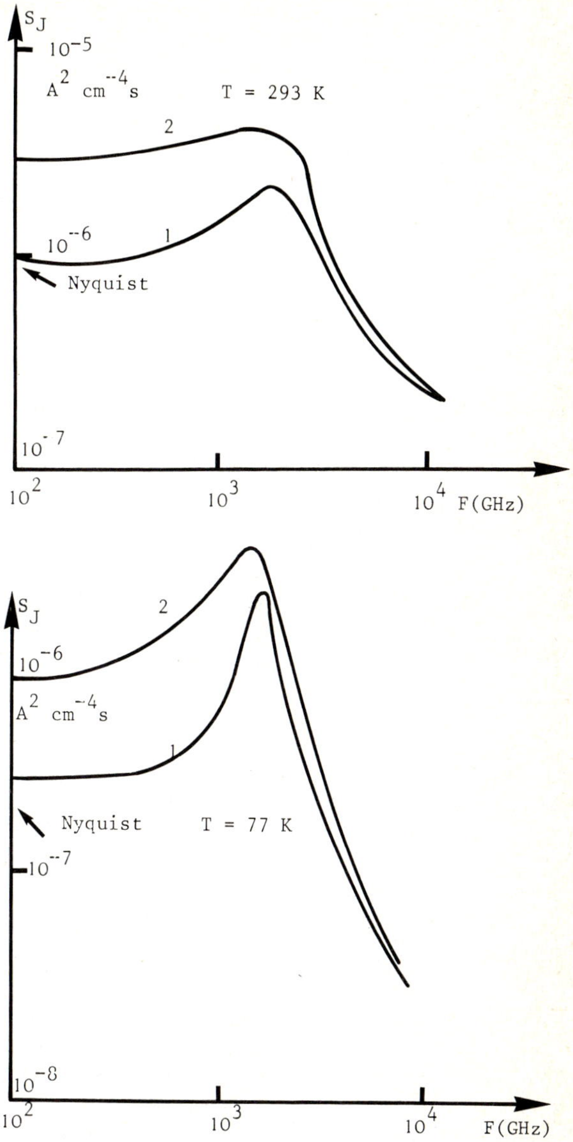

Figure IV : Current fluctuation noise spectra

1 V = 0. V.

2 V = 0.5 V.

n^+ i n^+ GaAs structure.

On the last figure, the current autocorrelation function computed in the same condition of temperature and bias voltage, for device (1) (n^+ i n^+ homostructure) and device (2) (n^+/i n^+ heterostructure) are compared. The total noise power (in whole spectrum) : $C_J(0)$ are of comparable values but noticeable differences appear in the time dependence of the $C_J(\tau)$'s, therefore in the frequency dependence of the $S_J(F)$.

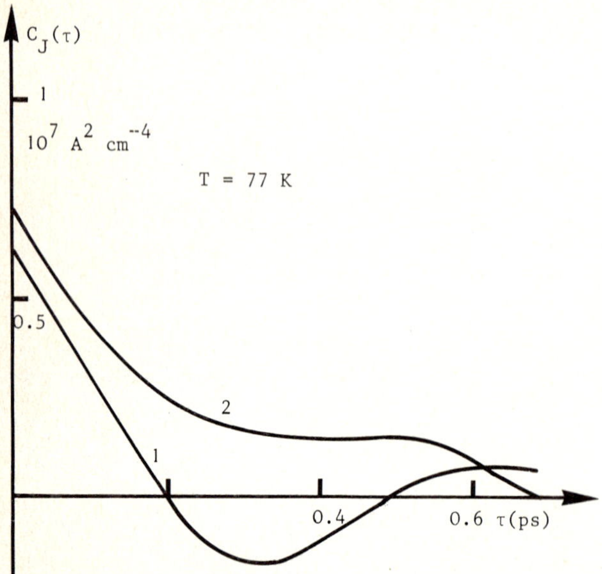

Figure V : Current autocorrelation functions for constant voltage biased structures

1 $n^+ i n^+$ GaAs structure (δt = 0.01 ps, δz = 0.01 μm)

2 $n^+ i n^+$ heterojunction (δt = 0.005 ps, δz = 0.005μm)

The rather complicated shape of the noise spectral densities (and of the current autocorrelation functions) can fondamentally be explained by three phenomena

- the effect of the dielectric relaxation time of the material
- the effect of the mobility cut-off frequency, depending on the momentum and energy relaxation times of the carriers and this effect is not the same all over the component
- lastly, the carrier transit time effect through the active region.

CONCLUSION

This study shows that the Monte Carlo simulation associated with Poisson's equation integration can be a powerful method to study noise properties of devices. But, some care is required in the choice of space mesh size and time step specially in submicron structures where strong space electric field variations are appearing.

The noise properties of the structures have been pointed out by observation and spectral analysis of current fluctuations through the component.

REFERENCES

[1] ZIMMERMANN, J. and CONSTANT, E., Application of Monte Carlo techniques to hot carrier diffusion noise calculation in unipolar semiconducting components, Solid-State Electron. 23(1980) 915-925

[2] GHIS, A., CONSTANT, E. and BOITTIAUX, B., Ballistic and overshoot electron transport in bulk semiconductors and in submicronic devices, J. Appl. Phys 54 (January 1983) 214-221

NOISE IN PHYSICAL SYSTEMS AND 1/f NOISE
M. Savelli, G. Lecoy and J-P. Nougier (eds.)
© *Elsevier Science Publishers B.V., 1983*

MONTE-CARLO PARTICLE MODELLING OF NOISE IN SEMICONDUCTORS

C Moglestue

GEC Research Laboratories Hirst Research Centre Wembley UK

In the Monte Carlo particle model the electric fields, particle velocities and free flight times have the same stochastic distribution as in the material simulated. The model therefore lends itself to the study of noise. However, due to approximations inherent in the model it will be necessary to make some corrections to the simulated noise to extract the physical noise. These will be discussed here.

1 INTRODUCTION

The Monte Carlo particle simulation of small semiconductor devices is now establishing itself as a way of modelling[1]. This method consists briefly of following individual carriers through the device to be modelled. Each carrier is accelerated by the local field for a time defined by a random number such that the times of free flights are distributed correctly. The subsequent scattering is also selected from many possible types by means of another random number such as the frequencies of the various scattering events will be the same as expected for the real semiconductor. We can therefore use the Monte Carlo particle model to simulate noise.

However, the nature of the technique does not allow the direct simulation of noise. The purpose of this paper is to study how the modelling influences the obtained noise, and to discuss how to extract the real noise, correcting for additional noise created by the practical approximations that have to be made. Fluctuations in computer simulated plasmas have been studied extensively by Langdon[2] and Hockney[3]. Langdon and his predecessors have studied the effect the model has on fluctuations in the charge density of the plasma. They were discussing the Fourier transform of the correlations in the density, but they do not give any explicit expression for the noise current. Tien and Moshman[4] and Pollack and Whinnery[5] have simulated the noise in a high frequency diode, using Monte Carlo techniques. They looked at the time correlation of the current, but, again, they did not present any explicit expression for it. Zimmermann and Constant[7] obtained analytical expressions for the noise spectrum based on the impedance method and the conventional transport equations. A semiconductor can be regarded as a plasma, but with lattice interactions, so that the techniques of calculating plasma correlations can be adapted. Below we shall regard a block of semiconductor material of uniform cross section and doping, through which a uniform current is flowing. The electric field is weak enough that interband transitions become insignificant.

2 THEORY

The correlation function for <u>one</u> particle is defined as

$$C(T) = \langle k_y(t)k_y(t+T)\rangle \qquad (1)$$

when k_y is the wave vector component in the direction of the current. The angular brackets indicate an averaging over all possible transport histories during the interval T correctly normalised as $C(0) = 1$. One particle represents a current[6] ev/s, s being the length of the block and v is the velocity of the particle. The fluctuation in this one-particle current is

$$S_1(T) = (e\hbar/sm^*)^2C(T) \qquad (2)$$

where e and m* represent the electronic charge and effective mass respectively.

Over a volume V with a particle density n the correlation function reads

$$S(T) = nVS_1(T) \qquad (3)$$

Because of the limited size of computer memory, it is only possible to follow N_s particles. Each simulated particle thus represents N/N_s real ones. For the purpose of solving Poisson's equation and estimating the currents, each simulated particle has to be attributed a charge Ne/N_s and an effective mass Nm^*/N_s. For all other purposes we use the single particle values for e and m*. The simulated correlation current will now be

$$S_s(T) = (N/N_s)S(T) \qquad (4)$$

thus exaggerated by a factor N/N_s. The

fluctuations one sees in the simulated current are thus exaggerated by a factor $\sqrt{N/N_s}$. If we could perform the summation involved in the angular brackets in (1) directly, we could then calculate the correct noise by dividing by the exaggeration factor. This, however, is not the end of the story, although it may be good enough as a first approximation.

Because the charges move, it is necessary to discretise Poisson's equation both in space and time. We introduce a two-dimensional uniform rectangular grid with a mesh size h_x, h_y along the x and y directions, respectively. There will be N_x cells in the x-direction and N_y cells in the main direction (y) of the current flow. (We have disregarded any changes in the third dimension, but a generalisation to three dimensions is straight forward. However, our calculations are three dimensional in k-space.) Furthermore, the potential distribution has to be recalculated at regular time intervals δT. Within each cell and time interval the potential stays uniform and fixed. This causes a certain 'graininess' both in time and space.

Each cell would contain ν particles if it were charge neutral. Assuming that the particles are moving independently of each other, which would be the case for most semiconductors, we can write down the distribution function for all possible distributions of the particles over the mesh. In this analysis the particles lose memory of their previous state on scattering. At any given time the statistical distribution of particles that have been flying freely for a time θ remains the same when the average electric field and temperature stay unaltered, but by the time T a given particle is scattered n times, namely at time $\{t_i\}_n$, i = 1,2,...n (n can have the value 0). These sets $\{t_i\}_n$ form a statistical distribution which we also have to consider in (4). C(T) is calculated using the analytical expressions for the scattering matrix and for the distributions of $\{t_i\}_n$. Actual measurements of noise take place over a long time compared to the free flights, so that we can consider the limit $T \to \infty$.

When $T \to \infty$, then $S_s(T)$ can be written in the form

$$S_s(T) = \frac{N}{N_s} S(T) + D(h_x h_y)^2 N_x B_y f(T)/\nu \qquad (5)$$

with

$$\lim_{t\to\infty} f(T) \equiv 2\tau_s\{\tau_s - \delta T[\exp(\delta T/\tau_s) - 1]^{-1}\} \qquad (6)$$

The last term of (5) represents the corrections due to the mesh and the finite time step δT. When $\delta T \to 0$ this term vanishes. D represents a coefficient containing material constants, τ_s is a relaxation time and B_y is a 'bulk factor'. It reflects the shape of the block to be simulated. B_y increases with N_x and N_y, at first, then it levels off for $N_x \approx N_y \approx 16$, indicating that reflections from the boundaries (surfaces) of the semiconductor lose their relative importance when increasing the mesh density. We notice from (5) that the correlation noise is proportional to $(h_x h_y)^2$ and N_x, and inversely proportional to ν but does not depend on N_y except through B_y. With $h_x = h_y$ we have simulated the noise current through a rectangular block of semiconducting GaAs and verified this dependency of S_s on the model parameters and on the bulk factor B_y.

The Fourier transform of S_s, the power spectrum, depends on the model parameters in the same way. It reads

$$\hat{S}_s(\omega) = \hat{A}\,\frac{\tau_s}{1+(\omega\tau_s)^2} + \hat{D}\,\frac{(h_x h_y)^2}{\nu} N_x B_y \left[\frac{\tau_s^2\,\delta T}{1+(\omega\tau_s)^2} \left(\frac{2e^{\delta T/\tau_s}}{e^{\delta T/\tau_s}-1} + \frac{1-(\omega\tau_s)^2}{1+(\omega\tau_s)^2} \right) + \left(\tau_s^2 - \frac{\tau_s\,\delta T e^{\delta T/\tau_s}}{e^{\delta T/\tau_s}-1} \right) \delta(\omega) \right]. \qquad (7)$$

$\hat{A}$ and $\hat{D}$ represent, like D, coefficients containing material constants. $\hat{S}_s(\omega)$ peaks at zero frequency. For the other frequencies $\hat{S}_s(\omega)$ is nearly constant for the very lowest ones, then falls off towards zero for high frequencies $(\omega \gg \tau_s^{-1})$. Around $\omega = \tau_s^{-1}$ the shape of $\hat{S}(\omega)$ resembles $1/\omega$ most, but for $\omega < \tau_s^{-1}$ the slope is less than $1/\omega$, and for $\omega > \tau_s^{-1}$. $\hat{S}(\omega)$ approaches zero faster than $1/\omega$.

3 CONCLUSION

We have derived an expression for the correlation current in a semiconductor as obtained by our Monte-Carlo particle model. This has been done by studying the correlation for an individual particle at time t and t+T, summing over all possible paths the particle could take during the time interval T, normalised by the probability of the a priori realisation of each path. During this interval T the particle has been scattered n times, with n varying from zero. During scattering the particle will have lost the memory of its previous state, which is the case for inter valley transitions and acoustic phonon scattering . Including a partial retention of the previous state, as is the case for polar optical phonon scattering, is straightforward, but does not significantly alter the form of expressions (5) and (7). The inclusion of non randomising scattering will show in expressions for A and D.

We find that our correlation current relates to the one expected for a real semiconductor through an exaggeration factor N/N_S, representing the impaired statistics due to considering only some of the carriers, and an additive term containing model parameters reflecting the choice of mesh and timesteps introduced to solve Poisson's equation. By actually simulating the current we have verified that our expression for the correlation is correct. The current chosen for this purpose was weak enough that the second part of (5) would dominate. In many practical applications the first part is the dominant one, making the corrections due to the choice of model of secondary importance. In this case a reduction of the simulated noise by the exaggeration factor would be sufficient to extract the real noise from the calculations.

The power spectrum (7) shown in Figure 1 represents the Fourier transform of (5). It depends on the model parameters and on the restricted number of particles followed during simulation in the same way as (5). It contains a relaxation time τ_S which is a measure for the average scattering frequency. Around this frequency the power spectrum resembles $1/\omega$ the most. When the first form (the one in A) dominates, the noise spectrum varies like ω^{-2} for $\omega >> \tau_S$.

4 REFERENCES

[1] Molgestue, C., Monte Carlo particle modelling of small semiconductor devices, Comp. Math. Appl. Mech. and Eng., 30 (1982) 173-208.

[2] Langdon, A.B., Effect of the spatial grid in simulation plasmas, J. Comp. Phys. 6 (1970) 247-267.

[3] Hockney, R.W.,Measurements of collision and heating times in a two dimensional thermal computer plasma, J. Comp. Phys. 8 (1971) 19-44.

[4] Tien, P.K. and Moshman, J., Monte Carlo calculation of noise near the potential minimum of a high frequency diode, J. Appl. Phys. 27 (1956) 1067-1078.

[5] Pollack, M.A. and Whinnery, J.R., Noise transport in the cross-field diode, IEEE Trans. Elect. Dev. ED-11 (1964) 81-89.

[6] v.d. Ziel, A. and Chenette, E.R., Noise in solid state devices, Adv. Electronics and Electr. Phys. 46 (1978) 313-383.

[7] Zimmermann, J. and Constant, E., Application of Monte Carlo techniques to hot carrier diffusion noise calculation in unipolar semiconducting components, Solid State Electronics, 23 (1980), 915-925.

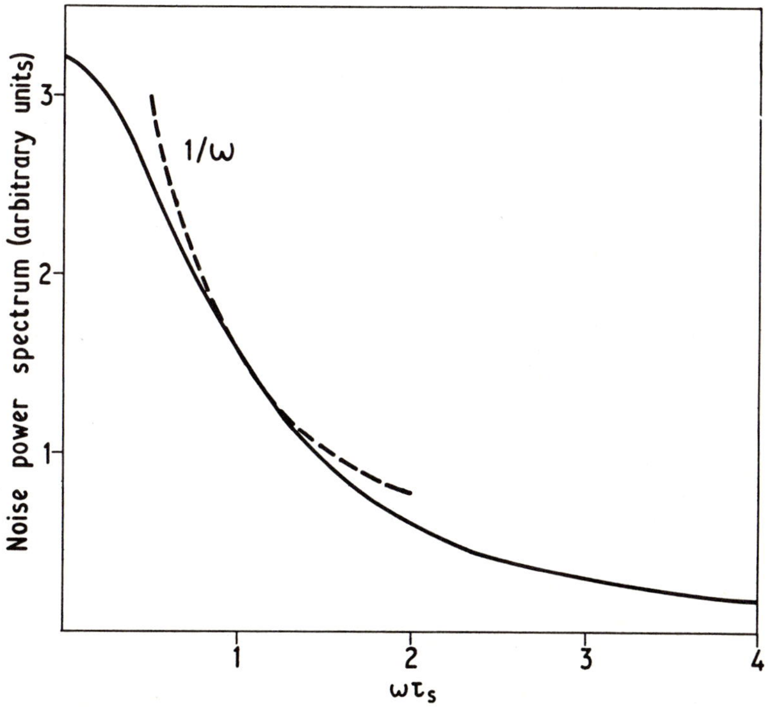

Figure 1 Power spectrum versus $\omega\tau_s$. Full curve : Eq 7. Dotted line : τ_s/ω

NOISE IN PHYSICAL SYSTEMS AND 1/f NOISE
M. Savelli, G. Lecoy and J-P. Nougier (eds.)
© Elsevier Science Publishers B.V., 1983

MASTER EQUATION APPROACH TO QUANTUM BROWNIAN MOVEMENT

Hermann Grabert* and Peter Talkner
Institut für Theoretische Physik, Universität Basel
Klingelbergstrasse 82, CH-4056 Basel, Switzerland

Some aspects of the problem of quantum Markov processes are investigated. Shortcomings of the conventional approach to describe damped quantum systems are outlined. A recently proposed new master equation for quantum Brownian motion is presented. The symmetries of quantum stochastic processes are emphasized.

1. INTRODUCTION

Within the framework of classical physics irreversible processes are frequently described in terms of Langevin equations or stochastically equivalent Fokker-Planck equations. These evolution equations can often be conceived from phenomenological considerations because their form is constrained by general features that are macroscopic manifestations of the underlying molecular process, such as the reciprocal relations and the fluctuation dissipation theorem (FDT). Since one avoids to study the molecular dynamics in detail, these semiphenomenological Fokker-Planck models are very useful in practice. If both thermal and quantal fluctuations are of importance, the Fokker-Planck equation (FPE) has to be replaced by a quantum master equation (QME). In this communication, we shall discuss some aspects of this approach using quantum Brownian motion by way of example.

2. CLASSICAL BROWNIAN MOTION

The motion of a Brownian particle of mass m moving in a potential V(q) is described on the deterministic level by the phenomenological equations

$$\dot{q} = \frac{1}{m} p \tag{1}$$

$$\dot{p} = -\partial V/\partial q - \gamma p ,$$

and on the more refined stochastic level by the FPE

$$\dot{\rho} = -\frac{\partial}{\partial q}\frac{p}{m}\rho + \frac{\partial}{\partial p}\left(\frac{\partial V}{\partial q} + \gamma p\right)\rho + k_B T m\gamma \frac{\partial^2}{\partial p^2}\rho \tag{2}$$

where $\rho(p, q, t)$ is the probability distribution of the Brownian particle, γ a damping constant and T the temperature of the environment.

* on leave from: Institut für Theoretische Physik, Universität Stuttgart, Pfaffenwaldring 57, D-7000 Stuttgart 80, Germany

Work supported by Swiss National Science Foundation

Using the Poisson bracket, the FPE (2) may also be written

$$\dot{\rho} = \{H,\rho\} + k_B T m\gamma \{q, \rho_{eq}\{q, \rho_{eq}^{-1}\rho\}\} \tag{3}$$

where

$$H = p^2/2m + V(q) \tag{4}$$

is the Hamiltonian, and

$$\rho_{eq} = Z^{-1}\exp(-H/k_B T) \tag{5}$$

the canonical equilibrium distribution. The first term on the RHS of (3) gives the reversible motion while the second term describes dissipation and thermal fluctuations. The form (3) of the FPE is a convenient starting point for the study of formal properties of the process. For instance, it is apparent that ρ_{eq} is the stationary solution of (3), and one easily shows that the first term on the RHS transforms odd (as $\dot{\rho}$) under the time reversal transformation while the second term transforms even.

The FPE governs the relaxation of initial nonequilibrium distributions. It can, however, also be used to calculate correlations of fluctuations about equilibrium. Introducing the short-hand notation

$$\dot{\rho} = L_{FP}\,\rho \tag{6}$$

where L_{FP} is the Fokker Planck operator, the equilibrium correlation of two variables X(p,q) and Y(p,q) reads

$$C_{XY}(t) = \langle X(t)Y\rangle = \underline{tr}\,(X \exp(L_{FP}t)\, Y \rho_{eq}) \tag{7}$$

where $\underline{tr}$ denotes the integration over p and q, that is the integral over the phase space of the Brownian particle, and where $\exp(L_{FP}t)$ is the time evolution operator for the probability, the kernel of which is the Green's function of the FPE, that is the conditional probability.

Furthermore, we may study the response to external forces by adding a perturbation to the

Hamiltonian (4) and investigating the solutions of the FPE in the presence of these forces. This way the classical response functions (dynamic susceptibilities) are found to read

$$\chi_{XY}(t) = -\theta(t)(1/k_BT)\underline{tr}(X\exp(L_{FP}t)L_{FP}Y\rho_{eq}) \quad (8)$$

The response functions are related to the correlation functions (7) by the FDT

$$k_BT\chi_{XY}(t) = -\theta(t)\,\dot{C}_{XY}(t) \quad (9)$$

Here $\theta(t)$ is the unit step function.

3. THE CONVENTIONAL APPROACH

In the quantal case the FPE is replaced by a quantum master equation (QME) describing the time evolution of the density matrix $\rho(t)$. The density matrix is an operator acting in the Hilbert space of the Brownian particle and it gives the mean value $< X(t) >$ of observables X by virtue of

$$< X(t) > = tr\,(X\,\rho(t)) \quad (10)$$

where tr denotes the trace over a complete orthonormal system of states in the Hilbert space.

Clearly, one expects that the QME is of the form

$$\dot{\rho} = -\,iL\rho = -(i/\hbar)\,[H,\rho] - iL_d\,\rho \quad (11)$$

where the first term describes the reversible dynamics of a quantum particle moving in a potential, and where the second term is supposed to describe dissipation. The super operator L is called the Liouvillian and L_d is its dissipative part.

In classical physics, the mean relaxation toward equilibrium is intimately connected with the decay of spontaneous fluctuations about equilibrium [cf.(7)] and there should be some relation in the quantal case too. Frequently, the calculation of equilibrium correlation functions from the QME is based on the so-called quantum regression hypothesis [1,2] saying that for a quantum Markov process the correlations of two variables X(t) and Y(0) are given by $(t \geq 0)$

$$\begin{aligned} <X(t)Y> &= tr(X\exp(-iLt)\,Y\,\rho_{eq}) \\ <YX(t)> &= tr(X\exp(-iLt)\,\rho_{eq}\,Y) \end{aligned} \quad (12)$$

Since the variables X(t) and Y do not commute in general, two differently time ordered correlations must be distinguished in the quantal case.

By virtue of the hypothesis (12), the quantum response function [3]

$$\chi_{XY}(t) = \Theta(t)\,(i/\hbar)< [X(t),Y] > \quad (13)$$

takes the form

$$\chi_{XY}(t) = \Theta(t)\,tr(X\,e^{-iLt}\,\frac{i}{\hbar}[Y,\rho_{eq}]) \quad (14)$$

Since in the classical limit $\hbar \to 0$ a commutator $(i/\hbar)[X,Y]$ is replaced by the Poisson bracket $\{Y,X\}$, we obtain for the classical response function

$$\chi_{XY}^{c\ell}(t) = \Theta(t)\,tr(X\,e^{L_{FP}t}\,\{\rho_{eq},Y\}) \quad (15)$$

where we have assumed that the classical time evolution is governed by the Fokker Planck operator L_{FP}. Now, since $\{\rho_{eq},Y\} = -(1/k_BT)\{H,Y\rho_{eq}\}$ we find that (15) coincides with the correct expression (8) only for reversible systems where L_{FP} is given by the first term on the RHS of (3).

Hence, the conventional theory of quantum Markov processes is not an extension to the quantum regime of the theory of classical stochastic processes. It can be shown [4] that irrespective of the particular form of the Liouvillian L_d, the quantum regression hypothesis (12) always violates the quantum version of the FDT, and that the symmetry of detailed balance cannot be satisfied simultaneously with the constraints following from Ehrenfest's theorem [5] . Since the hypothesis (12) holds for reversible systems, the errors can be small numerically for weakly damped systems. Hence, an approach using (12) is not necessarily meaningless. However, one would certainly like to develop the theory of quantum Markov processes in such a way that the properties of classical Markov processes are recovered in the limit $\hbar \to 0$. This problem will be addressed in the following sections.

4. EXACT TIME EVOLUTION OF THE DENSITY MATRIX

An exact non-Markovian master equation for the density matrix $\rho(t)$ of a quantum Brownian particle can be obtained by means of the projection operator technique in statistical mechanics [6] . One finds

$$\dot{\rho}(t) = \frac{i}{\hbar}[\rho_{eq},\hat{\chi}^{-1}\rho(t)] + \left(\frac{i}{\hbar}\right)^2 \int_0^t ds\,[q,m\Lambda(t-s)[q,\hat{\chi}^{-1}\rho(s)]] \quad (16)$$

Here the operator $\hat{\chi}$ determines the static susceptibilities by virtue of

$$\chi_{XY} = tr(\delta X\,\hat{\chi}\,\delta Y) \quad (17)$$

where $\delta X = X - <X>$ is the fluctuation about equilibrium. χ_{XY} gives the mean response of X to an external field coupling to Y. The operator $\hat{\chi}$ has an inverse since the equilibrium state is stable. The classical susceptibilities are given by

$$\chi_{XY}^{c\ell} = (1/k_BT)\langle\delta X\delta Y\rangle = \underline{tr}(\delta X\ \hat{\chi}^{c\ell}\ \delta Y) \tag{18}$$

Hence, the operator $\hat{\chi}$ reduces in the classical limit to a multiplication by $\hat{\chi}^{c\ell} = \rho_{eq}/k_BT$.

Noting that the first term on the RHS of the FPE(3) may be written $\{H,\rho\} = -k_BT\{\rho_{eq},\rho_{eq}^{-1}\rho\}$, we see that the QME(16) is formally obtained from the FPE(3) if (i) the Poisson brackets $\{X,Y\}$ are replaced by commutators $(-i/\hbar)[X,Y]$ as usual, (ii) $k_BT\rho_{eq}^{-1}$ is replaced by its quantum analogue $\hat{\chi}^{-1}$, and (iii) $\gamma\rho_{eq}$ is replaced by a time-retarded damping operator $\Lambda(s)$. This structure of the QME(16) makes sure that the correct classical behaviour is recovered in the limit $\hbar\to 0$.

Besides the mean relaxation toward equilibrium the QME(16) also governs the time evolution of fluctuations about equilibrium[6] . The result is stated conveniently in terms of the canonical correlation function

$$C_{XY} = tr(XG(t)Y) \tag{19}$$

where for $t \geq 0$ the super operator G(t) satisfies the QME

$$\dot{G}(t) = \frac{i}{\hbar}[\rho_{eq}, \hat{\chi}^{-1} G(t)] + \left(\frac{i}{\hbar}\right)^2 \int_0^t ds\,[q, m\Lambda(t-s)\,[q, \hat{\chi}^{-1} G(s)]] \tag{20}$$

with the initial condition $G(0) = k_BT\hat{\chi}$. The relation of $C_{XY}(t)$ to the frequently used symmetrized correlation function

$$S_{XY}(t) = (1/2)\ \langle X(t)Y + YX(t)\rangle \tag{21}$$

is expressed at its clearest in terms of the associated spectral functions [3]

$$S_{XY}(\omega) = (\hbar\omega/2k_BT)\coth(\hbar\omega/2k_BT)C_{XY}(\omega) \tag{22}$$

Both, the canonical correlation (19) and the symmetrized correlation (21) reduce in the limit $\hbar\to 0$ to the classical correlation function (7). However, a simple quantum regression theorem relating the mean regression from nonequilibrium states [cf.(16)] with the regression of fluctuations about equilibrium [cf.(19,20)] can only be given in terms of the canonical correlation functions. Further exact results related with the QME(16) are given in [4] and [6] .

5. A MARKOVIAN MODEL FOR QUANTUM BROWNIAN MOVEMENT

The exact QME(16) includes memory effects described by the retarded damping operator $\Lambda(s)$. Hence, the stochastic process is non-Markovian. However, Markovian behaviour can be expected if the Brownian particle is much heavier than the reservoir particles. In this section we shall discuss such a Markovian model for quantum Brownian movement [7] .

We assume that the static behaviour of the Brownian particle can be described in terms of an effective Hamiltonian H of the form (4). Then, the stationary density matrix ρ_{eq} is of the canonical form (5) and the static susceptibilities (17) are given in terms of the operator

$$\hat{\chi} = \beta K\ ,\quad \beta = 1/k_BT \tag{23}$$

where

$$KX = (1/\beta Z)\int_0^\beta d\alpha\ e^{-\alpha H}\ X\ e^{-(\beta-\alpha)H} \tag{24}$$

is the Kubo transformation. From (5) and (23) there follows $[\rho_{eq},\hat{\chi}^{-1}\rho(t)] = -[H,\rho(t)]$ so that the QME (16) takes the form (11) if memory effects are disregarded. Statistical mechanical considerations lead to the damping operator

$$\Lambda X = (1/\beta Z)\int_0^\beta d\alpha\ \gamma(\alpha)\ e^{-\alpha H}\ X\ e^{-(\beta-\alpha)H} \tag{25}$$

where $\gamma(\alpha)$ is given in terms of the correlation function of the force Γ exerted by the reservoir upon the Brownian particle

$$\gamma(\alpha) = (1/mk_BT)\int_0^\infty ds\langle\Gamma(s-i\hbar\alpha)\Gamma\rangle \tag{26}$$

Hence, the dissipative Liouvillian reads

$$L_d\rho = (k_BTm/i\hbar^2)\,[q,\Lambda[q,K^{-1}\rho]] \tag{27}$$

In the high temperature limit the QME (11,27) reduces to the FPE(3) with a damping constant given by Kirkwood's formula.

Some interesting properties of the QME (11,27) can easily be seen. For the mean relaxation we find

$$\begin{aligned}\langle\dot{q}(t)\rangle &= (1/m)\ \langle p(t)\rangle\\ \langle\dot{p}(t)\rangle &= -\langle\partial V(t)/\partial q(t)\rangle - \gamma\langle p(t)\rangle\end{aligned} \tag{28}$$

where

$$\gamma = (1/\beta)\int_0^\beta d\alpha\ \ \gamma(\alpha) \tag{29}$$

is a damping constant. This is in accordance with Ehrenfest's theorem.

The symmetries of the process are easier recognized if the Liouville operator L is written as

$$iL = k_BT\ R\ K^{-1} \tag{30}$$

where

$$RX = -(i/\hbar)\ [\rho_{eq},X] + (m/\hbar^2)\ [q,\ \Lambda[q,X]] \tag{31}$$

is a transport operator satisfying the reciprocity relation

$$\Pi\, R\, \Pi = R^{T} \qquad (32)$$

Here Π is the time reversal transformation and the superscript T denotes the transpose. The symmetry (32) is connected with the detailed balance symmetry of the canonical correlation functions (19). In the Markovian approximation the operator $G(t)$ is given by $t \geq 0$

$$G(t) = \exp(-iLt)K \ , \ G(-t) = G^{T}(t) \qquad (33)$$

Further, one can study the response to external forces and derive the FDT. Hence, the model incorporates the basic symmetries of the process correctly.

REFERENCES

[1] Lax, M., Phys. Rev. 172, 350 (1968)

[2] Haake, F., in: Springer Tracts in Modern Physics, Vol. 66, p. 98 (Springer, Heidelberg, 1973), and references cited therein

[3] Kubo, R., Rep. Progr. Phys. (London) 29, 255 (1966)

[4] Grabert, H., Z. Phys. B49, 161 (1982)

[5] Talkner, P., (unpublished)

[6] Grabert, H., Springer Tracts in Modern Physics, Vol. 95 (Springer, Heidelberg, 1982)

[7] Grabert, H., Talkner, P., Phys. Rev. Lett. 50, 1335 (1983)

NOISE IN PHYSICAL SYSTEMS AND 1/f NOISE
M. Savelli, G. Lecoy and J-P. Nougier (eds.)
© Elsevier Science Publishers B.V., 1983

WHY IS NATURE FRACTAL AND WHEN SHOULD NOISES BE SCALING?
(Invited Paper)

Benoit B. Mandelbrot and Richard F. Voss

Thomas J. Watson Research Center
International Business Machines Corporation
Yorktown Heights, NY 10598 USA

Both measurements on the natural world and the remarkable resemblance of simple fractal computer simulations to Nature support Mandelbrot's contention that the geometry of Nature is largely fractal. Fractal shapes share many common characteristics with the "scaling" or "1/f" noises. The 1/f noises, like many random fractals, are statistically self-similar: they "look the same" when viewed on different temporal or spatial scales. General arguments can be made for the prevalence of fractal shapes in Nature: they invoke, respectively, the limit theorems of probability theory, new singularities of partial differential equations in real physical space, and the fractal character of the "strange" attractors in phase space. These arguments will be summarized, and their relevance to 1/f noise discussed.

1. INTRODUCTION

Mountains are not cones, clouds are not spheres, and the natural shapes encountered in physics—especially in the study of condensed matter—are not typically representable by standard geometric shapes from Euclid. To represent them, Mandelbrot conceived, developed and applied widely a new approach he calls the *fractal geometry of nature.* A fractal's degree of roughness or irregularity can be characterized by a number called the *fractal dimension* D, which is strongly related to the intuitive notion of "dimension" yet need not be an integer.

As a preliminary, and before defining the term *fractal,* the thesis that the geometry of Nature is largely fractal can be conveyed with the help of Figure 1, generated and rendered by Voss. These clouds, mountains and noise records are *totally artificial* random surfaces or curves. Compare carefully each curve on the top row of diagrams with the top edge of the black "front wall" of the corresponding relief in the second row of diagrams. This curve and this edge are identical, which underlines the intimate relationship between the diagrams in each of the two columns.

Originally, the probabilistic rules underlying Figure 1 have *not* been selected as the outcome of a theories deduced from basic physical principles. They have been selected merely because they were the most convenient rules whose outcomes a) are continuous shapes and b) are scale invariant, that is, remain statistically unchanged when one moves closer to examine them in detail, or away to gain an overall view of them. The relevance of all this to noise theory is obvious. Indeed, one can legitimately argue that 1/f noises ought to be called "scaling noises", since they can be characterized as being noises whose rules remain unchanged as they are contracted or expanded in time. The point, the line and the plane are scaling geometric shapes and are of course very smooth. All the other scaling geometric shapes, however, are rough and Mandelbrot called them *fractals.* Actually, the mathematical notion of fractal is more general and more delicate, but for the moment the above will do.

The remarkable resemblance of Figure 1 and of other algorithmic "fractal forgeries" to the natural world is confirmed by diverse theories put forward in [1], by diverse conventional measurements reported in [1], for example of such naturally occurring shapes as the perimeters of coastlines, and by numerous more recent results concerning percolation clusters[2], fracture surfaces of metals[3] and cloud and rain areas[4]. With the benefit of perfect hindsight, one can also find many unrecognized fractal symptoms in papers written long ago (Appendix A). Granted these increasingly numerous and unquestionable confirmations, a broad question arises. No one had expected Nature to be fractal, and fractals started by being highly controversial. How can their ubiquity be explained, today?

It has been argued by Mandelbrot that in many cases scientists who clamor for explanations use this term as "code word" for very different wishes. First point: explanations can either bring the promise of control over nature, or merely reassure the engineer well in advance of any conceivable implementation of control. To give an extreme example, fluctuations in precipitation or temperature over the very long term (tens or hundreds of years) are scaling[5]; there can be no thought of controlling them, yet hydrologists would feel better if they were explained. Second point: when a theory founded upon phenomenology is criticized for being as yet unexplained, it may often be that what is criticized is the fact this theory has not yet been stated in familiar vocabulary and "broken in" by repeated successful use.

It must also be mentioned that the urge to explain is being repeatedly given a bad name by certain very common *circular*

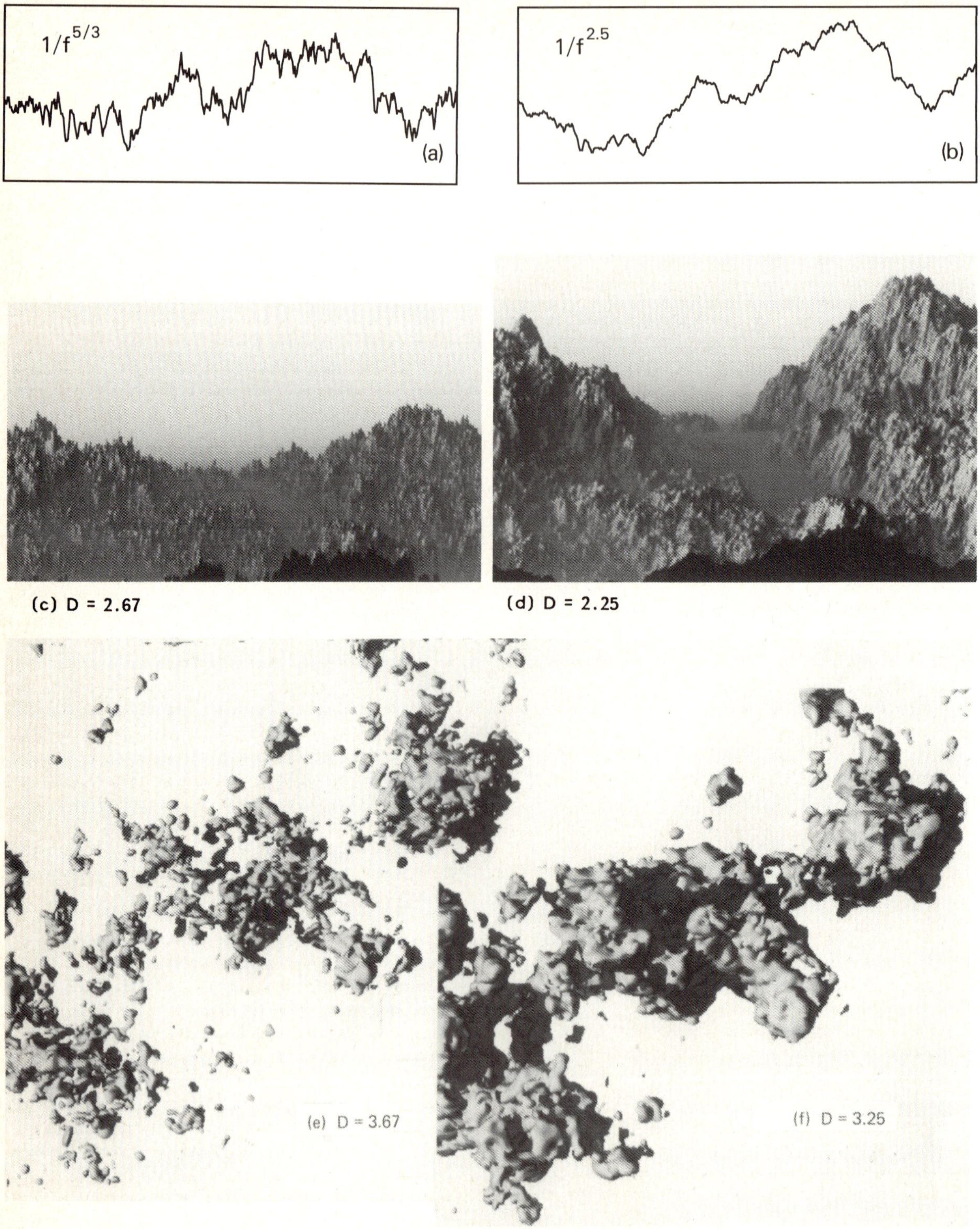

Figure 1. Samples of fractal noises, landscapes, and flakes with $1/f^B$ spectra along any direction: (a), (c), (e) B=1.67; (b), (d), (f) B=2.5.

pseudo-explanations. A telltale symptom of fractal behavior is often found in a hyperbolic function or probability distribution. It is child's play to construct black boxes that yield the desired hyperbolic output, if one is allowed to assume that their input is *also* hyperbolic. Other circular arguments, all dating to the mid 1800's but continuing to be published by (or at least submitted to) today's journals, are discussed in Appendix B. A common feature is that these models are linear, perhaps in a hidden fashion.

Despite the strictures in the preceding paragraphs, a full explanation of the existing fractal models *would be very important*. Two main lines of argument, each coming in two flavors, are presented in [1] and will now be summarized. The first line lays stress on linearity, while the second line of argument demands nonlinearities. An eclectic synthesis would claim that some nonlinearity is required somewhere to create fractal structures, but, once these structures have been created, they will not be destroyed by linear transformations.

For the specific case of scaling or 1/f noises, however, experimental attempts to determine if nonlinearities are important have been inconclusive[6]. Systems with spectral densities closest to an exact 1/f dependence show no evidence of macroscopic nonlinearities. Moreover, measurements of 1/f noise via the variance of Johnson noise[7] convincingly demonstrate that power dissipation is not necessary for 1/f noise and strongly suggest that 1/f noise can, in fact, be a thermal equilibrium phenomenon.

2. SCALE INVARIANCE TAKEN AS A PRINCIPLE OF NATURE

This argument is not very serious, but must be mentioned. In outline it runs as follows. Many analytical aspects of nature are now known to be scaling. For example, the theory of critical phenomena is built upon scaling relations (many of which have since been explained via renormalization group argument). Therefore, the corresponding geometric aspects (as evidenced, for example, in the shape of percolating clusters, [1] Chapter 13 and [2]) should also be expected to be scaling. Straight lines and planes are scaling, but they are rarely found in nature. Fractal geometry offers nature a far richer pallet.

For a would-be "explanation", this argument is peculiar: it amounts to declaring that the scale invariance should *not* in fact be viewed as something requiring further explanation, but as being itself a basic principle of Nature. Just as one need not "explain" stationarity, which is the same as "translation invariance", one need not explain "scaling", which is the same as "dilatation invariance". Stated differently: By a widely held viewpoint, mechanics did not become fully mature until the books ceased to be full of strings and pulleys and started instead to be full of invariance groups.

The weakness of this argument is that it proposes to enshrine linearity, while the few cases for which substantial progress have been achieved (see sections 4 and 5), the explanations are dominated by the role of nonlinearity, and the linearity implied by strict scaling only enters as an approximation to more complex forms of invariance.

3. AN ATTRACTIVE ARGUMENT FOR FRACTAL BEHAVIOR THAT IS UNFORTUNATELY OF KNOWN AND LIMITED SCOPE: LIMIT THEOREMS OF PROBABILITY THEORY

The sum of many independent random variables is "often" Gaussian, and the result of many independent displacements is often Brownian motion. These are called "generic" results, and physicists would say that the Gaussian and Brownian motion have broad "domains of universality", in the sense that the Gaussian and Brownian motion obtains in the limit under seemingly very undemanding conditions. Brownian motion's path is a fractal, and its spectral density is of course $1/f^2$. So here is a convincing argument, but it only applies to a fractal that is so well-known as to be without interest.

What about the other, "non-generic" limit theorems? Many are known, and the typical limits are also scaling and graphed by fractals. Unfortunately, the conditions of validity of any nongeneric limit theorem are extremely stringent: the input must be a slightly generalized form of the scaling invariance to be found in the theorem's output.

Thus, by and large, the central limit theorems express that addition (a linear operation) will not *destroy* a fractal form of scaling. But they are not capable of creating it except in the Brownian $1/f^2$ case. For most practical purposes, the limit theorem argument is a logically circular pseudo-explanation like those criticized in Section 1.

4. A POTENTIALLY CONVINCING ARGUMENT FOR FRACTAL BEHAVIOR. ANALOGY WITH THE MANDELBROT CONJECTURES CONCERNING TURBULENCE. TURBULENCE RESEMBLES 1/f NOISES, INSOFAR AS IT INVOLVES NUMEROUS SPECTRA OF THE FORM $1/f^B$

In outline, the prevalence of fractals can be accounted for directly, by real space arguments. Many basic equations of physics are scale-free: for example, the Euler equations of the flow of a nonviscous fluid. The solutions are dominated by the equations' singularities. These singularities in turn reflect the symmetry of the original equations and should be expected to be scaling geometric shapes. If turbulence is to be associated with nonstraight singularities, it is natural to trace it back to fractal singularities of fluid motion. This thesis has recently been strongly buttressed by extensive computer simulations described in Appendix C.

Now, let us elaborate somewhat. Many authors have been attracted by the notion that a $1/f$ noise is a turbulence-like behavior in a system ruled by a different set of nonlinear equations. Unfortunately, turbulence could be of little help because its theory was so undeveloped. For example, the "Kolmogorov" $f^{-5/3}$ spectrum of the velocity of homogeneous turbulence was *not* introduced by truly physical but by purely "dimensional" arguments. A physical derivation has been sought for forty years, but none is known, perhaps because homogeneous turbulence is not a physical notion. The dissipation of turbulence also has a $1/f^B$ spectrum, and a careful examination of its spatial distribution led Mandelbrot to conjecture (as described in Appendix C) that fully-developed turbulence prevails in a fluid when the solution of the equations of fluid motion—in real physical space—develop fractal singularities. The estimated dimension of singularities was predicted to be about 2.5 to 2.6. The onset of turbulence will be referred to momentarily, but the problems raised by fully-developed turbulence differ from those raised by the onset. Fully-developed turbulence may even look as a steady state (stationary) process, even though energy is dissipated somewhere, namely throughout shapes representable by fractals. The equation of fluid motion (Euler at zero viscosity, Navier-Stokes at positive viscosity) are nonlinear and very hard. Previous to the Mandelbrot conjectures, the search for singularities of motion was limited—without conscious decision to do so—to the standard geometric shapes: points, lines and planes. Now singularities are being found where the new conjectures said they should be. A broader form of the Mandelbrot conjectures is that other nonlinear equations have fractal singularities.

For further elaborations, see Appendix C.

5. A POTENTIALLY CONVINCING ARGUMENT FOR FRACTAL BEHAVIOR. ANALOGY WITH THE FACTS THAT THE SO-CALLED "STRANGE" ATTRACTORS OF DYNAMICS ARE FRACTAL, AND THAT THEIR STUDY IS RIFE WITH $1/f^B$ SPECTRA

In outline, the prevalence of fractals can also be accounted for indirectly for systems whose time evolution is traced in a phase space. In most cases, the attractor is "strange", which means that it is a fractal. This fact was first established by H. Poincaré, P. Fatou and G. Julia, for self-mappings of the complex plane and will be illustrated by exhibiting some fractal associated with the mapping $z \to z^2 - \mu$.

Now let us elaborate. The approach that is currently "in" in the study of dynamical systems arose from the revival of a remark by Poincaré. Having failed in his attempt to prove that the solar system is stable, and having introduced (informally) the notion of attractor for a system's trajectory in a phase space, Poincaré went on to observe that he could not even exclude attractors of extraordinarily bizarre shapes. When revived by Ruelle and Takens[8], they were labeled "strange". There is no generally accepted definition of "strangeness", but all these attractors are fractals and Chapter 20 of [1] suggests that strangeness be defined by being fractal. While Section 4 involves fractals in real physical space, strange attractors are fractals in a phase space.

Formally, a dynamical system is a mapping of a phase space upon itself: it transforms the system's position at time t into its position at time t+1. Hence, later positions are obtained by "iterating" this mapping.

It is well known that probability theory is dominated by the study of the binomial distributions (and of their "central limit", the Gaussian distribution). In the study of dynamical systems defined by iteration, the role of the binomial is played by the map $z \to z^2 - \mu$. The study of $z \to z^2 - \mu$ on the real line is due especially to J. Myrberg and M. Feigenbaum[9]. The systematic study of $z \to z^2 - \mu$ in the complex plane was initiated by B. B. Mandelbrot[10]. The fractals it involves are of two kinds. Given the value of μ, the $\mathcal{F}$-set in the z-plane (due to G. Julia and P. Fatou) is the boundary of the set that fails to converge to infinity by the iteration of the map $z \to z^2 - \mu$. And the $\mathcal{M}$-set in the μ-plane (due to B. B. Mandelbrot) is the set of μ such that the $\mathcal{F}$-set is is connected. The $\mathcal{M}$-set (which is illustrated as Figure 2) has a very distinctive shape: A heart-shaped central "atom" is surrounded by sprouts made of smaller atoms, each of which has lesser atoms and sprouts growing on it, and so on to infinity. Furthermore, all these sprouts have the same shape irrespective of size. Thus, the whole adds up to a "molecule", whose boundary is a fractal curve. In addition, the little dots floating around the big molecule are in fact small molecules, each a small-scale image of the big molecules. Furthermore, all these molecules are linked together.

6. GEOMETRIC CONNECTIONS BETWEEN FRACTALS AND RANDOM TIME SERIES OR NOISES WITH A SPECTRAL DENSITY OF THE FORM $1/f^B$

Against the background of the fractal geometry of nature, the mathematical similarity between fractal shapes and the "scaling" or "$1/f$" noises is obvious. Both possess a statistical invariance or scaling of various moments under changes of temporal or spatial scale. As already mentioned, this similarity is illustrated in Figure 1 which includes the extension of $1/f$ noises to higher dimensions. A fractal landscape $H(x,y)$ is random surface with the property that a cut along any direction gives a $1/f$ noise. Similarly, a fractal flake is a 3D function $T(x,y,z)$ (such as local temperature or water vapor density) with the same property. Mathematically, whenever a spectrum can be attached to a fractal (e.g. when the fractal is a vertical section through a relief or a distribution of galaxies), this spectrum is of the form $1/f^B$ and the exponent B is a linear function of the fractal's fractal dimension $D = d+(3-B)/2$. Thus, B can serve to measure the fractal's fractal dimension.

7. POSSIBLE PHYSICAL CONNECTIONS BETWEEN FRACTALS AND 1/f NOISES

The remaining (and by far the most interesting) issue is that of a possible physical connection. Here, the recent work on the reasons for the occurrence of fractal shapes may serve as a guideline for studies of 1/f noises. As mentioned above, a totally linear system would preserve a scaling, i.e. fractal, geometry, but could not generate it. However, recent work on turbulence and strange attractor theory demonstrates that nonlinearity, even in seemingly small amounts, can generate fractal structures.

A large number of "exotic" or "high level" systems (Nile river flood levels[5], musical melodies[11], traffic currents[12]), and motions of the geographic pole[13], are known to exhibit 1/f noise[14]. Although such observations, no doubt, fuel much of the speculation about the "universal" nature of 1/f noise, the "equations of motion" for these systems are unknown. Thus, the favored targets of the investigations of 1/f noise mechanisms have been the far simpler electronic systems such as resistors, individual transistors or diodes, or junction elements. Here, one begins (one hopes) with a reasonably complete understanding of the basic physical processes.

It is precisely in these simple electronic systems, however, where the 1/f noise exhibits several characteristics that are particularly difficult to reconcile with the known methods of generating fractals[15].

1) The observed 1/f noise shows no threshold. In simple resistors 1/f noise is observed in the voltage fluctuations with constant driving current. The amplitude of the spectral density varies smoothly with average voltage $\langle V \rangle$ as $\langle V \rangle^2$. Moreover, measurements of 1/f noise via the variance of Johnson noise[7] convincingly demonstrate that power dissipation is not necessary for 1/f noise and strongly suggest that 1/f noise can, in fact, be a thermal equilibrium phenomena. Such measurements cannot, however, distinguish metastable or slow mechanical non-equilibrium effects (such as annealing of defects) from true equilibrium.

2) Macroscopic I-V nonlinearities are not needed for 1/f noise; it can be found in the most linear of resistors. Experimental[6] attempts to determine, on a more microscopic scale, if nonlinearities are important in 1/f noise (by studying the characteristic generation and decay of fluctuations of different amplitudes) are overall inconclusive. Of primary importance here, however, is the finding that systems, such as carbon resistors with spectral densities closest to an exact 1/f dependence, show no evidence of nonlinearities. The rare fluctuations of large amplitude behave exactly the same as small excursions.

3) There is no evidence for a connection between 1/f noise and actual fractal shapes in the sample. 1/f noise can also be found in the most uniform of samples such as metal films or single crystal semiconductors. Even granular films or resistors show only small scale disorder; their properties become uniform of large scales. Although Marinari et al.[16] recently proposed that a 1 dimensional random walk in a random environment (only small scale disorder) could lead to 1/f noise, extensive computer simulations[17] fail to confirm their hypothesis and show, instead, a $1/f^{1.6}$ dependence.

The recent wide acceptance of the possible role of chaotic behavior in producing "noisy" systems has lead many researchers to consider (if only privately) the connection between 1/f noise and chaos. As mentioned above, chaotic systems are invariable associated with orbits that tend to fractal ("strange") attractors in phase space. Such "exact" or "deterministic" noise has (built-in) the long memories necessary for 1/f noise. The "intermittency" often found at the onset of chaos or turbulence seems remarkably like 1/f noise. In fact, the specific model studied by Manneville[18] can produce an extended "1/f" spectrum with appropriate parameters. Unfortunately the "noise" in all such models is extremely sensitive to the control parameters and the equations themselves are often not directly related to physical systems. All exhibit thresholds for chaotic behavior unlike the 1/f noise in real experimental systems and extended 1/f spectral densities are almost never observed. The "observation" of 1/f noise by Arecchi and Lisi[19] in an analog model of a driven two-well system is unfounded. Not only do extensive computer simulations[20] of the same system show no region of 1/f noise, but theoretical treatments[21] predict Lorentzian-like spectra.

Thus, while the mathematical similarity between fractals and 1/f noise remains clear, the physical connection remains obscure. It is likely that there is no "universal" mechanism for generating $1/f^B$ noises. In some cases (such as the "high level" or strongly driven systems) 1/f noise may be related to turbulence or chaotic behavior and nonlinearities. In other cases (like oxide junctions or semiconductor surfaces) the $1/f^B$ spectrum may, in fact, be due to a (oft proposed and intellectually unsatisfying) sufficiently wide and flat distribution of characteristic energies. Once again, the simple macroscopically-linear, electronic systems pose the greatest problem.

The lessons of fractal geometry, however, continually return to the importance of nonlinearities. If present, such nonlinearities must be at a sufficiently microscopic level that usual measurements, which average over a large number of independent elements, can not detect. These nonlinearities could be classical or quantum mechanical and experimental resolution is possible only with new experiments on the smallest possible samples involving few independent elements.

APPENDIX A. OLD EVIDENCE FOR SCALING (EXCERPT FROM [1], CHAPTER 41)

A1. Scaling in Elastic Silk Threads

The earliest empirical observation that can now be reinterpreted as evidence of scaling in a physical system was made, extraordinarily enough, *a hundred and fifty years ago*. On the urging of Carl Friedrich Gauss, Wilhelm Weber set out to investigate the torsion of the silk threads used to support moving coils in electric and magnetic instruments. He found that applying a longitudinal load provokes an immediate extension which is followed by a further lengthening with time. On removal of the load, an immediate contraction equal to the initial immediate extension takes place. This is followed by a gradual further decrease of length until the original length is reached. The aftereffects of a perturbation follow a law of the form $t^{-\gamma}$: they decay hyperbolically in time, not exponentially as everyone expected then, and expects to this day.

The next work on this topic is Kohlrausch in 1847, and the elastic torsion of glass fibers is further studied by William Thomson, later Lord Kelvin, in 1865, by James Clerk Maxwell in 1867, and by Ludwig Boltzmann, in a 1874 paper that Maxwell viewed as important enough to discuss in the ninth (1878) edition of *Encyclopaedia Britannica*.

These names and dates should be pondered carefully. They prove that, in order to make a problem worth studying, a show of interest by the likes of Gauss, Kelvin, Boltzmann, and Maxwell is not enough. A problem that fascinated but defeated them could fall into extreme obscurity.

A2. Scaling in Electrostatic Leyden Jars

Kohlrausch in 1854 found for the speed of discharge by the Leyden jar the same result as in his work on silk threads: the charge decays hyperbolically in time. Dielectrics other than glass are investigated in detail in the Ph.D. thesis of Jacques Curie (Pierre Curie's brother and his first collaborator), who finds that in some dielectrics the decay is exponential, but in others it is hyperbolic, with varying values of the exponent γ.

APPENDIX B. SCALING: DURABLE PANACEAS (EXCERPT FROM [1], CHAPTER 41)

Innumerable explanations of the scaling decays or noises are scattered over a hundred years of the most diverse journals. All make for sad reading. Their lack of success is consistent and monotonous, since dead-ends recognized in the 1800's keep being explored again and again, in different contexts and words.

B1. Hopkinson's Mixture Panacea

Faced with the hyperbolic decay of the charge of a Leyden jar, Hopkinson (a colleague of Maxwell's at Kings College, London) advances in 1878 the "rough explanation [that] glass may be regarded as a mixture of a variety of different silicates that behave differently." It would follow that a decay function that seems to be a hyperbola, is in fact a mixture of two or more different exponentials of the form $\exp(-s/\tau_m)$, each of them characterized by a different relaxation time τ_m. However, even the early data suffices to show that two to four exponentials do not suffice, and the argument is abandoned.

But it keeps popping out wherever data are not sufficiently abundant to disprove it.

B2. Distributed Relaxation Times Panacea.

When data cover many decades and cannot be fitted unless the mixture involves exponentials in ridiculous number, say 17 or 23, one is tempted to go all the way to a mixture of an *infinite* number of exponentials. The definition of Euler's gamma function yields

$$t^{-\gamma} = [\Gamma(\gamma)]^{-1} \int_0^\infty \tau^{-(\gamma+1)} e^{-t/\tau} d\tau.$$

This identity shows that, if the exponential relaxation time τ has the "intensity" $\tau^{-(\gamma+1)}$, the mixture is hyperbolic. However, this argument is logically circular. A scientific explanation's output is supposed to be less obvious a priori than its input, but $t^{-\gamma}$ and $\tau^{-(\gamma+1)}$ are functionally identical.

B3. TRANSIENT BEHAVIOR PANACEA

Upon hearing of the diverse symptoms of scaling listed in the preceding entry, a second near-universal first reaction is this: Surely, these hyperbolic functions $t^{-\gamma}$ are only transient complications that will be cut off exponentially when decays are observed long enough. The first systematic search for the cutoff was by von Schweidler in 1907[22]. He measured Leyden jars' decay, first at intervals of 100 seconds, then (less frequently) for a total time of 16 million seconds (200 days, through summer and winter!). The hyperbolic decay continues on the dot. More recent experiments on electric 1/f noises had started by lasting a few hours, then a night, then a weekend, then a short vacation. In surprisingly many cases, the 1/f behavior continues on the dot.

Other chapters of *The Fractal Geometry of Nature*[1], for instance the study of galaxy clusters, note that scientists can become so engrossed in the search for a cutoff as to neglect the need for describing and explaining the phenomena characteristic of the scaling range. Oddly, an over-involvement with the cutoff can be even stronger among engineers. To take an example, discussed in Chapter 27, many hydrologists hesitate to use my model[5] because it involves an infinite cutoff to scaling. In an engineering project, the finiteness of the cutoff is immaterial, nevertheless a finite cutoff is fervently desired by presumably practical people.

APPENDIX C: FRACTAL SINGULARITIES OF DIFFERENTIAL EQUATIONS (EXCERPT FROM [1], CHAPTER 11)

C1. A Split in Turbulence Theory

A major defect of the current theoretical study of turbulence is that it separates into at least two disconnected parts. One part includes the successful phenomenology put forth by Kolmogorov in 1941. And the other part includes the differential equations of hydrodynamics, due to Euler for nonviscous fluids, and to Navier (and Stokes) for viscous fluids. These two parts remain unrelated: If "explained" and "understood" mean "reduced to basic equations", the Kolmogorov theory is not yet explained or understood. And Kolmogorov has not helped solve the equations of fluid motion.

I assert (in Chapter 10 in *The Fractal Geometry of Nature*) that turbulent dissipation is not homogeneous over the whole space, only over a fractal subset. This may seem at first sight to make the gap even greater. *But I contended that the opposite is the case.* And there is increasing evidence in my favor.

C2. The Importance of Singularities

Let us review the procedure that allows an equation of mathematical physics to be solved successfully. Typically, one draws up a list that combines solutions obtained by solving the equation under special conditions, and solutions guessed on the basis of physical observation. Next, neglecting details of the solutions, one draws a list of elementary "singularities" characteristic of the problem. From then on, more complex instances of the equation can often be solved in the first approximation by identifying the appropriate singularities and stringing them together as required. This is how the student of calculus draws the graph of a rational function. Of course, the standard singularities are standard Euclidean sets: points, curves, and surfaces.

C3. Conjecture: the Singularities of Fluid Motion are Fractal Sets (Mandelbrot[23])

In this perspective, I interpret the difficulties experienced in deriving turbulence from the Euler and Navier-Stokes solutions as implying that *no* standard singularity accounts for what we perceive intuitively to be the characteristic features of turbulence.

I contend instead that the turbulent solutions of the basic equations involve singularities or "near singularities" of an entirely new kind. The singularities are locally scaling fractal sets, and the near singularities are approximations thereto.

An unspecific motivation for this contention is that, standard sets having proven inadequate, one may as well try the next best known sets. But more specific motivation is available.

C4. Nonviscous (Euler) Fluids

First Specific Conjecture. Part of my contention is that the singularities of the solutions of the Euler equations are fractal sets.

Motivation. This belief relies on the very old notion that the symmetries and other invariances present in an equation "ought" to be reflected in the equation's solution. Of course, preservation of symmetries is by no means a general principle of Nature, hence one cannot exclude the possibility of "broken symmetry" here. I propose, however, that one try the consequences of symmetry preservation. Since the Euler equations are scale-free, the equations' typical solutions should also be scale-free, and the same should hold of any singularities they may possess. If the failure of past efforts is taken as evidence that the singularities are not standard points or lines or surfaces, they must be fractals.

It may of course happen that a scale is imposed by the boundary's shape and the initial velocities. It is, however, likely that the solutions' local behavior is ruled by a "principle of not feeling the boundary." Hence the solutions should be locally scaleless.

Alexandre Chorin's Work. Chorin[24] provides strong support for my contention, by applying a vortex method to the analysis of the inertial range in fully developed turbulence. The finding is that the highly stretched vorticity collects itself into a body of decreasing volume, and of dimension $D \sim 2.5$ compatible with the conclusions in Chapter 10. The fractal correction to the usual Kolmogorov exponent 5/3 is compatible with experimental data. The calculations suggest that the solutions of Euler's equations in three dimensions blow up in a finite time. Subsequent work of Chorin comes even closer to experiment: $2.5 < D < 2.6$.

C5. Viscous (Navier-Stokes) Fluids

Second Specific Conjecture. Furthermore, I contended that the singularities of the solutions of the Navier-Stokes equations can only be fractals.

Dimension Inequalities. Furthermore, we have the intuitive feeling that the solutions of the Navier-Stokes equations are necessarily smoother, hence less singular, than those of the Euler equations. Hence the further conjecture that the dimension is larger in the Euler than in the Navier-Stokes case. The passage to zero viscosity is doubtless singular.

Near Singularities. A final conjecture in the implementation of my overall contention concerns the peaks of dissipation involved in the notion of intermittency: they are Euler singularities smoothed out by viscosity.

V. Scheffer's Work. The examination of my conjectures for the viscous case was pioneered by V. Scheffer, recently joined by others in studying in this light a finite or infinite fluid subject to the Navier-Stokes equations with a finite kinetic energy at $t=0$.

Assuming that singularities are indeed present, Scheffer[25] shows that they necessarily satisfy the following theorems. First, their projection over the time axis has at most the fractal dimension ½. Second, their projection on the space coordinates is at most a fractal of dimension equal to 1.

It turns out, after the fact, that the first of the above results is a corollary of a remark in an old and famous paper. Writing in 1934, Leray stops abruptly after a formal inequality of which Scheffer's first theorem is a corollary, in fact merely a restatement. But is it fair to say "merely"? Restating a result in more elegant terminology is (for sound reasons) rarely viewed as a scientific advance, but I think that the present instance is different. The inequality in Leray's theorem was nearly useless until the Mandelbrot-Scheffer corollary placed it in proper perspective.

The almost routine uses of Hausdorff Besicovitch dimension in recent studies of the Navier-Stokes equations can all be traced back to my conjecture.

C6. Singularities of Other Nonlinear Equations of Physics

The other phenomena which this Essay claims involve scaling fractals have nothing to do with either Euler or Navier and Stokes. For example, the distribution of galaxies is ruled by the equations of gravitation. But the symmetry preservation argument applies to all scaling equations.

More generally, the singularities' fractal character is likely to be traceable to generic features shared by many different equations of mathematical physics. Can it be some very broad kind of nonlinearity?

Figure 2. The fractal $\mathcal{M}$-set of the map $z \to z^2 = \mu$ of the complex plane, viewed as a dynamical system.

References

[1] Mandelbrot, B. B., *The Fractal Geometry of Nature* (Freeman, San Francisco 1982). This book supersedes and makes obsolete the same author's *Fractals: Form, Chance and Dimension* (1977) and his original *Objets fractal* (1975).

[2] Voss, R. F., Laibowitz, R. B. and Alessandrini, E. I. *Phys. Rev. Lett.* 49, 1441 (1982).

[3] Mandelbrot, B. B., Passoja, D. and Paullay, A. To appear.

[4] Lovejoy, S., *Science 216,* 185 (1982).

[5] Mandelbrot, B. B. and Wallis, J. R., *Water Resour. Res. 4,* 909 (1968); *5,* 321 (1969).

[6] Voss, R. F. *Phys. Rev. Lett. 40,* 913 (1978).

[7] Voss, R. F. and Clarke, J., *Phys Rev. Lett. 36,* 42 (1976); Beck, H. G. E. and Spruit, W. P., *J. Appl. Phys. 49,* 3384 (1978).

[8] Ruelle, D. and Takens, F., *Comm. Math. Phys. 20,* 167 (1971).

[9] Collet, P. and Eckmann, J. P., *Iterated maps on the intervals as dynamical systems* (Boston, Birkhauser, 1980).

[10] Mandelbrot, B. B. *Non-linear dynamics* (Ed. R. H. G. Helleman) *Ann. N.Y. Acad. Sci. 357,* 249 (1980); Mandelbrot *Physica D* (in the press).

[11] Voss, R. F. and Clarke, J., *Nature 258,* 317 (1975); *J. Accoust. Soc. Am. 63,* 258 (1978).

[12] Musha, T. and Higuchi, H., *Jap. J. Appl. Phys 15,* 1271 (1976).

[13] Mandelbrot, B. and McCamy K., *Geophysical J.* 21, 217 (1970).

[14] For general reviews of "exotic" 1/f noise sources, see Voss, R. F., *Proc. 33rd Ann. Symp. on Freq. Control,* (Atlantic City, NJ, 1979) p. 40; W. H. Press, *Comments on Astrophys. 4,* 103 (1978).

[15] For a review of general characteristics of 1/f noise in simple electronic systems, see Dutta, P. and Horn, P. M., *Rev. Mod. Phys. 53,* 497 (1981); van der Ziel, A. *Adv. Electronics and Electron Phys. 49,* 225 (1979); and Weissman, M. B. *Proc. 6th Int. Conf. on Noise in Phys. Systems,* (NBS Special Publication 614, Gaithersburg Md., 1981) p. 133.

[16] Marinari, E., Parisi, G., Ruelle, D. and Windey, P., *Phys. Rev. Lett. 50,* 1223 (1982).

[17] Voss, R. F. and Koch, R. H., to be published.

[18] Manneville, P., *J. de Physique 41,* 1235 (1980).

[19] Arecchi, F. T. and Lisi, F., *Phys. Rev. Lett. 49,* 94 (1982).

[20] Voss, R. F., *Phys. Rev. Lett. 50,* 1329 (1983).

[21] Beasley, M. R., D'Humières, D. and Huberman, B. A., *Phys. Rev. Lett. 50,* 1328 (1983).

[22] von Schweidler, E., *Ann. d. Phys. (4)24,* 711 (1907).

[23] Mandelbrot, B. B., *Comptes rendus* (Paris) *282A,* 119 (1976).

[24] Chorin, A., *Comm. in Pure Appl. Math. 34,* 853 (1981).

[25] Scheffer, V. *Turbulence and Navier Stokes equations* (Ed. R. Temam) *Lecture Notes in Math 565,* 174 (1976).

NOISE IN PHYSICAL SYSTEMS AND 1/f NOISE
M. Savelli, G. Lecoy and J-P. Nougier (eds.)
© Elsevier Science Publishers B.V., 1983

TRANSITION PHENOMENA INDUCED BY MULTIPLICATIVE NOISE IN NEMATIC LIQUID CRYSTAL SYSTEMS

(Invited Paper)

René LEFEVER

Chimie Physique II, C.P. 231
Université Libre de Bruxelles
1050 Brussels, Belgium

Werner HORSTHEMKE

Physics Department, Institut for Fusion Studies
and Center for Studies in Statistical Mechanics
University of Texas
Austin, Texas 78712

We analyze the influence of a nonlinear external multiplicative noise on the Freedericksz transition in a nematic liquid crystal layer. Exact results are reported for the system's stationary behavior as a function of the intensity and correlation time of the noise. It is found that with respect to noiseless conditions, the threshold of instability may be shifted to higher or lower values of the control parameter depending on the value of the correlation time of the noise. The relation with electrohydrodynamic instabilities in liquid crystal systems is discussed.

1. INTRODUCTION

Numerous recent studies are devoted to the properties of nonequilibrium systems submitted to a fluctuating environment (see e.g. (1-5) and the references cited therein). More particularly there is a widespread interest for the influence of external noise in systems which undergo a fork bifurcation as a function of a control parameter, say h (see Fig. 1).

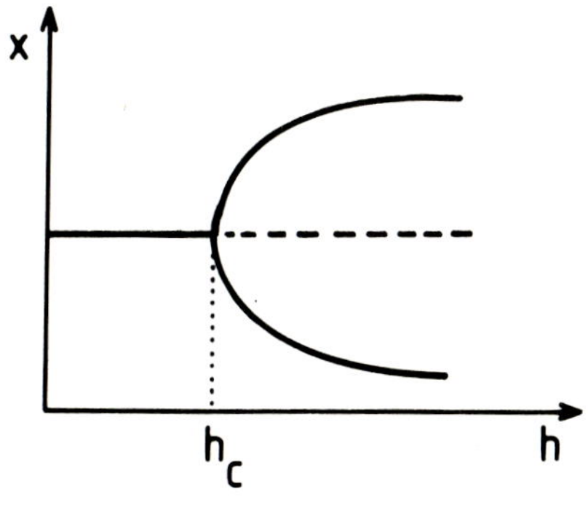

Figure 1: Fork bifurcation. At the critical point h_c the regime x_o becomes unstable and two new stable branches $x_{1,2}$ appear.

When this parameter fluctuates in time the question arises of what impact these fluctuations have on the bifurcation. Over recent years, this problem has been under intense investigation in such diverse fields as hydrodynamics (6-9), oscillatory electrical circuits (10), photochemical systems (11,12), nonlinear optics (13).

All the systems studied in the fields mentioned above have in common the property that their behavior near the threshold point is governed by a balance equation in which the environmental noise is multiplicative, i.e. the noise enters the equation multiplied by a function which depends on the state of the system. Typically the balance equation is of the form

$$\dot{x} = f(x) + \mu(h, z_t)g(x) \quad (1.1)$$

where f and g are functions of the amplitude x of some mode which becomes unstable at h_c and where μ depends on the average value h and noise z_t of the control parameter. One has:

$$h_t = h + z_t \text{ with } h = \text{constant and}$$
$$E\{z_t\} = 0. \quad (1.2)$$

It is now well known that multiplicative noise may give rise to a specific class of transition phenomena, usually called noise induced transitions. In the case of the fork bifurcation which interests us more particularly here, clearly the various branches of the fork modulate differently the intensity of the noise function μ. The remarkable result of this effect, demonstrated experimentally by several authors, is that it shifts the position of the fork bifurcation with respect to its position in the noiseless case. Remarkably also, the instability point remains sharply defined even for noises having a large variance while the magnitude of the shift is in first approximation proportional to the variance.

In this paper we will analyze a noise induced transition of this type taking place in a nematic liquid crystal layer perturbed by a fluctuating magnetic field. This system exhibits a fork bifurcation corresponding to a Freedericksz transition involving a twist mode instability. The behavior of this instability under fluctuating field conditions has not yet been the object of an experimental study. Nevertheless sufficient motivations exist for undertaking a theoretical study of this system and for making predictions on the outcome of possible experiments. Indeed:(i)The mechanism of the mechanism of the Freedericksz transition is well understood under constant environmental conditions and there exists a reliable theoretical description which can be taken as starting point for an analyzis of the influence of noise. Futhermore this instability is closely related with the electrohydrodynamic instabilities leading to the formation of Williams domains and experimentally studied by the groups of Kai (6) and Kawakubo (7). The mechanism of the Freedericksz transition is however much simpler and thus permits a less ambiguous interpretation of the effects induced by multiplicative noise. (ii)In all experiments made so far with multiplicative noise, the control parameter is perturbed by a broad-banded noise whose cut-off frequency is much greater than the normal relaxation time of the system. One works with a noise which in practice can be considered as δ-correlated (white noise). It is most desirable to explore completely the response of the system not only in term of the intensity of the noise but also in term of its correlation time. The system considered here is quite appropriate for such a study. This is done below, in the case of markovian dichotomous noise. This choice of noise is advantageous indeed because it allows for an exact determination of the stationary response of the system for any value of the noise intensity and correlation time; futhermore it can easily be generated experimentally. (iii)In the case of the electrohydrodynamic instabilities mentioned above as well as in the case of the Freedericksz transition, the noise function μ is nonlinear. This poses the question of finding the white noise limit of a nonlinear noise. As we report below this problem finds easily its answer in the case of dichotomous noise.

In the next section we will first review briefly the properties of the Freedericksz transition in a constant environment. This behavior will be compared in section 3 with the system subjected to a fluctuating magnetic field. The results are discussed in section 4.

2. THE FREEDERICKSZ TRANSITION

For the sake of clarity, let us first recall some properties of nematic liquid crystals (for more details see (14-16)). A nematic liquid crystal layer consists of elongated molecules, i.e. little rods, which on the average align parallel to a prefered direction. If the nematic layer is enclosed between two parallel plates, the prefered direction can be prescribed by treating the surface of the plates. Rubbing the surface along a fixed direction creates grooves and the nematic molecules will lie parallel to the rubbed direction. As far as the elastic energy of the material near the surface is concerned, this is the most favorable direction. The extra energy required to leave the alignment parallel to the grooves is quite appreciable and leads to a firm anchoring of the director near the surface. The director is a vector $\vec{n}$, choosen to represent the direction of prefered orientation of the molecules in the neighbourhood of any point.

If nematic molecules are submitted to a magnetic field $\vec{H}$, the magnetisation which is induced is given by

$$\vec{M} = \chi_\perp \vec{H} + (\chi_\parallel - \chi_\perp)(\vec{H}.\vec{n})\vec{n} \qquad (2.1$$

where $\chi_\parallel$ and $\chi_\perp$ are the magnetic susceptibility for $\vec{H}$ parallel or perpendicular to n. Usually the anisotropic susceptibility

$$\chi_a = \chi_\parallel - \chi_\perp \qquad (2.2$$

is positive and the free energy is minimum when the molecules are aligned parallel to the lines of force of the field. The magnetic torque produced by this anisotropy is

$$\vec{\Gamma}_M = \vec{M} \times \vec{H} = \chi_a(\vec{n}.\vec{H})\vec{n}\times \vec{H}. \qquad (2.3$$

Let us now examinate the situation depicted in Fig.2. The x-axis coincides with the direction along which two parallel plates (only the lower plate is represented) confining a nematic layer have been rubbed. In the absence of any external influence, the director $\vec{n}$ in the whole nematic layer is parallel to the x-axis: the angle θ of n with the x-axis is zero. If a magnetic field $\vec{H}$ is applied parallel to the y-axis, the torque (2.3) acting on the rodlike molecules will tend to align them parallel to $\vec{H}$. However, if the lower and upper plates,placed at z=-d/2 and z=d/2, have been properly treated the anchoring of the molecules there is quite firm and almost impossible to overcome by a magnetic field. Therefore the director will remain parallel to the x-axis near the end plates and we have

$$\theta(\pm d/2) = 0 \qquad (2.4$$

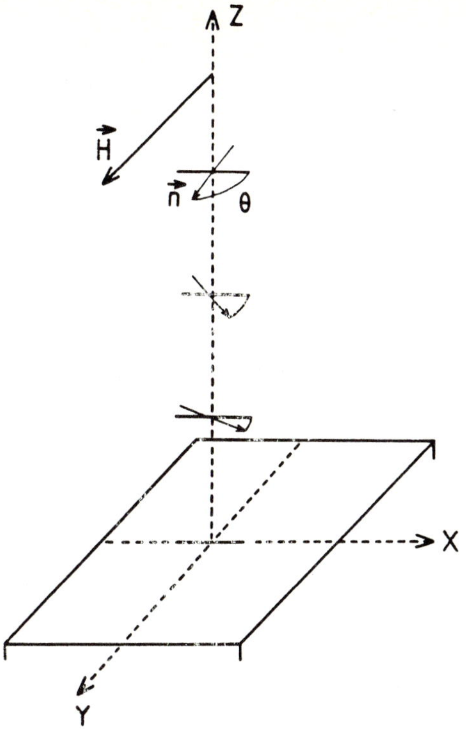

Figure 2: Twist mode instability.

for all values of $\vec{H}$. This has the consequence that in the layer the magnetic torque is counteracted by an elastic torque stemming from the preferential alignment of the molecules and a viscuous torque opposing any rapid rotation of the director. Balancing the torques, one obtains (14,15)

$$\lambda_1 \partial_t \theta = K_{22} \partial_{zz} \theta + \chi_a H^2 \sin\theta\cos\theta, \qquad (2.5)$$

where λ_1 is the twist viscosity and K_{22} is an elastic constant, namely the twist modulus.

It is easily seen that $\theta(z)=0$ is a steady state solution of (2.5) and it is the one which prevails at $\vec{H}=0$. The stability of this homogeneous steady state is easily determined by a linear normal mode analysis:

$$\theta(z,t) = \hat{\theta} e^{\omega t} \cos\frac{(2n+1)\pi z}{d}, \quad n = 0,1,..$$

It is found that the n-th mode becomes unstable, i.e. $\omega_n \geqslant 0$, for

$$H^2 \geqslant (K_{22}\pi^2)(2n+1)^2/(\chi_a d^2). \qquad (2.6)$$

This implies that at

$$H_c^2 = (K_{22}\pi^2)/(\chi_a d^2) \qquad (2.7)$$

the homogeneous steady state $\theta(z)=0$ becomes unstable against perturbations of the form $\cos(\pi z/d)$. This transition phenomenon is the Freedericksz transition.

Only if the value of the magnetic field is larger than the critical value H_c can the external field twist the director in the nematic layer away from the prefered direction imposed by the plates. The state of the system becomes spatially inhomogeneous in the z-direction. For $H_c<H<3H_c$ only the lowest spatial mode is unstable and will dominate the behavior in the vicinity of the critical point (2.7). To describe the state of the nematic layer, we can neglect higher spatial modes and write

$$\theta(z,t) = \hat{\theta}(t)\cos\frac{\pi z}{d}. \qquad (2.8)$$

For not too large values of $\hat{\theta}$, i.e. close to the critical point,(2.7) yields (dropping the caret)

$$\lambda_1 \dot{\theta} = -K_{22}(\pi^2/d^2)\theta(t) + \chi_a H^2(\theta - \theta^3/2) \qquad (2.9)$$

3. BEHAVIOR IN A FLUCTUATING MAGNETIC FIELD

Let us now consider that the magnetic field fluctuates. One has

$$H_t = H + I_t \qquad (3.1)$$

where I_t is the markovian dichotomous noise whose state space consists only of two levels $I_1 = -$, $I_2 =$. The waiting times in each state are exponentially distributed and equal on the average. The transition probability of I_t obeys the master equation

$$\frac{d}{dt}\begin{pmatrix} P(\Delta,t|I_0) \\ P(-\Delta,t|I_0) \end{pmatrix} = -\frac{\gamma}{2}\begin{pmatrix} 1 & -1 \\ -1 & 1 \end{pmatrix}\begin{pmatrix} P(\Delta,t|I_0) \\ P(-\Delta,t|I_0) \end{pmatrix}$$

$P(\pm\Delta,t|I_0)$ is the probability that $I_t = \pm\Delta$ given that $I_t = I_0$ at time $t=0$. The general solution is

$$\begin{pmatrix} P(\Delta,t|I_0) \\ P(-\Delta,t|I_0) \end{pmatrix} = \begin{pmatrix} (P(I_0=\Delta)-1/2)e^{-\gamma t}+1/2 \\ (P(I_0=-\Delta)-1/2)e^{-\gamma t}+1/2 \end{pmatrix}$$

Thus if I_t is started with equal probabilities for each state, then it is a stationary process with mean equal to zero and an exponentially decreasing correlation function

$$E\{I_t\} = 0, \quad E\{I_t I_{t+\tau}\} = \Delta^2 e^{-\gamma\tau}. \qquad (3.2)$$

The correlation time is given by

$$\tau_{cor} = \gamma^{-1} \qquad (3.3)$$

and the power spectrum is Lorentzian

$$S(\nu) = (\gamma\Delta^2/\pi)/(\nu^2 + \gamma^2). \qquad (3.4)$$

For

$$\gamma \to \infty, \ \Delta \to \infty \text{ such that } \Delta^2/\gamma = \sigma^2 \text{ (finite)} \qquad (3.5)$$

$S(\nu)$ is constant at all frequencies ν and $I_t \to \sigma\xi_t$, where ξ_t is gaussian white noise with $E\{\xi_t \xi_{t+\tau}\} = \delta(\tau)$ (20). If the control parameter appears nonlinearly in the phenomenological equation, the white noise limit has to be taken with circumspection, since nonlinear operation on white noise cannot be defined for mathematical reasons. As shown below however, our system can be written as a formally linear noise problem for which the white noise limit can be taken without any difficulty.

Indeed, if the magnetic field fluctuations are given by a dichotomous Markov process, then (2.9) reads

$$\lambda_1\dot{\theta}_t = -K_{22}(\pi^2/d^2)\theta_t + \chi_a(H+I_t)^2(\theta_t-\tfrac{1}{2}\theta_t^3).$$

The nonlinearity of the control parameter is quadratic so that for dichotomous noise $I_t^2=\Delta^2$. Thus the nonlinear multiplicative noise problem reduces to a formally linear noiseproblem

$$\lambda_1\dot{\theta}_t = -K_{22}(\pi^2/d^2)\theta_t + \chi_a(H^2+\Delta^2)\times(\theta_t-\theta_t^3/2) + 2\chi_a HI_t(\theta_t-\theta_t^3). \quad (3.6)$$

We call (3.6) a formally linear noise problem, since the equation is not obtained from (2.9) by writing $H_t^2=H^2+I_t$. (3.6) retains the full nonlinearity of the problem at hand. Proceeding according to (3.5), the white noise limit can be obtained straightforwardly (for a treatment of this limit in the case of a gaussian colored noise see (18,19). We will come back to the white noise limit somewhat later.

Introducing the zero field relaxation time τ_0

$$\tau_0 = (\lambda_1 d^2)/(K_{22}\pi^2) \quad (3.7)$$

and the reduced fields

$$h_t = H_t/H_c,\ i_t = I_t/H_c,\ h = H/H_c$$

$$\delta = \Delta/H_c \quad (3.7)$$

(3.6) can be written in the convenient form

$$\dot{\theta}_t = \tau_0^{-1}\theta_t((h^2+\delta^2-1) - \tfrac{1}{2}(h^2+\delta^2)\theta_t^2) + 2\tau_0^{-1}\theta_t h i_t(1-\tfrac{1}{2}\theta_t^2) = F(\theta_t,i_t). \quad (3.8)$$

The main theoretical advantage of using the dichotomous Markov noise in the theoretical analysis is that, despite the nonmarkovian character of θ_t, its stationary probability density $p_s(\theta)$ can be evaluated exactly (20).It is given by

$$p_s(\theta) = N\left(\frac{1}{F(\theta,\Delta)} - \frac{1}{F(\theta,-\Delta)}\right)\times \exp\left\{-\frac{\gamma}{2}\int^{\theta}\left(\frac{1}{F(\theta,\Delta)} + \frac{1}{F(\theta,-\Delta)}\right)d\theta\right\} \quad (3.9)$$

for θ in the interior of the support

$$U = (\theta_-,\theta_+) \text{ and } p_s(\theta) = 0 \text{ if } \theta \quad U.$$

N is a normalization constant. The boundaries of the support θ_- and θ_+ obey the the equation

$$0 = F(\theta_\pm,\pm\Delta) \quad (3.10)$$

If (3.10) admits multiple steady states for Δ and $-\Delta$, then U is an appropriate union of intervals between the stable steady states. One has

$$U = \{0\} \text{ if } h-\delta<1,\ h+\delta<1 \quad (a)$$
$$U = \{0,\theta_+\} \text{ if } h-\delta<1<h+\delta \quad (b) \quad (3.11)$$
$$U = \{\theta_-,\theta_+\} \text{ if } 1<h-\delta<h+\delta \quad (c)$$

where

$$\theta_\pm^2 = 2(1 - (h\pm\delta)^{-2}). \quad (3.12)$$

We only have to consider cases with $0<\delta<h$. If $h+\delta<1$ (see (3.11a)), $p_s(\theta)=\delta(\theta)$. We are interested in transitions in the steady state behavior of the layer taking place when $h+\delta>1$ (cf. (3.11a,b)). For results concernig the dynamical properties of noise induced transitions see the paper of Suzuki in this volume.

Extending the deterministic definition of a transition point to the fluctuating case we say that a noise induced transition occurs, if the number of extrema of $p_s(\theta)$ or their nature (minimum or maximum) changes. In other words the extrema are our indicator of transition, or <u>order parameter</u>. This choice of order parameter is in general more reliable than other indicators such as the moments. The latter indeed are obtained through an averaging procedure which washes out a lot of information (for a complete discussion of the justification and merits of this choice, see(1,18)). For instance, a smooth transition from a symmetric single humped probability density to a symmetric double humped probability density does not show up in the moments unless the influence of additional fast relaxating variables is explicitely taken into account (21) (see also the paper of Graham in this volume).

After lenghty but straightforward calculation one finds that the behavior of the extrema of $p_s(\theta)$ summarizes as follows (18)

(i) $\theta=0$ is always an extremum for $0<h-\delta<1<h+\delta$.

(ii) The transition point h_F at which the maximum at $\theta=0$ transforms into a minimum is given by

$$h_F^2=1+\delta^2-\tfrac{1}{2}\gamma\tau_0+(\tfrac{1}{4}(\gamma\tau_0)^2+(4-2\gamma\tau_0)\delta^2)^{1/2} \quad (3.13)$$

(iii) This point is also the point where $p_s(\theta)$ ceases to diverge near zero.

(iv) $p_s(\theta) \underset{\theta\to\theta_+}{\to} \infty$ if $\gamma\tau_0<4((h+\delta)^2-1)$ (3.14)

(v) $p_s(\theta) \underset{\theta\to\theta_-}{\to} \infty$ if $\gamma\tau_0<4((h-\delta)^2-1)$, $h-\delta>1$ (3.14)

Clearly, changes in the variance δ^2 of the external noise as well as in the correlation time $\tau_{cor} = \gamma^{-1}$ modify the threshold value of the Freedericksz transition. The value of the systematic part h_F necessary to induce the transition is represented in Fig. 3 as a function of the amplitude δ of the fluctuating part for different values of the correlation time.

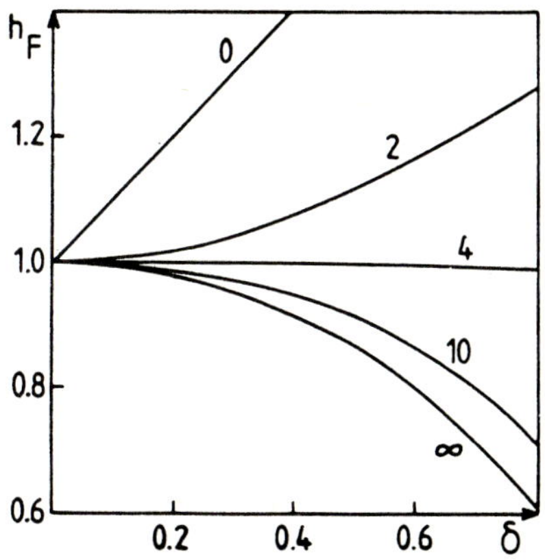

Figure 3: Average magnetic field at the Freedericksz transition point as a function of the amplitude of the noise and for different values of $\gamma\tau_0$.

We see that for small values of $\gamma\tau_0$, i.e. long correlation times, the threshold value increases with the variance of the noise, while for short correlation times, i.e. large values of $\gamma\tau_0$, the external noise lowers the threshold value. The change over in behavior can be found from (3.13) by differentiating with respect to δ^2

$$\left.\frac{\partial(h_F^2)}{\partial(\delta^2)}\right|_{\delta^2=0} = \frac{4-\gamma\tau_0}{\gamma\tau_0} . \qquad (3.15)$$

This implies that dichotomous external noise will lower the threshold value for the Freedericksz transition if its correlation time is shorter than one fourth of the zero field relaxation time. In other words rapid external noise destabilizes the system, whereas slow noise enhances the stability of the undeformed layer.

If a constant magnetic field is impressed upon the nematic layer, the Freedericksz transition is the only instability which occurs for $H<3H_c$. In a fluctuating magnetic field the behavior is much richer. The possibility of additional transition phenomena is evident from (3.13-3.14). Using this information permits to construct phase diagrams in term of the noise characteristics. In Fig. 4, we report a phase diagram which represents the propertis of the system in the $(\gamma\tau_0$-h)-plane for a fixed value of δ. In addition to hard

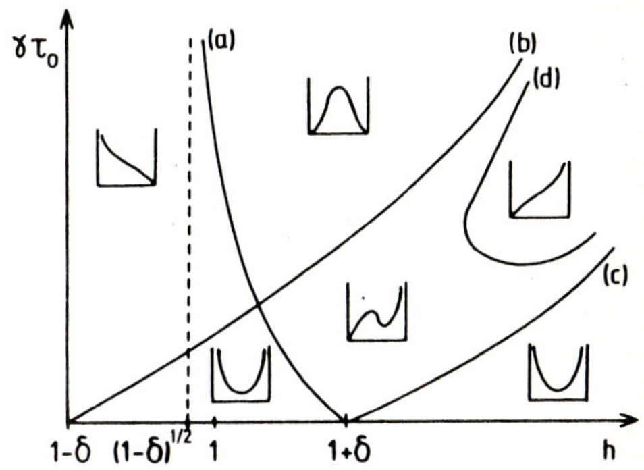

Figure 4: Qualitative sketch of the behavior in the $(\gamma\tau_0$-h)-phase plane. The shape of the probability density in each domain is represented.

transitions (first-order-like transitions) which are caused by changes in the boundary behavior and occur when the lines (a),(b),(c) are crossed, one also observes a pure noise induced soft transition corresponding to the line (c). For large values of $\gamma\tau_0$ (fixed) one observes the following sequence of events when h is increased. As the line (a) is crossed, the most probable value changes from zero value to nonzero value via a critical point. This is the Freedericksz transition whose threshold is shifted by the noise. Increasing the value of h further, we observe a noise induced hard transition as the line (b) is crossed. The system displays now bistable behavior, one state corresponds to the Freedericksz deformation, the other coincides with the upper boundary of the support, i.e. the steady state corresponding to $H+\Delta$. As the line (c) is crossed, the maximum and the minimum in the interior of the support, i.e. the stable Freedericksz state and an unstable noise induced state, coalesce via a saddle point. This is a soft transition to monotone behavior of $p_s(\theta)$. The noise induced bistability disappears and the behavior is dominated by the state θ_+. Interestingly, upon a further increase of h, the bistability reappears via a soft transition as the line (c) is crossed again. A further increase of h leads finally to the trivial behavior that the relaxation to the states θ_- and θ_+ becomes shorter than the waiting time of the noise in the states $\pm\Delta$. This leads to a divergence of the probability density at both boundaries of the support. The system follows the external noise.

4. DISCUSSION

Let us compare the behavior predicted here with the observations on the electrohydrodynamic instability associated with the formation of Williams domain (6,7). Therefore we consider the white noise limit which is appropriate to make contact with the situation investigated experimentally. One derives easily that in this limit (18)

$$h_F^2 = \frac{1 - \sigma^2/2}{1 - 2\sigma^2\tau_0^{-1}} \quad (4.1)$$

where $\sigma = (2\delta^2/\gamma)^{1/2}$ is defined in (3.5). For small white noise intensity, we have

$$h_F^2 = 1 - 2\sigma^2\tau_0^{-1} - \gamma\sigma^2/2 + O(\sigma^4). \quad (4.2)$$

The white noise idealization is meaningful if the bandwidth of the noise γ is much larger than any frequency of the deterministic system. In our case, this implies that $\gamma >> \tau_0^{-1}$. Thus we find

$$h_F^2 \simeq 1 - \gamma\sigma^2/2 = 1 - \delta^2. \quad (4.3)$$

This clearly implies that the systematic effect due to the nonlinearity of the noise, i.e. $E\{I_t^2\}$, dominates over the multiplicative noise effect. This phenomenon had already been noted by San Miguel and Sancho (19). This has the consequence that white noise shifts the Freedericksz threshold to lower values of the magnetic field.

The implications for the electrohydrodynamic instability are the following: If the theoretical description of Kai and coworkers is adequate, then the observed shift to higher values of the average voltage must be entirely due to the excitation of modes with very many different wavelengths. The wavenumber spectra of Kawakubo and coworkers, which do not show any appreciable broadening of the peak corresponding to the Williams domain, could be explained by the fact that once a prefered spatial mode is established in the system, all other modes are strongly damped by competition with the prefered mode. The other possibility is that the analogy between the electrohydrodynamic instability and the Freedericksz transition which works well with a constant electrical field becomes inadequate under fluctuating environmental conditions. It is not uncommon that models which show close agreement deterministic situations predict completely different behavior for fluctuating environmental conditions. Another approach has thus been proposed by Kawakubo (7). It is based on the theory of Dubois-Violette et al. (23) and takes into account the nonuniformity of the electric field and the formation of space charges due to the distortion of the layer. Though this approach predicts in the white noise limit a shift to higher values of the voltage, in apparent agreement with the experiment, the result is nevertheless not totally conclusive since it is based on a linearized description.

In the light of the results reported above and of these concluding remarks, the situation clearly calls for further experimental stidies. The purpose of such experiments would be threefold: a) to study the effect of external multiplicative nonlinear noise, b) to test the validity of the assumption that the only mode excited for $H+\Delta<3H_c$ is $\cos(\pi z/d)$ and c) to evaluate with less ambiguity the contribution of the various effects which participate in the formation of the Williams domain when the electric field fluctuates.

Acknowledgement

This work was supported by the U.S. Department of Energy, grant DE-AS05-81ER10947 and by the NATO Research grant 125.82.

REFERENCES

1.-See the papers by W. Horsthemke, R. Lefever, J. M. Sancho and M. San Miguel in "Stochastic Nonlinear Systems in Physics, Chemistry and Biology", Synergetic Series vol. 8 (Springer-Verlag)(1981)
-Horsthemke,W. and Lefever,R., Noise induced transition. Theory and applications in physics, chemistry and biology (to appear Springer-Verlag)
2. Sancho,J.M., San Miguel,M., Katz,S.L., and Gunton,J.D., Phys.Rev. A26,1589(1982)
3. Graham,R. and Schenzle,A., Phys. Rev. A25, 1731 (1982)
4. Suzuki,M., Kaneko,K., and Sasagawa,F., Prog. Theor. Phys. 65, 828 (1981)
5. Fujisaka,H. and Grossmann,S., Z. Phys. B43, 69 (1981)
6. Kai,S., Kai,T., Takata,M., Hirakawa,K., J. Phys. Soc. Japan 47, 1379 (1979)
7. Kawakubo,T., Yanagita,A., Kabashima,S., J. Phys. Soc. Japan 50,1451 (1981)
8. Gollub,J.P.,Steinman,J.F., Phys. Rev. Lett. 45,551 (1980)
9. Moss,F. and Welland,G.V., Phys. Rev. A25,3389 (1982)
10. Kabashima,S., Kogure,S.? Kawakubo,T., and Okada,T., J. Appl. Phys. 50, 6296 (1979)
11. de Kepper,P.and Horsthemke,W., C. R. Acad. Sci. Paris 287C, 251 (1978)
12. Lefever,R. and Horsthemke,W., Proc. Natl. Acad. Sci. USA 76, 2490 (1979)
13. Graham,R., Höhnerbach, Schenzle,A., Phys. Rev. Lett. 48, 1396 (1982)
14. de Gennes,P.G., The physics of liquid crystals, Clarendon Press (1974)
15. Guyon,E., Am. J. Phys. 43, 877 (1975)

16. Deuling,H.J., Solid State, Phys. Supp. 14 77 (1978)
17. Chandrasekhar,S., Liquid crystals, Cambridge University Press (1977)
18. Horsthemke,W.,Doering,C., Lefever,R. and Chi A.S.,(in preparation)
19. San Miguel,M., and Sancho,J.M., Z. Phys. B43, 361 (1981)
20. Kitahara,K., Horsthemke,W., Lefever, R.,and Inaba,Y., Prog. Theor. Phys. 64, 1233 (1980)
21. Graham,R. and Schenzle,A., Phys. Rev. A26, 1676 (1982)
22. Horsthemke,W. and lefever,R., Biophys. J. 35, 415 (1981)
23 Dubois-Violette,E., de Gennes,P.G., and Parodi,O., J. Phys. 32, 305 (1971)

NOISE IN PHYSICAL SYSTEMS AND 1/f NOISE
M. Savelli, G. Lecoy and J-P. Nougier (eds.)
© Elsevier Science Publishers B.V., 1983

ON THE DISTRIBUTION OF THE LEVEL-CROSSING TIME-INTERVALS OF RANDOM PROCESSES

T. Munakata and D. Wolf

Institut für Angewandte Physik der Universität Frankfurt a.M., FRG

A 6-states model is presented which enables to derive in explicit form approximative solutions for the probability density $P_0(\tau)$ of the level-crossing time-intervals τ and for the probability density $P_1(\tau)$ of the sum τ of two adjacent level-crossing intervals. The model has been applied to Gaussian and Rayleigh processes with different power spectra and with various level values. The theoretical results were verified by experiments. Throughout our approximations represent the experimental data with high accuracy and in almost all cases outperform the approximations known so far.

1. INTRODUCTION

Up to now the probability density $P_0(\tau)$ (or $P_1(\tau)$) of a random process for the first downward (upward) crossing of a certain level I within the infinitesimal time-interval $(t+\tau, t+\tau+d\tau)$ given an upward crossing of the level I in $(t,t+dt)$ have not been found. Since the famous work of Rice [1], who firstly introduced approximations for $P_0(\tau)$ and $P_1(\tau)$, called $Q(\tau)$ and $W(\tau)$, resp., implicit approximative solutions have been derived from integral equations [2] or by an excursion model [3]. Recently the authors have proposed a 4-states model which yields explicit approximative solutions [4].

In this paper the 4-states model is developed to a 6-states model by splitting two of the four states. This extended model leads to more precise approximations for $P_0(\tau)$ and $P_1(\tau)$.

2. THEORY

The 4-states model starts with the four signal states Z_i, $i=1,\dots,4$, characterized by their conditional probabilities S_i, resp., which are defined as follows:
Given an upward crossing of I in $(t,t+dt)$, then the process in

- Z_1 remains above the level I at least until the time $t+\tau$;
- Z_2 is found below the level I at the time $t+\tau$ after just one subsequent downward crossing;
- Z_3 is found above the level I at time $t+\tau$ after at least one subsequent upward crossing during the time-interval τ;
- Z_4 is found below the level I at time $t+\tau$ after at least two subsequent downward crossings during the time-interval τ.

With the assumptions (argument τ omitted)

$$P_0 = \frac{S_1 Q}{S_1+S_3} = \frac{QS_1}{P_+} \equiv g\cdot S_1 \tag{1}$$

and

$$P_1 = \frac{S_2 W}{S_2+S_4} = \frac{WS_2}{P_-} \equiv h\cdot S_2 , \tag{2}$$

where

$$P_+(\tau) = 1-\int_0^\tau [Q(t)-W(t)]dt \tag{3}$$

and

$$P_-(\tau) = 1-P_+(\tau) \tag{4}$$

the differential equations ($\dot{S}_i = dS_i/d\tau$)

$$\dot{S}_1(\tau) = -g(\tau)S_1(\tau) \tag{5}$$

and

$$\dot{S}_2(\tau) = g(\tau)S_1(\tau)-h(\tau)S_2(\tau) \tag{6}$$

are obtained, which lead to the explicit solutions

$$P_0^{(4)}(\tau) = g(\tau)\cdot\exp\{-G(\tau)\} \tag{7}$$

$$P_1^{(4)}(\tau) = h(\tau)\cdot\exp\{-H(\tau)\}\cdot \int_0^\tau g(t)\cdot\exp\{-G(t)+H(t)\}dt \tag{8}$$

with

$$G(\tau) = \int_0^\tau g(t)dt$$

and

$$H(\tau) = \int_0^\tau h(t)dt\,. \tag{10}$$

This 4-states model can be modified to a 6-states model assuming the states Z_3 and Z_4 to consist of two states each. The relaxation states Z_{31} and Z_{41} together with the transition states Z_{32} and Z_{42} now form the states Z_3 and Z_4, respectively. Thus

$$S_3 = S_{31} + S_{32} \tag{11}$$

and

$$S_4 = S_{41} + S_{42} \tag{12}$$

hold. As is illustrated by Figure 1 the process having arrived at a state Z_{31} or Z_{41} remains in this state for a certain time θ_3 or θ_4, resp., before the process proceeds to Z_{32} or Z_{42}. Generally θ_3 and θ_4 are random variables, for simplicity, however, they are replaced by mean values

$$\theta_3 = T_1 \ , \ \theta_4 = T_2 .$$

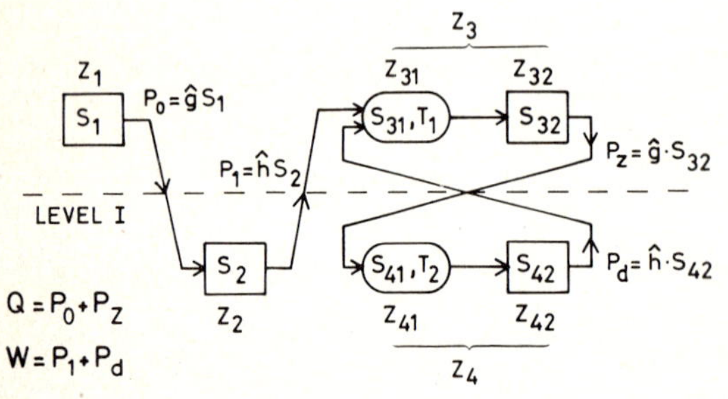

Figure 1 Scheme of the 6-states model

Then the relaxation probabilities can be written as

$$S_{31} = \int_{\tau-T_1}^{\tau} W(t)\,dt \tag{13}$$

and

$$S_{41} = \int_{\tau-T_2}^{\tau} [Q(t)-P_o(t)]\,dt. \tag{14}$$

Now replacing S_3 by S_{32} in eq. (1) and S_4 by S_{42} in eq. (2) the modified assumptions

$$P_o = \hat{g}S_1 \equiv \frac{Q}{P_+-S_{31}}\, S_1 = \frac{S_1 Q}{S_1+S_{32}} \tag{15}$$

and

$$P_1 = \hat{h}S_2 \equiv \frac{W}{P_--S_{41}}\, S_2 = \frac{S_2 W}{S_2+S_{42}} \tag{16}$$

are obtained leading to the differential equations

$$\dot{S}_1(\tau) = -\hat{g}(\tau)\, S_1(\tau) \tag{17}$$

and

$$\dot{S}_2(\tau) = \hat{g}(\tau)\, S_1(\tau) - \hat{h}(\tau)\, S_2(\tau) . \tag{18}$$

Eqs. (17) and (18) correspond to eqs. (5) and (6), resp.; therefore, the explicit solutions are

$$P_o^{(6)}(\tau;T_1) = \hat{g}(\tau;T_1)\exp\{-\hat{G}(\tau;T_1)\} \tag{19}$$

$$P_1^{(6)}(\tau;T_1,T_2) = \hat{h}(\tau;T_1,T_2)\exp -\hat{H}(\tau;T_1,T_2)\cdot \tag{20}$$

$$\cdot \int_o^{\tau} \hat{g}(t;T_1)\exp\{-\hat{G}(t;T_1)+\hat{H}(t;T_1,T_2)\}dt$$

with

$$\hat{G}(\tau;T_1) = \int_o^{\tau} \hat{g}(t;T_1)\,dt \tag{21}$$

and

$$\hat{H}(\tau;T_1,T_2) = \int_o^{\tau} \hat{h}(t;T_1,T_2)\,dt . \tag{22}$$

It should be noticed that the solution $P_o^{(6)}$ can be evaluated directly whereas $P_1^{(6)}$ is accessible via $P_o^{(6)}$ only.

Since the solutions $P_o^{(6)}(\tau;T_1)$ and $P_1^{(6)}(\tau;T_1,T_2)$ depend on the relaxation times T_1 and T_2, resp., other relationships must be known determining these parameters. For this purpose the moments of $P_o^{(6)}(\tau;T_1)$ and $P_1^{(6)}(\tau;T_1,T_2)$,

$$m_o(T_1) = \int_o^{\infty} \tau P_o^{(6)}(\tau;T_1)\,d\tau = \int_o^{\infty} \exp\{-\hat{G}(t;T_1)\}dt \tag{23}$$

and

$$m_1(T_1,T_2) = \int_o^{\infty} \tau\, P_1^{(6)}(\tau;T_1,T_2)\,d\tau$$

$$= m_o(T_1) + \int_o^{\infty} \exp\{-\hat{H}(t;T_1,T_2)\}\cdot \tag{24}$$

$$\cdot \int_o^{t} P_o^{(6)}(u;T_1,T_2)\exp\{\hat{H}(u;T_1T_2)\}du\,dt$$

are set equal to the first moments μ_o of $P_o(\tau)$ and μ_1 of $P_1(\tau)$ which - as is well known - are accessible analytically without any knowledge of $P_o(\tau)$ and $P_1(\tau)$, respectively. This can be achieved by varying T_1 and T_2 appropriately. By means of this the approximative solutions $P_o^{(6)}$ and $P_1^{(6)}$ in theory are completely defined.

Finally it should be noted that a simple relationship exists between T_1 and the crossing rate β_+ of upward crossings per unit time and the decay constant α of the exponential law which governs the asymptotic decrease of $P_o(\tau)$. This relationship

$$T_1 = \frac{\hat{g}(\infty)P_+(\infty)-\beta_+}{\hat{g}(\infty)\beta_+} \simeq \frac{\alpha P_+(\infty)-\beta_+}{\alpha\beta_+} \tag{25}$$

enables to test the solution.

Theoretical results have been evaluated numerically for Gaussian processes with unit power and for Rayleigh processes derived from them. The analysis included processes with Butterworth-type power spectra for various level values $|I| \leq 3$.

3. EXPERIMENTS

In the experiments the probability densities $P_0(\tau)$ and $P_1(\tau)$ of level crossing time-intervals were measured for Gaussian and Rayleigh random processes with various power spectra and for various values of the normalized level I. The power spectra of the Gaussian processes were shaped by Butterworth low-pass and band-pass filters of the orders 4, 7, and 8. The Rayleigh process was composed of two statistically independent Gaussian processes ξ and ζ according to

$$\eta = \sqrt{\xi^2+\zeta^2} \quad .$$

The experimental arrangement used is described in [5]. The main parts of this set-up were two noise generators, adjustable active filters, operational amplifiers, the level-clipper, the time-interval counter, a buffer memory for sequential storage of the time-interval lengths, and a digital computer for control and data processing. Additional experiments were performed by complete computer simulations. Details of the system are given in [6].

Some typical results for the densities $P_{0+}(\tau)$ and $P_{1+}(\tau)$ with Gaussian and Rayleigh processes are presented in Figures 2 and 3. For more precise notation the index "+" has been added indicating that the process crosses the level I in (t,t+dt) with positive slope. Results for $P_{0-}(\tau)$ of Rayleigh processes, crossing the level I in (t,t+dt) with negative slope, are given in [6]. The same notation is used for Q and W.

In Figure 2 the approximations $P_{0+}^{(6)}(\tau)$ and $P_{1+}^{(6)}(\tau)$ are shown together with the RICE functions Q and W in comparison with the measured data of a Gaussian process with 8th-order Butterworth low-pass spectrum. The level values I=-1,0,+1 are normalized with respect to the total power; τ is normalized with respect to the cut-off frequency ω_0. In all cases these results demonstrate clearly the coincidence of the theoretical 6-states-model-solutions and the measured values. For the level I=0 an example is shown of the asymptotic behaviour of our solutions. The solutions fit very well the observed

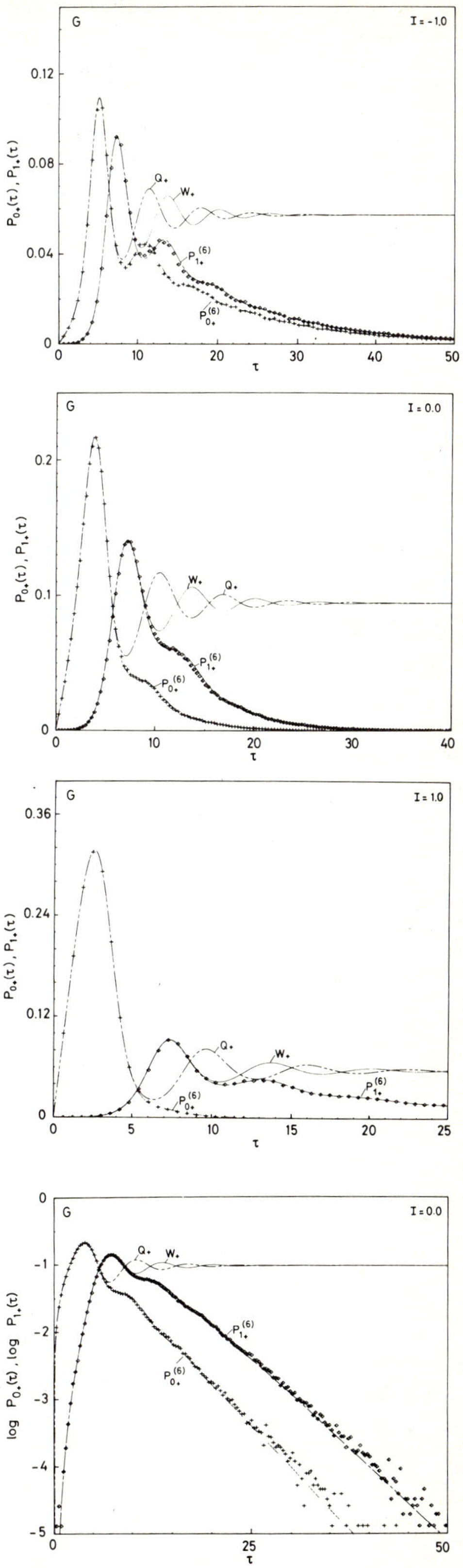

Figure 2 Probability densities $P_{0+}(\tau)$ and $P_{1+}(\tau)$ of Gaussian random processes with 8-th-order-Butterworth low-pass spectrum for levels I=-1,0,+1. Experimental values (+,◇) in comparison with approximations $Q_+, W_+, P_{0+}^{(6)}$, and $P_{1+}^{(6)}$.

decay with only small deviations. In the case of levels $|I|>1.0$ this fit becomes perfect.

In Figure 3 corresponding results are depicted which were obtained from a Rayleigh process composed of two Gaussian processes identical to that of Figure 2. Also in this case the 6-states model solutions coincide well with the experimental data.

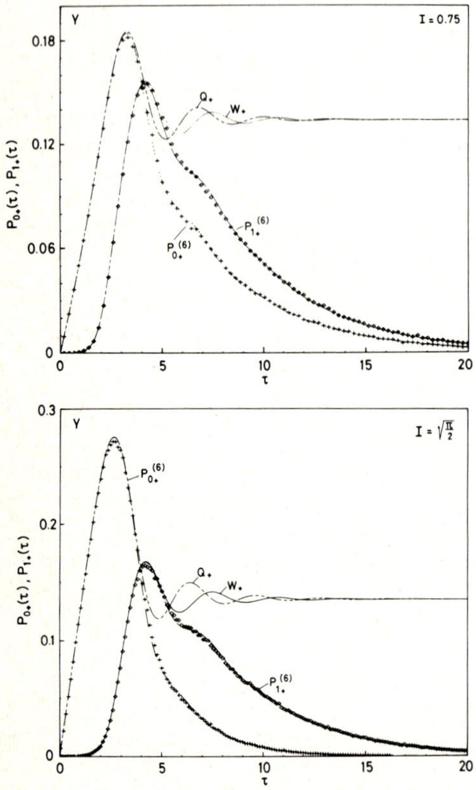

Figure 3 Probability densities $P_{0+}(\tau)$ and $P_{1+}(\tau)$ of Rayleigh random processes composed of two Gaussian processes (8th-order Butterworth low-pass) for levels $I=0.75$, $\sqrt{\pi/2}$. Experimental values (+,◊) in comparison with approximations $Q_+, W_+, P^{(6)}_{0+}$, and $P^{(6)}_{1+}$.

4. CONCLUSION

The solutions $P^{(6)}_0(\tau)$ and $P^{(6)}_1(\tau)$ obtained from the 6-states model have been verified excellently by the experiments. These approximations represent the level-crossing behaviour of Gaussian and Rayleigh random processes far better than previous approximations. Recent preliminary investigations applying the 6-states model to RICE processes indicate the representations to be of similar accuracy. Thus the 6-states model approach may be expected to become a powerful tool in solving level-crossing problems.

5. REFERENCES

[1] Rice, S.O., On the mathematical analysis of random noise, Bell Syst. Techn. J. 23 (1944) 282-332; 24 (1945) 46-156.

[2] Mc Fadden, J.A., The axis-crossing intervals of random functions, IRE Trans. Inf. Theory, IT-4 (1958) 14.

Rainal, A.J., Zero-crossing intervals of Gaussian processes, IRE Trans. Inf. Theory, IT-8 (1962) 372-378.

Rainal, A.J., Axis crossing intervals of Rayleigh-processes, Bell Syst. Techn. J. 44 (1965) 1219-1224.

Wolf, D. und Brehm, H., Die Verteilungsdichte der Zeitintervalle zwischen Nulldurchgängen bei Gaußschen stochastischen Signalen, AEÜ 27 (1973) 477-489.

Brehm, H. and Wolf, D., The distribution of level-crossing time-intervals of Gaussian random signals, AEÜ 29 (1975) 415-420.

[3] Mimaki, T. and Sato, H., Level-crossing intervals of Gaussian noise, Proc. VIth Int. Symposium on Noise in Physical Systems, Washington 1981.

[4] Munakata, T. und Wolf, D., Neue theoretische Lösungen für das Problem der Pegelüberschreitungen bei Gaußschen Prozessen, 4. Aachener Kolloquium "Theorie und Anwendung bei der Signalverarbeitung", RWTH Aachen 1981, 151-154

Munakata, T. and Wolf, D., A novel approach to the level-crossing problem of random processes, IEEE Int. Sympos. on Inform. Theory, Les Arcs (France) 1982.

[5] Wolf, D. and Brehm, H., Die Verteilungsdichte der Zeitintervalle zwischen Nulldurchgängen bei Gaußschen stochastischen Signalen, AEÜ 27 (1973) 477-489.

[6] Wolf, D., Munakata, T., und Wehhofer, J., Die Verteilungsdichte der Pegelunterschreitungszeitintervalle bei Rayleigh-Fadingkanälen, NTG-Fachberichte 84 (1983) 23-32.

The authors thank the Deutsche Forschungsgemeinschaft for financial support.

NOISE IN PHYSICAL SYSTEMS AND 1/f NOISE
M. Savelli, G. Lecoy and J-P. Nougier (eds.)
© Elsevier Science Publishers B.V., 1983

FLUCTUATIONS IN DISSIPATIVE STEADY STATES OF THIN METALLIC FILMS

A.-M. S. Tremblay and François Vidal

Département de physique, Université de Sherbrooke,
Sherbrooke, Québec J1K 2R1
Canada

It is shown that high-frequency current fluctuations induced by the application of a steady electric field to a thin metallic film provide, in certain cases, a way to determine an inelastic-scattering time in a regime where it is otherwise hardly accessible because the usual transport coefficients and equilibrium fluctuations are mainly determined by elastic scattering.

Since the following work has already appeared in print[1], we restrict ourselves, for the benifit of this audience, to a short summary of our motivation and main results and we briefly discuss one related new development.

In our opinion, there are two reasons which make the study of fluctuations in nonequilibrium systems interesting.

a) Firstly, the fundamental aspect. The measurement of fluctuations probably constitutes the most stringent test of Statistical Mechanics available. Yet there are very few experimental tests of fluctuations in nonequilibrium systems, and very often these experiments are not very well understood. 1/f noise is in such a category. One can then certainly claim that Nonequilibrium Statistical Mechanics (N.E.S.M.) rests on much less solid grounds than its equilibrium counterpart. We thus think it is important to suggest experiments which could check certain aspects of the theory.

b) Secondly, we think that if one takes the opposite point of view and assumes that N.E.S.M. is correct, then one can use the predictions of the theory to learn about properties of materials. Indeed, there is no simple fluctuation-dissipation theorem for fluctuations about nonequilibrium states (even steady-states) and hence in general, measurements of fluctuations about nonequilibrium states give results which cannot be predicted from transport measurements and the like. Again, we can take 1/f noise as an example. 1/f noise may be telling us something about trap disbributions, atomic diffusion or similar processes which are not well known because they have very little bearing on traditional measurements.

The work of ref. 1 addressed a few examples of fluctuations in nonequilibrium steady states. We summarize here the results we obtained for the problem of high frequency current fluctuations in a resistor subjected to a constant electric field. We refer to frequencies lower than the usual high frequency cutoffs but higher than those at which other physical effects tend to give 1/f fluctuations.

Our calculation applies to a metallic resistor in the residual resistivity regime where the resistance is determined by elastic (defect) scattering processes. We take the phonons into account explicitly. They provide a sink for Joule heat. Our results are obtained from a Boltzmann-Langevin type theory[2], which can also be justified with Green's function techniques. The extra nonequilibrium current fluctuations are given by,

$$S_I(\omega) \equiv \int_{-\infty}^{\infty} dt\; e^{i\omega t} \langle I(t)\, I(0)\rangle_{ne} \tag{1}$$

$$= \frac{2}{R} k_B T \left[0.54 \; \frac{eE\ell^*}{k_B T} \right]^2 \tag{2}$$

where k_B is Boltzmann's constant, T the temperature, R the sample resistance, e the electron charge, E the applied electric field and

$$\ell^* \equiv v_F^2 \, \tau_e \, \tau_o \tag{3}$$

is the inelastic mean free path. v_F is the Fermi velocity, τ_e the elastic collision time and τ_o an inelastic collision time. (See Eq.(3.24) of Ref.1 for a more precise definition). We assume that $\tau_e \ll \tau_o$, i.e. that the resistance is, to an excellent approximation, determined by τ_e and hence temperature independent.

One can show explicitly that it is possible to find experimental conditions such that phonon heating is negligible. Hence the above results apply, for example, in films of the order of 1000 Å thick and for temperatures of the order of 1 K. With $\tau_e \sim 10^{-14}$s, one could obtain a 10% effect in a field of the order of 10^{-2} V/cm. Experimental conditions are discussed in detail in Eqs. (4.22) to (4.24) of ref. 1 .

Let us now discuss how the above results bear upon the two aspects of nonequilibrium fluctuations mentioned at the beginning.

a) The result (2), if measured experimentally, would be a check of the theory of nonequilibrium fluctuations only if the result could be obtained with great precision and if the parameters entering Eq. (2) were also known independently from other sources and with a comparable precision. Indeed, the order of magnitude of the result (2) can be obtained from the following simple argument. Consider the equilibrium Nyquist-Johnson result,

$$S_I(\omega) = \frac{2}{R} k_B T \; . \tag{4}$$

In the nonequilibrium steady state, the electronic temperature will rise by an amount given roughly by the balance between the Joule heating rate and the rate at which electrons can loose their energy to the external world, this being limited by the electron-phonon inelastic collision time τ_o since the phonons are in very good contact with the thermal reservoir. More quantitatively we have

$$\frac{C_v \delta T}{\tau_o} = \frac{1}{\Omega} \frac{V^2}{R} \tag{5}$$

where C_v is the specific heat, δT the electronic temperature change, Ω the volume and V the applied voltage. Since R is temperature independent, the most simple-minded ideas about nonequilibrium fluctuations would predict that the nonequilibrium current fluctuations are given by

$$S_I(\omega) = \frac{2}{R} k_B \, \delta T \tag{6}$$

with δT determined by Eq.(5). That result is a factor 5.33 smaller than the true result. If one becomes a bit more sophisticated and tries to evaluate the temperature rise from a more realistic formula than Eq.(5) one must first solve the Boltzmann equation appropriate to this problem. One then finds that the steady-state distribution <u>cannot</u> be described by a temperature. If one insists in calculating a temperature from the formula

$$\delta T = \frac{1}{C_v} \int \frac{d^3k}{(2\pi)^3} \, \varepsilon_k \, \delta f_k \tag{7}$$

where ε_k is the energy in state k and δf_k the deviation from the equilibrium distribution function, one finds that the result Eq.(2) is a factor 1.27 larger than what may be obtained by substituting Eq.(7) in Eq.(6). It is in a sense this factor of 1.27 which is a real test of N.E.S.M. and, hence, one must measure the result Eq.(2) to at least 20% accuracy and know independently all the parameters entering that equation to claim that it is a check of N.E.S.M.

b) Assuming Statistical Mechanics to be correct and now considering Eq.(2) from the point of view of measuring material properties, we see that we have here a most interesting result. Indeed, Eq.(2) shows that measuring nonequilibrium current fluctuations gives a way of obtaining the inelastic relaxation time τ_o in a regime where it is not otherwise easily accessible because the usual transport measurements are sensitive to properties, when $\tau_e \ll \tau_o$, which are determined by elastic scattering. This inelastic time is relevant to at least the following three fields of Condensed Matter Physics:

i) In nonequilibrium superconductivity, the normal-state inelastic scattering time τ_o appears as a parameter in many of the results and hence it can be indirectly measured in the superconducting state[3]. Our calculation provides a way of obtaining this inelastic scattering time directly in the normal state.

ii) In the field of low-temperature refrigeration, it has been known for some time that in sufficiently small metallic systems, the thermal impedance between electrons and phonons becomes an important mechanism limiting the efficiency of heat transfer[4]. If we define the thermal resistance as $R_T = \delta T/\dot{Q}$, where $\dot{Q} = \sigma E^2 V$ (σ is the electrical conductivity) one finds that nonequilibrium current fluctuations provide a way of measuring δT, and hence the thermal resistance between electrons and phonons. Given the widespread use of metal sinters in the refrigeration process, this electron-phonon thermal resistance is very relevant for low temperature experimentation. The thermal resistance can also be measured by different techniques[4].

iii) Finally, it is interesting to speculate that if results analogous to those we have derived hold in the localized regime of two-dimensional metallic films, high-frequency nonequilibrium current fluctuations may provide a way to independently determine the inelastic scattering rate which arises as a parameter determining the logarithmic temperature dependence of the resistance in the theory of localization in two dimensions[5]. In view of the existence of a competing theory[6] for this logarithmic temperature dependence of resistance, it may be important to have this additional way to estimate the inelastic relaxation rate to remove one adjustable parameter in the theory.

To conclude, we should like to point out that the calculation we described has recently been extended non-perturbatively to the high-field limit ($eE\ell^*/k_BT \gg 1$) by M. Arai[7]. He finds that δT in Eq.(6) is proportional to $E^{2/5}$, a highly non-ohmic effect which would be most interesting to measure. Note that in the limit $T \to 0$ one is always in the high field limit and hence, as Arai points out, his result gives a fundamental noise limit to many electrical measurements.

ACKNOWLEDGMENTS

We should like to thank B. Pannetier for very useful conversations. Discussions with J.D.N. Cheeke, J.P. Harrison and T. Lemberger are also gratefully acknowledged. This work was supported by the National Science and Engineering Research Council through an individual operating grant and through the program "Attaché de recherche".

REFERENCES

[1] Tremblay, A.M.S., Vidal, F., Fluctuations in dissipative steady states of thin metallic films, Phys. Rev. B25 (1982) 7562.

[2] Gantsevich, S.V., Gurevich, V.L., Katilius, R., Theory of fluctuations in nonequilibrium electron gas, Riv. Nuevo Cimento 2 (1979) 1.

[3] Chang, J.J., Properties of Nonequilibrium Superconductors: A Kinetic Equation Approach. Gray, K.E. (ed), Nonequilibrium Superconductivity, Phonons and Kapitza Boundaries (Plenum, New York, 1981).

[4] Harrison, J.P., Review paper: Heat Transfer Between Liquid Helium and Solids Below 100 mK, J. Low Temp. Phys. 37 (1979) 467.

[5] Abrahams, E., Anderson, P.W., Licciardello, D.C., Ramakrishnan, T.V., Scaling Theory of Localization: Absence of Quantum Diffusion in Two Dimensions, Phys. Rev. Lett.42 (1979) 673. Momentum and energy relaxation rates should not, however, be confused.

[6] Altshuler, B.L., Khmel'nitzkii, D., Larkin, A.I., Lee, P.A., Magnetoresistance and Hall effect in a disordered two-dimensional electron gas, Phys. Rev. B22 (1980) 5142.

[7] Arai, M., A Fundamental Noise Limit for Biased Resistors at Low Temperatures, Appl. Phys. Lett. in press.

NOISE IN PHYSICAL SYSTEMS AND 1/f NOISE
M. Savelli, G. Lecoy and J-P. Nougier (eds.)
© *Elsevier Science Publishers B.V., 1983*

CALCULATION OF VACANCY NOISE IN MONOCRYSTALLINE METALS

Wolfgang Michel
Institut für Theoretische Physik, Universität Stuttgart, D-7000 Stuttgart

Hermann Grabert[1]
Institut für Theoretische Physik, Universität Basel, CH-4056 Basel

We study nonequilibrium voltage fluctuations in metal films caused by the coupling between the electric current and the vacancy diffusion mode which has a major impact on the excess noise close to the melting point. Our approach is based on a nonlinear Langevin model incorporating the conservation laws and the fluctuation-dissipation theorem. An expression for the nonequilibrium power spectrum in the presence of a steady electric current is obtained. Using boundary conditions appropriate to monocrystalline samples, the power spectrum is determined analytically for thin plates and numerically for samples with quadratic cross sections. The relation of the spectrum to vacancy properties is emphasized.

1. INTRODUCTION

Voltage fluctuations in a metal film carrying a constant electric current show excess noise typically enhanced at low frequencies. Since vacancies contribute to the electric resistivity and since their concentration increases exponentially with temperature, the excess noise of metal films at temperatures close to their melting point is strongly influenced by the coupling between the electric current and the vacancy diffusion mode. In the following we shall focus on this special type of excess noise referred to as vacancy noise, henceforth. The power spectrum of vacancy noise is particularly interesting, since it allows for a determination of vacancy properties, e.g., the formation and migration enthalpies [1,2].

A complete treatment of low frequency noise in a metal should include all slow modes of the system, e.g., the heat and sound modes. One may argue that the fast sound modes can be disregarded and that we have to study the influence of the slow diffusive modes only. In a dislocation-free metal these are the heat and vacancy modes. Disregarding the heat mode in an isothermal approximation, we arrive at a two variable model for the charge and vacancy densities. Such a model will be studied in this work.

The paper is organized as follows. In Sec. 2 we present a nonlinear Langevin model incorporating the conservation laws and the fluctuation-dissipation theorem. An expression for the nonequilibrium power spectrum in the presence of a steady electric current is derived. The spectrum consists of the familiar Johnson noise and a frequency dependent excess noise which can be expressed in terms of the vacancy correlation function.

Sec. 3 is concerned with the experimentally important case of a thin metal film. We obtain an analytical expression for the vacancy noise taking into account boundary conditions appropriate to monocrystalline thin metal plates. The connection of the power spectrum with vacancy properties is discussed. In Sec. 4 we show that samples with rectangular cross sections behave quite similar to thin plates. For the nontrivial case of a sample with quadratic cross section the calculation of the power spectrum is reduced analytically to one remaining integration which is evaluated numerically. Finally, Sec. 5 contains our conclusions.

2. NONLINEAR LANGEVIN MODEL

A phenomenological model describing low frequency excitations in metals can be obtained by following the lines of reasoning traced by Landau and Lifshitz in their theory of fluctuations in fluid dynamics [3]. For cubic crystals in the isothermal and isobaric approximations the charge density $\rho(\vec{x},t)$ and the vacancy density[2] $n(\vec{x},t)$ obey the conservation laws

$$\dot{\rho} = - \operatorname{div} \vec{j} \quad (1)$$

$$\dot{n} = - \operatorname{div} \vec{j}_n \quad (2)$$

where the electric current density $j(\vec{x},t)$ and the vacancy flux $\vec{j}_n(\vec{x},t)$ are given by the constitutive equations

$$\vec{j} = -\sigma \operatorname{grad} \phi + \vec{h} \quad (3)$$

$$\vec{j}_n = -\zeta \operatorname{grad} \mu + \vec{k} \quad (4)$$

Here ϕ is the electric potential related to the electric field by $\vec{E} = - \operatorname{grad} \phi$, μ is the chemical potential of the vacancies, σ is the electric conductivity, and ζ is related to the vacancy diffusion constant D_n by

$$D_n = \zeta \frac{\partial\mu}{\partial n} \tag{5}$$

Finally, $\vec{h}(\vec{x},t)$ and $\vec{k}(\vec{x},t)$ are Gaussian random forces with vanishing mean, and their correlations are given by

$$<h_i(\vec{x},t)h_j(\vec{y},s)> = 2k_B T\sigma\delta_{ij}\delta(\vec{x}-\vec{y})\delta(t-s)$$

$$<h_i(\vec{x},t)k_j(\vec{y},s)> = 0 \tag{6}$$

$$<k_i(\vec{x},t)k_j(\vec{y},s)> = 2k_B T\zeta\delta_{ij}\delta(\vec{x}-\vec{y})\delta(t-s)$$

Perfect screening of the system was assumed to arrive at (3). From (5) we see that (4) may be written

$$\vec{j}_n = - D_n \text{ grad } n + \vec{k} \tag{7}$$

Disregarding the random forces, the equations (1-3, 7) allow for a stationary solution with a constant electric current $\vec{j}_o$ and a constant vacancy density n_o. The solution satisfies the charge neutrality condition $\rho_o = 0$ and there is no steady vacancy flux[3]. In order to determine the fluctuations about the steady state, the Langevin equations (1-3,7) can be linearized in the deviations of the variables from their steady state values. Such a procedure is appropriate to systems which are not near critical points or instabilities [4,5] , and it leads to linear Langevin equations for the fluctuations of the form

$$\delta\dot{\rho} = \text{div } (\sigma \text{ grad } \delta\phi + \vec{j}_o\beta\delta n - \vec{h}) \tag{8}$$

$$\delta\dot{n} = D_n \, \Delta \, \delta n - \text{div } \vec{k} \tag{9}$$

where the coefficient $\beta = -(1/\sigma)\, \partial\sigma/\partial n$ characterizes the dependence of the electric conductivity σ upon the vacancy density n . Because of this state dependence of the transport coefficient σ , the charge fluctuations are coupled to the vacancy fluctuations. It will be noted that this coupling is a typical nonequilibrium effect since it vanishes with j_o. For low frequencies ($\omega << 4\pi\sigma$) the LHS of (8) can be neglected, leading to a quasi stationary model where

$$\text{div } \delta\vec{j} = -\text{div}(\sigma \text{grad } \delta\phi + \vec{j}_o\beta\delta n - \vec{h}) = 0 \tag{10}$$

This implies that the normal component of the fluctuating current has to vanish on a free metal surface, i.e., $\delta\vec{j}_\perp = 0$.

In the sequel we consider a sample of rectangular shape. The steady current j_o flows in the x direction. The quantity of interest is now the noise voltage $\delta V(t)$ between two metal cross sections at $x = 0$ and $x = L_x$ of area $A = L_y L_z$. Using

$$\delta V(t) = \overline{\delta\phi(x = L_x, y, z)} - \overline{\delta\phi(x = 0, y, z)}$$

$$= - \frac{L_x}{\Omega} \int_\Omega d^3x \, \delta E_x(\vec{x},t) \tag{11}$$

where the bar denotes an average over the metal cross section, and where $\Omega = L_x L_y L_z$ is the volume between the cross sections which are connected with the measuring device, the voltage correlation function may be written

$$<\delta V(t)\delta V(o)> = \frac{L_x^2}{\Omega^2} \iint_\Omega d^3x d^3x' <\delta E_x(\vec{x},t)\delta E_x(\vec{x}',0)> \tag{12}$$

To determine the electric field correlation function we start by noting that (10) gives upon averaging over the metal cross section

$$\frac{\partial}{\partial x}\overline{\delta j_x} = \frac{\partial}{\partial x}\left\{\sigma\overline{\delta E_x} - j_o\beta\overline{\delta n} + \overline{h_x}\right\} = 0 \tag{13}$$

where we have made use of the fact that $\delta j_\perp$ vanishes on a free surface. From (13) we find

$$\overline{\delta E_x} = \frac{j_o\beta}{\sigma}\,\overline{\delta n} - \frac{1}{\sigma}\overline{h_x} \tag{14}$$

which combines with (11) to yield

$$<\delta V(t)\delta V(o)> = \frac{L_x^2}{\Omega^2\sigma^2} \iint_\Omega d^3x d^3x'$$

$$\left\{<h_x(\vec{x},t)h_x(\vec{x}',0)> + j_o^2\beta^2<\delta n(\vec{x},t)\delta n(\vec{x}',0)>\right\}$$

$$= 2k_B TR\delta(t) + j_o^2 \frac{\beta^2}{A^2\sigma^2} <\delta N(t)\delta N(0)> \tag{15}$$

where we have used (6). Here $R = L_x/\sigma A$ is the resistance of the piece of metal of volume Ω between the contacts and

$$\delta N(t) = \int_\Omega d^3x \, \delta n(\vec{x},t) \tag{16}$$

is the deviation of the number of vacancies in the volume Ω from its steady state value. The first term on the RHS of (15) represents the usual Johnson noise while the second term gives the nonequilibrium vacancy noise.

By Fourier transforming (15), we obtain the power spectrum[4]

$$S_V(\omega) = 2k_B TR + S_V^{neq}(\omega) \tag{17}$$

where the nonequilibrium spectrum

$$S_V^{neq}(\omega) = (j_o\beta/A\sigma)^2 S_N(\omega) = \beta^2 V_o^2 \frac{1}{\Omega^2} S_N(\omega) \tag{18}$$

is given in terms of the spectrum

$$S_N(\omega) = \int_{-\infty}^{+\infty} dt \, e^{-i\omega t} <\delta N(t)\delta N(0)> \tag{19}$$

of vacancy number fluctuations. $V_o = R\, j_o A$ is the average voltage between the contacts.

Resistance fluctuations due to vacancy fluctuations are very small. In order to detect vacancy noise the current density has to be very high and the sample has to be small. These requirements are met with thin metal films on a substrat removing the Joule heat.

3. VACANCY NOISE IN THIN METAL FILMS

In this section we consider a thin monocrystalline metal film. Since there are no dislocations and grain boundaries, the only sinks and sources for vacancies are at the surface. If the thickness $L_y = d$ of the film is much smaller than the other dimensions, only the surfaces at $y = 0$ and $y = d$ must be taken into account and the problem becomes essentially one-dimensional.

The formation of vacancies is a microscopic process not directly accessible to our continuum model. Boundary conditions for the macroscopic diffusion process can however be set up in the following way. Assume that on the average within the infinitesimal time interval dt a surface of area a emits

$$dN = \nu a dt \tag{20}$$

vacancies which are created at a small distance ε from the surface. The vacancies diffuse through the metal until they reach a surface where they are absorbed. The creation rate ν will be determined below in such a way that the vacancy fluctuations become independent of ε if ε tends to 0.

For simplicity, let us first consider a single vacancy starting at $t = 0$ from its point of creation on the $y = \varepsilon$ plane. The probability $p(y,t)dt$ to find the vacancy at time t in the interal $[y,y+dy]$ is then given by the solution of the one-dimensional diffusion equation

$$\frac{\partial}{\partial t} p(y,t) = D_n \frac{\partial^2}{\partial y^2} p(y,t) \tag{21}$$

with initial condition $p(y,0) = \delta(y-\varepsilon)$ and absorptive boundary conditions $p(0,t)=p(d,t)=0$ at the surfaces. $p(y,t)$ can be determined by the method of images. In fact, we shall need only the probability

$$p(t) = \int_0^d dy\ p(y,t) \tag{22}$$

to find the vacancy anywhere in the sample. $p(t)$ is related to the lifetime distribution $w(\tau)$ of the vacancies by

$$w(\tau) = -\frac{d}{d\tau} p(\tau) \tag{23}$$

Now, the Laplace transform of $p(t)$, i.e.,

$$p(s) = \int_0^\infty dt\ e^{-st}\ p(t)\ , \tag{24}$$

is found to be

$$p(s) = \frac{1}{s}\left\{1 - \frac{\cosh[\sqrt{s/D_n}(\varepsilon-d/2)]}{\cosh[\sqrt{s/D_n}\ d/2]}\right\} \tag{25}$$

from which we obtain for the Laplace transformed lifetime distribution

$$w(s) = 1-sp(s) = \frac{\cosh[\sqrt{s/D_n}(\varepsilon-d/2)]}{\cosh[\sqrt{s/D_n}\ d/2]} \tag{26}$$

Hence, the average lifetime

$$\langle\tau\rangle = \int_0^\infty d\tau\ \tau\ w(\tau) = \int_0^\infty d\tau\ p(\tau) = p(s=0) \tag{27}$$

is given by

$$\langle\tau\rangle = \varepsilon(d-\varepsilon)/2D_n \tag{28}$$

Clearly, the same lifetime statistics is obtaines for vacancies created at $y = d - \varepsilon$.

Now, consider a thin plate of volume $\Omega = ad$. In a steady state the average number of vacancies created at the surface has to balance the average number of absorbed vacancies. Hence

$$2a\nu = \langle N\rangle/\langle\tau\rangle = \bar{n}\Omega/\langle\tau\rangle \tag{29}$$

where $\bar{n} = \langle N\rangle/\Omega$ is the average density of vacancies. Using (28), we obtain for small ε

$$\nu = \bar{n}\ D_n/\varepsilon \tag{30}$$

Clearly, in order to maintain a finite vacancy density $\bar{n}$, the creation rate ν has to increase in the limit $\varepsilon\to 0$, since most vacancies will be re-absorbed by the emitting surface immediately. The rate (30) is however independent of the sample size, as it should be since the formation of vacancies is a local microscopic process.[5]

We shall assume that the vacancies are created independently. Then the process is Poissonian and the variance of vacancy fluctuations is related to the average number of vacancies by

$$\langle(\delta N)^2\rangle = \langle N\rangle \tag{31}$$

To determine the decay of vacancy fluctuations we first note that a vacancy created at time $s < 0$ will still be in the sample at time $t > 0$ with probability $p(t-s)$. Hence, a vacancy present at time $t_0 = 0$ which has been created at an unknown earlier time s will still be present at time t with probability

$$\pi(t) = \frac{\int_{-\infty}^{0} ds\ p(t-s)}{\int_{-\infty}^{0} ds\ p(-s)} = \frac{1}{\langle\tau\rangle}\int_t^\infty d\tau\ \ p(\tau) \tag{32}$$

where (27) has been used. The function (32) determines the decay of vacancy fluctuations, and we obtain in view of (31)

$$S_N(t) = \langle\delta N(t)\delta N(0)\rangle = \frac{\langle N\rangle}{\langle\tau\rangle}\int_{|t|}^\infty d\tau\ p(\tau)$$

$$= \frac{<N>}{<\tau>} \int_{|t|}^{\infty} d\tau \ (\tau - |t|) \ w(\tau) \qquad (33)$$

where (23) has been used to arrive at the second line. The Fourier transform of (33) gives the spectrum of vacancy fluctuations

$$S_N(\omega) = \frac{<N>}{<\tau>} \frac{2}{\omega^2} \left\{ 1 - \int_0^{\infty} d\tau \cos(\omega\tau) \ w(\tau) \right\} \qquad (34)$$

Now, by virtue of (26) we have

$$\int_0^{\infty} d\tau \cos(\omega\tau) \ w(\tau) = \mathrm{Re} \ w(s = i\omega) \qquad (35)$$

$$= 2 \ \frac{\cosh \Delta \cosh\sqrt{\bar\omega}/2 \cos\sqrt{\bar\omega\varepsilon/d} + \sin\Delta \sin\sqrt{\bar\omega}/2 \cosh\sqrt{\bar\omega\varepsilon/d}}{\cosh\sqrt{\bar\omega} + \cos\sqrt{\bar\omega}}$$

where $\Delta = \sqrt{\bar\omega\varepsilon/d} - \sqrt{\bar\omega}/2$. We also have passed to the scaled dimensionless frequency $\bar\omega = \pi^2 \tau_o |\omega| / 2$, where $\tau_o = d^2/\pi^2 D_n$ is the time corresponding to the lowest eigenvalue of the diffusion equation (21). Inserting (28) and (35) into (34) gives in the limit $\varepsilon \to 0$

$$S_N(\omega) = \frac{<N>d^2}{D_n} \frac{F_-(\sqrt{\bar\omega})}{\bar\omega^{3/2}} \equiv \frac{<N>d^2}{D_n} \ I(\bar\omega) \qquad (36)$$

where we have introduced the function

$$F_{\pm}(x) = \frac{\sinh x \pm \sin x}{\cosh x + \cos x} \qquad (37)$$

Finally, by virtue of (18), we obtain the desired expression for the nonequilibrium noise spectrum[4]

$$S_V^{neq}(\omega) = \left(\frac{\pi\beta V_o}{\Omega}\right)^2 <N> \tau_o \ I(\bar\omega) \qquad (38)$$

We note that the same spectrum can be obtained in a more straight forward manner by solving the Langevin equation (9) for the vacancy density fluctuations δn with absorbing boundary conditions on the surface. The present calculation relates our phenomenological Langevin model to the method of vacancy pulses.

The spectrum is depicted in Fig. 1. It has a characteristic knee at a frequency ω_K . Defining the knee position as the intersection of the asymptotes for large and small frequencies, i.e.,

$$S_V^{neq}(\omega) = \left(\frac{\pi\beta V_o}{\Omega}\right)^2 <N> \tau_o \begin{cases} 1/6 & \text{for } \omega \ll \omega_K \\ \bar\omega^{-3/2} & \text{for } \omega \gg \omega_K \end{cases} \qquad (39)$$

We find for the knee frequency ω_K and the noise power S_K at the knee

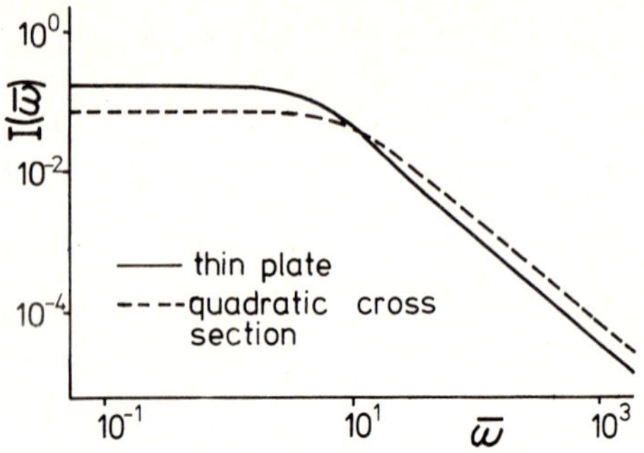

Fig. 1: Vacancy Noise $I(\bar\omega)$, scaled units

$$\omega_K = 2\sqrt[3]{36}/\pi^2 \tau_o$$
$$S_K = (\pi\beta V_o/\Omega)^2 <N> \tau_o/6 \qquad (40)$$

The temperature dependence of the knee position can be related to vacancy properties. The number of vacancies may be written [6]

$$<N> = \Omega n_A c(T) \qquad (41)$$

where n_A is the density of atoms and

$$c(T) = \exp(S^F/k_B) \ \exp(-H^F/k_B T) \qquad (42)$$

the concentration of vacancies. S^F and H^F are the formation entropy and enthalpy, respectively. For cubic crystals the vacancy diffusion constant may be written

$$D_n(T) = \ell^2 \nu^* \exp(-H^M/k_B T) \qquad (43)$$

where ℓ is the lattice spacing, ν^* an attempt frequency, and H^M the migration enthalpy of vacancies. Since $\tau_o \propto D_n^{-1}$, the temperature dependence of the knee frequency ω_K is dominated by the factor $\exp(-H^M/k_B T)$ which allows for a determination of H^M. The power spectrum S_K is proportional to $<N>/D_n$ leading to a temperature dependence according to $\exp[(H^M - H^F)/k_B T]$. This allows for a determination of $H^F - H^M$.

4. VACANCY NOISE IN SAMPLES WITH QUADRATIC CROSS SECTION

In rectangular samples the lifetime of vacancies is essentially determined by the smallest dimension of the sample since only few vacancies will have a chance to reach the other surfaces. This leads to the one-dimensional behaviour discussed previously. Two-dimensional behaviour can be expected if the cross dimensions L_y and L_z are of comparable size and, in particular, for quadratic cross sections where $L_y = L_z = d$. This case will be examined now.

We shall assume again that vacancies emitted by the surface are created at a small distance ε form the surface. Now consider a vacancy starting at time $t = 0$ from $y = \varepsilon$ and $z = z_o$ where $\varepsilon < z_o < d-\varepsilon$. The probability $\rho(\varepsilon, z_o, t)$ to find the vacancy at time t anywhere in the sample is then given by

$$\rho(\varepsilon, z_o, t) = p(\varepsilon, t)p(z_o, t) \tag{44}$$

where $p(\varepsilon, t)$ is the solution of the one-dimensional diffusion equation (21) with the initial and boundary conditions given there. The first factor on the RHS of (44) is the probability that the vacancy avoids absorption by the surfaces at $y = 0$ and $y = d$ while the second factor is the same quantity for the surfaces at $z = 0$ and $z = d$. Averaging over the z_o coordinate, we obtain the probability to find a vacancy emitted at $t = 0$ anywhere in the sample at time t , i.e.,

$$\rho(\varepsilon, t) = \frac{p(\varepsilon, t)}{d - 2\varepsilon} \int_{\varepsilon}^{d-\varepsilon} dz_o p(z_o, t) \ . \tag{45}$$

To proceed, we use basically the same method as in the previous section, however, the calculations are more involved now. (25) and (45) lead to an integral representation for the Laplace transform $\rho(\varepsilon, s)$, which by virtue of $\langle\tau\rangle = \rho(\varepsilon, s = 0)$ yields the average lifetime for small ε

$$\langle\tau\rangle = \varepsilon d/4D_n \ . \tag{46}$$

This is half of the average lifetime (28) in a plate of thickness d . Since the surface to volume ratio is now doubled, the balance equation (29) leads to the same vacancy creation rate ν given in (30).

Using the Laplace transform $\rho(\varepsilon,s)$, we find for the spectrum of vacancy fluctuations in the limit $\varepsilon \to 0$

$$S_N(\omega) = \frac{\langle N\rangle d^2}{D_n} I(\bar\omega) \tag{47}$$

where

$$I(\bar\omega) = \int_0^\infty \frac{dx}{\pi} \left\{ \frac{F_-(u)F_-(v) + \mathrm{sign}(x-\bar\omega/2)F_+(u)F_+(v)}{|x^2 - (\bar\omega/2)^2|^{3/2}} + \frac{F_+(u)u - \mathrm{sign}(x-\bar\omega/2)F_+(v)v}{\bar\omega|x^2 - (\bar\omega/2)^2|} \right\} \tag{48}$$

Here $\bar\omega = |\omega| d^2/2D_n$ is the same dimensionless frequency used earlier, $F_\pm$ are the functions defined in (37) and u,v are given by $u = |x - \bar\omega/2|^{1/2}$, $v = |x + \bar\omega/2|^{1/2}$.

The numerical evaluation of the integral (48) is depicted in Fig. 1. Compared with the spectrum of a plate, the knee frequency is shifted towards higher frequencies and the noise power is reduced at low frequencies. This is a consequence of the fact that both spectra satisfy the sum rule

$$\int_{-\infty}^{\infty} \frac{d\omega}{2\pi} S_N(\omega) = \langle(\delta N)^2\rangle = \langle N\rangle \ . \tag{49}$$

The temperature dependence of the spectrum is related to vacancy properties in precisely the same way as discussed earlier.

5. CONCLUSIONS

We have investigated vacancy noise in dislocation-free metals for different geometries using a phenomenological diffusion model supplemented with appropriate boundary conditions. The relation of this model to the method of single vacancy pulses has been explained. For rectangular samples the spectrum has always the same shape characterized by a knee. These findings can also be extended to samples with circular cross section.
It is hoped that the present approach proves useful for the study of other types of excess noise too.

ACKNOWLEDGEMENTS

We wish to thank Dr. H. Stoll for many helpful discussions on the problem of vacancy noise. One of us (H.G.) acknowledges partial support by the Swiss National Science Foundation.

NOTES

1. On leave from Institut für Theoretische Physik, Universität Stuttgart, D-7000 Stuttgart 80
2. Strictly speaking the hydrodynamic variable n is the density of vacancies minus the density of interstitial atoms. Vacancies and interstitials can equilibrate locally whereby the number difference is conserved. For metals, however, the concentration of interstitial atoms is negligible.
3. We could incorporate into our model a cross coefficient leading to a steady vacancy flux due to a preferred direction of vacancy diffusion in the presence of an electric field. Such an average drift does not affect the excess noise spectrum and will be disregarded.
4. We use double sided power spectra while many authors prefer single sided spectra defined

for positive frequencies only. The latter are obtained by multiplying our results by 2.

5. In the limit $\varepsilon \to 0$ the boundary conditions used here are in fact equivalent to the boundary condition $n = \bar{n}$, that means absorbing boundary conditions for the fluctuations δn.

REFERENCES

[1] Celasca, M., Fiorillo, F., Mazzetti, P., Phys. Rev. Lett. 36, 38 (1976)

[2] Stoll, H., J. Appl. Phys. 30A, 117 (1983); and article in this volume

[3] Landau, L., Lifshitz, E.M., Fluid Mechanics, (Pergamon Press, Oxford, 1958)

[4] Grabert, H., J. Stat. Phys. 26, 113 (1981)

[5] Tremblay, A.-M.S., Arai, M., Siggia, E.P., Phys. Rev. 23A, 1451, (1981)

[6] Seeger, A., Mehrer, H., in: Vacancies and Interstitials in Metals, Seeger, A., Schumacher, D., Schilling, W., Diehl, J. (eds.) (North Holland, Amsterdam, 1970)

NOISE IN PHYSICAL SYSTEMS AND 1/f NOISE
M. Savelli, G. Lecoy and J-P. Nougier (eds.)
© Elsevier Science Publishers B.V., 1983

LIFETIME OF A METASTABLE STATE AT WEAK NOISE

Peter Talkner, Dietrich Ryter

Institut für Physik, Klingelbergstr. 82
CH-4056 Basel

The mean time for a trajectory of a randomly perturbed system to leave a domain of attraction is determined in leading order of weak noise, Considerable simplifications are obtained if a WKB type expansion applies to the solution of the stationary Fokker Planck equation.

1. INTRODUCTION

In the study of various equilibrium and nonequilibrium phenomena such as first order phase transitions, chemical reactions, optical bistability and electronic systems, the lifetime of a metastable state plays a decisive role: e.g. it determines nucleation rates, chemical reaction rates, the quality of optical switches and of electronic systems. Often it determines the long time behaviour of the system.

The metastability of a state arises as an interplay of nonlinear deterministic and weak stochastic forces. Under the influence of the deterministic force alone the metastable state would be one of several stable states. The stochastic force is negligible during most of the time, except for those rare events where the force becomes large enough to drive the system into another state. It is quite evident that the rate of these transitions is just the inverse of the mean lifetime of the metastable state, and one may show that the lifetime is exponentially distributed at weak noise.

Such rates were evaluated by Kramers [1] for a Brownian particle moving in a one-dimensional potential. Generalizations to Brownian motion in higher dimensions are due to Landauer et al.[2] and Langer [3]. In the considered models the metastable states are always point attractors of the deterministic forces; moreover the systems are supposed to reach thermal equilibrium for $t \to \infty$.

Under even stronger restrictions, the path integral method was used to calculate transition rates [4,5]. An attemt to reformulate and generalize Kramers' rate calculation was recently presented by Gardiner [6].

A different approach is to calculate the mean lifetime of the metastable state rather than the rate. In this case the problem can be formulated without any special assumptions about the metastable state or about the probable transition path; even the exit points from the domain of attraction and the stationary distribution are not a priori restricted. The domain of attraction of the metastable state is conveniently imagined to be surrounded by an absorbing boundary. Then the mean time of absorption is just half the lifetime of the metastable state. For a continuous Markov process i.e. with stochastic forces proportional to Gaussian white noise, the mean time of absorption is the solution of a boundary value problem of an inhomogeneous second order differential equation [7,8]. In one dimension it can be solved exactly for an arbitrary strength of the stochastic force [7]. In higher dimensions analytical solutions are not usually known and the standard numerical procedures inappropriate for small noise. However, the smallness of the noise which is inherent in the notion of metastability can be utilized to reduce the problem to first finding the solution of the homogeneous differential equation on a thin boundary layer and second solving the stationary Fokker Planck equation on the domain of attraction of the metastable state [9]. The general solution of the boundary layer problem was given recently by the present authors [10].

In this paper the expression for the mean exit time is briefly rederived in Sect 2, and in Sect 3 more explicit results are obtained by use of a WKB-type solution of the stationary Fokker Planck equation. This kind of solution reduces all occurring integrals to the saddle-point type, moreover it indicates where the transitions preferably take place.

2. THE MEAN ABSORPTION TIME

We suppose that a set of first order autonomous differential equations

$$\dot{x}^i = K^i(\underset{\sim}{x}), \quad \underset{\sim}{x} = (x^1, x^2, \dots x^n) \in \Gamma \tag{1}$$

describes the deterministic motion of the system in configuration space Γ, and that this set of differential equations has a connected but otherwise arbitrary attractor with a domain of attraction Ω smaller than Γ. The boundary $\partial\Omega$ is supposed to be smooth. If the system is perturbed by white noise the duration of stay within Ω is generally finite, even if the noise is arbitrarily weak.

The perturbed motion is described by the Fokker Planck operator L

$$L = -\partial_i K^i(\underset{\sim}{x}) + \frac{1}{2}\varepsilon\,\partial_i\partial_j D^{ij}(\underset{\sim}{x}) \tag{2a}$$

with the adjoint

$$L^+ = K^i(\underset{\sim}{x})\,\partial_i + \frac{1}{2}\varepsilon D^{ij}(\underset{\sim}{x})\,\partial_i\partial_j \tag{2b}$$

where $\varepsilon D^{ij}(\underset{\sim}{x})$ is the diffusion matrix, $D^{ij}(\underset{\sim}{x})$ is bounded and ε measuring the strength of the noise is positive and small.

The mean time $t(\underset{\sim}{x})$ at which a trajectory starting at $\underset{\sim}{x} \in \Omega$ reaches the boundary $\partial\Omega$ for the first time is given by [8]

$$L^+ t = -1, \quad t(\underset{\sim}{x}) = 0 \quad \text{for } \underset{\sim}{x} \in \partial\Omega \tag{3}$$

By integrating eq.(3) with a solution w of the stationary Fokker Planck equation

$$Lw = 0 \tag{4}$$

which is integrable on Ω one obtains by the Gaussian theorem

$$\frac{\varepsilon}{2}\int_{\partial\Omega} dS_i\, w\, D^{ij}\,\partial_j t = -\int_{\Omega} d^n x\, w \tag{5}$$

where $d\underset{\sim}{S}$ is the oriented surface element on $\partial\Omega$. Note that there is no boundary condition imposed on w and that consequently w is not uniquely specified.

For small noise ($\varepsilon \to 0$) a trajectory starting within Ω will typically first approach the attractor and stay within its neighbourhood for a long time compared with time constants of the deterministic motion, until an occasional fluctuation drives it to the boundary. Hence, the mean absorption time $t(\underset{\sim}{x})$ assumes the same large value T everywhere in Ω, except for a thin layer $\Delta\Omega$ along the boundary where the small noise is still sufficient to cause a direct exit. Accordingly, one may define a function $f(\underset{\sim}{x})$ which is unity in the inner part of Ω:

$$t(x) = Tf(x), \quad f(x)=0 \text{ for } x\in\partial\Omega \text{ and } f(x)\approx 1 \text{ for } \underset{\sim}{x} \in \Omega - \Delta\Omega \tag{6}$$

Since T is exponentially large in ε^{-1} [11] and since clearly $\Delta\Omega$ shrinks to $\partial\Omega$ for $\varepsilon \to 0$ (a quantitative estimate will be given below) the inhomogeneity in the equation for $f(\underset{\sim}{x})$ following from eq.(3) becomes negligible on the boundary layer $\Delta\Omega$:

$$L^+ f(\underset{\sim}{x})\approx 0 \text{ for } \underset{\sim}{x}\in\Delta\Omega,\ f(\underset{\sim}{x})=0 \text{ for } \underset{\sim}{x}\in\partial\Omega \text{ and } f(\underset{\sim}{x})\approx 1 \text{ for } \underset{\sim}{x}\in\Omega-\Delta\Omega \tag{7}$$

With eq.(5) T may be expressed in terms of w and of the gradient of f on $\partial\Omega$:

$$T = -\int_{\Omega} d^n x w \Big/ \frac{\varepsilon}{2}\int_{\partial\Omega} dS_i\, w\, D^{ij}\,\partial_j f \tag{8}$$

To solve eq.(7) we make for $f(\underset{\sim}{x})$ the ansatz

$$f(\underset{\sim}{x}) = \sqrt{\frac{2}{\pi\varepsilon}}\int_0^{\rho(\underset{\sim}{x})} dz\ \exp\{-z^2/2\varepsilon\}. \tag{9}$$

The function $\rho(\underset{\sim}{x})$ introduced thereby satisfies

$$K^i\partial_i\rho - \frac{1}{2}D^{ij}(\partial_i\rho)\,(\partial_j\rho)\rho = 0 \tag{10}$$

as follows from (7) in leading order of $\varepsilon \to 0$. Since on $\partial\Omega$ the normal component of the drift field $\underset{\sim}{K}$ vanishes, eq.(10) admits a $\rho(x)$ vanishing on $\partial\Omega$. Next we suppose that in a coordinate system in the boundary with one axis (r) along $d\underset{\sim}{S}$ on $\partial\Omega$ (r=0 on $\partial\Omega$ and $r > 0$ in Ω) and with all other axes lying in $\partial\Omega$, the r-component of the drift vanishes liearly in r

$$K^r = g\,r, \quad g > 0 \tag{11}$$

which respresents the typical behaviour of a force field near a separatrix. Then near $\partial\Omega$ the function $\rho(\underset{\sim}{x})$ is given by

$$\rho = ar \tag{12}$$

where a is a function on $\partial\Omega$ which obeys the equation

$$ga + K^\alpha(\partial_\alpha a) - \frac{1}{2}D^{rr}a^3 = 0 \tag{13}$$

with $\alpha = 1,\ldots n-1$ denoting coordinates on $\partial\Omega$. Note that at the stationary points on the separatrix a is simply given by

$$a = \sqrt{2g/D^{rr}}. \tag{14}$$

The characteristic system of the partial differencial equation (13) consists of the deterministic system (1) restricted to the separatrix and the equation

$$\dot{a} = \frac{1}{2}D^{rr}a^3 - ga \tag{15}$$

Since the gradient of f on $\partial\Omega$ is proportional to a, the constant part T of the absorption time becomes

$$T = \sqrt{2\pi/\varepsilon}\,\int_{\Omega} d^n x w \Big/ \int_{\partial\Omega} dS_r\, w D^{rr} a\,. \tag{16}$$

It is thus expressed in terms of the solution w of the stationary Fokker Planck equation, the boundary $\partial\Omega$, and the function a essentially determined by the deterministic dynamics on $\partial\Omega$. We recall that w need not fulfill any boundary conditions. Our next aim will be to utilize this for the further evaluation of eq. (16).

3. WKB EXPANSION OF THE STATIONARY FOKKER PLANCK EQUATION

For $\varepsilon \to 0$ the solution of the stationary Fokker Planck equation with natural boundary conditions will tend to δ-functions concentrated at the attractors of the deterministic motion. If near a point attractor, the deterministic motion can be linearized this solution will locally become a Gaussian with variance proportional to $\sqrt{\varepsilon}$. We assume that this ε dependence is still obeyed for nonlinear systems in leading order

$$w(\underset{\sim}{x}) = z(\underset{\sim}{x},\varepsilon)\, e^{-\frac{\phi(\underset{\sim}{x})}{\varepsilon}} \tag{17}$$

where $\phi(\underset{\sim}{x})$ is independent of ε and $z(\underset{\sim}{x},\varepsilon)$ can be expanded in a power series in ε, from which in the following only the lowest order term will be kept. For a process obeying detailed balance this is exact. Note the analogy to the WKB expansion in quantum mechanics: The Schrödinger and the Fokker Planck equations are both second order differential equations, which in the classical and deterministic limit, respectively, have vanishing coefficients of the second order differential operators. Moreover (17) is the precise analogue of the WKB wave function.

From eq.(4) we find that in leading order in ε $\phi(\underset{\sim}{x})$ and $z(\underset{\sim}{x}) = z(\underset{\sim}{x},0)$ obey first order differential equations [12]

$$K^i \partial_i \phi + \frac{1}{2} D^{ij} (\partial_i \phi)(\partial_j \phi) = 0 \tag{18}$$

$$[K^i + D^{ij}(\partial_j \phi)]\partial_i z + [(\partial_i K^i) + (\partial_j D^{ij})(\partial_i \phi) + \frac{1}{2} D^{ij}(\partial_i \partial_j \phi)]\; z = 0 \tag{19}$$

Since D^{ij} is a nonnegative matrix from (18) it follows that ϕ decreases along the trajectories of the deterministic dynamics (1) [13]

$$\left.\frac{d\phi}{dt}\right|_d = -\frac{1}{2} D^{ij}(\partial_i \phi)(\partial_j \phi) \leqslant 0 \tag{20}$$

Therefore ϕ is a Lyapunov function of the deterministic motion. This shows that for $\varepsilon \to 0$ the ansatz converges to δ-functions at the deterministic attractors as desired.

Near a point attractor, at which the drift is supposed to vanish linearily, as

$$K^i(\underset{\sim}{x}) = B^i_{\;j}\, x^j , \tag{21}$$

ϕ varies quadratically

$$\phi(\underset{\sim}{x}) = \phi_0 + \frac{1}{2} \phi_{ij}\, x^i x^j . \tag{22}$$

It follows from (1) that the matrix ϕ_{ij} of the second derivates is determined by

$$B^k_{\;i}\phi_{kj} + B^k_{\;j}\phi_{ki} + D^{k\ell}\phi_{ki}\,\phi_{\ell j} = 0 \tag{23}$$

where $D^{k\ell}$ is taken at the attractor. The inverse of the matrix ϕ_{ij} obeyes a linear inhomogeneous equation which may be solved by standard methods. In the same way the curvature of ϕ may be determined at a hyperbolic point of the drift field $\underset{\sim}{K}$ which corresponds to a saddle-point of ϕ because of the Lyapunov property (20). For a limit cycle a similar method applies for the calculation of the matrix ϕ_{ij} of second derivatives transvers to the limit cycle [14] . This matrix determines ϕ in a linear neighbourhood of the limit cycle. To calculate ϕ it is most convenient to interprete ϕ as the action of a reversible system with the Hamiltonian

$$H = \frac{1}{2} D^{ij}\, p_i p_j + K^i\, p_i \tag{24}$$

in a 2n-dimensional phase space.
Clearly, with

$$p_i = \partial_i \phi \tag{25}$$

eq.(16) is recovered as the Hamilton Jacobi equation of a system with the Hamiltonian (24) moving on the "energy" -hypersurface H = 0. Consequently, the characteristic system of eq.(18) is given by the canonical equations of the Hamiltonian (24):

$$\dot{x}^i = \frac{\partial H}{\partial p_i} = D^{ij}\, p_j + K^i \tag{26}$$

$$\dot{p}_i = -\frac{\partial H}{\partial x^i} = -(\partial_i D^{kj})\, p_k p_j - (\partial K^i) p_j \tag{27}$$

Note that the action Φ grows on a solution of the canonical equations:

$$\left.\frac{d\Phi}{dt}\right|_c = \frac{1}{2} D^{ij} p_i p_j \geq 0 . \tag{28}$$

Together with the Lyapunov property (20) of Φ it follows that the solutions with H = 0 go away from the attractors of the drift field $\underset{\sim}{K}$. Since H is invariant under the "time"-reversal symmetry

$$p_i \to -p_i , \quad K^i \to -K^i \tag{29}$$

the solutions of the "time"-reversed eqs. (26, 27) go away from unstable points and finally approach an attractor of $\underset{\sim}{K}$. In this way one can easily find trajectories connecting e.g. hyperbolic points with attractors. For the difference of the action it follows

$$\Delta\Phi = \int_0^\infty dt\; \dot{x}^i(t) p_i(t) . \tag{30}$$

Some comments are in order. Since both the drift field $\underset{\sim}{K}$ and the momentum $\underset{\sim}{p}$ vanish at a hyperbolic point one has to start the trajectory in the linear neighbourhood. Assuming that Φ is of the form (22) eq. (25) yields the initial momentum. Since it needs an infinite amount of "time" to reach the attractor, one must terminate the trajectory at a point in the linear neighbourhood of the attractor. If Φ is there too of the form (22) one can correct the error madein the action. However, the assumtions about the action near the stationary points of the drift field $\underset{\sim}{K}$ need not be consistent with each other. In order to check the consistency one has to compare the momenta following from the trajectory with those derived analytically from Φ in the linear neighbourhood of the attractor.

Next we discuss eq. (19) for the prefactor z(t). On a trajectory of the Hamiltonian system (26,27) eq. (19) reduces to

$$\dot{z}+\{(\partial_i K^i)+(\partial_j D^{ij})p_i+\tfrac{1}{2}D^{ij}(\partial_i\partial_j\Phi)\}z = 0 \qquad (31)$$

where $\partial_i K^i$ and $(\partial_j D^{ij})p_i$ are known functions of t. Since the second derivative of Φ can too be determined from the eqs. (26,27) linearized about the considered trajectory as functions of t, z obeys an ordinary differential equation.

Finally we will apply the WKB expansion to the evaluation of the mean absorption time T given by eq. (16). First we notice that for the volume integral only the absolute minimum of Φ in Ω prevails which due to the Lyapunov property (20) is taken on the attractor. For a point attractor Φ is typically given by eq. (22), and since at the attractor the prefactor $z(\underset{\sim}{x})$ is a slowly varying function compared with $\exp(-\Phi/\varepsilon)$ it can be put constant and only a Gaussian integral has to be performed. For attractors,limit cycles,and higher dimensional invariant tori again typically a quadratic approximation of Φ transvers to the attractor will be sufficient.

For the surface integral only the absolute minima of Φ on $\partial\Omega$ prevail, which due to the Lyapunov property (20) are taken at the attractors of the drift field $\underset{\sim}{K}$ restricted to the separatrix. Hence, isolated minima are taken at the hyperbolic points of $\underset{\sim}{K}$ and, thus, easy to find. At these points the separatrix can be replaced by its tangential plane and the function a is given by eq. (14). Since again the nonexponential part of the integrand varies slowly compared with exp($-\Phi/\varepsilon$) it can be put constant, and assuming that Φ is quadratic there a Gaussian integral remains. In this case the difference of the action at the saddle point and the attractor can be determined by eq. (30) using a "time"-reversed path.

If the lowest minimum of Φ on $\partial\Omega$ is assumed at a limit cycle, the eq. (13) determining the function a reduces to an ordinary first order differential equation which can easily be integrated. If moreover Φ is a quadratic form in local coordinates transvers to the limit cycle the mean time T can be calculated in the same way as discussed for the hyperbolic point.

For a d_a-dimensional attractor and a d_s-dimensional minimum of Φ on $\partial\Omega$ the ε dependence of the mean time T becomes

$$T \sim \varepsilon^{(d_s-d_a)/2} \exp(-\Delta\Phi/\varepsilon), \quad \Delta\Phi > 0 \qquad (32)$$

if again Φ is quadratic in a local coordinate system transvers to the attractor on $\partial\Omega$.

4. CONCLUSIONS

In this paper we investigated the mean lifetime of a trjectory in the domain of attraction of a deterministically stable state which is weakly perturbed by a random force. As a function of the initial state this time is constant within the domain of attraction and decreases to zero like a steep error function at the boundary. The constant part of the mean lifetime depends on a solution of the stationary Fokker Planck equation, which is trivially found if the system obeys detailed balance. E.g. Kramers' results for Brownian motion in multistable potentials [1,2], and Langers nucleation rates are recovered [15]. Another application is given in [16].

A possible noise induced drift proportional to ε can be included in eq. (2): it merely modifies the function w, more specifically the prefactor z in eq. (17).

References

[1] H.A. Kramers, Physica 7 (1940) 284

[2] R. Landauer, and J.A. Swanson, Phys. Rev. 121 (1961) 1668

[3] J.S. Langer, Ann. Phys. (NY) 54 (1969) 258

[4] U. Weiss, and W. Häffner, in Functional Integration, ed. J.P. Antoine, E. Tirapegui (1980)

[5] B. Caroli, C. Caroli, and B. Roulet, J. Stat. Phys. 28 (1982) 757

[6] C.W. Gardiner, J. Stat. Phys. 30 (1983) 157

[7] R.L. Stratonovitch, Topics in the theory of random noise, Gordon Breach, NY (1963)

[8] E.B. Dynkin, Markov processes, Springer NY (1965)

[9] Z. Schuss, Theory and applications of stochastic differential equations, J. Wiley NY (1980)

[10] P. Talkner, and D. Ryter, Phys. Lett. 88A (1982) 162

[11] A.D. Ventsel, and M.I. Freidlin, Russ. Math. Surveys 25 (1970) 1

[12] D. Ludwig, SIAM Rev. 17 (1975) 605

[13] R. Graham, in Stochastic nonlinear systems in physics, chemistry, and biology, ed. L. Arnold, R. Lefever, Springer, Berlin (1981)

[14] P. Talkner, D. Ryter, and P. Jordan, to be published

[15] D.Ryter, and P. Talkner, to be published

[16] D. Ryter, P. Talkner, and P. hänggi, Phys. Lett. 93A (1983) 447

NOISE IN PHYSICAL SYSTEMS AND 1/f NOISE
M. Savelli, G. Lecoy and J-P. Nougier (eds.)
© Elsevier Science Publishers B.V., 1983

INSTABILITIES CONTROLLED BY A NOISY PARAMETER

R. Graham, A. Schenzle

Universität Essen, Fachbereich Physik
W. Germany

Systems undergoing continuous symmetry breaking bifurcations are considered for the case where the bifurcation parameter is perturbed by colored noise. Using the Lorenz model and a subharmonic oscillator as examples, we determine the Markoff process which governs the dynamics asymptotically for weak noise and close to the bifurcation point. We also determine in the weak noise limit the dependence of the threshold on the bandwidth and the intensity of the noise.

1. INTRODUCTION

The behavior of systems exhibiting continuous symmetry breaking instabilities under the influence of noise has attracted much interest, both experimentally [1-4] and theoretically [5-15]. In a recent paper [16] using Fokker Planck methods, we have shown for a simple model that fluctuations of the control parameter can stabilize a system in the symmetrical state by shifting the threshold of instability. The efficiency of stabilization, measured by the shift of the threshold, was found to depend on the noise intensity and the bandwidth of the noise.

In the present paper using direct methods of stochastic calculus [1)], we wish to study the effect of fluctuations of the control parameter for two physically more realistic models.

First we consider the effect of such fluctuations in the Lorenz model [17]. This model is e.g. useful for describing the onset of thermal convection in small convection cells, and it is also closely related to a model of a single mode laser [18]. We show that the Lorenz model is stabilized in the symmetrical state in precisely the same way as the simpler model of ref. [16]. As a second example we consider a subharmonic oscillator, cf. [19]. Here, we find that the threshold of the subharmonic oscillator is insensitive to the noise of the pump. Both examples are found to reduce to the same Markoff process in the limit of small noise. Analytical and numerical solutions of the limit process have been given in the literature [11-13].

2. LORENZ MODEL

The Lorenz model satisfies the three equations

$$\begin{aligned} \dot{x} &= -\sigma(x-y) \\ \dot{y} &= -y - x\cdot z + (r+u)x \\ \dot{z} &= -bz + x\cdot y \end{aligned} \tag{1}$$

x, y, z are the dynamical variables of the model in dimensionless units. In the context of convection they are the amplitudes of the streaming velocity and the first two spatial Fourier components of the temperature profile, respectively. The Prandtl number σ and the geometrical factor b of order 1 are fixed parameters. The instantaneous value of the Rayleigh number is r + u. At the usual threshold of instability of the trivial symmetrical state x = y = z = 0 we have r = 1, u = 0. We now suppose that the Rayleigh number has a fluctuating component u with noise-intensity $\langle u^2(t)\rangle = q$ and bandwidth $\Delta\omega$. For simplicity we take u as a stationary Ornstein Uhlenbeck process, which satisfies

$$du = -\Delta\omega\, u\, dt + \sqrt{2\Delta\omega q}\, dW_t \tag{2}$$

Here dW_t is the standard white noise or Wiener process, which satisfies

$$\langle dW_t\rangle = 0 \quad , \langle dW_t^2\rangle = dt \tag{3}$$

Eqs. (1), (2) form a set of stochastic differential equations. In the following we chose to work with the Ito calculus for reasons which become clear below. We note, however, that the Stratonovich interpretation and the Ito interpretation of eqs. (1), (2) coincide.

We now introduce

$$\varepsilon = \sqrt{\frac{2q\sigma^2}{\Delta\omega(1+\sigma)^3}} \tag{4}$$

as a small parameter and scale the variables and parameters by

$$z = \frac{(1+\sigma)^2}{\sigma}\left(\varepsilon^2 X_\varepsilon^2 + \varepsilon Z_\varepsilon\right)$$
$$X = \varepsilon\sqrt{2(1+\sigma)^2}\, X_\varepsilon$$
$$u = \varepsilon\frac{(1+\sigma)^2}{\sigma}\, u_\varepsilon \tag{5}$$
$$t = \frac{1}{(1+\sigma)\varepsilon^2}\,\tau$$
$$a = (\gamma-1)\cdot(1+\sigma)^2\Delta\omega/2q\sigma$$
$$\lambda_1 = \frac{\Delta\omega}{1+\sigma}, \quad \lambda_2 = \frac{b}{1+\sigma}, \quad \lambda_3 = \frac{2\sigma-b}{1+\sigma}$$

Then the basic equations are

$$dX_\varepsilon = \frac{1}{\varepsilon}\,p_\varepsilon\, d\tau \tag{6}$$
$$dp_\varepsilon = \frac{1}{\varepsilon}\left(aX_\varepsilon - X_\varepsilon^3\right)d\tau - \frac{1}{\varepsilon^2}\left(p_\varepsilon - X_\varepsilon u_\varepsilon + X_\varepsilon Z_\varepsilon\right)d\tau \tag{7}$$
$$dZ_\varepsilon = -\frac{\lambda_2}{\varepsilon^2}\, Z_\varepsilon\, d\tau + \frac{\lambda_3}{\varepsilon}\, X_\varepsilon^2\, d\tau \tag{8}$$
$$du_\varepsilon = -\frac{\lambda_1}{\varepsilon^2}\, u_\varepsilon\, d\tau + \frac{1}{\varepsilon}\lambda_1\, dW_\tau \tag{9}$$

We are interested in the process obtained from eqs. (6)-(9) by letting $\varepsilon \to 0$.

$$X(t) = \lim_{\varepsilon\to 0} X_\varepsilon(t) \tag{10}$$

From eqs. (6), (7) we obtain

$$dX_\varepsilon = \left(aX_\varepsilon - X_\varepsilon^3\right)d\tau + \frac{1}{\varepsilon}X_\varepsilon\left(u_\varepsilon - Z_\varepsilon\right)d\tau + O(\varepsilon) \tag{11}$$

Here and in the following we make use of the fact that a stochastic differential in the sense of Ito multiplied by ε is negligible for $\varepsilon \to 0$ and formally $O(\varepsilon)$. This condition must be satisfied if the limit process for $\varepsilon \to 0$ indeed exists, as we assume here. The Ito differential then represents the actual infinitesimal stochastic increment of a process which exists and becomes independent of ε for $\varepsilon \to 0$. It may be useful to note at this point that the Ito calculus is distinguished in this respect, since it describes the actual stochastic increments, and that a stochastic differential in the sense of Stratonovich need not have a limit for $\varepsilon \to 0$, even if the Ito differential has such a limit. E.g. the Ito differential

$$du_\varepsilon^2 = 2u_\varepsilon\, du_\varepsilon + \frac{1}{\varepsilon}\lambda_1^2\, d\tau$$

differs by the last term on the right hand side from the corresponding Stratonovich differential, and this term does not exist for $\varepsilon \to 0$.

In the following we describe a systematic procedure which leads to the elimination of u_ε and Z_ε from eq. (11) in favour of X_ε to formal order ε.

Using the basic equations we compute

$$dX_\varepsilon u_\varepsilon = X_\varepsilon du_\varepsilon + u_\varepsilon dX_\varepsilon$$
$$= -\frac{\lambda_1}{\varepsilon}X_\varepsilon u_\varepsilon d\tau + \frac{\lambda_1}{\varepsilon}X_\varepsilon dW_\tau + \frac{1}{\varepsilon}u_\varepsilon p_\varepsilon d\tau \tag{12}$$

and therefore

$$\frac{1}{\varepsilon}X_\varepsilon u_\varepsilon d\tau = \frac{1}{\lambda_1}u_\varepsilon p_\varepsilon d\tau + X_\varepsilon dW_\tau + O(\varepsilon) \tag{13}$$

Similarly we obtain

$$\frac{1}{\varepsilon}X_\varepsilon Z_\varepsilon d\tau = \frac{1}{\lambda_2}Z_\varepsilon p_\varepsilon d\tau + \frac{\lambda_3}{\lambda_2}X_\varepsilon^3 d\tau + O(\varepsilon) \tag{14}$$

In order to eliminate $u_\varepsilon p_\varepsilon$ and $Z_\varepsilon p_\varepsilon$ we continue along these lines and obtain in the next step

$$u_\varepsilon p_\varepsilon d\tau = \frac{1}{1+\lambda_1}\left(X_\varepsilon u_\varepsilon^2 - X_\varepsilon Z_\varepsilon u_\varepsilon\right)d\tau + O(\varepsilon) \tag{15}$$
$$Z_\varepsilon p_\varepsilon d\tau = \frac{1}{1+\lambda_2}\left(-X_\varepsilon Z_\varepsilon^2 + X_\varepsilon Z_\varepsilon u_\varepsilon\right)d\tau + O(\varepsilon) \tag{16}$$

Repeating the procedure to eliminate

$X_\varepsilon u_\varepsilon^2, X_\varepsilon Z_\varepsilon u_\varepsilon$, and $X_\varepsilon Z_\varepsilon^2$

we have to evaluate in Ito calculus

$$dX_\varepsilon u_\varepsilon^2 = u_\varepsilon^2 dX_\varepsilon + 2X_\varepsilon u_\varepsilon du_\varepsilon + X_\varepsilon\frac{\lambda_1^2}{\varepsilon^2}d\tau \tag{17}$$

The last term on the right hand side is peculiar to the Ito calculus. This term later on turns out to have direct physical significance since it is responsible for the stabilizing effect of the noise. Continuing as above we obtain

$$X_\varepsilon u_\varepsilon^2 d\tau = \frac{\lambda_1}{2}X_\varepsilon d\tau + O(\varepsilon)$$
$$X_\varepsilon u_\varepsilon Z_\varepsilon d\tau = O(\varepsilon) \tag{18}$$
$$Z_\varepsilon^2 X_\varepsilon d\tau = O(\varepsilon)$$

Collecting all results we are left with

$$dX_\varepsilon = \left(a + \frac{1}{2(1+\lambda_1)}\right)X_\varepsilon d\tau - \left(1 + \frac{\lambda_3}{\lambda_2}\right)X_\varepsilon^3 d\tau + X_\varepsilon dW_\tau + O(\varepsilon) \tag{19}$$

Taking the limit $\varepsilon \to 0$ we obtain the limiting Markoff process

$$X(\tau) = \lim_{\varepsilon\to 0} X_\varepsilon(\tau)$$

governed by the Ito equation

$$dX = \left(a + \frac{1}{2(1+\lambda_1)}\right)X d\tau - \left(1 + \frac{\lambda_3}{\lambda_2}\right)X^3 d\tau + X dW_\tau \tag{20}$$

The process (20) has been analyzed in great detail in the literature to which we refer for properties like moments, correlation functions and probabilities densities. For our present purpose it is sufficient to remark that the stationary moment $\langle X^2\rangle$ is given by

$$\langle X^2\rangle = \begin{cases} 0 & a \le \frac{1}{2}\frac{\lambda_1}{1+\lambda_1} \\ \left(a - \frac{\lambda_1}{2(1+\lambda_1)}\right)\frac{\lambda_2}{\lambda_2+\lambda_3}, & a > \frac{1}{2}\frac{\lambda_1}{1+\lambda_1} \end{cases} \tag{21}$$

Thus, there is a sharp threshold at

$$\gamma = \gamma_{th} = 1 + \frac{q\sigma}{(1+\sigma)^2(1+\Delta\omega+\sigma)} \tag{22}$$

which depends on the intensity q and the bandwidth $\Delta\omega$ of the noise. We note that this dependence, expressed in scaled variables, is exactly the same as in the simpler model analyzed in [16]. We conclude that the Lorenz model is stabilized by small noise in the Rayleigh number. The shift of the threshold of instability is proportional to the noise intensity. For given total noise intensity q the stabilization is more effective if the bandwidth of the noise is small compared to the sum of the viscous and the thermal relaxation times.

3. SUBHARMONIC OSCILLATOR

As a second example we consider the subharmonic oscillator

$$\dot{\alpha}_1 = -\varkappa_1\alpha_1 - g\alpha_2^2 + F_0 + u(t) \tag{23}$$

$$\dot{\alpha}_2 = -\varkappa_2\alpha_2 + 2g\alpha_2^*\cdot\alpha_1 \tag{24}$$

with the complex amplitudes α_1 at the fundamental frequency and α_2 at the subharmonic frequency, respectively. $\varkappa_1$ and $\varkappa_2$ are damping constants, g is the coupling constant, F_0 + U(t) is the pump at the fundamental frequency with the constant amplitude F_0 and the noisy component u. Again u is supposed to be given by eq. (2). Eqs. (23), (24) can be analyzed by the same methods as before. The scaling

$$\begin{aligned} &\varepsilon = \frac{g}{\varkappa_1}\sqrt{\frac{2q}{\Delta\omega\varkappa_1}} \\ &t = \frac{1}{\varkappa_1\varepsilon^2}\tau \quad , \alpha_1 = \frac{\varkappa_1\varepsilon}{g}x_\varepsilon + \frac{1}{\varkappa_1}F_0 \\ &\alpha_2 = \varepsilon\frac{\varkappa_1}{g}y_\varepsilon \quad , u = \frac{\varkappa_1^2}{g}\varepsilon u_\varepsilon \\ &a = \frac{1}{q}\Delta\omega\varkappa(F_0 - \frac{1}{2g}\varkappa_1\varkappa_2) \\ &\lambda_1 = \Delta\omega/\varkappa_1 \quad , \lambda_2 = \varkappa_2/\varkappa_1 \end{aligned} \tag{25}$$

reduces the basic equations to the form

$$dx_\varepsilon = -\frac{1}{\varepsilon^2}(x_\varepsilon - u_\varepsilon)d\tau - \frac{1}{\varepsilon}y_\varepsilon^2 d\tau \tag{26}$$

$$dy_\varepsilon = -\frac{\lambda_2}{\varepsilon^2}(y_\varepsilon - y_\varepsilon^*)d\tau + \frac{2}{\varepsilon}x_\varepsilon y_\varepsilon^* d\tau + a y_\varepsilon^* d\tau \tag{27}$$

$$du_\varepsilon = -\frac{\lambda_1}{\varepsilon^2}u_\varepsilon d\tau + \frac{\lambda_1}{\varepsilon}dW_\tau \tag{28}$$

As before we use the Ito calculus. Our aim is to find the equation of motion which governs the limit process

$$y(\tau) = \lim_{\varepsilon\to 0} y_\varepsilon(\tau) \tag{29}$$

assuming that the limit process exists. From eqs. (26), (27) we immediately see that

$$(x_\varepsilon - x_\varepsilon^*)d\tau = O(\varepsilon), (y_\varepsilon - y_\varepsilon^*)d\tau = O(\varepsilon) \tag{30}$$

i.e., x_ε and y_ε may be considered as real with corrections of order ε. It is then straight forward to eliminate x_ε and u_ε to order ε from the equation for $(y_\varepsilon + y_\varepsilon^*)$ by the procedure described above, with the result

$$dy_\varepsilon = (a+2)y_\varepsilon d\tau - 2y_\varepsilon^3 d\tau + 2y_\varepsilon dW + O(\varepsilon) \tag{31}$$

The limiting process $y(\tau) \lim_{\varepsilon\to 0} y_\varepsilon(\tau)$

therefore satisfies the Ito equation

$$dy = \left[(a+2)y - 2y^3\right]d\tau + 2y\,dW \tag{32}$$

which has the same form as eq. (20). We read off the threshold condition $a \geq 0$, i.e.

$$gF_0 \geq \frac{1}{2}\varkappa_1\varkappa_2 \tag{33}$$

Thus, the threshold condition in the limit of weak noise, eq. (33), turns out to be independent of the noise. It is easy to understand this result directly from eqs. (23), (24), if one accepts the fact (30) that the process $\alpha_2(t)$ becomes real for $\varepsilon \to 0$ (we note that this conclusion breaks down in the case of detuning): The threshold is defined by the condition that $\alpha_2(t) \neq 0$ in the statistical steady state becomes possible for the first time. Dividing then by $\alpha_2(t) = \alpha_2^*(t)$ and taking the average in the statistical steady state we obtain the condition

$$\varkappa_2 = 2g\langle\alpha_1\rangle = -\frac{2g^2}{\varkappa_1}\langle\alpha_2^2\rangle + \frac{2g}{\varkappa_1}F_0$$

Here we used the fact that $\langle u\rangle = 0$, and that averages of time derivatives vanish in the statistical steady state. For

$$gF_0 < \frac{1}{2}\varkappa_1\varkappa_2$$

this condition cannot be satisfied, hence $\alpha_2(t) \equiv 0$ is the only solution in the statistical steady state. For

$$gF_0 > \frac{1}{2}\varkappa_1\varkappa_2$$

on the other hand, $\alpha_2(t) \neq 0$ is a possible solution, since then

$$\langle\alpha_2^2\rangle = \frac{1}{g^2}(gF_0 - \frac{1}{2}\varkappa_1\varkappa_2) > 0$$

Therefore, the threshold is completely determined by the averaged equation of the fundamental mode and must be independent of any noise in the pump. The same conclusion is still valid if the noise of the pump is complex. Then x_ε is complex to zero order in ε, but only its real part is coupled to $(y_\varepsilon + y_\varepsilon^*)$ to lowest order in ε. Hence, only the real part of the noise of the pump can matter, but it leaves the the threshold unshifted due to the argument presented before.

Footnote:

1) We are indebted to Laurie Davies for pointing out to us the efficiency of this method for the model of ref. [16].

References

1. S. Kabashima, S. Kogure, T. Kawakubo, and T. Okada, Appl. Phys. 50, 6296 (1979)
2. S. Kai, T. Kai, M. Takata, and K. Hirakawa, J.Phys. Soc. Jpn. 47, 1379 (1979)
3. P. De Kepper and W. Horsthemke, C.R. Acad. Sci., Ser. C287, 251 (1978)
4. T. Kawakubo, A. Yanagita, and S. Kabashima, J. Phys. Soc. Jpn. 50, 1451 (1981)
5. W. Horsthemke and M. Malek Mansour, Z. Phys. B24, 307 (1976)
6. W. Horsthemke and R. Lefever, Phys. Lett. 64A, 19 (1977)
7. L. Arnold, W. Horsthemke, and R. Lefever, Z.Phys. B 29, 367 (1978)
8. H. Mori, T. Morita, and K.T. Mashiyama, Prog. Theor. Phys. 63, 1865 (1980)
9. T. Morita, H. Mori, and K.T. Mashiyama, Prog. Theor. Phys. 64, 500 (1980)
10. H. Brand and A. Schenzle, J. Phys. Soc. Jpn. 48, 1382 (1980)
11. A. Schenzle and H. Brand, Phys. Rev. A20, 1628 (1979)
12. R. Graham and A. Schenzle, Phys. Rev. A 25, 1731 (1982)
13. R. Graham, Phys. Rev. A25, 3234 (1982)
14. H. Fuijisaka and S. Grossmann, Z. Phys. 43, 69 (1981)
15. H. Brand, A. Schenzle, and G. Schröder, Phys. Rev. A25, 2324 (1982)
16. R. Graham, A. Schenzle, Phys. Rev. A26, 1676 (1982)
17. E.N. Lorenz, J. Atmos. Sci. 20, 130 (1963)
18. H. Haken, Phys. Lett. 53A, 77 (1975)
19. R. Graham, Springer Tracts in Mod. Phys. 66, 1 (1973)

THEORY 1/f NOISE

NOISE IN PHYSICAL SYSTEMS AND 1/f NOISE
M. Savelli, G. Lecoy and J-P. Nougier (eds.)
© Elsevier Science Publishers B.V., 1983

LONG-TIME TAIL RELAXATION NOISE
(Invited Paper)

Masuo SUZUKI

Department of Physics, Faculty of Science
University of Tokyo, Bunkyo-Ku
Tokyo 113, Japan

A phenomenon of noise-induced long-time tail relaxation is explained intuitively and it is extended to fractional stochastic processes. General conditions on the noise-induced long-time tail are obtained and are applied to simple multiplicative stochastic processes. The noise-induced long-time tail occurs due to the balance (or cancellation) between the deterministic (drift) term and noise term.

1. INTRODUCTION

The purpose of the present paper is to explain a very interesting situation in which the existence of noise induces a new phenomenon, namely noise-induced long-time tail.(1-3) In almost all ordinary cases, physical quantities relax exponentially to their stationary states. At the critical point (if it exists), however, they decay in a power-law (i.e., $\sim t^{-\psi}$ for large time t).

First we explain a very simple example of the form

$$\frac{dx}{dt} = \gamma x - gx^3;\ g > 0. \tag{1.1}$$

The solution of this equation is easily given by

$$x(t) = x(0)e^{\gamma t}\{1+gx^2(0)(e^{2\gamma t}-1)/\gamma\}^{-1/2}, \tag{1.2}$$

as is well-known.(4) The critical point of the equation (1.1) is $\gamma=0$.(4) The relaxation of $x^p(t)$ shows a power-law decay of the form

$$x^p(t) = x^p(0)\{1+2gx^2(0)t\}^{-p/2} \sim t^{-p/2} \tag{1.3}$$

at the critical point for $p > 0$. That is, the long-time exponent ψ_Q defined by

$$Q(t) \sim t^{-\psi_Q} \tag{1.4}$$

depends on Q in the above simple example. This is always true for deterministic systems such as (1.1). However, a new situation appears in the following multiplicative stochastic processes

$$\frac{dx}{dt} = a(x) + b(x)\eta(t), \tag{1.5}$$

where $\eta(t)$ is a random force which is, for simplicity, assumed to be Gaussian and white, namely

$$\langle\eta(t)\ \eta(t')\rangle = 2\varepsilon\delta(t-t'). \tag{1.6}$$

The above type of multiplicative stochastic processes have been studied by many authors.(5-10) In particular, the following multiplicative stochastic processes

$$\frac{dx}{dt} = \gamma x - gx^m + x^n\eta(t) \tag{1.7}$$

are quite interesting in many respects. The case $n=1$ is called "random growing-rate model", because it is rewritten as

$$\frac{dx}{dt} = (\gamma + \eta(t))x - gx^m. \tag{1.8}$$

In the case $1 < m < 2n-1$, the relaxation is essentially deterministic as in (1.1) and we have

$$\langle x^p(t)\rangle \sim t^{-p/(m-1)} \tag{1.9}$$

for $p > 0$, at the critical point $\gamma = 0$.(1-3) That is, in this deterministic case $1 < m < 2n-1$ the noise term $x^n\eta(t)$ in (1.7) is irrelevant and the relaxation of $\langle x^p(t)\rangle$ is governed only by the deterministic part in (1.7).

On the other hand, in the case $m > 2n-1 \geq 1$, a new situation appears, namely all the moments $\langle x^p(t)\rangle$ for $p > 0$ show the following same power-law decay

$$\langle x^p(t)\rangle \sim t^{-1/2} \tag{1.10}$$

at the critical point. Of course, the numerical factor in front of $t^{-1/2}$ depends on the parameters ε, g, m, n and p. This is a quite remarkable situation in contrast to the deterministic case. That is, the existence of the noise term $x^n\eta(t)$ in (1.7) is essential in this case. Thus, this is called "noise-induced long-time tail". This was found first by Brenig and Banai (8) in a special case $m=3$ and $n=1$ for the second moment $\langle x^2(t)\rangle$. The universality that the long-time tail exponent ψ_p for $\langle x^p(t)\rangle$ does not depend on p for any positive value of p was obtained first by Suzuki et al.(1) together with some intuitive explanation of the appearance of the noise-induced long-time tail, as will be discussed again in the present paper.

2. NOISE-INDUCED LONG-TIME TAIL IN A SIMPLE EXACTLY SOLUBLE MODEL

In the present section we explain the mechanism of appearance of noise-induced long-time tail in a very simple example.

We consider here the random growing-rate model

(1.8). This can be solved rigorously and consequently the moment $\langle x^p(t)\rangle$ has been shown explicitly to have the following asymptotic form

$$\langle x^p(t)\rangle \simeq \{\frac{(m-1)\varepsilon}{g}\}^{p/(m-1)} \frac{\Gamma(p/(m-1))}{(m-1)(\pi\varepsilon t)^{1/2}} \qquad (2.1)$$

at the critical point $\gamma = 0$ in the case $p > 0$ for large t and for $g > 0$, and $m > 1$.(2)

Before discussing a general case, we study here how the above noise-induced long-time tail appears. One of the simplest explanations will be to make use of the following moment equation

$$\frac{d}{dt}\langle x^p(t)\rangle = p\gamma\langle x^p(t)\rangle - pg\langle x(t)^{m+p-1}\rangle + p^2\varepsilon\langle x^p(t)\rangle , \qquad (2.2)$$

which is easily derived from (1.8). If the most dominant singular part of the deterministic term $\langle x(t)^{m+p-1}\rangle$ cancels that of the noise term $\langle x^p(t)\rangle$ in (2.2) and consequently the next singular part is ballanced by the left-hand side in(2.2), then the noise-induced long-time tail appears at the critical point $\gamma = 0$.

Another explanation of the noise-induced long-time tail will be given by the simplification of (1.8) into the follwoing additive stochastic differential equation

$$\frac{dy}{dt} = -ge^{(m-1)y} + \eta(t) \qquad (2.3)$$

by the transformation of variable

$$y = \log x \quad \text{or} \quad x = e^y . \qquad (2.4)$$

Then, the natural boundary $x = 0$ of the stochastic process (1.8) corresponds to $y = -\infty$ in the new variable. As t goes to infinity, the stochastic variable y goes to $-\infty$, in average, (i.e., $\langle y(t)\rangle \to -\infty$) and consequently the deterministic part $-ge^{(m-1)y}$ in (2.3) becomes exponentially smaller and smaller for larger t. Then, the deterministic part in (2.3) may be replaced by the average of it. Thus, distribution function P(y,t) is qualitatively described by that corresponding to the following simple Wiener process

$$y(t) = \langle y(t)\rangle + \int_0^t \eta(t')dt' \qquad (2.5)$$

for large t. The average $\langle y(t)\rangle$ may be determined approximately by the following self-consistent equation

$$\frac{d}{dt}\langle y(t)\rangle = -g\exp\{(m-1)\langle y(t)\rangle\} . \qquad (2.6)$$

That is, we have

$$\langle y(t)\rangle \sim -\frac{1}{m-1}\log t \qquad (2.7)$$

for large t. The corresponding probability distribution function P(y,t) for (2.5) is given by

$$p(y,t) = \frac{1}{(4\pi\varepsilon t)^{1/2}} \exp[-\frac{(y-\langle y(t)\rangle)^2}{4\varepsilon t}] \qquad (2.8)$$

Since $\langle y(t)\rangle^2 \ll t$ for large t, the distribution function (2.8) can be reduced asymptotically to

$$P(y,t) \sim \frac{1}{(4\pi\varepsilon t)^{1/2}} \qquad (2.9)$$

for a fixed value of y. Thus, the moment $\langle x^p(t)\rangle$ for $p > 0$ takes the following asymptotic form

$$\langle x^p(t)\rangle \sim \int_{-\infty}^{\delta} e^{py} P(y,t)dy \sim \frac{1}{(4\pi\varepsilon t)^{1/2}} \int_{-\infty}^{\delta} e^{py}dy \sim t^{-1/2} \qquad (2.10)$$

for some appropriate finite constant δ. The average of a more general function $Q(x(t))$ takes the same universal behavior

$$\langle Q(x(t))\rangle \sim \int_{-\infty}^{\delta} Q(e^y)P(y,t)dy \sim \frac{1}{(4\pi\varepsilon t)^{1/2}} \times \int_{-\infty}^{\delta} Q(e^y)dy \sim t^{-1/2} \qquad (2.11)$$

for large t. That is, the universal noise-induced long-time tail $\sim t^{-1/2}$ appears due to the asymptotic behavior of the Wiener process or due to the normalization factor of the probability distribution function. This is the reason why it is called noise-induced long-time tail. Thus, the noise plays an essential role in our problem. However, the noise-induced long-time tail does not appear unless the nonlinear term $-gx^m$ $(m > 1)$ exists, because otherwise x (or y) does not approach zero (or $-\infty$) as t goes to infinity. Namely, the noise-induced long-time tail is a quite interesting phenomenon which occurs as a synergetic effect of the noise and nonlinearity of the system.

Although our system discussed in the present section is very simple, the above explanation of the mechanism of the noise-induced long-time tail is essential and quite universal, and consequently it is extended to more general situations, as in the next section.

3. NOISE-INDUCED LONG-TIME TAIL IN GENERAL SITUATIONS

Up to now, we have explained the mechanism of the noise-induced long-time tail by using a very simple exactly soluble model (1.8). Now we discuss a general situation described by the following stochastic process

$$\frac{dx}{dt} = a(x) + b(x)\eta(t), \qquad (3.1)$$

where $\eta(t)$ is a Gaussian and white noise satisfying

$$\langle\eta(t)\,\eta(t')\rangle = 2\varepsilon\delta(t-t') \qquad (3.2)$$

The speed measure dm(x) and scale measure ds(x) for the process (3.1) are given by

$$dm(x) = \frac{1}{b(x)} \exp(\frac{1}{\varepsilon}\int_{x_0}^{x} \frac{a(\xi)}{b^2(\xi)} d\xi)dx \equiv \hat{P}_{st}(x)dx, \qquad (3.3)$$

and

$$ds(x) = \frac{1}{b(x)} \exp\left(-\frac{1}{\varepsilon}\int_{x_0}^{x} \frac{a(\xi)}{b^2(\xi)} d\xi\right) dx \qquad (3.4)$$

for the initial value $x_0 > 0$, respectively.(11) Without loss of generality, we confine our arguments into the following situation that $x=0$ is a natural boundary and $x=\infty$ is an entrance boundary, and consequently that the region $[0,\infty)$ is relevant in our problem. For this, we have, at least, that

$$a(x) = o(x) \text{ and } b(x) = O(x) \text{ or } o(x) \text{ near } x = 0, \qquad (3.5)$$

as in the example (1.8), and that $m(x)$ and $s(x)$ are monotonically increasing functions of x. This last condition is assured by

$$b(x) > 0 \quad \text{for} \quad x > 0, \qquad (3.6)$$

as is seen from (3.3) and (3.4). Furthermore, we assume that

$$a(x) < 0 \quad \text{for} \quad x > 0 \qquad (3.7)$$

so that y may go, in average, to $-\infty$ as t goes to infinity.

Now it is convenient to make a change of variable as

$$y = \int_{x_0}^{x} \frac{1}{b(\xi)} d\xi \equiv y(x), \qquad (3.8)$$

then (3.1) is transformed into

$$\frac{dy}{dt} = \alpha(y) + \eta(t), \qquad (3.9)$$

where

$$\alpha(y) = a(x(y))/b(x(y)) \qquad (3.10)$$

and $x = x(y)$ is the inverse function of $y = y(x)$ defined by (3.8). By taking into account the condition (3.5) of $b(x)$ together with (3.8), we find that the boundary $x=0$ corresponds to $y=-\infty$ (natural boundary).

As is easily expected from our previous arguments in Section 2 for a simple example, the condition for the appearance of the noise-induced long-time tail is that Eq. (3.9) is asymptotically reduced to the Wiener process (2.5) for $y \to -\infty$ namely that

$$\lim_{y\to-\infty} y\alpha(y) = 0, \qquad (3.11)$$

i.e., $\alpha(y)$ approaches zero faster than $1/y$ as y goes to $-\infty$. This condition is easily understood from the scaling property of the following Fokker-Planck equation

$$\frac{\partial P(y,t)}{\partial t} = \left(-\frac{\partial}{\partial y}\alpha(y) + \varepsilon\frac{\partial^2}{\partial y^2}\right)P(y,t), \qquad (3.12)$$

which corresponds to the Langevin equation (3.9). That is, the condition (3.11) yields that the drift term $(\partial/\partial y)\alpha(y)$ in (3.12) becomes less dominant than the diffusion term $\partial^2/\partial y^2$ in (3.12) for $y \to -\infty$. The above condition (3.11) is equivalent to the following condition that

$$\int_{-\infty}^{\delta} \alpha(y)dy = \int_0^{\delta'} \frac{a(x)}{b^2(x)}dx < \infty \qquad (3.13)$$

for finite δ and δ'. This is simply expressed by the condition that

$$dm(x) \sim ds(x) \quad \text{near} \quad x = 0. \qquad (3.14)$$

Thus, the condition of the appearance of the noise-induced long-time tail is that the speed measure and scale measure have the same behavior near $x=0$, together with the conditions that $x=0$ is a natural boundary and that $x=\infty$ is an entrance boundary. As is well-known, the condition that $x=0$ is a natural boundary is expressed by

$$\int_0^{\delta} dm(x) = \infty \text{ and } \int_0^{\delta} ds(x) = \infty \qquad (3.15)$$

for a positive δ. This yields

$$\int_0^{\delta} \frac{1}{b(x)} dx = \infty \qquad (3.16)$$

from (3.3) and (3.4) together with (3.13). The above condition is reduced to the second condition in (3.5). Furthermore, the condition that $x=\infty$ is an entrance boundary is expressed by

$$\int_{\delta}^{\infty} dm(x) < \infty \quad \text{and} \quad \int_{\delta}^{\infty} ds(x) = \infty. \qquad (3.17)$$

This condition is equivalent to

$$\lim_{x\to\infty} xa(x)/b^2(x) = -\infty \qquad (3.18)$$

Under the above conditions, the original stochastic differential equation (3.1) may be asymptotically reduced to the following simple Wiener process

$$y(t) = \langle y(t)\rangle + \int_0^t \eta(t')dt' \qquad (3.19)$$

for large t. Here, the average $\langle y(t)\rangle$ may be determined approximately by the solution of the deterministic equation

$$\frac{d}{dt} y(t) = \alpha(y(t)). \qquad (3.20)$$

Thus, the probability distribution function $P(x,t)$ takes the following asymptotic form

$$P(x,t) \simeq \frac{1}{(\pi\varepsilon t)^{1/2}} \exp\left(-\frac{(y(x)-\langle y(t)\rangle)^2}{4\varepsilon t}\right) \frac{1}{b(x)} \qquad (3.21)$$

near $x=0$ for large t. The asymptotic solution of $P(x,t)$ in the whole region $[0,\infty)$ of x will be obtained by replacing $1/b(x)$ in (3.21) by $\hat{P}_{st}(x)$ in (3.3), namely

$$P(x,t) \simeq \frac{1}{(\pi\varepsilon t)^{1/2}} \exp\left(-\frac{(y(x)-\langle y(t)\rangle)^2}{4\varepsilon t}\right)\hat{P}_{st}(x) \qquad (3.22)$$

for large t.

It is easily shown from (3.11) and (3.20) that

$$\langle y(t)\rangle^2 \ll t \qquad (3.23)$$

for large t. Therefore, the distribution function $P(x,t)$ in (3.22) can be reduced asymptotically to

$$P(x,t) \simeq \frac{1}{(\pi\varepsilon t)^{1/2}} \hat{P}_{st}(x) \qquad (3.24)$$

for $t \to \infty$ and for a fixed value of x.

Therefore, the average of an arbitrary quantity $Q(x(t))$ takes the following asymptotic form

$$\langle Q(x(t))\rangle = \int_0^\infty Q(x)P(x,t)dx$$

$$\simeq \frac{1}{(\pi\varepsilon t)^{1/2}} \int_0^\infty Q(x)\exp\left(-\frac{(y(x)-\langle y(t)\rangle)^2}{4\varepsilon t}\right)\hat{P}_{st}(x)dx$$

$$\simeq \frac{1}{(\pi\varepsilon t)^{1/2}} \int_0^\infty Q(x)\hat{P}_{st}(x)dx$$

$$= \left(\frac{1}{\pi\varepsilon t}\right)^{1/2} \int_0^\infty Q(x)dm(x)$$

$$\simeq c[Q]/t^{1/2} \qquad (3.25)$$

for large t. This is our required general formula for the noise-induced long-time tail.

The above formula can be obtained alternatively if we assume the last form in (3.25), namely

$$\langle Q(x(t))\rangle \simeq c[Q]/t^{1/2} . \qquad (3.26)$$

That is, the expression of the numerical factor $c[Q]$ is easily obtained as follows. The Fokker-Planck equation corresponding to (3.1) is given by

$$\frac{\partial P}{\partial t} = -\frac{\partial}{\partial x}(a(x)+\varepsilon b(x)b'(x))P + \varepsilon\frac{\partial^2}{\partial x^2}b^2(x)P. \qquad (3.27)$$

This yields the following equation of motion for the average $\langle Q(x(t))\rangle$:

$$\frac{d}{dt}\langle Q(x(t))\rangle = F(Q(x(t))) \equiv F_{drift}(Q(x(t))) + F_{dif}(Q(x(t))) \qquad (3.28)$$

where

$$F_{drift}(Q(x(t))) = \langle Q'(x(t))(a(x(t)) + \varepsilon b(x(t))b'(x(t)))\rangle \qquad (3.29)$$

and

$$F_{dif}(Q(x(t))) = \varepsilon\langle Q''(x(t))b^2(x(t))\rangle . \qquad (3.30)$$

As was discussed in Section 2, the noise-induced long-time tail appears due to the cancellation of the most dominant parts of the drift and diffusion terms in (3.28). Thus, we obtain many simultaneous equations

$$c[Q'(x)(a(x)+\varepsilon b(x)b'(x))] + c[\varepsilon Q''(x)b^2(x)] = 0 \qquad (3.31)$$

for the coefficient functional $c[Q]$ defined by (3.26), which is linear in Q. On the other hand, the stationary property of (3.28) for $t \to \infty$ gives the following relation

$$\int_0^\infty Q'(x)(a(x)+\varepsilon b(x)b'(x))dm(x) + \int_0^\infty \varepsilon Q''(x)b^2(x)dm(x) = 0 \qquad (3.32)$$

for any quantity $Q(x)$, as far as the above integrals are defined. By comparing (3.31) with (3.32), we find that

$$c[Q] \propto \int_0^\infty Q(x)dm(x) . \qquad (3.33)$$

The proportionality coefficient of (3.33) will be found easily by studying some special case and consequently we arrive again at the formula (3.25) for the noise-induced long-time tail.

This derivation clarifies the essence of the noise-induced long-time tail.

The result (3.25) suggests another derivation, namely to make use of the speed measure $m(x)$ itself as a stochastic variable and to linearize the relevant stochastic equation expressed by $m(x)$.

For this purpose, we introduce the following new variable $z(t) = m(x(t))$ (essentially equivalent to (3.3) except the lower bound of the integral)

$$z(t) = \int_{x_0}^{x(t)} \frac{1}{b(x')} \exp\left(\frac{1}{\varepsilon}\int_0^{x'} \frac{a(\xi)}{b^2(\xi)} d\xi\right)dx'. \qquad (3.34)$$

Owing to the condition (3.13), the integral in the exponential of (3.34) is convergent. Then, the equation of $z(t)$ is given by

$$\frac{dz(t)}{dt} = \frac{dz}{dx}\frac{dx(t)}{dt} = \beta(x(t))+\gamma(x(t))\eta(t), \qquad (3.35)$$

where

$$\beta(x) = \frac{a(x)}{b(x)}\gamma(x) \qquad (3.36)$$

and

$$\gamma(x) = \exp\left(\frac{1}{\varepsilon}\int_0^x \frac{a(\xi)}{b^2(\xi)} d\xi\right) \qquad (3.37)$$

Since we are interested in the long-time behavior, we now make the following asymptotic evaluation

$$z(t) = \int_0^t \beta(x(t'))dt' + \int_0^t \gamma(x(t'))\eta(t')dt'$$

$$\simeq \int_0^t \langle\beta(x(t'))\rangle dt' + \langle\gamma(x(t))\rangle\int_0^t \eta(t')dt'$$

$$\simeq \langle z(t)\rangle + \langle\gamma(x(t))\rangle\int_0^t \eta(t')dt' \qquad (3.38)$$

for large t. Here, the average $\langle z(t)\rangle$ may be approximately determined by the solution of the deterministic equation

$$\frac{dz(t)}{dt} = \beta(x(t)) , \qquad (3.39)$$

namely $z(t) = \hat{z}(x(t))$ and

$$\frac{dx(t)}{dt} = a(x(t)). \tag{3.40}$$

The average $\langle\gamma(x(t))\rangle$ is also given, in the above approximation, by

$$\langle\gamma(x(t))\rangle \simeq \gamma(\langle x(t)\rangle) \tag{3.41}$$

with the solution $\langle x(t)\rangle$ of (3.40). The above asymptotic evaluation (3.38) corresponds, in spirit, to the so-called Ω-expansion on stochastic processes.(13-16)

Now it is easily shown that

$$\lim_{t\to\infty} \langle\gamma(x(t))\rangle = 1. \tag{3.42}$$

Therefore, (3.38) is reduced again asymptotically

$$z(t) = \langle z(t)\rangle + \int_0^t \eta(t')dt' \tag{3.43}$$

for large t. Thus, the distribution function $\hat{P}(z,t)$ for (3.43) is given by

$$\hat{P}(z,t) = N(t)\exp\{-\frac{(z-\langle z(t)\rangle)^2}{4\varepsilon t}\} \quad . \tag{3.44}$$

Here, the normalization $N(t)$ is calculated as

$$N(t)^{-1} = \int_{-\infty}^{\hat{z}(\infty)} \exp\{-\frac{(z-\langle z(t)\rangle)^2}{4\varepsilon t}\}dz$$

$$= 2\sqrt{\varepsilon t}\int_{-\infty}^{c(t)} e^{-v^2}dv$$

$$= \sqrt{\pi\varepsilon t}\{1 + \frac{2}{\sqrt{\pi}}\int_0^{c(t)} e^{-v^2}dv\} \ , \tag{3.45}$$

where

$$c(t) = \frac{\hat{z}(\infty)-\langle z(t)\rangle}{2\sqrt{\varepsilon t}} \quad . \tag{3.46}$$

As before, from (3.11), (3.20) and (3.40), we can show that

$$\lim_{t\to\infty} \langle z(t)\rangle^2/t = 0 \quad \text{i.e.,} \quad \langle z(t)\rangle^2 \ll t \tag{3.47}$$

for large t, which corresponds to (3.23). Therefore, we have

$$\lim_{t\to\infty} c(t) = 0. \tag{3.48}$$

Consequently, the second term in (3.45) vanishes, as t goes to infinity. Thus, we get

$$N(t) \simeq 1/\sqrt{\pi\varepsilon t} \quad . \tag{3.49}$$

We arrive finally at

$$P(x,t)dx \simeq \frac{1}{\sqrt{\pi\varepsilon t}} \exp\{-\frac{(\hat{z}(x)-\langle z(t)\rangle)^2}{4\varepsilon t}\}d\hat{z}(x) \tag{3.50}$$

for large t. This yields again (3.25) by the help of (3.47).

4. FRACTIONAL BROWNIAN MOTION AND LONG-TIME TAIL

In the present section, we extend our arguments in preceding sections to the following fractional brownian processes

$$\frac{dx}{dt} = a(x) + b(x)\xi(t), \tag{4.1}$$

where $\xi(t)$ is a Gaussian and fractional random noise with the correlation

$$\langle\xi(t)\xi(t')\rangle \equiv \gamma(t-t') \sim \frac{1}{|t-t'|^\alpha} \tag{4.2}$$

for large time difference $|t-t'|$. In Mandelbrot's notation (17), we have

$$\langle B(s)B(t)\rangle = \frac{\varepsilon}{2}\{|s|^{2H} + |t|^{2H} - |s-t|^{2H}\}, \tag{4.3}$$

where

$$B(t) = \int_0^t \xi(t')dt' \ . \tag{4.4}$$

The case $\alpha = 1$, namely $H = \frac{1}{2}$ corresponds to the ordinary Wiener process. If the correlation of the noise $\xi(t)$ is exponential as

$$\langle\xi(t)\xi(t')\rangle = 2\varepsilon e^{-\lambda|t-t'|} \ , \tag{4.5}$$

then the long-time tail is essentially the same as that for a Gaussian white noise, as was discussed already by the present author.(3) Thus, we discuss here the fractional random noise (4.2).

Almost all arguments in preceding sections are still valid except the expression of the variance

$$\langle B^2(t)\rangle = \int_0^t dt'\int_0^t dt''\gamma(t'-t'') \sim t^{2H}, \tag{4.6}$$

where (18)

$$2H = 2-\alpha;\ 0 < \alpha < 1 \tag{4.7a}$$

and

$$2H = 1\ ;\ 1 \le \alpha \ . \tag{4.7b}$$

Thus, we have

$$\frac{1}{2} \le H < 1 \ . \tag{4.8}$$

As before, we transform here (4.1) into the following additive stochastic process

$$\frac{dy}{dt} = \alpha(y) + \xi(t), \tag{4.9}$$

with use of the change of variable (3.10). If the deterministic part (or drift term) $\alpha(y)$ is asymptotically more dominant than the remaining noise term $\xi(t)$, then the deterministic long-time tail appears.(1-3) The condition for this is given by

$$|\alpha(y)| \gg y^{(H-1)/H} \quad \text{for } y \to -\infty, \tag{4.10}$$

because the scaling of $\xi(t)$ is specified by t^{H-1} and consequently because $\alpha(y)$ should be more dominant than t^{H-1}(i.e., $y^{(H-1)/H}$).

Conversely, if

$$|\alpha(y)| \lesssim y^{-(1-H)/H} \quad \text{for } y \to -\infty, \tag{4.11}$$

then the noise-induced long-time tail appears

under the conditions that $a(x) < 0$ and $b(x) > 0$ for $x > 0$ and that $b(x) \sim O(x)$ or $o(x)$.

The noise-induced long-time tail exponent ψ defined by

$$\langle\langle Q(x(t))\rangle\rangle \sim t^{-\psi} \tag{4.12}$$

is now given by $\psi = H$, namely

$$\psi = 1 - \frac{1}{2}\alpha \quad \text{for } 0<\alpha<1, \tag{4.13a}$$

and

$$\psi = \frac{1}{2} \quad \text{for } \alpha \geq 1. \tag{4.13b}$$

This gives a generalization of our previous arguments on stochastic equations with the Wiener process.(1-3)

In particular, when we consider the following multiplicative stochastic processes

$$\frac{dx}{dt} = \gamma x - gx^m + x^n\xi(t), \tag{4.14}$$

with the fractional random noise $\xi(t)$, the condition of the appearance of the noise-induced long-time tail is given by

$$m > (n + H - 1)/H \text{ and } n > 1 \tag{4.15}$$

from the condition (4.11) with the relation (3.11), because

$$\alpha(y) = \frac{a(x(y))}{b(x(y))} \sim x^{(m-n)} \sim |y|^{-(m-n)/(n-1)}. \tag{4.16}$$

From (4.8) and (4.15), we find that

$$m - n > 0 \quad . \tag{4.17}$$

Thus, $\alpha(y(x))$ vanishes, as x goes to zero. This is also easily seen from (4.11).

It will be interesting to note here an invariant property of the inequality (4.15) for the following nonlinear transformation

$$y = x^q \quad \text{for } q > 0. \tag{4.18}$$

By applying (4.18) to (4.14), we obtain the following transformed equation

$$\frac{dy}{dt} = \gamma' y - g' y^{m'} + y^{n'}\xi'(t) \tag{4.19}$$

with $\gamma' = q\gamma$, $g' = qg$, $\xi'(t) = q\xi(t)$ and

$$m' = (m-1+q)/q, \; n' = (n-1+q)/q. \tag{4.20}$$

The quantity Q_1' defined by $Q_1' \equiv q\{m'-(n'+H-1)/H\}$ is invariant for the transformation (4.18), that is,

$$Q_1' \equiv q\{m'-(n'+H-1)/H\} = m-(n+H-1)/H \equiv Q_1 \; . \tag{4.21}$$

Thus, the above condition (4.15) for the appearance of the noise-induced long-time tail is consistent for any representation of stochastic variable as far as the transformation (4.18) is concerned.

By the way, the quantity Q_2 defined by $Q_2 = (m-n)/(n-1)$ is also invariant under the transformation (4.18).

The case $H = 1/2$ corresponds to the previous situation and (4.15) is reduced to

$$m > 2n-1 > 1, \tag{4.22}$$

as was already discussed in Section 3.

The above arguments can be easily extended to the case $0 < H < \frac{1}{2}$ with similart results, namely $\psi = H$, using the parameter H defined by (4.3).

5. LONG-TIME TAIL OF GEOMETRICAL OBJECTS

Fractional Brownian motions are quite often seen in nature. For example, the relative diffusion in turbulence and clumps are all described by generalized Brownian motions with large correlations.(19) A simple example of fractional Brownian motions will be given by a string animal introduced by the present author.(19) This is a geometrical object in which each segment can move in a stochastic way with the condition that the total length of string is conserved.

The gravitational center $x(t)$ of a string animal is shown to take the following asymptotic form

$$\langle x^2(t)\rangle \sim t^{\psi} \tag{5.1}$$

for large time t. The exponent ψ is about 2/3 in two dimensions, which is fractional.(20,21)

There are many other topological systems showing fractional long-time tails, and they will be discussed in future elsewhere.

6. MONTE CARLO SIMULATIONS OF LONG-TIME TAIL

It is easy to perform Monte Carlo simulations on stochastic processes (3.1) with Gaussian and white noise. In fact, Hamada and Muto (10), Sasagawa, Miyashita and Suzuki (22), and Okada (23) have calculated explicitly several moments in the multiplicative stochastic processes (1.7), for example, with $m = 3$ and $n = 1$, and with $m = 5$ and $n = 2$. Their results agree very well with our theoretical predictions on the noise-induced long-time tail described in the present paper, including the coefficients of the long-time tail.

It will be also quite interesting to make Monte Carlo simulations on the long-time tail for the fractional stochastic processes (4.14) with the fractional random noise $\xi(t)$. For this purpose, the following representation by Mandelbrot and Van Ness(24) on the fractional Brownian motion

$$B(t) = N\{\int_{-\infty}^{0}((t-\xi)^{H-1/2}-(-\xi)^{H-1/2})dW(\xi) + \int_0^t(t-\xi)^{H-1/2}dW(\xi)\}, \tag{6.1}$$

will be useful (25), where N is the normalization

factor and W(ξ) denotes the ordinary Wiener process.

Monte Carlo simulations on fractional stochastic processes will be quite useful to understand the asymptotic behavior of their solutions, and to confirm the long-time tail.

7. SUMMARY AND DISCUSSION

In the present paper, we have explained intuitively the mechanism of the appearance of the noise-induced long-time tail in stochastic processes with Gaussian and white noise, and also with Gaussian and fractional random noise. That is, the noise-induced long-time tail occurs due to the balance (or cancellation) between the deterministic part and noise part in multiplicative stochastic processes. Thus, the noise plays an essential role for this interesting phenomenon. We may also say that the noise-induced long-time tail occurs owing to the "asymptotic self-similarity" of the relevant stochastic process at the critical point.

Quite recently N. Minami (26) has given, to our intuitive general arguments on the noise-induced long-time tail, a mathematicl justification by extending Kasahara's theorem (27) on the asymptotic behavior of stochastic processes for compact supports to those for non-compact supports.

Monte Carlo simulations and experiments on fractional stochastic processes will be expected to be performed in the near future in order to confirm the present theoretical predictions.

ACKNOWLEDGEMENTS

The author would like to thank Professor N. Ikeda, Dr. Y. Ogura, and Mr. N. Minami for useful discussions. This work is partially financed by Yoshida Foundation for Science and Technology, and also by the Scientific Research Fund of the Ministry of Education.

REFERENCES:

[1] Suzuki, M., Kaneko, K. and Takesue, S., Prog. Theor. Phys. 69 (1982) 1756-1775.
[2] Suzuki, M., Takesue, S., and Sasagawa, F., Prog. Theor. Phys. 68 (1982) 98-115.
[3] Suzuki, M., Prog. Theor. Phys. 68 (1982) 1917-1934.
[4] Suzuki, M., and Kubo, R., J. Phys. Soc. Jpn. 24 (1968) 51.
[5] Horsthemke, W., and Malek-Mansour, M., Z. Phys. B24 (1976) 307
[6] Horsthemke, W., and Lefever, R., Phys. Lett. 64A (1977) 19.
[7] Schenzle, A. and Brand, H., Phys. Rev. 20B (1979) 1628.
[8] Brenig, L. and Banai, N., Physica 5D (1982) 208.
[9] Graham, R. and Schenzle, A., Phys. Rev. A25 (1982) 1731.
[10] Hamada, Y. and Muto, K., Prog. Theor. Phys. 69 (1983) 451 and references cited therein.
[11] Ikeda, N. and Watanabe, S., Stochastic Differential Equations and Diffusion Processes, Kinokuniya co. 1982.
[12] Suzuki, M., Asymptotic Behavior of Nonlinear Brownian Motion near the Instability Point. Proceedings of the Taniguchi International Symposium on Stochastic Analysis, ed. Ikeda, N. (Springer).
[13] Van Kampen, N.G., Can. J. Phys. 39 (1961) 551.
[14] Kubo, R., Matsuo, K., and Kitahara, K., J. Stat. Phys. 9 (1973) 51.
[15] Suzuki, M., Prog. Theor. Phys. 53 (1975) 1657.
[16] Suzuki, M., Adv. in Chem. Phys. 46 (1981) 195-278.
[17] Mandelbrot, B.B., The Fractal Geometry of Nature, W.H. Freeman Company, San Fransisco, 1982.
[18] The last inequality $1 < \alpha < 2$ in (2.12) of Ref.3 should read $0 < \alpha < 1$.
[19] Suzuki, M., Dynamics of Topological Disorder —Brownian Motion with Geometrical Restriction— Proceedings of the Taniguchi International Symposium on Topological Disorder in Condensed Matter. (Springer).
[20] Suzuki, M., Application of Fractal Analysis to Phase Transitions and Other Phenomena, Proceedings of U.S.-Japan Workshop on Statistical Physics and Chaos in Fusion Plasmas, ed. Horton, W., and Reichl, L.(Springer).
[21] Suzuki, M., and Sasagawa, F., to be published.
[22] Sasagawa, F., Miyashita, S. and Suzuki, M., Prog. Theor. Phys. to be submitted.
[23] Okada, T., to be published.
[24] Mandelbrot, B.B., and Van Ness, J.W., SIAM Review, 10 (1968) 422.
[25] Mandelbrot, B.B., and Taqqu , M.S., Robust R/S Analysis of Long Run Serial Correlation, IBM Research Report, 1979.
[26] Minami, N., Master thesis, 1983.
[27] Kasahara, Y., Spectral theory of generalized second order differential operators and its applications to Markov processes, Japan J. Math. 1 (1975) 67-85.

NOISE IN PHYSICAL SYSTEMS AND 1/f NOISE
M. Savelli, G. Lecoy and J-P. Nougier (eds.)
© Elsevier Science Publishers B.V., 1983

FLICKER NOISE IN NON-ELECTRIC SYSTEMS

Bruno Pellegrini

Istituto di Elettronica e Telecomunicazioni
Università di Pisa, Via Diotisalvi 2
56100 Pisa, Italy

The island theory of the flicker, burst and generation-recombination noises of the electric devices is extended to non-electric systems, even non-physical. Such a result is achieved by omitting the electric phenomena in the previous general model and by retaining the diffision processes occurring in many large element-ensembles. The noise spectra and the conclusions so obtained are analogous to the ones holding true for the electric systems and they appear able to explain the ubiquity of the flicker noise.

1. INTRODUCTION

Many physical, biological and economic systems produce fluctuations with $1/f^{\gamma}$ power spectral density ($0.7<\gamma<1.4$) that is they have a spectrum which diverges when the frequency f tends to zero.

The systems which, for intrinsic interest and measurement simplicity, have been and are more studied, both theoretically and experimentally, are the electric systems (ES). For them we have proposed the island model [1-4] which puts the origin of the low-frequency noises in zones, called islands, of any nature and size that are able to capture, to store and to release electric carriers. Such a model yields an unitary theory of the generation-recombination, burst and flicker noises [1,2], together with the diffusion-noise theory it removes any difference between the models of the carrier-number and -mobility fluctuations [3] and, finally, it shows that a pure 1/f noise does not exist physically [4].

In this paper we show that the model may be extended to non-electric systems (NES) too. This result is achieved from the general analysis developed in the Ref. 2 by omitting properly the electric drift and effects and by retaining, instead, the diffusion phenomena and capture, storage and release processes from the islands which occur in many types of large element-ensembles.

2. RELAXATION TIME

In a non-electric system let n_e, $\underline{J}_e$ and μ be the densities of the particles and of their current, and the chemical potential, respectively. Moreover let N_e and i_e be the number of the particles belonging to an island and the net current of those which enter it, respectively. Such quantities are related to the corresponding ones of the electric systems by the relationships [2]

$$n_e=n, \quad \underline{J}_e=-\underline{J}/q, \quad \mu=-qu \quad , \tag{1}$$

$$N_e=-Q/q=\sum_j N_j f_j \quad , \quad i_e=-i/q \quad , \tag{2}$$

where q and Q are the electron and island charge, respectively, N_j and f_j are the state number and the occupation factor of the jth energy level.

In the NES, instead, the electric field and potential $\underline{\xi}$ and v are null everywhere, i.e.

$$\underline{\xi}=0 \quad , \quad v=0 \quad , \tag{3}$$

By taking into account (1)-(3) in the analysis of the Ref. 2, we obtain that the fluctuation Δi_e of the current entering the island becomes

$$\Delta i_e=G_e(b\Delta\mu_E-\Delta\mu_I) \quad , \tag{4}$$

where

$$G_e=i_{eo}/kT=\sum_j \bar{e}_j N_j \bar{f}_j/kT \quad , \tag{5}$$

is the island conductance, b is a proper coefficient [2], μ_E and μ_I are the chemical potential outside and inside the island, respectively, e_j is the emission coefficient of the jth level and i_{eo} is the average value of the currents entering and leaving the island.

On the other hand, according to (2), the N_e fluctuation is given by

$$\Delta N_e=C_{Ie}\Delta\mu_I \quad , \tag{6}$$

where

$$C_{Ie}=\sum_j N_j\bar{f}_j(1\pm\bar{f}_j)/kT=\bar{N}_I/kT \quad , \tag{7}$$

is the island capacitance (the - and + signs holding true for fermions and bosons, respectively) k is the Boltzmann constant and T is the absolute temperature.

Since the aim of this first part of the analysis is to compute the island relaxation time

$$\tau \equiv -\Delta N_e/\Delta i_e \quad , \qquad (8)$$

and the fluctuations

$$\Delta i_A = \int_A d\underline{A}.\Delta\underline{J}_e \quad , \qquad (9)$$

of the current through a given probe surface A generated by the island variations ΔN_e, we must take into account also the transport and continuity phenomena outside the island which lead to the relationships [2]

$$\Delta\underline{J}_e = b(\underline{J}_{eo}\Delta\mu - D\bar{n}_e\underline{\nabla}\Delta\mu)/kT \quad , \qquad (10)$$

$$\nabla^2\Delta\mu = 0 \quad , \qquad (11)$$

where $\underline{J}_{eo}$ is the direct current density of the particles and D is their diffusion constant.

The solution of (11) is of type [2]

$$\Delta\mu = \Delta\mu_E\alpha(\underline{r}) = \Delta\mu_E r_I/r \quad , \qquad (12)$$

where the third member holds true if the island, of radius r_I and surface A_I, and the boundary conditions are spherical.

The third equation which we are looking for in order to compute τ follows from (10) and (12) in the form

$$\Delta i_e = -G_{Ee}\Delta\mu_E \quad , \qquad (13)$$

where

$$G_{Ee} = -\frac{bD\bar{n}_e}{kT}\int_{A_I} d\underline{A}.\underline{\nabla}\alpha = \frac{bD\bar{n}_e A_I}{kTr_I} \quad , \qquad (14)$$

is the conductance of the medium surrounding the island.

Finally, from (4)-(8), (13) and (14), we obtain

$$\tau = \frac{\bar{N}_I}{i_{eo}}\left(1+\frac{i_{eo}r_I}{D\bar{n}_e A_I}\right) \simeq \frac{\bar{N}_I}{i_{eo}} \simeq \frac{1}{\bar{e}_j} \quad , \qquad (15)$$

where the third member holds true for

$$i_{eo} \ll D\bar{n}_e A_I/r_I \quad , \qquad (16)$$

whereas the fourth one, according to (5) and (7), is verified when the levels belong to an interval smaller than kT and $\bar{f}_j \ll 1$. We observe that in the last two cases τ does not depend on the medium surrounding the island and, in the form $\tau = 1/\bar{e}_j$, it is independent also of the element statistics $\bar{f}_j$.

3. FLUCTUATION PICKING UP

In the NES the island variations ΔN_e can not generate, of course, any voltage fluctuations between two probe points.

There is another difference between the ES and the NES. In fact in the ES the electric field makes $\Delta\underline{J}_e$ an exponential function of the distance r from the island so that its current-dipole vector $\Delta\underline{\dot{P}}_E = \int \Delta\underline{J}_e d^3x$ is finite. In the NES, instead, since $\Delta\mu$, according to (12), is inversely proportional to r, $\Delta\dot{P}_e$ is not finite and it can not be used as a probe for picking up the island fluctuations ΔN_e.

They may be probed, instead, both in the NES and in the ES, by means of the fluctuations Δi_A, induced by ΔN_e, of the current across a given probe surface A.

In fact from (8), (10) and (12)-(14) it follows that

$$\Delta\underline{J}_e = \frac{bD\bar{n}_e}{G_{Ee}}\left[\frac{\alpha(r)}{D\bar{n}_e}\underline{J}_{eo} - \underline{\nabla}\alpha(r)\right]\frac{\Delta N_e}{\tau} \quad , \qquad (17)$$

which for spherical symmetry and

$$J_{oe} \gg D\bar{n}_e/r \quad , \qquad (18)$$

becomes

$$\Delta\underline{J}_e = \frac{\underline{J}_{eo}\Delta N_e}{4\pi D\bar{n}_e r\tau} \quad , \qquad (19)$$

Then, by indicating with

$$i_A \equiv \int_A d\underline{A}.\underline{J}_{eo} \quad , \qquad (20)$$

the direct current across A and with R the mean distance between the island and A itself, from (9) and (19) we obtain finally

$$\Delta i_A = \frac{i_A}{4\pi D\bar{n}_e R\tau}\Delta N_e \quad . \qquad (21)$$

i.e. Δi_A is proportional to i_A. However, according to (9) and (17), when the condition (18), for instance for $J_{eo}=0$, is not verified, we have a fluctuation $\Delta i_A \propto \Delta N_e$ also for $i_A=0$. This is another difference between NES and ES.

4. NOISE SPECTRA

Let us compute now the power spectral densities of the fluctuations due to ΔN_e.

The entity conservation in the island leads, according to (8), to the relaxation equation

$$\frac{d\Delta N_e}{dt} = -\frac{\Delta N_e}{\tau} + \eta \quad , \qquad (22)$$

where η is the stochastic component of the current crossing the island surface. If such crossing is a Poisson process, the spectrum of η becomes

$$S_\eta = 4\, i_{eo} \quad , \qquad (23)$$

so that the Langevin method applied to (22) leads

to the spectrum

$$S_N = \frac{4<(\Delta N_e)^2>_t \tau}{1+\omega^2\tau^2} \quad , \quad (24)$$

of ΔN_e, where according to (15) and (23), its variance $<(\Delta N_e)^2>_t$ is given by

$$<(\Delta N_e)^2>_t = \bar{N}_I(1+\frac{i_{eo}^r I}{D\bar{n}_e A_I}) \simeq \bar{N}_I \quad , \quad (25)$$

in which the last term holds true when (16) is verified. By summing the contributes of all the islands, of density D_I, the total spectrum S_I of Δi_A, from (21), (24) and (25), becomes

$$S_I = \int \frac{i_A^2 \ \bar{N}_I}{4(\pi D\bar{n}_e R)^2 \tau(1+\omega^2\tau^2)} D_I D_{\bar{N}_I} \cdot D_\tau d^3x d\bar{N}_I d\tau \quad , \quad (26)$$

where $D_{\bar{N}_I}$ and D_τ are the distribution functions of $\bar{N}_I$ and τ, respectively.

From (26) it follows that S_I is independent of the frequency f when $D_\tau(\tau)$ is constant with respect to τ, whereas it is $S_I \propto 1/f$ for $D_\tau \propto \tau$. More in general, according to the general method and results of the Ref. 2, the spectrum S_I becomes of flicker type $S_I \propto 1/f^\gamma$, with an exponent γ independent of f and about equal to 1 on several f decades, when (D_τ/τ) has broad maximum for high values of τ.

These results for the NES differ from those of the ES for which it is $S_I \propto$ constant, $S_I \propto 1/f$ and $S_I \propto 1/f^\gamma$ for $D_\tau \propto 1/\tau^2$, $D_\tau \propto 1/\tau$ and (τD_τ) with a broad maximum, respectively. That is the flicker noise existence needs, both in the NES and in the ES, a wide dispersion of τ and, about, $D_\tau \propto \tau$ and $D_\tau \propto 1/\tau$ for the NES and the ES, respectively.

In the NES, as well in the ES, S_I is proportional to the square steady current i_A^2. In particular if, in the considered system, i_A has a.c. component at a frequency f_A, the corresponding amplitude modulation of ΔN_e given by (21) leads to an intermodulation part of the spectrum S_I dispersed around f_A.

The model has been developed on the assumption that the particle distribution f_j is given by the Fermi-Dirac or Bose-Einstein statistics. However it is to be observed that the spectrum S_I given by (26), with τ and N_I given by the last term of (15) and (25), respectively, holds true more in general for any diffusive element ensemble containing collection, storage and emission islands characterized by dispersed, or not, memory times τ. Moreover the system may be in a gaseous or liquid state, or in a whichever condition of the mobile entities.

5. EXAMPLES AND CONCLUSIONS

We think that the fluctuations phenomena in many physical, biological and economic systems may be studied and explained by means of the proposed approach.

In this way we could probably explain the fluctuations of the generation and of the flow of organic liquids, such as the insulin and the hormones, in the living organisms, of the monetary fluxes or of other variables in the economic systems, and so on.

The model should justify the flicker noise in the neutron flux in the terrestrial magnetosphere [5], the flow rate of sand in hourglass [5], the flux of cars on an expressway [6], the flow rate of a river (e.g. of the Nile over the last 2000 years [5]) and so on. In the last case the system to be considered is the river drainage-basin, fed by the rains, with its water collection-emission elements; the probe quantity i_A is the flow per second of the river itself. So in the traffic example i_A is the car flux on the expressway and the system is constituted by the entire national traffic structure, where the towns are the storage islands.

In conclusions the island model of the low-frequency noises, and its extension, here proposed, from the electric systems to the non-electric ones, seem to find the "universal" flicker-noise origin and mechanism behind its various manifestations and its ubiquity, in the unavoidable existance, in almost all the systems of mobile entities, of storage elements, with highly dispersed hold times, which, by interacting with the transport process of the entities themselves, determine just the typical spectrum $1/f^\gamma$, with $\gamma \simeq 1$, of their flow fluctuations.

All such various systems, apart from their particular nature, have in common, for the fluctuations, similar mathematical models and analogous statistic structures and results.

REFERENCES

[1] Pellegrini, B., New theory of flicker noise, Phys.Rev. B22 (1980) 4684-4691.

[2] Pellegrini, B., One model of flicker, burst and generation-recombination noises, Phys. Rev. B24 (1981) 7071-7083.

[3] Pellegrini, B., Diffusion, mobility fluctuation, and island models of flicker-noise, Phys.Rev. B26 (1982) 1791-1797.

[4] Pellegrini, B., Saletti, R., Terreni, P., Prudenziati, M., $1/f^{\gamma}$ noise in thick-film resistors as an effect of tunnel and thermally activated emissions, from measures versus frequency and temperature. Phys.Rev. B27 (1983) 1233-1243.

[5] Handel, P.H., Quantum approach to 1/f noise, Phys.Rev. A22 (1980) 745-756.

[6] Musha, T., Higunchi, H., Jap.J.Appl.Phys. 15 (1976) 1271-1275.

NOISE IN PHYSICAL SYSTEMS AND 1/f NOISE
M. Savelli, G. Lecoy and J-P. Nougier (eds.)
© Elsevier Science Publishers B.V., 1983

THEORY OF 1/f NOISE IN METALS

Yoshimasa Isawa

Research Institute of Electrical Communication
Tohoku University, Katahira, Sendai, Japan

We propose a microscopic model to make possible to generate 1/f fluctuations in metal films. The 1/f fluctuations in energy spectral density of conduction electrons are caused by the scattering of diffusely propagating conduction electrons due to surface phonon modes accompanying the displacements of randomly distributed defects. The energy fluctuation leads to a mobility fluctuation which results in 1/f noise for current or voltage fluctuations. The details of the model and the results obtained are compared with the experimental ones and the empirical formulas.

1. INTRODUCTION

We proposed a new microscopic model[1] to explain 1/f noise in metals at relatively high temperature above Debye temperature but below the Fermi energy. The model is based on the scattering of diffusely propagating conduction electrons due to surface acoustic modes accompanying the displacement of randomly distributed defects. An analysis of 1/f noise in metal films based on the model together with its details is presented in this paper.

2. MODEL FOR 1/f NOISE

The dimension of metal films is assumed to be length L, width L, thickness d and volume $V=L^2d$ (L>>d). An electron is free to move in x- and y-directions parallel to the surface of films, while the motion of an electron normal to the surface can be described by standing wave modes. The surface phonon modes are wavelike in x- and y-directions, but decay with a distance of the order of wavelength away from the surface. We only treat the surface modes with wavelength much larger than d. Hence the mode amplitudes are regarded as constant with respect to z-direction. The defects are supposed to be point defects (like vacancies, interstitial atoms or substitutional atoms), dislocations and grain boundaries. The discussions, however, are not detailed enough to distinguish between these defects. The defects are introduced here to destroy the periodic structure of the crystal. They lead not only to an elastic scattering of conduction electrons but also to an inelastic (electron-surface phonon) scattering accompanied by the displacement of defect atoms.

The electron-surface phonon interaction accompanying the displacement of defects is given by, based on the rigid ion model[2],

$$-\frac{1}{L}\sum_{\substack{s,q,\lambda\\ p,p'}} e^{iq\cdot r_s}\int dr\,\psi_p^*(r)Q_\lambda(q)\cdot\nabla v_s(r-r_s)\psi_{p'}(r)c_p^+c_{p'} \tag{1}$$

where $Q_\lambda(q)$ is the displacement operator for q, λ-th surface mode, c_p^+ and c_p are the creation and annihilation operators of an electron, r_s denotes the position of the s-th defect, $v_s(r-r_s)$ is the s-th defect potential. The $\psi_p(r)$ is the Bloch function given by

$$\psi_p(r)=\sqrt{\frac{2}{V}}\,e^{ip_\perp\cdot r_\perp}\sin(p_z z)u_p(r_\perp), \tag{2}$$

where $p_z=\frac{m\pi}{d}$ (m=1,2,--), $p_\perp$ is the momentum in the plane parallel to the surface and $u_p(r_\perp)$ has the periodicity of the Bravais lattice. In the following only one kind of defects is assumed, for simplicity, and the scattering processes in which the momentum normal to the surface is conserved are taken into account, i.e., $p_z=p_z'$, since other processes ($p_z\neq p_z'$) do not contribute to 1/f noise. If the defects distribute randomly and the average distance between the neighboring defects is much smaller than the surface phonon wavelength and also the thickness of the film, then the homogeneity is recovered with respect to long wavelength behaviors. Hence the electron-surface phonon interaction hamiltonian accompanying the displacement of defects is written as follows,

$$H_{e-sp}=\frac{1}{L}\sum_{p,q,\lambda}v_s(p,q)\cdot Q_\lambda(q)c_p^+c_{p-q}\ . \tag{3}$$

The coupling constant $v_s(p,q)$ is given by

$$v_s(p,q)=-n_s\int dr\,e^{-iq\cdot r}u_p(r+\Delta r_s)\nabla v_s(r)u_{p-q}(r+\Delta r_s), \tag{4}$$

where n_s is the density of defects and Δr_s denotes the position of the s-th defect within the unit cell. The H_{e-sp} has a nonzero contribution proportional to $Q_\lambda(q)$ even when q is approached to zero since $v_s(p,0)\neq 0$.

In the absence of defects, an electron-phonon interaction in crystal vanishes for $q\to 0$ if Umklapp processes are ignored. This is because it is proportional to

$$(\psi_p|[H_c,p]|\psi_p)=0, \quad (5)$$

where H_c is the one electron Hamiltonian with periodic potential and ψ_p is the Bloch state with momentum p. However eq.(5) is not correct when the periodicity of the lattice is broken. This is the reason why electron-surface phonon interactions do not disappear at q=0. The $v_s(p,0)$ term means that the uniform displacement of the total atoms gives rise to a net energy change of the system, which is, of course, unphysical. In order to remove this difficulty, H_{e-sp} should be changed as follows,

$$H'_{e-sp} = \frac{1}{L}\sum_{p,q,\lambda} v_s(p,q)\cdot Q_\lambda(q)(c_p^+c_{p-q}-\delta_{q,0}\langle c_p^+c_p\rangle), \quad (6)$$

where $\langle c_p^+c_p\rangle$ is the average density of electrons with momentum p. The additional term in eq.(6) indicates that the redefinition of the displacement operator due to the introduction of defects has to be made. However the additional term has no effects on the noise. Therefore we ignore this term and adopt eq.(3) as the electron-surface phonon interactions in this paper.

The total hamiltonian of the system is given by

$$H = H_0 + H_{e-sp}, \quad (7)$$

where H_0 includes the scattering of the conduction electrons due to static part of defect potential and the interactions with bulk acoustic phonon modes.

Since the electron-phonon interaction due to piezo electric coupling has the same form as eq.(3), the following arguments can also be applied to piezo electric substances.

3. SPECTRAL DENSITY FUNCTION

The auto-correlation function of X is defined as follows

$$\phi_X(t)=\frac{1}{2}\mathrm{Tr}\,\rho[X(t+t_0)X(t_0)+X(t_0)X(t+t_0)], \quad (8)$$

where ρ is the density matrix of the system, X(t) is the Heisenberg operator of X. Then the spectral density function for X is given by

$$S_X(f)=4\mathrm{Re}\int_0^\infty dt\, e^{i\omega t}\phi_X(t) = i(P^R(\omega)-P^A(\omega))\coth\frac{\omega}{2k_BT}, \quad (9)$$

where $f=\omega/2\pi$, and $P^R(\omega)$ and $P^A(\omega)$ are the retarded and the advanced Green's function defined by

$$P^R(\omega)=-i\int_{-\infty}^{\infty}dt\,\mathrm{Tr}\rho[X(t),X(0)]\theta(t)e^{i\omega t}, \quad (10.a)$$

$$P^A(\omega)=i\int_{-\infty}^{\infty}dt\,\mathrm{Tr}\rho[X(t),X(0)]\theta(-t)e^{i\omega t}. \quad (10.b)$$

These are obtained from the analytic continuation of thermal Green's function, $\mathcal{P}(\omega)$ [3]. Equation (9) indicates that 1/f fluctuation is found for the systems where $(P^R(\omega)-P^A(\omega))$ takes a nonzero constant value proportional to -isign(ω) as ω becomes smaller.

The observed quantity in 1/f noise is the spectral density function for current or voltage fluctuation. We do not, however, derive 1/f term from the direct analysis at the present time. This is because we have to do an intricate calculation of a 4-point function since 1/f noise might be proportional to I^2 or V^2 where I and V are the external dc current and voltage, respectively. Hence we first calculate the spectral density function for energy fluctuation ($X=H_{e-sp}$) of the conduction electrons, and then transform it into the one for current or voltage fluctuations by using a phenomenological relation.

For simplicity, the temperature is assumed to be in the range;

$$E_F \gg k_BT \gg k_B\Theta, \quad (11)$$

where E_F is the Fermi energy and Θ is the Debye temperature. The lowest order process of H_{e-sp} for $\mathcal{P}(\omega)$ is shown in Fig.1. The vertex correction shown in Fig.2 is dominated by the multiple scattering of the conduction electrons due to defects and electron-bulk phonon interaction which leads to a diffusive motion to electrons.

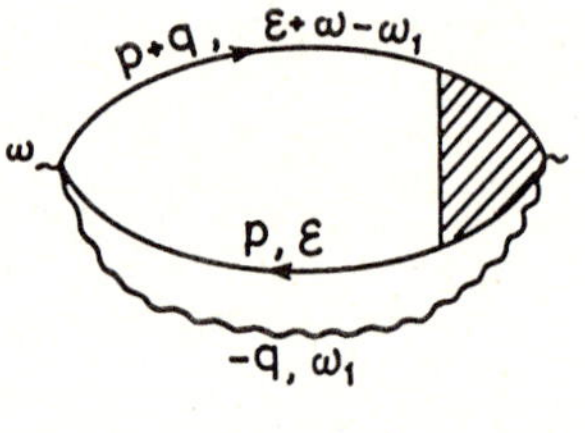

Fig.1. The lowest order diagram of electron-surface phonon interaction for $\mathcal{P}(\omega)$. The wavy line is a surface phonon Green's function. Real line means the electron Green's function. The parallel hatched triangle is the vertex correction defined in Fig.2. The momentum normal to the surface is given by $p_z=m\pi/d$. The sum of m is limited by $p_z< p_F$. The phonon momentum q is 2-dimensional vector parallel to the surface.

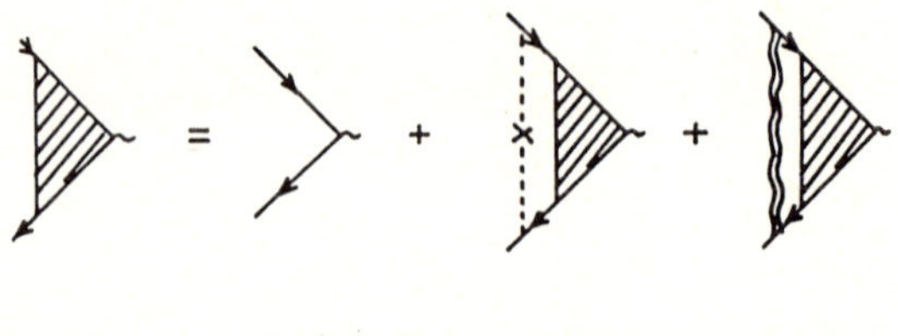

Fig.2. The equation for the vertex correction. The dotted line with a cross means a scattering of the conduction electron due to a static part of defect potential. The double wavy line denotes the bulk phonon Green's function.

The thermal Green's function is given by, on the assumption of $|\omega_1-\omega|<1/\tau$ and $Dq^2<1/\tau$,

$$\mathcal{P}(\omega)=-2N(0)\frac{k_BT}{\rho_0L^2}\sum_{q,\lambda,\omega_1}\frac{1}{\omega_{q,\lambda}^2+\omega_1^2}\frac{Dq^2}{Dq^2+|\omega_1-\omega|}v_s^2, \quad (12)$$

where ρ_0 is the surface mass density, $N(0)=L^2 d\times\frac{mp_F}{2\pi^2}$ is the density of states per spin at the Fermi energy with effective mass, m, and the Fermi momentum, p_F, $\omega_{q,\lambda}$ is the surface phonon energy, $v_s(p,0)$ denotes the electron-surface phonon interaction, and $D=\frac{1}{2}v_F^2\tau$ is the diffusion constant for conduction electrons propagating freely in 2-dimensional area with the Fermi velocity, v_F. In Ref[1] we used $D=\frac{1}{3}v_F^2\tau$ since the conduction electrons were supposed to propagate freely in 3-dimensional area.

The scattering time, τ, due to defects and electron-bulk phonon interaction is given by, in Born approximation shown in Fig.3,

$$1/2\tau=\pi N(0)n_s\overline{u(p-p')^2}+\pi k_BTN(0)\overline{g(p-p')^2}, \quad (13)$$

where $u(p-p')$ is the static part of defect potential, $g(p-p')$ is the electron-bulk phonon interaction and $\overline{A(p-p')}$ means the average over solid angle of A on the Fermi surface, $|p|=|p'|=p_F$. The large momentum transfer processes are dominant in Fig.2 and Fig.3. Thus the discreteness of p_z can be ignored in eq.(13).

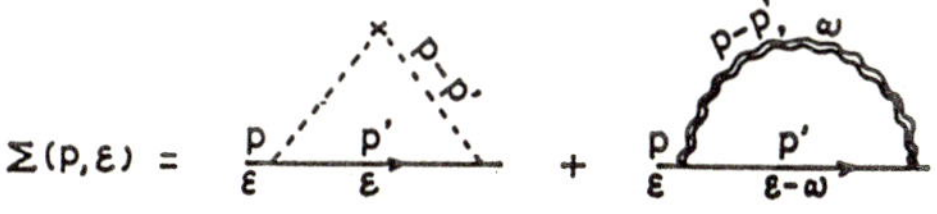

Fig.3. The self-energy part, $\Sigma(p,\varepsilon)$, of the electron Green's function in Born approximation. The relaxation time, τ, is given by $1/2\tau=\mathrm{Im}\Sigma^A(p_F,\varepsilon=0)$.

If a small momentum part of bulk phonon is taken into account, the vertex correction coming from the multiple scattering of electrons due to randomly distributed defects in metals has to be considered. However we can show that these corrections are cancelled each other, and the diffusive nature of conduction electrons is preserved since particle-hole scattering processes do not lead to relaxation time.

The main contributions in eq.(12) is approximately written as

$$\mathcal{P}(\omega)=-2N(0)\frac{k_BTv_s^2}{\rho_0L^2}\sum_{q,\lambda}\left[\frac{1}{\omega_{q,\lambda}^2+\omega^2}+\frac{Dq^2}{\omega_{q,\lambda}^2(Dq^2+|\omega|)}\right]. \quad (14)$$

By analytic continuation, $P^R(\omega)$ is obtained as follows,

$$P^R(\omega)=-2N(0)\frac{k_BTv_s^2}{\rho_0L^2}\sum_{q,\lambda}\left[\frac{1}{\omega_{q,\lambda}^2-(\omega+i\delta)^2}+\frac{Dq^2}{\omega_{q,\lambda}^2(Dq^2-i\omega)}\right]. \quad (15)$$

There may be some lower cut off momentum, q_m, for surface phonon momentum determined by the dimension and the geometry of the sample or the circuit. The q_m is supposed to be the order of 1/L. The noise aspects can be classified into the following three cases dependening on the values of q_m and ω.

(i) When q_m is smaller than $|\omega|/v_\lambda(\omega_{q,\lambda}=v_\lambda q)$ and $\sqrt{|\omega|/D}$, two terms at the right hand side of eq.(15) give the same contribution. Then,

$$P^R(\omega)-P^A(\omega)=-i\sum_\lambda\frac{2N(0)}{\rho_0v_\lambda^2}k_BTv_s^2\,\mathrm{sign}(\omega). \quad (16)$$

(ii) When $|\omega|/v_\lambda<q_m<\sqrt{|\omega|/D}$, the second term contributes to 1/f noise, and we obtain

$$P^R(\omega)-P^A(\omega)=-i\sum_\lambda\frac{N(0)}{\rho_0v_\lambda^2}k_BTv_s^2\,\mathrm{sign}(\omega). \quad (17)$$

(iii) When $q_m>\sqrt{|\omega|/D}$,

$$P^R(\omega)-P^A(\omega)\propto -i\omega. \quad (18)$$

From eq.(9),(16) and (17), we can see that $S_{He-sp}(f)$ has a 1/f term in the frequency range of (i) and (ii).

The spectral density function for current or voltage can easily be obtained from the one for resistance. The latter is deduced from $S_{He-sp}(f)$ as follows. The local resistivity fluctuation, $\delta\rho(r,t)$, may be expressed as, on the assumption of local equilibrium,

$$\delta\rho(r,t)=V\left(\frac{\delta\rho}{\delta E}\right)_0\delta\varepsilon(r,t)+V\left(\frac{\delta\rho}{\delta N}\right)_0\delta n(r,t), \quad (19)$$

where the suffix, $_0$, means the value in an equilibrium state, $\delta\varepsilon(r,t)$ and $\delta n(r,t)$ are the fluctuations in energy and number density of conduction electrons and E and N are the total energy and number of electrons.

Owing to our theoretical analysis, the fluctuation spectrum of carrier number does not give rise to 1/f noise. Hence we ignore the second term in eq.(19). From the thermodynamic formula, the first term in eq.(19) is written as

$$\frac{1}{\rho}\left(\frac{\delta\rho}{\delta E}\right)_0=\frac{1}{\rho C_V}\left(\frac{\delta\rho}{\delta T}\right)_0+\frac{1}{\rho}\left(\frac{\delta V}{\delta E}\right)_T\left(\frac{\delta\rho}{\delta V}\right)_0, \quad (20)$$

where C_V is the specific heat of conduction electrons at constant volume given by $C_V=\frac{2}{3}\pi^2k_B^2T\times N(0)$ for $k_BT\ll E_F$, and $(\delta V/\delta E)_T=\frac{5}{3}\frac{1}{nE_F}$ (n: the density of conduction electron) is obtained in free electron approximation. The second term in eq.(20) is much smaller than the first one in typical metals. Hence the spectral density function for resistance is given by

$$S_R(f)/R^2=\left(\frac{\gamma}{C_V}\right)^2S_{He-sp}(f), \quad (21)$$

where R is the resistance and $\gamma=\frac{1}{R}\left(\frac{\delta R}{\delta T}\right)$ is the temperature coefficient of R.

At low frequencies, we obtain the following results;

(i) When $q_m < |\omega|/v_\lambda$ and $\sqrt{|\omega|/D}$,

$$S_R(f)/R^2 = \frac{9}{4\pi^5}\left(\frac{\gamma}{k_B}\right)^2 \frac{1}{N(0)} \sum_\lambda \frac{2v_s^2}{\rho_0 v_\lambda^2} \frac{1}{f} . \qquad (22)$$

(ii) When $|\omega|/v_\lambda < q_m < \sqrt{|\omega|/D}$,

$$S_R(f)/R^2 = \frac{9}{4\pi^5}\left(\frac{\gamma}{k_B}\right)^2 \frac{1}{N(0)} \sum_\lambda \frac{v_s^2}{\rho_0 v_\lambda^2} \frac{1}{f} . \qquad (23)$$

(iii) When $q_m > \sqrt{|\omega|/D}$, the noise spectrum is no longer 1/f-like and flattened at low frequencies.

4. DISCUSSION

We made use of collective modes with low excitation energy in materials as extended sourses of 1/f noise. The proposed model is summarized as follows. The conduction electrons undergo the energy fluctuations due to electron-surface phonon interaction H_{e-sp} which is the origin of 1/f noise. The energy fluctuations bring about the mobility fluctuations through its temperature dependence. Hence we could also interpret the origin of 1/f noise in our model as due to mobility fluctuations of the conduction electrons[4,5].

The system size dependence of $S_R(f)$ is $1/V = 1/L^2d$ since $N(0) \propto V$, which is consistent with the empirical formula by Hooge[6]. However eq.(22) and eq.(23) do not inversely proportional to the total number of mobile charge carrier but to the density of states at the Fermi energy. In this sense, our results are consistent with "1/V" description proposed by Weissman[7].

The temperature coefficient of resistance, γ, is proportional to (μ/μ_{latt}) at constant temperature where μ and μ_{latt} are the total mobility and the one that would be found if only lattice scattering were present. Therefore, eq.(22) and eq.(23) are proportional to $(\mu/\mu_{latt})^2$. This result agrees with Hooge and Vandamme[8].

The electron-surface phonon interactions, v_s, is proportional to the density of defects. There may be two types of defects. The one is native defects. The other is activated by temperature rise, electron bombardment, irradiation of light or any other changes of the external conditions. When an activation type of defect with activation energy, E_g, is dominant, $S_R(f)/R^2$ is proportional to $\exp(-E_g/k_BT)/T^2$ since $\gamma \propto 1/T$. Hence we can expect a peak value of 1/f noise as a function of temperature. This is in qualitative agreement with the recent experimental results of Eberhard and Horn[9].

The noise spectrum being flattened at low frequencies may be related to the experimental results of Ketchen and Clarke[10] for freely suspended metal films, although their sample dimension and the temperature range are not the same as the ones treated in this paper.

REFERENCES

[1] Y.Isawa: J.Phys.Soc.Japan.52,726(1983).
[2] See for example, J.M.Ziman: Electrons and Phonons(Oxford University Press,1960),pp183.
[3] A.A.Abrikosov, L.P.Gorkov and E.Dzyaloshinskii: Methods of the Quantum Theory of Fields in Statistical Physics (Prentice-Hall, Englewood Cliffs,New Jersey, 1963).
[4] F.N.Hooge: Physica.60,130(1972).
[5] T.G.M.Kleinpenning: Physica.77,78(1974).
[6] F.N.Hooge:Phys.Lett.29A,139(1969).
[7] M.B.Weissman: Physica.100B,157(1980).
[8] F.N.Hooge and L.K.J.Vandamme: Phys.Lett.66A ,315(1978).
[9] J.W.Eberhard and P.M.Horn: Phys.Rev.Lett.39 ,643(1977).
[10]M.B.Ketchen and J.Clarke: Phys.Rev.B17,114 (1978).

NOISE IN PHYSICAL SYSTEMS AND 1/f NOISE
M. Savelli, G. Lecoy and J-P. Nougier (eds.)
© *Elsevier Science Publishers B.V., 1983*

UNIFIED MODEL FOR 1/f NOISE IN SEMICONDUCTORS AND METALS

Hisanao Sato

Central Research Laboratory, Matsushita Electric
Industrial Co., Ltd., Moriguchi, Osaka 570, Japan

The noise current spectral density is calculated by assuming that the number fluctuation of lattice defects causes the Fermi level fluctuation. For semiconductors both volume and surface effects are analyzed. The spectral density is found to decrease with increase of the carrier density in qualitative agreement with experimental results.

1. INTRODUCTION

The fluctuation of the number of lattice defects has been shown to be a cause of the 1/f noise in metals. (1) For semiconductors it has been suggested that the number of phonons fluctuates with a 1/f spectrum, since the noise decreases with increase in the impurity scattering.(2)

The defect fluctuation mechanism has advantage over phonon fluctuation model in obtaining long characteristic times observed in semiconductors.(3) Moreover, in contrast with the trap model (4), it can exhibit either the bulk or the surface effects, since the lattice defects may be generated at surfaces, grain boundaries, and dislocations as well as in homogeneous crystals.

Therefore, we propose a model which assumes that the fluctuation of the Fermi energy, which is caused by the defect fluctuations, is an origin of the conductivity fluctuations in metals and semiconductors. It is allowable to define the Fermi energy at every time, since the defect fluctuations which we consider are slow processes in comparison with electronic redistribution among the defect states. The 1/f spectrum is attributed to the number fluctuation of defects, whose spectral density is taken to be

$$S_t(f)=AN_t/f. \qquad (1)$$

Here, N_t is the number of the defects and $A=kTD(-kT\ln(2\pi f\tau_0))$ when the activation energy E in the relaxation time $\tau=\tau_0\exp(E/kT)$ has a distribution D(E). (1) If the distribution of the relaxation time is given as $1/\tau$ in a range from τ_1 to τ_2, A becomes $1/\ln(\tau_2/\tau_1)$. (5)

In this paper we calculate the noise current spectral density with the model and discuss the carrier density dependence of the spectral density.

2. VOLUME EFFECT IN SEMICONDUCTORS

We consider a non-degenerate semiconductor containing carrier traps of an acceptor type, the number of which is assumed to fluctuate with the distribution of relaxation times as described in the previous section.

The Fermi energy E_F is determined by the neutrality condition

$$N+N_A^-+N_t^-=P+N_D^+, \qquad (2)$$

where N, P, N_A, N_D, and N_t are the numbers of electrons, holes, shallow acceptors, shallow donors, and trap centers, and the superscripts ± indicate the charged centers. Then, the Fermi energy fluctuation ΔE_F is derived from the same condition concerning the fluctuations

$$\Delta N+\Delta N_A^-+\Delta N_t^-=\Delta P+\Delta N_D^+. \qquad (3)$$

We assume that the number of the traps fluctuates independently of its charged states, so that

$$\Delta N_t^-=f_t\Delta N_t+N_t(\partial f_t/\partial E_F)\Delta E_F, \qquad (4)$$

where the Fermi function f_t for the trap state E_t is written as

$$f_t=[1+\exp(E_t-E_F)/kT]^{-1} \qquad (5)$$

If the state densities of the conduction and valence bands are not influenced by the traps, carrier number fluctuations can be represented in terms of the Fermi level fluctuation:

$$\Delta N/N=-\Delta P/P=\Delta E_F/kT. \qquad (6)$$

This gives the fluctuation ΔI of the steady current I in the form

$$\Delta I/I=[(\mu_n N-\mu_p P)/(\mu_n N+\mu_p P)](\Delta E_F/kT), \quad (7)$$

where changes in the mobilities, μ_n and μ_p, due to the variation of the trap density are neglected.

Denote probabilities that the donor and acceptor states are filled with electrons by f_D and f_A, respectively, and we have

$$\Delta N_D^+=-N_D(\partial f_D/\partial E_F)\Delta E_F, \quad (8)$$

$$\Delta N_A^-=N_A(\partial f_A/\partial E_F)\Delta E_F. \quad (9)$$

Substituting Eqs.(4), (6), (8), and (9) into Eq.(3) leads to

$$\Delta E_F/kT=-f_t\Delta N_t/(N+P+X), \quad (10)$$

$$X=kT\partial(N_Df_D+N_Af_A+N_tf_t)/\partial E_F. \quad (11)$$

Accordingly, the noise current spectral density $S_I(f)$ is obtained with combining Eqs.(1), (7), and (10).

For n-type semiconductors we find

$$S_I(f)/I^2=AN_tf_t^2/(N+X)^2f, \quad (12)$$

which indicates that $S_I(f)$ is inversely proportional to the volume of specimens and that it depends on the relative magnitude of E_F to E_t. In the case $E_D>E_F>E_t$, where E_D is the donor state energy, X can be neglected compared with N and Eq. (12) becomes

$$S_I(f)/I^2=A(n_t/n)/Nf, \quad (13)$$

where n_t and n are the densities of the traps and conduction electrons.

In the case $E_D>E_t>E_F>E_i$, where E_i is the intrinsic Fermi level, X may approximately be put equal to N_tf_t, so that for $N>N_tf_t$

$$S_I(f)/I^2=AN_tf_t^2/N^2f, \quad (14)$$

and for $N_tf_t>N$

$$S_I(f)=A/N_tf. \quad (15)$$

Since f_t/n is nearly independent of the Fermi level position, here $S_I(f)$ does not depend on the carrier density.

Fig.1 is a schematic representation of the carrier density dependence of the noise current spectral density for the acceptor type traps, where curve A is the case $E_t>E_i$ and curve B is for $E_i>E_t$ as indicated by arrows. The trap density is assumed to be smaller than n at $E_F=E_t$ in the curve A, and to be smaller than p at $E_F=E_t$ in the curve B. Main features in $S_I(f)$ are as follows: S_I consists of three parts, each of which behaves in A as p^{-4}, constant, and n^{-2}, and in B as p^{-4}, p^{-2}, and n^{-2}, respectively. Even when the trap densities become high, those features remain unchanged.

For donor type traps we have a factor $1-f_t$ instead of f_t in $S_I(f)$, so that $S_I(f)$ vs E_F curves become the mirror images with regard to the line at $E_F=E_i$ of those represented in Fig.1.

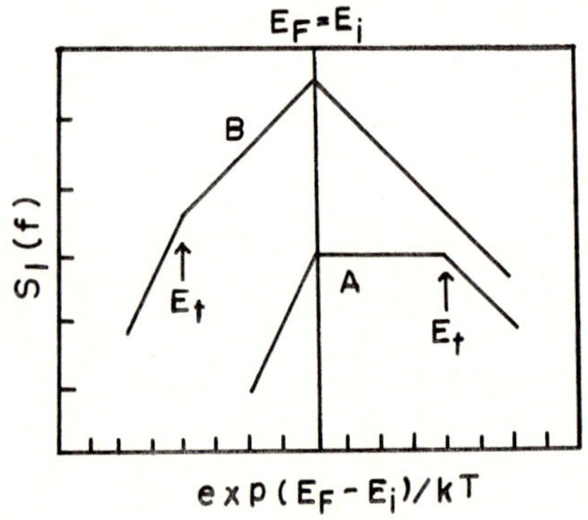

Fig.1. The noise current spectral density $S_I(f)$ vs $\exp(E_F-E_i)/kT$ for acceptor type traps, where E_F is the Fermi level and E_i the intrinsic Fermi level.

3. SURFACE EFFECT IN SEMICONDUCTORS

The noise current due to fluctuations in the surface states is discussed by assuming the depletion layer at the surface of an n-type semiconductor film. The energy diagram at the surface of the film is shown in Fig.2, where E_0 is the neutral level at the surface, E_F is the Fermi level, E_c is the bottom of the conduction band, E_{VL} is the vacuum level, and φ_s and φ_b are defined by

$$\varphi_s=E_{VL}(0)-E_0, \quad (16)$$

$$\varphi_b=E_{VL}(\infty)-E_F. \quad (17)$$

From these definitions we have a relation

$$\varphi_s-\varphi_b=E_{VL}(0)-E_{VL}(\infty)+E_F-E_0. \quad (18)$$

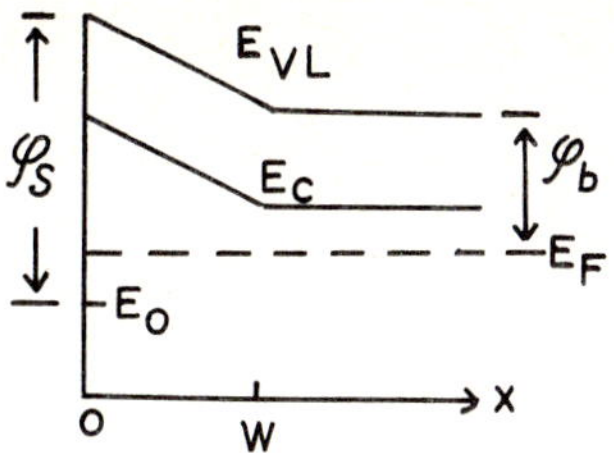

Fig.2. Energy diagram at the surface of n-type semiconductors.

In Eq.(18) $E_{VL}(0)-E_{VL}(\infty)$, the potential energy difference between the surface and the bulk, is related to the depletion layer width W by

$$E_{VL}(0)-E_{VL}(\infty)=(2\pi e^2 n_D/\varepsilon)W^2, \quad (19)$$

where n_D is the donor density and ε is the static dielectric constant. The Fermi energy difference, E_F-E_O, is written from the balance of the charge density between the surface and the depletion layer as

$$E_F-E_O=n_D W/D_S, \quad (20)$$

where D_S is the surface state density per unit energy. From Eqs.(18), (19), and (20) we obtain

$$W=(w_b^2+w_s^2)^{1/2}-w_s, \quad (21)$$

where $w_b=(\varepsilon(\varphi_S-\varphi_b)/2\pi e^2 n_D)^{1/2}$ and $w_s=\varepsilon/4\pi e^2 D_S$.

Consider the case that the steady current flows in the direction of the length L of the film, whose thickness and width are denoted by a and b, respectively. Then, the number of electrons which contribute to the current is the electron density multiplied by abL-2(a+b)LW. The fluctuation ΔD_S in the surface states induces the fluctuation ΔW, resulting in the current fluctuation of the form

$$\Delta I/I=-2(a+b)\Delta W/ab. \quad (22)$$

We put in Eq.(22)

$$\Delta W=\gamma W\Delta D_S/D_S=\gamma W\Delta N_S/N_S, \quad (23)$$

where $\gamma=d\ln W/d\ln D_S$ and N_S is the number of the surface states in the band gap, and obtain

$$S_I(f)/I^2=4A\gamma^2((a+b)/ab)^2W^2/N_S f. \quad (24)$$

This indicates that the carrier density dependence of $S_I(f)$ is the same as that of W^2, which is found from Eq.(21) with $n_D=n$ to be proportional to 1/n for small n and to $1/n^2$ for large n.

Provided that $a \ll b$, we rewrite Eq.(24) in the form

$$S_I(f)/I^2=4A\gamma^2 C/Nf, \quad (25)$$

$$C=W^2N/a^2N_S. \quad (26)$$

Fig.3 shows the carrier density dependence of W^2n, that is of C, with taking following values: $\varepsilon=12$, $D_S=10^{12}/cm^2eV$, $E_C-E_O=0.8eV$, and the effective density of states in the conduction band is $2.7\times10^{19}/cm^3$. Thus, we find that $S_I(f)$ due to the surface effect is written to be in the form (25), in which C is independent of n for small n and decreases with increasing n. The magnitude of C may be appreciable for some values of sample parameters.

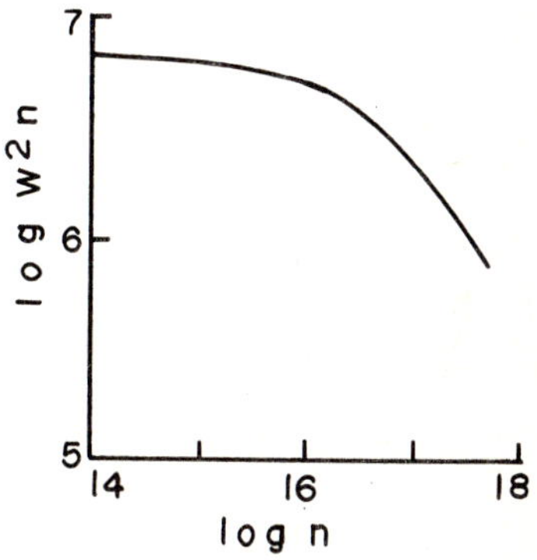

Fig.3. Carrier density dependence of the depletion layer width W.

4. METALS

For degenerate n-type semiconductors the noise current spectral density due to the volume effect is given by Eq.(13), if the effect of traps on the conduction band state density is neglected. For the volume effect in metals, however, we must assume that the state density of the band is changed with the lattice defects.

The electrical conductivity σ of metals is written as

$$\sigma=\int B(E)D(E)(-\partial f/\partial E)dE, \qquad (27)$$

where B(E) is a function of an electron energy E, D(E) is the state density in the band, and f(E) is the Fermi distribution function. The Fermi energy is determined by

$$n=\int D(E)f(E)dE, \qquad (28)$$

which leads to

$$\Delta E_F=-\Delta D/D(E_F), \qquad (29)$$

where ΔD is the fluctuation of the electronic states below the Fermi level per unit volume

$$\Delta D=\int \Delta D(E)f(E)dE. \qquad (30)$$

The conductivity fluctuation $\Delta\sigma$ is derived from Eq.(27)

$$\Delta\sigma=(dBD/dE)_{E_F}\Delta E_F+B(E_F)\Delta D(E_F). \qquad (31)$$

If the last term in Eq.(31) can be neglected:

$$\Delta D(E_F)/D(E_F)\ll \Delta D/E_F D(E_F), \qquad (32)$$

we have

$$\Delta\sigma/\sigma=\Delta E_F/E_F. \qquad (33)$$

ΔD is assumed to be connected with the defect fluctuation ΔN_t by

$$\Delta D/E_F D(E_F)=\eta\,\Delta N_t/N, \qquad (34)$$

where η is a numerical factor of the order unity. Then, combining Eqs.(29), (33), and (34), we obtain

$$S_I(f)/I^2=A\eta^2(n_t/n)/Nf. \qquad (35)$$

This gives too much smaller values than the observed ones for metal films, so that there the surface effects may be dominant.

5.SUMMARY

A model is proposed, which explains the 1/f noise in metals and semiconductors in terms of fluctuations in the electronic states associated with the lattice defects.

For the volume effect in semiconductors the noise current spectral density $S_I(f)$ is found to depend strongly on the relative magnitude of the Fermi energy to the electron energy in trap states. For n-type semiconductors with acceptor type traps $S_I(f)$ is independent of the carrier density n for E_t E_F E_i, as shown in Fig.1. When the Fermi level goes above the trap state E_t, $S_I(f)$ is proportional to $1/n^2$.

In the case of surface state fluctuations in semiconductors, $S_I(f)$ is proportional to the square of the depletion layer width, so that it varies as 1/n for low doping levels and as $1/n^2$ for high doping levels.

The magnitudes and the carrier density dependence of $S_I(f)$ calculated for semiconductors appear to be consistent with the experiment.(2)

For metal films the noise power expected from the volume effect is very small, so that the effects of surfaces or grain boundaries should play the dominant role.

REFERENCES

(1) Dutta, P., Dimon, P., and Horn, P.M., Phys. Rev. Lett. 43 (1979) 646.
(2) Hooge, F.N., and Vandamme, L.K.J., Phys. Lett. 66A (1978) 315.
(3) Caloyannides, M.A., J. Appl. Phys. 45 (1974) 307.
(4) McWhorter, A.L., Semiconductor Surface Physics, ed. Kingston, R.H. (University of pennsylvania, Philadelphia, 1957) p.207.
(5) Van Der Ziel, A., Fluctuation Phenomena in Semiconductors, (Butterworths, London, 1959).

NOISE IN PHYSICAL SYSTEMS AND 1/f NOISE
M. Savelli, G. Lecoy and J-P. Nougier (eds.)
© *Elsevier Science Publishers B.V., 1983*

DERIVATION OF THE NYQUIST-1/F NOISE THEOREM

Peter H. Handel
McDonnell Douglas Research Laboratories, St. Louis, Missouri 63166, USA

C. M. Van Vliet
University of Montreal, Canada, and University of Florida-Gainesville, USA

A. Van Der Ziel
University of Minnesota, Minneapolis and University of Florida-Gainesville, USA

A more general first principles derivation of the Nyquist theorem, which includes quantum 1/f noise in a natural way is obtained by performing infrared radiative corrections on the basic matrix elements and by using the delete 1/f noise approach. This derivation shows how the infrared-divergent coupling to various systems of infraquanta causes thermal noise to differ in its statistical properties from classical Nyquist noise.

1. INTRODUCTION

In order to derive the Nyquist theorem, also known as fluctuation-dissipation theorem,we use the method of Callen and Welton [1] but include the effect of infrared radiative corrections. After defining the system in Section 2 and the infrared-divergent coupling in Section 3, we calculate the dissipation in Section 4 and the fluctuations in Section 5. Comparison of the calculated fluctuations with the rate of dissipation yields the Nyquist theorem in Section 6.

2. THE RESISTOR

We describe a resistor of conductance $g(\omega)$ with an applied voltage $V(t) = V_0 \cos\omega t$ as a system with a Hamiltonian

$$H = H_0(q_k, p_k) + V(t)\, Q(q_k, p_k). \tag{2.1}$$

Here $H_0(q_k, p_k)$ is the Hamiltonian in the absence of the applied potential and $V(t)\, Q(q_k, p_k)$ is a peturbation:

$$Q = \sum_{j=1}^{N} e_j x_j/L, \tag{2.2}$$

where e_j and x_j are the charges and coordinates (along the circuit) of the N carriers present in the resistor of length L.

The eigenfunctions ψ_n of the unperturbed Hamiltonian are defined by

$$H_0 \psi_n = E_n \psi_n, \tag{2.3}$$

where E_n are the corresponding eigenvalues. The evolution of the state of the system is given by the Schrödinger equation

$$H_0 \psi + V_0 \sin\omega t\, Q\psi = i\hbar \frac{\partial\psi}{\partial t}. \tag{2.4}$$

According to Dirac's time dependent perturbation theory the expansion

$$\psi = \sum_n a_n(t)\, \psi_n \tag{2.5}$$

leads to a set of differential equations for the coefficients $a_n(t)$ which can be integrated easily. If the system was initially in the state ψ_n, calculation of the coefficients $a_m(t)$ yields the transition rates for absorption and emission of a quantum of energy $\hbar\omega$. The total transition rate out of the state n is

$$\dot{P}_n = (\pi V_0^2/2\hbar)\, \{|\langle E_n + \hbar\omega|Q|E_n\rangle|^2 \rho(E_n + \hbar\omega)$$

$$+ |\langle E_n - \hbar\omega|Q|E_n\rangle|^2 \rho(E_n - \hbar\omega)\}. \tag{2.6}$$

Here $\rho(E)$ is the density of (final) states at the energy E and we have used the Dirac notation for the matrix elements of Q

$$\langle E_{n'}|Q|E_n\rangle = \int \Psi_{n'}^{*}\, Q\psi_n\, dv$$

$$= [1 + \sum_{\nu=1}^{\Lambda/\varepsilon_o} b_{\nu,n'} e^{i\nu\varepsilon_o t + i\gamma_\nu} (\nu\varepsilon_o)^{-1/2}]$$

$$\times \int \psi^*_{n'} Q\psi_n \, dv. \qquad (2.7)$$

The special form of the final state $\Psi_{n'}$ and the rectangular bracket will be explained in detail below, and will ultimately yield 1/f fluctuations.

3. INFRARED-DIVERGENT COUPLING

In Equation (2.7) the matrix element for the transition from the initial state n to the final state n′ contains an additional factor, written in rectangular brackets, which takes into account the infrared-divergent coupling of the current carriers to infraquanta and which generates the 1/f fluctuations. The infraquanta are massless particles of arbitrarily small energy such as low-frequency (down to small fractions of 1 Hz) photons, phonons, electron-hole pairs at the Fermi surface of a metal, spin waves, etc. Due to the spontaneous emission of infraquanta accompanying the transition, the final state will contain also some components with small energy losses $\hbar\varepsilon = \nu\varepsilon_o\hbar$

$$\Psi_{n'} = [1 + \sum_{\nu=1}^{\Lambda/\varepsilon_o} b_{\nu n'} e^{i\nu\varepsilon_o t + i\gamma_\nu} (\nu\varepsilon_o)^{-1/2}] \psi_{n'}. \qquad (3.1)$$

Here $(\nu\varepsilon_0)^{-1/2} b_{\nu n'}$ is the (real) amplitude and $e^{i\gamma_\nu}$ the phase of the component with energy loss $\nu\varepsilon_o\hbar$. The energy losses are integer multiples of $\hbar\varepsilon_o = 2\pi\hbar/T$, where T is the duration of the experiment. The rectangular bracket corresponds to a series expansion over the duration T of the experiment, with energy losses only. There are no gains from spontaneous emission into the vacuum state of the field which describes the infraquanta, e.g., the electromagnetic field. The expansion is truncated at an energy $\Lambda\hbar$ which is of the order of the energy E_n of the carriers. It is well known that the coefficients are roughly proportional to $\varepsilon^{-1/2}$ if the coupling is infrared-divergent, such as it is for the infraquanta mentioned above. Indeed, this corresponds to a matrix element $\sim\varepsilon^{-1}$, i.e., an infrared divergence in the total process rate $\sim \int d\varepsilon/\varepsilon$. Therefore $b_{\nu n'}$ is simply a constant coupling parameter

$$b_{\nu n} \simeq (\alpha A_n)^{1/2} \varepsilon_o^{1/2}, \qquad (3.2)$$

independent of ν or of ε in first approximation, if higher-order processes are neglected. If the coupling parameter is small ($\alpha A_n \ll 1$), $b_{\nu n'}$ will remain close to the constant given by Equation (2.2) even if the higher-order contributions from multiple emission of infraquanta are summed up and are all included:

$$b_{\nu n} = (\alpha A_n)^{1/2} \nu^{\alpha A_n/2} \varepsilon_o^{1/2}. \qquad (3.3)$$

This shows how the inclusion of all higher-order processes, known as infrared radiative corrections, turns the infrared divergence into a integrable singularity $\sim\varepsilon^{\alpha A-1}$ which is, for small αA, very close to the initial divergent ε^{-1} singularity. For convenience we have dropped the subscript n.

The total energy-loss spectrum is the sum of energy-loss spectra caused by different systems of infraquanta, and therefore the coupling parameters are additive

$$\alpha A = \sum_i \alpha^{(i)} A^{(i)}. \qquad (3.4)$$

Coupling parameters αA and coupling constants α are easy to calculate for some systems of infraquanta with infrared-divergent coupling to the current carriers.

The form of the radiative corrections provided in rectangular brackets in Equations (2.7) and (3.1) is simplified in two ways. First, there is in fact a factor of this form for every charge carrier in the system with independent random phases $\gamma_{\nu j}$; we will try to relax this simplification in the final results. Second, the summation over ν is a substitute for a more complicated summation over all modes of the electromagnetic field, i.e., over the wavevectors and polarization states of the emitted quanta, which yield the same results.

4. THE CONDUCTANCE

In order to derive the Nyquist theorem, we have to calculate the rate of dissipation; this calculation will implicitely define the electric conductance, or the resistance, of our system.

The transition probabilities which determine the rate of dissipation are determined by squaring the matrix elements in Equation (2.7)

$$|\langle E_n|Q|E_n\rangle|^2 =$$

$$[1 + 2\sum_{\nu=1}^{\Lambda/\varepsilon_o} b_{\nu n'} \cos(\nu\varepsilon_o t + \gamma_\nu)(\nu\varepsilon_o)^{-1/2} + O(\alpha A)]$$

$$\times \left|\int \psi_{n'} Q \psi_n \, dv\right|^2. \qquad (4.1)$$

The rate of absorption of energy for the system in the state n will be

$$p_n = (\pi/2)V_0^2\omega\ \{|\langle E_n + \hbar\omega|Q|E_n\rangle|^2\ \rho(E_n + \hbar\omega)$$

$$- |\langle E_n - \hbar\omega|Q|E_n\rangle|^2\ \rho(E_n - \hbar\omega)\}\ . \qquad (4.2)$$

Here $\rho(E)$ is the system's density of states. Equation (4.2) is an expression of the "Golden Rule."

If we assume that the system is close to equilibrium at the temperature T, the occupation of the state n is given by a Boltzmann factor

$$f(E_n) \sim e^{-E_n/kT}\ . \qquad (4.3)$$

The power dissipated in the system is then

$$p = (\pi/2)\ V_0^2 \int_0^\infty \rho(E)\ f(E)$$

$$\times\{|\langle E + \hbar\omega|Q|E\rangle|^2\ \rho(E + \hbar\omega)$$

$$- |\langle E - \hbar\omega|Q|E\rangle|^2\ \rho(E - \hbar\omega)\}\ dE. \qquad (4.4)$$

The electric current $I = dQ/dt \equiv \dot{Q}$ allows us to define the impedance $Z(\omega) = V/\dot{Q}$ and the conductance $g(\omega) = \mathrm{Re}\ [1/Z(\omega)]$. The dissipated power is then

$$p = (1/2)\ V_0^2\ g(\omega) \qquad (4.5)$$

and from Equation (4.4) we obtain

$$g(\omega) \equiv \pi\omega\ \text{"A"} = \pi\omega \int_0^\infty \rho(E)\ f(E)$$

$$\times\{|\langle E + \hbar\omega|Q|E\rangle|^2\ \rho(E + \hbar\omega)$$

$$- |\langle E - \hbar\omega|Q|E\rangle|^2\ \rho(E - \hbar\omega)\}\ dE. \qquad (4.6)$$

Here the notation "A" has been introduced for later reference. According to Equations (2.7) and (4.1) the conductance given by Equation (4.6) contains small fluctuations in time, with a 1/f spectral density:

$$g(\omega) = g_0(\omega)\ [1 + 2 \sum_\nu \overline{b_\nu}\ \cos(\nu\varepsilon_0 t + \gamma_\nu)$$

$$\times(\nu\varepsilon_0)^{-1/2}]\ . \qquad (4.7)$$

Indeed, the spectral density is proportinal to the squared amplitude of $\cos\ (\nu\varepsilon_0 t + \gamma_\nu)$

$$(1/2)[2\ \overline{b_\nu}\ (\nu\varepsilon_0)^{-1/2}]^2 = 2\alpha\bar{A}\nu^{\alpha A-1}$$

$$\simeq 2\alpha\bar{A}/\nu, \qquad (4.8)$$

the last approximation being valid for $\alpha\bar{A} \ll 1$. This represents a spectral density of the fractional conductivity flucuations of $2\alpha\bar{A}/f$. The bar represents an average over all energies, i.e., over n.

5. CURRENT FLUCTUATIONS

So far we have considered the stationary state close to equilibrium. Let us consider now the current fluctuations in equilibrium (V = 0) at temperature T. The mean squared current in the state n is

$$\langle E_n|\dot{Q}^2|E_n\rangle = \sum_m \langle E_n|\dot{Q}|E_m\rangle\langle E_m|\dot{Q}|E_n\rangle$$

$$= -\hbar^{-2} \sum_m \langle E_n|H_0Q - QH_0|E_m\rangle\ \langle E_m|H_0Q - QH_0|E_n\rangle$$

$$= \hbar^{-2} \sum_m (E_n - E_m)^2|\langle E_m|Q|E_n\rangle|^2$$

$$= \hbar^{-2} \int_0^\infty (\hbar\omega)^2|\langle E_n + \hbar\omega|Q|E_n\rangle\ \rho(E_n + \hbar\omega)\ \hbar\ d\omega$$

$$+ \hbar^{-2} \int_0^\infty (\hbar\omega)^2|\langle E_n - \hbar\omega|Q|E_n\rangle\ \rho(E_n - \hbar\omega)\ \hbar\ d\omega$$

$$= \int_0^\infty \mathcal{S}_{\dot{Q}_n}(\omega)\ d\omega\ . \qquad (5.1)$$

Here we have successively expanded the square of the operator $\dot{Q}$, used the Heisenberg equations of motion (or Schrödinger's equation) used the eigenfunction property of the states ψ_n, written $E_m - E_n = \pm\ \hbar\omega$, switched to an integral and defined the spectral density of the current fluctuations in the state n. In general the spectral density of current fluctuations in equilibrium at temperature T will be

$$S_{\dot{Q}}(\omega) = \sum_n f(E_n)\ S_{\dot{Q}}(\omega)$$

$$= \sum_n f(E_n)\hbar\omega^2\{|\langle E_n + h\omega|Q|E_n\rangle|^2\rho(E_n + h\omega)$$

$$+ |\langle E_n - \hbar\omega|Q|E_n\rangle|^2 \rho(E_n - \hbar\omega)\} \equiv \hbar\omega^2 \text{"B"}$$

$$= \hbar\omega^2 \int_0^\infty \rho(E)\, f(E)\, \{|\langle E + \hbar\omega|Q|E\rangle|^2 \rho(E + \hbar\omega)$$

$$+ |\langle E - \hbar\omega|Q|E\rangle|^2 \rho(E - \hbar\omega)\}\, dE. \qquad (5.2)$$

We observe that this integral differs from that in Equation (4.6) only by a sign. By using the notations "A" and "B" we can simply write

$$S_{\dot{Q}}(\omega) = (\hbar\omega/\pi)\, g(\omega)\, \text{"B"/"A"} \qquad (5.3)$$

6. THERMAL NOISE

Equation (5.3) is a relation between a thermal equilibrium property of the system and the kinetic coefficient $g(\omega)$ which characterizes the response of the system to an applied voltage, i.e., the stationary state close to equilibrium. In order to simplify this equation we note that $\langle E - \hbar\omega|Q|E\rangle = 0$ for $E < \hbar\omega$ because this would lead to an energy below the energy of the ground state. Therefore,

$$\int_0^\infty f(E)|\langle E - \hbar\omega|Q|E\rangle|^2 \rho(E - \hbar\omega)\, \rho(E)\, dE$$

$$= \int_0^\infty f(E' + \hbar\omega)|\langle E'|Q|E' + \hbar\omega\rangle|^2 \rho(E')$$

$$\rho(E' + \hbar\omega) dE'$$

$$= e^{-\hbar\omega/kT} \int_0^\infty f(E)|\langle E + \hbar\omega|Q|E|^2 \rho(E)\rho(E + \hbar\omega) dE \qquad (6.1)$$

The last integral can thus be factored out both from "A" and "B", and can be simplifed. We obtain

$$S_{\dot{Q}}(\omega) = \frac{\hbar\omega}{\pi}\, g(\omega)\, \frac{1 + \exp(-\hbar\omega/kT)}{1 - \exp(-\hbar\omega/kT)}$$

$$= \frac{2}{\pi}\, g(\omega)\, \{\frac{\hbar\omega}{2} + \frac{\hbar\omega}{\exp(\hbar\omega/kT) - 1}\}. \qquad (6.2)$$

This is the Nyquist theorem. In curly brackets we recognize the expectation value of the energy of a harmonic oscillator in equilibrium at temperature T. As a result of our derivation, $g(\omega)$ is not the expectation value $g_o(\omega)$ of the conductance, but the physical value $g(\omega)$ given by Equation (4.7) and fluctuating in time with a 1/f spectral density. If we now relax the simplification introduced earlier and allow for independent random phases $\gamma_{\nu j}$ for each current carrier's 1/f noise part, we obtain

$$g(\omega) = \pi\omega \int_0^\infty f(E)\rho(E)\rho(E + \hbar\omega)$$

$$*[1 - \exp(-\hbar\omega/kT)]$$

$$\times\, |\sum_j \langle E + \hbar\omega|e_j \frac{x_j}{L} [1 + \sum_{\nu=1}^{\Lambda/\varepsilon_o} b_\nu(E)$$

$$\times \exp(-i\nu\varepsilon_o t - i\gamma_{\nu j})(\nu\varepsilon_0)^{-1/2}]|E|^2 dE. \qquad (6.3)$$

When we calculate the spectral density of the 1/f noise fluctuations, the presence of N sets of independent random phases leads to a factor 1/N in the resulting 1/f noise. This is equivalent to the inclusion of a factor $1/\sqrt{N}$ into the 1/f noise amplitude of Equation (4.7) for a resistor with N current carriers, and allows us to write the Nyquist - 1/f Noise Theorem in the form

$$S_{\dot{Q}}(f) = 4\, g(f)\, \hbar f\, [\frac{1}{2} + \frac{1}{\exp(\hbar f/kT) - 1}]$$

$$[1 + (4/N)^{1/2} \sum_\nu \bar{b}_\nu \cos(\nu\varepsilon_o t + \gamma_\nu)(\nu\varepsilon_o)^{-1/2}] . \qquad (6.4)$$

This result shows that the physical phenomenon of "thermal noise" is more complicated than the simple Nyquist model which describes it as a Gaussian process. As a result of Equation (6.4) thermal noise is a non-Gaussian process with an amplitude distribution given by an integral which can not be expressed in terms of elementary functions.

REFERENCES

[1] Callen, H. B. and Welton, T. E., Irreversibility and generalized Noise, Phys. Rev. 83 (1951) 34-40.

NOISE IN PHYSICAL SYSTEMS AND 1/f NOISE
M. Savelli, G. Lecoy and J-P. Nougier (eds.)
© Elsevier Science Publishers B.V., 1983

ANY PARTICLE REPRESENTED BY A COHERENT STATE EXHIBITS 1/f NOISE

P. H. Handel, Physics Department, University of Missouri-St. Louis and McDonnell Douglas Research Laboratories, St. Louis, MO, USA

Starting from the known description of the physical electron in terms of coherent states of the electromagnetic field, this paper proves that any charged particle must exhibit 1/f noise. The proof follows from the uncertainty present in the energy of any coherent state, which causes fluctuations in time.

1. INTRODUCTION

The electromagnetic field of a free electron is best described by a coherent state.[1] However, a coherent state does not have a definite energy, and is therefore nonstationary. Physically this means that there will be quantum fluctuations in the probability distribution and in the current distribution. The main purpose of the present paper is to prove that the spectrum of these fundamental fluctuations, which are part of the definition of a charged particle, is a 1/f spectrum with no lower frequency limit. This proof yields an equivalent formulation of the quantum 1/f noise theory.[2-10]

The proof is composed of three parts. In the first part we prove that the autocorrelation function for the probability distribution in time of a single field mode $\vec{q}$ in a coherent state of amplitude $z_{\vec{q}}$ with $|z_{\vec{p}}|^2 \ll 1$ is

$$A(\tau) = \frac{\pi}{\sqrt{2}} \{1 + 2|z_{\vec{p}}|^2 \cos\omega\tau\} ,$$

where $\omega = cq$ is the frequency of the mode considered.

In the second part the amplitude $z_{\vec{q}}$ of the coherent field of the electron is found to be proporational to $q^{-3/2}$ from lowest-order (in $\alpha = e^2/hc$) perturbation theory.

In the third part, the autocorrelation function for the electron with its field is calculated by taking the product of all autocorrelation functions calculated for individual field modes, again with $|z_{\vec{q}}|^2 \ll 1$, as calculated in the second part. The resulting autocorrelation function is essentially the Fourier transform of 1/f multiplied by $\alpha/\pi = (137\cdot\pi)^{-1}$. As in the diffraction of particles, quantum 1/f noise is determined by the behavior of a single particle, but can be observed only with a large number of particles.[5]

2. AUTOCORRELATION FOR A SINGLE FIELD MODE

Consider a mode of the electromagnetic field characterized by the wave vector $\vec{q}$ and the polarization λ. Denoting $\vec{q}$, λ simply by q in the labels of the states, we write the coherent state [1],[11] of amplitude $|z_q|$ and phase arg z_q in the form

$$|a_q\rangle = e^{-(1/2)|z_q|^2} e^{z_q a_q^+} |0\rangle$$

$$= e^{-(1/2)|z_q|^2} \sum_{n=0}^{\infty} z_q^n (n!)^{-1/2} |n\rangle , \quad (2.1)$$

where z_q is a complex number and both the energy eigenstates and the coherent states are normed to unity

$$\langle n|n\rangle = 1; \ \langle z_q|z_q\rangle = 1 , \quad (2.2)$$

while a_q^+ is the creation operator for a photon in the state q.

Let us choose the representation of the energy states in terms of Hermite polynomials $H_n(x)$. With

$$|n\rangle \rightarrow (2^n n! \sqrt{\pi})^{-1/2} e^{-x^2/2} H_n(x) e^{in\omega t} \quad (2.3)$$

we obtain for the coherent state $|z_q|\rangle$ the representation

$$\psi_q(x) = e^{-(1/2)|z_q|^2} e^{-x^2/2} \sum_{n=0}^{\infty} \frac{(z_q e^{-i\omega t})^n}{n!(2^n\sqrt{\pi})^{1/2}} H_n(x)$$

$$= e^{-(1/2)|z_q|^2} e^{-x^2/2} \exp\left(-z^2 e^{-2i\omega t} + 2xze^{i\omega t}\right) \quad (2.4)$$

The last form has been obtained by setting $r = ze^{i\omega t}$ in the generating function of the Hermite polynomials

$$\exp\left(-r^2 - 2rx\right) = \sum_{n=0}^{\infty} (n!)^{-1/2} r^n H_n(x) \ . \tag{2.5}$$

The probability density corresponding to Equation (2.4) in the space of field values x is

$$|\phi_q(x)|^2 = \exp\left(-2|z_q|^2 - x^2\right)$$

$$\times \exp\left(-z_q^2 e^{-2i\omega t} - z_q^{*2} e^{2i\omega t}\right)$$

$$\times \exp\left[2x\left(ze^{-i\omega t} + z^{*} e^{i\omega t}\right)\right] \ . \tag{2.6}$$

In order to calculate the autocorrelation function of the probability density in time, we first consider the product

$$P_q(t,\tau,x) = |\phi_q(x)|_t^2 |\phi_q(x)|_{t+\tau}^2$$

$$= \exp\left\{-2x^2 - 2|z_q|^2[1 + \cos(2\omega t + 2\phi)\right.$$

$$\left. + \cos(2\omega t + 2\omega\tau + 2\phi)]\right\}$$

$$\times \exp\left\{4x|z_q|[\cos(\omega t + \phi) + \cos(\omega t + \omega\tau + \phi)]\right\} \ . \tag{2.7}$$

If we factorize the cosines we obtain

$$P_q(t,\tau,x) = \exp\left\{-2x^2 + 8x|z_q|\cos(\omega t + \omega\tau/2 + \phi\right.$$

$$\times \cos(\omega\tau/2) - 2|z_q|^2[1 + 2\cos(2\omega t$$

$$\left. + \omega\tau + 2\phi)\cos\omega\tau]\right\} \ . \tag{2.8}$$

With the assumption $|z_q| \ll 1$, which will be justified later, the exponential can be expanded in a series, and this yields

$$P_q(t,\tau,x) = \left\{1 + 8x|z_q|\cos(\omega t + \omega\tau/2\right.$$

$$+ \phi)\cos(\omega\tau/2) + 32x^2|z_q|^2\cos^2(\omega t + \omega t/2 + \phi)$$

$$\cos^2(\omega\tau/2) - 2|z_q|^2 - 4|z_q|^2\cos(2\omega t + \omega\tau + 2\phi)$$

$$\left.\cos\omega\tau + O(z_q^3)\right\}e^{-2x^2} \ . \tag{2.9}$$

Here $O(z_q^3)$ is the rest of the series which is small, of the order z_q^3 and negligible. The autocorrelation function for this mode is obtained by averaging over time, or equivalently, over the random phase of the mode, $\phi(= \phi_q)$, and then integrating over x from $-\infty$ to ∞

$$P_q(\tau,x) = \langle|\phi_q|_t^2|\phi_q|_{t+\tau}^2\rangle$$

$$= \left\{1 + 8x^2|z_q|^2(1 + \cos\omega\tau) - 2|z_q|^2\right\}e^{-2x^2} \ ; \tag{2.10}$$

$$A(\tau) = \pi(2)^{-1/2}\left\{1 + 2|z_q|^2\cos\omega\tau\right\} \ . \tag{2.11}$$

The last form shows that the probability density contains a constant background with superposed small oscillations of frequency ω which is the frequency of the mode q.

3. THE COHERENT FIELD OF A CHARGED PARTICLE

It is well known that a bare, electrically charged, particle interacts with the electromagnetic field modes and thereby acquires its own electromagnetic field z_q. Therefore, in order to determine the complex amplitudes z_q of the own field of the electron, we let the particle get dressed by itself, with the interaction Hamiltonian

$$H' = A_\mu j^\mu = -\frac{e}{c}\vec{v}\vec{A} + e\Phi \ , \tag{3.1}$$

where $\vec{A}$, Φ are the vector potential and electric potential Φ, i.e., components of the four-vector A^μ, while j^μ are the components of the current four-vector, with summation over repeating greek indices implied. The first-order perturbation theory yields for the state of a physical elementary charge

$$|\vec{k}, 0\rangle^{(1)} = |\vec{k}, 0\rangle^{(o)} + \sum_q |\vec{k} - \vec{q}, q\rangle^{(o)}$$

$$\times \frac{\langle\vec{k} - \vec{q}, q|H'|\vec{k}, 0\rangle}{\varepsilon_{\vec{k}} - \varepsilon_{\vec{k}-\vec{q}} - \omega_q} \ , \tag{3.2}$$

where $|\vec{k}, 0\rangle^{(o)}$ is the bare electron of momentum $\vec{k}$ and the electromagnetic vacuum with no interactions as indicated by the superscripts,(o) and $\varepsilon_{\vec{k}}$ is the bare electron's energy at momentum $\vec{k}$, while ω_q is the energy of a photon of wave number $\vec{q}$. The matrix elements have the well known form[1], [11]

$$\langle\vec{k} - \vec{q}, q|H'|\vec{k}, 0\rangle = \frac{e}{m}\frac{k^\mu E_\mu(q)}{V^{1/2}(2q)^{1/2}}, \qquad (3.3)$$

where m is the mass of the electron, $E_\mu(q)$ the polarization vector of the state $q(=\vec{q},\lambda)$, while q in the demominator is just $|\vec{q}|$. V is the volume of a normalization box. From Equation (3.2), we identify the coefficient of the field state q as being

$$z_q = \frac{(e/m)\, k^\mu E_\mu}{(k^2/2m) - (\vec{k} - \vec{q})^2/2m - \omega_q} \times \frac{1}{V^{1/2}(2q)^{1/2}}$$

$$= \frac{ek^\mu E_\mu}{\vec{k}\cdot\vec{q} - m\omega_q}\,\frac{1}{hc\, V^{1/2}(2q)^{1/2}}. \qquad (3.4)$$

By taking the square of the absolute value and calculating the scalar product we obtain

$$|z_q|^2 = (e^2/q^2)(2hcq)^{-1}V^{-1}. \qquad (3.5)$$

4. AUTOCORRELATION OF THE DENSITY OF PARTICLES

Considering now all modes of the electromagnetic field, we obtain from the single-mode result of Equation (2.11)

$$A(\tau) = C\prod_q \{1 + 2|z_q|^2 \cos \omega_q\tau\} \cong$$

$$C\{1 + 2\sum_q |z_q|^2 \cos \omega_q\tau\}$$

$$= C\{1 + 2(V/2^3\pi^3)\int d^3q\, |z_q|^2 \cos \omega_q\tau\}. \qquad (4.1)$$

Here C is a constant, and we have again used the smallness of $|z_q|^2$. Using Equation (3.5) we obtain

$$A(\tau) = C\{1 + 2(V/2^3\pi^3)(4\pi/V)(e^2/2hc)$$

$$\times \int(dq/q)\cos \omega_q\tau\}$$

$$= C\{1 + 2(\alpha/\pi)\int\cos(\omega\tau)d\omega/\omega\}. \qquad (4.2)$$

Here $\alpha = e^2/hc$ is the fine structure constant close to 137^{-1}. The first term in curly brackets is unity and represents the constant background, or the d.c. part. The autocorrelation function for the relative, or fractional density fluctuations, or for current density fluctuations in the beam of charged particles is obtained therefore by dividing the second term in curly brackets by the first term. The constant C is drops out when the fractional fluctuations are considered. According to the Wiener-Khintchine theorem, the coefficient of cos ωτ is the spectral density of the fluctuations, $S_{|\phi|^2}$, or S_j for the current density $\vec{j} = e(\vec{k}/m)|\phi|^2$

$$\frac{S_{|\phi|^2}}{\langle|\phi|^2\rangle^2} = \frac{S_j}{\langle j\rangle^2} = 2\,\frac{\alpha}{\pi}\,\frac{1}{f} \cong 4.6\cdot10^{-3}f^{-1}. \qquad (4.3)$$

This result is related to the well known quantum 1/f noise result[1-8]. If a beam of charged particle is scattered, passes from one medium into another medium (e.g., at contacts), is accelerated, or is involved in any kind of transitions, the amplitudes z_q which describe its field will change. Then, even if the initial state was prepared to have a well-determined energy, the final state will have an indefinite energy, with an uncertainty determined by the difference between the new and old z_q amplitudes, Δz_q. This, however, is just the bremsstrahlung amplitude Δz_q. We thus regain the quantum 1/f noise result, according to which the small energy losses from bremsstrahlung of infra-quanta yield a final state of indefinite energy, and therefore lead to fluctuations of the process rate, or cross section, of the process in which the electrons have participated, and which has occasioned the bremsstrahlung in the first place. The calculation of piezoelectric 1/f noise[10] although it deals with phonons as infraquanta, is phrased in terms of the coherent field amplitudes z_q for the first time. It has α substituted by the piezoelectric coupling constant g.

5. DISCUSSION

The final result in Equation (4.3) appears as a universal result valid for any beam of physical charged particles described by coherent states of their own field. The question is to what extent are the carrier prepared with coherent field states in an actual 1/f noise measurement. Is it possible to consider the energy of the bare electron uncertain in such a way that it compensates exactly the energy fluctuations of the coherent field?

We are used to describe an isolated system always as a system with a well defined energy. But how well can the electrons in a 1/f noise measurement be isolated from other influences? The screening efforts of the observer just substitute the interaction with the screen for the high-frequency part of the universe's field modes. Depending on the quality of the screening, the high-frequency part may start at frequencies as low as 1 Hz or below.

From Equations (3.5) and (4.2) we see that the smallness of the z_q-amplitudes is insured by the small value of the fine structure constant. This justifies the approximations which we have made in the derivation. However, for some infraquanta the coupling constant turns out to be large. In this case we remember that we are, in fact, interested in the 1/f noise of the whole macroscopic sample. Due to the independence of the phases $\psi_q^{(j)}$ corresponding to different electrons $j = 1,2,...N$, the effective coupling constant for the sample should contain a factor N in the denominator, and is proportional to the low-frequency radiation-resistance of the sample. If divided by N, the quantities $|z_q|^2$ are usually very small compared to unity, even for strong coupling cases. Finally, if the coupling parameter remains large up to the end, i.e., even for the whole sample, no 1/f noise contributions should be expected from the system of infraquanta involved.

Finally, we mention again that the presence of the fourth power of Ψ in Section 2 expresses the fact that the autocorrelation function is determined by the two-particle wave function, in terms of which our result is just bilinear, as required by the postulates of quantum mechanics. As we had pointed out earlier, and as was shown in detail recently[5], the 1/f noise obtained in Section 4 is a one-particle effect which can, however, be noticed only in the presence of many particles, or by repeating the experiment many times, just as in the well known case of a diffraction pattern.

In conclusion, in spite of the simplicity of the final result in Equation (4.3) and of the connection between coherent states and 1/f noise established in this paper, we expect most quantum 1/f noise to be of the Δz_q type, i.e., bremsstrahlung type, eventually somehwere between the z_q and Δz_q case, depending on the details of the experiment and on the type of infraquanta yielding the larger 1/f noise contributions.

REFERENCES

[1] Chung, V., Infrared divergence in quantum electrodynamics, Phys. Rev. 140B (1965) 1110-1122.

[2] Handel, P. H., 1/f Noise - an "infrared" phenomenon, Phys. Rev. Lett. 34 (1975) 1492-1495.

[3] Handel, P. H., Quantum approach to 1/f noise, Phys. Rev. A22 (1980) 745-757.

[4] Handel, P. H., in Van Vliet, K. M. (ed.) Proc. II Internat. Symposium on 1/f Noise (University of Florida Gainesville Press, 1980).

[5] Sherif, T. S. and Handel, P. H., Unified treatment of diffraction and 1/f noise, Phys. Rev. A26 (1982) 596-602.

[6] Handel, P. H. and Sherif, T. S., Direct calculation of the Schrödinger field which generates quantum 1/f noise (in this volume).

[7] Handel, P. H. and Wolf, D., Characteristic functional of quantum 1/f noise, Phys. Rev. A26 (1982) 3727-3730.

[8] Van Vliet, K. M., Handel, P. H., and van der Ziel, A., Superstatistical emission noise, Physica 108A (1981) 511-530.

[9] Widom. A., Pancheri. G., Srivastava, Y., Megaloudis, G., Clark, T. D., Prance, H., and Prance, R. J., Quantum electrodynamic circuit soft-photon renormalization of the conductance in electronic shot-noise devices, Phys. Rev. B27 (1982) 3412-3417.

[10] Handel, P. H. and Musha, T., Quantum 1/f noise from piezoelectric coupling (in this issue).

[11] Jauch, J. M. and Rohrlich, F., The Theory of Photons and Electrons (Springer, Heidelberg, 1976).

NOISE IN PHYSICAL SYSTEMS AND 1/f NOISE
M. Savelli, G. Lecoy and J-P. Nougier (eds.)
© Elsevier Science Publishers B.V., 1983

QUANTUM 1/f NOISE FROM PIEZOELECTRIC COUPLING

P. H. Handel, University of Missouri-St. Louis, 63121
and McDonnell Douglas Research Laboratories, St. Louis
U.S.A. and T. Musha, Department of Applied Electronics
Tokyo Institute of Technology, Nagatsuta, Yokohama 227, Japan

Quantum 1/f noise resulting from direct (e.g. piezoelectric) electron-phonon coupling is calculated explicitly for the first time. The results of this calucation predict a giant 1/f noise for piezoelectric substances, exceeding the Hooge parameter of $2\cdot 10^{-3}$ by a factor of the order of 10^5. This is found to be in agreement with experimental observations.

1. INTRODUCTION

Quantum 1/f noise[1-4] is a self-interference effect caused by very small energy losses present in the Schrödinger-field which describes the current carriers, due to infrared-divergent coupling to various infraquanta. Such infraquanta are, for example, photons, phonons, spin waves, electron-hole pairs on the Fermi-surface of a metal, various hydrodynamic excitations, and gravitons.

The coupling with photons is present for any type of charge carriers and has been considered as an example in constructing the theory. Electron-phonon interactions have been considered similar to electron-photon interactions, and have therefore not been treated explicitly, although the piezoelectric coupling constant was listed.[2-3] It turns out that the usual deformation-potential coupling contains an additional wave-vector as factor in the matrix element, and therefore will lead to infrared-divergent coupling only in special cases, to be discussed in a different paper. However, the piezoelectric coupling is very similar to the electrodynamic interaction and will be discussed here in some detail.

2. PIEZOELECTRIC COUPLING

The interaction Hamiltonian can be written in the form [5]

$$H' = (2\pi g/qV)^{1/2}\, \hbar v_s \sum_{\vec{k},q} c^+_{\vec{k}+\vec{q}}\, c_{\vec{k}}\, (a_{\vec{q}} - a^+_{-\vec{q}}). \quad (1)$$

Here $c_{\vec{k}}$ is the annihilation operator for electrons and $a_{\vec{q}}$ the same for phonons; v_s is the speed of sound and V the volume of the sample. The dimensionless coupling constant g replaces here the fine structure constant α present in the electrodynamic coupling, and can be calculated in terms of an appropriately averaged piezoelectric constant Q

$$g = \pi C^2 \frac{e^2\, Q^2}{\varepsilon \hbar v_s\, \varepsilon \rho v_s^2}\ . \quad (2)$$

Here e is the charge of the electron, ε the static dielectric constant, ρ the density of the crystal, and C a numerical constant of order unity depending on the crystal structure. For example [5] for zincblende structures $C = (4/5)^{1/2} \simeq 0.9$.

Examples of piezoelectric coupling constants for a few substances are listed in the following table:

Substance	g-value
ZnS	0.1482
CdTe	0.018
CdS	4.1

Table 1: Piezoelectric coupling constants

Due to the piezoelectric electron-phonon coupling defined by Eq. (1), the electron will be "dressed" with a cloud of phonons, as we can readily see by constructing the first-order approximation in perturbation theory:

$$\Psi_{\vec{k}} = |\vec{k},0\rangle + \sum_{\vec{q}} |\vec{k}-\vec{q},\vec{q}\rangle \frac{\langle \vec{k}-\vec{q},\vec{q}|H'|\vec{k},0\rangle}{E_{\vec{k}-\vec{q}} + \hbar\omega_{\vec{q}} - E_{\vec{k}}}\ \cdot \quad (3)$$

Here $|\vec{k},\vec{q}\rangle$ is an unperturbed momentum eigenstate corresponding to one electron of wavevector $\vec{k}$ with energy E_k and one phonon of wavevector $\vec{q}$ (of energy $\hbar\omega_{\vec{q}}$). The amplitude of the phonon state $|\vec{q}\rangle$ is thus

$$A_{\vec{q}}(\vec{k}) = \frac{(2\pi g/qV)^{1/2} v_s}{q\left(\frac{\hbar q}{2m} - \frac{\hbar \hat{q}\cdot\vec{k}}{m} + v_s\right)}; \quad \hat{q} = \vec{q}/q. \tag{4}$$

When the electron is scattered from the state $\vec{k}$ to the state $\vec{k}'$ due to any type of interaction, the phonon cloud will be characterized by different amplitudes $A_{\vec{q}}(\vec{k}')$. The difference $A_{\vec{q}}(\vec{k}')-A_{\vec{q}}(\vec{k})$ is the amplitude of the emitted phonon-bremsstrahlung. Indeed, each phonon mode can be compared with a harmonic oscillator. If the force causing the constant displacement from the equilibrium position is suddenly changed, the oscillator will start oscillating with an amplitude exactly equal to the difference of the displacements, i.e.,

$$A_{\vec{q}} = A_{\vec{q}}(\vec{k}')-A_{\vec{q}}(\vec{k}). \tag{5}$$

The probability of phonon emission is given by the squared absolute value of this amplitude, i.e., for small q,

$$P_{\vec{q}} = (2)|A_{\vec{q}}|^2 \simeq (2)\frac{2\pi g}{qV}\,\frac{v_s^2}{q^2}\left(\frac{1}{v_s-\hbar\hat{q}\cdot\vec{k}'/m} - \frac{1}{v_s-\hbar\hat{q}\cdot\vec{k}/m}\right)^2 = (2)\frac{2\pi g}{q^3 V}\,\frac{[\hat{q}\cdot(\vec{k}'-\vec{k})/k_s]^2}{(1-\vec{k}'\cdot\hat{q}/k_s)^2\,(1-\vec{k}\cdot\hat{q}/k_s)^2}, \tag{6}$$

where $k_s = mv_s/\hbar$ and the factor (2) is included in front when two transversal acoustical modes are considered. The total number of piezoelectric phonons emitted leads to an infrared-divergent integral:

$$N_{ph} = \sum_{\vec{q}} P_{\vec{q}} = \frac{V}{(2\pi)^3}\int P_{\vec{q}}\, d^3q = \frac{(2)g}{(2\pi)^2}\int_0^{q_D}\frac{dq}{q}$$

$$\int \frac{[\hat{q}\cdot(\vec{k}'-\vec{k})/k_s]^2\, d\Omega_q}{(1-\vec{k}'\cdot\hat{q}/k_s)^2\,(1-\vec{k}\cdot\hat{q}/ks)^2}. \tag{7}$$

3. BREMSSTRAHLUNG OF PHONONS

Here we have used again the Debye model with q_D as the Debye-wavenumber. The angular integration can be performed by using standard techniques, and is facilitated by using some of Feynman's integration tricks.[6] The result is

$$N_{ph} = \frac{(2)g}{3\pi k_s^2}(\vec{k}'-\vec{k})^2\int_0^{g_D}\frac{dq}{q} = \frac{(2)g}{3\pi}(\Delta\vec{\beta})^2\int_0^{q_D}\frac{dq}{q};$$

$$\Delta\vec{\beta} = \frac{\hbar(\vec{k}'-\vec{k})}{m\,v_s}. \tag{8}$$

This result can be obtained from the corresponding non-relativistic electrodynamic result[2] by replacing the fine-structure constant α by g, and by interpreting β as the velocity of the carriers divided by the speed of sound.

Note that the energy emitted is finite, and can be obtained by including a $\hbar\omega_q$-factor in the integral of Eq. (8):

$$W = \frac{(2)g\hbar v_s}{3\pi k_s^2}(\vec{k}'-\vec{k})\int_0^{q_D} dq = \frac{(2)g\hbar v_s}{3\pi k_s^2}\, q_D(\vec{k}'-\vec{k})^2$$

$$\equiv gA\hbar v_s q_D. \tag{9}$$

Here we have introduced the notation $2(\vec{k}'-\vec{k})^2/3\pi k_s^2 = A$, similar[1-4] to the electromagnetic case.

4. QUANTUM 1/f NOISE

Due to the simultaneous emission of soft phonons, the electron scattering cross section σ, i.e. the rate of the $\vec{k}\to\vec{k}'$ electron transition, will exhibit low-frequency fluctuations with a 1/f spectral density as we explain below. This spectral density corresponds in the well-known quantum 1/f noise formalism[1-4] to an autocorrelation function $A(\tau)$ obtained by including a 2 cos ωt factor in the integral present in Eq. (8):

$$\langle\sigma\rangle^{-2}\,\langle(\delta\sigma)^2\rangle_f = A(\tau)$$

$$= (2)\,\frac{2g}{3\pi}\left(\frac{\vec{k}'-\vec{k}}{k_s}\right)^2\int_0^{\omega_D}\cos\omega t\; d\omega/\omega$$

$$= 2gA\int_0^{\omega_D}\cos\omega\tau\; d\omega/\omega. \tag{10}$$

According to the Wiener-Khintchine theorem this implies a spectral density of 2gA/f for the fractional fluctuations of the scattering cross section.

The derivation [1-4] of the autocorrelation function (10) is based on an evaluation of the beat current between the main, non-bremsstrahlung, part and the various bremsstrahlung parts with energy loss $\varepsilon = \hbar\omega$.

The main, non-bremsstrahlung, part is in fact the amplitude-sum of all "bremsstrahlung" parts with energy losses below the detection threshold $\hbar\omega_0=\varepsilon_0= 2\pi\hbar/T$ where T is the duration of the experiment. In this treatment the stochastic Schrödinger field which describes the current carriers makes abstraction of other, much larger, energy losses of different nature present in the system, which are statistically independent and therefore irrelevant. The form of the stochastic field used in Infra-Quantum-Physics since the beginning[1] of this discipline is, in our case, for the scattered electrons,

$$\psi = a e^{i\vec{k}\cdot\vec{r}-iEt/\hbar}$$

$$\times \left[1+(gA)^{1/2} \sum_{n=1}^{\omega_D/\omega_0} n^{gA/2} n^{-1/2} e^{in\omega_0 t+i\gamma_n}\right]. \quad (11)$$

Here $\hbar\omega = \hbar n\omega_0$ is the very small energy loss, γ_n is a random phase, and a is a constant. The factor $n^{gA/2}$ is close to unity for $gA \ll 1$ and arises from infrared radiative corrections. This simplified form has been proven to yield correct results by a recent direct calculation[7] starting from first principles.

The expectation value of the scattering cross section is proportional to the phase (or time) average of $|\psi|^2$ which can be calculated from Eq. (11):

$$\langle\sigma\rangle = C\left[1+gA \int_{\omega_0}^{\omega_D} (\omega/\omega_0)^{gA}\, d\omega/\omega\right]. \quad (12)$$

Here C is $|a|^2$ times hk'/m divided by the incoming flux of electrons, and we have replaced the sum by an integral in the final form. The first term represents the non-bremsstrahlung part of the cross-section, and the second is the well-known expression of the bremsstrahlung part. The autocorrelation function of the cross-section fluctuations $\delta\sigma=\sigma-\langle\sigma\rangle$ is then obtained from Eqs. (11) and (12) in the form

$$\langle\delta\sigma(t)\delta\sigma(t+\tau)\rangle = C^2\, 2gA \int (\omega/\omega_0)^{gA} \cos\omega\tau\; d\omega/\omega. \quad (13)$$

Here a small "noise of noise" term proportional to g^2A^2 has been neglected. If we divide by the square of $\langle\sigma\rangle$, we obtain the exact form of Eq. (10)

$$\langle\sigma\rangle^{-2} S_\sigma(f) = \frac{2gA(f/f_0)^{gA}\, f^{-1}\, N^{-1}}{1 + gA \int_{f_0}^{f_D} (f/f_0)^{gA}\, df/f}$$

$$= 2gA\left(\frac{f}{f_D}\right)^{gA} \frac{1}{fN} \simeq \frac{2gA}{fN}, \quad (14)$$

where N is the number of carriers in the sample, yielding independent noise contributions. The last form is a good approximation for $gA \ll 1$. Eq. (14) is equivalent to Eq. (10) by virtue of the Wiener-Khintchine theorem.

5. DISCUSSION

Using even the smallest piezoelectric coupling constant g=0.018 from table 1, and a speed of sound of the order of 10^5 cm/s, we obtain

$$gA \simeq 1.53\cdot10^{-12}(s^2/cm^2)v^2 \sin^2(\theta/2) \quad (15)$$

If we take 10^7 cm/s for the thermal speed of the electrons v, we obtain

$$gA = 150 \sin^2(\theta/2) \quad (16)$$

Consequently, the condition $gA < 0.1$ will be satisfied only for small-angle scattering and for slow, subthermal electrons. However, we observe that the facter 1/N of Eq.(14) will be present also in the expression of the spectrum of the low-frequency power radiated by the sample, i.e. in the radiation resistance. Therefore, the effective coupling parameter or the infrared exponent, should be gA/N rather than gA, if we are interested in the noise of the whole sample. This parameter will always be much smaller than unity, and, by replacing in the exponents of Eq.(14) gA by gA/N, we always obtain the final result 2gA/fN. Therefore we expect unusually large 1/f noise from piezoelectric coupling, compared to the Hooge formula with the coeffieient $2\text{-}10^{-3}$. Indeed from Eq. (16) we expect, by taking $\sin^2(\theta/2) = 1/2$ in average, a Hooge coefficient of 75, i.e. about $4\cdot10^4$ times larger. For the other substances included in Table 1 the expected noise is even larger, in proportion to the g-values.

For barium strontium titanate ($Ba_{.597}Sr_{.4}La_{.003}TiO_3$) the piezoelectric coupling constant g = 0.02 is close to that for CdTe. Therefore we expect the experimental 1/f noise coeffiecient to be also $4\cdot10^4$ times larger than the reference value of $2\cdot10^{-3}$. Measurements by Mytton and Benton [8] as well as by Agarwal, Ambrozy and Hartnagel [9,10] yield a 1/f noise coeffient of barium strontium titanate which is more than 10^5 times the reference value of $2\text{-}10^{-3}$. This is close to the prediction from piezoelectric coupling. Finally, noise in $Hg_{1-x}Cd_xTe$ photodiodes also appears to be very large [11], in rough agreement with piezoelectric coupling noise from CdTe. In conclusion, the experimental evidence agrees in principle with the calculations, and seems to support our calculations of quantum 1/f noise from piezoelectric coupling of current carries to phonons in some solids in which this type of coupling is present.

REFERENCES

[1] Handel, P.H., 1/f Noise-an "infrared" phenomenon, Phys. Rev. Lett. 34 (1975) 1492-1495

[2] Handel, P.H., Quantum approach to 1/f noise, Phys. Rev A 22(1980) 745-757.

[3] Handel, P.H. in Van Vliet, K.M. (ed) Proc. II Internat. Symposium on 1/f Noise (University of Florida-Gainesville Press, 1980)

[4] Sherif, T.S. and Handel, P.H., Unified treatment of diffraction and 1/f noise, Phys. Rev. A26(1982) 596-602.

[5] Duke, C.B. and Mahan, G.D., Phonon-broadened imurity spectra. I. Density of states, Phys. Rev. 139A(1965)1965-1982.

[6] See, e.g., Björken, G.D., and Drell, S.D., Relativistic Quantum Mechanics (McGraw-Hill, New York, 1969) p.126

[7] Handel, P.H. and Sherif , T.S., Direct calculation of the Schrödinger field which generates quantum 1/f noise (in this issue)

[8] Mytton, R. J., and Benton, R.K., High 1/f noise anomaly in semiconducting barium strontium titanate, Physics Letters 39A(1972) 329-330

[9] Agarwal, R.P., Ambrozy, A. and Hartnagel, H.L., Excess noise in Ba Sr TiO_3, IEEE Trans. on Electon Devices ED-26(1979) 1937-1941

[10] Ambrozy, A., A model for excess noise of semiconduting Ba Sr TiO_3, IEEE Trans. on Electron Devices ED-26(1979) 1368-1369

[11] Reine, M.B., Sood, A.K. and Tredwell, T.J., Photovoltaic infrared detectors, in Semiconductors and Semimetals, Vol 18 (Academic Press, 1981)

NOISE IN PHYSICAL SYSTEMS AND 1/f NOISE
M. Savelli, G. Lecoy and J-P. Nougier (eds.)
© Elsevier Science Publishers B.V., 1983

SOFT PHOTON NOISE IN ELECTRICAL CIRCUITS

A. Widom, G. Megaloudis, G. Pancheri and Y. Srivastava

Physics Dept., Northeastern Univ., Boston, Mass., U.S.A.

and

T.D. Clark, R.J. Prance, H. Prance and J.E. Mutton

Physics Dept., Univ. of Sussex, Brighton, Sussex, England.

The copious emission of soft photons, which takes place whenever an electron passes through a linear passive impedance, (i) increases the spectral concentration of the noise in the low frequency regime, and (ii) induces energy correlations towards the formation of quantized bundles of magnetic flux. The noise can be viewed as the incoherent disintegration of the magnetic flux bundles.

1. INTRODUCTION

The notion that soft photon radiation is important for understanding noise in electrical circuits is due to Handel (1), who is reporting further considerations at this conference. For specific electronic devices, it is not difficult to carry out the Feynman diagram summations required to understand the effect [using the Schwinger source functional methods in the context of quantum electrodynamic circuit theory (2)]. This has been illustrated by the authors (3,4).

In what follows, the soft photon emission for electronic current impulses will be computed for the purpose of understanding two physical phenomena: (i) the concentration of the noise spectrum towards the low frequency regime, and (ii) the formation of virtual quantized bundles of Faraday magnetic flux.

The circuit element chosen for consideration is merely a simple linear passive impedance. The phenomena to be discussed are purely quantum electrodynamic in nature, and would not be present in a classical electromagnetic theoretical view of circuits.

2. SOFT PHOTONS

Consider a linear passive impedance $Z(\zeta)$. Suppose that a single electron passes through the impedance as a current impulse $e\delta(t)$. From a classical circuit viewpoint, the energy of heating is given by

$$\Delta E = (e^2/\pi)\int_0^\infty d\omega \; \mathrm{Re}Z(\omega+i0^+). \qquad (1)$$

If the impedance is in a cold environment, then the heating energy ultimately radiates with a mean number $dN(\omega)$ of photons in a bandwidth $d\omega$, i.e. from the viewpoint of quantum electrodynamics Eq. (1) reads

$$\hbar\omega \; dN(\omega) = (e^2/\pi)\mathrm{Re}Z(\omega+i0^+)d\omega. \qquad (2)$$

Hence, if the coupling strength

$$\beta = (e^2/\pi\hbar) \lim_{\omega\to 0} \mathrm{Re}Z(\omega+i0^+) \qquad (3)$$

is finite, then $dN(\omega) \to \beta(d\omega/\omega)$ as $\omega \to 0$ so that a simple resistor exhibits the full "infrared catastrophy" well known in soft photon quantum electrodynamics.

3. RENORMALIZATION

The catastrophy can be rendered harmless by concentrating on the physically important total radiation energy, $\hbar\Omega$, summed over the very many photons emitted whenever an electron passes through the impedance. The probability $dP(\Omega)$ that the radiation energy is in the interval $\hbar d\Omega$ defines a finite characteristic function $\chi(s)$ via

$$\int_0^\infty e^{-i\Omega s} \; dP(\Omega) = e^{-\chi(s)}, \quad \mathrm{Im}\; s < 0. \qquad (4)$$

The Bloch-Nordsieck summation of soft photon processes amounts to the assertion of Poisson counting statistics for the radiation. Mathematically, this is expressed as

$$\chi(s) = \int_0^\infty (1 - e^{-i\omega s})dN(\omega). \qquad (5)$$

From Eqs. (2)-(5) it is not difficult to prove the "$(1/\Omega)^{(1-\beta)}$ noise" theorem, i.e.

$$dP(\Omega) \sim d\Omega/\Omega^{(1-\beta)}, \quad \Omega \to 0, \qquad (6)$$

where β is easily found from engineering considerations, i.e. with

$$\alpha = (e^2/\hbar c) \qquad (7)$$

as the usual quantum electrodynamic coupling, and

$$R = (4\pi/c) \qquad (8)$$

as the vacuum radiation impedance,

$$\beta = (4\alpha/R) \lim_{\omega \to 0} \mathrm{Re}Z(\omega+i0^+) \qquad (9)$$

defines the renormalized coupling strength when the vacuum impedance, R ≃ 377 Ohms, is "replaced" by a condensed matter impedance $Z(\zeta)$. Non-linear circuit elements require a somewhat more subtle photon emission amplitude analysis, but the concentration of noise power to the low frequency part of the spectrum, as in Eq.(6), can be easily understood for weak coupling.

4. SHOT NOISE

Suppose that a random sequence of electronic impulses passes through the impedance,

$$I_{shot}(t) = e \sum_j \delta(t-t_j), \qquad (10)$$

at a mean frequency ν,

$$\langle I_{shot} \rangle = e\nu \qquad (11)$$

At first glance it would appear that under the combined action of random current pulses and random photon emissions the noise would be quite chaotic. This view perhaps overlooks the duality in quantum electrodynamics between electric flux and magnetic flux, i.e. the quantisation of charge in units of 0, ± e, ± 2e, ... leads to a correlation energy for the quantization of magnetic flux bundles spaced in units of 0, $\pm \phi_0$, $\pm 2\phi_0$, ... where

$$\phi_0 = (2\pi\hbar c/e). \qquad (12)$$

For electronic shot noise, this is most easily proved by Schwinger source functional actions. A detailed treatment is then required.

5. COMPLEX LAGRANGIANS

The action corresponding to a current source $I(t)$, forced through a linear passive impedance $Z(\zeta)$, is defined via the Feynman circuit photon propagator

$$D(t) = i(c/\pi) \int_0^\infty (d\omega/\omega) e^{-i\omega|t|} \, \mathrm{Re}Z(\omega+i0^+), \qquad (13)$$

i.e. the action is given by

$$S(I) = \frac{1}{2c} \int dt_1 \int dt_2 \, I(t_1) \, D(t_1-t_2) \, I(t_2). \qquad (14)$$

If, in addition to the current source $I(t)$, there is shot noise of the form in Eq. (10), then the renormalized action $W(I)$ is simply found from

$$e^{iW(I)/\hbar} = \langle e^{iS(I+I_{shot})/\hbar} \rangle \qquad (15)$$

where the average is taken over the random times for which electronic current impulses occur. The computation of such averages is standard electrical engineering theory and leads here to the result

$$e^{iW(I)/\hbar} = e^{\frac{i}{\hbar}\int \mathcal{L}'(\Phi)dt} \, e^{iS(I)/\hbar}, \qquad (16)$$

where Φ in Eq.(16) is to be interpreted as a functional derivative $-i\hbar c\delta/\delta I(t)$, and the complex interaction Lagrangian is given by

$$\mathcal{L}'(\Phi) = i\hbar\nu \, [1 - \exp(ie\Phi/\hbar c)]. \qquad (17)$$

Let us now turn to the physical meaning of the complex Lagrangian in Eq.(17) as a function of magnetic flux (viewed as a c-number) producing a voltage across the impedance in accordance with Faraday's law

$$V = -(d\Phi/cdt) \qquad (18)$$

6. CORRELATION ENERGIES

The "potential energy" for forming bundles of magnetic flux is obtained from the real part of the complex Schwinger interaction Lagrangian $U = -\mathrm{Re}\mathcal{L}'$, i.e.

$$U = \hbar\nu \, \sin(e\Phi/\hbar c), \qquad (19)$$

and this energy has a set of minima spaced apart by the flux quantum of Eq.(12). Each flux bundle which cuts across the impedance gives rise to a voltage signal via Eq.(18). The lifetime of the flux bundle can be obtained from the imaginary part of the Schwinger interaction Lagrangian, i.e. the transition rate for the dissipative dissociation of the bundle $\Gamma = (2/\hbar)\,\mathrm{Im}\mathcal{L}'$ is given by

$$\Gamma = 2\nu \, [1 - \cos(e\Phi/\hbar c)]. \qquad (20)$$

Thus, in a shot noise device, the energy width of the bundle is comparable to the correlation energy required to produce the bundle and a somewhat chaotic disintegration into soft photons (concentrating the spectral power into the low frequency noise regime) is the final result.

7. OTHER SYSTEMS

There are systems in which the magnetic flux bundles are quite stable against breakup into soft photons. For example, in some superconducting weak links, the energy gap which induces the gauge symmetry breaking of quantum electrodynamics within the bulk superconductor (5), can also effectively forbid the dissipative breakup of the magnetic flux bundles. In those Josephson junctions wherein the low frequency noise is copious, the tunnelling of flux bundles would be hard to analyse. This is not true for some less noisy constricted region weak links. The Schwinger Lagragian treatment has been discussed elsewhere for the Josephson effect (6).

Superconductivity is not a requirement for the observation of stable quantum electrodynamic flux bundles. Our view is that the flow of such stable bundles can be observed on the gate region of a field effect transistor as the so-called "quantum Hall effect" (7).

In this regard, we do not view as coincidental the fact that the most accurate experimental determinations of the quantum electrodynamic coupling strength $\alpha = (e^2/\hbar c)$ have been made using the devices to which we here refer. This coupling is simply the ratio of "electric charge strength" to "magnetic flux strength", i.e. $\alpha = 2\pi(e/\phi_0)$.

8. CONCLUSIONS

There is little doubt that the present sophistication of electronic devices depends on a theoretical framework in which electrons are viewed as quantum mechanical objects. Although Faraday was able to use purely classical reasoning to obtain the laws capable of describing the electro-mechanical engineering of crucially important industrial machines, purely classical electronic current views would be incapable of describing the micro-circuits which are now much more delicate. Although the circuit currents are made up of many flowing electrons, the quantum mechanics of the individual electrons is presently central to engineering design.

Similarly, although the macroscopic Maxwell fields, upon which the notion of voltage depends, describe many photons, there exist noise sources which depend on the quantum mechanics of the individual photons. Yet, the engineering designs used to eliminate unwanted noise in micro-circuits rarely use the long understood fact that (in a logical sense) it is not consistent to quantize the motion of electronic currents and not also quantize the electromagnetic field.

Hence, in the use of quantum theory for the engineering design of circuits, until the quantum electrodynamics of the voltage (Maxwell field) is routinely included along with the quantum mechanics of the current (electronic motions) we cannot go the full route in exploring the technical uses of that theory of the electromagnetic interaction which has long been understood.

REFERENCES

[1] Handel, P., Phys. Rev. Letters 34 (1975) 1495-1498.

[2] Schwinger, J., Quantum Kinematics and Dynamics (Benjamin, New York, 1970).

[3] Widom, A., Pancheri, G., Srivastava, Y., Megaloudis, G., Clark, T.D., Prance, H. and Prance, R.J., Soft photon emission from electron pair tunnelling in small Josephson junctions. Phys. Rev. B 26 (1982) 1475.

[4] Widom, A., Pancheri, G., Srivastava, Y., Megaloudis, G., Clark, T.D., Prance, H. and Prance, R.J., Quantum electrodynamic circuit soft photon renormalisation of the conductance in electronic shot noise devices, Phys. Rev. B 27 (1983) 3412-3417.

[5] Aitchison, I.J.R. and Hey, A.J.G., Gauge theories in particle physics (Adam Hilger, Bristol 1982), Chapter 9.

[6] Widom, A., Megaloudis, G., Clark, T.D., Prance, R.J. and Prance, H., Quantum electrodynamic formulation of the Josephson tunnelling theory, J. Phys. A. Math. Gen. 16 (1983) L.27.

[7] Widom, A. and Clark, T.D., Thermodynamic equations for gate charge storage on a field effect transistor, J. Phys. D. Appl. Phys. 15 (1982) L.181.

NOISE IN PHYSICAL SYSTEMS AND 1/f NOISE
M. Savelli, G. Lecoy and J-P. Nougier (eds.)
© *Elsevier Science Publishers B.V., 1983*

DIRECT CALCULATION OF THE SCHRÖDINGER FIELD WHICH GENERATES QUANTUM 1/F NOISE

P. H. Handel and T. S. Sherif

Department of Physics
University of Missouri-St. Louis
St. Louis, MO 63121, USA
and
McDonnell Douglas Research Laboratories
St. Louis, MO USA

A derivation of the starting points of quantum 1/f noise theory from first principles is given by direct calculation of the outgoing wave of a scattering experiment which involves particles coupled to infraquanta. This coupling is assumed to be infrared-divergent, and leads to 1/f noise in the outgoing beam and in the scattering cross section. The result of the calculation is a stochastic outgoing Schrödinger field with random phases reflecting the random initial phases of the electromagnetic field oscillators, or of the infra-quantal modes in general. This result is similar to the starting expression used so far in quantum 1/f noise, and gives the same 1/f noise, although the new expression contains a three-dimensional lattice of random phases. Although only the case of photons has been treated explicitely, the results apply to any type of infraquanta with infrared-divergent coupling to the current carriers.

1. INTRODUCTION

Quantum 1/f noise(1-6) is a low-frequency fluctuation process present in any current of particles which exhibit infrared-divergent coupling to a system of massless infra-quanta. It will be present in electric currents whenever the current is carried by a small number of carriers, because the carriers have infrared-divergent coupling to photons, transversal phonons,electron-hole pairs on the surface of the Fermi sphere, etc. It will also be present at low frequencies in large currents of neutral matter, such as currents of water in hydroelectric storage plants in oceanic currents, or in astrophysical jets of matter, because of the coupling to gravitons,(1,5) and to other infraquanta. All infrared-divergent couplings manifest themselves as generalized bremsstrahlung with a number of emitted infraquanta inversely proportional to their frequency.

The starting point of the quantum 1/f noise approach is the description of the particle current in terms of a stochastic Schrödinger field, caused by random energy losses due to bremsstrahlung of infraquanta. Questions have been raised about the validity of this fundamental expression, on whether the random phases should be present in the expression, etc. The justification given for the starting expression was based on physical intuition at the beginning,(1) and a more detailed discussion was given in conference papers(2) which clarify the meaning of the coherent non-bremsstrahlung part in terms of the resolution limit or threshold $\varepsilon_0 = 2\pi\hbar/T$ where T is of the same order as, or smaller than the duration of the 1/f noise measurement. Equivalent later reformulations of the quantum 1/f noise approach by other authors were also based on the same starting point, i.e. on the same expression of the outgoing Schrödinger field. Finally, we mention that our recent two-particle density matrix formulation(6) also rests on this starting expression.

The purpose of this paper is to derive the correct form of this basic expression of the Schrödinger field from first principles, and to check whether it yields the same 1/f current fluctuations and the same noise spectrum as the simple expression used so far. The derivation will be carried out on the example of photons as infraquanta but can be directly applied to other infraquanta such as phonons, spin waves, electron-hole pairs at the Fermi surface, gravitons, etc. by replacing the electromagnetic coupling αA with the corresponding coupling constant.

2. ELECTROMAGNETIC MODES OF FREE SPACE

It is most convenient to describe the electromagnetic field in terms of plane waves. The vector potential is taken in the radiation gauge as

$$\vec{A}(\vec{r},t) = \sum_{\vec{k},\sigma} (\hbar^2 c/\Omega\omega_{\vec{k}})^{1/2}\, \vec{u}_{\vec{k},\sigma} \times [a_{\vec{k},\sigma}(t)\, e^{i\vec{k}.\vec{r}} + a^*_{\vec{k},\sigma}(t)\, e^{-i\vec{k}.\vec{r}}]. \qquad (2.1)$$

The polarization vectors $\vec{u}_{\vec{k},1}$ and $\vec{u}_{\vec{k},2}$ are mutually orthogonal unit vectors perpendicular to $\vec{k}$.

3. DEVELOPMENT OF THE SCHRÖDINGER FIELD UNDER THE INFLUENCE OF ONE ELECTROMAGNETIC FIELD MODE

The Schrödinger equation for an electron moving in a vector potential A and scattering potential V is

$$(1/2m)[-i\hbar\vec{\nabla} - e\vec{A}/c]^2\ \Psi + V\Psi = i\hbar\dot{\Psi}\ . \qquad (3.1)$$

A dot has been used to indicate the time-derivative. The electromagnetic field is treated as a classical field at this point. In order to eliminate the A^2 term from Eq. (3.1), we write

$$\Psi \equiv \exp\left((-i/\hbar)\int^t (e^2/2mc^2)\ A^2 dt'\right)\Phi\ . \qquad (3.2)$$

Thus, Eq. (3.2) is reduced to

$$\left((-\hbar^2/2m)\ \nabla^2 + (ie\hbar/mc)\ \vec{A}\cdot\vec{\nabla} + V\right)\Phi = i\hbar\dot{\Phi}. \quad (3.3)$$

It is convenient to consider first the influence of a single electromagnetic mode, i.e. a single term from Eq. (2.1). Therefore, we take $\vec{A} = \vec{a}\cos(\omega t+\gamma)$, where γ is an initial phase constant, and we treat $V\Phi$ as a perturbation source term. The solution for Eq. (3.1) is an incoming plane wave plus scattered waves given by the integral equation

$$\Phi_{\vec{k}_o}(\vec{r},t) = \phi_{\vec{k}_o} - \int d^3x' \int_{-\infty}^{t} dt' G V(\vec{r}')\Phi_{\vec{k}_o}(\vec{r}',t'). \qquad (3.4)$$

Here $\phi_{\vec{k}_o}$ is the solution of the homogeneous equation, i.e. with V = 0, and can be written in the form

$$\phi_{\vec{k}_o} = e^{i\vec{k}_o\cdot\vec{r}}\exp\left[-(i\hbar/2m)\int^t (k_o^2 - 2e\vec{k}_o\cdot\vec{A}/c\hbar)dt\right]\ . \qquad (3.5)$$

G is the Green's function which satisfies the equation

$$\left[(-\hbar^2/2m)\nabla^2 + (ie\hbar\ \vec{A}\cdot\vec{\nabla}/mc) - i\hbar\partial/\partial t\right]G = \delta(\vec{r}-\vec{r}')\delta(t-t'). \qquad (3.6)$$

Given $\vec{A} = \vec{a}\cos(\omega t+\gamma)$, G can be found to be

$$G = i/(2\pi)^3\hbar\int d^3k\ e^{i\vec{k}.(\vec{r}-\vec{r}')} \qquad (3.7)$$
$$\times \exp\left[-(i\hbar/2m)\left(k^2 t - 2e\vec{k}\cdot\vec{a}\ \sin(\omega t+\gamma)/\hbar c\omega\right)\right]$$
$$\times \exp\left[(i\hbar/2m)(k^2 t' - 2e\vec{k}\cdot\vec{a}\ \sin(\omega t'+\gamma)/\hbar c\omega)\right]$$

In the first Born approximation we set $\Phi_{\vec{k}_o}(\vec{r}',t') = \phi_{\vec{k}_o}(\vec{r}',t')$ in the integral present in Eq. (3.4) and obtain for the scattered wave

$$[i/(2\pi)^3\hbar]\int d^3x'\int^t dt' V(\vec{r}')\int d^3k e^{i\vec{k}(\vec{r}-\vec{r}')}$$
$$\times \exp\left[-(i\hbar/2m)\left(k^2 t - 2e\vec{k}\cdot\vec{a}\ \sin(\omega t+\gamma)/\hbar c\omega\right)\right]$$
$$\times \exp\left[(i\hbar/2m)(k^2 - k_o^2)t'\right]$$
$$\times \{\exp\left[ie(\vec{k}_o - \vec{k})\cdot\vec{a}\ \sin(\omega t'+\gamma)/2mc\omega\right]e^{i\vec{k}_o\cdot\vec{r}'}\}.$$

Using the relation

$$e^{i\beta\sin(\omega t'+\gamma)} = \sum_{n=-\infty}^{\infty} J_n(\beta)e^{in(\omega t'+\gamma)}\ , \qquad (3.8)$$

where $J_n(\beta)$ is the nth order Bessel function, we expand the expression contained in curly brackets in Fourier series. Then Eq. (3.4) takes the form

$$\Phi_{\vec{k}_o,n}(\vec{r},t) - \phi_{\vec{k}_o,n}(\vec{r},t)$$
$$= \left(-i/(2\pi)^3\hbar\right)\int d^3x'\int_{-\infty}^{t} dt' V(\vec{r}')\int d^3k e^{i\vec{k}.(\vec{r}-\vec{r}')}$$
$$\times \exp\left[-(i\hbar/2m)\left(k^2 t - 2e\vec{k}\cdot\vec{a}\ \sin(\omega t+\gamma)/\hbar c\omega\right)\right]$$
$$\times \exp\left[(i\hbar/2m)(k^2 - k_o^2)t'\right]$$
$$\left(\sum_{n=-\infty}^{\infty} J_n(\beta)e^{in(\omega t'+\gamma)}\right)e^{i\vec{k}_o\cdot\vec{r}'}. \qquad (3.9)$$

After performing the integration over t' we use a contour integration method for k. Then Eq. (3.9) is reduced to

$$\Phi_{\vec{k}_o,n}(\vec{r},t) - \phi_{\vec{k}_o,n}(\vec{r},t) = (-m/(2\pi)\hbar^2)$$
$$\times \sum_{n=-\infty}^{\infty} J_n(\beta)e^{in\gamma}\left(e^{ik(n)r}/r\right)$$
$$\times \exp\{i\hbar[k^2(n)t - 2e\vec{k}(n)\cdot\vec{a}\ \sin(\omega t+\gamma)/\hbar c\omega]/2m\}$$
$$\times \int d^3x' e^{i\vec{k}(n).\vec{r}'}\ V(\vec{r}')e^{i\vec{k}_o\cdot\vec{r}'}\ , \qquad (3.10)$$

where

$$\beta = -e(\hbar\vec{k}-\hbar\vec{k}_o)\cdot\vec{a}/mc\hbar\omega$$
$$= -e\ \vec{Q}\cdot\vec{a}/mc\hbar\omega\ , \qquad (3.11)$$

and $\vec{Q}$ is the momentum transfer. In Eq. (3.10) k(n) is defined by

$$(\hbar k(n))^2/2m = (\hbar k_o)^2/2m - n\hbar\omega, \qquad (3.12)$$

Then the total scattered wave can be written as

$$\Psi_s = [-m/(2\pi)\hbar^2 r]\sum_{n=-\infty}^{\infty} e^{ik(n)r}\exp\{-i\hbar[k^2(n)t$$
$$- 2e\vec{k}(n)\cdot\vec{a}\ \sin(\omega t+\gamma)/\hbar c\omega]/2m\}.$$
$$\times V_{\vec{k}(n),\vec{k}_o} J_n(\beta)e^{in\gamma}, \qquad (3.13)$$

where $V_{\vec{k}(n),\vec{k}_o} = \int e^{-i\vec{k}(n).\vec{r}'} V(\vec{r}')e^{i\vec{k}_o.\vec{r}'} d^3x'$ is the scattering matrix element calculated without consideration of the interaction with the electromagnetic field oscillators.

Since we are interested in the spontaneous emission part only, we consider the electromagnetic oscillators in the ground state and take only the terms with $n \geq 0$.

We approximate now Eq. (3.13); terms $|n| > 1$ should be neglected due to the smallness of β. We obtain for small β, up to the power β^2, with $k(1) \simeq k(0) \equiv k$.

$$|\Psi_s|^2 \cong (m/2\pi\hbar^2)^2 |V_{\vec{k},\vec{k}_o}|^2 \tag{3.14}$$

$$\times\ [J_o^2(\beta) + 2J_o(\beta)J_1(\beta)\cos(\omega t+\gamma) + J_1^2(\beta)]/r^2 \ .$$

Since $J_o(\beta) \cong (1 - \beta^2/4)$ and $J_1(\beta) \cong \beta/2$, by

substituting these values into Eq. (3.14) we get

$$|\Psi_s|^2 \cong (|a|^2/r^2)[1 + \beta\cos(\omega t + \gamma)] \ , \tag{3.15}$$

where

$$|a|^2 = (m/2\pi\hbar^2)^2 \ |V_{\vec{k},\vec{k}_o}|^2. \tag{3.16}$$

Then the autocorrelation function $A(\tau)$ for the total scattered wave is

$$A(\tau) = \langle |\Psi_s|^2{}_t |\Psi_s|^2{}_{t+\tau}\rangle \cong (|a|/r)^4[1+\beta^2(\cos\omega\tau)/2] \ . \tag{3.17}$$

The average is defined as an average over the random phase γ, or over t.

4. FORM OF THE SCHRÖDINGER FIELD WITH ALL ELECTROMAGNETIC MODES CONSIDERED

We will consider now the actual case of interaction with all electromagnetic field modes. The classical vector potential is given by

$$\vec{A} = \sum_{\vec{k}',s} \vec{a}_{\vec{k}',s} \cos(\omega_{\vec{k}'}t + \gamma_{\vec{k}',s}) \ , \tag{4.1}$$

where s labels the two polarization states. Using Eq. (4.1) to substitute in Eq. (3.7), we can write the Green's function as

$$G = (i/(2\pi)^3\hbar)\int d^3k e^{i\vec{k}.(\vec{r}-\vec{r}')}$$

$$\times\ \exp[\sum_{\vec{k}',s} -i\hbar(k^2t-2e\vec{k}\cdot\vec{a}_{\vec{k}',s}\sin(\omega_{\vec{k}'}t+\gamma_{\vec{k}',s})/\hbar\omega_{\vec{k}'})/2m]$$

$$\times\ \exp[\sum_{\vec{k}',s} i\hbar(k^2t' - 2e\ \vec{k}\cdot\vec{a}_{\vec{k}',s}\ \sin(\omega_{\vec{k}'}t'+\gamma_{\vec{k}',s})$$

$$/\hbar\omega_{\vec{k}'})/2m]. \tag{4.2}$$

The Born approximation has been used in the integral Eq. (3.4) as before. Using Eq. (3.8) for each mode and each polarization state, the scattered wave can be put in the final form as

$$\Psi_s = (-m/(2\pi)\hbar^2)\ \prod_{\vec{k}',s} \sum_{n_{\vec{k}',s}=-\infty}^{\infty} (e^{ik(n_{\vec{k}',s})r}/r)$$

$$\exp\{-i\hbar[k^2(n_{\vec{k}',s})t - 2e\ \vec{k}(n_{\vec{k}',s})$$

$$\cdot\vec{a}_{\vec{k}',s}\sin(\omega_{\vec{k}'}t + \gamma_{\vec{k}',s})/h\omega_{\vec{k}'}]/2m\}$$

$$\times\ V_{\vec{k}(n_{\vec{k}',s}),\vec{k}_o} J_{n_{\vec{k}',s}}(\beta_{\vec{k}',s})\ e^{in_{\vec{k}',s}\gamma_{\vec{k}',s}} \ . \tag{4.3}$$

Here the definitions given in Eq. (3.11), and Eq. (3.12) have been used again. If we limit ourselves again to the terms $n_{\vec{k}',s} = 0$ and 1, we obtain

$$\Psi_s \cong [-m/(2\pi)\hbar^2 r]$$

$$\times \prod_{\vec{k}',s} e^{ikr} \exp[-i\hbar(k^2t-2e\vec{k}\cdot\vec{a}_{\vec{k}',s}$$

$$\times\ \sin(\omega_{\vec{k}'}t+\gamma_{\vec{k}',s})/\hbar\omega_{\vec{k}'})/2m].$$

$$\times\ [J_o(\beta_{\vec{k}',s})+e^{i(\omega_{\vec{k}'}t + \gamma_{\vec{k}',s})}J_1(\beta_{\vec{k}',s})]\ V_{\vec{k},\vec{k}_o} \ . \tag{4.4}$$

In Eq. (4.4) we have again used $k(0) \cong k(1) \cong k$.

5. CALCULATION OF 1/f NOISE AND COMPARISON WITH PREVIOUS RESULTS

In this section we will calculate again the probability density and its autocorrelation in the outgoing beem. The probability density in the outgoing Schrödinger-field given by Eq. (4.4) has the form:

$$|\Psi_s|^2 \cong (m/(2\pi)\hbar^2)^2(|V_{\vec{k},\vec{k}_o}|^2/r^2)$$

$$\times \prod_{\vec{k}',s} [J_o^2(\beta_{\vec{k}',s})+2J_o(\beta_{\vec{k}',s})$$

$$\times\ J_1(\beta_{\vec{k}',s})\cos(\omega_{\vec{k}'}t+\gamma_{\vec{k}',s}) + J_1^2(\beta_{\vec{k}',s})]. \tag{5.1}$$

We substitute the approximate values of Bessel's function as before and neglect all powers of $\beta_{\vec{k},s}$ greater than one in Eq. (5.1). This yields

$$|\Psi_s|^2 = (|a|^2/r^2)[1 + \sum_{\vec{k},s} \beta_{\vec{k},s}\cos(\omega_{\vec{k}}t + \gamma_{\vec{k},s})]. \tag{5.2}$$

The autocorrelation function $A(\tau)$ for the scattered wave can be calculated as

$$|\Psi_s|^2_t |\Psi_s|^2_{t+\tau} = (|a|^4/r^4)\ \{1 +$$

$$\sum_{\vec{k},s} \beta_{\vec{k},s}\ [\cos(\omega_{\vec{k}} t + \gamma_{\vec{k},s}) +$$

$$\cos(\omega_{\vec{k}}(t+\tau) + \gamma_{\vec{k},s})]$$

$$+ \sum_{\vec{k},s} \sum_{\vec{k}',s'} \beta_{\vec{k},s}\ \beta_{\vec{k}',s'}\ \cos(\omega_{\vec{k}} t + \gamma_{\vec{k},s}) \qquad (5.3)$$

The ensemble average over the random phases is now considered for Eq. (3.6) with $\langle e^{i\gamma_{\vec{k},s}}\rangle = 0$ and $\langle e^{i(\gamma_{\vec{k},s} - \gamma_{\vec{k}',s'})}\rangle = \delta_{\vec{k},\vec{k}'}\ \delta_{s,s'}$.

We obtain

$$A(\tau) = \langle|\Psi_s|^2_t\ |\Psi_s|^2_{t+\tau}\rangle,$$

$$A(\tau) = (|a|/r)^4\ \{1 + 1/2 \sum_{\vec{k},s} |\beta_{\vec{k},s}|^2 \cos\omega_{\vec{k}}\tau\} \ . \qquad (5.4)$$

By expressing the summation in terms of the integral $(\Omega/(2\pi)^3)\int d^3k$ in Eq. (5.4), the autocorrelation function becomes

$$A(\tau) = (|a|/r)^4\ \{1 + 1/2 \sum_{s=1}^{2} (\Omega/(2\pi)^3)$$

$$\int |\beta_{\vec{k},s}|^2 \cos\omega_{\vec{k}}\tau d^3k\} \ . \qquad (5.5)$$

Note that in Eq. (4.1) the expansion coefficients also depend on the volume Ω of the normalization box

$$\vec{a}_{\vec{k},s} = \vec{\varepsilon}_s\ c(\hbar/\omega\Omega)^{1/2} \ . \qquad (5.6)$$

From Eq. (3.11) one finds then

$$\beta_{\vec{k},s} = -(e/m\hbar\omega)(\hbar/\omega\Omega)^{1/2}\ \vec{Q}\cdot\vec{\varepsilon}_s. \qquad (5.7)$$

Substituting into Eq. (5.5), we obtain

$$A(\tau) = (|a|/r)^4\ [1 + 2\alpha A \int_{\omega_0}^{\omega_1} \cos\omega\tau\ d\omega/\omega], \qquad (5.8)$$

where the fine structure constant $\alpha = e^2/4\pi\hbar c$ has been introduced and

$$\alpha A = \frac{2\alpha}{3\pi}\frac{Q^2}{m^2c^2} \qquad (5.9)$$

is the infrared exponent present both in the infrared radiative corrections and in the spectral density of the fractional bremsstrahlung cross section $2\alpha A/f$. The lower integration limit $\omega_0 > 2\pi T^{-1}$, where T is the duration of the experiment, defines the frequency resolution. The upper limit $\omega_1 < E/\hbar$ is given by the energy E of the particles in the beam. According to the Wiener-Khintchine theorem, the spectral density of the density fluctuations $S_n(f)$ is the Fourier transform of the Autocorrelation function $A(\tau)$

$$S_n(f)\ (|a|/r)^4\ [\delta(f) + 2\alpha A/f] \qquad (5.10)$$

Consequently, the spectral density of fractional density fluctuations $S_{\delta n/n}$ will be

$$S_{\delta n/n}(f) \equiv \langle n\rangle^{-2}\ \langle(\delta n)^2\rangle_f = 2\alpha A/f \qquad (5.11)$$

This final result and Eqs. (5.8) - (5.11) coincide with the well-known quantum 1/f noise formula(5),(1-4), with the infrared radiative correction factor $(f/f_0)^{\alpha A}$ approximated by unity.

If we consider the current density $j = (e/m)\hbar k|\psi_s|^2$ in Eqs. (5.1) - (5.4) rather than $|\psi_s|^2$, we obtain in a similar way

$$S_{\delta j/j} = S_{\delta\sigma/\sigma} = S_{\delta\mu/\mu} = 2\alpha A/f, \qquad (5.12)$$

which indicates the presence of fluctuations in the current, the cross section σ, and (for currents in solids) also mobility, with the same spectral density of fractional fluctuations.

There is a detail which was obtained for the first time in the present paper. This is the three-dimensional lattice-sum and polarization-states sum character of the incoherent summation of bremsstrahlung parts in Eqs. (5.2) - (5.5). However, this detail is not reflected in the final result, since the random phases drop out anyway in the calculation of the autocorrelation function and of the spectral density. Therefore, the simple starting form can still be used for simplicity in a less detailed presentation of quantum 1/f noise theory.

REFERENCES

[1] Handel, P.H., 1/f Noise-an infrared phenomenon, Phys. Rev. Lett. 34 (1975) 1492-1494.

[2] Handel, P.H., Low frequency fluctuations in electronic transport phenomena, in Devreese, J.T. and van Doren, V.E. (eds.), Linear and Nonlinear Electron Transport in Solids (Plenum, New York, 1976).

[3] Handel, P.H. and Eftimiu, C., Survival of the long time correlations in 1/f noise, in T. Musha (ed.), Proceedings of the Symposium on 1/f Fluctuations (Tokyo Inst. of Technology Press, Tokyo, 1977).

[4] Handel, P.H., Quantum 1/f noise in the presence of a thermal radiation background, in van Vliet, C.M. (ed.), Proceedings of the Second International Symposium on 1/f Noise (Univ. of Florida Press, Gainesville, FL, 1980).

[5] Handel, P.H., Quantum approach to 1/f noise, Phys. Rev. A 26 (1980) 745-757.

[6] Sherif, T.S. and Handel, P.H., Unified treatment of diffraction and 1/f noise, Phys. Rev. A 26 (1982) 596-602.

CHAOTIC SYSTEMS

NOISE IN PHYSICAL SYSTEMS AND 1/f NOISE
M. Savelli, G. Lecoy and J-P. Nougier (eds.)
© Elsevier Science Publishers B.V., 1983

NOISE IN CHAOTIC FLUID SYSTEMS
(Invited Paper)

A. LIBCHABER

Ecole Normale Supérieure
24 rue Lhomond
75231 PARIS CEDEX 05
France

We illustrate, in a Rayleigh-Benard experiment, the various types of noise appearing in dynamical systems. Intermittent noise, power law spectrum, exponential noise may appear depending on the scenario leading to a chaotic behavior.

This has been already published in the two review papers given in the references below.

References :

1. J.P. Eckmann, Rev. Mod. Physics 53, 643 (1981)
2. A. Libchaber, S. Fauve, "melting, localization, chaos", Kalia and Vashishta Ed., Elsevier 1982.

NOISE IN PHYSICAL SYSTEMS AND 1/f NOISE
M. Savelli, G. Lecoy and J-P. Nougier (eds.)
© Elsevier Science Publishers B.V., 1983

CHAOS AND NOISE IN JOSEPHSON TUNNEL JUNCTIONS

(Invited Paper)

John Clarke, Robert F. Miracky, and John Martinis

Department of Physics
University of California
Berkeley, California 94720

and

Materials and Molecular Research Division
Lawrence Berkeley Laboratory
Berkeley, California 94720

and

Roger H. Koch

IBM Thomas J. Watson Research Center
Yorktown Heights, New York 10598

The current-voltage characteristics of Josephson tunnel junctions shunted by a conductance with substantial self-inductance exhibit regions of stable negative resistance. At certain values of bias current, the junctions exhibit chaos, which is manifested as a low frequency voltage noise equivalent to a noise temperature of about 10^3K. At other bias points, switching between subharmonic modes or between a subharmonic mode and a chaotic regime produces low frequency noise with a noise temperature often exceeding 10^6K. Analog simulations indicate that, at least under some conditions, the switching is induced by intrinsic thermal noise. The analog has also been used to show that switching induced by thermal noise is responsible for the noise rise observed experimentally in three-photon Josephson parametric amplifiers.

1. INTRODUCTION

There has recently been considerable interest in nonlinear systems that exhibit chaos.(1) Such systems are governed by deterministic equations, but for an appropriate choice of parameters the solutions appear to fluctuate randomly, although in fact this behavior is completely determined by the equations of motion and the given set of initial conditions. An attractive system for studying chaos is provided by the Josephson (2) junction, which has well-established equations of motion so that realistic comparisons can be made between the observed behavior of real junctions and the predictions of analog or digital simulations. Huberman, Crutchfield, and Packard (3) were the first to point out that a resistively shunted Josephson junction with appropriate parameters should exhibit chaos when driven by an external radio frequency field. This problem has been studied in detail by Kautz (4) and D'Humieres et al.(5) using computer simulations; as emphasized by D'Humieres et al.,(5) this problem is equivalent to the forced pendulum. Apart from the inherent interest in the chaotic behavior of these systems, these ideas may by very relevant to the design of devices involving the Josephson effect. For example, it has been suggested that the large levels of excess noise observed in Josephson parametric amplifiers (6-16) were due to chaos, although there is hardly widespread agreement on this point.(13-16) Unfortunately, there has been relatively little experimental work on chaos in Josephson junctions under conditions where the experimental parameters are sufficiently well known that one can make meaningful comparisons with theoretical predictions; in particular it is very difficult to estimate the magnitude of the rf voltage across the junction. Because of this problem, we have chosen to study chaotic effects in a different system, namely a Josephson tunnel junction with self-capacitance that is shunted by a resistance with substantial self-inductance.(17)

We first describe the model system, and obtain the equations of motion. We next describe the junction configurations used, outline the measurement techniques, and present the experimental data. We discuss solutions to the equations of motion obtained on an analog computer, and show that all the features observed experimentally can be explained with the analog. We find from simulations that there are period-doubling sequences to chaos, followed by intermittency and odd- and even-period windows. In addition, we find large amounts of low frequency noise arising from switching induced by thermal noise between stable subharmonic modes or between a stable mode and a chaotic regime. As an example of the importance of these switching processes in a practical device, we describe the results of an analog simulation of a three-photon parametric amplifier. We conclude with a brief summary.

2. EQUATIONS OF MOTION

The Josephson junction that we have investigated is shown schematically in Fig. 1. The junction

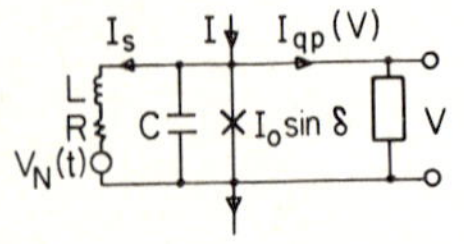

Figure 1 : Schematic representation of a Josephson tunnel junction with critical current I_o and self-capacitance C shunted with an external resistance R which has a self-inductance L. The junction is also shunted with the quasiparticle tunneling conductance, $1/R_J$.

has a critical current I_o and is shunted by its self-capacitance, C, and the quasiparticle tunneling conductance, $1/R_J$, through which the quasiparticle current I_{qp} flows. The external shunt has a resistance R and a self-inductance L, through which a current I_s flows; the resistance R generates a Nyquist voltage noise, $V_N(t)$, with a spectral density

$$S_v(f) = 4k_BTR, \tag{1}$$

where T is the temperature and f is the frequency. The coupled equations of motion can be written in the form

$$I = I_o\sin\delta + \hbar C\ddot{\delta}/2e + I_s + I_{qp}, \tag{2}$$

and

$$\hbar\dot{\delta}/2e = I_sR + \dot{I}_sL + V_N(t). \tag{3}$$

Here, I is the fixed bias current, δ is the phase difference across the junction, and we have set $\dot{\delta} = 2eV/\hbar$, where the dot implies the differentiation with respect to time. It is convenient to rewrite these equations in a dimensionless form by introducing a dimensionless time variable $t/(\Phi_o/2\pi I_oR)$, where $\Phi_o \equiv h/2e$ is the flux quantum. We obtain

$$i = \sin\delta + \beta_C\ddot{\delta} + i_s + i_{qp}, \tag{4}$$

and

$$\dot{\delta} = i_s + \beta_L\dot{i}_s + v_N. \tag{5}$$

In these equations, we have set $i = I/I_o$, $i_s = I_s/I_o$, $i_{qp} = I_{qp}/I_o$, $v_N = V_N/I_oR$, $\beta_C = 2\pi I_oR^2C/\Phi_o$, $\beta_L = 2\pi LI_o/\Phi_o$, and $\Gamma = 2\pi k_BT/I_o\Phi_o$. The three parameters that determine the behavior of the system are β_L (the reduced inductance), β_C (the reduced capacitance), and i (the reduced bias current); for a given junction i serves as the control parameter.

3. EXPERIMENTAL TECHNIQUES

We have studied two types of junction, illustrated in Fig. 2. The small area junctions (10×10 μm^2), shown in Fig. 2(a), were fabricated in

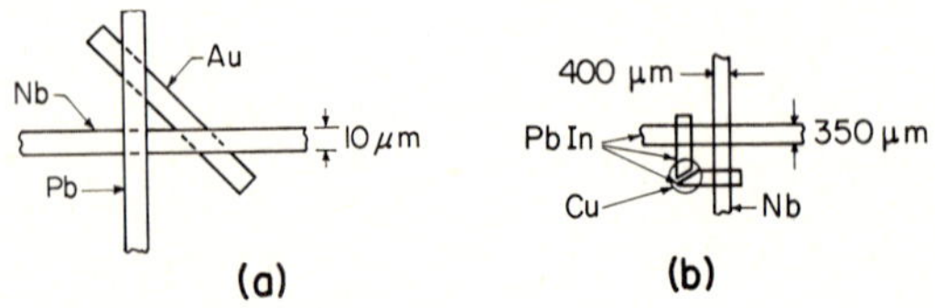

Figure 2 : (a) Small and (b) large area Josephson tunnel junctions shunted by an external resistance with substantial self-inductance. In (b), the loop is overlaid with a superconducting ground plane, insulated from the loop with a layer of SiO.

nine batches of six junctions on a 50 mm diameter Si wafer, using photolithographic lift-off techniques. First, a gold strip about 10 μm wide and 160 nm thick was deposited, followed by a Nb strip about 200 nm thick. The lift-off for the Nb strip was performed, and the resist patterned for the PbIn (5 wt.%In) strip. The wafer was diced to give nine individual substrates, each with six junctions; each substrate was processed individually from this point. The surface of the Nb was cleaned by ion-milling in Ar, and oxidized with a radio frequency discharge in an Ar-O_2 mixture. The PbIn film, about 300 nm thick, was then deposited and lifted off. Typical parameters for these junctions were I_o = 0.5 mA, C = 4 pF, R = 0.4 Ω, and L = 4 pH. The large area junctions (400×350 μm^2), shown in Fig. 2(b), were fabricated by depositing films through metal shadow masks. First, a disk of Cu about 300 nm thick was deposited, followed by two strips of PbIn that overlap the Cu. A Nb strip about 400 μm wide and 300 nm thick was sputtered and oxidized thermally. The junction was completed by depositing a 170 nm-thick PbIn strip. The sample was covered with an insulating SiO layer about 130 nm thick, and a PbIn ground plane was evaporated to reduce the inductance of the loop. Typical parameters were I_o = 1 mA, C = 6 nF, R = 2 mΩ, and L = 3 pH.

Several different measurements were made. First, we obtained a current voltage (I-V) characteristic by slowly sweeping the current, and recording the resulting voltage. Second, we measured the low frequency voltage noise across the junction by amplifying the noise with a cooled resonant tank circuit, with a resonant frequency, f_t, of about 100 kHz, coupled via a room temperature preamplifier to a mean square voltmeter. For the large area junctions, the bandwidth, B, of the measurement was determined by the Q ($\approx$ 480) of the tank circuit: $B = \pi f_T/2Q$. For the small area junctions, a lower Q ($\approx$ 60) was used, and the bandwidth was limited by a low pass filter in the subsequent signal processing to about 250 Hz. Third, in the case of the large area junctions, we connected the junctions directly to a low noise high frequency preamplifier, with a bandwidth of about 1 GHz, and examined the voltage on a spectrum analyzer.

4. EXPERIMENTAL RESULTS

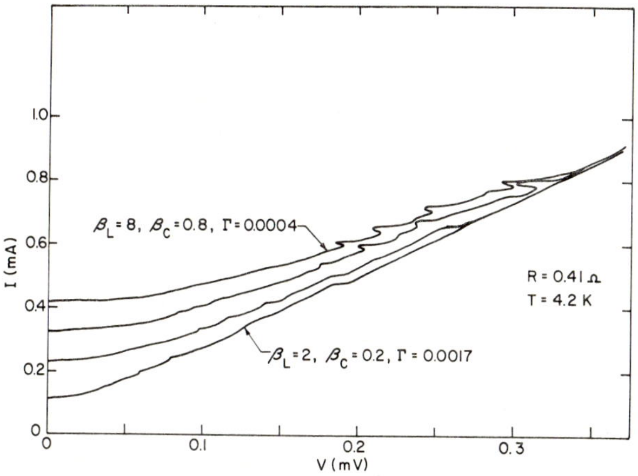

Figure 3 : Current-voltage (I-V) characteristics of a small area junction for four values of critical current.

Figure 3 shows the I-V characteristic of a small area junction for four values of critical current; the critical current was reduced by trapping magnetic flux in the junction. The characteristics of the two higher values of critical current exhibit stable regions of negative dynamic resistance. The I-V characteristics of the two lower values of critical current show structure on the I-V characteristics, but the dynamic resistance is always positive.

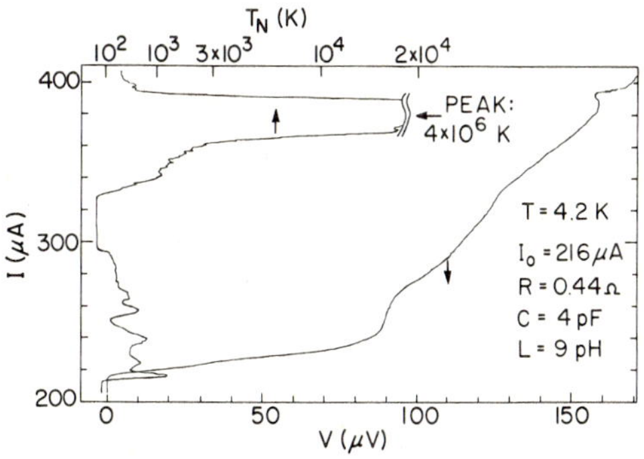

Figure 4 : I-V characteristic and voltage noise temperature, T_N, at 97 kHz for a small area junction with β_C = 0.5, β_L = 6, and $\Gamma = 8.1 \times 10^{-4}$. T_N was measured at a reduced frequency of 2.1×10^{-6}.

Figure 4 shows the I-V characteristic of a different small area junction, and, in addition, the voltage noise measured at a frequency of 97 kHz with a tank circuit Q of 57. This frequency corresponds to a reduced frequency of 2.1×10^{-6} in units of the characteristic frequency I_0R/Φ_0. The noise has been characterized by a noise temperature, T_N, defined by

$$T_N = \langle V_N^2\rangle/4k_BRB, \tag{6}$$

where $\langle V_N^2\rangle$ is the mean square voltage across the junction. The noise temperature, T_N, as a function of bias current generally falls into one of three ranges. An example of the first occurs for the bias region near 300 μA. Here the noise is independent of bias current, with a magnitude of about 40K. This noise arises from the preamplifier and from Nyquist noise associated with losses in the tank circuit rather than from the junction itself. The second range of noise level appears, from example, in the intervals from 220 to 295 μA and 330 to 360 μA. Here we observe noise temperatures varying from several hundred to one thousand Kelvin. Finally, there is an enormous peak that is off-scale on this plot. In a separate measurement we found that its noise temperature was about 4×10^6K. This peak occurs just below the negative resistance region on the I-V characteristic.

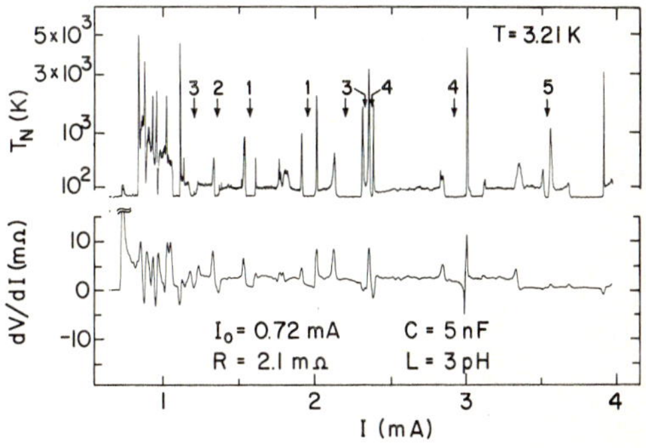

Figure 5 : dV/dI vs. I and T_N vs. I for large area junction with β_C = 0.05, β_L = 6.6, and $\Gamma = 1.9 \times 10^{-4}$. The numbers on the upper trace refer to the subharmonic number. T_N was measured at a reduced frequency of 1.6×10^{-4}.

Figure 5 shows the dynamic resistance, dV/dI, vs. I and T_N vs. I for a large area junction. There is considerable structure, including numerous values of current for which the dynamic resistance is negative. The noise temperature again shows three types of regions: Regions of about 32K, where the noise arises from the tank circuit and the preamplifier, broad regions of a hundred Kelvin, and large spikes where the noise is several thousand Kelvin. These spikes are sometimes but by no means always associated with local maxima in the dynamic resistance. Several of the regions where the noise is low have been labelled with the subharmonic number of the oscillations. This number was obtained by amplifying the signal across the junction with the wideband amplifier, determining the fundamental frequency on a spectrum analyzer, and dividing this frequency into the Josephson frequency 2eV/h. It is evident that both even and odd subharmonics are present.

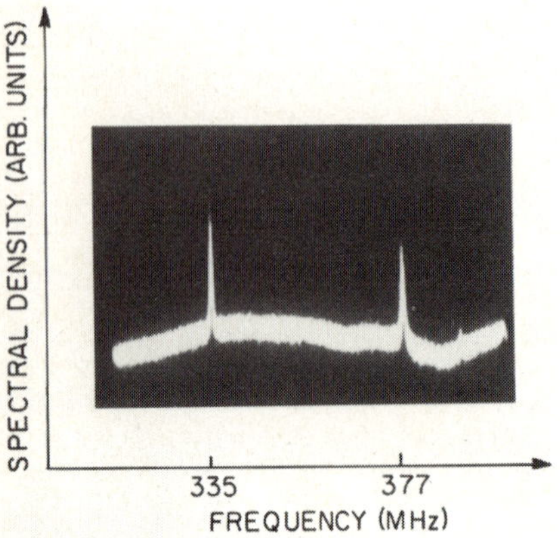

Figure 6 : Power spectrum of voltage across a large area junction at 4.2K with I_o = 2.35 mA, I = 6.68 mA, C = 5 nF, R = 1.7 mΩ, L = 3 pH, β_C = 0.10, β_L = 21, and Γ = 7.5 × 10^{-5}.

Finally, we examined the voltage across a large area junction when it was biased at a low frequency noise spike, and displayed the result on a spectrum analyzer. The result of such a measurement is shown in Fig. 6, where we observe two well-defined peaks at 335 MHz and 377 MHz. As one sweeps the bias current through the region where the noise spike occurs, one observes first one peak, then the growth of the second peak as the first one shrinks, and finally the disappearance of the first peak. Thus, the junction switches between two subharmonic modes, giving rise to large levels of noise at frequencies below the characteristic switching frequencies. Switching between a subharmonic mode and a chaotic regime can also occur (see below): A manifestation of this behavior is the appearance of a low frequency noise spike at the boundary between noise free and moderately noisy regions. Examples of this appear in Fig. 5 at about 1.11 mA and 2.02 mA.

In order to shed more light on the behavior observed experimentally, we have performed extensive digital and analog simulations. Some of the results of the analog simulation are now briefly reported.

5. ANALOG SIMULATIONS

We have simulated the junction using an electronic analog of a Josephson junction involving a phase-locked loop (Philipp Gillette and Associates, model JA-100). To achieve large inductances (1 - 10H) with low loss, we used an active circuit for the inductance of the shunt.

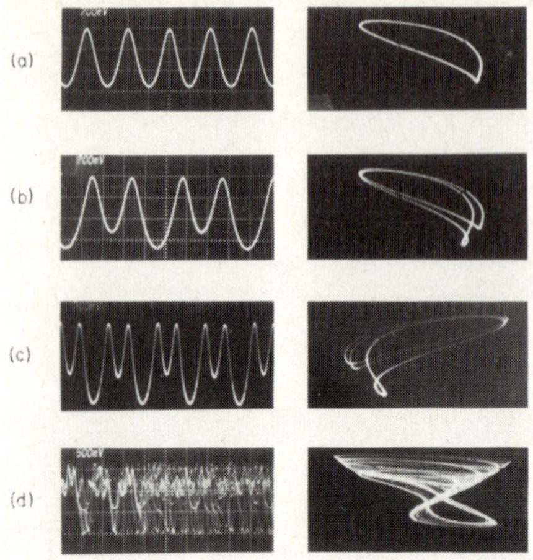

Figure 7 : Voltage vs. time (left-hand column) and $\dot{\delta}$ vs. sinδ (right-hand column) for the analog simulator with β_C = 0.25, β_L = 8.0, and $\Gamma \approx$ 0: (a) period 1, (b) period 2, (c) period 4, and (d) chaos.

In Fig. 7 we show a typical set of solutions for β_C = 0.25, and β_L = 8.0. The left-hand column shows the voltage across the junction vs. time for four values of bias current, while the right-hand column shows the corresponding phase portrait, $\dot{\delta}$ vs. sinδ. In (a), the solution is the Josephson oscillation with period 1, and the corresponding phase portrait is a single closed loop that repeats each time δ evolves through 2π. In (b), a bifurcation has occurred to a period 2 solution, and the phase portrait contains two loops. In (c), a second bifurcation has occurred to a period 4 solution, while in (d) the system has become chaotic. This is an illustration of a Feigenbaum (18) period-doubling sequence to chaos.

As the current is further reduced, the system exhibits Pomeau-Manneville (19) intermittency, followed by tangent bifurcations to limit cycles of even or odd periodicity. In the theory of chaotic behavior for one-dimensional mappings with single quadratic maxima, there is an explicit sequence in which stable limit cycles of period n should first appear as the control parameter is varied monotonically.(20) The behavior we observe in our junctions as the bias current is reduced does not appear to fit this simple picture, suggesting that a reduction of this third-order system to a one-dimensional mapping may not be possible. The lack of order in the appearance of periodic windows also indicates that the basins of attraction are probably quite complicated, and that the observed behavior may depend crucially on the amount of external noise present.

In Figs. 8 and 9 we show two current-voltage characteristics obtained from the simulator with

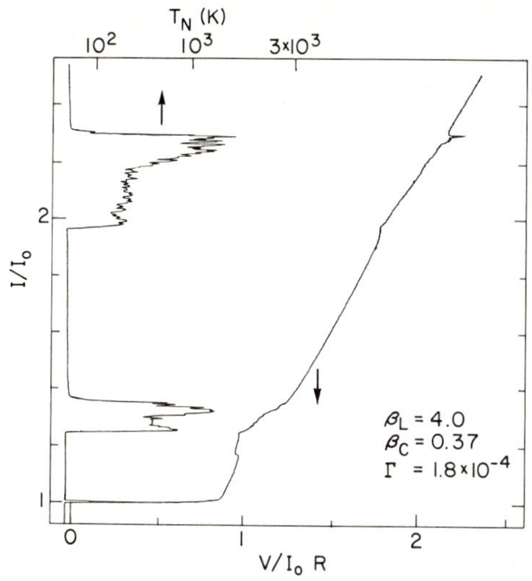

Figure 8 : I/I_o vs. V/I_oR and T_N vs. I/I_o for analog simulator with parameters chosen to approximate those of the junction in Fig. 4. T_N was measured in the reduced frequency range 0.016 to 0.079.

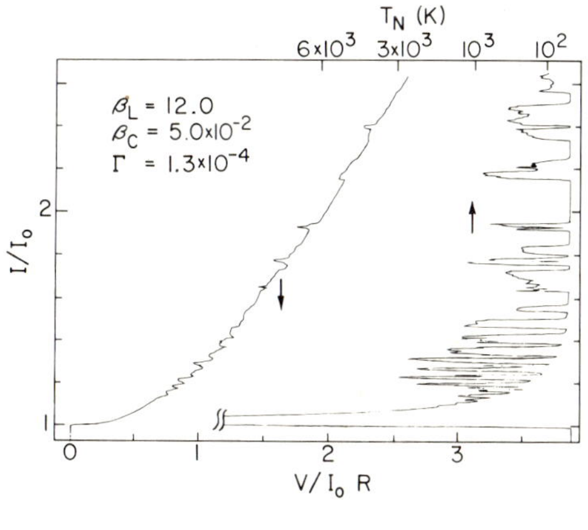

Figure 9 : I/I_o vs. V/I_o R and T_N vs. I/I_o for analog simulator with parameters chosen to approximate those of the junction in Fig. 5. T_N was measured in the reduced frequency range 0.016 to 0.079.

values of β_C, β_L, and Γ chosen to approximate those of the real junctions in Figs. 4 and 5. We observe the same general structure on the characteristics, including regions of stable negative resistance. Also shown in Figs. 8 and 9 are the corresponding noise temperatures, measured at frequencies between 10 and 50 Hz, corresponding to reduced frequencies of 0.016 and 0.079. In the region of low noise, the junction is in a stable limit cycle, and the noise is from the measurement system. There are also relatively broad regions of current where the noise ranges from a few hundred Kelvin to perhaps 10^3K: In these regions the junction is chaotic. Thus, for the experimental junctions, we can identify the values of bias current where the noise is of the same order of magnitude as regions in which chaos occurs.

Notice that the very large noise temperatures ($\gtrsim 10^5$K) measured in the real junctions are not present in the bandwidth of the noise measurements in Figs. 8 and 9. For the real junctions, the noise measurements in Figs. 4 and 5 were performed at reduced frequencies, $f/(I_oR/\Phi_o)$, of 2.1×10^{-6} and 1.6×10^{-4}, respectively, while for the simulations the reduced frequency of the measurements spans the range 0.016 to 0.079. The lack of a large noise peak in this portion of the analog spectrum suggests than switching noise is greater at much lower reduced frequencies.

In order to observe this switching behavior on the analog, we measured the power spectrum of the noise down to dimensionless frequencies equivalent to those of the measurements on the large-area junctions. In Fig. 10, we show the spectral density of the voltage noise at a bias

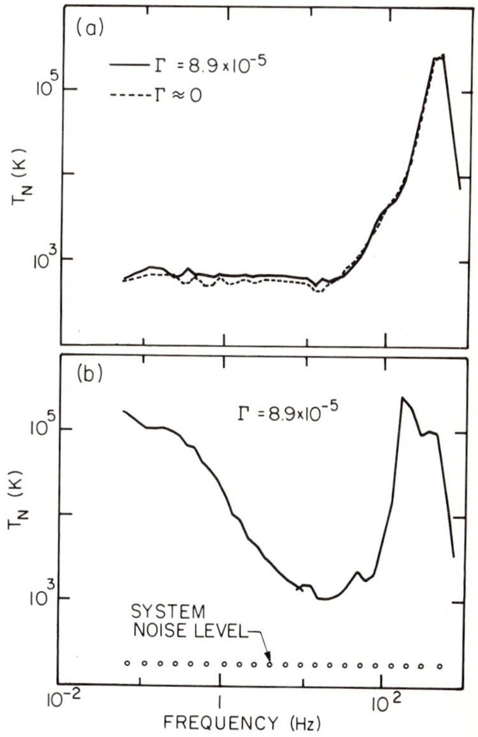

Figure 10 : Noise temperature (for I_o = 1.49 mA) for analog with β_C = 0.105 and β_L = 12.0 (a) in chaotic regime and (b) in switching regime, with and without thermal noise. The horizontal scale spans the normalized frequency range 1.55×10^{-5} to 1.55.

point where the system is chaotic, and at a point where switching occurs between a subharmonic mode and a chaotic regime. In the chaotic regime, we see a residual, broadened subharmonic peak at about 500 Hz, and at lower frequencies white noise with a noise temperature of about 700K. We notice that this behavior is essentially independent of the presence or absence of injected thermal noise equivalent to 3.18K. In the absence of injected noise, the intrinsic noise temperature of the analog was $\lesssim$ 7 mK. In Fig. 10(b), the noise temperature of the junction in the absence of thermal noise is below the noise temperature of the measurement system. In the presence of thermal noise equivalent to 3.18K, however, the noise temperature increases dramatically, with a power spectrum of approximately 1/f at frequencies below 10 Hz to a value of about 10^5K at 0.1 Hz. The 1/f nature of this noise has also been observed in real junctions. Thus, it appears that this type of behavior is due to thermal noise currents that can induce the system to "hop" between two different regimes, for example, two subharmonic modes that would be stable in the absence of thermal noise, or between a subharmonic mode and a chaotic region. Somewhat related behavior has been discussed by Arrecchi and Lisi (21) for a non-linear electronic oscillator. However, these authors claim that the 1/f noise is intrinsic to the deterministic equation of motion, without the addition of thermal noise. Hopping between modes has also been discussed by Kautz,(4) D'Humieres et al.,(5) and Ben-Jacob et al.,(22) but they do not appear to have observed a 1/f power spectrum.

6. JOSEPHSON PARAMETRIC AMPLIFIERS

To illustrate the importance of some of the ideas described above to practical devices, we consider briefly the Josephson parametric amplifier. The inductance of a Josephson junction for $I < I_0$ is non-linear, and may be written in the form (2)

$$L_J = \Phi_0/2\pi I_0\cos\delta = \Phi_0/2\pi(I_0^2 - I^2)^{\frac{1}{2}}. \qquad (7)$$

This non-linearity may be used for parametric amplification of electromagnetic signals, and amplifiers of this kind have been extensively studied. There are two principal modes of operation, the singly-degenerate three-photon mode, (9-12) and the doubly-degenerate four-photon mode.(6-8) We shall confine ourselves to the former, in which the pump frequency, ω_p, the signal frequency, ω_s, and the idler frequency, ω_i, satisfy the relation

$$\omega_p = \omega_s + \omega_i, \qquad (8)$$

where $\omega_s \approx \omega_i \approx \omega_p/2$. The junction is biased with a current below the critical current, and an appropriate level of pump power is applied. Both the bias current and pump amplitude change the plasma frequency of the junction according to the relation

$$\omega_0(\eta, I/I_0) = \omega_0(0,0)[J_0(\eta)]^{\frac{1}{2}}[1-(I/I_0)^2]^{\frac{1}{4}}, \qquad (9)$$

where

$$\eta = 2\pi V_p/\Phi_0\omega_p. \qquad (10)$$

Here,

$$\omega_0(0,0) = (I_0/2\pi\Phi_0 C)^{\frac{1}{2}} \qquad (11)$$

is the maximum plasma frequency, and V_p is the voltage across the junction at the pump frequency. Thus, one adjusts the bias current and pump amplitude until the plasma frequency is equal to the signal frequency. As the pump power is increased, the signal amplification increases, but, unfortunately one finds that the input noise temperature also increases. This so-called "noise rise" has prevented the Josephson parametric amplifier from becoming important as a high frequency amplifier. There have been several attempts to explain the noise rise,(13-16) but, to our knowledge, none of these explanations has been in satisfactory agreement with the observed phenomena.

We (23) have simulated the three-photon Josephson parametric amplifier using the circuit (16) shown in Fig. 11. The junction has capacitance

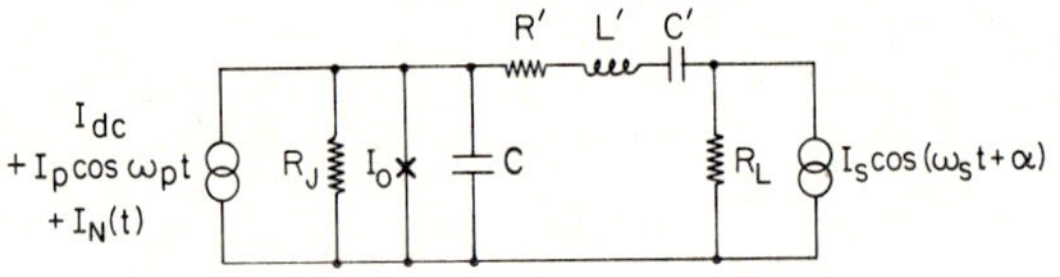

Figure 11 : Circuit for analog simulation of three-photon Josephson parametric amplifier.

C and is shunted with a resistance R_J representing the quasiparticle tunneling resistance; associated with this resistance is a Nyquist noise current, $I_N(t)$. In practice, the signal is coupled in by means of a circulator, which we represent by a series resonant circuit with inductance L', capacitance C', and resistance R'. The resistance of the signal source is R_L. The values of L' and C' were chosen so that $\omega_s = (L'C')^{-\frac{1}{2}}$. To optimize the performance, we set $R_L = R'$ and maintained $R_L + R' \approx R_J$. The Q of the resonant circuit, $(L'/C')^{\frac{1}{2}}/(R' + R_L)$ was typically 20. The pump was represented by a current source, $I_p\cos\omega_p t$. To establish parametric amplification, we first applied low levels of pump and signal power and adjusted I_{dc} until signal gain and the idler were observed. We then increased the pump amplitude, adjusting I_{dc} to achieve maximum gain.

Figure 12 shows typical results for four levels of pump power with $I_{dc}/I_0 = 0.74$ and $\Gamma = 2.32 \times 10^{-5}$; this value of Γ corresponds to a temperature of 0.55K for a junction with I_0 = 1 mA. The frequency has been normalized to $\omega_0(0,0)$. The lowest trace is for $I_p/I_0 = 0$, so that a re-

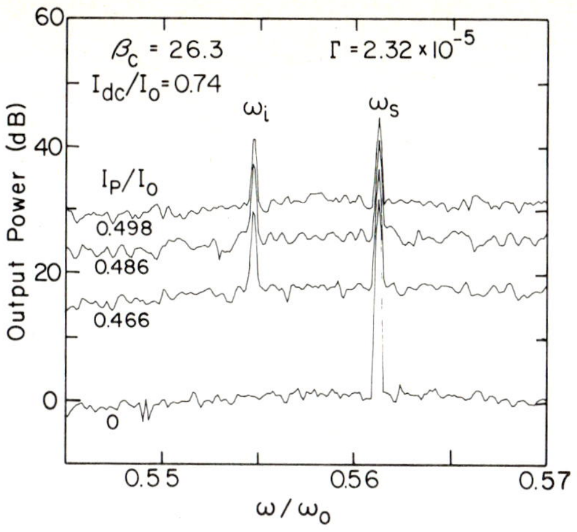

Figure 12 : Power spectral density of voltage across R_L for 4 values of I_p/I_0. Reference level of power is arbitrary. Frequency ω has been normalized to the maximum plasma frequency $\omega_0(0,0)$.

sponse is seen only at the signal frequency, ω_s. The remaining three traces show the results for increasing levels of I_p/I_0, the largest level corresponding to the maximum gain that could be achieved with the given circuit parameters. For larger values of I_p or I_{dc}, a bifurcation occurs to an oscillation at $\omega_p/2$. This behavior, which is also seen in real devices,(24) is a useful means of recognizing that the condition for maximum gain has been achieved. Once the bifurcation has occurred, the gain drops sharply. We see in Fig. 12 that while the signal gain increases as the pump power increases, the level of the noise increases much more dramatically. Thus, when the pump power is increased from zero to its value for maximum gain, the signal gain is 12.8 dB, while the noise increases by 30 dB. To illustrate the much more rapid rise in noise than in signal gain, in Table I we have listed the signal gain, G_S, the noise gain, G_N, and

Table I : Signal gain, G_S, at $\omega = \omega_s$, noise gain, G_N, for $\omega \approx \omega_s$, and G_N/G_S for the amplifier represented in Fig. 12.

I_p/I_0	Signal Gain G_S (dB)	Noise Gain G_N (dB)	G_N/G_S (dB)
0.466	4.7 ± 0.1	16.7 ± 0.4	12.0 ± 0.4
0.475	6.9 ± 0.1	17.7 ± 0.4	10.8 ± 0.4
0.486	9.3 ± 0.1	24.9 ± 0.7	15.6 ± 0.7
0.498	12.8 ± 0.1	30.0 ± 0.4	17.2 ± 0.4

their ratio G_N/G_S for 4 values of I_p/I_0. The gains G_S and G_N (measured at frequency ω_s) are the ratios of the signal and noise powers, respectively, to their values at $I_p/I_0 = 0$. Notice that G_N/G_S, which is proportional to the noise temperature T_N, increases from 12 dB to 17 dB as the signal gain is increased from 4.7 dB to 12.8 dB. (By way of comparison, for an amplifier with no noise rise we would have $G_N/G_S = 1$, independent of G_S.) The observed noise rise is such that G_N/G_S, and hence T_N, rise faster than G_S for low values of G_S. This result is consistent with data obtained by Mygind et al. (10) for a real amplifier when they varied the applied pump power.

We have also performed simulations for $\Gamma \approx 6.1 \times 10^{-7}$. In this limit, substantial amounts of signal gain were achieved, without an accompanying noise rise. However, other simulations suggest that the threshold level of the thermal noise is quite low: We observed a noise rise for $\Gamma = 4.2 \times 10^{-6}$ corresponding to a temperature of 0.1K for a junction with I_0 = 1 mA.

From these results, we can draw two important conclusions concerning noise in the three-photon parametric amplifier. First, the noise cannot arise from chaos: To achieve gain, one always operates the amplifier with I_{dc} and I_p below the threshold for a period-doubling bifurcation of the pump frequency. Second, the observed noise rise requires the presence of thermal noise. We believe the noise rise is due to hopping between a bias point in the high gain region (that would be stable in the absence of thermal noise) and an unstable point in the bifurcated region. Since an increase in gain necessarily implies that the device is closer to the bifurcation threshold, one would expect that the probability of hopping and hence the noise produced would increase as the gain is increased.

Finally, we emphasize that the amplification process in the four-photon Josephson parametric amplifier is quite different from that just described, and the device is not operated near a period-doubling bifurcation. Thus, we believe that the mechanism we have described here does not account for the noise in the four-photon amplifier.

7. SUMMARY

We have described the properties of Josephson tunnel junctions shunted by a resistance with non-negligible self-inductance and biased with a constant current. The general features observed for the experimental junctions are explained very adequately by analog simulations. The I-V characteristics usually exhibit regions of stable negative resistance that can be reduced in extent and made to disappear by the progressive reduction of the critical current in an external magnetic field. When the junction is in a stable limit cycle, which may or may not be bifurcated, the voltage noise is below the level of the measurement system. In the chaotic regimes, the noise temperature of the junction is typically a few hundred Kelvin. At certain values of the bias current the noise temperatures may become very large, often exceeding 10^6K. These noise temperatures, which increase as the measurement frequency is decreased, arise from switching between two subharmonic modes or between a subharmonic mode and a chaotic regime. The analog simulations indicate that, at least under certain conditions, this switching is induced by relatively low levels of intrinsic thermal noise; in the absence of thermal noise, the switching did not occur. However, we would not rule out other situations in which the switching is intrinsic to the equations of motion, and is not induced by thermal noise.

As an example of the importance of the switching mechanism, we have used the analog simulator to study noise in the three-photon Josephson parametric amplifier. The amplifier gain increases as the pump amplitude is increased until the system is about to bifurcate to the half-harmonic of the pump frequency. At that point, the gain drops dramatically. The simulations show no noise rise in the absence of thermal noise. However, a small level of thermal noise is sufficient to produce a dramatic increase in the noise in the amplifier. We believe that this noise rise is produced by switching between bias points in the bifurcated and unbifurcated regions.

ACKNOWLEDGMENTS

We are indebted to Henry Abarbanel, Raymond Chiao, John David Crawford, Bernardo Huberman, Edgar Knobloch and Kurt Wiesenfeld for very helpful discussions. T. D. Van Duzer kindly lent us the analog simulator. We are indebted to the Micro Electronics Facility in the Electronics Research Laboratory of the Electrical Engineering and Computer Science Department of the University of California at Berkeley for the use of their facilities. This work was supported by the Director, Office of Basic Energy Sciences, Materials Sciences Division of the U. S. Department of Energy under Contract Number DE-AC03-76SF00098.

REFERENCES

[1] For reviews, see, for example, May, R.M., Simple mathematical models with very complicated dynamics, Nature 261 (1976) 459-467; Eckmann, J.-P., Roads to turbulence in dissipative dynamical systems, Rev. Mod. Phys. 53 (1981) 643-654.

[2] Josephson, B.D., Possible new effects in superconductive tunneling, Phys. Lett. 1 (1962) 251-253; Supercurrents through barriers, Adv. in Phys. 14 (1965) 419-451.

[3] Huberman, B.A., Crutchfield, J.P., and Packard, N.H., Noise phenomena in Josephson junctions, Appl. Phys. Lett. 37 (1980) 750-752.

[4] Kautz, R.L., The ac Josephson effect in hysteretic junctions: Range and stability of phase lock, J. Appl. Phys. 52 (1981) 3528-3541; Chaotic states of rf-biased Josephson junctions, J. Appl. Phys. 52 (1981) 6241-6246; Chaos in Josephson circuits, IEEE Trans. Mag. (1983) to be published.

[5] D'Humieres, D., Beasley, M.R., Huberman, B. A., and Libchaber, A., Chaotic states and routes to chaos in the forced pendulum, Phys. Rev. A 26 (1982) 3483-3496.

[6] Feldman, M.J., Parrish, P.T., and Chiao, R. Y., Parametric amplification by unbiased Josephson junctions, J. Appl. Phys. 46 (1975) 4031-4042.

[7] Taur, Y. and Richards, P.L., Parametric amplification and oscillation at 36 GHz using a point contact Josephson junction, J. Appl. Phys. 48 (1977) 1321-1326.

[8] Wahlsten, S., Rudner, S., and Claeson, T., Arrays of Josephson tunnel junctions as parametric amplifiers, J. Appl. Phys. 49 (1978) 4248-4263.

[9] Mygind, J., Pedersen, N.F., and Soerensen, O.H., X-band singly degenerate parametric amplification in a Josephson tunnel junction, Appl. Phys. Lett. 32 (1978) 70-72.

[10] Mygind, J., Pedersen, N.F., Soerensen, O.H., Dueholm, B., and Levinsen, M.T., Low-noise parametric amplification at 35 GHz in a single Josephson tunnel junction, Appl. Phys. Lett. 35 (1979) 91-93.

[11] Levinsen, M.T., Pedersen, N.F., Soerensen, O.H., Dueholm, B., and Mygind, J., Externally pumped millimeter-wave Josephson junction parametric amplifier, IEEE Trans. Elec. Dev. ED-27 (1980) 1928-1934.

[12] Soerensen, O.H., Dueholm, B., Mygind, J., and Pedersen, N.F., Theory of the singly quasidegenerate Josephson junction parametric amplifier, J. App.. Phys. 51 (1980) 5483-5494.

[13] Chiao, R.Y., Feldman, M.J., Peterson, D.W., Tucker, B.A., and Levinsen, M.T., Phase instability noise in Josephson junctions, in Future Trends in Superconductive Electronics, A.I.P. Conf. Proc. 44 (1978) 259-263.

[14] Feldman, M.J. and Levinsen, M.T., Theories of the noise rise in Josephson PARAMPS, IEEE Trans. Magn. MAG-17 (1981) 834-837.

[15] Pedersen, N.F. and Davidson, A., Chaos and noise rise in Josephson junctions, Appl. Phys. Lett. 39 (1981) 830-832.

[16] Levinsen, M.T., Even and odd subharmonic frequencies and chaos in Josephson junctions: Impact on parametric amplifiers?, J. Appl. Phys. 53 (1982) 4294-4299.

[17] Miracky, R.F., Clarke, J., and Koch, R.H., Chaotic noise observed in a resistively

shunted self-resonant Josephson tunnel junction, Phys. Rev. Lett 50 (1983) 856-859.

[18] Feigenbaum, M.J., Quantitative universality for a class of nonlinear transformations, J. Stat. Phys. 19 (1978) 25-52; The universal metric properties of nonlinear transformations, J. Stat. Phys. 21 (1979) 669-706.

[19] Manneville, P. and Pomeau, Y., Intermittency and the Lorenz model, Phys. Lett. 75A (1979) 1-2; Pomeau, Y. and Manneville, P., Intermittent transition to turbulence in dissipative dynamical systems, Commun. Math. Phys. 74 (1980) 189-197.

[20] Metropolis, N., Stein, M.L., and Stein, P. R., On finite limit sets for transformations on the unit interval, J. Combinatorial Theory 15A (1973) 25-44.

[21] Arecchi, F.T. and Lisi, F., Hopping mechanism generating 1/f noise in nonlinear systems, Phys. Rev. Lett. 49 (1982) 94-98.

[22] Ben-Jacob, E., Goldhirsch, I., Imry, Y., and Fishman, S., Intermittent chaos in Josephson junctions, Phys. Rev. Lett. 49 (1982) 1599-1602.

[23] Miracky, R.F. and Clarke, J., Simulation of the noise rise in three-photon Josephson parametric amplifiers, (1983) submitted to Appl. Phys. Lett.

[24] Pedersen, N.F., Soerensen, O.H., Dueholm, B., and Mygind, J., Half-harmonic parametric oscillations in Josephson junctions, J. Low Temp. Phys. 38 (1980) 1-23.

NOISE IN PHYSICAL SYSTEMS AND 1/f NOISE
M. Savelli, G. Lecoy and J-P. Nougier (eds.)
© *Elsevier Science Publishers B.V., 1983*

1/f SPECTRA IN NONLINEAR SYSTEMS WITH MANY ATTRACTORS

Fortunato Tito Arecchi (*), Remo Badii and Antonio Politi

Istituto Nazionale di Ottica
Firenze, Italy

Power spectra of multistable systems show a low frequency component corresponding to jumps among independent attractors. This situation is modeled with a one-dimensional cubic map disturbed by noise.

Recent experiments [1,2] on nonlinear driven systems yield power spectra with a low frequency divergence $f^{-\alpha}$, α being around 1 whenever the following conditions are fulfilled: i) the system is multistable, that is, it has 2 or more attractors; ii) the attractors are near to be destabilized or they have just become unstable; and iii) the system is "open" to external fluctuations, i.e., the presence of white noise is essential to yield jumps between different basins of attraction. For the nonlinear electronic oscillator of Ref.1, we report in fig.1 the positions of subharmonic bifurcations, chaotic regions and 1/f region versus amplitude A and frequency f of the driving term.

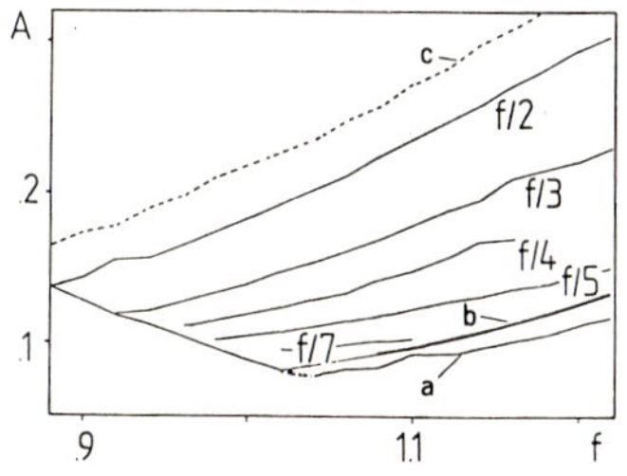

Figure 1 : Experimental parameter space for the system of Ref.1. Under line a the system has stable limit cycles. Above line b there are chaotic attractors covering two valleys, mixed with "islands" of periodic motion as indicated by f/5, f/4 and so on. Line c sets an upper limit due to electronics. 1/f noise is present within the narrow under the thick part of line b.

Usually our power spectra extend over less than two decades, at frequencies much less than the driving frequency. For the CO_2 laser [2], we have an $f^{-0.6}$ spectrum between 1 and 100 Hz, while the driving frequency is around 60 KHz.

To understand the above features we have investigated numerically the dynamical equations of the systems of Ref.1 and 2. An integration of the Duffing equation (Ref.1, Eq.(1)) for a suitable choise of parameters allowing for the simultaneous coexistence of four attractors, two period-4 and two period-7, leads to a spectrum (fig.2) showing a power law region extended over about 2 decades with a slope around 1.3. However, numerical integration of a differential system is a lengthy procedure requiring a few hundred steps per period, hence we have preferred to describe the dynamics by a recursive map as done in some approaches to chaotic phenomena [3].

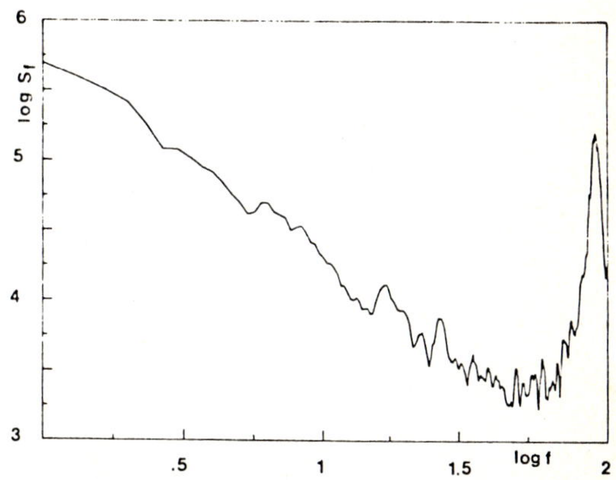

Figure 2 : Power spectrum for the driven Duffing oscillator with k = 0.154, A = 0.114, ω = 1.22 and spectral density of added noise 6×10^{-6}.

The above listed three conditions show that we are in presence of a phenomenon which occurs beyond the usual approach to chaos by either one of the current scenarios [4]. A simple jump between two attractors, as introduced in recent models [5], is not sufficient to explain the phenomenology. Indeed experiments [1,2] show that, when leaving an attractor, the representative point in phase space has a long erratic motion before landing onto another attractor. This "transient" regime is made of motions among repulsive orbits.

A dynamics in terms of a recursive map must allow for at least two independent attractors. The simplest one-dimensional map with two attractors must have two extrema [6]. Hence we study a cubic map in the interval (-1, 1)

$$x_{n+1} = (a-1)x_n - ax_n^3. \qquad (1)$$

To account for item iii) the map will be disturbed by white noise with r.m.s. between 10^{-7} and 10^{-5}. Up to a value $a = \bar{a} = 3\sqrt{3}/2 + 1 = 3.598\ 076 \ldots$, the motion is confined either on the interval (-1,0) or (0,1) with qualitative features alike the well known logistic map. For $a = \bar{a}$, we may still have two independent attractors, whose domains, however, are interlaced in complicated ways over the interval (-1,1). For a = 4, even this new structure becomes unstable and there are no longer attractors.

The simplest stable pair of attractors A,B above $\bar{a}$ is a pair of period-3 attractors which are superstable for $a_s = 3.981\ 797\ 394 \ldots$. These period-3 attractors disappear for $a = \tilde{a} = 3.982\ 000\ 642 \ldots$. For $a_s < a < \tilde{a}$ (fig. 3) the

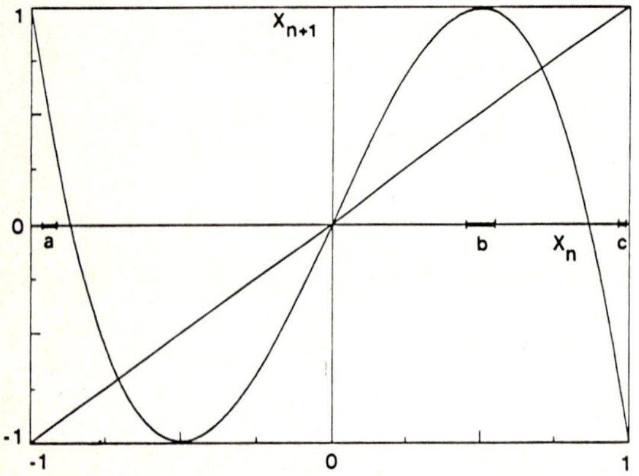

Figure 3: Cubic map for $\tilde{a}$. a,b, and c show (not in scale) the intervals covered by one of the two period-3 attractors.

presence of a small amount of noise makes it easy to leave one attractor and jump toward the other one. Before landing onto the other attractor, the representative point wanders on the available space through a long transient, because of the complex structure of the two basins of attraction. The corresponding low frequency power spectra, for different noise levels, are given in fig. 4.

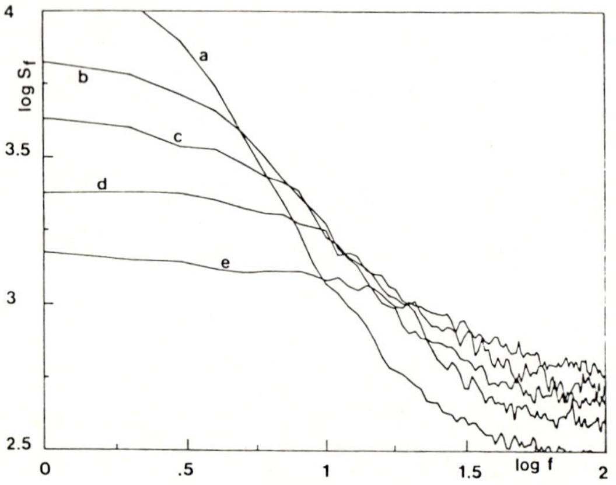

Figure 4 : Power spectra for $a = \tilde{a}$, and for increasing noise levels σ, that is, a) 5×10^{-7}, b) 10^{-6}, c) 2×10^{-6}, d) 4×10^{-6}, e) 10^{-5}. They can be fitted by $f^{-\alpha}$, with α decreasing from 1.5 for a) to 0.5 for e).

These spectra show a power law region extending over about one decade with a slope between 0 and 2. They appear qualitatively in agreement with the experimental spectra of Refs.1 and 2.

Fig. 5 shows the experimental dependence of the Lyapunov exponent on the r.m.s. σ of the applied noise for different values of the control parameter a. The relevant λ range is confined between the value $\lambda_0 = (\ln 2)/3$ where the two period-3 attractors disappear and $\lambda_T = \ln 3$ where the whole map becomes unstable. The added noise allows to escape from an attracting region whenever the phase point is away from the boundary by less than about 3σ. For any a value, the corresponding $\sigma = \sigma_0(a)$ for which the λ curve crosses $\lambda = \lambda_0$ is approximately one third the minimum distance between the border of the region covered by the attractor and the frontier of the immediate attraction domain [7]. Experimentally the dependence is fitted by

$$\sigma_0(a) = \sigma_M \left[1 - \left(\frac{a - a^0}{\tilde{a} - a^0}\right)^{\gamma}\right], \qquad (2)$$

where $\sigma_M = 1.45 \times 10^{-5}$, $a^0 = 3.98184$, $\gamma = 1.40 \pm 0.1$, $a^0 < a < \tilde{a}$. In other words, the jumps between attractors appear as a noise-induced destabilization. σ_T is the noise value for which the whole map becomes unstable, hence all the λ

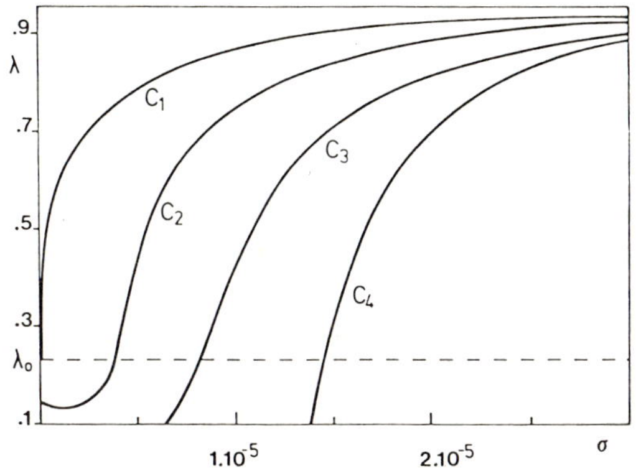

Figure 5 : Lyapunov exponents λ vs noise σ for different a values. C_1:a=$\tilde{a}$, C_2 :a=3.98197, C_3:a=3.98193, C_4:a=a^0

curves cluster at λ_T for $\sigma = \sigma_T$ with an empirical scaling law $\bar{\lambda} = F(a, \sigma)$ where

$$\bar{\lambda} = (\lambda - \lambda_0)/(\lambda_T - \lambda_0). \tag{3}$$

During the motion, when the phase point is on an attractor, the Lyapunov exponent is practically equal to λ_0, and when it is in the transient region, we can approximately take the asymptotic value λ_T. We can then split the sum defining λ into separate contributions. Calling N_A, N_T the number of steps on the attractor and on the transient respectively, we get

$$\lambda = \lim_{N \to \infty} \left[2 \sum_{i=1}^{N_A} \ln|f'(x_i)| + \sum_{i=1}^{N_T} \ln|f'(x_i)|\right]/N = \tag{4}$$

$$=2k\lambda_0 + k'\lambda_T ,$$

where $k = \lim_{N \to \infty} N_A/N$ and $k' = \lim_{N \to \infty} N_T/N$, with

$2k + k' = 1$, are the fractional times spent on the attractor and on the transient, respectively. Notice from eqs.(3) and (4) that the fractional time k' spent on the transient coincides with the reduced Lyapunov exponent $\bar{\lambda}$ given in the scaling law. Inspection of the date of fig.4 and fig.5 shows that the slopes α of the $f^{-\alpha}$ regions scale linearly with $\bar{\lambda}$ as $\alpha = 2(1 - \bar{\lambda})$, from $\alpha = 2$ (single Lorentzian), when $\bar{\lambda} = 0$ (no transient), to $\alpha \to 0$ (flat spectrum) for $\bar{\lambda} = 1$ (all transient).

A model explanation of the above spectra was given[8] in terms of jump processes among three regions of phase space (the two attractors and the intermediate transient), by taking the jump probabilities decorrelated from the internal motions within the three regions. This leads to a power spectrum made of three Lorentzians which fit well the spectra shown in fig.4.

In the particular case of equal transition probabilities and nearest neighbors jumps between attractors, we obtain a low frequency spectrum which approaches 1/f. Our hopping mechanism does not explain in general 1/f noise in linear equilibrium situations. It opens, however, a new area of investigation, namely that of noise induced interactions among many attractor domains.

(*) Also with Phys. Dept., University of Florence, I-50125 Florence, Italy.

[1] F.T. Arecchi and F. Lisi, Phys. Rev. Lett. 49 (1982) 94. The slopes in the spectra of this paper must be multiplied by 2, for a miscalibration of the spectrum analyzer.

[2] F.T. Arecchi, R. Meucci, G. Puccioni and J. Tredicce, Phys. Rev. Lett. 49 (1982) 1217.

[3] E. Ott, Rev. Mod. Phys. 53 (1981) 643.

[4] J.P. Eckmann, Rev. Mod. Phys., 53 (1981) 743.

[5] E. Ben Jacob, I. Goldhirsch, Y. Imry and S. Fishman, Phys. Rev. Lett. 49 (1982) 1599.

[6] I. Gumowski and C. Mira: dynamique chaotique (Cepadues Editions, Toulouse, 1980).

[7] The immediate basin of attraction is defined (Ref. 6) as the neighborhood of an attractor within which the distance from the latter shrinks monotonically.

[8] F.T. Arecchi, R. Badii and A. Politi, to be published.

QUANTUM NOISE

NOISE IN PHYSICAL SYSTEMS AND 1/f NOISE
M. Savelli, G. Lecoy and J-P. Nougier (eds.)
© *Elsevier Science Publishers B.V., 1983*

INFLUENCE OF NOISE ON THE VOLTAGE VERSUS CURRENT CHARACTERISTICS OF AC-BIASED SQUID MAGNETOMETER

J.E. Mutton, R.J. Prance and T.D. Clark

Physics Dept., Univ. of Sussex, Brighton, Sussex, England

and

A. Widom

Physics Dept., Northeastern Univ., Boston, Mass., U.S.A.

We show that the well known voltage steps in the voltage versus current characteristics of high frequency (AC) biased niobium point contact SQUID ring magnetometers are an artifact of external noise. We provide a simple "voltage pulse" model to describe the actual dynamical behaviour of such magnetometers.

1. INTRODUCTION

It has been conventional practice to describe the dynamical behaviour of a thick superconducting ring containing a Josephson weak link (a "SQUID" ring) by the nonlinear equation of motion (1)

$$C\cdot\ddot{\Phi} + \frac{1}{R}\cdot\dot{\Phi} = -\frac{\partial u(\Phi,\Phi_x)}{\partial\Phi} \qquad (1)$$

Equation (1) describes the motion of the "SQUID particle", with effective mass the weak link capacitance C, in the potential

$$U(\Phi,\Phi_x) = \frac{(\Phi-\Phi_x)^2}{2\Lambda} - \frac{I_c\Phi_0}{2\pi}\cos(2\pi\Phi/\Phi_0) \qquad (2)$$

where Φ and Φ_x are, respectively, the external applied flux and included flux in the ring, I_c is the weak link "critical" current, R is a phenomenological dissipative element associated with the weak link [the "shunt resistor" in the resistively shunted junction (RSJ) model of a weak link (2)] , Λ is the geometric inductance of the ring and $\Phi_0 = h/2e$.

Equations (1) and (2) have been used to calculate the behaviour of a weak link ring in the "SQUID magnetometer configuration", i.e. a SQUID ring coupled to a parallel inductor/capacitor tank circuit, as depicted in figure 1. Here, the tank circuit is driven on resonance, at frequency $\omega_R/2\pi$, by an external high frequency current I_{IN}. The response of the coupled system to I_{IN} is given by the high frequency voltage V_{OUT}, within the passband of the tank circuit. In the coupled situation depicted in figure 1, the quasiclassical equations (1) and (2), which include a flux velocity dependent damping term, yield V_{OUT} versus I_{IN} characteristics for a linearly ramped current amplitude which contain a set of "SQUID steps". In figure 2, we show, as an example, characteristics for two DC bias flux states [$\Phi_{XDC} = N\Phi_0$ and $(N+\frac{1}{2})\Phi_0$, N integer] calculated by us numerically from equations (1) and (2) and the circuit equations for the tank circuit (3). Here, we have used circuit parameters typical of a practical very high performance AC-biased SQUID magnetometer based on a point contact weak link ring (4), i.e. $I_c = 10\mu A$, $R = 1\Omega$, $C = 10^{-14}$ F, $\Lambda \doteq 5\times10^{-10}$ H, $L = 4\times10^{-8}$ H, $M = 5\times10^{-10}$ H (ring-tank circuit mutual inductance), $\omega_R/2\pi = 4.3\times10^8$ Hz, and "Q" (ring-tank circuit) = 40. Within this model the so-called step slope parameter α (figure 2) is set by the "intrinsic" current (flux) noise generated by

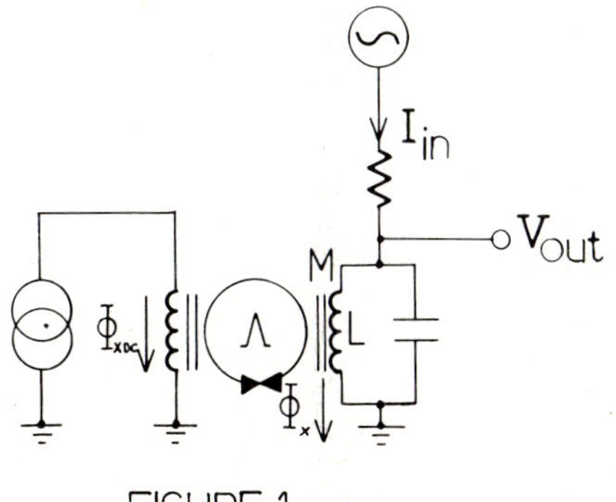

FIGURE 1

FIGURE 2

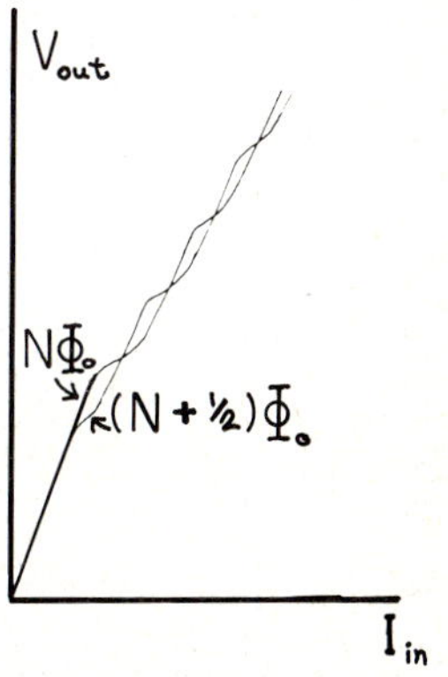

the weak link shunt resistor. In terms of the potential expressed in equation (2) this flux noise is given by (5,6)

$$\langle\Delta\Phi^2_{IN}\rangle^{\frac{1}{2}} \simeq 1.56\,\frac{\Lambda I_c^{1/3}}{(\omega_R/2\pi)^{1/3}}\left(\frac{k_B T}{\Phi_0}\right)^{2/3}\left[1-\left(\frac{\Phi_0}{2\Lambda I_c}\right)^2\right]^{1/6} \qquad (3)$$

with, in the limit of $\omega_R/2\pi \ll R/\Lambda$,

$$\alpha \simeq 1.4\,(\omega_R/2\pi)^{\frac{1}{2}}\,\langle\Delta\Phi^2_{INT}\rangle^{\frac{1}{2}}/\Phi_0 \text{ Hz} \qquad (4)$$

Here, T is the temperature of the ring, I_c is the weak link ring critical current defined as the value of the screening supercurrent (I_S) at which magnetic flux can first enter the ring, and R/Λ is the assumed "RSJ" model relaxation time of the ring after flux entry. We note that according to equations (3) and (4) for $T > 0$, α must always be finite and positive.

2. WEAK LINK RING DYNAMICS IN THE VOLTAGE PULSE MODEL

Although much effort has been devoted to predicting the behaviour of AC-biased SQUID magnetometers through numerical computations based on equations (1) and (2) it is by no means apparent that this nonlinear description is adequate. In this context it is pertinent to consider the simplest possible model of such magnetometers. In this model we view the weak link as a switch which allows flux to enter the ring in units of Φ_0 at an external flux (Φ_{XC}) corresponding to a precise level of screening current flowing in the ring, i.e. at $I_s = I_c$. There is no weak link shunt resistor or ring relaxation time Λ/R; rather, we introduce a relaxation time τ for the ring-tank circuit combination since clearly this is what governs the externally observed behaviour of a SQUID magnetometer. In our model, the magnetic energy storage in the ring is $W_n = (n\Phi_0-\Phi_x)^2/2\Lambda$ where $n = 0, \pm 1, \pm 2, \ldots$ etc., and we assume that when $\Phi_x = \Phi_{XC}$ (and $I_s = I_c$) only "nearest neighbour" flux transitions occur, such that $n \to n \pm 1$.

In figure 3(a) we show a set of W_n branches in Φ_x-space; for each n branch there is an included flux $n\Phi_0$ in the ring. If we assume that $I_c > \Phi_0/\Lambda$, which is the operating condition for a "hysteretic-regime" AC-biased SQUID magnetometer (5), then we can generate SQUID trajectories in Φ_x-space, as shown in heavy lines in figure 3(a). We note that at a certain value of W_n(at $I_s = I_c$) there is a discontinuous (catastrophic) change in the flux state of the ring. These fast quantised flux changes in the ring will generate back emf voltage pulses at the tank circuit which will then decay over time τ. In figure 3(b) we show [for the $I_c > \Phi_0/\Lambda$ value of 3(a)] a typical succession of voltage pulse event times over a single AC cycle of the external flux Φ_x generated in the tank circuit inductor.

It is easy to calculate V_{OUT} versus I_{IN}, at frequency $\omega_R/2\pi$, in the voltage pulse model provided we assume that between fast flux transitions the

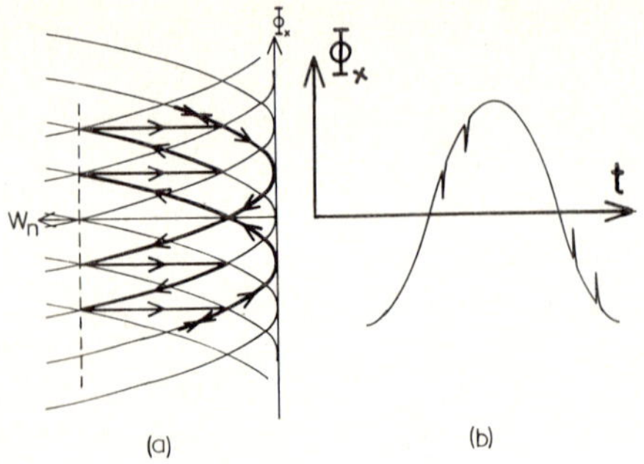

FIGURE 3

flux in the tank circuit inductor $\propto I_{IN}$. This will be valid as long as the flux transition time $\ll 2\pi/\omega_R$. On resonance, and in the hysteretic SQUID regime, the voltage response to a succession of quantised flux changes is given by (7)

$$V_{OUT} = Z_0 I_{IN} \qquad (5)$$

for no flux transitions ($I_s < I_c$), and

$$V_{OUT} = Z_1 I_{IN} - \frac{\omega_R}{2\pi}\cdot\frac{M}{\Lambda}\cdot\Phi_0\cdot\beta(\omega_R\tau)\times\left\langle\sum_{\substack{\text{pulses over}\\ \text{one cycle}\\ j}}\sigma_j e^{i\omega_R t_j}\right\rangle \qquad (6)$$

when flux transitions are taking place.

Here, Z_0 is the tank circuit impedance in the absence of flux transitions, Z_1 is this impedance when such transitions are taking place, renormalised to account for feedback between the tank circuit and the ring, $\sigma_j = \pm 1$, depending on the direction (sign) of the flux transition, the t_j's are the times of the flux transitions, the brackets represent averaging over every AC cycle of I_{IN} [i.e. Φ_x(AC)] and $\beta(\omega_R\tau) = (1 - i\omega_R\tau)^{-1}$ is a damping parameter used in this simplest case to define an exponential relaxation time τ for the ring-tank circuit combination.

Obviously, the calculational scheme of equation (6) and figure 3, requires a precise knowledge of the relative voltage pulse times t_j. In electronic terms this means that the relative phase of V_{OUT} to I_{IN} must be accurately defined. In figure 4 we show the calculated V_{OUT} versus I_{IN} characteristics [V_{OUT} in phase (zero phase) with I_{IN}] for $\Phi_{XDC} = N\Phi_0$ and $(N+\frac{1}{2})\Phi_0$ using the circuit parameters of figure 2 with, in addition, $\omega_R\tau = 1$ and $Z_0/Z_1 = 1$. We see that these voltage pulse model characteristics consist of a series of "pull down" features which start at a value of I_{IN} corresponding to $I_s = I_c$. These triangles split and invert when Φ_{XDC} is changed by $\pm \Phi_0/2$. In figures 5a and b we show the experimental in phase V_{OUT} versus I_{IN} characteristics of a 430 MHz-biased, very high performance, stabilised niobium point contact SQUID magnetometer with the same circuit

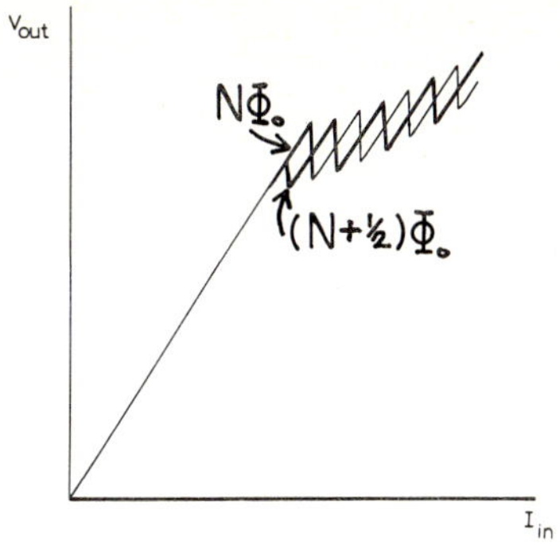

FIGURE 4

parameters used to generate figures 2 and 4. The timing reference is provided by a very stable 430 MHz phase sensitive detector system (bandwidth ≃ 35 MHz). In this system we make use of cryogenically cooled GaAs FET preamplifiers for the UHF signal, V_{OUT}. We see that the characteristics of figure 5 match closely those generated by the voltage pulse model (figure 4).

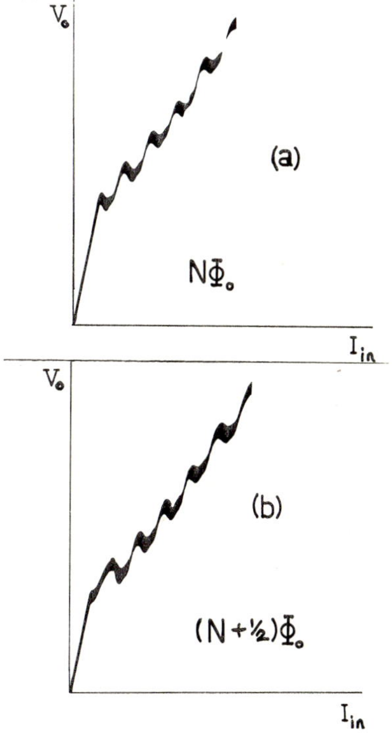

FIGURE 5

It is evident that if the succession of voltage pulses is not recorded accurately in time then the sharply defined features of figure 4 must deteriorate. In figure 6 we show the effect of introducing timing (i.e. phase) jitter on the sequences of pulses required to generate the V_{OUT} versus I_{IN} characteristics of figure 4. This jitter is created by adding to the input drive flux (i.e. I_{IN}) each AC cycle a constant

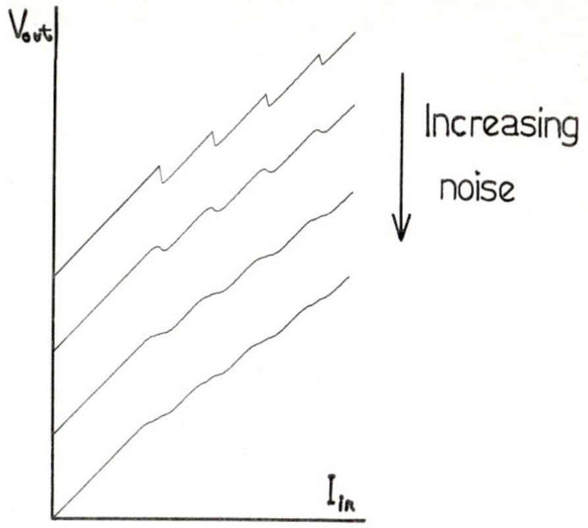

FIGURE 6

noise amplitude. This flux noise acts to change the origin (and hence the relative pulse times) in figure 3a. The noise, which is essentially low frequency ($\lesssim \omega_R/2\pi$), has an amplitude for each AC cycle which is chosen randomly. The relative frequency of occurrence of any particular amplitude is governed by a Poisson distribution, its width and mean (μ) having been predetermined. In figure 7 we show the experimental effect of increasing low frequency noise

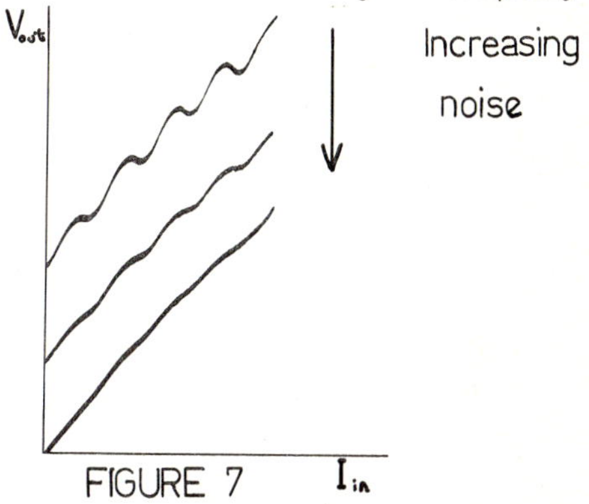

FIGURE 7

amplitude on the triangular features shown in figure 5a. For practical reasons we have arranged for this noise to be bandstop filtered within the bandpass of the tank circuit (roughly 415 to 445 MHz). We see from both figures 6 and 7 that simply by adjusting the noise amplitude injected into a phase sensitive detected SQUID magnetometer system the triangular features, typical of the voltage pulse model, can be changed continuously into the rounded steps associated with the quasiclassical nonlinear description of a SQUID magnetometer (figure 2). For our 430 MHz system we find that the noise dependent characteristics shown in figure 7 can be created by flux noise typically in the frequency range from 10 to 1000 MHz.

3. CONCLUSIONS

It is clear from the above discussion and results that the voltage pulse model provides an accurate description of the dynamical behaviour of hysteretic AC-biased niobium point contact SQUID ring magnetometers. It is also apparent that the shunt resistor (RSJ) model of a weak link ring, used to generate the finite α SQUID steps of figure 2, is unrealistic - these steps being an artifact created by external noise induced jitter in the SQUID magnetometer electronics.

REFERENCES

[1] Leggett, A.J., Quantum tunnelling and noise in SQUIDs, Sixth Int. Conf. on Noise in Physical Systems - 1981, NBS Special Publication 614 (Libr. Congr. Number 81-600084), 355-358.

[2] McCumber, D.E., Effect of a.c. impedance on d.c. voltage current characteristics of superconductor weak-link junctions, Jrnl. Applied Physics 39 (1968) 3113-3118.

[3] Simmonds, M.B. and Parker, W.H., Analog computer simulation of weakly connected superconducting rings, Journal of Applied Physics 42 (1971) 38-45.

[4] Long, A.P., Clark, T.D. and Prance, R.J., Varactor tuned ultrahigh frequency SQUID magnetometer, Rev. Sci. Instrum. 51(1) (1980) 8-13.

[5] Jackel, L.D. and Buhrman, R.A., Noise in the r.f. SQUID, Jrnl. Low Temp. Phys. 19 (1975) 201-246.

[6] Kurkijärvi, J. and Webb, W.W., Thermal fluctuation noise in a superconducting flux detector, IEEE Pub. No. 72CHO 682-5-TABSC (IEEE, New York, 1972), 581-587.

[7] Scott, R.E., Linear circuits (Addison-Wesley, Reading, Mass. 1960), 321-425.

NOISE IN PHYSICAL SYSTEMS AND 1/f NOISE
M. Savelli, G. Lecoy and J-P. Nougier (eds.)
© *Elsevier Science Publishers B.V., 1983*

QUANTUM LIMIT CONSTRAINTS ON DC SQUID AMPLIFIERS

C. D. Tesche

I.B.M. Thomas J. Watson Research Center
P.O. Box 218
Yorktown Heights, New York 10598

A model for a class of dc SQUID linear amplifier circuits is described in which the entire circuit is treated in a consistent fashion. The amplifier noise temperature, gain, effective input impedance and noise current generated in the input circuit is determined as a function of the SQUID and input circuit parameters, and the well known isolated SQUID characteristics. As a result, quantum noise limit constraints are obtained for the isolated characteristics.

1. Introduction

DC SQUID linear amplifiers fabricated with Josephson tunnel junctions appear to have the potential of approaching the quantum noise limit for an arbitrary linear, phase preserving amplifier.[1] In addition, the dc SQUID amplifier is a particularly attractive candidate for use in quantum nondemolition measurement schemes.[2,3,4] However, in contrast to other more conventional amplifiers, the detailed behavior of the dc SQUID in even the simplest amplifier circuit depends not only on the isolated device parameters, but also on the parameters of the input circuit. As a result, the analysis of a circuit containing a dc SQUID should begin with a careful consideration of the entire circuit. A pattern of analysis for a certain class of linear amplifiers is illustrated in this paper. Particular attention is paid to the back action of the SQUID on the input circuit, and the effect of the input circuit on the operation of the SQUID.

2. ISOLATED DC SQUID

We begin with a brief description of the dc SQUID. The isolated SQUID is not a linear amplifier, but rather a flux-to-voltage transducer. The device consists of a superconducting loop of inductance L interrupted by two Josephson junctions. The voltage V_o at the device output is determined by oscillations of the quantum mechanical phase drops δ_1 and δ_2 across the junctions,

$$V_o = (\Phi_o/2\pi)(d\delta_1/dt + d\delta_2/dt) \qquad (1)$$

In an idealized device with no noise sources, the fundamental frequency of these oscillations occurs in the GHz range. The time averaged voltage $<V_o>$ is a function of the dc bias current I and signal flux Φ_a linking the loop inductance. For low frequency applications (f<<GHz), the device is characterized by a forward transfer function $V_\phi = d<V_o>/d\Phi_a$, and by the value of the white, low frequency output voltage noise spectral density, S_V, determined by the intrinsic noise sources. These terms can be combined into an energy factor referred to the output with dimensions of energy/Hz, $S_{OE} = S_V/2LV_\phi^2$. Although the output energy factor is an important figure of merit for the SQUID, it does not, in general, correspond to a minimum energy resolution per unit bandwidth, because the isolated SQUID is not being operated as a detector, but rather as a flux-to-voltage transducer.

Another important figure of merit is related to the noise component of the circulating current, J, around the SQUID loop. For example, the SQUID may be incorporated into an untuned linear amplifier circuit (Fig. 1). A signal current generated by the voltage source V_i is coupled into the SQUID via the mutual inductance M. In addition, noise components in J are necessarily coupled back into the input circuit, thus altering the total output noise. We define the energy factor referred to the input as $S_{IE} = LS_J/2$, where S_J is the low frequency component of the circulating current noise spectral density. Again, note that S_{IE} does not correspond to a minimum energy resolution per unit bandwidth be-

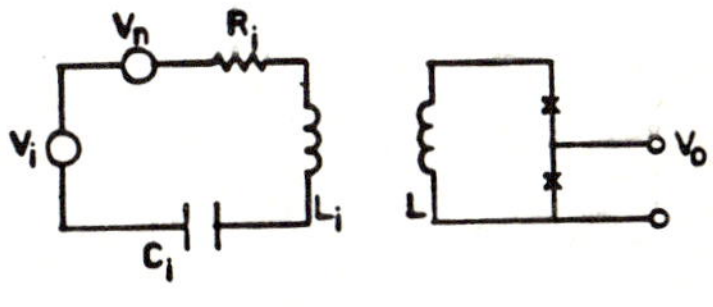

Fig. 1. Untuned dc SQUID Linear Amplifier.

cause this device is being operated, not as a detector, but rather as a linear voltage amplifier. C. M. Caves has shown that the noise temperature, T_N, of such an amplifier operated with large gain at signal frequency f is subject to a constraint imposed by the uncertainty principle, $T_N > \hbar f/k_B \ln 3$.[1] Since the noise temperature is a function of both S_{OE} and S_{IE} this relationship can be used to derive a constraint on the energy factors of an isolated SQUID, $(S_{IE}S_{OE})^{1/2} > \hbar$.[5]

Finally, note that since the time averaged circulating current, $<J>$, is a function of the applied flux, the SQUID will react back on the source of the applied field. This reaction is described by the function $J_\phi = d<J>/d\Phi_a$. However, since $<J>$ is generally less than 0.05 Φ_o for a SQUID with $2LI_o = \Phi_o$, this back action can be neglected in most applications of an isolated SQUID as a transducer. In addition, we may define a reverse transfer function $J_I = d<J>/dI$, which describes the degree to which noise components in the bias current introduced by elements at the SQUID output are transmitted back into the SQUID input. Again, because the circulating currents are fairly small, this is a dimensionless number on the order of a few percent.

3. DC SQUID WITH A SUPERCONDUCTING INPUT COIL

As an illustration of the general procedure which will be followed in the rest of the paper, we analyze a circuit containing a SQUID inductively coupled to a superconducting input circuit which consists of an input coil L_i and a coupling coil to the SQUID L_c. The input signal, Φ_a, and output voltage, V_o, are coupled into and out of the device through the input inductance, L_i, and the output voltage leads, respectively. Thus the device can be viewed as a black box with a coil at the input and two voltage leads at the output (in addition to an adjustable dc current and flux bias). The behavior at the output is described by a set of differential equations isomorphic to those which describe the behavior of the isolated SQUID. The solutions to the isolated SQUID equations are well known.[6,7,8] The nonlinear equations of motion for the phase drops across the junctions are[6]

$$(\Phi_o/2\pi R)d\delta_1/dt = I/2 - J - I_o \sin \delta_1 + I_1 \quad (2)$$

and

$$(\Phi_o/2\pi R)d\delta_2/dt = I/2 + J - I_o \sin \delta_2 + I_2 \quad (3)$$

where I_o is the junction critical current, and I_1 and I_2 are the Johnson noise sources associated with the junction shunt resistances (it is assumed that the device is operated at a frequency such that the the 1/f noise is negligible). Quantization of the fluxoid around the SQUID loop yields

$$D = LJ + MJ_i + \Phi_b \quad (4)$$

where $D = (\Phi_o/2\pi)(\delta_1 - \delta_2)$, the mutual inductance $M^2 = \alpha^2 LL_c$, the bias flux to the SQUID is Φ_b, and the current in the input coil is J_i. Quantization of the fluxoid in the input coil yields

$$0 = (L_i + L_c)J_i + MJ + \Phi_a, \quad (5)$$

where Φ_a is the signal flux applied to the input coil. Combining Eqns. 3 and 4,

$$D = L(1-\alpha^2 r)J - M\Phi_a/(L_i + L_c) + \Phi_b \quad (6)$$

where $r = L_c/(L_i + L_c)$. Eq. 6 is identical to the quantization condition for an isolated SQUID with inductance $L_e = L(1-\alpha^2 r)$ and signal flux $\Phi_e = -M\Phi_a/(L_i + L_c)$. Note that the loop inductance does not appear in Eqns.1,2 or 3. Thus, the phase drops, circulating current and output voltage develop identically in both the isolated SQUID with inductance L_e and the coupled device as expected.

The figures of merit for the black box device can now be expressed in terms of those of the isolated device. The forward transfer function is $V_\phi^C = -MV_\phi/(L_i + L_c)$ where V_ϕ is calculated for an isolated SQUID with loop inductance L_e. The output energy factor, S_{OE} is determined by the voltage noise spectral density, the forward transfer function, and the "black box" input inductance, L_i. Thus $S_{OE}^c = S_{OE}/\alpha_s^2$, where $\alpha_s^2 = \alpha^2 rL_i/(L_i + L_c)$. The input energy factor is computed from the noise components in J_i, and the input inductance, L_i. Thus, $S_{IE}^C = \alpha_s^2 S_{IE}$, where both S_{OE} and S_{IE} are computed for an isolated SQUID with loop inductance L_e. Note that the product $S_{IE}^C S_{OE}^C = S_{IE}S_{OE}$, even though the two device input inductances, and forward transfer functions are distinct. It can be shown that this implies that the noise temperature of a SQUID amplifier containing these two devices is the same.

Finally, from Eqn. 5 we see that the back action term $J_\phi^C = -MJ_\phi/(L_i + L_c)$, and the reverse transfer function $J_I^C = -MJ_I/(L_i + L_c)$. Thus, these quantities may be reduced by an appropriate chioce of the mutual and input inductances.

4. UNTUNED DC SQUID LINEAR AMPLIFIER

A lumped circuit element model for the simplest possible inductively coupled dc SQUID voltage amplifier circuit is shown in Fig. 1. The voltage source V_i and source resistance R_i are connected in series with the input inductance L_i. The mutual inductance $M^2 = \alpha^2 LL_i$. A careful analysis of this circuit is useful because it illustrates a pattern appropriate for many devices in which the SQUID is inductively coupled to a linear input circuit. Of course, it is possible that a particular attempt to realize this circuit in the laboratory may result in significant stray capacitance, particularly within and between the coils of the input and SQUID inductances. As a result, resonance behavior in the input circuit and/or significant variations in the SQUID characteristics may be observed. Such a situation does not imply that this analysis is incorrect, rather that the analysis for the circuit fabricated must include all relevant circuit elements.

The equations for the untuned linear amplifier are identical to those for the SQUID coupled to a superconducting input circuit (Eqn.1-4), except for the input circuit quantization condition, Eqn. 5, which becomes

$$L_i dJ_i/dt + MdJ/dt + R_i J_i = V_i + V_n \qquad (7)$$

where V_n is the noise source associated with the source resistance. Numerical integration of Eqns. 1-4,7 indicate that the detailed behavior of the phase drops δ_1 and δ_2 in the amplifier are altered by the presence of the input circuit.[5] In particular, the IV characteristics and current and voltage noise spectral densities closely resemble those of a SQUID coupled to a superconducting input circuit with resultant effective inductance $L_e = L(1-\alpha^2)$, even though the input circuit in the amplifier is lossy, provided that the time constant $\tau_i = L_i/R_i >> \Phi_o/<V_o>$. Motivated by this observation, we eliminate the input current J_i and replace L by $L_e/(1-\alpha^2)$ in Eqns. 4 and 7 to obtain

$$dX/dt + X/\tau_i = (M/L_i)(V_i + V_n + MJ/\tau_i) \qquad (8)$$

where $X = D - L_e J - \Phi_b$. Notice that the forcing term on the right hand side includes the input voltage sources V_i and V_n and the term MJ/τ_i. Thus, when the screening of the SQUID inductance is explicitly displayed as in Eq. 8, the effect of the circulating current J on the system is equivalent to that produced by an input voltage source MJ/τ_i, even though the EMF produced in the input circuit is MdJ/dt (Eq. 7). This pattern is observed in the analysis of other linear input circuits, with the qualification that the forcing term depends on the detailed structure of the input circuit analyzed. For example, the forcing term for the tuned linear amplifier with capacitance C_i in series with inductance L_i has Fourier components $MJ(\omega)/\tau_i - jMJ(\omega)/\omega L_i C_i$.[9]

Up to this point, no approximations have been made to the original lumped circuit element equations. In many cases of interest, however, all the characteristic times associated with the input circuit are long compared to those associated with the oscillations of the phase drops δ_1 and δ_2. Thus the effects of the higher frequency components of the forcing terms are reduced by a factor $1/\omega\tau_i$ and can be neglected. The low frequency noise components of the input source resistance, V_n^o, and of the SQUID circulating current, J^o, and the low frequency signal, V_i, appear in the approximate integral of Eqn. 8 as components of an effective applied flux. That is

$$D = L_e J + \Phi_e + \Phi_a \qquad (9)$$

where

$$\Phi_e = (M/Z_i)(V_i + V_n^o + MJ^o/\tau_i)$$

and $Z_i = R_i + j\omega L_i$.

Eqns. 1-4,9 which approximate the behavior of the untuned linear amplifier are identical to those of an isolated SQUID with loop inductance L_e and signal flux Φ_e. The voltage signal at the output is $(MV_\phi/Z_i)V_i$, and the voltage noise at the output is $(MV_\phi/Z_i)(V_n^o + MJ^o/\tau_i) + V^o$ where V^o is the low frequency component of the output voltage noise computed at fixed applied flux (small signal limit) for an isolated SQUID with loop inductance L_e. This is in disagreement with the corresponding expression in ref. 10, in which the current noise produces a contribution to the output voltage noise of $-j\omega M^2 V_\phi J^o/Z_i$. In addition, the noise factors and forward transfer function in ref. 10 are computed for a SQUID with unscreened loop inductance.

We have now obtained the voltage signal and voltage noise at the output of the linear amplifier in terms of the input voltage source, the input and SQUID circuit parameters, and the well known solutions for $V(t)$ and $J(t)$ for an isolated SQUID. At this point, computation of the noise temperature as a function of the source resistance is straightforward.[5] In addition, since the solution for $J(t)$ is identical to that of the SQUID imbedded in the amplifier circuit, we may now return to Eqn. 7 and solve for the component of the input current, J_i, generated by the SQUID current J. That is, the back action of the SQUID on the input circuit may now be determined. First, note that when the input voltages are zero, the noise current in the input, $J_i = -(M/Z_i)dJ/dt$, is generated by the intrinsic SQUID noise sources. In addition, note that the components of the current J_i near the Josephson frequency are approximately equal to $(M/L_i)J$. This is similar to the result obtained for a superconducting input circuit (Eqn. 5) and consistent with the numerical result that the induced current in the input circuit produces screening of the SQUID loop inductance.

The change in the input current, ΔJ_i, due to a small signal, ΔV_i, at frequency ω is calculated as follows. The source produces an effective applied flux in the equivalent isolated SQUID of $\Phi_e = (M/Z_i)\Delta V_s$, and a change in the screening current in the SQUID loop of $\Delta J = J_\phi \Phi_e$. Again, since the solution for J is the same for both the isolated and coupled SQUIDs, the expression for the induced current can be substituted into Eqn. 7. Thus

$$L_i dJ_i/dt + (M^2 J_\phi / Z_i)\Delta V_s + R_i J_i = \Delta V_s \quad (10)$$

Solving for the effective impedance, $\Delta V_s/\Delta J_i$, to lowest order in J_ϕ, the effective SQUID impedance in series with the input coil inductance is

$$R_{SQUID} = -R_i(\omega M^2 J_\phi)^2/(R_i^2 + \omega^2 L_i^2)$$

and

$$L_{SQUID} = M^2 J_\phi .$$

Note that J_ϕ is negative. Thus both the effective resistance and inductance of the input circuit are reduced by the presence of the SQUID. In the low frequency limit, R_{SQUID} falls to zero. Maximum screening of the input coil occurs in the limit J_ϕ = -1, when the phase drops across the junctions become insignificant.

5. CONCLUSIONS

The effect of the input circuit on the SQUID forward transfer function and output voltage noise has been obtained by a careful analysis of the entire circuit. In addition, the effective impedance in the input circuit produced by the screening currents in the SQUID loop, and the noise current generated in the input by the intrinsic noise sources in the SQUID have been obtained in a consistent fashion. Thus, a simplified model for the SQUID has been constructed, the elements of which include both the isolated SQUID parameters and noise factors, and the input circuit parameters. This model is now in a form appropriate for the discussion of more general problems, such as those presented by quantum non-demolition measurement schemes.

References

1. C. M. Caves, Quantum limits on noise in linear amplifiers, Phys. Rev. D 26 (1982) 1817-1839.
2. C. M. Caves, K. S. Thorne, R. W. P. Drever, V. Sandberg, and M. Zimmermann, On the measurement of a weak classical force coupled to a quantum mechanical oscillator. 1. Issues of principle, Rev. Mod. Phys. 52 (1980) 341-392.
3. W. W. Johnson and M. Bocko, Approaching the quantum "limit" for force detection, Phys. Rev. Lett. 47 (1981) 1184-1187.
4. M. F. Bocko and W. W. Johnson, Surpassing the amplifier limit for force detection, Phys. Rev. Lett. 48 (1982) 1371-1374.
5. C. D. Tesche, Optimization of dc SQUID linear amplifiers and the quantum noise limit, Appl. Phys. Lett. 41 (1982) 490-492.
6. C. D. Tesche and J. Clarke, DC SQUID: Noise and optimization, J. Low Temp. Phys. 29 (1977) 301-331.
7. J. J. P. Bruines, V. J. de Waal, and J. E. Mooij, Comment on "DC SQUID: Noise and optimization" by Tesche and Clarke, J. Low Temp. Phys. 46 (1982) 383-386.
8. C. D. Tesche and J. Clarke, Current noise in the dc SQUID, J. Low Temp. Phys. 37 (1979) 397-403.
9. C. D. Tesche, Analysis of strong inductive coupling on SQUID systems, to appear in IEEE Trans. Magn.
10. J. Clarke, C. D. Tesche, and R. P. Giffard, Optimization of dc SQUID voltmeter and magnetometer circuits, J. Low Temp. Phys. 37 (1979) 405-420.

NOISE IN PHYSICAL SYSTEMS AND 1/f NOISE
M. Savelli, G. Lecoy and J-P. Nougier (eds.)
© Elsevier Science Publishers B.V., 1983

NOISE SPECTROSCOPY OF A WEAK LINK CONSTRICTION RING

R.J. Prance, J.E. Mutton, H. Prance and T.D. Clark

Physics Dept., Univ. of Sussex, Brighton, Sussex, England

and

A. Widom and G. Megaloudis

Physics Dept., Northeastern Univ., Boston, Mass., U.S.A.

We show that superconducting circuits, incorporating Josephson weak link devices, can act as macroscopic quantum objects with energy levels observable by a new technique - double derivative noise spectroscopy. We demonstrate that such circuits provide macroscopic ($\sim$ 1 cm spatial scale) examples of gauge symmetry breaking systems with properties previously considered only in the context of high energy physics.

1. PHOTONS IN SUPERCONDUCTORS

The essential coherence property characterising a superconductor is that a condensed state is formed from paired electrons which all have the same centre of mass momentum. Thus, below the transition temperature T_c the superconducting state comprises a coherent plasma of all these paired electrons. It is now well established (1,2) that the transverse screening provided by this electron pair condensate ($<\psi\uparrow\psi\downarrow> = 0$) leads to photons (with energy < the superconducting gap energy) "growing a mass" in a superconductor (3). This process (the Meissner effect) has an associated screening length (the "London" penetration depth λ_L) which, in appropriate units, is inversely proportional to the mass of the photon in a superconductor. Now, it can be demonstrated (3) that massless vector fields (for photons outside a superconductor) can have only two (transverse) states of polarisation. However, a true massive vector field (for photons in a superconductor) must have three independent states of polarisation, two transverse and one longitudinal. The extra longitudinal degree of freedom is the phase of the pair condensate wave function (3). In a superconductor this longitudinal component is manifest as a longitudinal oscillation of the coherent pair condensate plasma, i.e. as a plasma oscillation. We shall see that the concept of a photon growing a mass in a superconductor, together with its physical embodiments [transverse screening + phase (plasma) oscillations] will allow us to construct a realistic model of the canonical superconducting circuit, viz. a thick superconducting ring containing a Josephson weak link constriction device.

2. THE WEAK LINK RING

It is well understood that the current carrying states of a thick superconducting ring can be described in terms of the electron pair condensate. In such current carrying states the principal quantum number will be the "winding number" $n = 0,\pm1,\pm2, \ldots$ of the electron pair condensate wave function along a path around the ring inside the bulk superconductor. When the ring is subjected to an external applied flux (Φ_x) the energy associated with the winding number state n is just $W_n(\Phi_x) = (n\Phi_0-\Phi_x)^2/2\Lambda$, where $\Phi_0 = h/2e$, Λ is the geometric inductance of the ring and $n\Phi_0$ is the flux trapped in the ring by the transverse screening processes.

Obviously, for a thick superconducting ring there exists no mechanism for changing the winding number while the superconductivity of the ring is maintained. However, the inclusion in the ring of a weak link constriction of cross-sectional dimensions $\sim$ few $\times\ \lambda_L$ allows for the possibility of n-changes by the coherent (quantum) transfer of flux bundles (Φ_0) across the constriction. Such a process can be considered conjugate (4) to the coherent transfer of electron pairs through a weak link, i.e. the Josephson effect.

We assume, in the simplest quantum mechanical model, nearest neighbour winding number transitions (4,5) [$n \to n \pm 1$] with each n-change ($\pm$ 1) corresponding to a $\pm\ \Phi_0$ flux change in the weak link ring. For small flux transfer frequencies $\Omega/2$ [i.e. in a tight binding approximation model (5)] we can write down a Schrodinger equation to describe all possible nearest neighbour n-transitions of the form (6)

$$W_n(\Phi_x)D_n - (\hbar\Omega/2)[D_{n+1} + D_{n-1}] = ED_n \qquad (1)$$

where D_n is the amplitude for the ring to be in the winding number state n.

Equation (1) provides a limited treatment of the quantum mechanical behaviour of a weak link ring. A full description requires the introduction of quantised phase (plasma) oscillation modes arising from the longitudinal component of the massive vector field for photons in superconductors. For a weak link ring the frequency ω_0 of the photon mode arising from this plasma oscillation is controlled by the weak link. Taken alone, the plasma oscillation frequency of a weak link is given in the conventional theory (7) by the "Josephson" plasma frequency $\omega_0^2 = 1/\Lambda_{eff}C$, where Λ_{eff} and C are, respectively, the effective inductance and geometric capacitance of the weak link. With the weak link incorporated in a superconducting ring $1/\Lambda_{eff} = 1/\Lambda + 1/\Lambda_{kin}$, where $\Lambda_{kin} = \hbar/4e^2\nu_J$ is the kinetic inductance of the weak link (7) and ν_J is the electron pair transfer frequency through the weak link - here termed the "Josephson" frequency. Effectively, the plasma oscillation frequency is controlled, via ν_J, by the minimum value of the superconducting energy gap in the ring which will occur in the region of the weak link (8).

We can include electromagnetic oscillations of energy $m\hbar\omega_0$, where m = 0, 1,2,3..., in the Schrödinger equation for the interacting pair condensate - photon condensate (i.e. $\langle\bar{B}\rangle = 0$ in the ring) by introducing the energy $W_{n,m}(\Phi_x) = (n\Phi_0 - \Phi_x)^2/2\Lambda + m\hbar\omega_0$. The Schrödinger equation which yields the weakly excited spectrum of energy levels of the weak link ring can then be written, in the winding number (n) representation, as

$$W_{nm}D_{nm} + \sum_{m'} \frac{\hbar}{2} \Omega_{mm'} [D_{n+1,m'} + D_{n-1,m'}] = ED_{nm} \qquad (2)$$

We can write the amplitude for the pair condensate to rotate by angle θ (with screening supercurrent in the ring $\propto \dot{\theta}$), in the presence of energy quanta $\hbar\omega_0$ of electromagnetic field oscillation energy, as $\psi(\theta,m)$. By this we mean as $n \to n \pm 1$, there can also be a transition m (initial) $\to$ m (final) in the number of electromagnetic field oscillation quanta in the region of the weak link. By using the usual angular transformation $\psi(\theta,m) = \sum_n D_{nm}e^{in\theta}$ (9) we can write equation (2) in the form

$$1/2\Lambda[-i\Phi_0 (\frac{\partial}{\partial\theta}) - \Phi_x]^2\psi(\theta,m) + m\hbar\omega_0\psi(\theta,m) + \hbar \sum_{m'} \Omega_{mm'} \cos\theta\psi(\theta,m') = E\psi(\theta,m) \qquad (3)$$

which yields Φ_x-dependent energy levels (ground state + weakly excited) of the weak link ring, periodic in Φ_0, such that $E_\kappa(\Phi_x + \Phi_0) = E_\kappa$, where κ = 1,2,3,...etc. Each band with "good" quantum number κ comprises a coherent superposition of different n and m states, depending on the band number κ. Here, $\Omega_{mm'}$, which is the matrix element for m' photons to go into m photons during a transition between neighbouring winding numbers, can be expressed as

$$\Omega_{mm'} = \Omega \sum \sqrt{m!m'!}\,[r!(m-r)!(m'-r)!]^{-1}\lambda^{(m+m'-2r)}$$

with r going from zero to the minimum of m and m' and λ acting as a coupling parameter linking winding and photon transitions (10).

3. DOUBLE DERIVATIVE NOISE SPECTROSCOPY

The technique we use to observe the eigenbands (E_κ versus Φ_x) of a weak link ring [solutions of equation (3)] is in principle very simple although its practical realisation requires the use of state of the art low noise high frequency (in our case 430 MHz centre frequency) electronics. Let us first consider the UHF receiver system, shown in block form in figure 1, in which a weak link ring is coupled to a parallel resonant (LC) tank circuit, with quasistatic flux

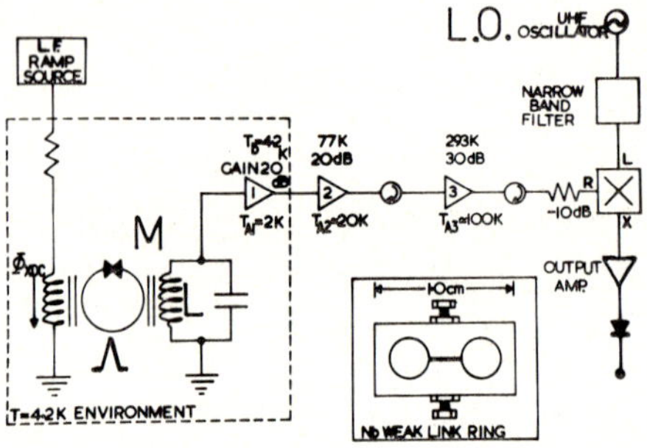

FIGURE 1

bias (Φ_{XDC}) for the ring provided by a separated coil. With no coupling to the ring the voltage fluctuations $\langle\Delta V^2\rangle$ across the tank circuit, resonant frequency ω_R, can be expressed in terms of the flux fluctuations $\langle\Delta\Phi^2\rangle$ in the tank circuit inductor as $\langle\Delta V\rangle \simeq \omega_R^2\langle\Delta\Phi^2\rangle$, where $\langle\Delta\Phi^2\rangle = k_BT_NL$ for tank circuit coil inductance L at noise temperature T_N. With mutual inductance M between the ring and tank circuit, such that $K^2 = M^2/\Lambda L$, the total classical flux noise in the tank circuit coil is just (11)

$$\langle\Delta\Phi^2\rangle = k_BT_NL\,[1 + K^2\chi] \qquad (4)$$

where $\chi = \Lambda(d^2E_\kappa/d\Phi_x^2)$ is the magnetic moment polarizability of the ring treated as macroscopic quantum object.

It is clear from equation (4) that if the weak link ring acts as a macroscopic quantum object plots of $\langle\Delta V^2\rangle$ versus Φ_{XDC} (i.e. in the quasistatic limit) will yield $d^2E_\kappa/d\Phi_x^2$ for the κth energy level $E_\kappa(\Phi_x)$ of the weak link ring, as solutions of equation (3). For $\hbar\omega_R \ll \hbar\Omega$ or $\hbar\omega_0$ we can look on K^2 as arising from elastic photon scattering from the tank circuit to the weak link ring back to the tank circuit, i.e. χ is formally the quantum electrodynamic scattering amplitude involved.

The actual receiver used in our experiments has a centre frequency of 430 MHz and a 3 dB bandwidth of 35 MHz. The approximate noise temperatures and gains of the various UHF amplifiers, are shown in figure 1. The principal features of this receiver are (i) an extremely stable local oscillator (few parts in 10^{10} frequency stability) (ii) very broad band matching of the intermediate frequency amplifier beyond the mixer &(iii) massive electromagnetic shielding. To observe χ_κ we simply adjust all the UHF (+IF) amplifiers for minimum noise temperature and plot $<\Delta V^2>$ versus Φ_{XDC}, integrated over the tank circuit bandpass. All the experimental data reported here were taken at T = 4.2K in a x-t output recorder bandwidth from DC to a few Hz.

4. THE WEAK LINK RING

In the experiments reported in this paper we used a standard design "Zimmerman" two hole SQUID device (12) incorporated a preoxidised, mechanically stabilised niobium point contact weak link in a niobium SQUID block (6) - see inset, figure 1. The point contact was adjusted at 4.2K to yield a critical current = 12 μA as measured on a 20 MHz SQUID magnetometer rig (6), utilising a relatively high excess noise temperature preamplifier ($T_N \simeq$ 100K). The cross-sectional dimensions and capacitance of such well made point contacts appear to be $\lesssim$ 1000Å square and $\lesssim 10^{-15}$ F, respectively (6).

5. POINT CONTACT RING POLARIZABILITIES

In figure 2(a) we show $<\Delta V^2>$ versus Φ_{XDC} with the receiver set for minimum attainable system noise temperature. The plot consists of a series of noise spikes on a k_BT_NL background, accurately (better than 1%) spaced by Φ_0 and showing no position hysteresis when the Φ_{XDC} sweep is reversed, over $\Phi_{XDC} > \pm 20\Phi_0$. The time taken for this recording is $\sim$ 1 minute. The pattern of figure 2(a) is just the $d^2E_\kappa/d\Phi_x^2$ versus Φ_{XDC} plot for the lowest (κ = 1) weak link ring band. In figure 2(b) we show χ versus Φ_{XDC} calculated from the κ = 1(i.e. m = 0) solution of equation (3), taking $\hbar\Omega$ from the half power width of the experimental noise spikes in figure 2(a) [i.e. for best fit $\hbar\Omega = 0.05\ \Phi_0^2/\Lambda$ so that $\hbar\omega_R << \hbar\Omega$]. The χ-pattern in figure 2(b) [and all subsequent χ-patterns] have been suitably scaled vertically to match the experimental data.

We have found that we can make the weak link ring jump stochastically between the lowest lying bands by making small variations in the gate voltage [$V_G(2)$] on the 2nd stage UHF amplifier. These variations make negligible difference to the inband (430 ± 15 MHz) noise temperature seen by the ring (< ± 0.1K) but changes the extremely high frequency ($\gtrsim$ 50 GHz) noise generated in the GaAs FET device. Small changes in $V_G(2)$ [even if reset afterwards to the initial value] have the effect of generating a burst of extremely high frequency noise photons which act to kick the ring from band to band. We emphasise that on the time scale of our observations ($\sim$ few minutes) the ring stays in one band, so considerations of Boltzmann distributions for the bands would be incorrect. In figure 3(a) we show $<\Delta V^2>$ versus Φ_{XDC} for the ring kicked into the κ = 2 (m = 1) band. The characteristic splitting of the original κ = 1 noise spikes arises from the presence of plasma mode oscillations of the pair condensate in the ring. The $\chi(\kappa$= 2) versus Φ_{XDC} plot for the second band solution is shown in figure 3(b) for $\hbar\Omega = 0.05\Phi_0^2/\Lambda$ [taken from figure 2(a)] with $\hbar\omega_0$ set at $0.078\Phi_0^2/\Lambda$ for best fit and λ = 0.5 (the line shape of figure 3(b) is not strongly dependent on λ, i.e. within $\lambda = 0.5 \pm 0.2$).

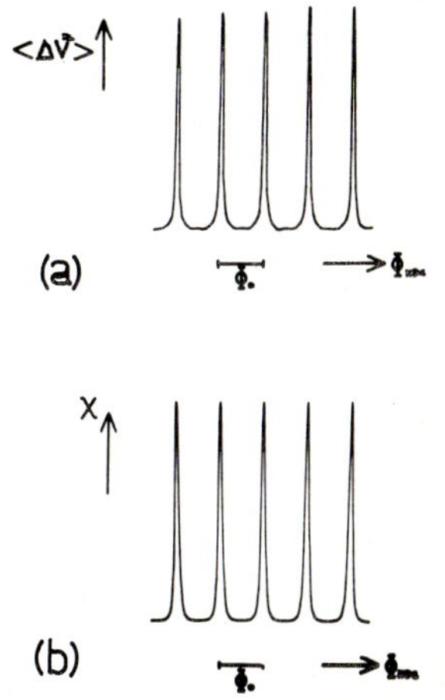

FIGURE 2

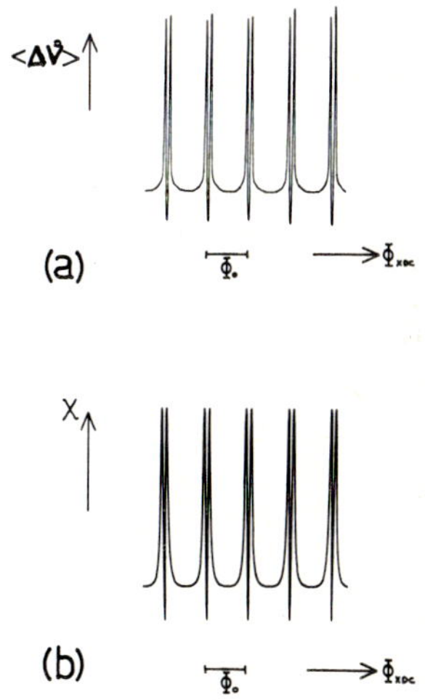

FIGURE 3

With the values of $\hbar\Omega$ and $\hbar\omega_0$ set from figures 2(a) and 2(b) we have calculated the first three band solutions of equation (3) with λ = 0.5. These are shown in figure 4. The spacing between these bands is $\hbar\omega_0$ at Φ_{XDC} = 0. We note that the κ = 3(m = 2) band generates a very distinctive second derivative (χ versus Φ_{XDC}) pattern, as shown in figure 5(a). In figure 5(b) we show the experimental $<\Delta V^2>$ versus Φ_{XDC} pattern for the ring kicked into the third (m = 2) band.

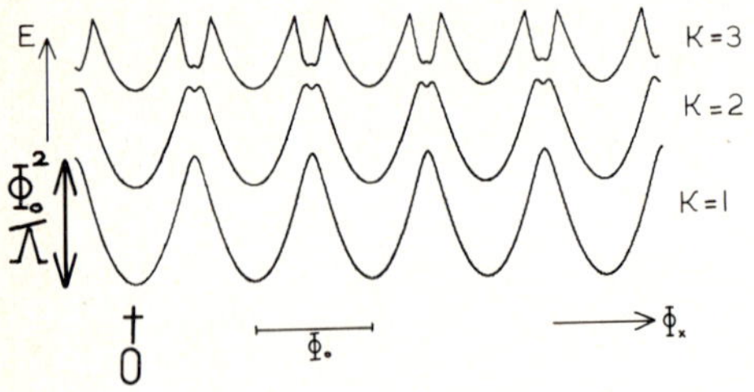

FIGURE 4

Again, as for figures 2(a) and 3(a) no hysteresis is observed in this pattern when Φ_{XDC} is reversed.

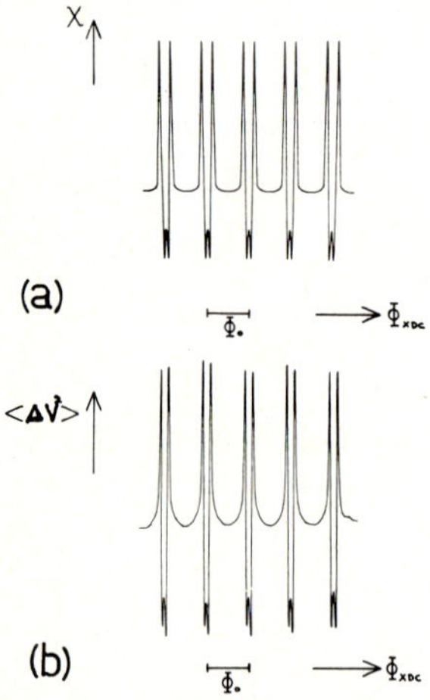

FIGURE 5

5. CONCLUSIONS

We have shown by means of a radically new technique - double derivative noise spectroscopy - that a weak link ring acts as a macroscopic quantum object on length scales $\sim$ 1 cm, i.e. this ring can exist in an amplitude superposition of macroscopically different states. The form of this superposition, as manifest in the non-invasive $d^2E_K/d\Phi_X^2$ ($<\Delta V^2>$ versus Φ_{XDC}) plots, determines that the weak link ring acts as a gauge symmetry breaking system for the electromagnetic gauge field.

REFERENCES

[1] Anderson, P.W., Coherent Excited States in the theory of superconductivity: gauge invariance and the Meissner Effect, Phys. Rev. 110 (1958) 827-835.

[2] Anderson, P.W., Plasmons, gauge invariance and mass, Phys. Rev. 130 (1963) 439-442.

[3] Aitchison, I.J.R. and Hey, A.J.G., Gauge theories in particle physics, (Adam Hilger, Bristol, 1982) chapter 9.

[4] Widom, A., Megaloudis, G., Sacco, J.E. and Clark, T.D., Theory of quantum electrodynamic flux tunnelling in a superconducting ring with a Josephson weak link, Il Nuovo Cimento 61B (1981) 112-122.

[5] De Gennes, P.G., Superconductivity of metals and alloys. (W.A. Benjamin, New York, 1968) 118-121.

[6] Prance, R.J., Long, A.P., Clark, T.D., Widom, A., Mutton, J.E., Sacco, J., Potts, M.W., Megaloudis, G. and Goodall, F., Macroscopic quantum electrodynamic effects in a superconducting ring containing a Josephson weak link, Nature 289 (1981) 543-549.

[7] Solymer, L., Superconductive tunnelling and applications (Chapman Hall, London) 176.

[8] Likharev, K.K., Superconducting weak links, Rev. Mod. Phys. 51 (1979) 101-159.

[9] Itzykson, C. and Zuber, J.B., Quantum field theory (McGraw Hill, 1980) 571.

[10] Feynman, R.P., Mathematical Formulation of the Quantum Theory of Electromagnetic Interaction, Phys. Rev. 80 (1950) 440-457.

[11] Widom, A., Megaloudis, G., Clark, T.D. and Prance, R.J., Schwinger renormalisation group equations of state in circuits with classical electrodynamic noise, Il Nuovo Cimento 69A (1982) 128-132.

[12] Zimmerman, J.E. and Frederick, N.V., Miniature Ultrasensitive superconducting magnetic gradiometer and its use in cardiography and other applications, Appl. Phys. Lett. 19 (1971) 16-19.

HOT CARRIERS AND SUBMICRON DEVICES

NOISE IN PHYSICAL SYSTEMS AND 1/f NOISE
M. Savelli, G. Lecoy and J-P. Nougier (eds.)

© Elsevier Science Publishers B.V., 1983

THERMAL, SHOT, DIFFUSION AND 1/f NOISE IN GaAs AND GaAs DEVICES
(Invited Paper)

A. van der Ziel

Electrical Engineering Department
University of Minnesota
Minneapolis, Minnesota 55455
U.S.A.

A summary is given of thermal, shot, diffusion and 1/f noise in GaAs and GaAs devices. The thermal noise is interpreted in terms of velocity fluctuation noise: hot electron effects are also discussed. The characteristic and the noise: of short GaAs diodes are discussed; in some cases the noise can be interpreted as shot noise that leads to near-thermal noise. Diffusion noise is discussed and applied to GaAs current limiters. Finally 1/f noise is discussed and deviations from the 1/f spectrum are dealt with. The 1/f noise in short n^+-n^--n^+ diodes turn out to be very small. 1/f noise in GaAs MODFETs is large but can probably be controlled now that its noise source has been located.

1. THERMAL NOISE OF GaAs DEVICES.

Thermal noise in GaAs resistors and devices such as MESFETs and MODFETs (Hemts) is esentially velocity fluctuation noise. That is, if $\overline{\Delta v_x(t)\Delta v_x(t+s)}$ is the autocorrelation function of the velocity fluctuation $\Delta v_x(t)$, then the spectral intensity of the short-circuited noise current in a section of cross-section $\Delta y\Delta z$ and length Δx is[1,2]

$$S_I(f)=4q^2n(x)(\Delta y\Delta z/\Delta x)\int_0^\infty \overline{\Delta v_x(t)\Delta v_x(t+s)}\cos\omega s\,ds \quad (1)$$

where n(x) is the carrier density and q the electron charge.

According to Kubo[3] the integral is the real part of the h.f. complex diffusion constant D, since D is the Fourier transform of $\overline{\Delta v_x(t)\Delta v_x(t+s)}$, or

$$D=\int_0^\infty \overline{\Delta v_x(t)\Delta v_x(t+s)}\exp(-j\omega s)\,ds \quad (2)$$

If $\overline{\Delta v_x(t)\Delta v_x(t+s)} = \overline{\Delta v_x^2}\exp(-s/\tau)$, where τ is the average time between collisions, then

$$D = D_0/(1+j\omega\tau)\ ;\ Re(D) = D_0/(1+\omega^2\tau^2) \quad (2a)$$

where D_0 is the low frequency diffusion constant. Hence

$$S_I(f) = 4q^2Re(D)n(x)\Delta y\Delta z/\Delta x \quad (3)$$

The low-frequency form of this equation was first derived by Becking[4].

If now the extended Einstein relation

$$qRe(D)=kT\,Re(\mu) \quad (4)$$

is valid, we may rewrite (3) as

$$S_I(f) = 4kT[qn(x)Re(\mu)\Delta y\Delta z/\Delta x] = 4kT/\Delta R \quad (5)$$

where $\Delta R = \Delta x/[qRe(\mu)n(x)\Delta y\Delta z]$ is the resistance of the box $\Delta x\Delta y\Delta z$. Equation (5) corresponds to Nyquist's theorem for the resistance ΔR.

At very high frequencies quantum effects occur and kT in (5) must be replaced by the average energy $\overline{E} = 1/2hf + hf/[\exp(hf/kT) - 1]$. Van Vliet and van der Ziel[5] have extended above discussion for the case that quantum effects are important. At very low temperatures carrier freeze-out can occur and the resistor or device may not operate properly.

The velocity fluctuation approach also holds for high fields since $\overline{\Delta v_x(t)\Delta v_x(t+s)}$ can be defined for high fields. Therefore the expression[2]

$$S_I(f) = 4q^2D_0n(x)\Delta y\Delta z/\Delta x \quad (6)$$

remains valid at high fields, provided that it is borne in mind that D_0 is now a function $D_0(F)$ of the field F.

In devices like MESFETs, MODFETs and current limiters one introduces the electron temperature T_e by the definition

$$qD_0 = kT_e\mu_0 \quad (6a)$$

Substituting into (6) yields

$$S_I(f) = 4kT_e[q\mu_0n(x)\Delta y\Delta z/\Delta x] = 4kT_eg_{dc} \quad (6b)$$

where $g_{dc} = I/V$ is the d.c. conductance of the

box $\Delta x \Delta y \Delta z$. At high fields T_e can be considerably larger than the device temperature T (hot electron effect).

In resistors it is customary to put

$$qD_0' = kT_e'\mu_0' \quad (7)$$

where $\mu_0' = du_d/dF$ is the differential mobility, $u_d = \mu_0 F$ is the drift velocity and T_e' is an alternate electron temperature.
Substituting into (6) yields

$$S_I(f) = 4kT_e' \,[q\mu_0' n(x)\Delta y\Delta z/\Delta x] = 4kT_e' g_{ac} \quad (7a)$$

where $g_{ac} = dI/dV$ is the a.c. conductance of the box $\Delta x \Delta y \Delta z$. At high fields μ_0 tends to saturate and $T_e' > T_e > T$.

When GaAs MESFET devices are cooled down to 20 °K, T_e decreases considerably, even though T_e/T increases with decreasing T[6]. As a consequence the noise temperature of a MESFET receiver decreases considerably. This is important for space communication.

When the length of a GaAs MESFET device is decreased, the transit time of the carrier decreases, and as a consequence the microwave operation improves and the noise figure decreases. Devices with gate length of 0.25 μm[7] and 0.15 μm[8] are in the experimental stage. These devices allow operation well beyond 30 GHz.

Recently developed GaAs MODFETs show good microwave noise temperatures up to 10 GHz[9]; these were devices with a gate length of 1.5 μm. The good properties come from the fact that the electrons in the channel behave as a two-dimensional electron gas and have a very high mobility (8,000 cm^2/volt sec at room temperature, above 100,000 cm^2/volt sec at 77 °K)[10]. When these devices will be scaled down to 0.25-0.15 μm gate length, operation above 30 GHz should be feasible.

With the very short channels and high mobility, one ultimately comes to the point where carrier collisions become rare events. The associated noise problem has not yet been solved; it deserves careful attention.

2. SHOT NOISE IN SHORT GaAs DIODES

The diodes referred to here are of the n^+-n^--n^+, n^+-p^--n^+, p^+-p^--p^+ and p^+-n^--p^+ varieties; the length of the middle region is typically 0.40 μm but shorter and longer devices have been made.

In each device considerable spillover occurs from the heavily doped end regions into the middle region[11]. As a consequence the current flow in the n^+-p^--n^+ device should be due to electrons and the current flow in the p^+-n^--p^+ device should be due to holes. Because of the spillover effect the potential distribution in the n^--region shows a potential minimum of depth $-V_m$, that decreases in depth and moves toward the cathode when the anode voltage V_a is raised. This has important consequences for the characteristic.

In the n^+-n^--n^+ device the device has a linear characteristic up to about 0.30 V per diode[12]. This can be explained by neglecting collisions and treating the device as a Langmuir-Fry type planar vacuum diode consisting of two cathodes at different potential. The first try gave a nearly linear characteristic[13]; a more accurate numerical calculation by Bosman[14], based on the modified Langmuir-Fry theory, gave indeed a practically linear characteristic up to 0.3V that was nearly independent of temperature. The reason why the method works is that only collisions very close to the potential minimum can affect the characteristic, and these collisions are very rare events. The ballistic approximation of neglecting the collisions altogether is therefore a good one.

The primary noise in these devices is shot noise of the incremental current dI_v with an emission velocity between v_x and $v_x + dv_x$; this produces a fluctuation δV_m in the depth of the potential minimum that partly compensates the primary fluctuation. By integrating over the velocity distribution one obtains the total noise[15]. Experimentally one observes thermal noise of the conductance I/V[12]; the shot noise approach gives this result at low bias and probably a result close to it at larger bias. According to thermodynamics the device should show thermal noise at zero bias.

It would be important to find out for what device lengths the ballistic approach becomes inaccurate. This is being worked on at the University of Florida.

Short n^+-p^--n^+ GaAs diodes have a strongly non-linear characteristic[12]. At low bias voltage V_a the characteristic is linear and the current is about a factor 100 smaller than for an n^+-n^--n^+ device at comparable bias. For $V_a > 0.2V$ the characteristic is also linear and the current is only about a factor 3-4 smaller than for an n^+-n^--n^+ device at comparable bias. In between the d.c. conductance I/V increases rather rapidly from a low value at low bias to a much higher value at high bias; the linear (I,V) characteristic in the high bias regime extends to at least 0.5V. Preliminary calculations by Bosman[14] seem to indicate that the ballistic approximation is valid at large bias ($V_a > 0.2V$). The reasons are the same for the n^+-n^--n^+ device.

The h.f. noise at low forward bias is thermal noise as expected for any linear resistor. The situation is not quite clear at intermediate and large forward bias because of a large amount of 1/f noise present[12]. Theoretically one would expect thermal noise of the d.c. conductance I/V in the linear high bias regime.

The causes for the low conductance at low forward bias are not quite clear. In view of the large difference in mobility ($\mu_n \simeq 7000$ cm^2/volt sec for electrons and $\mu_p \simeq 350$ cm^2/volt sec for holes) we first thought about ambipolar effects as a possibility. We now consider it more likely that the low current is caused by a strong carrier recombination near the potential minimum due to holes trapped in that region; this effect gradually diminishes with increasing forward bias and finally disappears, resulting in the strong non-linear characteristic.

Of great interest would be the p^+-p^--p^+ and p^+-n^--p^+ device diodes. The p^+-p^--p^+ is important to see whether the ballistic approach is also valid here, or whether the characteristic is determined by collision-limited space-charge-limited current flow. The p^+-n^--p^+ device is important because the ambipolar effect and the recombination effect give a radically different a.c. conductance at low bias.

Permeable base transistors[16] are in fact solid state triodes and therefore should have near-thermal noise at high frequencies. Reasonable noise figures were observed at 10 GHz[16]. The limit of low-noise operation probably lies considerably higher, especially if the device dimensions can be shrunk still further.

3. DIFFUSION NOISE[17]

When carriers or charged particles diffuse out of a small active volume into a larger non-active volume, the spectrum shows typical spectra that are different for one, two and three dimensions[17]. In one-dimensional diffusion noise the spectrum varies as $1/f^{1/2}$ at low frequencies and as $1/f^{3/2}$ at high frequencies, with a short transition region in between. In two-dimensional diffusion noise the spectrum varies as $\ln f$ at low frequencies, and as $1/f^{3/2}$ at high frequencies, with a transition region in between, whereas in three-dimensional diffusion noise the spectrum is constant at low frequencies and varies as $1/f^{3/2}$ at high frequencies with a transition region in between.

We discuss one-dimensional noise in detail, since it seems to occur in short GaAs current limiters, as evidenced by the $1/f^{1/2}$ and $1/f^{3/2}$ branches of the spectrum[18].

According to van Vliet and Fassett[17] the fluctuation ΔN in the number N of particles due to a one-dimensional diffusion process has a spectrum

$$S_{\Delta N} = \frac{\overline{\Delta N^2} L^2}{D\Theta_1^3}[1-\exp(-\Theta_1)(\cos\Theta_1 + \sin\Theta_1)] \quad (8)$$

where L is the effective length, D the diffusion constant, ω the angular frequency and $\Theta_1 = L(\omega/2D)^{1/2}$. The $1/f^{-1/2}$ branch occurs for $\Theta_1 \ll 1$ and has

$$S_{\Delta N}(f) = \frac{2^{1/2}\overline{\Delta N^2}L}{\omega^{1/2}D^{1/2}}, \quad (8a)$$

the $f^{-3/2}$ branch occurs for $\Theta_1 \gg 1$ and has

$$D_{\Delta N}(f) = \frac{2^{3/2}\overline{\Delta N^2}D^{1/2}}{\omega^{3/2}L} \quad (8b)$$

whereas the turn-over frequency is located at the point where the two curves cross. This occurs at Θ=1, or

$$f_t = \frac{D}{\pi L^2} \quad (8c)$$

Since for GaAs current limiters at room temperature $f_t \simeq 10^3$Hz, give or take a factor 10, it follows that $D \simeq 3\times10^{-5} cm^2/sec$[18]. The diffusing particles are thus impurity atoms or ions. Furthermore the low frequency branch increases strongly with decreasing temperature, the high frequency branch decreases strongly with decreasing temperature and f_t decreases strongly with decreasing temperature. This would be expected if D is governed by an activation energy E_a as

$$D = D_o(\exp(-qE_a/kT) \quad (9)$$

Finally the turn-over frequency increases rather rapidly with increasing voltage V[18]. This would be expected for diffusion of impurity ions governed by an activation energy E_a that decreases with increasing field F = V/L because of the Poole-Frenkel effect; according to the theory this yields

$$E_a = E_{ao} - [qV/(\pi\varepsilon\varepsilon_o L)]^{1/2} \quad (10)$$

where ε is the relative dielectric constant and ε_o the MKS conversion factor.

This discussion shows how Eq. (8) can be used to unravel very complicated spectra from a few general features. While the GaAs is not a one-dimensional device, a one-dimensional process would occur if the impurity ions diffuse along dislocation lines[18].

What is measured is not the fluctuation ΔN in N but the resulting fluctuation ΔI in the current I. It comes about by two fully correlated processes:

a) Carrier number fluctuations. Since the impurity ions act as donors or acceptors, ΔN gives fluctuations in the number of free carriers.

b) Carrier mobility fluctuations. Since the impurity ions scatter the carriers, ΔN gives rise to fluctuations $\Delta\mu$ in the mobility μ.

With the help of these simple considerations the l.f. noise in GaAs current limiters can be explained as onedimesional diffusion noise[18].

4. 1/f NOISE

1/f noise is often interpreted in terms of a distribution of time constants[19]. In some cases the noise is due to carrier number fluctuations. If the carrier fluctuations have a single time constant τ then the fluctuation ΔN in the number of carriers has a spectrum

$$S_{\Delta N}(f) = 4\overline{\Delta N^2} \frac{\tau}{1 + \omega^2\tau^2} \tag{11}$$

For a distribution $g(\tau)\,d\tau$ in time constants with

$$\int_0^\infty g(\tau)d\tau = 1 \text{ (normalization)} \tag{11a}$$

one has instead

$$S_{\Delta N}(f) = 4\,\overline{\Delta N^2} \int_0^\infty \frac{\tau g(\tau)d\tau}{1 + \omega^2\tau^2} \tag{11b}$$

In the particular case that[19]

$$g(\tau)d\tau = [\ln(\tau_1/\tau_0)]^{-1}\, d\tau/\tau \tag{12}$$

for $\tau_0 < \tau < \tau_1$ and zero otherwise, one obtains

$$S_{\Delta N}(f) = \frac{2}{\pi} \frac{\overline{\Delta N^2}}{f \ln(\tau_1/\tau_0)} [\tan^{-1}(\omega\tau_1) - \tan^{-1}(\omega\tau_0)] \tag{12a}$$

which has a wide 1/f regime if $\tau_0 \ll \tau_1$ for $1/\tau_1 < \omega < 1/\tau_0$. Usually τ_1 is so large that the turn-over at $\omega\tau_1 = 1$ cannot be observed and τ_0 is often so small that the turnover at $\omega\tau_0 = 1$ is not observed either.

Suh and van der Ziel[20] found that the current noise spectrum of some NEC GaAs FETs could be accurately described by the formula

$$S_I(f) = (A/f)\,[1 - (2/\pi)\tan^{-1}(\omega\tau_0)] \tag{12b}$$

which follows from (12a) for $\omega\tau_1 \gg 1$. This indicated that the devices in question showed number fluctuation noise.

Hooge[21] has shown that in many cases 1/f noise in semiconductor resistors is a bulk effect and that the current noise satisfies the equation

$$S_I(f)/I^2 = \alpha/(fN) \tag{13}$$

where N is the number of carriers in the sample and α is Hooge's parameter, often of the order of 2×10^{-3}.

This interpretation may well be correct for bulk GaAs resistors, but does not seem to hold for GaAs MESFETs. The devices are made by epitaxially growing an n-type active layer on a semi-insulating substrate. The 1/f noise of the device can be substantially reduced, by at least one order of magnitude, by giving the interface between the semi-insulating substrate and the active layer the following treatments[22]:

1. The semi-insulting layer was etched before the active layer was applied.
2. An undoped buffer layer was deposited between the semi-insulating layer and the active layer.
3. The semi-insulating substrate was first etched and then a buffer layer was applied.

Each treatment gave improvement, but the combined treatment (3) gave the best results. This clearly indicated that most of the 1/f noise is generated at the interface with the semi-insulating substrate. It is therefore a surface effect, most likely caused by generation-recombination processes.

To test whether the 1/f noise in bulk GaAs is of the mobility fluctuation type, the following experiment should be performed: A good GaAs sample with two ohmic contacts is taken and its 1/f noise is measured. Then the surface should be sputtered in an argon atmosphere to clean it and the 1/f noise should be remeasured. If the 1/f noise decreases sustantially, this is an indication that it is due to a surface effect.

Zheng et al[23] carried out the experiment in HgCdTe. They found an A/f spectrum throughout before cleaning, and a spectrum

$$\frac{2A}{\pi f} \tan^{-1}(\omega\tau_2) \tag{14}$$

after sufficient sputtering; here τ_2 was of the order of 10^{-3} sec, or orders of magnitude smaller than τ_1. Apparently all the long time constants of the 1/f noise were eliminated by the sputtering process. This clearly indicates that here the 1/f noise was caused by a surface mechanism, leading to number fluctuations.

We now turn to the 1/f noise in MODFETs. The composition of the MODFET (modulation doped FET) is shown in Fig. 1. Under a Schottky barrier gate is a layer of n^+-GaAlAs separated from the undoped GaAs active layer by a thin layer of undoped GaAlAs. The n^+-GaAlAs layer is depleted of electrons and some of these are located on the gate side of the GaAs layer, where they are form a conducting channel (two-dimensional electron gas).

The n^+ - GaAlAs layer has a large number of deep-lying traps that can change the characteristic. As a consequence the (I,V) characteristic of a MODFET often shows a hysteresis loop that can be removed by shining light on the device.

The GaAs MODFET shows a large amount of 1/f noise[24].

Experiments by Duh on ungated structures[25] have shown that the 1/f noise is generated in the space-charge region between gate and channel, probably by the modulation of the channel by the fluctuating occupancy of the traps, and is not caused by the conducting channel. This means that it should be possible to reduce the 1/f noise of MOSFETs appreciably by reducing the trap density in the n^+ -GaAlAs layer.

The n^+-n^--n^+ structures have very little 1/f noise[12,26]. This would be expected if the 1/f noise is either a surface effect or a bulk effect. The device diameter is 100 μm and the width of the n^- -region is of the order of 4000 Å so that there is very little surface area and hence very little surface 1/f noise. Moreover, there are very few collisions in the n^- -region and hence there is little mobility flucutation 1/f noise.

If it is assumed that the noise is of the mobility fluctuation type, one can evaluate Hooge's parameter α. Neglecting spillover effects, the effective number of carriers is $N = ALN_d$ where A and L are the area and the length of the n^- -region. In that case one finds $\alpha = 5\times10^{-8}$, which must be compared with Hooge's value[27] $\alpha = 6\times10^{-3}$ for bulk GaAs. The introduction of the Hooge paramter in these thin samples is a doubtful procedure, because the device is not a uniform resistor.

The n^+-p^--n^+ structure has several orders of magnitude larger noise[12]. The nature of this 1/f noise is not quite clear; it could be recombination 1/f noise in the region around the potential minimum. More experiments and new ideas are needed to explain this phenomenon.

The permeable-base transistor (PBT) is a solid state triode and its 1/f noise should be similar to that of an n^+-n^--n^+ diode, i.e. it should be very small. No data are available at present to test this prediction.

Work supported by NSF grant

REFERENCES

[1] N. Wiener, Acta. Math, 55, 117, 1930; A Khintchine, Math. Ann., 109, 604, 1934.

[2] A. van der Ziel, Solid State Electronics, 23, 1035, 1980.

[3] K. Kubo, J. Phys. Soc. Japan, 12, 570, 1957.

[4] A. G. Th. Becking, unpublished: his derivation of (3) is discussed by A. van der-Ziel, Noise, Scources, Characterization, Measurements, Prentice Hall Inc., 1970, Chapter 5.

[5] K. M. van Vliet and A. van der Ziel, Solid State Electronics, 20, 931, 1977: Physica, 99A, 337, 1979.

[6] S. Weinreb, 1982 IEEE MTT-S Digest.

[7] H. F. Cooke, Varian Associates, Private Communication

[8] W. Fichtner, R. K. Watts, D. B. Fraser, R. Johnston and S.M. Sze, IEEE El. Dev. Lett., EDL-3, 412, 1982.

[9] M. Laviron et al, Electron. Lett., 17, 536, 1981.

[10] R. Dingle et al, Appl. Phys. Lett, 33, 665, 1978.

[11] A. van der Ziel et al, IEEE Trans. El. Devices, ED-30, 128, 1983.

[12] R. R. Schmidt, Ph. D. Thesis, U. of Florida, 1983; R. R. Schmidt et al, Solid State Electronics, 1983, in the press.

[13] A. J. Holden and B. T. Debny, Electron. Lett., 18, 558, 1982.

[14] G. Bosman, to be published.

[15] D. O. North, RCA Review, 4, 441, 1940; 5, 106, 1941.

[16] C. O. Bolzer and G. D. Alley, IEEE Trans. El. Dev., ED-27, 1128, 1980; Proc. IEEE, 20, 46, 1982; C. O. Bolzer, private communication.

[17] K. M. van Vliet and J. R. Fassett, Fluctuation Phenomena in Solids, (Ed. R. Burgess), Acad. Press, Inc., New York, 1965, Chapter 7.

[18] A. Peczalski, A. van der Ziel and R. Zuleeg, Solid State Electronics, 26, 1983, in the press.

[19] A. van der Ziel, Physica, 16, 359, 1950; F. K. duPré, Phys. Rev. 78, 615, 1940.

[20] C. H. Suh and A. van der Ziel, Appl. Phys. Lett., 37, 565, 1980.

[21] F. N. Hooge, Phys. Lett., A29, 139, 1969.

[22] M. K. Ahmed and H. Beneking, Semi-Insulating III-V Materials, Nottingham, England, April 1980, Conf. Rept., 346, 1980; C. Tsirionis, Private Communication.

[23] K. Zheng, K. H. Duh and A. van der Ziel, Physica, 1983, in the press.

[24] K. H. Duh, A. van der Ziel and H. Morkoc, IEEE El. Dev. Lett., EDL-4, 10, 1983.

[25] K. H. Duh et al, to be published.

[26] A. Peczalski, A. van der Ziel and M. A. Hollis, IEEE Trans. El. Dev. ED-29, 1837, 1982.

[27] F. N. Hooge, T. G. M. Kleinpenning and L. K. van Damme, Repts. Progress in Physics, 44, 479, 1971.

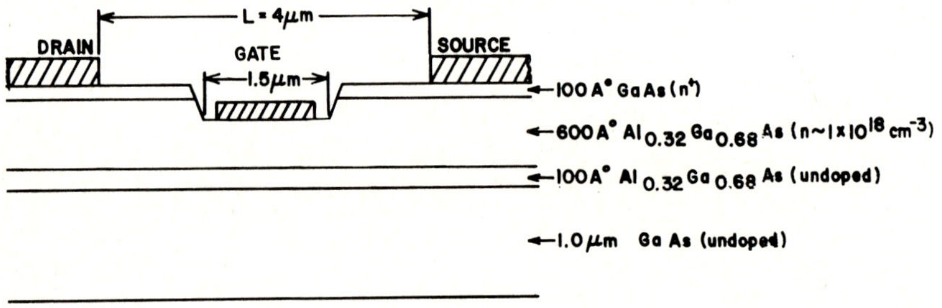

Fig. 1. Cross-section of MODFET

NOISE IN PHYSICAL SYSTEMS AND 1/f NOISE
M. Savelli, G. Lecoy and J-P. Nougier (eds.)
© Elsevier Science Publishers B.V., 1983

NOISE OF SUBMICRON DEVICES

Invited paper

J.P. Nougier

Centre d'Electronique de Montpellier
Université des Sciences et Techniques du Languedoc, Pl. E. Bataillon
34060 Montpellier Cedex, France*

In this paper, are reviewed the main problems encountered in modeling the noise of submicron devices, most of which are yet unsolved : change of the nature of the noise, non stationnary effects, space correlations, tridimensional effects, non instantaneous collisions, discretization, ...

1. INTRODUCTION

I shall focus here on theoretical aspects, experiments being reviewed in the paper of A. Van der Ziel [1]. In a recent paper [2], were reviewed the main aspects of theories used for modeling noise of devices of classical dimensions (i.e. larger than a few micrometers) : these theories are more or less well established, understood and reliable, and have been applied to a variety of devices [2]. In the conclusion of that paper, the question of the formulation of theories applicable for short devices was raised.
The present paper thus appears as a natural continuation of the previous one [2]. Since the few last years, a great effort has been devoted to the understanding of the behaviour of submicron devices and systems. However, as concerning the theory of noise, although some progress have been performed, much remains to be done. As a consequence, the present paper is not a review gathering results is a coherent synthesis, as was due in the previous one [2], but rather is a survey of the main problems encountered in modeling submicron devices, most of which are yet unsolved. It is hoped that such a presentation will raise curiosity and interest of searchers for this large area of physics yet uncleared.

2. UNEXPECTED INFLUENCE OF SOME PARAMETERS

When reducing the dimensions of a device, some parameters which had insignificant effects in long devices may get a great importance in short devices. In that case, the short device is considered as the limit of the long one, when the size is scaled down, the theory used for modeling the long device is supposed to apply also to the short one.

<u>Example 1</u> : n^+nn^+ or p^+pp^+ silicon diodes.

We suppose here that the current is carried by the majority carriers (single injection). For a given current density, the electric field profile is given figure 1, where it can be clearly seen that the space charge extends over a distance d_s.
If the length d of the diode (that is the thickness of the n (or p) region) is $d \gg d_s$, then the electric field in the device can be considered as being uniform, the sample is a "homogeneous bar": many experiments are performed on such bars, where the field is well controlled. If the current density is large enough, the field is large and one measures hot electron phenomena, for both I-V and noise characteristics.

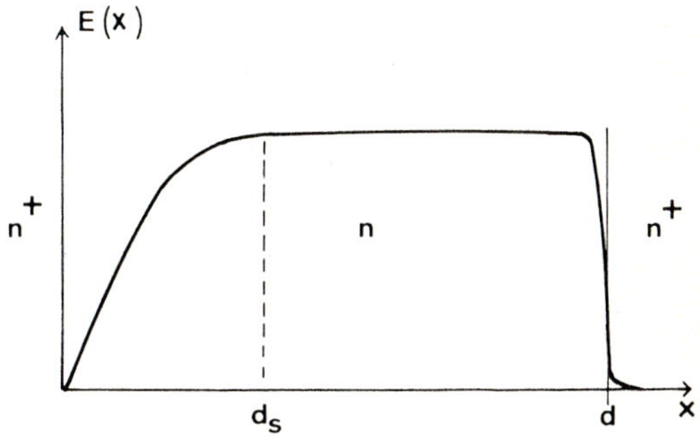

Fig.1 - Electric field profile in a n^+nn^+ single injection diode.

If $d \lesssim d_s$, one deals with a space charge limited current diode, where a mixture of ohmic and hot electron behaviours are involved, since the field is strongly inhomogeneous : such a behaviour has been studied [3] [4] [5] and is completely different from that of bars.
The value of d_s depends on the dopings of the n^+ and n regions, it can reach hundred microns for pure n dopings ([3] to [5]) and only a few microns for doped materials of high mobility [6].

* Associé au CNRS, Greco Microondes.

Example 2 : GR noise in n type channel silicon junction gate field effect transistor. The noise current in a JG FET, due to generation recombination (GR) noise sources in the channel, can be expressed as [7] :

$$S_{IGR} = \frac{4q\mu_0}{L^2(1+V_D/E_cL)^2} \frac{\alpha\tau}{(1+\omega^2\tau^2)} I_D V_D \qquad (1)$$

where I_D and V_D are the drain current and the drain voltage, α and τ are constants at 300K, L is the length of the transistor between source and drain, and E_c is the characteristic field, $E_c \simeq 10$kV/cm at 300K in n-type silicon. Usually the transistor is operated at $V_D \lesssim 5$V.
For long transistors, $L \simeq 20$ µm, then $E_cL \simeq 20$V, V_D/E_cL is small and $S_{IGR} \propto I_D V_D$, that is roughly proportional to the d.c. power in the transistor. For short transistors, $L \simeq 0.5$ µm, then $E_cL \simeq 0.5$V, and $V_D/E_cL \gg 1$, so that $S_{IGR} \propto I_D/V_D$. This shows that the behaviour of short transistors may be quite different from the behaviour of long transistors.

3. DEFINITION OF A SUBMICRON DEVICE

In spite of interesting or unexpected behaviour as was just mentionned in the two previous examples, the purpose of this paper is not to investigate short devices but submicron devices.
What are submicron devices ? They are not merely short devices. Short devices are devices which differ from long (usual) ones only by reducing the dimensions and to which the theories used for long devices still apply. Submicron devices are devices so short that usual theories of not apply any more : their length is of the order of a few mean free paths or even less. Let L be the mean free path, that is the average distance between two collisions, and d be the length of the device. In a long device, $d \gg L$,a given carrier undergoes many collisions inside the device, and thus reaches a steady state:the time required for the carrier to pass through the device is d/v, where v is the (steady state) drift velocity. We are dealing with a device in a steady state regime were each carrier is in a steady state. In a submicron device, $d \simeq L$, one also gets a steady state when a dc. bias is applied. However, a given carrier undergoes too few collisions so as to be in a steady state, the time required for this carrier to pass through the device is d/v', where v' differs from the steady state velocity. We are now dealing with a device in a steady state regime where each carrier is not in a steady state.
A device can be schematically compared with a column of atmosphere between the ground and an helicoptere steady in the air. The length d of the device is the altitude of the helicoptere, the d.c. field is represented by the gravity field, the carriers are represented by parachutists dropping from the helicoptere at random instants but with an average constant rate (Poissonian emission of carriers) : this will be called in the following the "para-device". When the parachutist number i has dropped from the helicoptere, it takes a certain time τ_i for the parachute to open : during this time, the parachutist is accelerated, its velocity reaches a maximum value (velocity overshoot), then decreases, when the parachute opens, so as to reach a stationary value v_i at time of the order of τ_i : the parachutist has then travelled over a distance L_i. This is about the same for a carrier in a device. The mean free path L is the average of the values L_i of all the parachutists, the drift (steady state) velocity v is the average of the values v_i. When the helicoptere is at a high altitude, $d \gg L$, we get a "long para-device" : for a constant flow of dropping parachutists, we get a steady state current where each person is in a steady state regime, the time of each flight is in the average d/v. Conversely, when $d \simeq L$ or even smaller,one gets a "submicron para-device" : a constant flow of dropping parachutists also gives a steady current, but most of the parachutist have not yet reached their steady state velocity since the parachutes had not yet opened. The time of flight is $d/v' < d/v$.
We will now investigate the main problems encountered in submicron devices.

4. CHANGE OF THE NATURE OF THE NOISE

In submicron devices, the nature of the noise sources themselves change. Experiments have just been reported in order to try to bring evidence of such a change [1] [8], and are based on the following idea : let us consider for example a n^+nn^+ diode of length d between the electrodes. If $d \gg L$ (L is the mean free path), each electron, while travelling through the diode, undergoes many collisions, the noise source is diffusion noise. If $d \ll L$, the electrons are not scattered in the diode, and one deals with shot noise. If $d \simeq L$, one should observe some "damped diffusion noise". It is not easy to bring evidence of such a phenomenon, since the noise actually observed takes also into account the "parasitic" regions (n^+n transition regions in particular). These experiments will not be discussed here since they are reported extensively elsewhere [1]. In particular, it appears that even the 1/f noise is modified in submicron devices [8].
An other experiment was recently suggested, which could perhaps show this phenomenon [9]. The basic idea is to use the gate of a field effect transistor as a probe which then allows sensoring the noise sources inside the channel. In a n-type channel silicon J.G FET, the noise current induced in the gate, by the noise sources inside the channel, is expected to be [9] :

$$S_{Igate} = \frac{4k_B\omega^2q^2N^3_D\mu_0}{I^2_D} \int_0^L T_n(x)A(x)[A(L)-A(x)]^2dx \qquad (2)$$

where $A(x)$ is the area of a cross section of the channel at point x, $A(L)$ being the cross section near the drain or near the source. Since $A(x) \simeq A(L)$ outside the channel, eq.(2) brings a confirmation of the basic idea that the gate behaves as a probe. Thus S_{Igate}, through $T_n(x)$, should be different for long FETS than for FETS with submicron gate length.

5. NON STATIONARY EFFECTS

For studying fast transient regimes, of the order of a fraction of picosecond, in a semiconductor where an electric field step is applied, one should solve the Boltzmann equation (B.E.) : one may then use iterative or direct numerical solutions of the Boltzmann equation, or Monte Carlo methods. These methods provide exact solutions of the B.E. Unfortunately, they require big computers (important memory occupancy and long computation time), and are therefore expensive. It was recently shown [10] that the dynamic equations [11] were approximate solutions of the transient behaviour of transport coefficients, indeed departing by less than 10% from exact transient regimes for a variety of semiconductors [10] and in a wide range of initial conditions [12]. These equations are :

$$\begin{cases} \dfrac{d[m(\varepsilon)v]}{dt} = qE - \dfrac{m(\varepsilon)v}{\tau_m(\varepsilon)} \\ \\ \dfrac{d\varepsilon}{dt} = qEv - \dfrac{\varepsilon-\varepsilon_0}{\tau_\varepsilon(\varepsilon)} \end{cases} \qquad (3)$$

in eqs.(3), v and ε depend on time, and the momentum and energy relaxation times τ_m and τ_ε as well as the effective mass m, are supposed to be funtions of ε only, and therefore can be determined in steady state regime. If v and ε are known at a given time, the right band sides of eqs.(3) is known, which gives dv/dt and $d\varepsilon/dt$, and thus $v(t+\Delta t)$ and $\varepsilon(t+\Delta t)$ (for example, $v(t+\Delta t) = v(t) + \Delta t\, dv/dt$).

Eqs.(3), coupled with the Poisson equation, can be used for modeling static and dynamic characteristics of submicron devices [13]. Obviously, in such devices, the drift velocity at point x, where the d.c. electric field is E, is steady. However, it is not equal to the value it would have in a long device in the same electric field, because, as was mentionned earlier, each carrier undergoes too few collisions to be in a steady state (in the "submicron para-device", the velocity is strongly varying along the device, although the driving force is homogeneous ; in the "long para-device", the velocity is uniform, except over the distance L which is negligible compared with the length of the device).

Since v and ε have not their stationary value, it can be expected that the diffusion coefficient D also takes a nonstationary value. It was assumed by Carnez et al [14] that $D = D(\varepsilon)$ is a function of ε . This results in a non stationary noise source.

$$K(x) = 4q^2 n(x)\, D[\varepsilon(x)] \qquad (4)$$

This noise source, carried in the impedance field formula, gives the noise voltage :

$$S_V = \int_0^L K(x) |\nabla Z(x)|^2 A(x)dx \qquad (5)$$

It was shown [14] that these non stationary relaxation effects could have an important influence on the noise properties of short GaAs MES FETs: the noise figure of short gate transistors is much lower when taking into account non stationary effects then when using the usual steady state noise source term, as is shown in figure 2, taken from ref. [14]. Hence, non stationary effects are important and should be taken into account.

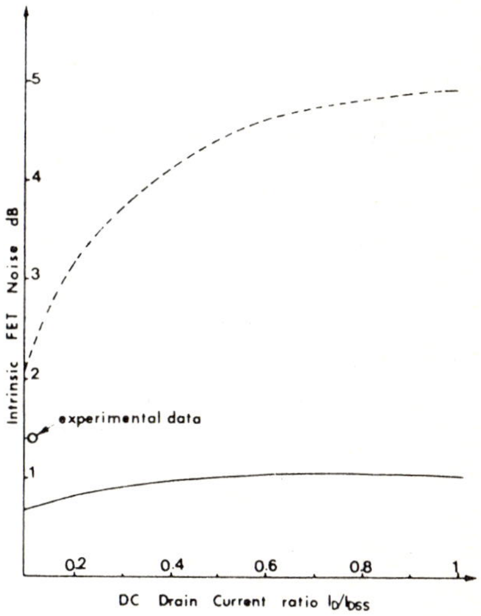

Figure 2 : Intrinsic GaAs MES FET noise figure, length of the gate : 0.4 μm (taken from ref.[14]).

6. SPACE CORRELATIONS OF NOISE SOURCES

Up to now, the noise sources located at two different points were supposed to be uncorrelated:

$$S_j(\vec{r}, \vec{r}', \nu) = K(\vec{r}, \nu)\, \delta(\vec{r}' - \vec{r}) \qquad (6)$$

Indeed this is not true, and correlations do occur over distances of the order of the mean free path. This was proved very recently [15], the basic idea is the following : suppose that no electric field is applied (ohmic regime) ; an electron is located at time t at point r within d^3r, with a velocity $\vec{v}(\vec{r},t) = \vec{v}$. At time $t+\theta$, if this electron has not been scattered, its velocity is still $\vec{v}$, and its position is $\vec{r}' = \vec{r} + \vec{v}\theta$ (fig.3).

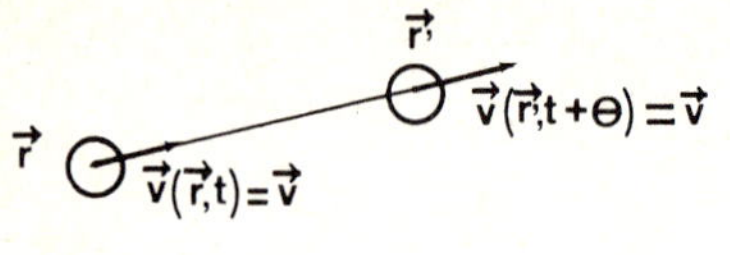

Figure 3

Therefore, the correlation $C_{\alpha\beta}(\vec{r},\vec{r}',\theta)$ of the velocities over the time interval θ, between the points $\vec{r}$ and $\vec{r}'$, along the directions α and β is equal to $v_\alpha v_\beta$ with the probability PP'. P is the probability that the electron has not been scattered during the time interval θ, that is $P = \exp[-\theta/\tau(k)]$ where $\vec{k}$ is the state of the electron. P' is the probability that $\vec{r}$, $\vec{r}'$ and θ are related by $\vec{r}' = \vec{r}+\vec{v}\theta$, that is $P' \propto \delta(\vec{r}'-\vec{r}-\vec{v}\theta)$ This leads to the following expression of the correlation of the current density in non polar semiconductors, with $\vec{R} = \vec{r}' - \vec{r}$:

$$C_{j\alpha\beta}(\vec{R},\theta) = q^2 n \int d^3k f_o(\vec{k}) v_\alpha v_\beta \delta(\vec{R}-\theta\vec{v}) \exp-[\theta/\tau(\vec{k})] \quad (7)$$

where $f_o(\vec{k})$ is the thermal equilibrium distribution function. For distances $\vec{R}$ such that $|\vec{R}| \gg v_{max} \times \theta_{max}$, where θ_{max} equals a few relaxation times $\tau(\vec{k})$, and $v_{max} = v(\vec{k}_{max})$ such that $f_o(\vec{k}_{max}) \simeq 0$, then $\delta(\vec{R}-\theta\vec{v}) \simeq \delta(\vec{R})$, the correlation function is a Dirac function of space, and the Fourier transform of eq.(6) gives :

$$\begin{cases} S_{j\alpha\beta}(\vec{R},\nu) \simeq \delta(\vec{R})\, 4q^2 n\, D_{\alpha\beta}(\nu) \\ D_{\alpha\beta}(\nu) = \int d^3k\, f_o(\vec{k})\, v_\alpha v_\beta \tau(\vec{k})/[1+\omega^2\tau^2(\vec{k})] \end{cases} \quad (8)$$

this is the usual expression of the diffusion noise source term and of the diffusion coefficient, to be used in long devices, of dimensions larger than $\theta_{max} v_{max}$. For submicron choices, eq.(7) should be taken into account. Figure 4 shows the Fourier transform $S_{j//}(\vec{R},\nu)$ of $C_{j\alpha\beta}(R,\theta)$ given by eq.(7), along the direction R, for various values of the frequency ν, in p-type silicon: the correlation length is equal to a few hundreds angströms. Therefore, space correlation should be taken into account in p-type silicon devices much shorter than 0.1 μm. In GaAs, the mean free path is $\simeq$ 1000Å instead of $\simeq$ 100Å in p-type Si, and furthermore the preeminent scattering mechanisms, which are polar optical scatterings, are not randomizing : this means that several collisions are needed in order to decorrelate the velocities of the carriers : one may then expect correlation lengths of the order of or even larger than 0.1 μm, so that space correlations of noise sources should likely be taken into account for submicron GaAs devices. Of course, these correlations are also involved in hot electron regimes [16].

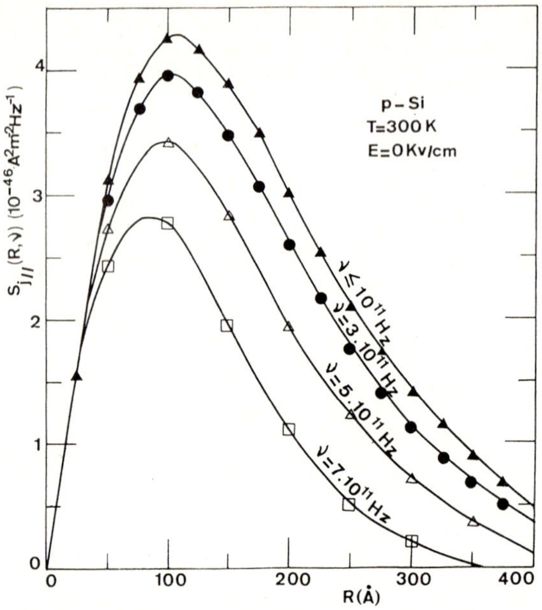

Figure 4 : Noise source term $S_{j//}(R,\nu)$, along the direction R, for various values of ν, in p-type silicon (taken from [15]).

It should be pointed out that the delta function $\delta(\vec{r}-\vec{r}')$, involved in the usual expression, of the correlation $C_j(\vec{r},\vec{r}',\theta)$ or of the noise source $S_j(\vec{r},\vec{r}',\nu)$ (eq.(6)), is an approximation of eq.(7) and never appears as some limiting case of this equation. The reason is that this delta function was empirically stated, and was never demonstrated. Indeed eq.(7) shows that this statement is wrong, although leading to a good approximation for long devices. This can be explained as follows : the presence of the delta function $\delta(\vec{r}'-\vec{r})$ in $C_j(\vec{r},\vec{r}',\theta)$ would mean that the current density $\vec{j}(\vec{r},t)$ at point $\vec{r}$ and time t is correlated with the current density $\vec{j}(\vec{r},t+\theta)$ at the same point and at time θ later. However, all the carriers at point $\vec{r}$ at time t, having a non zero velocity, which contribute to $\vec{j}(\vec{r},t)$, will have gone away from point $\vec{r}$ later on, so that $\vec{j}(\vec{r},t+\theta)$ is due to carriers, different from those giving $\vec{j}(\vec{r},t)$, which are obviously uncorrelated with them : hence $\vec{j}(\vec{r},t)$ and $\vec{j}(\vec{r},t+\theta)$ should not be correlated and one should have $C_j(\vec{r},\vec{r}',\theta)=0$ and $S_j(\vec{r},\vec{r}',\theta)=0$ if $\vec{r}'=\vec{r}$: this result is consistant with eq.(7), but is in contradiction with the usual empirical expression (eq.(6)) of the noise source term.

7. TRIDIMENSIONAL EFFECTS

In most cases, usual devices can be considered as being one dimensional. This does not mean that the conduction is performed along a line, like in some polymers for example, but that the conduction equations depend only on one space variable, x for example. An other way to say this is to consider that the equipotential surfaces are parallel planes. The noise measured at the electrodes, which are generally taken as the bias electrodes, is then given by the impedance field formula of Shockley et al. [17], giving the noise voltage $S_V(\nu)$ at the terminals, through the local noise source $K(x,\nu)$ and the impedance field $\nabla Z(x,\nu)$ (see eq.(4)). This formula can be extended to the three dimensional case [2], and is no more valid when space correlations are taken into account, as they should be in submicron devices. The correct expression is then, when $S_j(\vec{r},\vec{r}',\nu)$ is not proportional to $\delta(\vec{r}-\vec{r}')$ [18] :

$$S_V(\nu)=\iint \vec{\nabla} Z(\vec{r},\nu)\vec{\vec{S}}_j(\vec{r},\vec{r}',\nu)\vec{\nabla} Z(\vec{r}',\nu)d^3rd^3r' \qquad (9)$$

As was shown in section 6 above, $\vec{\vec{S}}_j(\vec{r},\vec{r}',\nu)$ actually depends on $\vec{R}=\vec{r}-\vec{r}'$, and does not depend only on x-x'.

Figure 5 shows that point M is equally correlated with all the points at the same distance, here taken as the mean free path L from M, which are not all at the same abscissa x'. The correlation between M and M' is not equal to the correlation between M and M". As a consequence, eq.(9) cannot be reduced to an integration over a single couple of variables x and x' : one deals with tridimentional effects, even if the equipotential surfaces are parallel planes.

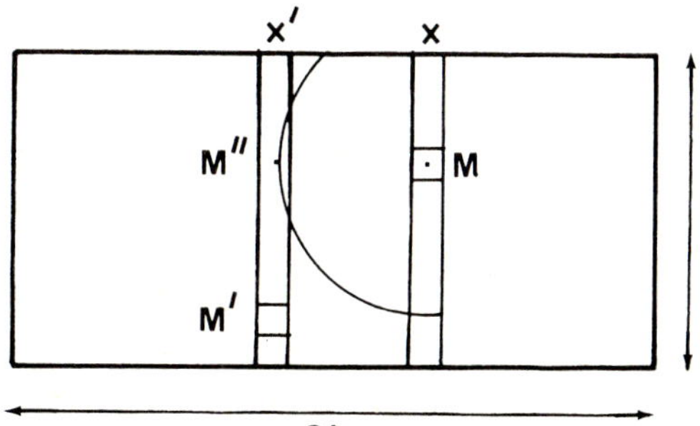

Figure 5 : example of a submicron device, with dimensions of the order of the mean free path L.

8. NON INSTANTANEOUS COLLISIONS

Usually, the collisions are supposed to be instantaneous : this means that the position of a carrier undergoing a scattering mechanism is kept unchanged during the collision, and that the relation between the $\vec{k}$ vectors before and after the collision (and thus the transition rates) does not depend on the electric field. The Boltzmann equation includes this hypothesis.

A few years ago [19] [20], a new formulation was given, taking into account collisions duration, leading to a modified Boltzmann equation, which has not yet been solved.

Indeed as collision is said to be instantaneous if its duration is small compared with the free flight between two collisions. This duration is not easy to define, but may be estimated as follows [21] : at time t=0, an electron, which is in the state $\vec{k}$ of energy ε, undergoes a collision with, say a phonon of energy $\hbar\omega_0$; since then, its state evolves, and at time t the electron is in the stake $\vec{k}'$ of energy ε'. Indeed $\vec{k}'$ and ε' are not known, but one knowns the probability $\mathcal{P}_t(\vec{k},\vec{k}')$ to find the electron in the state $\vec{k}'$ at time t, starting from $\vec{k}$ at time O :

$$\begin{cases} \mathcal{P}_t(\vec{k},\vec{k}') = \dfrac{\pi|V_{kk'}|^2}{\hbar^2} \; t \; \dfrac{\sin^2\alpha t}{\pi\alpha^2 t} \\ \alpha = (\varepsilon'-\varepsilon_f)/2\hbar \\ \varepsilon_f = \varepsilon \pm \hbar\omega_0 \end{cases} \qquad (10)$$

V is the interaction potential between the electron and the phonon. If $t\to\infty$, $\sin^2\alpha t/\pi\alpha^2 t\to\delta(\alpha)$, which means that the energy of the electron tends towards its final value $\varepsilon_f = \varepsilon \pm \hbar\omega_0$ (the + and - signs refer to absorption or emission of the phonon) : strictly speaking, the duration of the collision is infinite. In practice, one may say that the collision is finished when $\alpha \simeq 0$, that is when ε' is close to ε_f. Since in most semiconducting materials $\hbar\omega_0$ is about 50 to 100meV, we may decide that the collision is finished when ε' departs from ε_f by less than, say, 5meV. Therefore, we estimate the duration of the collision as being the time such that the probability to get $|\varepsilon'-\varepsilon_f| > 5$meV becomes negligible, compared with the probability that $|\varepsilon'-\varepsilon_f| \leq 5$meV. Figure 6, taken from [21], shows that this is obtained for $t \simeq 0.5$ps, which is of the same order of magnitude as the free flight duration. Hence a collision may not be considered as being instantaneous. The two main consequences are that:

* the processes are not Markovian : the next state depends both on the present state and on the history,
* many body problems are involved, since one collision begins before the previous one was finished.

Of course this effect does not depend on the size of the device, but one may expect its influence to be more important in submicron devices than in long devices, since then each carrier undergoes only few collisions.

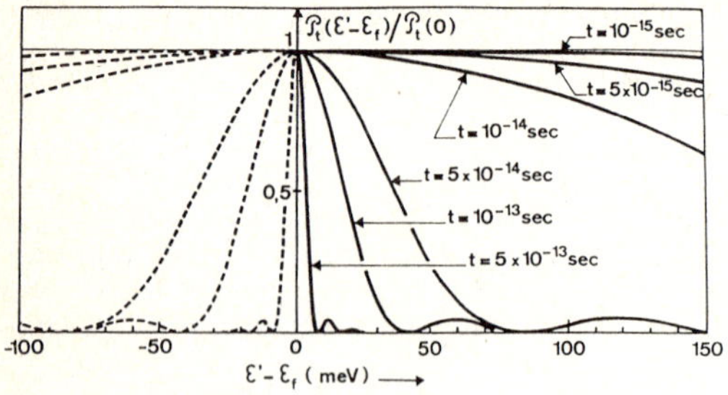

Figure 6 : Transition probabilities, versus energy $\varepsilon'-\varepsilon_f$, at various times (after [21]).

9. DISCRETIZATION

Till now, devices were considered to be continuous media. This may be no longer true in small systems. The noise depends then widely on the detailed structure of the device. This was clearly shown by the study of the noise in hopping process through a limited number of potential wells [22] supposed to represent the motion of ions through pores. Independently from the fact that this model is or not realistic as applied to this specific problem, the noise spectrum of the current was shown to depend much strongly on the detailed variation of the potential V(x). Figure 7 shows that, modifying only the height of the central potential barrier, gives quite different shapes of the noise current spectrum.

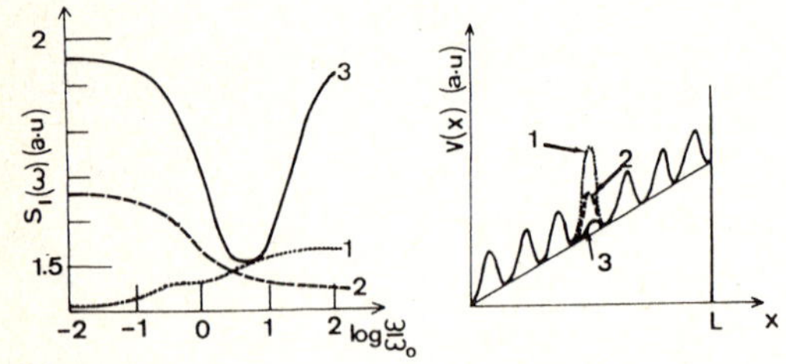

Figure 7 : Noise current spectrum in hopping process, for three values of the central potential barrier height.

One would expect that tunnel emission of electrons through a barrier containing traps would lead to similar results.
Also it is well recognized that defects or radiation changes may have drastic effects in submicron devices, although they may have minor importance in usual devices.
Another typical example of discretization due to dimension scaling down is that of thin films, inversion layers, and superlattices, which are quantized systems where the theory of continuous media obviously do not apply.

10. INFLUENCE OF STATISTICS

Finally one would point out that one usually deals with Gaussian statistics due to the large amount of electrons involved in devices, and one could think that scaling down the size would lead to electron number so low that Gaussian statistics no longer applies. Indeed, apart from exceptionally scarse cases, this is not expected to happen in the next future in devices, as is shown by the following simple reasoning : the d.c. current flowing through the device is : I=envA=evN/L, so that the total number of conducting electrons is N=IL/eV. The d.c. current will be a few milliamps, while the drift velocity is $\simeq 10^7$ cm/s in short devices ; hence, with I=1.6mA, one gets N < 30 for L < 3Å, we are far from such device lengths !

11. CONCLUSION

Throughout this paper, one could see that much more problems remain to be solved than have been solved till now : a lot of work remains to be done :
* experimentaly, one should make and study submicron devices : diodes [8], field effect transistors, etc...
* theoretically, one should improve and even reformulate device modeling concepts and compare these new theoretical models, suited for submicron devices, with experimental results. For this purpose it is necessary to give correct expressions of the noise sources, taking into account space correlations [15] [16] for various mechanisms, and to reformulate the impedance field method, then to make simulations by solving coupled Poisson and transport equations : this was shown to be possible either by Monte Carlo techniques [23], or by using balance equations [14] [24], which are less rigorous but much faster techniques.

REFERENCES

[1] A. Van der Ziel,
1/f, diffusion and thermal noise in GaAs devices, Proc. 7th. Int. Conf. Noise in Physical Systems, invited paper, Montpellier (France) (1983).

[2] J.P. Nougier,
Proc. 6th. Int. Conf. Noise in Physical Systems, Washington (USA), 1981, National Bureau of Standards, Special Publication n° 614, p.397 (1981).

[3] A. Gisolf and R.J.J. Zijlstra,
Solid St. Electron 17, 839 (1974).

[4] G. Bosman and R.J.J. Zijlstra, Phys. Lett. 71A, 464 (1979).

[5] J.P. Nougier, D. Gasquet, J.C. Vaissière and H.R. Bilger. Proc. 5th. Int. Conf. Noise Physical Systems, Springer Series in Electrophysics 2, Springer Verlag ed., p.110 (1978).

[6] D. Gasquet, M. Fadel, J.P. Nougier Noise of hot electrons in indium phosphide. Proc. 7th. Int. Conf. Noise in Physical Systems, Montpellier (France) (1983).

[7] J.P. Nougier, J.C. Vaissière and D. Gasquet, Proc. 6th. Conf. Noise in Physical Systems, Washington (USA), 1981, National Bureau of Standards, Special Publication n° 614, p. 42 (1981).

[8] R.R. Schmidt, G. Bosman, C.M. Van Vliet, A. Van der Ziel, L.F. Eastman and M. Hollis. Noise in near ballistic n^+nn^+ and n^+pn^+ GaAs submicron diodes. Proc. Int. Conferences on Noise, Montpellier (1983).

[9] P. Hesto, J.C. Vaissière, D. Gasquet, R. Castagné, and J.P. Nougier J. Phys. (Paris), C7, 235 (1981).

[10] J.P. Nougier, J.C. Vaissière, D. Gasquet, J. Zimmermann and E. Constant. J. Appl. Phys. 52, 825 (1981).

[11] M. Shur Solid St. Electron 20, 389 (1977).

[12] L. Reggiani, J.C. Vaissière, J.P. Nougier, and D. Gasquet. J. Phys. (Paris), C7, 357 (1981).

[13] B. Carnez, A. Carpy, A. Kaszinski, E. Constant, and G. Salmer. J. Appl. Phys. 51, (1980).

[14] B. Carnez, A. Cappy, R. Fauquembergue, E. Constant and G. Salmer. IEEE Trans. El. Dev. ED 28, 784 (1981).

[15] J.P. Nougier, C. Gontrand, and J.C. Vaissière, Microscopic spatial correlation at thermal equilibrium in non polar semiconductors. Int. Noise Conf. Montpellier (1983).

[16] J.P. Nougier, J.C. Vaissière, and C. Gontrand to be published.

[17] W. Shockley, J.A. Copeland and R.R. James, in Quantum theory of atoms, molecules and the solid state. P. Lowdin ed., Acad. Press, New York, (1966).

[18] J.P. Nougier, J.C. Vaissière and C. Gontrand to be published.

[19] J.R. Barker, Sol. State Electron 21, 197 (1978).

[20] D.K. Ferry, and J.R. Barker, Solid State Commun. 30, 361 (1979).

[21] J.P. Nougier, J.C. Vaissière, and D. Gasquet J. Phys. (Paris), C7, 283 (1981).

[22] E. Frehland. Biophys. Struct. Mech. 7, 1 (1980).

[23] B. Boittiaux, E. Constant, and A. Ghis. Simulation of diffusion noise in a device. Proc. Noise Conf. Montpellier (1983).

[24] E. Allamando, G. Salmer, E. Constant and B. Carnez. A new noise model of submicrometer dual gate MESFET. Proc. Noise Conf., Montpellier, (1983).

NOISE IN PHYSICAL SYSTEMS AND 1/f NOISE
M. Savelli, G. Lecoy and J-P. Nougier (eds.)
© Elsevier Science Publishers B.V., 1983

CURRENT NOISE IN MULTIVALLEY SEMICONDUCTORS*

A. Chatterjee and P. Das

Electrical, Computer, and Systems Engineering Department
Rensselaer Polytechnic Institute
Troy, New York 12181

Current noise in multivalley semiconductors can be attributed to fluctuation of the valley populations (intervalley noise) in addition to velocity fluctuations of carriers within the valley (Nyquist noise). Intervalley noise for GaAs is calculated using a two valley model and the Nyquist noise is obtained from the linearized transport equations. The intervalley component is significant under hot electron conditions, indeed it dominates the noise spectrum at low frequencies (< 1 GHz). Field dependence of the noise is studied and the maxima in the negative differential resistance region is discussed for both low and high frequency noise. A comparison is made between features of noise in GaAs and silicon. Calculations on the small-signal microwave mobility and reflection coefficient are also presented. The effect of finite collision duration was also included. However, the results were insensitive to the collision duration time.

*Partially supported by NSF Grant No. ECS-82-19070

I. INTRODUCTION

Consider a conductor of cross-sectional area A and length L maintained in thermal equilibrium at a temperature T. The random motion of charge carriers caused by thermal agitation results in current fluctuations, whose spectrum is given by Nyquist's formula

$$S_I(\omega) = 2nek_BT(A/L)\mathrm{Re}\{\mu(\omega)\} , \qquad (1)$$

where n is electron density, e is electronic charge, k_B is Boltzmann's constant and μ is the complex small-signal mobility. As a bias electric field is applied, the electron gas is energized and thermalizes to an effective temperature different from the ambient lattice temperature. For high carrier concentrations, typical of electronic devices, the energy is shared by electron-electron interactions so that the distribution function is well approximated by a displaced Maxwellian:

$$f(\vec{k}) = C \exp[-\varepsilon(\vec{k}-\vec{k}_d)/k_BT_e] , \qquad (2)$$

where $\vec{k}$ is the wavevector of an electron, ε is energy, $\vec{k}_d$ is the displacement due to drift, and T_e is effective temperature. In a single valley semiconductor or if there are many equivalent valleys with spherical energy surfaces, all the valleys are heated to the same temperature. In this case, if we make a linear approximation to the nonlinear conductor about a bias electric field, the noise spectrum is given by Nyquist's formula with T_e as the noise temperature [1]. Thus the macroscopic current response is related to the dynamics of microscopic current fluctuations, thereby establishing a unique connection between the small-signal mobility, velocity autocovariance and velocity overshoot [2]. The possibility of experimental determination of relaxation times and collision duration time from hot electron noise measurements has also been explored [3].

The conduction band of GaAs has a light mass central valley and heavy mass satellite valleys at 0.4 eV above the conduction band minimum and at high electric fields the electron distribution is characterized by distinct electron temperatures for the lower and upper valleys. Fluctuation of the valley populations (due to random transitions of electrons between valleys) would cause the average mobility and hence the conductivity to fluctuate resulting in electrical noise when a net current is flowing. The intervalley noise is conceptually similar to generation-recombination noise where transitions are between the band and trap levels [4]. Presence of intervalley noise precludes a simple application of Nyquist's theorem to calculate the noise spectrum for GaAs [2,5-7]. In the next section we present the nonlinear balance equations and the linearization scheme to obtain small-signal microwave mobility, reflection coefficient and Nyquist noise. Section 3 exhibits and analyzes the noise spectrum and its field dependence. Other applications of intervalley noise and a comparison between noise in silicon and GaAs is given in the concluding section.

2. SMALL-SIGNAL ANALYSIS AND NYQUIST NOISE

A hierarchy of balance equations can be generated from the Boltzmann transport equation by multiplying it by an arbitrary function $\phi(\vec{k})$ and integrating over $\vec{k}$ assuming a Maxwellian distribution in each valley. The populations, drift velocities and effective temperatures of each valley are evaluated from the number, momentum and energy balance equations obtained by choosing $\phi(\vec{k})$ equal to unity, $\hbar\vec{k}$ and $\frac{\hbar^2k^2}{2m}$ (parabolic band with effective mass m), respectively. The nonlinear balance equations for a homogeneous material are [8,9]

$$\frac{\partial n_i}{\partial t} + n_i\Gamma_{ni}(T_i) - n_j\Gamma_{nj}(T_i) = 0 \quad \text{(number balance)} \tag{3}$$

$$m_i \frac{\partial v_i}{\partial t} + m_i v_i \Gamma_{mi}(T_i) = eF \quad \text{(momentum balance)} \tag{4}$$

and

$$\frac{3}{2}\frac{\partial}{\partial t}\left(k_B T_i\right) + k_B T_i \Gamma_{ei}(T_i) - \frac{n_j}{n_i} k_B T_j \Gamma_{ij}(T_j) - ev_i F = 0 \quad \text{(energy balance)} \tag{5}$$

Here n_i, v_i and T_i denote, respectively, the population, drift velocity and effective temperature of the ith valley and the Γ's are the various relaxation rates. The subscripts would run as i = 1,2 and j = 2,1 to generate a total of six equations for a two valley semiconductor. If the applied electric field F(t) is a small a.c. field $F_1 e^{j\omega t}$ superimposed on a large d.c. field F_o, a first order perturbation analysis is valid. This gives linearized balance equations which can be solved for the first order response, once the zeroth order solutions, n_{io}, v_{io} and T_{io}, are known. The linearized balanced equations are [1]:

$$(j\omega + \Gamma_{nio})n_{i1} - \Gamma_{njo}\, n_{j1} + n_{io}\Gamma'_{ni}T_{i1} - n_{jo}\Gamma'_{nj}T_{j1} = 0 \tag{6}$$

$$(j\omega + \Gamma_{mio})v_{i1} + v_{io}\Gamma'_{mi}\, T_{i1} = eF_1/m_i \tag{7}$$

and

$$\left(\frac{3}{2} j\omega + \Gamma_{eio} + T_{io}\Gamma'_{ei}\right) k_B T_{i1} - \frac{n_{jo}}{n_{io}}(\Gamma_{ijo} + T_{jo}\Gamma'_{ij})k_B T_{j1} - \frac{n_{jo}}{n_{io}}\left(\frac{n_{j1}}{n_{jo}} - \frac{n_{i1}}{n_{io}}\right) k_B T_{jo}\Gamma_{ijo} - ev_{i1}F_o = ev_{io}F_1 , \tag{8}$$

where the prime indicates a derivative with respect to temperature evaluated at the bias value. To include non-zero collision duration Γ's are to be replaced by $\Gamma/(1+j\omega\tau_c)$ with τ_c as the collision duration time [10].

The small-signal mobility,

$$\mu(\omega) = \frac{\sum_i n_{io}v_{i1}}{F_1 \sum_i n_{io}} \tag{9}$$

is readily obtained by solving equations (6)-(8). Figures 1 and 2 show the real part of the complex small-signal mobility plotted for different frequencies and different bias fields in GaAs*. The Gunn effect or negative differential conductivity due to a field induced transfer of electrons from the fast central valley to the slow upper valley is apparent. Since the speed of response to an oscillating field is limited by the relaxation rates, it is expected that there would be a high frequency limit to the Gunn effect [11]. We see that there is no negative differential resistance (ndr) for frequencies beyond, approximately, the population relaxation rate Γ_n. The reflectivity of a plane wave depends on the small-signal conductivity and is a useful experimental tool to probe hot electron transport in semiconductors [12]. Reflectivity is given by [10]

$$|R| = \left|\frac{\kappa^{1/2} - 1}{\kappa^{1/2} + 1}\right| , \tag{10}$$

where
$$\kappa = \kappa_s\left(1 - j\,\frac{ne\mu(\omega)}{\omega\varepsilon_o\kappa_s}\right) , \tag{11}$$

κ_s is the dielectric constant of the semiconductor and ε_o is the permittivity of free space. For perpendicular polarization (that is the oscillating field normal to the bias field) the electron temperatures are constant to a first order. The small-signal mobility, in this case, obtained by setting the T_{i1}'s to zero in the balance equations, is different from that for parallel polarization. In Figure 3 we have plotted reflectivity in GaAs for parallel polarization. The plasma edge is not strongly dependent on electric field and the shift observed in silicon [10], in the non-zero collision duration regime, is not seen in GaAs.

The mean time between intervalley transitions is long compared to the time scale on which electron's velocity fluctuates within a valley. Thus, intravalley fluctuations are essentially uncorrelated to the relatively slow fluctuation of the number of electrons in the valley. Consequently, the current noise spectrum can be written as [7] the sum of the thermal noise, where the valley populations are constant, and intervalley noise, where all electrons in a valley move with a constant velocity equal to the corresponding drift velocity. To calculate the thermal noise spectrum we, therefore, set the perturbation in the valley population, n_{i1}, to zero in the balance equations and use Nyquist's formula to get

$$[S_I(\omega)]_{th} = (2eA/L) \sum_i n_{io}k_B T_{io}\, \mathrm{Re}(\mu_i(\omega))_{n_i=n_{io}} \tag{12}$$

Although, in the hot electron regime, Nyquist noise has polarized spectral features, the total noise spectrum is dominated by the isotropic intervalley noise for most regions of interest.

*using the constants given in references (9) and (14)

3. THE NOISE SPECTRUM

The number fluctuation is modeled as that of a two level system [4] and the corresponding electrical noise spectrum is [7]

$$[S_I(\omega)]_{int} = (2e^2AF^2/L)(\mu_1-\mu_2)^2 \frac{n_1n_2}{(n_1+n_2)}$$

$$\times \frac{\tau_{eff}}{(1+\omega^2\tau_{eff}^2)} \ . \qquad (13)$$

To determine intervalley noise spectral density the semiconductor is considered to be otherwise noiseless, that is, the electron velocity is the d.c. drift velocity v_{io}. Hence, μ_1 and μ_2 in (13) are the corresponding chordal mobilities in valley 1 and 2. The characteristic relaxation time, τ_{eff}, is the effective time constant with which the valley populations relax when disturbed from their equilibrium values. From equation (3), noting that $n_1 + n_2 = n$, constant, we can identify

$$1/\tau_{eff} = \Gamma_{n10} + \Gamma_{n20} \ . \qquad (14)$$

Intervalley noise is a nonequilibrium phenomenon since it is proportional to F^2 and, like generation-recombination noise, it appears only when a net current is flowing through the conductor. Also the noise power becomes weaker as the difference in mobilities get small, reducing to the single valley case when $\mu_1 = \mu_2$. In Figure 4 the low frequency ($f << 1/\tau_{eff}$) noise spectral density is plotted as a function of the bias electric field. For low fields almost all the electrons are in the lower valley whereas for extremely high fields most of the electrons are transferred to the upper valley. In both cases one valley is almost empty and hence intervalley noise is small. For intermediate fields $n_1 \approx n_2$ making the product n_1n_2 large with a consequent dramatic rise in the noise power in the ndr region. Figure 5 shows the spectral density plotted against frequency. Since τ_{eff} is large compared to momentum and energy relaxation times, the high frequency noise ($f >> 1/\tau_{eff}$) is predominantly thermal noise. Figure 6 shows plots of noise spectral density vs. electric field for different frequencies. The peak in the high frequency noise in the ndr region is due to two competing effects. As the field is increased, the electron temperatures rise tending to increase thermal noise. On the other hand, the mobility falls, due to population transfer and velocity saturation, tending to reduce current fluctuations.

4. CONCLUSION

On comparing these results with earlier calculations [2,3] on hot electron noise in silicon, we observe certain pertinent differences. In GaAs the mechanism primarily responsible for velocity overshoot is repopulation. Although there is enhancement of the conductivity at high frequencies for each valley, the real part of the mobility calculated by forcing all changes in the valley populations to zero, has a flat spectrum. Thus, unlike silicon, there is no peak in the noise spectral density vs. frequency. The noise is mostly isotropic, whereas in silicon the polarized spectral features are significant. The Kramer-Kronig relations between the real and imaginary parts of conductivity allow us to relate the small-signal step response of velocity to noise spectral density in a simple and direct way when the noise is given by Nyquist's formuala [2]. Thus, in GaAs, there is no simple relationship between velocity overshoot and the noise spectrum.

Ellipsoidal constant energy surfaces near the conduction band minima in silicon and germanium cause unequal heating of the valleys which, ultimately, is responsible for the Sasaki effect [13]. Thus, intervalley noise would be observed with the electric field applied in an arbitrary direction in silicon. By rotating the field in different crystal directions this intervalley noise should change giving rise to a noise analogue of the Sasaki effect. Population fluctuation could also be important in quantized inversion layers, adversely affecting the noise performance of small geometry FETs.

REFERENCES

[1] P. Das, D.K. Ferry and H. Grubin, J. Physique 42, C7-227 (1981).

[2] A. Chatterjee and P. Das, Solid-St. Electron. 26, 227 (1983).

[3] A. Chatterjee and P. Das, J. Appl. Phys. 53, 5289 (1982).

[4] K.M. van Vliet and J.R. Fassett, Fluctuation Phenomena in Solids, R.E. Burgess Ed., Academic Press (1965).

[5] P.J. Price, J. Appl. Phys. 31, 949 (1960).

[6] W. Shockley, J. Copeland and R. James, Quantum Theory of Atoms, Molecules and Solid State, pp. 537-563, Academic Press (1966).

[7] P.H. Handel and A. van der Ziel, Solid-St. Electron. 25, 541 (1982).

[8] J.S. Moore and P. Das, J. Appl. Phys. 50, 8082 (1979).

[9] J.S. Moore, Ph.D. Thesis, Rensselaer Polytechnic Institute, Troy, NY (1981).

[10] P. Das, D.K. Ferry and H. Grubin, Solid-St. Comm. 38, 537 (1981).

[11] P. Das and R. Bharat, Appl. Phys. Lett. 11, 386 (1967).

[12] S.J. Allen, D.C. Tsui, F. Derosa, K. Thornbar and B.A. Wilson, J. Physique 42, C7-369 (1981).

[13] E.M. Conwell, High Field Transport in Semiconductors, Academic Press (1967).

[14] M.A. Littlejohn, J.R. Hauser and T.H.Glisson, J. Appl. Phys. 48, 4587 (1977).

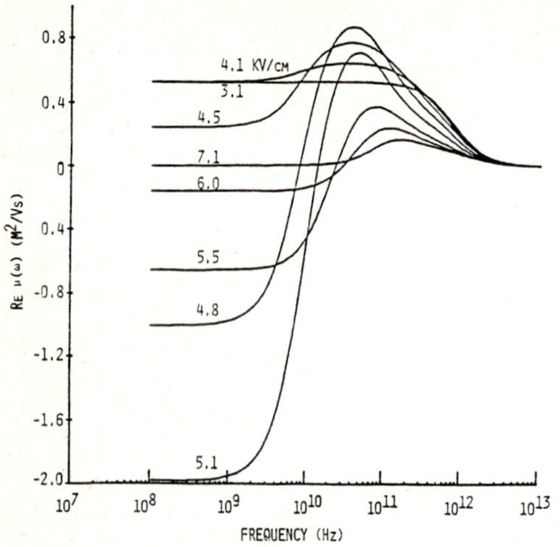

Fig. 1 The real part of small-signal mobility vs. frequency in GaAs.

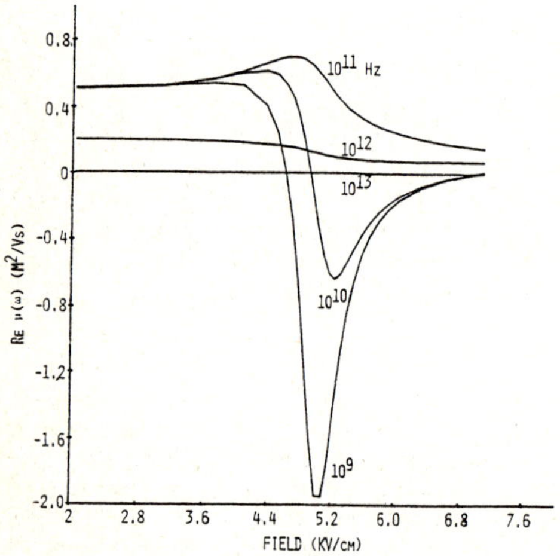

Fig. 2 The real part of small-signal mobility vs. electric field in GaAs.

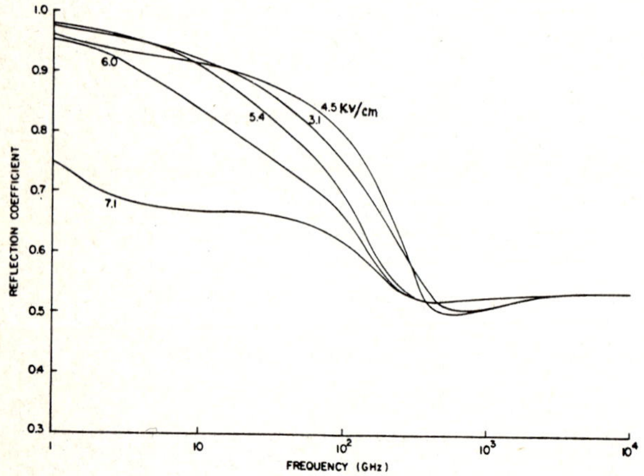

Fig. 3 Reflectivity vs frequency in GaAs for parallel polarization.

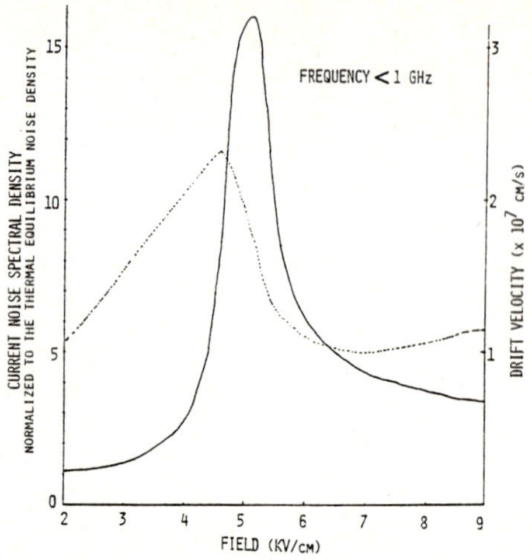

Fig. 4 Low frequency noise vs electric field in GaAs. The dotted curve shows the drift velocity.

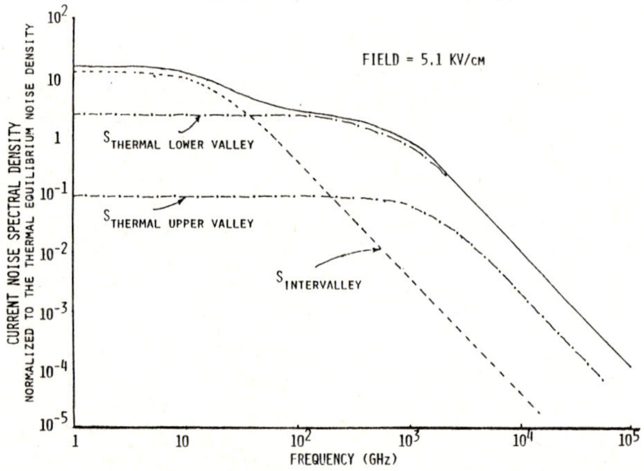

Fig. 5 Current noise spectral density in GaAs with a bias field corresponding to the ndr region.

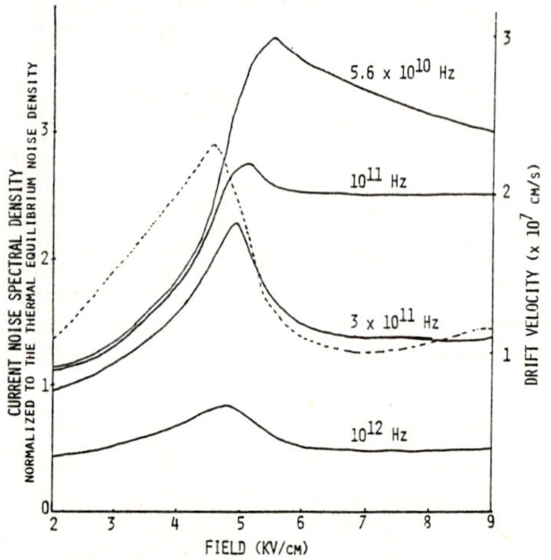

Fig. 6 Intermediate and high frequency noise vs electric field in GaAs. The dotted curve shows the drift velocity.

NOISE IN PHYSICAL SYSTEMS AND 1/f NOISE
M. Savelli, G. Lecoy and J-P. Nougier (eds.)
© Elsevier Science Publishers B.V., 1983

DIFFUSION AND GENERATION RECOMBINATION OF HOT ELECTRONS IN SILICON AT 77K

D. Gasquet, H. Tijani, J.P. Nougier, A. Van der Ziel

Centre d'Electronique de Montpellier - Université des Sciences et Techniques du Languedoc
Pl. E. Bataillon - 34060 Montpellier-Cedex - France

University of Minnesota, E.E. Dept., 123 Church Street S.E.
Minneapolis MN 55455 - U.S.A.

The excess noise previously observed for hot electrons in silicon à 77K is investigated versus frequency and electric field strength. This allows a separate experimental characterization of the diffusion noise and of generation recombination (G.R) noise. One is then able to determine experimentally, versus the electric field, the G.R. lifetime τ(E), the free carrier density n(E) and the diffusion coefficient D(E).

1. INTRODUCTION

A large excess noise of hot electrons was observed in n type silicon at 77K in bars and FETs, and was accounted for by generation recombination (G.R.) due to Poole Frenkel effect [1]. The purpose of the present work is to investigate experimentally the diffusion and the G.R. contribution to this noise. Since the cut-off frequency of the observed noise is higher than 1GHz, we performed measurements at 10.5GHz, using a pulsed technique described previously [2].

2. EXPERIMENTAL RESULTS

We measured the low frequency I-V characteristic, thus giving $\sigma(E)=j/E=(I/A)$ / (V/L) and the differential conductivity $\sigma'(E)=dj/dE$, versus the electric field (E//<111>), so as the longitudinal excess noise temperature $T_n(E)-T_o$ at frequencies 220MHz, 350MHz, 460MHz, 650MHz, 850MHz, 10.5GHz (T_o is the lattice temperature, T_o = 77K. In order to avoid Joule heating, the measurements were performed during pulses (as short as 500ns at higher fields, 10kV/cm at 300K and 4kV/cm at 77K). Unfortunately we had no available measurements between 1 and 10.5GHz and we are now building an apparatus at 4GHz for further experiments.The samples used were bars of n-type silicon 0.5mm length $N_D=2.75\times10^{14}$ cm^{-3} , with ohmic contacts at the ends obtained by metallic deposition af Ag over Ag - Si eutectic described prevlously [3].
Due to the great length the field is uniform in the sample.
Figure 1 shows results obtained at 295K : the noise is white up to 10kV/cm, only hot electron diffusion noise is present.
Figure 2 shows the chordal longitudinal relative conductivity $\sigma(E)/\sigma_o$ measured at 77K.

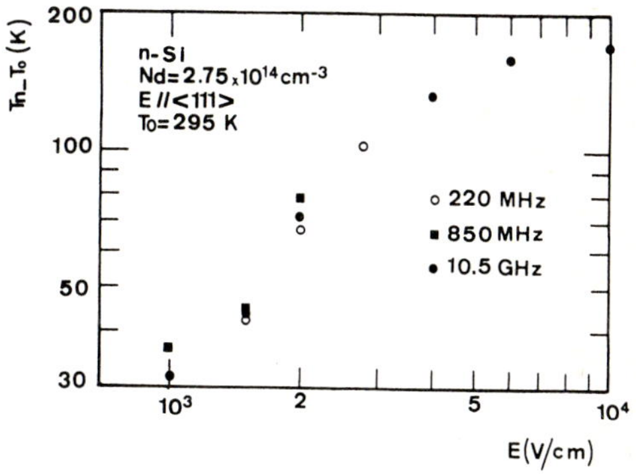

Figure 1 : Excess noise temperature versus electric field in n-type Si, at 300K.

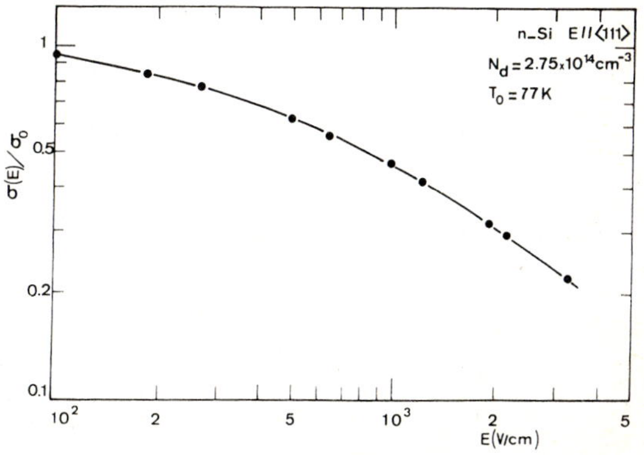

Figure 2 : Low frequency chordal relative conductivity versus electric field in n-type Si , at 77K.

Figure 3 gives, at 77K, the excess noise temperature $T_n(E) - T_o$, versus E, at the frequencies indicated above : obviously the noise is not white, indicating that, contrarily to what happens at 300K, an extra noise source is superimposed to diffusion.

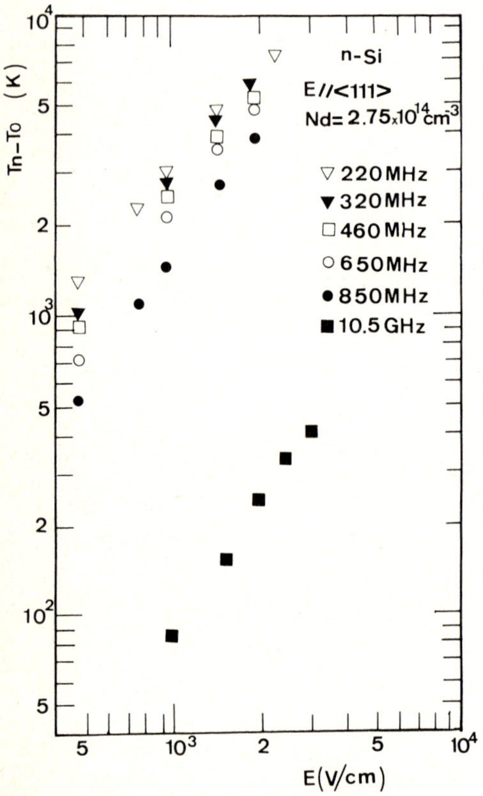

Figure 3 : Excess longitudinal noise temperature versus electric field at 77K.

3. INTERPRETATION

At 77K only part of the impurities are ionized [4] ; when an electric field is applied, the free carriers become more energetic and ionize part of the neutral impurities either by impact ionization or by Poole Frenkel effect [1] : this results in electron transfers between the impurity level and the conduction band, thus producing G.R. noise. The total noise $S_I(f)$ is then the sum of the diffusion noise S_{Idiff} and of the G.R. noise S_{IGR}, that is for a two level process [6] [1] [5]:

$$S_{Idiff} = \frac{4q^2AN_D}{L} u\, D(E) \quad (1)$$

$$S_{IGR} = \frac{4I^2}{N_DAL} \frac{\gamma(1-u)}{[\gamma^2(2-u)^2 + \omega^2u^2]} \quad (2)$$

$$u = u(E) = n(E)/N_D \quad (3)$$

$$\gamma = \gamma(E) = u(E)/[\tau(E)(2-u(E))] \quad (4)$$

n(E) is the density of free electrons (u is thus the fraction of ionized impurities), D(E) is the diffusion coefficient, and $\tau(E)$ is the G.R. lifetime. L and A are the sample length and cross section, I is the bias current, N_D the impurity concentration, $q = 1.6 \times 10^{-19}$ Cb.

The procedure is then the following. At a given electric field strength E, one measures the I-V characteristic, which gives :

$\sigma(E) = (I/A)/(V/L)$ (see fig.2), and the excess noise temperature $T_n - T_o$ versus frequency (see fig. 3), which gives :

$$S_I = S_{Idiff} + S_{IGR} = 4k_BT_n / (dV/dI) \quad (5)$$

Note that the static characteristic I-V can be used in eq.(5) at the frequencies studied. The three unknown parameters γ, u and D are determined by a least square fitting of $S_I(f) = S_{Idiff} + S_{IGR}$, given by eq.(1) and (2), to the experimental results determined by eq.(5); n and τ are then deduced from eqs.(3) and (4). Finally, from the measurement of $\sigma(E)$, one gets the mobility $\mu(E) = \sigma(E)/qN_D\, u(E)$, and thus v(E).

Figure 4 shows the variation of $u(E) = n(E)/N_D$ versus E : at thermal equilibrium, 75% of the impurities are found to be ionized, which is in reasonnable agreement with the value 82% given in the litterature [4] for the same impurity concentration.

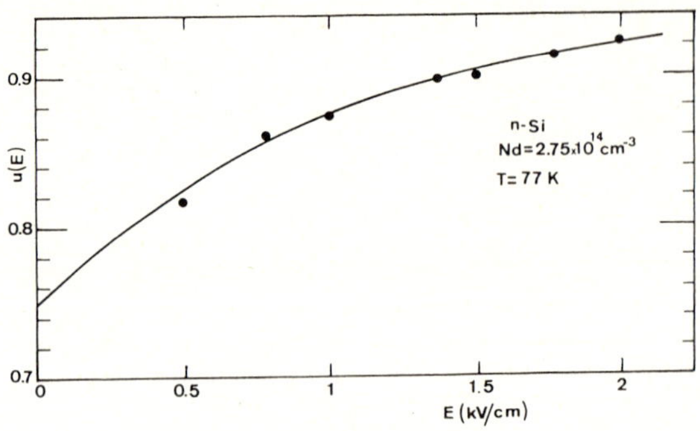

Figure 4 : fraction of ionized impurities versus electric field deduced from experiment (figs. 2 and 3).

From the ohmic resistance $R_c = 23\,\Omega$ and the dimensions of the sample (L=510 μm, A=0.38mm²), one gets the ohmic mobility μ_o=17 500cm²/Vs), in excellent agreement with previous results [4].
On figure 5 are plotted the G.R. liftime versus E, as well as the cut-off frequency :
$f_c(E) = 1/2\pi\tau(E)$: $\tau(E)$ is a decreasing function of E, and is very short since it is in the range 0.1-0.5ns, while the cut-off frequency is about one GHz, as can also be seen on fig.3.

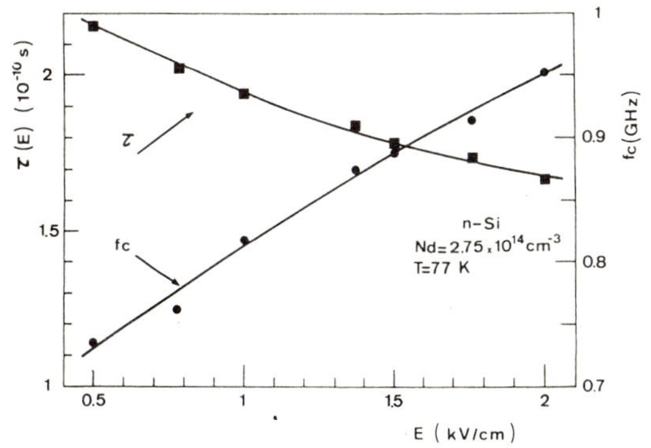

Fig.5 : GR lifetime $\tau(E)$ and cut-off frequency f_c (E) versus E.

Finally we report on fig.6 the variation of the diffusion coefficient $D(E)/D_o$ versus the electric field, D_o being given by the Einstein relation $k_BT/q = D_o/\mu_o$ which gives D_o=116cm²/s.

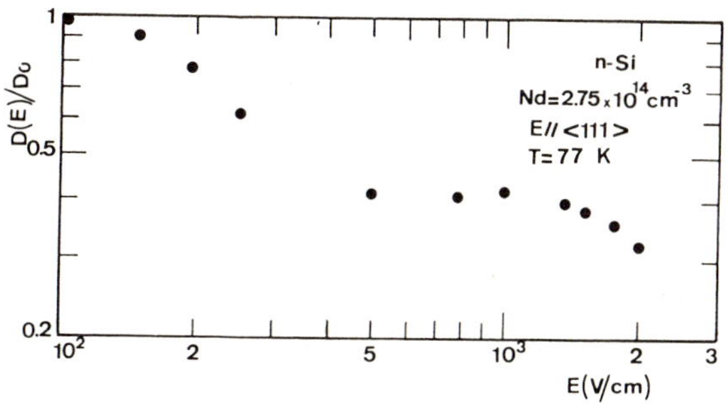

Figure 6 : Longitudinal diffusion coefficient versus electric field.

4. CONCLUSION

This paper brings, for the first time, the experimental proof that part of the noise of hot electrons in silicon at 77K is due to generation recombination induced by the electric field. We could determine the contribution of G.R. noise and diffusion noise to the hot electron noise, and derive the main transport parameters versus the electric field. In particular, we report here the first determination of the proportion of ionized impurities $u(E)=n(E)/N_D$ versus the electric field, and we could derive the true diffusion coefficient D(E), which is different from the value one could get by low frequency (< 1GHz) measurements, since when G.R. noise occurs, of course the importance of G.R. noise increases with increasing impurity concentration, but it should be noted that even when almost all the impurities are ionized, G.R. noise can be present [7] and produce a non negligible error.

REFERENCES

[1] A. Van der Ziel, R. Jindal, S.K. Kim, H. Park, J.P. Nougier, Solid State Electron. 22, 177 (1979).

[2] D. Gasquet, J.C. Vaissière, J.P. Nougier, Proc. 6th Int. Conf. Noise in Phys. Systems, Washington, USA, National Bureau of Standards, Special publication n° 614, p.305 (1981).

[3] M. Rolland, J.P. Nougier, D. Gasquet, R. Alabedra, Solid State Electron. 20, 323 (1977).

[4] J.C. Nash and J.W. Holm-Kennedy, Phys. Rev. B16, 2834 (1976).

[5] J.P. Nougier, Proc. 6th. Int. Conf. Noise in Physical Systems, Washington, USA, National Bureau of Standards, Special publication n° 614, p. 397 (1981).

[6] A. Van der Ziel, Noise sources, characterization, measurement, Prentice Hall, Englewood Cliffs, New Jersey (1970).

[7] D. Gasquet, M. Fadel, J.P. Nougier, Hot Carriers noise in n-type InP, 7th. Int. Conf. on Noise in Physical Systems, Montpellier, 17-20 may 1983.

NOISE IN PHYSICAL SYSTEMS AND 1/f NOISE
M. Savelli, G. Lecoy and J-P. Nougier (eds.)
© Elsevier Science Publishers B.V., 1983

NOISE OF HOT ELECTRONS IN INDIUM PHOSPHIDE

D. Gasquet, M. Fadel, J.P. Nougier

Centre d'Electronique de Montpellier
Université des Sciences et Techniques du Languedoc, Pl. E. Bataillon
34060 Montpellier-Cedex, France*

Noise measurements in InP at 300K exhibit diffusion noise and G.R. noise. The GR noise involves less than 1% of the carriers, is electric field independent, its cut-off frequency is 500MHz (τ=3.2ps). The diffusion coefficient D(E) of hot electrons is for the first time experimentally determined and is in very good agreement with Monte Carlo simulation.

1. INTRODUCTION

The cut-off frequency of devices is increased by decreasing their length and/or by increasing the mobility of the semiconducting material which they are made of. In that respect, indium phosphide is a promising material.

The purpose of this paper is to study the noise of n-type InP, at 300K, as a function of the electric field, in order to get hot electron transport coefficients and especially the diffusion coefficient D(E), which had not yet been obtained experimentally till now.

Noise measurements were performed, using a method previously described [1], between 220 and 850MHz, and at 10.5GHz.

The sample used were mesa n^+nn^+ diodes, provided by the Laboratoire Central de Recherches, Thomson CSF, France. An experimental active layer of 5 μm, $N_D=N_D^o=2.7\times10^{15}\,cm^{-3}$, $\mu_o = 4300cm^2/Vs$, $\rho = 0.51\Omega\,cm$, was grown on a n^+ doped wafer, $N_D = N_D^+ = 5\times10^{17}\,cm^{-3}$. A second epitaxial layer was then grown, n^+ doped, $N_D = N_D^+ = 5\times10^{17}\,cm^{-3}$. The mesa diodes were obtained by etching, and ohmic contacts were attached at both ends. The diodes we studied had comparable behaviours. Because of these low resistances (between 1.3 and 3.6Ω according to their area), the experiments had to be performed in pulses of 500ns duration, at a repetition rate of 1Hz to 0.1Hz according to the electric field intensity, in order to avoid joule heating.

2. MODELING OF THE DIODE

Because of the shortness of the diode (5 μm), in spite of its relatively low resistivity (0.51 Ωcm), it may happen that an important space charge is injected, resulting in a non uniform electric field. It is then necessary to determine the electric field map inside the diode. This was performed by numerically solving the coupled one dimensional conduction and Poisson's equations :

$$\begin{cases} j = -qn(x)\, v[E(x)] & (1) \\ \dfrac{dE(x)}{dx} = -\dfrac{q}{\varepsilon}\,[n(x) - N_D(x)] & (2) \end{cases}$$

where $N_D(x)$ is the doping profile, ε is the dielectric constant. The diffusion current and the displacement current were not taken into account, since a similar simulation performed on n-type silicon [2] proved their effect to be negligible. Non stationnary effect were also neglected. Deducing n(x) from eq.(2), and carrying it into eq.(1), gives :

$$j = v[E(x)]\, dE(x)/dx - qN_D(x)\, v[E(x)] \qquad (3)$$

In order to reduce the computation time, $N_D(x)$ and v(E) where approximated by analytic functions: $N_D(x)$ was deduced from erfc functions :

$$\frac{dN_D(x)}{dx} = \frac{N_D^+-N_D^o}{\sqrt{\pi}}\,\exp[-\alpha^2(L-\beta-x)^2]-\exp[-\alpha^2(x-\beta)^2] \qquad (4)$$

which gives :

$$N_D(x)=N_D^o+\frac{N_D^+-N_D^o}{2}\left\{\mathrm{erfc}[\alpha(x-\beta)]+\mathrm{erfc}[\alpha(L-\beta-x)]\right\} \qquad (5)$$

β is such that $N_D(\beta) = (N_D^+ + N_D^o)/2$, α determines the depth of the n^+n transition region. We used the following values : $N_D^+ = 2\times10^{17}\,cm^{-3}$; $N_D^o=2.7\times10^{15}\,cm^{-3}$; L = 7μm ; β = 0.7μm, so that the length of the active n layer is $\simeq$ 5μm; $\alpha = 5\times10^6\,m^{-1}$.

The drift velocity can be represented by [3]:

$$\begin{cases} v(E)=v_s\left\{1+(E/E_c-1)/[1+A(E/E_c)^\gamma]\right\} \\ A = (E_p/E_c-v_p/v_s)/[(E_p/E_c)^\gamma(v_p/v_s-1)] & (6) \\ \gamma = (E_p/E_c)/(E_p/E_c-v_p/v_s) \end{cases}$$

where $v_s=0.58\times10^7$ cm/s is the saturation velocity, $E_c=v_s/\mu_o=1.35$kV/cm, E_p=12kV/cm is the threshold field, beyond which we observed Gunn oscillations, and $v_p=v(E_p)=2.56\times10^7$ cm/s is the peak velocity. For a given value of j, the initial field inside the n^+ contact, x=0, was defined as :
$E(o)=j/q\,\mu_o N_D(o)=j/q\,\mu_o N_D^+$, and the first order differential equation (3) was numerically solved by a standard fourth order predictor corrector technique. Once E(x) is obtained, n(x) is determined by derivation (see eq.(2)), and the potential V(x) by integration.

* **This work was supported by the Direction Générale de la Recherche Scientifique et Technique, France, contrat n° 79.7.0726.**

Figure 1 shows a comparison between the experiment and the stimulation performed according to the previous equations. The agreement is quite satisfactory, taking into account uncertainties on both physical and technological parameters.

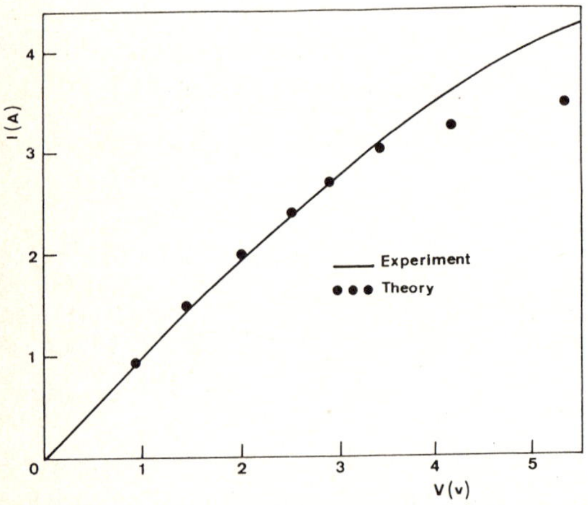

Figure 1 : static I-V characteristics of a n^+nn^+ InP diode : comparison between experiment (—) and theory (●).

Figure 2 shows electric field profiles for different current values of the diode studied. It can be seen that up to 5V, one may consider the field as being uniform E=V/L, with an approximation better than 10%.

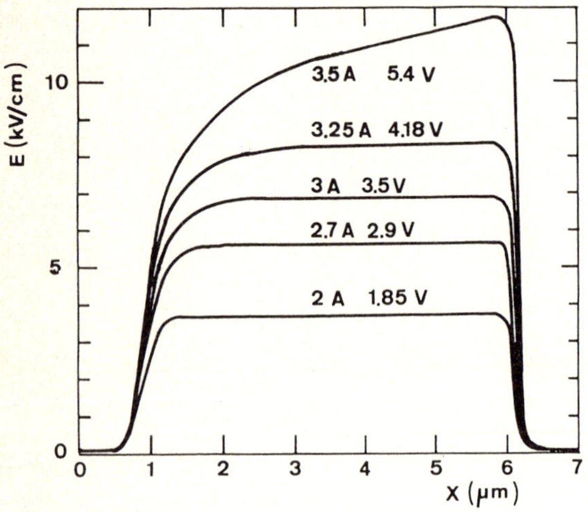

Figure 2 : Electric field profiles in the n^+nn^+ InP diode at various bias currents.

3. NOISE BEHAVIOUR

On figure 3 are plotted noise temperatures measured versus the electric field at various frequencies. We can clearly see that the noise is not white.

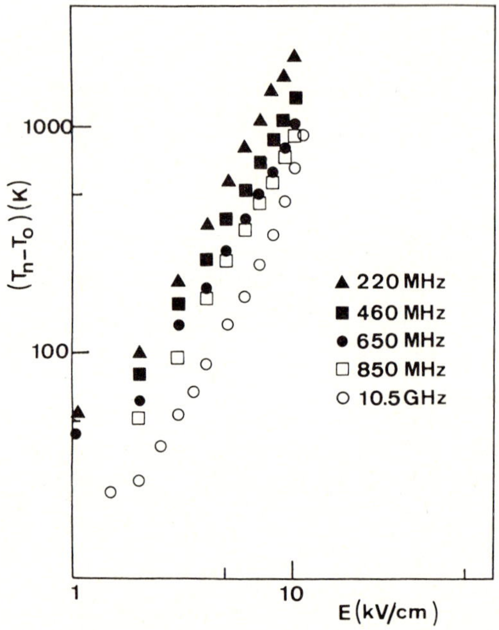

Figure 3 : Excess noise temperatures versus electric field at various frequencies, in n-type InP : $N_D=2.7\times10^{15}cm^{-3}$, $T_o=300K$.

A plot of noise temperatures versus frequency, at various electric fields, deduced from figure 3, is reported on figure 4, and indicates that no 1/f noise is present in the frequency range studied, but obviously generation recombination (G.R.) noise is superimposed to the diffusion noise.
The curves, passing through the experimental points in figure 4, are least squares fittings using the theory described in a previous paper [4] and allow determining the fraction of ionized impurities, $u = n(E)/N_D$, the G.R. lifetime τ (E) and the diffusion coefficient D(E). $u = n(E)/N_D$ was found to be constant, independant on E, equal to 0.99. This shows that the average population of the trapping level is not modified by the electric field intensity (no impact ionization, contrary to what was observed in n-type silicon at 77K [4]). It is interesting to note that a trapping level, containing in the average less than 1% of the carriers, can produce G.R. noise comparable to the diffusion noise. Also the G.R. lifetime τ is independant on E and was found equal to τ = 3.2ps, corresponding to a cut-off frequency of 500MHz (see figure 4).

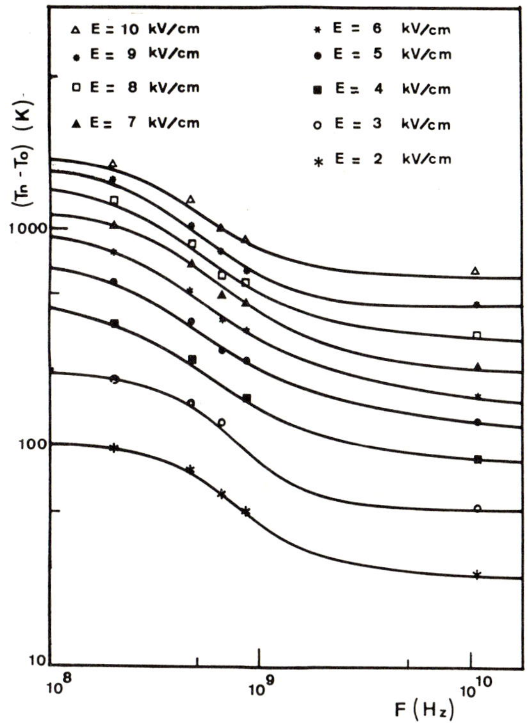

Figure 4 : Excess noise temperature versus frequency at various electric fields, in n-type InP, $N_D=2.7\times10^{15}cm^{-3}$, $T_o=300K$.

The diffusion coefficient D(E) can be, in that case, directly derived from the noise measurement at 10.5 GHz, since this frequency lies in the white diffusion noise plateau. D(E) is given figure 5, and compared with the theoretical Monte Carlo curve given by Hill and Robson [5] : the agreement between both results is quite satisfactory. This is up to our knowledge, the first experimental report on diffusion of hot electrons in Indium phosphide.

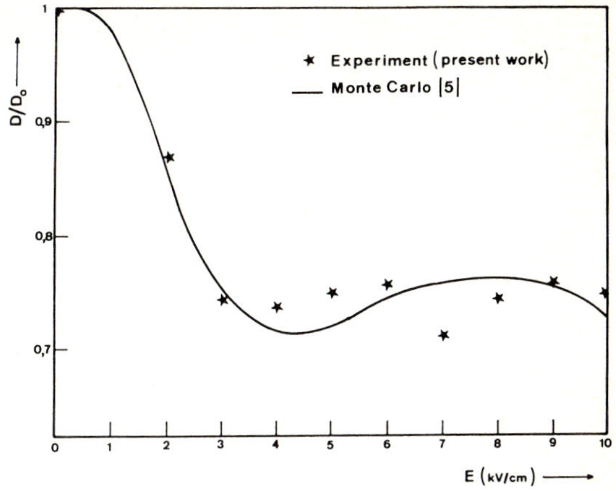

Figure 5 : Diffusion coefficient versus electric field in n-type InP, $N_D=2.7\times10^{15}cm^{-3}$, at 300K. Comparison between present experimental results (★) and Monte Carlo simulation [5] (——). D_o is the ohmic value $D_o=116cm^2/s$.

Acknowledgments : The authors are indebted to Dr. Convert, Dr. Linh and Dr. Armand, from LCR Thomson CSF, France, for providing InP mesa diodes.

REFERENCES

[1] D. Gasquet, J.C. Vaissière, J.P. Nougier, Proc. 6th International Conference on Noise in Physical Systems, Washington, 1981, National Bureau of Standards, Special Publication n° 614, p. 305 (1981).

[2] A. Moatadid, Thèse de 3e cycle, Montpellier, 1979.

[3] M. Shur, Solid State Electronics, 20, 389 (1977).

[4] D. Gasquet, H. Tijani, J.P. Nougier, 7th International Conference on noise in physical system, Montpellier, 1983.

[5] G. Hill, P.N. Robson, W. Fawcett, J. Appl. Phys. 50 356,(1979).

NOISE IN PHYSICAL SYSTEMS AND 1/f NOISE
M. Savelli, G. Lecoy and J-P. Nougier (eds.)
© *Elsevier Science Publishers B.V., 1983*

NOISE IN NEAR-BALLISTIC n^+nn^+ AND n^+pn^+ GALLIUM ARSENIDE SUBMICRON DIODES

R.R. Schmidt, G. Bosman, C.M. van Vliet and A. van der Ziel
Department of Electrical Engineering
University of Florida, Gainesville, FL 32611, USA

L.F. Eastman and M. Hollis
School of Electrical Engineering
Cornell University, Ithaca, NY 14853, USA

D.c. characteristics and noise measurements in the range 1 hz - 25 khz are reported for n^+nn^+ and n^+pn^+ near-ballistic devices, with n regions (p regions) of 0.4 μm (0.47 μm), fabricated by molecular beam epitaxy at Cornell University. The n^+nn^+ mesa structures show very low 1/f noise. This very low noise is attributed to the near absence of electron-phonon collisions. The thermal (-like) noise above 1 khz is equal to Nyquist noise at all currents. The n^+pn^+ noise, on the contrary, is quite high. It seems to be associated with electron traps in the bulk of the devices. The importance of noise measurements for confirming ballistic or near-ballistic behavior is discussed.

INTRODUCTION

Submicron gallium arsenide structures are of great current interest, since they permit ballistic electron flow, which in turn leads to carrier velocities that far exceed the saturation velocity in collision-dominated conduction, thus enabling the design of picosecond switching devices and other novel applications. The fabrication of submicron devices has been made possible by modern MBE techniques, electron lithography, etc. It is generally assumed that for the case of GaAs, near-ballistic behavior requires that the distance to be traveled by the injected electrons is less than or of the order of 0.7 μm. Eastman et al. report[1] that the mean free path for phonon emission into optical polar modes at room temperature is 0.1 μm for electrons of 0.05 eV, and 0.2 μm for electrons of 0.5 eV. Phonon absorption has a longer mean free path and can be neglected for the devices reported here, having thicknesses of 0.4 μm (thickness of n layer in n^+nn^+ devices) and 0.47 μm (thickness of p layer in n^+pn^+ devices). At higher electron energy intervalley scattering becomes important, thus limiting the near-ballistic range to about 0.5 eV of electron energy. In a sample of 0.4 μm thickness about two phonon emissions may occur. These involve, however, small angle deflections only (5° - 10°) and have little effect on the d.c. carrier characteristics, according to Ref. [1].

In the last few years a variety of physical models have appeared in literature.[2,3,4,5,6] Each model explained a particular aspect of ballistic charge transport in submicron structures only. The I-V characteristics predicted by these models were in poor agreement with the experimental data. In 1982 Holden and Debney[7] presented a straightforward approach, based on ideas from the well-known vacuum diode discussion of Fry.[8] Their results indicate that no fixed power law can be stated. We will use their model to explain our measured I-V characteristics.

In this paper we describe very accurate low-frequency and high-frequency noise measurements on near-ballistic devices. Such measurements serve a threefold purpose. First, from a practical point of view noise data reveal the practical performance limitations of the novel high-speed devices. As we will indicate, the noise of the n^+nn^+ devices is extremely low; the n^+pn^+ devices, however, fare much worse. Secondly, noise measurements at audio and sub-audio frequencies shed much light on the 1/f noise problem. According to most recent theories, such noise is thought to be caused by mobility fluctuations (see, e.g., Hooge et al.[9] and van der Ziel[10]). If collisions in the near-ballistic regime are rare, one expects the 1/f noise to be very low and ultimately, in "pure" ballistic devices, to be absent. Our work in n^+nn^+ devices indicates that this could be correct. Third, and not least, we believe that the high-frequency noise (thermal, velocity-fluctuation, or diffusion noise) will shed much light on the mode of operation of near-ballistic devices. To date, no full-fledged theory for such noise exists; we only have some preliminary computations by van der Ziel and Bosman.[11, 12] However, once this noise is understood, we will have a powerful means of substantiating or amending the various theories on near-ballistic behavior.

EXPERIMENTAL

The near-ballistic diode (NBD) is a sandwiched mesa structure of five lightly doped p or n layers, alternating with heavily doped n^+ layers, see Fig. 1. The doping densities of the various regions are 10^{18} cm^{-3} for the n^+

regions, approximately 2×10^{15} cm^{-3} for the n regions, and approximately 10^{15} cm^{-3} for the p regions. The diameter of the mesas is 100 μm. The devices were manufactured by molecular beam epitaxy at the Cornell University Submicron Research Facility. The mesas were provided with very low ohmic Au-Ge contacts.

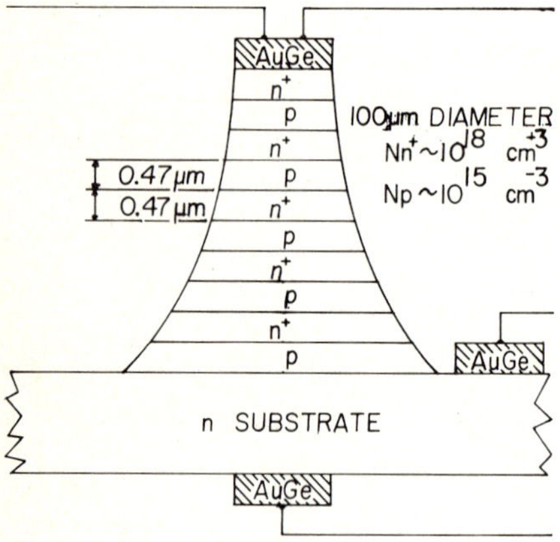

Figure 1. p-type NBD mesa structure

The bulk resistance of a single n^+nn^+ structure was 0.15Ω, comparable with the resistance of the golden bonding wires we used. To discriminate between the voltage drop across the device and across the bonding wires, we used a four-terminal configuration. Two terminals were used to supply the current, while the other two were used to measure the voltage and the voltage noise open circuited. The bulk resistance of the n^+pn^+ mesa was 90Ω at room temperature and 1 mA. The parasitic resistances in this case were negligible.

The characteristics of the two types of devices are quite different. The noise measurement of the n NBD's, in particular, was a challenge. A Hewlett Packard 3582 spectrum analyzer, featuring a dual-channel fast-Fourier transform method, was employed.[13] By measuring the coherence (square of the correlation) between the two channels, noise levels significantly below the noise level of the preamplifier could be detected. For the preamplifiers we used five common emitter transistors GE82 in parallel. This resulted in a 3Ω noise resistance for frequencies above 20 hz. The equivalent noise resistance of the cross-correlation setup was found to be less than 0.3Ω, thus enabling us to accurately measure the thermal noise of the very low ohmic n^+nn^+ devices.

The d.c. I-V characteristic of an n-type device at 300 K and 77 K is shown in Fig. 2. For the higher voltages, low duty-cycle pulsed measurements were made. The highest voltage over each layer is about 0.2 volts, well below the occurrence of intervalley scattering. We notice that within the experimental errors, the charac-

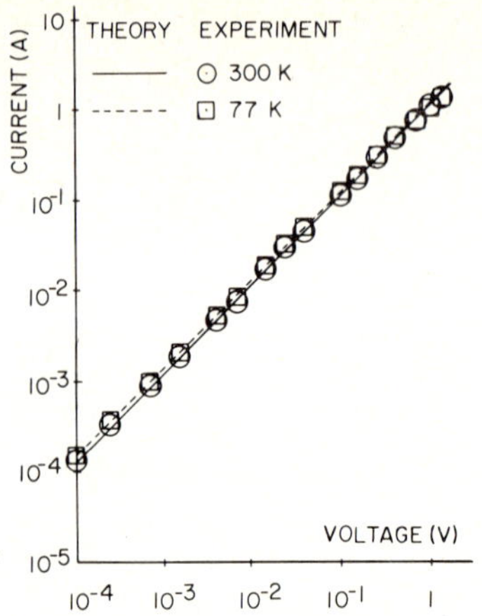

Fig. 2. I-V characteristic of an 0.4 μm n-NBD

teristic is linear for both temperatures except possibly at extreme bias. We note that no evidence of ballistic transport is observed. The I-V characteristic of an n^+pn^+ device is shown in Fig. 3 for two different temperatures. For high voltages the current is independent of temperature. For lower voltages the current decreases with decreasing temperature between 300 K and 77 K. For $12\,K \leq T \leq 77\,K$ no temperature dependence was observed.

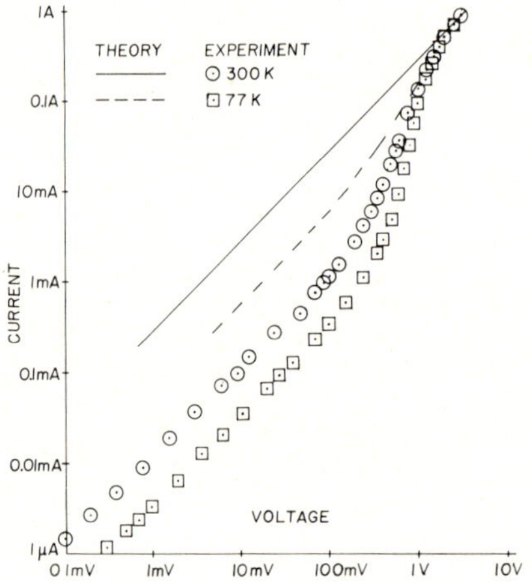

Fig. 3. I-V characteristic of an 0.47 μm p-NBD

The magnitude of the current noise of an n^+nn^+ device for four different currents at 300 K is shown in Fig. 4. Thermal levels and excess 1/f noise are seen. The magnitude of 1/f noise is proportional to I^2.

In Fig. 5 the current noise spectra of an n^+pn^+ device are shown for several bias currents at room temperature. The excess low-frequency noise of this device is orders of magnitude larger than for the n-type device. Another notable feature is the frequency dependence, which shows a slope of $f^{-0.7}$ to $f^{-0.8}$. Extrapolating to the corner frequency above which thermal noise dominates gives a value of over 100 Mhz for even the lowest (100 μA) bias current.

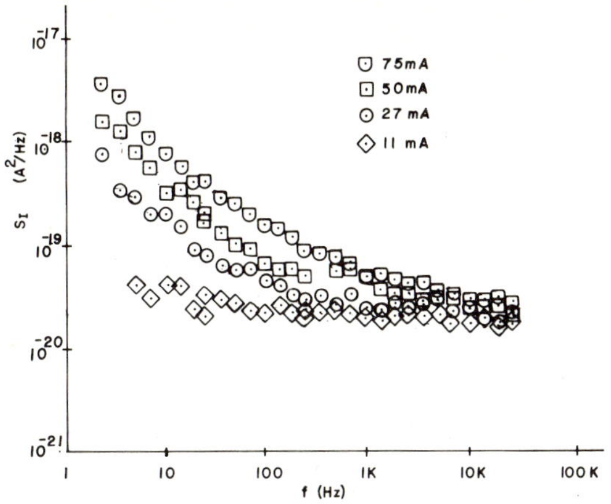

Fig. 4. Current noise of n-NBD at T = 300 K

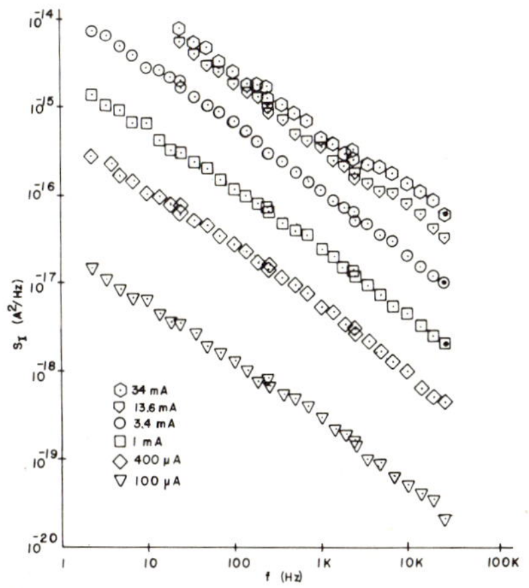

Fig. 5. Current noise of p-NBD at T = 300 K

At 300 K the magnitude of the l.f. noise level at 100 Hz is proportional to I^2 up to 1 mA. At higher currents the noise increases less fast and finally becomes independent of the bias current.

The temperature dependence of the noise in the low-bias regime was investigated. The low-frequency, small-bias (100 μA) noise spectra were measured for several temperatures (300, 250, 200, 150, 77 and 12 K). The magnitude increases with decreasing temperature down to about 200 K. Thereafter, further lowering of the temperature does not affect the magnitude, but the slope becomes more closely -1.

DISCUSSION

n^+nn^+ Devices: 1/f Noise

In 1969 Hooge developed the following empirical formula for 1/f noise:

$$S_{\Delta I}/I^2 = \alpha_H/fN ,$$

where f is frequency, N is total number of carriers in the sample, and α_H is the Hooge parameter. Recently, Kilmer and coworkers[14] found that the value of N in the preceding equation is not equal to the total number of carriers, but instead represents the number of carriers being scattered inside the sample. It is evident that these numbers are the same for nondegenerate, collision-dominated semiconductor samples. In submicron devices, however, carriers may contribute to the overall current without being scattered. Hence it is difficult to derive a value for N and consequently for α_H. To circumvent this problem, we introduce the parameter "noisiness," defined as $(S_{\Delta I}/I^2)f$. Substituting measured values gives for T = 300 K, noisiness = 1.6×10^{-15}, and for T = 77 K, noisiness = 1.4×10^{-14}.

These values can be compared with the noisiness of a collision-dominated GaAs device having an identical geometrical size and doping profile. Using $\alpha_H = 6 \times 10^{-3}$,[9] we find that the noisiness of the latter device is five orders of magnitude larger. This indicates that the number of collisions inside our sample is indeed very much reduced. Moreover, if Handel's theory of 1/f noise is valid,[15] very low noise can be expected from those collisions which still occur, involving polar phonon emission. As we noticed, the deflection angle θ for such processes is very small, whereas in Handel's theory of quantum 1/f noise the magnitude goes as $\sin^2 1/2\,\theta$.

Thermal Noise

The designation "thermal noise" is used here for the thermal-like noise observed at high frequencies. In a collision-limited device this noise is due to the diffusion-noise source, which by Einstein's relation transforms to a thermal-noise source for cold electrons. In the space-charge limited injection operation (Mott Gurney law), the noise becomes then, $8\,kT/R_x$, see [12]. In a pure ballistic device, on the other hand, this noise is due to shot noise. Lacking a detailed theory, van der Ziel and Bosman[11] indicated, nevertheless, that subthermal noise, $\theta 4kT/R_x$ with $\theta < 1$, can be expected. This is not corroborated by the results of Fig. 4, though it is very unlikely that the collision-limited case applies in view of the low 1/f noise reported above. A Langevin equation patterned after the momentum and energy balance equations of Shur and Eastman,[3,4] but with velocity dispersion as in the theory of Holden and Debney,[7] should give accurate predictions.

I-V Characteristics

The generalized Fry model we used accounts for: 1) injection of electrons from both n^+ contacts into the n region; 2) velocity dispersion of the injected electrons (a Maxwell-Boltzmann distribution is assumed for the electrons in the n^+

contacts); 3) space charge caused by ionized impurities; 4) presence of a dielectric . The model neglects collisions of electrons. The solid and dashed lines in Fig. 2 indicate the results obtained with this model. Note the very good agreement between experiment and theory. We conclude that although the contact spacing is larger than the electron mean free path, collisions do not affect the I-V characteristic of the devices. This can be explained as follows: The current measured at the device terminals consists of the superposition of two opposing currents, each current made up of electrons able to pass the potential minimum located in the n region. The value of the potential minimum controls the magnitude of the terminal current and therefore the I-V characteristic. This value is determined by the space-charge profile inside the device. Apart from the fixed ionized, shallow donor contribution, the space-charge density ρ is proportional to 1/v, where v is the electron velocity. Collisions cause v to change and therefore ρ according to $\Delta\rho = -1/v^2 \cdot \Delta v$. This shows that ρ is most affected by collisions in a region where v is small, i.e., in the direct vicinity of the potential minimum. We conclude that it is not the contact spacing, but the width of the potential minimum, that determines the I-V characteristic of a submicron device. Preliminary measurements on 1.1 μm n^+nn^+ GaAs devices support this conclusion.

In addition, Fig. 4 shows that we do not make a serious error by using a Maxwell-Boltzmann velocity distribution function for the degenerate contacts. This is obvious since fast electrons play a more important role than the slow ones.

n^+pn^+ Device; l.f. Excess Noise

The n^+pn^+ device shows much larger noise levels than the n-type device. The slope and the magnitude of the spectra seem to indicate the presence of generation-recombination processes. Carbon and oxygen are likely to be present in these devices. These elements introduce a set of discrete energy levels in the bandgap through which the Fermi level may move as a function of temperature, causing a redistribution of trap occupancies and related time constants. This would explain the temperature dependency of the slope of the noise spectra.

I-V Characteristic

The solid and dashed lines in Fig. 3 indicate the results obtained with the generalized Fry model. Note that for high bias voltages the agreement is quite satisfactory. For low bias, however, the calculated current values are too high. We assume that this discrepancy can be explained by taking electron trapping into account. A detailed calculation is underway.

CONCLUSIONS

The n^+nn^+ submicron device exhibits extremely low 1/f noise, indicating near-ballistic charge transport. The thermal noise is Nyquist noise for the current levels observed. The I-V measurements can be explained using a generalized Fry model. This demonstrates clearly that the magnitude of the measured 1/f noise of a submicron device is a good indicator for ballistic operation.

The n^+pn^+ device shows very large low-frequency noise with a temperature dependent slope between -0.85 and -1.0, indicating the presence of electron traps. The I-V measurements show that charge transport is ballistic for high bias voltages, whereas for lower bias voltages electron trapping dominates charge transport.

ACKNOWLEDGEMENT

The research in Florida was sponsored by AFOSR contract #80-0050.

REFERENCES

[1] Shur, M.S. and Eastman, L.F., Electronics Lts. 16 (1980) 522; also Eastman, L.F., Stall, R., Woodard, D., Dandekar, N., Wood, C.E.C., Shur, M.S. and Board, K., Electronics Lts. 16 (1980) 524.

[2] Shur, M.S. and Eastman, L.F., IEEE Trans. Electr. Dev. ED-26 (1979) 1677.

[3] Shur, M.S., IEEE Trans. Electr. Dev. ED-28 (1981) 1120.

[4] Shur, M.S. and Eastman, L.F., Solid St. Electr. 24 (1981) 11.

[5] Rosenberg, J.J., Yoffa, E.J. and Nathan, M.I., IEEE Trans. Electr. Dev. ED-28 (1981) 941.

[6] Cook, R.K. and Frey, J., IEEE Trans. Electr. Dev. ED-28 (1981) 951.

[7] Holden, A.J. and Debney, B.T., Electronics Lts. 18 (1982) 558.

[8] Fry, T.C., Phys. Rev. 17 (1921) 441.

[9] Hooge, F.N., Kleinpenning, T.J.G., and van Damme, L.K., Reports Progress in Physics 44 (1981) 479.

[10] van der Ziel, A., Advances in Electronics and Electron Physics (Martin, ed.) 49 (Academic Press, New York, 1979) 479.

[11] van der Ziel, A. and Bosman, G., Phys. Status Solidi (a) 73 (1982) K87.

[12] van der Ziel, A. and Bosman, G., Phys. Status Solidi (a) 73 (1982) K93.

[13] Schmidt, R.R., Bosman, G., Van Vliet, C.M., Eastman, L.F. and Hollis, M., Noise in near-ballistic n^+nn^+ and n^+pn^+ Gallium Arsenide submicron diodes, Solid-St. Elec. (in press).

[14] Kilmer, J., Bosman, G., Van Vliet, C.M. and van der Ziel, A., 1/f noise in metal films, submitted to Physica.

[15] Handel, P.H., Phys. Rev. A22 (1980) 745.

NOISE IN PHYSICAL SYSTEMS AND 1/f NOISE
M. Savelli, G. Lecoy and J-P. Nougier (eds.)

© *Elsevier Science Publishers B.V., 1983*

A NEW NOISE MODEL OF SUBMICROMETER DUAL GATE MESFET

E. ALLAMANDO, G. SALMER, E. CONSTANT, N.E. RADHY, A. CAPPY ★

and

B. CARNEZ ★★

A novel and simple model giving the noise behaviour and the high frequency performance of the submicrometer Ga As Dual-Gate MESFET is proposed.
This model takes into account the nonstationary electron-dynamicsand gate-edge effects that are important in submicrometer gate devices. It is considered that, in the microwave frequency range the noise is mainly due to the fluctuations in the carrier velocity whithin the channel.
The validity of this model is shown by comparison with experiments and them some new theoretical results are given.

I. INTRODUCTION

The dual gate Ga As MESFET is the most convenient device for many microwaves applications, such as a low noise gain controlled amplifiers, for instance. The most important improvement of the frequency response can be mainly achieved by reducing the gate length down to the submicrometer range. The design of such devices to minimize the noise figure needs a very accurate model. In the models available up to now, the dual gate MESFET was considered as two single gate FET's in cascade configuration. Such models cannot provide accurate theoretical predictions, because they do not take into account two main physical effects, which take a great importance in submicrometer dual gate FET's :

- the correlations between the noise current sources in gate 1 and 2 and in the drain circuits.
- the non stationnary electron dynamics effects during the electron transit in the channel.

2. NOISE MODEL

The model which we propose takes these effects into account. The numerical resolution of carrier transport current continuity and Poisson equations within the channel can be performed by means of a desktop computer. The carrier transport equations used are momentum and energy relaxation equations obtained by integration of the Boltzmann equation (1).

The most important assumption is that the equipotential lines in the channel are perpendicular to the source to drain axis, but the model takes into account edges effects on both sides of the gate and current injection into the buffer layer.

Noise is due to the carriers velocity fluctuations in the channel : for each section of width Δx in the channel, their spectral density is proportional to the local electronic temperature T, and it is assumed that T is only dependent on the average carrier energy. (2)

The contributions of these fluctuations to the gate and drain noise currents are numerically added and correlation coefficients are evaluated. The knowledge of the noise sources allows us to evaluate the noise figure by using classical methods and taking into account the influence of parasitic elements (source and gate resistances for instance). (3)

3. PHYSICAL BEHAVIOUR

Figure 1 presented the comparative contribution of each elementary section of the channel to the equivalent external noise current sources. Curve a refers to the fist gate, curve b to the second gate and curve c to the drain. This figure points out the predominant contribution of the channel zone under the first gate to the drain noise current source besides the second gate noise current source appears to be more important than the first gate one.

★ Centre Hyperfréquences et Semiconducteurs - L.A. C.N.R.S. n° 287
U.E.R. d'I.E.E.A. - Bât. P4
Université des Sciences et Techniques de Lille I
59655 VILLENEUVE D'ASCQ CEDEX (France).

★★ THOMSON-C.S.F., D.C.M. - 91401 ORSAY (France).

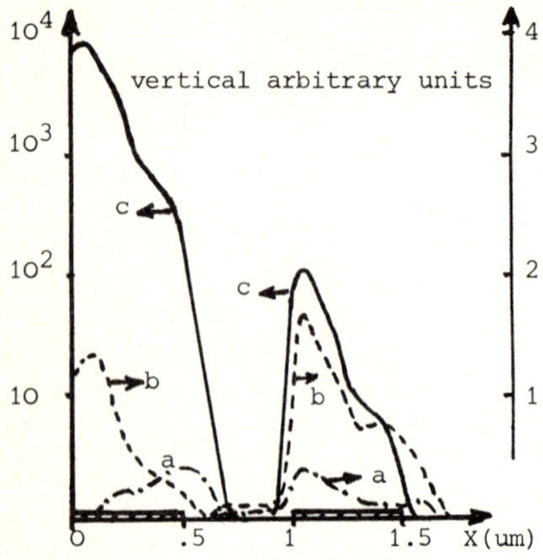

FIGURE 1

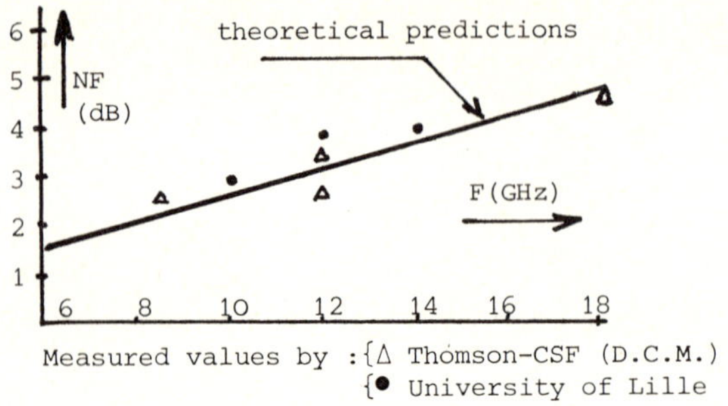

FIGURE 2

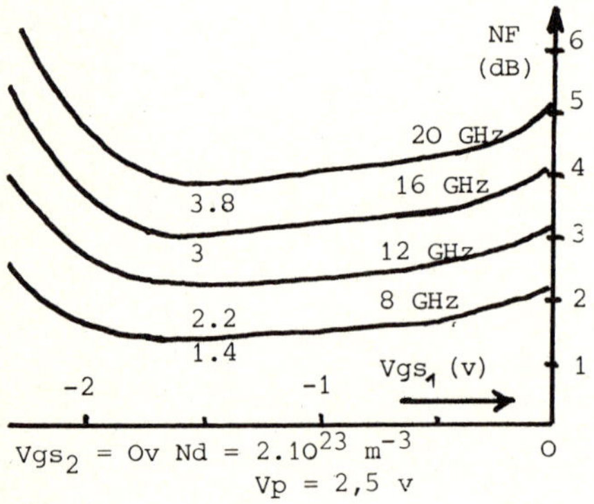

FIGURE 3

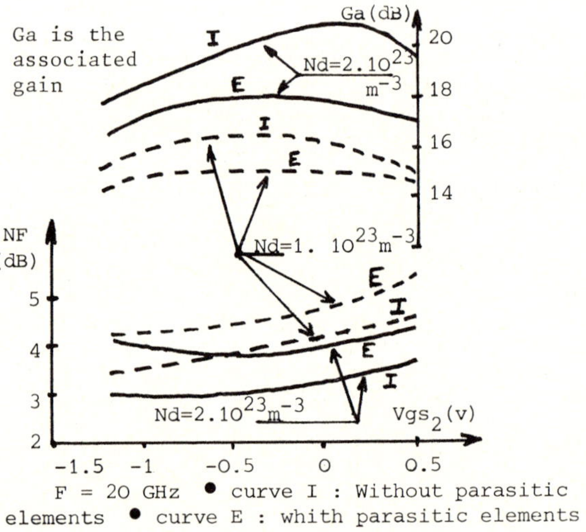

FIGURE 4

In order to prove the validity of our model, theoretical prediction were systematically compared with experimental results obtained for several different various dual gate FET's operating in the 6 to 18 GHz frequency range. A satisfactory agreement is achieved as shown in fig. 2 for submicrometer gate devices. These devices, used in prelimary studies, were fabricated by the D.C.M. department of Thomson-CSF

4. MAIN RESULTS

Fig. 3 shows the dependence of the noise figure on the first gate biasing condition and on the operating frequency. It is clear that, for submicrometer gate lengths, the noise figure does not strongly depend on the gate bias as it has been previously shown for single gate FET's (2). Moreover, we note the fast increase of the noise figure with increasing frequency.

The noise figure dependence upon technological parameters (doping level, epitaxial layer thickness, gate length...) has been accurately evaluated. As an example, Fig. 4 illustrates the theoretical evolutions of the noise figure with the second gate biasing conditions for two doping levels (Nd = 1 and 2.10^{23} m^{-3}).

This suggests to use structures with low parasitic elements and high doping level for microwave applications.

This work was supported by the D.R.E.T. under contract no 81/330.

REFERENCES

(1) A. Cappy : "Sur un nouveau modèle de transistor à effet de champ à grille submicronique", Thesis, Lille Univ. (june 1981)

(2) B. Carnez and al : "Noise modeling in submicrometer-gate FET's", IEEE Trans. Electron. Dev., 7, vol. ED-28 (1981) 784-789

(3) H. Rothe and W. Dahlke :"Theory of noisy fourpoles" Proc. IRE, vol 44 (1956) 811.

1/f NOISE IN RESISTORS AND THIN FILMS

NOISE IN PHYSICAL SYSTEMS AND 1/f NOISE
M. Savelli, G. Lecoy and J-P. Nougier (eds.)

© Elsevier Science Publishers B.V., 1983

IS THE 1/f NOISE PARAMETER α A CONSTANT?
(Invited Paper)

L.K.J. Vandamme

Department of Electrical Engineering
Eindhoven University of Technology
Eindhoven, Netherlands

Recent experimental studies on 1/f noise are reviewed. Experimental evidence in support of Hooge's bulk hypothesis has grown substantially since 1969. The relative and normalised 1/f noise parameter α is of the order of 10^{-3} or lower. α depends on scattering mechanism and temperature.

1. INTRODUCTION

1/f Noise has been the subject of a number of papers. The reviews on 1/f noise by Van der Ziel {1}, Voss {2} and Bell {3} give a complete picture of the state of the art till 1980. A survey of all recent controversial 1/f theories was given by Weissman {4} at the Noise Conference in 1981. In the review by Dutta and Horn {5} special attention was paid to the temperature dependence of the 1/f noise in metal films.

The review article by Hooge, Kleinpenning and Vandamme {6} is mainly concentrated on the validity of an empirical relation. Its application to electronic devices proved rather successful. The experimental results were confronted with two theories: Mc Whorther's surface-state theory and Clarke and Voss's theory of local temperature fluctuations. The applicability of either theory turns out to be very limited. In the latter review it is shown that 1/f noise obeying the empirical relation is a fluctuation in the mobility.
Here we deal with a survey of experimental work on 1/f noise. We shall try to put order in the experimental facts. The validity of the empirical relation will be investigated.

2. THE EMPIRICAL RELATION APPLIED TO INHOMOGENEOUS SITUATIONS

Voss and Clarke {7} and Beck and Spruit {8} have demonstrated that the conductance fluctuates even without applying any current. The current through the sample only serves to measure the conductance noise already existing. Therefore in ohmic samples the power spectral density in the noise voltage scales with the current as $S_V \propto I^2$. In ohmic samples the noise due to conductance fluctuations can often be summarised by a relative noise $C_{1/f}$ as follows

$$\frac{S_I}{I^2} = \frac{S_V}{V^2} = \frac{S_R}{R^2} = \frac{S_G}{G^2} = \frac{C_{1/f}}{f^\gamma} \tag{1}$$

where all symbols have the usual meanings. Often γ equals 1 ± 0.1 and is constant over many decades in frequency. Small contributions of burst noise and temperature fluctuations are very effective in spoiling a 1/f spectrum at low frequencies leading to $\gamma > 1.1$ {9,10}. Here we confine ourselves to γ close to 1 and $S_V \propto I^2$. Then $C_{1/f}$ is independent of polarisation and frequency and the dimensionless relative noise number is calculated from experimental results as

$$C_{1/f} = \frac{f\,S_V}{V^2} \tag{2}$$

$C_{1/f}$ was introduced to normalise results measured at different currents and frequencies. Hooge {11} proposed further normalisation to compare the noise from different samples as $C_{1/f} \propto 1/N$. The reason for his normalisation is the following: The conductance is proportional to the total number of free charge carriers N in the sample. If the free carrier contributes to the conduction and the noise uncorrelated from the others, then the fluctuations in G are $\langle(\Delta G)^2\rangle \propto N$ and $\langle(\Delta G/G)^2\rangle \propto N^{-1}$. Hooge's {11} empirical relation for the 1/f noise in homogeneous samples submitted to uniform fields is

$$C_{1/f} = \frac{\alpha}{N} = \frac{\alpha}{n\Omega} \tag{3}$$

The dimensionless proportionality factor α is of the order of 10^{-3} or lower, n is the free carrier concentration, and Ω is the volume of the sample. In view of uncorrelated fluctuations per charge carrier, the total number N is a trivial factor. It certainly does not mean that the 1/f noise is caused by number fluctuations.
In ohmic samples submitted to inhomogeneous fields N and Ω in eq. (3) become reduced effective values. Although the noise source is assumed to be distributed homogeneously in the volume, it seems to be concentrated in an effective volume Ω_{eff} submitted to a higher current density. The relation for the voltage between a pair of electrodes becomes

$$V = \frac{1}{I}\int_\Omega \rho J^2 \, d\Omega \tag{4}$$

By introducing the spatial cross-correlation function for the fluctuations in the conductivity or resistivity, the cross-correlation function for the voltage ϕ_V can be calculated using eq. (4) {12}. Its Fourier transform leads to S_V. The volume integral of the spatial cross-correlation spectral density function of a conductivity fluctuation at place $\underline{r}$ becomes $\alpha\sigma^2/n(\underline{r})f$. For calculating the resistivity fluctuation $\alpha\rho^2/n(\bar{r})f$ is to be used. S_V in a two-electrode arrangement becomes {12,13}

$$S_V = \frac{1}{I^2}\int_\Omega \frac{\alpha\rho^2}{nf} \cdot J^4 \, d\Omega \qquad (5)$$

If the medium is homogeneous with respect to the statistical noise properties, α/f can be put in front of the integral in eq. (5). For homogeneous samples, n and ρ are constant which simplifies eqs. (4) and (5) further. The integrals must be taken over the whole sample except the electrodes. From eqs. (4), (5) and $J \propto I$ follows the trivial facts $V \propto I$ and $S_V \propto I^2$. These eqs. were applied to 3- and 2-dimensional contact configurations {13-15}. The calculated results show agreement with the experimentally observed noise at contacts which were free from interface problems. For such constriction dominated contacts the relative noise $C_{1/f}$ scales like a volume effect for which holds

$$C_{1/f} \overset{\text{def}}{=} \frac{\alpha}{n.\Omega_{eff}} \qquad (6)$$

Using eqs. (4)-(6) the effective volume for a two electrode configuration can be written as

$$\Omega_{eff} = \left[\int_\Omega J^2 d\Omega\right]^2 \left[\int_\Omega J^4 d\Omega\right]^{-1} \qquad (7)$$

For a uniform current density in the sample Ω_{eff} has its largest value corresponding to the sample volume as can be seen from eq. (7). The noise due to conductance fluctuations in ohmic samples can best be measured by passing a constant current through a pair of driver electrodes and by observing the voltage fluctuations across a separate pair of sensor electrodes. The use of a constant current on a 4-electrode arrangement minimises the influence of unwanted contributions of an interface to the contact resistance and noise. In special cases the S_V across sensors equals $I^2.S_R$ where S_R is the spectral power density in the fluctuations in the resistance R between the sensors. The general relations for 4-electrode arrangements are given by Vandamme and van Bokhoven {16,17}:

$$S_V = \frac{1}{I^2}\int_\Omega \frac{\alpha\rho^2}{nf} |J.\tilde{J}|^2 \, d\Omega \qquad (8)$$

The integral must be taken over the whole sample volume except the electrodes. Relation (8) is quite similar to (5) except that J^4 has been replaced by $|J.\tilde{J}|^2$ which is the square of the scalar product of the current density J caused by the current I through the drivers and the adjoint current density $\tilde{J}$. In a thought experiment the adjoint current is the current that flows when the current source has been switched from drivers to the sensors. A number of analytical and numerical results are given in {16, 18, 19}. The adjoint current approach of eq. (8) provides a possibility to find the areas of high and low contribution to the noise. The noise across small sensor probes stems from the neighbourhood around the sensors where $J.\tilde{J}$ is large. Regions of high current density contribute heavily in eq. (8) if at least the densities J and $\tilde{J}$ are not perpendicular. When the sensors and drivers coincide, then $J=\tilde{J}$ and the four-terminal eq. (8) reduces to two-terminal eq. (5). The comparison between experimentally observed noise ratios and calculated ratios in ref. {18} provides a way to check if extraneous contact noise is important. When the sensors are placed so that the average voltage across them is zero, then the so-called transverse noise will be measured. Under the condition of negligible contact interface noise, the transverse noise does not change when the current source I at the drivers is replaced by a voltage source passing the same current I. The transverse noise arrangement is very convenient when using DC pre-amplifiers to measure spectra below 1 Hz. Cross-shaped samples observed in transverse noise condition show large areas of low noise contribution near the contacts {18}. With such geometry the contact interface noise contribution can be minimised.

3. SCALAR OR DYADIC NOISE SOURCES?

Weissman {20} has adopted the cross-shaped geometry and the general approach. He extended the calculations for two applied current sources. In order to separate current carrying contacts and noise sensors he ends up with an 8-terminal arrangement in a four-arm bridge. The scalar or tensor character of the conductivity fluctuations has a significant effect on the ratios of measurable fluctuations. The comparison between the cross-spectrum of resistance fluctuations on two nearly orthogonal paths with the spectrum of the transverse noise gives a quantity which is sensitive to how scalar the resistivity fluctuations are. The discrepancy between the scalar and dyad-model predictions is at most a factor 2 for an 8-contact arrangement with noise free contact interfaces. Weissman proved {20} that the geometrical factor appearing in Hall-effect noise measurements can be measured from the same geometry but in absence of a magnetic field and assuming the fluctuations as scalars and ignoring noisy contact interfaces. The advantage of multiprobe measurements lies in the reduction of contact interface effects and the possibility of observing a scalar or dyadic noise source. Much work has been done by Weissman et al. {21-24} to find out whether the noise source is dyadic or scalar. From the analysis of experimental results obtained from carbon, gold and chromium films followed that the conductivity fluctuations are cleary non-scalar {21}. This experimental evidence was obtained from granular films which are non-isotropic on a grain boundary scale but homogeneous and isotropic on a macro scale. The typical granule radius for the carbon film was 0.55 μm and the film thickness of the gold and chromium layers were 10 nm and 20 nm. The grain size is of the same order of magnitude of the film thickness for such thin films. Rather high values for α where observed pointing to the likely presence of local inhomogeneities. Weisman et al. found $\alpha=5\times10^{-3}$ in the chromium and $\alpha=5\times10^{-2}$ in gold. From experiments on lightly doped n-type silicon the conductivity fluctuations were spatially uncorrelated and approximately scalar {23}. The values of α ranged from 10^{-6} up to 10^{-4}. The mobility model was ruled out in ref. {23} because the noise source in n-Si was found to be scalar. However, Kleinpenning {25} has never stated that mobility fluctuations implied dyadic fluctuations as suggested in ref. {23}. In Kleinpenning's mobility fluctuation model a noise source is used with uncorrelated fluctuations in space and energy of the free carriers, not in $\bar{k}$ but in $|k|$. The integral over energies and space of the cross-correlation spectral density function of a conductivity fluctuation in $\underline{a}$ subband at energy E and place $\underline{r}$ becomes $\alpha\sigma^2/n(\bar{r},E)f$. In a recent paper {24} on mobility fluctuations and 1/f noise in Si, Weissman et al. agree that mobility fluctuation models can give either scalar or dyadic noise, depending on the model. In {24} the noise in the Hall coefficient was found to be smaller than predicted from carrier number fluctuations as was observed by Kleinpenning in 1977 {26,27}. In the SOS samples used by Weissman et al. {24}, the electron concentration was about $5\times10^{16}cm^{-3}$ and the mobility 400 cm^2/Vs. The observed noise was found to be scalar with an α-value of 10^{-4}. From the correlation coefficient between the Hall-voltage and the resistance fluctuations they conclude that primary noise sources must be number fluctuations. This doubtful conclusion is based on an analysis that does not take into account the change in μ due to the magnetoresistive effect.

4. 1/f NOISE AS A BULK EFFECT

In 1969, Hooge {11} collected data published on 1/f noise in semiconductors and gold. The survey made of them suggested that the relative noise was inversely proportional to the total number of mobile charge carriers N in the sample. This excluded surface effects as the origin of 1/f noise. The relative noise in continuous gold films was of the same order as in semiconductors for the same number of free charge carriers. The idea for a common physical origin of the 1/f noise in metals and semiconductors was born. The survey did not provide absolute evidence that α is the same for all semiconductors. In fact, α-values of 10^{-4} for n-Ge were registered in the survey, and an α-value of even 10^{-6}, unusually low at that time, was observed in rutile by Kleinpenning {28}. Fig. 1 presents a number of results of the relative noise as a function of N. All the gold layers were made with the same photo mask. Only the thickness was varied by using different evaporation times. Similar α-values were found by Voss and Clarke {31} and by Black et al. {32} in metal films. In most metals N equals the number of atoms N_c and this is used here in fig. 1. The hole concentration in p-type GaSb was $p=1.5\times10^{17}cm^{-3}$ and $2\times10^{13}<N<10^{14}$. N was varied by thinning the sample by etching and by keeping the contact regions unaffected. The

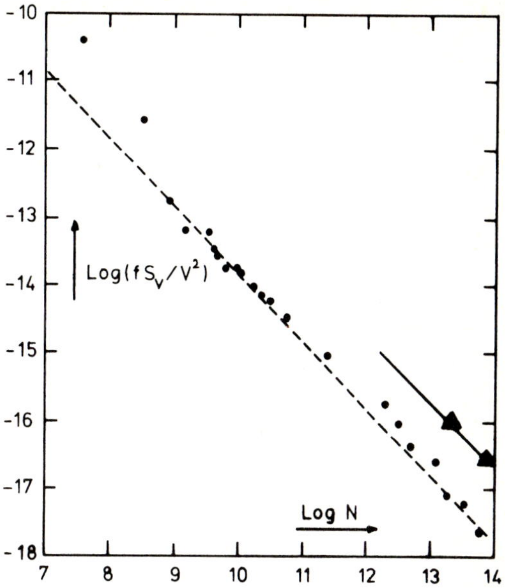

Fig. 1 : Relative noise $C_{1/f}=fS_V/V^2$ as a function of N. The dots represent the lowest $C_{1/f}$ for a given N observed in Pt by Fleetwood et al. {29}. The solid line represents results obtained with continuous Au layers {30} and the triangles are fron p-GaSb samples. Both lines represent the empirical relation with $\alpha=1.2\times10^{-4}$ for Pt and $\alpha=2\times10^{-3}$ for Au and GaSb.

scaling for the relative noise as described by eq. (3) seems well established. For platinum the relative noise is inversely proportional to N over at least 6 decades. This is strong evidence for the bulk origin of the noise. The deviations from a 1/N dependence for the low N platinum samples are caused by grainboundary problems which are typical of very thin films.
In general, the grain size of evaporated or sputtered layers is proportional to the layer thickness below 100 nm. Very thin layers (10 nm) often show island structures. The thin contacts in such granular structures cause local increases in the electric field. Such experimental results from samples with ill-defined non-uniform fields on a microscopic scale cannot be interpreted by eq. (3) in terms of α. Andrushko et al. {33} investigated the effects of grain size and inherent stresses on the excess noise for thin films of aluminium. Specimens with various structures were obtained by varying the condensation rate. Condensation was stopped when the film attained a thickness of 100 nm. Their experiments showed a relative noise increasing exponentially over more than 2 decades with the condensation rate in the range of 3 nm/s to 9 nm/s. Applying eq. (3) on these results obtained from grainy films, resulted in apparent α-values of at least 1. Their results proved that high density of fine grains corresponds to

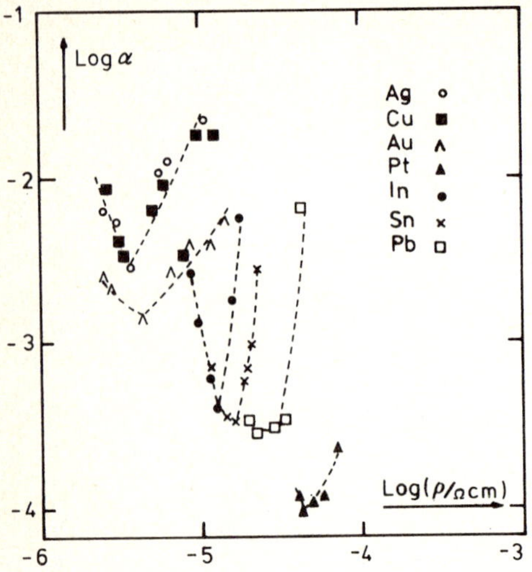

Fig. 2 : $\alpha = fS_V N/V^2$ as a function of resistivity for several metals. The data from Fleetwood et al. {34} are reduced by presenting only the lowest α at a given resistivity for each metal. The dotted line serve only as visual guides.

a large number of areas with distorted lattice, large-angle grain boundaries, elevated vacancy concentrations and more 1/f noise. In fig. 2 some data reduction has been applied to present "α-values" obtained from different metals by Fleetwood et al. {34}. The dotted lines represent the trends common to all metal layers of different thickness. Above a critical resistivity of the layer the relative noise increases strongly with ρ. This is due to the dominant role played by the grain-boundaries in very thin films. Ignoring the local increases in the electric field at the grain boundaries and using eq. (3) instead of eqs. (6) and (7), too high apparent α-values are obtained {35}. The trend on the left of each α-minimum is consistent with experiments, showing a reduction of α for samples with a reduced mobility due to scattering mechanisms other than lattice scattering {36,37}. This will be discussed in section 6.

5. POINT CONTACTS

For a well-defined single constriction in the current path, the electric field is known and the effective volume Ω_{eff} is calculable with eq. (7). For such a local increase in current density, the relative noise is interpretable in an α-value. A constriction-dominated contact is a possible arrangement to check the bulk hypothesis.

The equipotentials around a contact area on a 3-dimensional conductor can be approximated by hemispheres. Then the current density J equals $I/2\pi x^2$, where x is the radius of the hemisphere centered around the middle of the contact area with radius a. Introducing J in eq. (7), taking $d\Omega = 2\pi x^2 dx$, and integrating from a to ∞ leads to $\Omega_{eff} = 10\pi a^3$. The contact resistance can be obtained from V/I using eq. (4), which results in $R = \rho/2\pi a$. The relative noise scales as $(n\Omega_{eff})^{-1}$ or as a^{-3}, the contact resistance scales as a^{-1}, so the relative noise $C_{1/f} \propto R^3$. Vandamme {13} calculated the noise of a contact with the more realistic ellipsoidal equipotentials and found the same scaling for the noise.
Especially if $C_{1/f}$ is expresses as a function of resistance R, the relation $C_{1/f}$ for a double constriction calculated with the hemispherical equipotentials {38}

$$C_{1/f} = \frac{\alpha \pi^2 R^3}{20 n \rho^3} \tag{9}$$

is a surprisingly good approximation.
Even channel-like constrictions with a fixed ratio between contact diameter and rounding-off radius show the bulk-like scaling with $C_{1/f} \propto R^3$ {13}. It has been pointed out previously {39,40} that the noise from a bulk noise source which is proportional to $\int_\Omega J^4 d\Omega$ (5) would have a logarithmic divergence, if the field singularity at the rim of the contact spot were not truncated. In practice, a finite rounding-off at the rim of a contact due to the atomic structure of solids cause a breakdown of the macroscopic equations and the logarithmic divergence {13}.
Experimental work on crossed cylindrical metal bars was carried out {38} to investigate the bulk hypothesis. The resistance can easily be varied by a factor 20. When there are no complications, then $C_{1/f} \propto R^3$. In this way α-values of metals have been determined avoiding the problems of very thin films and poor thermal contact with the substrate. By using point contacs between crossed rods of p-type Ge with $\rho = 14.6\ \Omega$cm, we were able to vary Ω_{eff} and N_{eff} over 4 decades by changing the pressure on the bars {41}. The relative noise was proportional to the cube of the contact resistance and again additional experimental evidence for the bulk hypothesis was found.
From this type of experiments follows that constriction-dominated contact noise is not a special type of noise but is physically the same as bulk 1/f noise.
Owing to an interface on contacting members, the contact resistance and the noise can be dominated by a surface term, and one never finds a $C_{1/f} \propto R^3$. If contact technology fails for metal semiconductor contacts, there will be a non-negligible contribution of the contact interface noise that scales as $1/S^3$, where S is the area of the noisy contact interface. The relative noise then scales as 1/S. This was observed on contacts on 3- and 2-dimensional conductors {41,42}.
With 3-dimensional constrictions serious difficulties may arise from oxide films between the contact members. Such films may determine the resistance and the noise of a contact. If they do, the contact is callled film-dominated; otherwise it is constriction dominated, in which case eq. (9) applies. A film with constant resistivity

ρ_{film} and a constant thickness t may cause two situations {41}.

constriction dominated contact	film-dominated contact
$a\rho_{bulk} >> t.\rho_{film}$	$a\rho_{bulk} << t.\rho_{film}$
$R \propto a^{-1} \propto F^{-1/3}$	$R \propto a^{-2} \propto F^{-2/3}$
$C_{1/f} \propto R^3$	$C_{1/f} \propto R$

F is the force on the two contacting members. Which situation occurs depends on the ratio of the film resistance to the constriction resistance. The presence of the film is demonstrated simultaneously in the scaling of contact resistance with the force on the contact and in the scaling of the relative noise with the contact resistance. Reliable α-values can be found from crossed-bar experiments and eq. (9) only if $C_{1/f}$ remains proportional to R^3 over at least 3 decades in $C_{1/f}$ {43}.
Weissman {44} attempted to show that surface fluctuations produce a term in the conduction noise of a contact that shows the same scaling with contact radius as does bulk noise. In Weissman's highly artificial model the noise contribution stems from a surface source at the rim of the contact spot where a field singularity occurs. First of all, in real contacts there is no field singularity due to the rounding-off at the contact edges. So real contacts have a non divergent relative noise as calculated in refs. {13,14}. Secondly, he neglects the contribution of the surface layer on the contact resistance. Depending on the ratio $a\rho_{bulk}/t\rho_{film}$ this contribution cannot be ignored as was previously observed {41}. Thirdly, the surface noise source is concentrated in a torus of minor radius d_0 and major radius a at the rim of the contact spot. Of course if one prefers an artificial model where the surface noise source is concentrated in a subvolume while keeping $R \propto a^{-1}$, one will find that $C_{1/f} \propto R^3$ even for "surface" noise. This scaling should not be taken seriously and is not an argument against the bulk hypothesis.
From Vandamme's experimental results it has been demonstrated that the surface contribution can be seen in the scaling of $R \propto a^{-2}$ as well as in $C_{1/f} \propto R$, and the experimental results with $C_{1/f} \propto R^3$ are still a proof in favour of the bulk hypothesis.
Further, the influence of a change in oxide thickness was investigated on InSb point contacts and the influence of a changing ρ_{bulk} due to a temperature change {43}. It turned out that it was possible to change the contact noise from surface dominated to bulk dominated depending on the ratio $a\rho_{bulk}/t\rho_{film}$ for the same pair of InSb bars with the same oxide layer on it.

6. DOES α DEPEND ON SCATTERING MECHANISMS?

We have observed α-values at room temperature for semiconductors in the range of 10^{-7} to 10^{-2}. Samples with noisy contact interfaces often lead to a relative noise that scales inversely with the contact area. If, by ignoring the interface contribution, one described the relative noise as pure bulk noise with eq. (3), then one would find an apparent α often larger than 10^{-2}. A first systematic reduction in α was observed by Hooge and Vandamme {36}. In samples at room temperature with different mixtures of lattice and impurity scattering we found that α is proportional to $(\mu/\mu_{latt})^2$ where μ is the mobility of free charge carriers and μ_{latt} the mobility that would be present without impurities.
The experimental results were obtained from ohmic metal-semiconductor contacts on three p-type Ge crystals with $\rho=3.0\times10^{-2}\Omega cm$, $3.7\times10^{-3}\Omega cm$ and $4.5\times10^{-4}\Omega cm$ and on an n-type GaAs sample with $\rho=2.7\times10^{-3}\Omega cm$. The mobility in these samples is lower than μ_{latt}, the value found in high-ohmic material because of the additional impurity scattering. The results are presented in fig. 3.

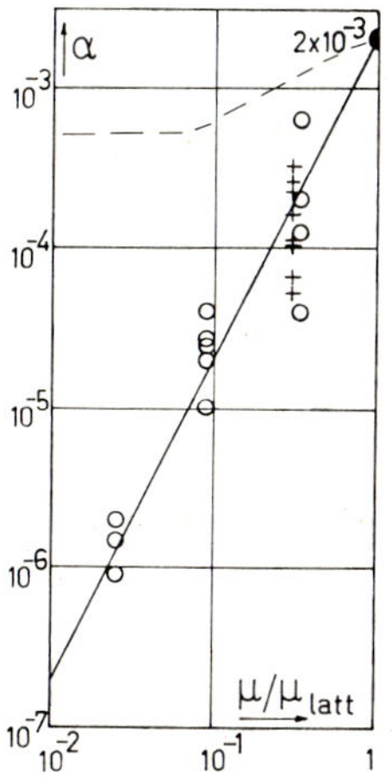

Fig. 3 : α as a function of μ at 300 K. α_{latt} and μ_{latt} are found in samples not affected by impurity scattering or surface scattering. Circles: p-type Ge. Crosses: n-type GaAs. Solid line: theory, eq. (11). Broken line: calculated for number fluctuations eq. (16).

If the lattice and impurity scattering are independent we find

$$1/\mu=1/\mu_{latt}+1/\mu_{imp} \qquad (10)$$

where μ_{latt} and μ_{imp} are the mobilities that would be found if only lattice or impurity scattering were present. We shall show that the experiments can be described well by assuming 1/f noise in the lattice scattering and little or no 1/f noise in the impurity scattering.
If $d\mu_{imp}=0$, after differentiating eq. (10) we find $d\mu/\mu^2 \cong d\mu_{latt}/\mu^2_{latt}$, and for α

$$\alpha=(\mu/\mu_{latt})^2\alpha_{latt} \qquad (11)$$

A large reduction of the mobility by influence of impurity scattering will occur only in highly doped material. The carrier concentration will then be high, which makes it necessary to use

very small samples. Even then the noise is low. This is why in all previous measurements of α low-doped material was chosen and the systematic reduction was not observed. The experimental evidence for (11) gives additional evidence in favour of mobility fluctuations as proved by Kleinpenning {45,27,46,47}. None of his proofs presented there were contradicted in the literature. Weissman {48} objects the relation (11) to be consistent with the spatial and energy independent mobility noise source presented as by Kleinpenning. In a reply, Kleinpenning {25} demonstrates that eq. (11) holds when the deviations in the temperature dependence of the impurity mobility from the Cornwell-Weiskopf model are taken into account. For impurity (ionized or neutral) scattering the collision time τ and thus μ_{imp} scales as E^s and μ_{imp} as T^s. Values for s close to 0 instead of 3/2 are very common in Si and Ge.
Another objection against eq. (11) pointing to a mobility source was given by Van der Ziel {1}. He claimed that the experimental results could be explained as number fluctuations. To discriminate between number fluctuations and mobility fluctuations we have to study a quantity which depends on μ and N in different ways {6}. The relation between μ and N is given by Hilsum as $\mu=\mu_{latt}(1+(N/N_o)^{\frac{1}{2}})^{-1}$ {49}. In our experiment the conductance G is a function of μ_{latt} and N as follows

$$G\propto\mu(\mu_{latt},N).N\propto N.\mu \tag{12}$$

A number fluctuation will lead to $\delta G=(dG/dN)\delta N$ and to

$$S_G=(NdG/dN)^2 S_N/N^2 \tag{13}$$

and a $\delta\mu_{latt}$ model will lead to $\delta G=(dG/d\mu_{latt})\delta\mu_{latt}$ and to

$$S_G=(\mu_{latt}dG/d\mu_{latt})^2 S_{\mu_{latt}}/\mu^2_{latt} \tag{14}$$

In this experiment we can discriminate between $\delta\mu_{latt}$ and δN because $|NdG/dN|\neq|\mu_{latt}dG/d\mu_{latt}|$. The lattice mobility fluctuations lead to

$$S_G/G^2=\left[\frac{\mu_{latt}}{G}\cdot\frac{dG}{d\mu_{latt}}\right]^2\cdot\frac{S_{\mu_{latt}}}{\mu_{latt}^2}=\left[\frac{\mu}{\mu_{latt}}\right]^2\frac{\alpha_{latt}}{Nf} \tag{15}$$

The number fluctuations result in

$$\frac{S_G}{G^2}=\left[\frac{NdG}{GdN}\right]^2\cdot\frac{S_N}{N^2}=h(N)\frac{\alpha}{Nf} \tag{16}$$

If the mobility does not depend on the number of free carriers, we have a reduction factor h(N)=1. In the other limit for $N>N_o$ we find $h(N)=\frac{1}{4}$. The experimental results show the most direct evidence so far in favour of the lattice mobility fluctuation hypothesis.
Further evidence for the reduction factor $(\mu/\mu_{latt})^2$ was found by Palenskis and Shoblitskas {50}. They demonstrated that the reduction in α proportional to $(\mu/\mu_{latt})^2$ is independent of the way in which the ratio μ/μ_{latt} was realised either with the help of temperature 77K<T<400K or by changing the amount of impurities. In their experiments α_{latt} was 9×10^{-4} for p-type Si and 2×10^{-3} for n-type in the whole temperature range. The α-value of 1.2×10^{-4} for Pt of Fleetwood et al. {29} as presented in fig. 1 can be interpreted with eq. (11) as $\alpha_{latt}=2\times10^{-3}$ by using their observed mobility ratio $\mu/\mu_{latt}=\frac{1}{4}$.
Another demonstration of α-dependence on scattering mechanism was given by noise measurements on Bi films at room temperature {37,51}. In thin layers the mobility is reduced by grain boundary scattering and near-inelastic surface scattering. Thin samples having a reduced mobility show reduced α-values. Bisschop and de Kuijper {52} found agreement between calculations based on eq. (11) and experimentally observed α-values for Bi films of different thicknesses in the temperature range of 77K-300K. Now we are also able to clarify the decrease in α with increasing film resistivity as observed by Fleetwood et al. {34} and presented in fig. 2. The trend left from each α minimum in fig. 2 shows some proportionality with μ. Since in metals we have charge carriers in different bands we cannot calculate α more precisely than within a factor 2. Only for Cu, Ag, Au and W there is agreement about $N=N_A$ (the number of atoms). Taking into account the large scatter up to factor 10 in α and the uncertainty in N, the experimental results in ref. {34} are no arguments against the $(\mu/\mu_{latt})^2$ reduction factor.
At strong electric fields the electrons are scattered mainly by creating optical phonons. This hot electron effect reduces the mobility and the α-values in a way similar to eq. (11){6}. An attempt to narrow down the number of possible noise sources may be the conjecture that the noise depends on the ratio of accoustical phonon scattering to the total scattering rate. The decrease in α with electric field strength was observed by Bosman et al., too {53}.

7. DOES α DEPEND ON TEMPERATURE?

Voss and Clarke {31} have interpreted the observed noise in metals in terms of temperature fluctuations that modulate the film resistance. The temperature coefficient of the film resistance togehter with spatial correlations of the noise play a central part in their model. The temperature hypothesis was tested from an experimental point of view {5,6,31,32,54,55}.
At room temperature the 1/f noise in metal films, Si, Ge and GaP lacks spatial correlation {5,6, 32,55}. Adkins and Koch {54} demonstrated that the relative noise was not proportional to the square of the temperature coefficient of the resistance of an inversion layer between 1.5 K and 4.2K. There was no correlation between the noise generated in different sections of the channel, yet the temperature fluctuation model predicts correlation over lenghts much greater than the size of the specimen.
Metal films often show a mixture of temperature induced 1/f like noise and 1/f noise at frequencies in the neighbourhood of the thermal relaxation

time {6}. The noise was found to depend strongly upon the type of substrate used and the strength of the adhesion between the film and the substrate {56}. Tin films prepared on sapphire with a larger thermal diffusivity as compared to glass showed 3 times less noise than tin films on glass. Samples prepared under proper conditions were nearly two orders of magnitude quieter than reported in ref. {31}. These high quality samples showed 1/f noise with an α-value of about 10^{-3}.

To check the theory of Clarke and Voss {31} we investigated the α of manganin with a negligible temperature coefficient {43}. From our experiments on manganin point contacts the usual value of α was obtained although the temperature coefficient of manganin is at least 2 decades smaller than that of other metals.
All these experimental results demonstrate that α noise is not a temperature induced noise. Cross correlations found in films with poor thermal contact to the substrate or at superconducting transitions {31} are clearly due to temperature fluctuations which do not have a good 1/f spectrum ($\gamma>1.2$).
As to the question of temperature influence, there are no unambiguous answers. We observed a strong temperature dependence of α for p-type Si of 30 Ωcm, although we assume mainly lattice scattering in the whole temperature range. The same trend of decreasing α-values with decreasing temperature were also observed by Dutta and Horn {5}. In the temperature range from 190 to 300K we observed $\mu_{latt}\propto T^{-2.3}$ and $\alpha\propto T^{4.6}$. This means that $<(\Delta\mu)^2>$ is approximately temperature independent for 190K<T<300K. Similar results were observed on p-Ge of 13 Ωcm {58}. There are other examples where α is weakly temperature dependent, if at all. {6,62,64}.

8. SUMMARY OF EXPERIMENTAL RESULTS

Owing to eq. (11) α is not constant when several scattering mechanisms are present simultaneously. So in the present situation we face two problems:
(i) Is α_{latt} a constant at roomtemperature?
(ii) How does α_{latt} depend on temperature?
The answer to the first question is that α_{latt} is often of the order of 10^{-3} at room temperature. This can be seen from the α-values in the table. The table is not complete, but omissions are not intentional. α-values were obtained from samples with an observed spectrum $S_V\propto V^{\beta}.f^{-\gamma}$ with $\beta=2+0.1$ and $\gamma=1\pm0.1$. That $fS_V/V^2=C_{1/f}$ is constant in a range of V and f is a prerequisite to accept a calculation of α. To indicate the sample and measuring conditions, the following code is used in the table. C: means the noise parameter α was obtained from a constriction-dominated arrangement using eq. (9). The contacts were ohmic metal-semiconductor point contacts of different diameter or semiconductor-semiconductor or metal-metal point contacts.
H: denotes homogeneous samples submitted to homogeneous fields.
T: α was calculated from the observed transverse noise on cross-shaped samples using eq. (8). This method implies a good suppression of possible contact interface noise.
L: means that the scattering mechanism was mainly lattice scattering. The colomn of reference numbers is headed by Refs. E refers to our unpublished results.

SEMICONDUC. AND METALS	CONDITION	α-VALUE OR RANGE	ρ(ΩCM) OR $n(CM^{-3})$	TEMP. (K)	REF.	COMMENT
P-Ge,N-Ge	C,H,L	$2x10^{-3}$	ρ>0.1	300	E	-
P-Ge	C,L	$2x10^{-3}$	13.2	300	58	
P-Ge	C,L	$2x10^{-3}$	5.9	300	58	
N-Ge	H.L	$4x10^{-3}$	3.5	300	57	a)
P-SI	-	$2x10^{-7}/9x10^{-5}$	10^{17}	77-400	50	$\alpha_L=9x10^{-4}$
N-SI	-	$3x10^{-5}/4x10^{-4}$	10^{13}	77-400	50	$\alpha_L=2x10^{-3}$
P-SI	H,L	$4x10^{-3}$	59	300	57	a)
	H,L	$3x10^{-3}$	14	77	57	a)
N-SI	H,L	$10^{-5}/3x10^{-3}$	7	300	58	b)
N-SI	T,L	$2x10^{-6}/10^{-4}$	30, 100	300	23	c)
N-SI	T	10^{-4}	$5x10^{16}$	300	24	d)
P-SI	H,L	$10^{-5}/2x10^{-4}$	10^{4}	300-77	53	e)
N-SI	H,L	$2x10^{-4}/10^{-3}$	$2.2x10^{3}$	300-77	53	
N-CHANNEL	INVERSION	$5x10^{-3}$	R SAMPLE	1.5	54	f)
	LAYER		$10^{5}Ω-10^{7}Ω$	4.2	54	
N-InSb	H	$4x10^{-4}$	10^{14}	76	59	g)
	H	$2x10^{-5}$	10^{14}	76	59	
N-InSb	H	10^{-4}	$1.9x10^{13}$	123	60	h)
N-InSb	C,H,L	$3x10^{-3}$	$1.6x10^{16}$	300	62	
N-InSb	C,H,L	10^{-3}	$1.6x10^{14}$	77	62	
P-InSb	C	$2x10^{-3}$	10^{15}	77	43	
N-InP	H,L	$2x10^{-4}$	$1.7x10^{15}$	300	E	i)
N-GaSb	H	$2x10^{-3}$	$1.5x10^{17}$	77	43	FIG. 1
N-GaAs	C,H,L	$6x10^{-3}$	$2.3x10^{16}$	300	62	
P-GaAs	H,L	$3x10^{-3}$	$2.3x10^{16}$	300	62	
N-GaP	H,L?	$9x10^{-3}$	$2.9x10^{16}$	300	62	
$Hg_{1-x}Cd_xTe$	H,L?	$3x10^{-3}$	$2x10^{16}$	300	63	j)
	H,L?	$5x10^{-3}$	$10^{15}-10^{16}$	77	63	j)
AU	C,H,L	$2x10^{-3}$	$5.9x10^{22}$	300-77	11	FIG. 1,2
W	C,L	10^{-3}	$6x10^{22}$	300	43	-
BI	C,H,T	$2x10^{-3}/2x10^{-4}$	10^{19}		37	
MANGANIN	C	$7x10^{-4}$	$3.5x10^{22}$	300	43	k)
Ga	C	$7x10^{-4}/10^{-2}$	$4x10^{22}$	77-600	62	l)
Hg	C	10^{-3}	$5x10^{22}$	300	57	l)
Sn	H	10^{-3}	$3.6x10^{22}$	300	56	m)

COMMENTS IN TABLE OF α-VALUE

a) After bending the crystal, the Hall mobility is reduced. The observed α-value did not change after bending the bars to a curvature radius as low as 5 mm. These experiments were carried out to investigate the influence of deformations and distortions on α. The bar ends remained unbent. Each straigth end was about 15% of the total length of the bar.
b) The samples were planar devices all having a volume of $2x10^{-4}cm^3$. Fifty-seven samples were investigated. Al contacts on n^+ layers were used. The distance between the contacts was about 250 µm. The α-values were distributed uniformly in the range 10^{-5}-$3x10^{-3}$.
c) Weissman et al. observed a spectrum close to 1/f in the 1 to 10 Hz range but less steep in the 10 to 100 Hz range. From our experimental results on oxidisedp-Si of 1 Ωcm we, too, observed extremely low α-values in the range of 10^{-7}

to 10^{-6}.

d) Measurements on n-type silicon-on-sapphire indicate significant impurity scattering (μ_H=400 cm^2/Vs) from surface and bulk. Taking into account the mobility reduction factor, we find $\alpha_{latt} \cong 10^{-3}$.

e) These samples were $p^+\pi p^+$ and n^+nn^+ planar devices with a cross-sectional area of 1 mm^2. The contact spacing for the phosphor-doped device was 78 μm and the boron doped device 40μm.

f) At constant N, α is found to increase with reduction of temperature from 4.2 to 1.5K. The results cannot be explained by trapping of carriers, and the thermal fluctuation model has proved not to apply. The relative noise is proportional to the channel resistance over more than 2 decades.

g) They found that α and the frequency dependence of the spectrum depend on the methods of sample preparation. Our calculated α-values from their experimental results are about a factor 6 to 120 lower than $2x10^{-3}$. We used the relation $fS_V/V^2 = q\mu R/l^2$. The structure in their spectra can be explained by a superposition of a 1/f and two generation-recombination spectra with characteristic frequencies of 10^2 and $3x10^3$Hz. A superposition of a $(1/f)^{1.3}$ spectrum with one generation-recombination spectrum also results in a nice fit of their experimental results. The difference in noise results were obtained on samples with different contact preparation. After etching, the relative noise was reduced by a factor 20.

h) Strong influence of etching. The geometrical mean α-value is presented. The largest α-values ($8x10^{-3}$) were observed in symmetrically etched samples {60}. Van de Voorde and Love {61} observed that a light surface etch usually reduces α.

i) The mesa structure has a height of 5μm and a diameter of about 200μm. A reduction in α due to the beginning of hot electron effects is possible. The resistance is increased by 15% of the low-field value.

j) α depends on surface treatment. This was also observed in {24,59,60,61}.

k) There was no detectable 1/f noise in the manganin films {31}. This fact was used to prove that the noise in the metal films arises from temperature modulation of the resistance. However, from our experiments on manganin point contacts the usual value of α was obtained. Our result demonstrated that the temperature coefficient does not play a major part in 1/f noise.

l) These experiments were carried out to investigate the dependence of α of the solid and liquid states. The α for solid, normal and undercooled liquid Ga is about $2x10^{-3}$.

m) The noise was found to depend upon the type of substrate used and the strength of the adhesion between film and substrate.

9. CONCLUSIONS

Hooge proposed a bulk noise source based on mobility fluctuations as the origin of 1/f noise {65}. Experimental evidence in support of this bulk hypothesis has grown substantially since 1969.

Considering more recent results, α seems not to be a universal constant. When lattice scattering prevails, $\alpha = \alpha_{latt}$ and it is often of the order of 10^{-3} at room temperature. Samples where lattice scattering is mixed with some other scattering mechanism show that only lattice scattering contributes to the 1/f noise. In this case $\alpha = \alpha_{latt}(\mu/\mu_{latt})^2$.

At the moment it is not clear if there are other mechanisms that may reduce α even in samples with pure lattice scattering.

$\alpha_{latt} \leq 10^{-6}$ goes hand in hand with spectra showing some generation-recombination noise or another structure.

α-values lower than 10^{-3} have no bearing on the question if the 1/f noise is a bulk phenomenon or is caused by mobility fluctuations.

Apparent α-values larger than 10^{-2} often go hand in hand with noisy contact interfaces or excessive grain boundaries in films. Small grain sizes are often seen in ultra thin films or in samples prepared at too high an evaporation rate. Irreproducible and high α-values are observed on metal films with poor adhesion to a substrate with low thermal conductivity.

As to the question of temperature influence there are no unambiguous answers.

α-values decreasing with decreasing temperature were observed in metal films {5}, Si and Ge {58} although lattice scattering can be assumed.

There are other examples where α is only weakly temperature dependent, if at all {62,50,30}.

The author wishes to express his sincere gratefulness towards F.N. Hooge and T.G.M. Kleinpenning for valuable suggestions.

REFERENCES

{1} van der Ziel, A., Flicker noise in electronic devices, Advances in Electronics and Electron Physics 49 (Ed. L.Marton,and C. Marton,Academic Press N.Y.) 1979) 225-297.

{2} Voss, R.F., 1/f (flicker) noise: a brief review, Proc. 33rd annual frequency control Symp. Atlantic City, N.J. 30 May-1 June (1979) 40-46.

{3} Bell, D.A., A survey of 1/f noise in electrical conductors, J. Phys. C: 13 (1980) 4425-4437.

{4} Weissman, M.B., Survey of recent 1/f noise theories, Proc. 6th Int. Conf. on Noise in Physical Systems NBS special publ. no. 614 (1981) 133-142.

{5} Dutta, P. and Horn, P.M., Low-frequency fluctuations in solids: 1/f noise, Rev. Mod. Phys. 53 (1981) 497-516.

{6} Hooge, F.N., Kleinpenning, T.G.M. and Vandamme, L.K.J., Experimental studies on 1/f noise, Rep. Prog. Phys. 44 (1981) 479-532.

{7} Voss, R.F. and Clarke, J., 1/f noise from systems in thermal equilibrium, Phys. Rev. Lett. 36 (1976) 42-45.

{8} Beck, H.G.E. and Spruit, W.P., 1/f noise in the variance of Johnson noise, J. Appl. Phys.

49 (1978) 3384-3385.
{9} Zaklikiewicz, A.M., Influence of burst noise on noise spectra, Solid-State Electron. 24 (1981) 1-3.
{10} Rahal, S., and Chovet, A., Variance noise and temperature fluctuations in semiconductors, Electron. Lett. 15 (1979) 271-272.
{11} Hooge, F.N. 1/f noise is no surface effect, Phys. Lett. 29A (1969) 139-140.
{12} Butterweck, H.J., Noise voltages of bulk resistors due to random fluctuations of conductivity. Philips Res. Rep. 30 (1975) 316*-321* (spec. issue).
{13} Vandamme, L.K.J. On the calculation of 1/f noise of contacts, Appl. Phys. 11 (1976) 89-96.
{14} Coppus, G.W.M. and Vandamme, L.K.J., Spreading resistance and 1/f noise of embedded ellipsoidal electrodes in a conductor, Appl. Phys. 20 (1979) 119-123.
{15} Vandamme, L.K.J. and Groot, J.C.F., 1/f noise and resistance between circular electrodes, Electron. Lett. 14 (1978) 30-32.
{16} Vandamme, L.K.J. and van Bokhoven, W.M.G., Conductance noise investigations with four arbitrarily shaped and placed electrodes. Appl. Phys. 14 (1977) 205-215.
{17} van Bokhoven, W.M.G., Calculation of noise in distributed conductive elements, due to stochastic conductivity fluctuations, Arch. Elektron. Uebertragungstechn. 32 (1978) 349-352.
{18} Vandamme, L.K.J. and de Kuijper, A.H., Conductance noise investigations on symmetrical planar resistors with finite contacts, Solid-State Electron. 22 (1979) 981-986.
{19} Vandamme, L.K.J. and Kamp, L.P.J., Transverse and longitudinal noise. J. Appl. Phys. 50 (1979) 340-342.
{20} Weissman, M.B., Relations between geometrical factors for noise magnitudes in resistors, J. Appl. Phys. 51 (1980) 5872-5875.
{21} Black, R.D., Snow, W.M. and Weissman, M.B., Non scalar 1/f conductivity fluctuations in carbon, gold and chrome films, Phys. Rev. B 25 (1982) 2955-2958.
{22} Weissman, M.B., Black, R.D. and Snow, W.M., Calculations of experimental implications of tensor properties of resistance fluctuations. J. Appl. Phys. 53 (1982) 6276-6279.
{23} Black, R.D., Weissman, M.B. and Restle, P.J., 1/f noise in silicon wafers. J. Appl. Phys. 53 (1982) 6280-6284.
{24} Black, R.D., Restle, P.J. and Weissman, M.B., Mobility fluctuation 1/f noise in silicon, to be published.
{25} Kleinpenning, T.G.M., Scattering mechanisms and 1/f noise in semiconductors. Physica, 103B (1981) 345-347.
{26} Vaes, H.M.J. and Kleinpenning, T.G.M., Hall-effect noise in semiconductors. J. Appl. Phys. 48 (1977) 5131-5134.
{27} Kleinpenning, T.G.M., 1/f noise in Hall effect: Fluctuations in mobility, J. Appl. Phys. 51 (1980) 3438.
{28} Kleinpenning, T.G.M., Current noise in some transition-metal compounds, Physica 59 (1972) 370-378.
{29} Fleetwood, D.M., Masden, J.T. and Giordano, N., 1/f noise in platinum film and ultrathin platinum wires: Evidence for a common, bulk origin, Phys. Rev. Lett. 50 (1983) 450-453.
{30} Hooge, F.N. and Hoppenbrouwers, A.M.H., 1/f noise in continuous thin gold films, Physica 45 (1969) 386-392.
{31} Voss, R.F. and Clarke, J., Flicker (1/f) noise:Equilibrium temperature and resistance fluctuations, Phys. Rev. B13 (1976) 556-573.
{32} Black, R.D., Weissman, M.B. and Fliegel, F.M. 1/f noise in metal fimlslacks spatial correlation, Phys. Rev. B24 (1981) 7454-7456.
{33} Andrushko, A.F., Bakshi, I.S. and Zhigal'skii, G.P., Effects of Structural Factors on the 1/f noise of aluminium films, Radiophys. & Quantum Electron. 24 (1981) 343-346. Transl. of "Izv. VUZ radio frz."
{34} Fleetwood, D.M., and Giordano, N., Resistivity dependence of 1/f noise in metal films, Phys. Rev. B27 (1983) 667-671.
{35} Vandamme, L.K.J., Criteria of low-noise thick film resistors, Electrocompon. Sci.&Technol. 4 (1977) 171-177.
{36} Hooge, F.N. and Vandamme, L.K.J., Lattice scattering causes 1/f noise, Phys. Lett. 66A (1978) 315-316.
{37} Hooge, F.N., Kedzia, J. and Vandamme, L.K.J., Boundary scattering and 1/f noise, J. Appl. Phys. 50 (1979) 8087-8089.
{38} Hoppenbrouwers, A.M.H. and Hooge, F.N., 1/f noise of spreading resistances, Philips Res. Rep. 25 (1970) 69-80.
{39} Honig, E.P., 1/f noise of bodies of arbitrary shape and of point contacts, Philips Res. Rep. 29 (1974) 253-260.
{40} Weissman, M.B., Simple model for 1/f noise. Phys. Rev. Lett. 35 (1975) 689-692.
{41} Vandamme, L.K.J., 1/f noise of point contacts affected by uniform films J. Appl. Phys. 45 (1974) 4563-4565.
{42} Vandamme, L.K.J. and Douib, A., Specific contact resistance and noise in contacts on thin layers, Solid-State Electron. 25 (1982) 1125-1127.
{43} Vandamme, L.K.J., On 1/f noise in Ohmic contacts. PhD. Thesis (1976), Eindhoven Univ. of Technology, Netherlands.
{44} Weissman, M.B., Surface 1/f noise in contacts, in press.
{45} Kleinpenning, T.G.M., 1/f noise in thermo EMF of intrinsic and extrinsic semiconductors, Physica, 77 (1974) 78-98 and Physica, 77 (1974) 102.
{46} Kleinpenning, T.G.M., 1/f noise of hot carriers in N-type silicon, Physica 103B (1981) 340-344.
{47} Kleinpenning, T.G.M., On 1/f noise of hot electrons in silicon, Physica 113B (1982) 189-194.
{48} Weissman, M.B., Implications of mobility-fluctuation descriptions of 1/f noise in semiconductors, Physica B, 100 (1980) 157-162.
{49} Hilsum, C. Simple empirical relationship between mobility and carrier concentration, Electron. Lett. 10 (1974) 259-260.
{50} Palenskis, V. and Shoblitskas, Z., Origin of 1/f noise, Solid State Commun. 43 (1982) 761-763.
{51} Vandamme, L.K.J. and Kedzia, J., Concentration, mobility and 1/f noise of electrons and holes

in thin bismuth films, Thin Solid Films, 65 (1980) 283-292.
{52} Bisschop, J. and de Kuijper, A.H., 1/f noise in thin bismuth films, to be published.
{53} Bosman, G., Zijlstra, R.J.J. and van Rheenen, A., 1/f noise of thermal and hot charge carriers in silicon, Physica 112B (1982) 188-196.
{54} Adkins, C.J. and Koch, R.H., Noise in inversion layers near the metal-insulator transition, J. Phys. C. 15 (1982) 1829-1839.
{55} Kilmer, J., Chenette, E.R., van Vliet, C.M. and Handel, P.H., Absence of temperature fluctuations in 1/f noise correlation experiments in silicon, Phys. Stat. Sol. (a) 70 (1982) 287-294.
{56} Fleetwood, D.M. and Giordano, N., Experimental study of excess low-frequency noise in tin, Phys. Rev. B 25 (1982) 1427-1430.
{57} Stroeken, J.T.M. and Kleinpenning, T.G.M., 1/f noise of deformed crystals, J. Appl. Phys. 47 (1976) 4691-4692.
{58} Bisschop, J., Ph.D. Thesis, Dep. of Elect. Eng. Eindhoven University of Technology, to be published (1983).
{59} Van de Voorde, P., Iddings, C.K., Love, W.F., and Halford, D., Structure in flicker-noise power spectrum of n-InSb, Phys. Rev. B 19 (1979) 4121-4124.
{60} Luk'yanchikova, N.B., Garbar, N.P., Malyutenko, V.K. and Teslenko, G.I., Fluctuations of the conductivity of indium antimonide in crossed electric and magnetic fields, Sov. Phys. Semicond. 15 (1981) 193-198, Transl. of "Fiz. & Tekh. Poluprovodn.".
{61} Van de Voorde, P. and Love, W.F., Magnetic effects on 1/f noise in n-InSb, Phys. Rev. B 24 (1981) 4781-4786.
{62} Vandamme, L.K.J., 1/f noise in homogeneous single crystals of III-V compounds, Phys. Lett. 49A (1974) 233-234.
{63} Hanafi, H.I. and van der Ziel, A., Flicker noise in $Hg_{1-x}Cd_xTe$, Physica 94B (1978) 351-356.
{64} Kedzia, J. and Vandamme, L.K.J., 1/f noise in liquid and solid gallium, Phys. Lett. 66A, (1978) 313-314.
{65} Hooge, F.N., 1/f Noise, Physica 83B (1976) 14-23.

NOISE IN PHYSICAL SYSTEMS AND 1/f NOISE
M. Savelli, G. Lecoy and J-P. Nougier (eds.)
© *Elsevier Science Publishers B.V., 1983*

1/f NOISE IN CERMET AND METANET RESISTORS

A. Van Calster, L. Van Den Eede, S. De Molder and A. De Keyser [1]

Laboratory of Electronics, Ghent State University
Sint-Pietersnieuwstraat 41, B-9000 Gent, Belgium.

1/f noise measurements are reported on screen printed cermet and metanet resistors in the frequency range of 0.1 Hz to 1 KHz. The dependence of the noise on the DC current, on the resistor paste and on the paste-metal contact are investigated. The link between the noise behaviour and the transport properties is examined.

1. INTRODUCTION

Cermet and metanet resistors are of great commercial importance, and are the basis of extremely reliable thick film resistor networks. The cermet thick film resistors are screen printed resistors with Dupont 1300 pastes, while the metanet resistors are made of special designed Sprague pastes. The main difference between both kind of pastes is the thickness. A cermet is typical about 10 μm thick, a metanet is only a few hundred Å thick. Although these are commercial available resistors, one does not understand well the electrical behaviour. This is mainly due to the non-homogeneous structure of the fired resistor : conductive particles embedded in a glassy matrix. Noise measurements combined with temperature coefficient (TCR) measurements are designed to get a better insight in the electrical behaviour of the resistors.

2. MEASUREMENTS

Because metanets and cermets are extremely low noise resistors extra care had to be taken to the noise measurement equipment, especially to the grounding and shielding. A schematic view of the measurement equipment is shown in figure 1. The noise of resistor R_T is amplified by a JFET amplifier. The amplified noise is filtered and sampled. The noise spectrum is digitally calculated using a fast fourier transform. For low resistor values a bridge method is used and an impedance transformer is placed in front of the amplifier. For extremely low frequencies a digital cross correlation technique is used [1]. With this equipment noise measurements could be carried out for resistors varying from 5 Ω to 10^4 K Ω in the frequency range of 10^{-2} to 10^4 Hz.

Our noise measurements can be divided into two parts. First we examined the presence of contact noise. In the second place we verified the 1/f behaviour of the noise spectrum and the square dependence of the noise spectrum on the applied DC voltage.

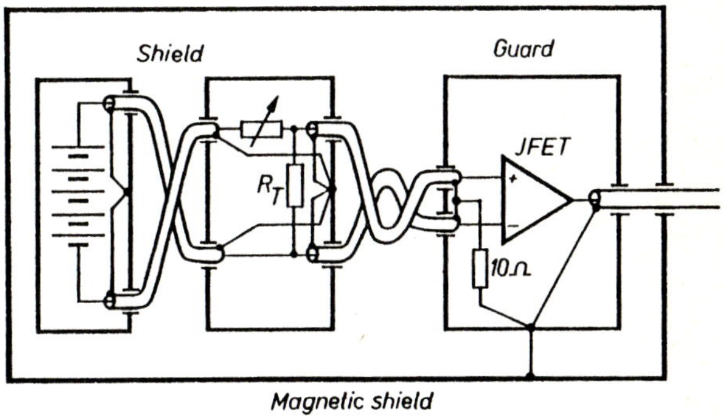

Figure 1. The noise measuring circuit

The presence of contact noise was examined on the test structures T1 and T2, shown in figure 2.

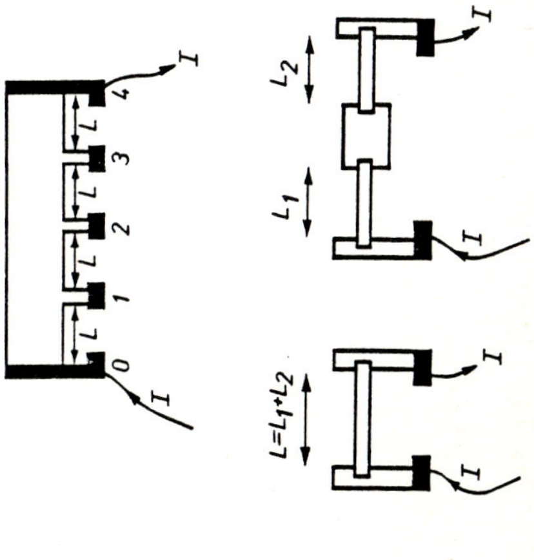

Figure 2. Teststructure T1 Teststructure T2

By comparing the noise spectra of the two resistors of T2 or by comparing the noise voltages between the tabs 01, 12, 23 and 34 of T1 one can detect the presence of contact noise. Neither in metanets, nor in cermets contact noise was measured. The measured noise spectra may be associated with the paste. The 1/f behaviour of the noise and the square dependence of the noise voltage on the applied DC voltage was verified by fitting the relation :

$$\frac{S_V}{V^2} = \frac{C}{f} , \qquad (1)$$

where S_V the measured noise spectrum, V the applied DC voltage, f the frequency and C a constant expected to be independent of V. A typical spectrum of a 57 kΩ metanet resistor is shown in figure 3.

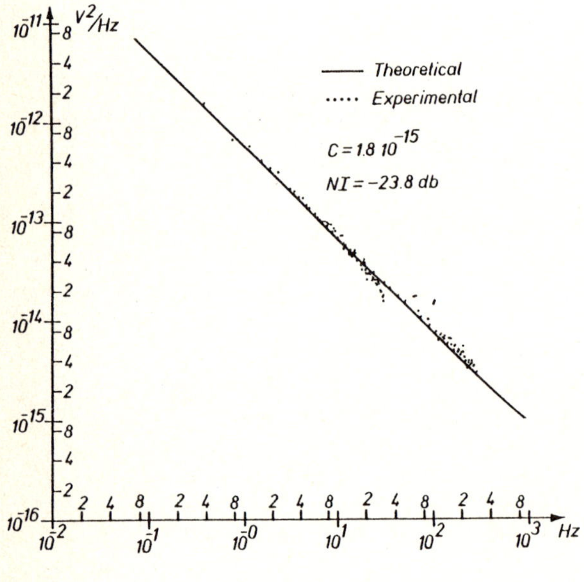

Figure 3. The noise spectrum of a 57KΩ metanet resistor.

From figure 3 it is seen that the noise shows a 1/f dependence. An example of the C values derived for different values of V is shown in table 1. NI represents the noise index.

V	C	NI (db)
27.3 V	1.810^{-15}	- 23.8
36.6 V	1.810^{-15}	- 23.8
63.7 V	1.710^{-15}	- 24.1

Table 1.

From table 1 it follows that C is constant, and thus that the noise fits the square law. The conclusions mentioned above also apply to cermets.

Although both pastes behave similar, there is a remarkable difference in noise level. For example a thin metanet resistor (400 Å) of 60 KΩ shows a noise index NI of -24 db, while a thick cermet resistor (15 μm) of 60 KΩ has a NI of -18 db. 1/f noise in thick film resistors is believed to be a volume effect [2], and normally leads to a NI which is inversely proportional to the volume of the resistor. Thus one could expect the cermet to be less noisier. This is not the case because cermets and metanets are made of different materials. On the other hand both pastes have noise indices which are inversely proportional to the resistor area, and thus both pastes agree with the expected dependence of 1/f noise on paste area [2].

3. DISCUSSION

Until now no physical model is available to explain 1/f noise in screen printed resistors. As 1/f noise in thick film is believed to be a volume effect, we tried to link 1/f noise with the bulk transport properties.
Generally it is assumed that the transport properties are determined by the amorphous (glassy) layer between the conductive particles. One normally assumes a variable range hopping mechanism, leading to the following characteristic TCR behaviour :

$$TCR \sim T^{1/2} \exp(T_o/T)^{1/4} , \qquad (1)$$

which shows a minimum at temperature T_{min} equal to $T_o/16$. T_o is inversely proportional to the density of localised states at the Fermi level per unit energy in the amorphous layer. The TCR of the metanets as well as the TCR of the cermets agreed very well with expression (1). But as can be seen from figure 4, the metanets showed a much smaller T_{min} (163 K) than the cermets (270 K).

Thus it may be concluded that most probably the current mechanism in both pastes is the same. Furthermore if one assumes that 1/f noise is related to the bulk transport properties, although this is perhaps less the case in the thin metanets, it may be concluded that also the noise mechanism in both pastes is the same. A possible link between the noise and the current mechanism is to relate the 1/f noise to the tunnel process involved in variable range hopping, which gives rise to a broad spectrum of time constants.

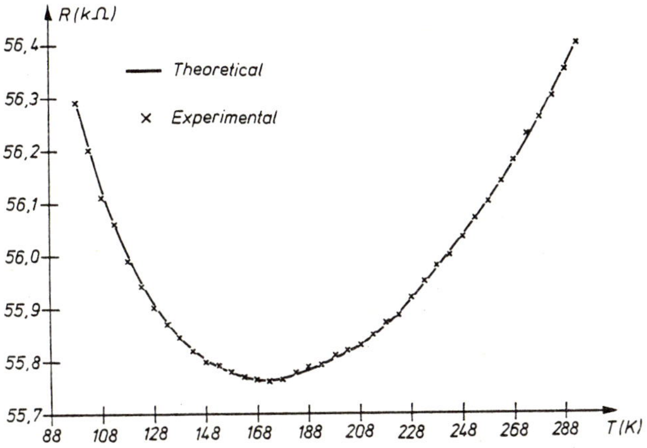

Figure 4. TCR of a metanet

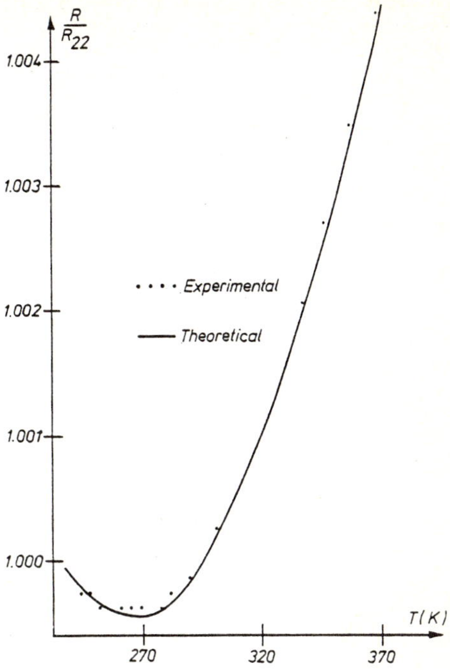

TCR of a cermet

4. CONCLUSIONS

Nonhomogeneous materials, such as thick film pastes, are far more difficult to investigate than homogeneous materials. Even in homogeneous materials 1/f noise is still a topic of a comprehensive research. In the case of thick film pastes there are even still some questions about the conduction mechanism, so the sources of 1/f noise are even more difficult to characterize. Nevertheless we believe that both investigated pastes (cermets as well as metanets) can be described by the variable range hopping conduction mechanism of the amorphous (glassy) layer between the conductive particles. From the point of 1/f noise it seems reasonable to link the noise with this bulk conduction process by the tunnel process involved, although surface phenomena may have some importance in the case of metanets.
A second fact is that although metanets pastes are much thinner than cermet pastes, metanets are less noisier than cermets. This makes metanets more attractive from the point of view of noise index.

5. REFERENCES

[1] S. DEMOLDER, M. VANDENDRIESSCHE and A. VAN CALSTER, "The measuring of 1/f noise of thick and thin film resistors", J. Phys. E, 13 (1980), p. 1323.

[2] S. DEMOLDER, A. VAN CALSTER and M. VANDENDRIESSCHE, "Current Noise in thick and thin film resistors", Proceedings of the European Hybrid Micro Electronics Conference, Avignon (France) 1981, p. 19.

1. DEKEYSER is with Sprague Electromag, De Merodestraat 2, B-9600 Ronse, Belgium.

NOISE IN PHYSICAL SYSTEMS AND 1/f NOISE
M. Savelli, G. Lecoy and J-P. Nougier (eds.)
© *Elsevier Science Publishers B.V., 1983*

1/f NOISE IN SILICON-ON-SAPPHIRE

M. B. Weissman, R. D. Black and P. J. Restle

Physics, Department, University of Illinois at Urbana-Champaign
1110 West Green Street, Urbana, IL 61801 USA

Measurements of 1/f noise in silicon-on-sapphire (SOS) resistors show that it is Gaussian and approximately scalar. The relative magnitude of the Hall coefficient noise and the resistivity noise is temperature dependent, but at 100 K, 140 K and 300 K these two variables are approximately completely correlated. Temperature dependence of spectral features shows that the kinetics are thermally activated. The results nicely fit a model in which the depth of fast interface traps depends on slow, thermally activated transitions of lattice defect states.

INTRODUCTION

The origin of 1/f noise in semiconductors and metals remains a topic for much dispute.[1-4] Among the question raised are:

1) To what extent is the noise a surface effect?
2) To what extent is the fluctuating variable mobility or carrier number?
3) What is the mechanism by which the fluctuating variable is affected?
4) What causes the spectrum to be nearly 1/f?

In this paper we present evidence which at least partially answers all of these questions for a simple system - SOS. Since some of this is being published elsewhere[5,6], we shall emphasize here the newest results, including some very suggestive results on other systems which indicate that the answer to our fourth question may have general applicability.

TECHNIQUES

The samples consisted of 8-contact crosses of ~ 0.6 μm thick SOS, doped n-type to room-temperature carrier concentration ~ $5 \cdot 10^{16} cm^{-3}$, with inner dimensions of the crosses being ~ 25 μm. Except when otherwise specified the samples were oxidized about 0.1 μm thick. More detailed descriptions of how the samples were prepared are being given elsewhere.[6] Our measuring apparatus has also been described elsewhere.[6,7,8]

Noise statistical properties were measured by taking histograms of integrated noise power per octave in batches of single discrete-transform records. Matrices representing normalized variances and covariances between these noise powers were also computed.

RESULTS

On all SOS samples the noise statistics as determined either by the histograms or by the covariance matrices were consistent with Gaussian statistics. Any envelope to the noise power represented a modulation of less than one part in 50, so that if the noise arose from many highly non-Gaussian sources (e.g. two-state systems) well over 10^3 would be involved. The noise magnitude differed by roughly a factor of three between samples, but on each sample the multi-lead measurements indicated that the noise arose about equally in different parts of the plane, also consistent with a large number of distributed noise sources.

Etching away the thick oxide, leaving only room-temperature thermal oxide, dramatically changed the spectral shape and temperature dependence of the noise. These properties were otherwise essentially unchanged between samples. The use of wet or dry oxide and the implantation of argon ions in the oxide caused little or no change in these quantities. We found slow equilibrations of the noise magnitudes after temperature changes only in thin-oxide samples, suggesting a slow equilibration of some surface contaminants. Treatment of the surface with fluorcarbons also leads to temporary changes in the noise magnitude. The Hooge parameter $S_V(f) \cdot N \cdot f/V^2$ for these samples at room temperature was ~ 10^{-4}, nearly two orders of magnitude larger than we have found in silicon wafers with about two orders of magnitude smaller surface-to-volume ratios.[7] We conclude that the noise in SOS is predominantly a surface effect, with the silicon-oxide interface being particularly important, the exterior oxide surface being important when the oxide layer is thin, and the surface potential affecting the noise level in all samples, reminiscent of previous observations.[9,10]

Cross-correlation measurements of noise on different paths[8,11] showed that the (two-dimensional) resistivity fluctuations were scalars or nearly so at 100 K, 140 K, and 300 K. (The noise in a thick silicon wafer was also scalar.[7]) When a magnetic field was introduced, Hall voltage fluctuations were found. Their magnitude, in comparison with the resistivity fluctuations was ~ 1.5, ~ 1.8, ~ 0.6 at 100 K, 140 K, and 300 K, where the normalization used would give a value of 1.0 for simple number fluctuations. The value of 0.6 in particular showed that some mobility fluctuations were present. It was found that shining a light on the sample changed both carrier concentration and mobility, which is precisely what would be expected for changing occupancy of traps in these samples, which affects Coulomb scattering as well as the number of free electrons.[6] The relative sign of the mobility fluctuations and number fluctuations would require that the traps be acceptors, and these are in fact the predominant active interface states on n-type silicon.[12]

The relatively large Hall noise at low temperature can be easily accounted for if carrier number fluctuations occur primarily near the surface, where the mobility is particularly low, especially at the lower temperatures where phonon scattering has decreased and the number of charged surface sites has increased.

If fluctuating trap occupancy caused both the mobility and number fluctuations, one would expect not only that the fluctuations would be scalars but also that the Hall coefficient fluctuations and resistivity fluctuations be highly correlated since they are due to fluctuations in the same parameter. (A slight non-correlation would appear if the locations of the traps were not all similar or if there were a few donors.) A correlation coefficient of 0.9 ± 0.15 was found at each temperature.

We have reported previously[5] that the spectra were not exactly 1/f and that the spectral features shifted in frequency as a function of temperature. The results are qualitatively consistent with the behavior predicted for activated kinetics with a range of activation energies[1] over the temperature range 100 K - 320 K and the frequency range 5 Hz - 10 kHz for all samples. (No tests were run outside this range.) Due to the wide frequency range and relatively narrow spectral frequency range and relatively narrow spectral features, the thermal activation is even more evident[5] than that found in metal films.[1] The noise magnitudes have no very strong temperature dependence other than that associated with the shifting kinetic features.

It is particularly interesting that while the thin-oxide sample showed noise (at room temperature) the turned from ~ f^{-1} below Hz to ~ $f^{-0.7}$ up to 10 kHz, the expected relation between the spectral shape and the temperature dependence[1] still held. In the ~ $f^{-0.7}$ range the noise magnitude was about proportional to T^{-5}, while it was nearly temperature-independent in the f^{-1} regime, all measured in the temperature range 276 K - 333 K in a dry N_2 atmosphere.

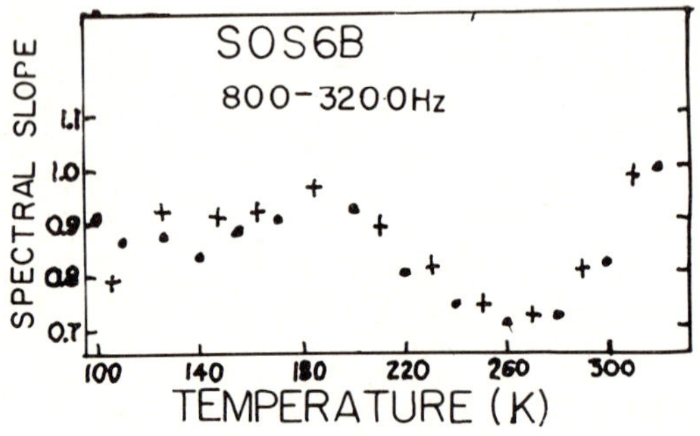

Figure 1 The logarithmic slope of the noise spectrum (dots) is compared with that computed from the temperature dependence of the noise power, according to a relation derived in reference 1 for noise with thermally activated kinetics. A temperature independent prefactor in the rate equation of 10^9 Hz was used. The absence of a vertical shift between the two plots shows that there is no strong temperature dependence (i.,e., sharper than $T^{\pm 2}$) other than that caused by the kinetic shifts. The sample has thick oxide.

CONCLUSION

The results point very strongly toward models in which the conductivity fluctuations are directly caused by fluctuating occupancy of the interface acceptor states. However, the thermal activation energies whose range determines the 1/f spectrum cannot correspond to the depths of these relatively shallow traps. The spectral form together with the temperature dependence cannot be reconciled with such a model.[5,6] The most plausible models would involve slow, thermally activated motions of the atoms which affect the depth of the electronic states. The fluctuations of the occupancy of the fixed states themselves is too fast to show up significantly in the observed frequency range. The range of activation energies found would then reflect the amorphous nature of the interface and the neighboring glassy oxide and would be reminiscent of the range of energies for two-level systems characteristic of the transitions which dominate the low-temperature heat capacity of amorphous materials.[13,14]

We emphasize that although our samples, thin semiconductors coated with oxide, are ideal model systems in which to see McWhorter noise, it was not the dominant noise source. The observed thermally activated kinetics are not compatible with the elegant distribution-of-tunneling-distances explanation of the spectrum. Although the mechanism by which the carrier number and mobility fluctuate seems to involve changing occupancy of traps, the evidence points away from theories which try to tie that mechanism too closely to the explanation of the spectral shape, as becomes even clearer when comparisons are made with other materials.

RELATED STUDIES OF OTHER MATERIALS

Given a reasonable physical model accounting for some rather detailed, reproducible experimental results in SOS, an obvious question is the extent to which the results generalize to other systems. The detailed mechanism by which the conductivity is affected cannot be very similar in metal films, for which the conductivity fluctuations are not close to scalars.[8,15] In fact, despite the qualitative similarity in thermal activation behavior in various metals, including gold and the semi-metal bismuth[1], the mechanism by which their conductivities fluctuate cannot be too similar, since chrome[8] and bismuth[15] have significantly more (locally) anisotropic fluctuations than does gold.[8] Number fluctuations in metal films are at any rate unimportant by magnitude arguments.[3] In metals there appear to be both "bulk" (the films are actually full of non-equilibrium defects) effects and smaller surface effects.[1] Therefore of our four initial questions we expect at most the fourth to have a general answer.

Good evidence for 1/f noise from thermally activated kinetics for transitions in ensembles lacking any very strong temperature dependence has now been found for gold[1], copper[1], silver[1], bismuth[1,15], thick-film resistors[16], ordinary carbon-composition resistors[6], the one-dimensional conductor NMP-phenazine-TCNQ[17], and Josephson junctions.[18] From the metal-to-metal variation in the spectrum of activation energies, it appears likely that in metals the activated process also involves some defect motion.[1]

In SOS we found that a solid-gas interface near the conducting region provided an additional source of noise characterized by a distribution of activation energies strongly weighted toward small energies. This is reminiscent of previous results on silver[1], in which a very thin sample appeared to have disproportionately many low-activation energy contributions.

Our data on bismuth[15] show that in addition to the temperature dependence of the noise magnitude attributable to kinetic shifts there is, near 300 K, a roughly T^{-3} dependence. (Such behavior is easier to detect in metals because the features are very broad, so that the error in determining the kinetic effects on the temperature dependence of the spectrum is small.) Together with the highly non-scalar fluctuations[8,15], this fact suggests the possibility that some of the impurity scattering includes locally anisotropic terms with time dependence caused by the motion (either constrained or diffusive) of defects. As the temperature is increased, this scattering becomes masked by relatively quiet phonon scattering.

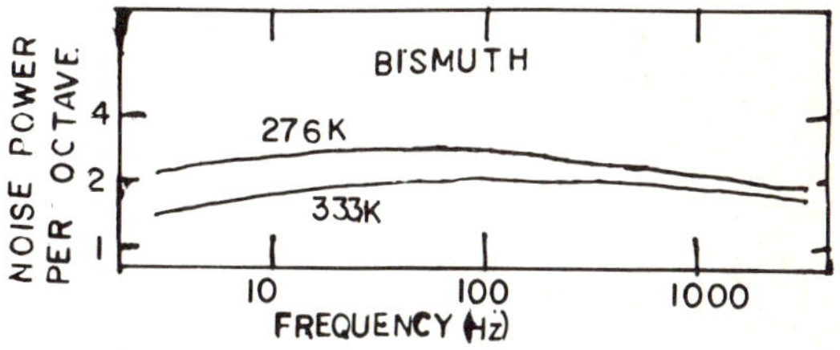

Figure 2 Noise power, normalized by squared DC voltage, per octave at two temperatures in a 30 nm thick Bi sample at two tempertures. Although the absolute scale is not calibrated, the comparison of the noise powers at the two temperatures is properly normalized. It is apparent that even in the flat ($f^{-1.0}$) part of the spectra, there is a decrease in noise at higher temperatures, which cannot be attributed to kinetic shifts.

The overall picture then seems to be that whenever there are sufficiently amorphous regions, the apparently universal presence of low-energy excitations[13,14] with smeared-out distributions of activation energies gives rise to a set of weakly temperature dependent processes, with temperature dependent, 1/f-like kinetics. The mechanisms by which such processes affect electrical properties vary as a function of material and temperature.

If our interpretation is correct, there is an important practical implication. To the extent that the lattice defects created by implantation of dopants can be annealed out of a semiconductor without allowing too much diffusion of the dopants, it should be possible to make devices in which 1/f noise is simply absent.

ACKNOWLEDGEMENTS

We thank J. Gilliand of Hewlett-Packard for the SOS, and G. Anner for use of micro-electronics facilities. This work was supported by NSF Grant DMR 80-07057 and, through facility usage at the Materials Research Laboratory by NSF Grant DMR 80-20250.

REFERENCES

[1] Dutta, P. and Horn, P. M., Rev. Mod. Phys. 53 (1981) 497.

[2] Hooge, F. N., Kleinpenning, T. G. M., and Vandamme, L. K. J., Rep. Proc. Phys. 44 (1981) 479.

[3] Weissman, M. B. in Meijer, P. H. E., Mountain, R. D. and Soulen, R. J., Proc. Sixth Int. Conf. on Noise in Phys. Syst. (Dept. of Commerce, Washington, D.C., 1981) p. 133.

[4] Van der Ziel, A. in Marton, L. and Marton, C., Adv. in Elect. and Elect. Phys. V49 (Academic, N.Y., 1979) p. 225.

[5] Weissman, M. B., Black, R. D., Restle, P. J., and Ray, T., Phys. Rev. B27 (1983) 1428.

[6] Black, R. D., Restle, P. J., and Weissman, M. B., submitted to Phys. Rev. B.

[7] Black, R. D., Weissman, M. B., and Restle, P. J., J. Appl. Phys. 53 (1982) 6280.

[8] Black, R. D., Snow, W. M., and Weissman, M. B., Phys. Rev. B25 (1982) 2955.

[9] Macrae, A. U., J. Appl. Phys. 33 (1962) 2570.

[10] Amberiadis, K., Van der Ziel, A., and Rucker, L. M., J. Appl. Phys. 52 (1981) 6989.

[11] Weissman, M. B., Black, R. D., and Snow, W. M., J. Appl. Phys. 53 (1982) 6276.

[12] Gray, P. V. and Brown, D. M., Appl. Phys. Lett. 8 (1966) 31.

[13] Anderson, P. W., Halperin, B. I., and Varma, C. M., Philos. Mag. 25 (1972) 1.

[14] Phillips, W. A., J. Low Temp. Phys. 7 (1972) 351.

[15] Restle, P. J., Black, R. D., and Weissman, M. B., unpublished results.

[16] Pellegrini, B., Saletti, R., Terreni, P., and Prudenziatti, M., Phys. Rev. B27 (1983) 1233.

[17] Rommelman, H., Epstein, A. J., Restle, P. J., Black, R. D., and Weissman, M. B., unpublished results.

[18] Koch, R. H., Bull. Ann. Phys. Soc. 28 (1983) 570.

NOISE IN PHYSICAL SYSTEMS AND 1/f NOISE
M. Savelli, G. Lecoy and J-P. Nougier (eds.)
© *Elsevier Science Publishers B.V., 1983*

1/f NOISE IN METAL FILMS: RESISTIVITY DEPENDENCE AND SAMPLE-TO-SAMPLE VARIATIONS

D. M. Fleetwood and N. Giordano

Department of Physics
Purdue University
West Lafayette, Indiana 47907, U. S. A.

The 1/f noise of a number of different types of metal films has been studied at room temperature. We find that the noise magnitudes of nominally identical samples can vary by more than a factor of 10; however, the minimum noise level of a given material is a fairly well-defined quantity. The resistivity dependence of the minimum noise level cannot be accounted for by the empirical formula of Hooge. Recent experiments imply that at least some of the sample-to-sample variations of the noise are due to very slow, nonstationary processes, perhaps associated with the relaxation of strain within the sample.

It has been widely accepted that the 1/f noise of continuous metal films at room temperature is in order-of-magnitude agreement with Hooge's empirical formula [1]:

$$S_V = \frac{\gamma_H V^2}{N_c f} \quad . \tag{1}$$

Here S_V is the excess voltage noise power spectral density, V is the average voltage across the resistor, N_c is the number of free charge carriers, and $\gamma_H \simeq 2 \times 10^{-3}$. Hooge and Vandamme have shown that (1) must be modified to account for the noise of thin bismuth films and heavily doped semiconductors.[2] In these cases they find:

$$S_V = \left(\frac{\mu}{\mu_{ph}}\right)^2 \frac{\gamma_H V^2}{N_c f} \quad , \tag{2}$$

where μ is the total mobility, and μ_{ph} is the contribution to the mobility due to electron-phonon scattering. Hence they conclude that the 1/f noise of these systems is related to electron-phonon scattering. Although neither (1) nor (2) is consistent with the temperature dependence of the noise in metal films observed by Eberhard and Horn , [3] many theories of the noise have been proposed in which the Hooge formula is an exact or approximate limiting form.[4] It is therefore important to determine the nature and extent of any limitations of (2) as a description of the noise at room temperature. We have performed an extensive study of the 1/f noise of several different types of metal films.[5] We find that the magnitude of the noise of nominally identical samples can vary by more than a factor of 10, which is well outside the estimated (factor of 2) uncertainty of the quantities that, according to (1) and (2), are thought to be relevant in determining the noise. Recent work that we have performed implies that at least some of the observed sample-to-sample variations are due to very slow, nonstationary processes, associated perhaps with the relaxation of strain. Despite these variations, we find that the minimum level of 1/f noise is a fairly well-defined quantity. Moreover, the minimum noise level exhibits a systematic resistivity dependence which cannot be accounted for by (1) or (2).

Details of the techniques used to fabricate the samples and perform the noise measurements have been given in previous publications.[5,6] In Fig. 1 we show the normalized noise magnitude, $\gamma \equiv S_V Nf/V^2$ at f = 10 Hz, where N is the number

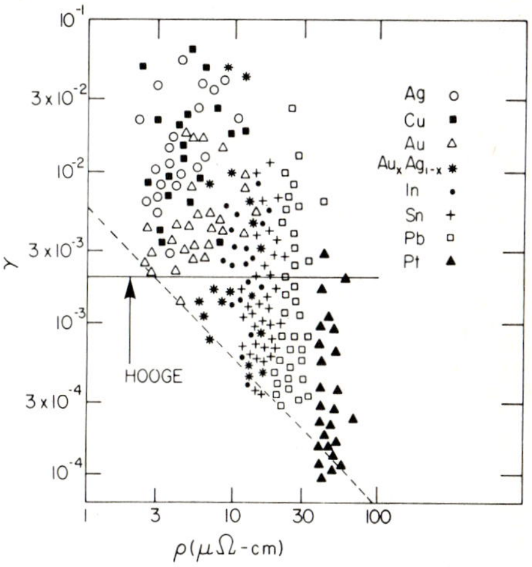

Figure 1: $\gamma \equiv S_V Nf/V^2$ at f = 10 Hz as a function of ρ for several metals. Overlapping points have been omitted for clarity. Here N is the number of atoms in the sample. The solid line is the prediction of (1) with $\gamma = \gamma_H$. The dashed line represents (3), with $\rho_0 = 6 \times 10^{-3}$ $\mu\Omega$-cm.

of atoms in the sample,[7] as a function of the (measured) sample resistivity, ρ. There is a large sample-to-sample variation of γ, even for "nominally identical" samples made of the same material. We shall discuss these variations below. The minimum noise observed for a given metal, however, is a fairly well-defined quantity, and exhibits a systematic dependence on ρ. It is seen in Fig. 1 that the variation of the minimum noise level of the pure metals in which electron-phonon scattering is dominant (all but Pt, as discussed below) is not consistent with (1), although some of this variation may be attributable to the differences between N and N_c.[5] Also, the noise of Pt and the Au_xAg_{1-x}, in which elastic scattering is important, cannot be accounted for by (2). Since $\mu/\mu_{ph} \simeq 1/4$ for Pt, and $\mu/\mu_{ph} \simeq 1/8$ for $Au_{0.5}Ag_{0.5}$, (2) predicts that $Au_{0.5}Ag_{0.5}$ should be about four times quieter than Pt. Instead, it is about four times noisier than Pt. Thus, we find no evidence from these data that the 1/f noise of metal films is associated with electron-phonon scattering. Rather, the results for all of these materials display a simple dependence on ρ, regardless of the dominant scattering mechanism. The dashed line in Fig. 1 is proportional to $1/\rho$, and the minimum noise levels are seen to be quite consistent with such a dependence. Incorporating this into (1), we find that the minimum noise levels are well-described by the relation

$$S_{V,min} \simeq \left(\frac{\rho_0}{\rho}\right) \frac{V^2}{Nf^\alpha} \quad , \tag{3}$$

where $\rho_0 \simeq 6 \times 10^{-3}$ $\mu\Omega$-cm.[5,8] We can offer no theoretical basis for (3), and it is not yet clear when or why one should expect to observe this minimum noise level experimentally. However, we feel that (3) represents an improved benchmark for the noise magnitudes of metal films, and may lead to insight into the origin of the noise.

Let us now consider the sample-to-sample variations of the noise observed in Fig. 1. All of the samples were checked for contact noise prior to the noise measurements, and it was found to be negligible. Also, the film was not significantly heated during the measurement. Thus, the noise observed is characteristic of the particular sample – if not of the metal itself. In addition to these sample-to-sample variations, we have found that occasional samples exhibit striking time-dependent, i.e. nonstationary, behavior. In Fig. 2 we show the noise magnitudes, γ, for a Ag and a Pt film as a function of time.[9,10] It can be seen that in both cases γ decreased systematically, which indicates that the process which causes the increased noise is itself relaxing and/or disappearing with time. One process which might account for such behavior is strain relaxation. We have performed experiments to determine the effect of stress/strain on the noise of metal films. In Fig. 3 we show γ as a function of time for a Pt film.[11] At A no external stress had been applied. The substrate was deformed at B by clamping its edges to a metal block, and raising a nylon screw from below. It can be seen that γ increased by a factor of 4, and then decreased

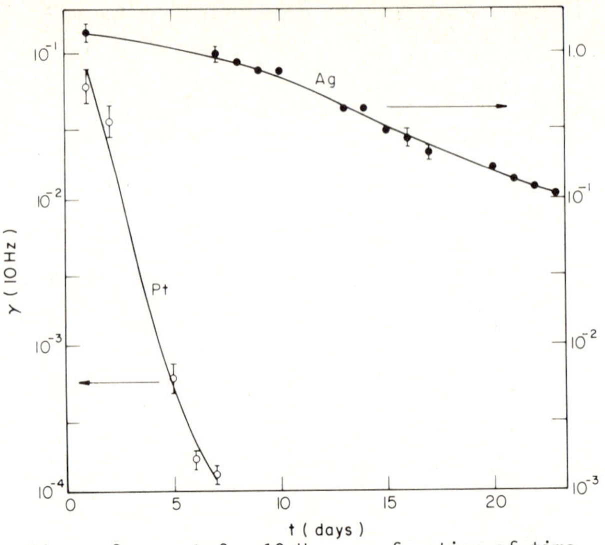

Figure 2: γ at f = 10 Hz as a function of time for a Ag film (solid symbols) and a Pt film (open symbols). The curves shown are guides to the eye. The left-hand scale refers to the Pt film, and the right-hand scale to the Ag film.

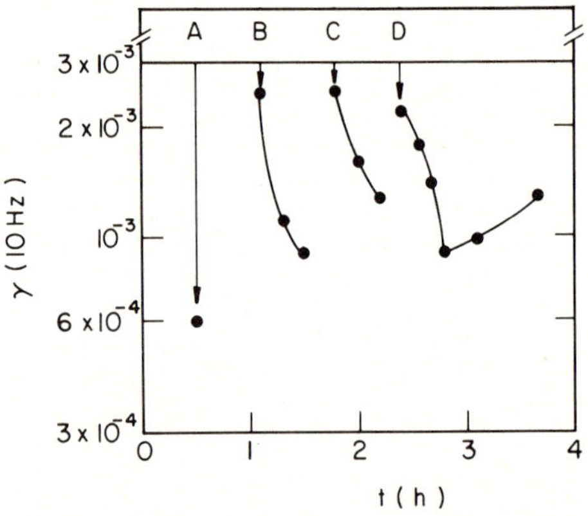

Figure 3: γ at f = 10 Hz as a function of time for a Pt film. Conditions under which the noise was measured at the various points (A, B, etc.) are given in the text. The curves shown are guides to the eye.

with time.[10] At C the deforming force was removed from the substrate. Again the noise initially increased, and then decreased with time. This indicates that, while there is no longer any stress applied to the substrate, the sample experiences a strain since it had relaxed to accommodate the previously deformed substrate. A strain similar to that applied at B (as determined by monitoring the sample resistance) produced a comparable increase in γ at D; hence, the behavior is qualitatively reproducible. Although the time scales are quite different,[12] we note that the decrease in γ with time for films in which strain had been induced is similar to that observed in Fig. 2. Consequently, we remeasured several films whose noise is shown in Fig. 1. We find that some Au_xAg_{1-x} films, which initially had the largest noise, now (six months to one year later) exhibit noise near the minimum level! Moreover, the noise magnitudes of films which were initially near the minimum level have not changed significantly. This strongly suggests that very slow, nonstationary processes, perhaps associated with the relaxation of strain within the sample, cause at least some of the sample-to-sample variations observed in Fig. 1.[13] In this regard we should mention that many processes associated with stress/strain relaxation have activation energies[14] similar to those which Dutta and Horn[15] have shown to generate 1/f noise, for the case in which the spread in energies is greater than k_BT. It is therefore worthwhile to consider at least two possibilities: 1) Strain relaxation, or a related process, increases the noise of the samples above a level that is characteristic of the metal film; or 2) All of the noise observed in these samples could be caused by a strain-sensitive process. While our results appear to be more consistent with the first interpretation, further work is needed to distinguish between these possibilities.

In summary, we have performed an extensive study of 1/f noise in a number of metal films. The noise magnitudes of nominally identical samples can vary by more than a factor of 10. Some of the sample-to-sample variations reported are evidently due to very slow, nonstationary processes, such as strain relaxation. Despite these sample-to-sample variations, the minimum noise level of a given material is a fairly well-defined quantity, which varies inversely with the resistivity of the sample. Such a variation does not appear to be consistent with Hooge's empirical formula, and certainly warrants further study.

We thank J. Clarke, S. J. Hruska, R. F. Voss, and M. B. Weissman for stimulating discussions, P. H. Keesom for the use of his screened room, and J. T. Masden and D. E. Beutler for assistance in sample fabrication. This work was supported in part by University and David Ross fellowships (to D.M.F.), by an Alfred P. Sloan Foundation Research Fellowship (to N.G.), and by the NSF-MRL program through Grant No. DMR80-20249.

REFERENCES:

[1] Hooge, F. N. and Hoppenbrouwers, A. M. H., 1/f noise in continuous thin gold films, Physica 45 (1969) 386-392; Hooge, F. N., Discussion of recent experiments on 1/f noise, Physica 60 (1972) 130-144. The frequency dependence of the noise is usually found not to be exactly 1/f, but is of the form $1/f^{\alpha}$, where $0.9 < \alpha < 1.4$.

[2] Hooge, F. N. and Vandamme, L. K. J., Lattice scattering causes 1/f noise, Phys. Lett. 66A (1978) 315-316; Hooge, F. N., Kedzia, J., and Vandamme, L. K. J., Boundary scattering and 1/f noise, J. Appl. Phys. 50 (1979) 8087-8089.

[3] Eberhard, J. W. and Horn, P. M., Excess (1/f) noise in metals, Phys. Rev. B18 (1978) 6681-6693.

[4] See, for example, Hooge, F. N., Kleinpenning, T. G. M., and Vandamme, L. K. J., Experimental studies on 1/f noise, Rep. Prog. Phys. 44 (1981) 479-532, and references therein.

[5] Fleetwood, D. M. and Giordano, N., Resistivity dependence of 1/f noise in metal films, Phys. Rev. B27 (1983) 667-671.

[6] Fleetwood, D. M. and Giordano, N., Experimental study of excess low-frequency noise in tin, Phys. Rev. B25 (1982) 1427-1430.

[7] A number of workers believe that N_c in (1) should be replaced by N. We shall use the latter quantity since it is readily determined; whereas, it is difficult to obtain a realistic estimate of the number of carriers in these samples.

[8] Note that if the units of ρ_0 are $\mu\Omega$-cm, then (3) is not dimensionally correct when α is not equal to 1. The dependence of $S_{V,min}$ on (V^2/ρ) suggests that for metal films $S_{V,min} \sim (I^2R)R_0$ instead of $S_{V,min} \sim V^2 = I^2R^2$.

[9] The contacts to the film were not altered during the course of these measurements.

[10] The frequency dependence of the noise, α, was relatively constant during these measurements.

[11] This particular film was on a glass substrate. Similar experiments have been performed on Au, Ag, Sn, and Pb films on mylar substrates (Fleetwood, D. M. and Giordano, N., to be published) with qualitatively similar results. Hence, we expect that these results should be quite general.

[12] This is not really surprising, owing to the very different manners in which stresses/strains are induced within the samples.

[13] Not all of the samples whose noise was significantly above the minimum level — in particular, Ag and Au films — had decreased noise when remeasured. Indeed, the noise of a few samples increased by up to a factor of 5. Hence, more work needs to be done to understand the variations in Fig. 1 fully, and experiments are underway in an effort to do so.

[14] See, for example, Nowick, A. S. and Berry, B. S., Anelastic Relaxation Processes in Crystalline Solids (Academic Press, New York, 1972).

[15] Dutta, P. Dimon, P., and Horn, P.M., Energy scales for noise processes in metals, Phys. Rev. Lett. 43 (1979) 646-649; Dutta, P. and

Horn, P. M., Low-frequency fluctuations in solids: 1/f noise, Rev. Mod. Phys. 53 (1981) 497-516.

NOISE IN PHYSICAL SYSTEMS AND 1/f NOISE
M. Savelli, G. Lecoy and J-P. Nougier (eds.)
© *Elsevier Science Publishers B.V., 1983*

1/f NOISE IN METAL FILMS OF SUBMICRON DIMENSIONS

J. Kilmer, C.M. Van Vliet, G. Bosman and A. van der Ziel

Department of Electrical Engineering
University of Florida, Gainesville, FL 32611, USA

Photonic quantum 1/f noise has been identified in Au metal films below the Debye temperature. The low values of the Hooge parameter predicted by Handel's theory (i.e., $\alpha_{true} \simeq 10^{-6}$ to 10^{-8}) are arrived at by realizing that only the fraction $3kT/2\Delta\mathcal{E}_F$ of the total number of carriers are available for scattering at a given temperature.

In recent years considerable progress has been made in the understanding of 1/f noise. It is now well established that in many cases there is fundamental 1/f noise caused by mobility fluctuations, in particular by fluctuations in the scattering cross section of scattering of electrons by phonons.[1][2]

The only general theory of 1/f noise which can explain such fluctuations was given by Handel in 1975.[3][4]. However, until recently experimental evidence verifying the theory did not exist. Specifically, Handel's quantum 1/f noise theory was questioned as the source of 1/f noise in electronic circuits because of the low value of the Hooge parameter, α_H, calculated from his theory. Briefly, the theory states that the interference between the part of the carrier's wave function which suffers losses due to an inelastic or "bremsstrahlung" scattering under the emission of infraquanta and the part of the wave function which does not suffer losses produces very low energy beats which translate themselves ($\Delta\mathcal{E} = hf$) as 1/f noise. Handel's theory predicts that scattering involving Umklapp processes (U-processes) provides the largest source of 1/f noise in metals since the photon infraquanta coupling constant, αA, is given by[4]

$$\alpha A = \frac{2\alpha}{3\pi}\left(\frac{\Delta p}{mc}\right)^2 \qquad (1)$$

where α is the fine structure constant $(137)^{-1}$, perhaps modified by the dielectric constant in the metal, c is the velocity of light in the metal, and $\Delta\vec{p}/\hbar = \Delta\vec{k}$ is the change in wave vector. Since the U-process gives the largest $\Delta\vec{k}$, we expect them to be the largest contributor to 1/f noise in metals. Though the dielectric constant of metals is not well known, and may be complex, one easily sees that the corrections in α and c cancel, so that we can further take the free space values.

Gold thin-film resistors (2,000 Å thick) were prepared for us by Dr. E. Wolf and R.A. Buhrman of the National Research and Resource Facility for Submicron Structures at Cornell University. The length of the resistors is close to 800 µm, and we have measured the 1/f noise in 1 µm-width samples. These dimensions give a resistance of a few hundred ohms and the noise spectrum can be readily measured, after amplification, by an HP 3582 Spectrum Analyzer. By incorporating a calibrated noise source, the absolute magnitude of the resistor's current noise spectrum, S_I, can be directly calculated by simply comparing the relative spectra of device on, device off, and calibration source on; it can be shown that the amplifier's parameters cancel out. The gold films were mounted to the cold head of a CTI Cryogenics Model 21 liquid He closed-cycle refrigerator capable of maintaining a stable temperature (i.e., ± 0.1 K over the duration of a low-frequency noise measurement) anywhere between 300 K and 8 K.

The results of the experiment give current spectra proportional to $1/f^{\gamma}$ between 1 and 100 Hz. Below 1 Hz we have cryostat noise, and above 100 Hz we have device noise competing with the amplifier's noise and thermal noise. The slope shows $\gamma \simeq 1.2$ from 300 K to about the Debye temperature (Θ_D = 165 K for gold) in agreement with Fleetwood and Giordano.[5] Below the Debye temperature the slope evens off to $\gamma \simeq 1$, indicating a more "pure" 1/f noise present at the lower temperatures. Next, we characterize the magnitude of the 1/f noise by calculating the dimensionless Hooge parameter, α_H, according to the formula

$$\frac{S_I(f)}{I^2} = \frac{\alpha_H}{fN} \qquad (2)$$

where N is the total number of available electrons in the metal; for those spectra where γ was $\neq 1$, we took f = 10 hz. In Figure 1 we have used N = nV where we set $n \approx 10^{22}$ cm^{-3} and V is the sample volume (800 µm × 1 µm × 0.2 µm). Figure 1 shows an interesting dependence of the magnitude of α_H below the Debye temperature corresponding to the occurrence of the more "pure" 1/f noise.

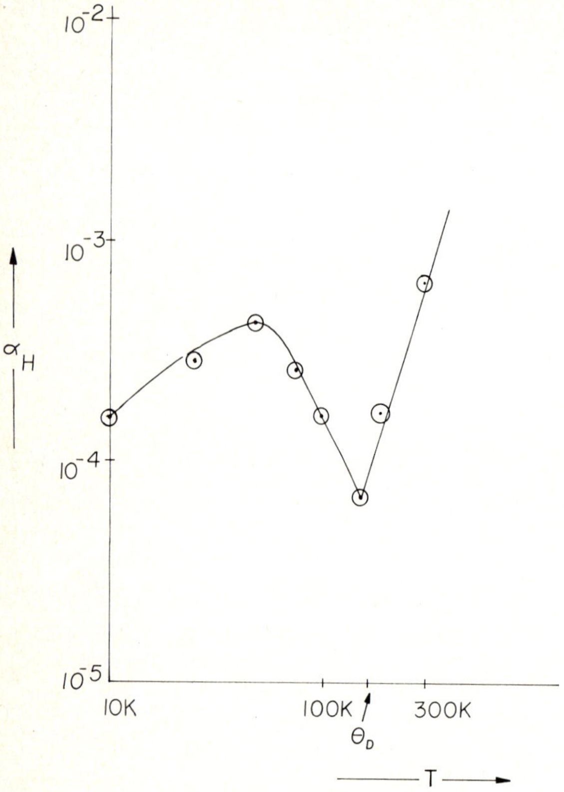

Figure 1. α_H as a function of temperature

We believe, however, that (2) is not a proper characterization of noise in metals. As was pointed out by Van Vliet and Zijlstra,[6] the basic formula for the mobility fluctuations for the scattering of a single carrier is

$$\frac{S_{\mu_i}(f)}{\mu_i^2} = \frac{\alpha_{true}}{f} \qquad (3)$$

where we subscripted the α-value as α_{true}. To obtain the fluctuations in the band mobility, or for that matter of the current I, we must sum over the scattering fluctuations of all carriers in the band.[7][8] For a nondegenerate semiconductor this provides a factor 1/N in the denominator, see (2). In metals, however, most of the carriers are "frozen" in the Fermi sea, a fact also noted by Dutta and Horn in their review paper.[9] Therefore, as we pointed out elsewhere, in connection with thermal noise from metals and the Einstein relation,[10] the result of the summation must be multiplied by $\langle \Delta N^2 \rangle_\infty / \langle N \rangle_\infty$, where the subscript "$\infty$" refers to the grand canonical ensemble.[10, sec. 3] From statistical mechanics the above factor is $kT\partial(\log N)/\partial \mathcal{E}_F$ where $\mathcal{E}_F$ is the Fermi energy. Explicitly, we have

$$\frac{kT\partial(\log N)}{\partial \varepsilon_F} = \frac{\mathcal{F}_{\frac{1}{2}}[(\mathcal{E}_c - \mathcal{E}_F)/kT]}{\mathcal{F}_{-\frac{1}{2}}[(\mathcal{E}_c - \mathcal{E}_F)/kT]} \qquad (4)$$

where $\mathcal{E}_c$ is the bottom of the conduction band, and $\mathcal{F}_k$ is the Fermi integral of order k. For total degeneracy, $\mathcal{F}_k(\eta) = \eta^{k+1}/\Gamma(k+2)$. Thus the ratio (4) becomes $2\Delta\mathcal{E}_F/3kT$ with $\Delta\mathcal{E}_F = \mathcal{E}_F - \mathcal{E}_c$. Consequently, eq. (3) followed by the proper statistical summation leads to

$$\frac{S_I(f)}{I^2} = \frac{\alpha_{true}}{N}\,\frac{2\Delta\mathcal{E}_F}{3kT} = \frac{\alpha_{true}}{fN^*}, \qquad (5)$$

indicating that the number of carriers available for scattering is $N^* = N(3kT/2\Delta\mathcal{E}_F)$. This is also intuitively obvious: the Fermi function differs only appreciably from 1 or 0 in a slice of order kT. That such a reduction in noise must occur in metals was perhaps first pointed out in a classic paper by Brillouin[11] on the first noise observations in metals, by Bernamont.[12] Comparing now (5) with (3) we find that the "true" Hooge parameter is related to the observed Hooge parameter α_H by

$$\alpha_{true} = \frac{3kT}{2\Delta\mathcal{E}_F}\,\alpha_H. \qquad (6)$$

With $\Delta\mathcal{E}_F = 5.5$ eV,[13] the values of α_{true} were computed to yield the data of Fig. 2.

As is noted, we now obtain α-values low enough to become in the ballpark expected from the quantum theory of 1/f noise. In the latter theory, the α_{true} of eq. (3) is just twice the infrared exponent, i.e.,

$$\alpha_{true} = 2\alpha A. \qquad (7)$$

With $\Delta p/m = 2v_F \sin\phi/2$, where v_F is the Fermi velocity (1.39×10^8 cm/sec) and ϕ is the scattering angle ($\approx 150°$ for U-processes), we obtain from (1) $2\alpha A \approx 2.4 \times 10^{-7}$. This value is approximate since N is not exactly known and since more correctly we must take into account the detailed geometry of the Fermi surface, being a sphere with eight "necks" (see Ziman[14]). However, this value comes close to the observed value of $(\alpha_{true})_{max}$ in Fig. 2, being 4.9×10^{-7}.

Qualitatively, we believe that the observed data of Fig. 2 can be well understood. Above the Debye temperature Θ_D, region C, some non-fundamental $1/f^\gamma$ noise occurs, similar to the "type B" noise observed by Dutta and Horn.[9] Below Θ_D we have for the first time a clear indication of the occurrence of quantum 1/f

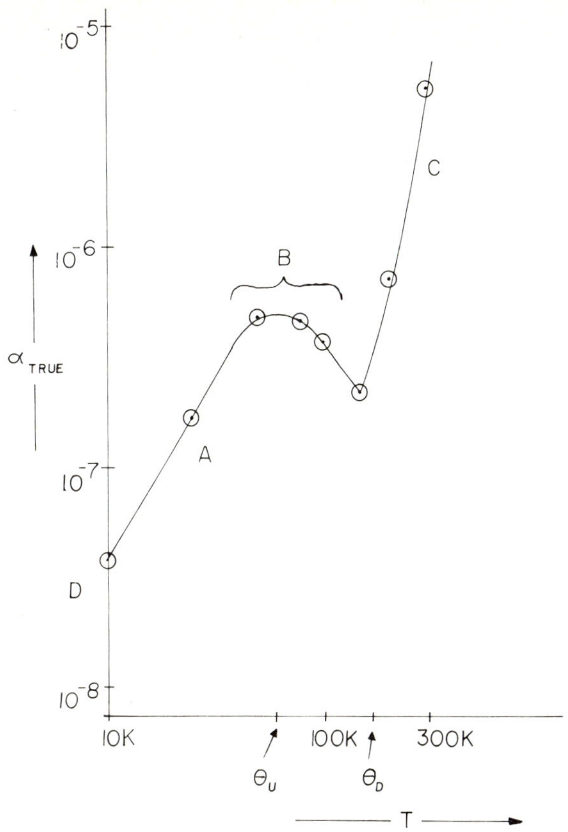

Figure 2. α_{true} as a function of temperature

noise. U-processes dominate in the region B. In the region A, U-processes freeze out and normal phonon processes (N-processes) take over. Finally, in a region D (not yet observed), ionized impurity scattering may give rise to a floor at very low temperatures.

A quantitative theory has not yet been fully developed. However, with U-processes dominating the noise, and N-processes and ionized impurity scattering dominating the resistance, one expects the temperature dependence to be of the form

$$\alpha_{true} \propto \left[\frac{C + e^{-\Theta_U/T}}{C_1 + C_2 T}\right]^2 \tag{8}$$

where Θ_U is the Umklapp temperature, $\Theta_U = hq_o u_o/k_B$, in which u_o is the transverse velocity of sound in gold, k_B is Boltzmann's constant, and q_o is a phonon vector associated with the length of the necks between adjacent Fermi surfaces in the extended zone scheme. We computed $q_o = 6.3 \times 10^7$ cm^{-1}, while $u_o = 1.2 \times 10^5$ cm/sec. This yields $\Theta_U = 57.6$ K. The observed maximum in the noise occurs at about 60 K. Though many details need fuller consideration, we believe that the observed noise can be reasonably well explained by the proposed processes. Development of a full theory and detailed 1/f noise measurements in metals may give much insight into the nature of the phonon processes undergone by the Fermi surface electrons.

We finally note that another type of confirmation of the theory of quantum 1/f noise was recently provided by α-particle decay statistics.[15]

ACKNOWLEDGEMENTS

We acknowledge support from an AFOSR contract, #82-0226, and from a user contract with NRRFSS at Cornell University.

REFERENCES

[1] Hooge, F.N., Van Damme, L.K.J. and Kleinpenning, T.G.M., Reports Progress Physics 44 (1981) 481.

[2] Hooge, F.N. and Van Damme, L.K.J., Phys. Letters 66A (1978) 315.

[3] Handel, P.H., Phys. Rev. Letters 34 (1975) 1492 and 1495.

[4] Handel, P.H., Phys. Review A22 (1980) 745.

[5] Fleetwood, D.M., Masden, J.T. and Giordano, N., "1/f Noise in Platinum Films and Ultrathin Platinum Wires: Evidence for a Common Bulk Origin," to be published.

[6] Van Vliet, K.M. and Zijlstra, R.J.J., Physica 111B (1981) 321.

[7] Hooge, F.N., Physica 114B (1982) 391.

[8] van der Ziel, A., Van Vliet, C.M., Zijlstra, R.J.J. and Jindal, R., Physica B, in press.

[9] Dutta, P. and Horn, P.M., Rev. Mod. Phys. 53 (1981) 497.

[10] Van Vliet, K.M. and van der Ziel, A., Solid State Electr. 20 (1977) 931.

[11] Brillouin, L., Helv. Physica Acta 7, suppl. 2 (1934) 47.

[12] Bernamont, J., Comptes Rendues de L'Ac. francaise 198 (1934) 1755 and 2144.

[13] Kittel, C., Introduction to Solid State Physics, 5th Ed. (McGraw-Hill, New York, 1976) 154.

[14] Ziman, J.M., Electrons and Phonons (Oxford University Press, London, 1967) 113.

[15] Gong, J., Van Vliet, C.M. and Ellis, W., Jr., "Observations of a Flicker Noise Floor in α-Particle Counting Statistics from $_{95}Am^{241}$," to be submitted to Physical Review.

NOISE IN PHYSICAL SYSTEMS AND 1/f NOISE
M. Savelli, G. Lecoy and J-P. Nougier (eds.)
© Elsevier Science Publishers B.V., 1983

LOW FREQUENCY BROADBAND NOISE IN $NbSe_3$

T.A. Davis, R.S. Lear, M.J. Skove, E.P. Stillwell and T.M. Tritt

Department of Physics and Astronomy
Clemson University, Clemson, S.C. 29631 U.S.A.

J.W. Brill

Department of Physics and Astronomy
University of Kentucky, Lexington, KY 40506 U.S.A.

We have measured the noise spectral density $S_V(f)$ below 50 Hz in $NbSe_3$ as a function of temperature T and average sample voltage V_s. The temperature was varied from 85 K to 300 K so as to include the introduction of the upper charge density wave CDW at T_u = 144 K. The range of sample voltages included the critical voltage for exciting motion of the CDW. For 108 K $\leq$ T $\leq$ 134 K we found large increases in the noise magnitude and in the frequency dependance when V_s reached a critical value. The exponent α in the relation $S_V(f) \sim f^{-\alpha}$ changed from 0.7 ± 0.1 to 2 ± 0.2 in this region.

I. INTRODUCTION

The structure of $NbSe_3$ has been determined by Hodeau,[1] et al. It consists of selenium triangular prisms, a niobium atom near the center of each, stacked on top of each other to form a chain. The chains are linked together to form a monoclinic structure with the chains along the b or two fold axis. $NbSe_3$ crystallizes in fibrous forms with the b axis parallel to the fiber axis.

Fig. 1 is a plot of the resistance versus temperature for a typical sample of $NbSe_3$. The longitudinal resistivity at 300 K is $\sim$0.6 x 10^{-5} Ω-m and the resistance ration r $\equiv$ R (300K)/R(4.2K) $\simeq$ 60.[2] The resistivity perpendicular to the chains is about five times that along the chains. At T_u = 144 K and T_L = 59 K, $NbSe_3$ undergoes distinct incommensurate charge density wave (CDW) transitions.[3] In the CDW state the electrical resistance is non-ohmic.[4,5] The effect of sample voltage V_s on the resistance near T_u is shown in Fig. 2. The differential resistance is constant up to a critical $V_s = V_{cr}$ then decreases with V_s to a value between its value in the CDW state and a value obtained by extrapolation of its metallic R vs. T curve. This reduction in resistance is due to motion of the CDW, not to its destruction. To compare this effect from sample to sample we compare longitudinal critical fields $E_{cr} = V_{cr}/\ell$ where ℓ is the sample length between potential contacts. E_{cr} is sample dependent (as is the resistance ratio r) and has been shown to depend upon the amount of impurity in the sample.[6] E_{cr} is also temperature dependent as is shown in Fig. 3.

In this paper we report on the low frequency (<50 Hz) wide band noise associated with the upper transition T_u.

II. EXPERIMENTAL

Thin whisker-like sample of $NbSe_3$ (cross section area A $\sim$ 10^{-7} to 10^{-8} cm^2, length 1 to 3 mm between potential contacts) were mounted with silver paint across four #36 Cu wires. The distance between each potential lead and its current lead was $\sim$0.3 mm and the distance between the two potential leads could be varied to insure that the sample was strain free. The dV_s/dI vs. V_s curves were made using a PAR model HR-8 lock-in amplifier, and the noise spectra were made using a PAR model 113 low noise amplifier and a Nicolet model 440 spectrum analyser. The samples were maintained in He gas, and surrounded by a copper shield the temperature of which was controlled by an electrical heater and a flow of cold N or He gas. Sample temperatures were held constant to about 1 part in 10^3 over the range of interest in this experiment.

III. EXPERIMENTAL RESULTS

a. 175 K $\leq$ T $\leq$ 300 K. $S_V(f)$ is well characterized by the Hooge relation $S_V(f) = k\, V_s^2/N f^{\alpha}$ with α = 1 ± 0.1 and k $\sim$ 1-10. Taking N = 10^{21} cm^{-3} as the carrier density we find k $\sim$ 1 for our samples at 300 K. k shows no change in this range.

b. 140 K < T < 175 K. α falls to 0.7 near 140 K and there is a small but clear failure of the noise to scale with V_s^2.

c. 108 K $\lesssim$ T $\lesssim$ 134 K. We find large increases in the noise at all frequencies, and rapid changes of α with V_s. For $V_s \sim 0.11$ volt each spectrum has segments with $\alpha > 1$.

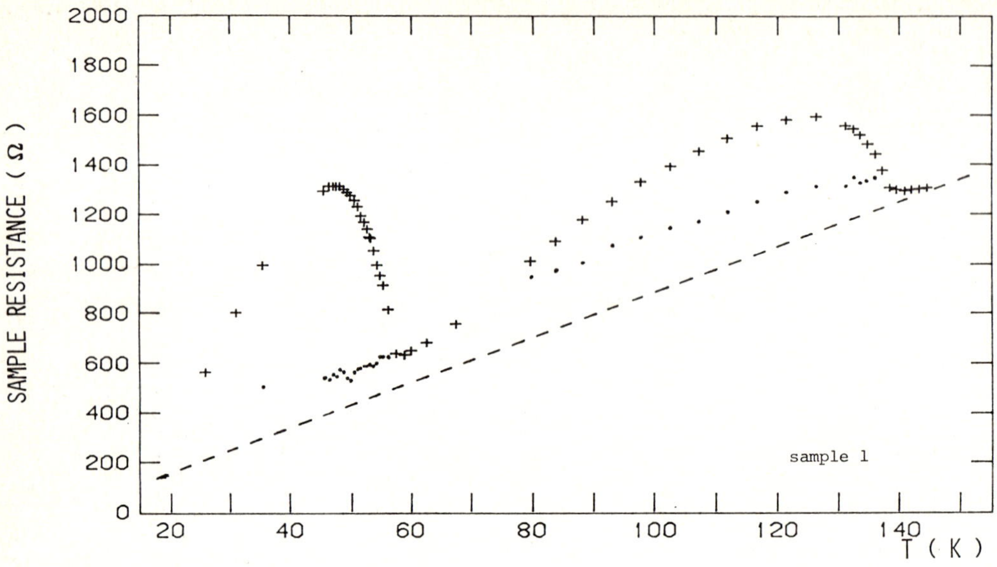

Figure 1. Plot of differential resistance vs. temperature for $E < E_{cr}$ (+) and $E > E_{cr}$ (•). The dashed line is an extrapolation of the metallic resistance of the sample.

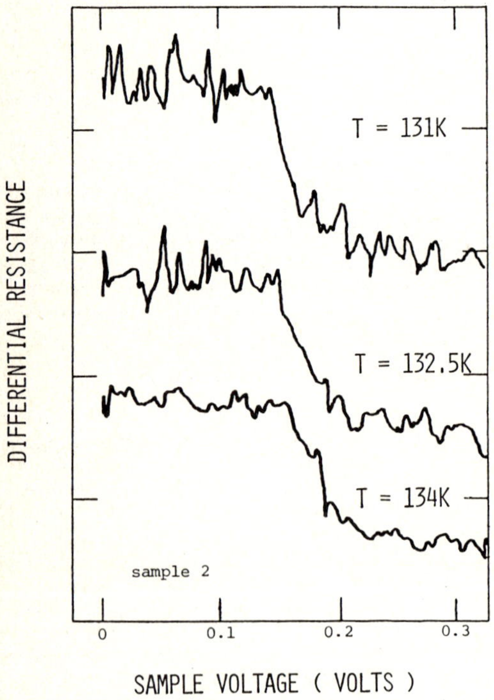

Figure 2. Differential resistance (dV/dI) as a function of sample voltage with $T \leqq T_u$ = 144K. (arbitrary units)

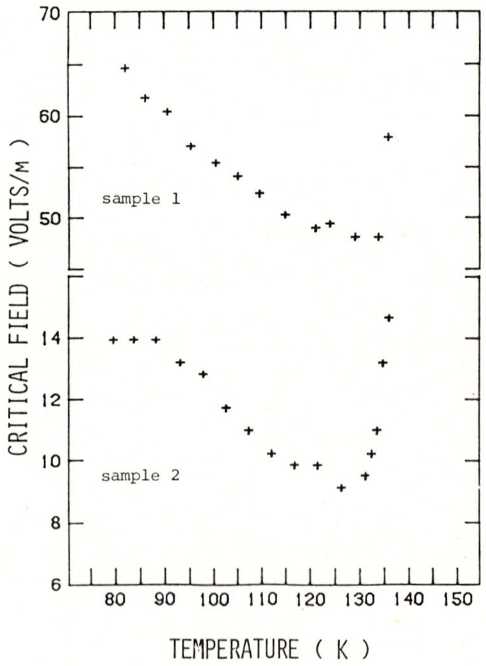

Figure 3. Critical field (V_s/ℓ) as a function of temperature for 2 samples of $NbSe_3$. The length of sample 1 was 2.38 mm, while that of sample 2 was 1.36 mm.

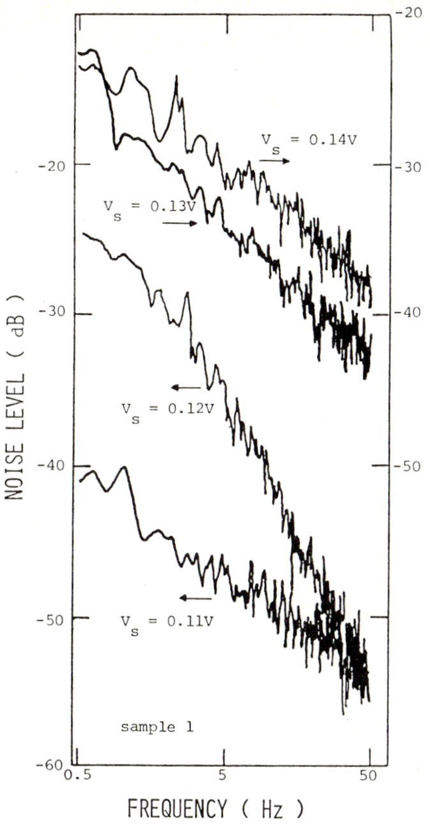

Figure 4. A log-log plot of the noise level (dB) vs. frequency at various sample voltages. Nyquist noise level is ≃ -65 dB. (T=114K)

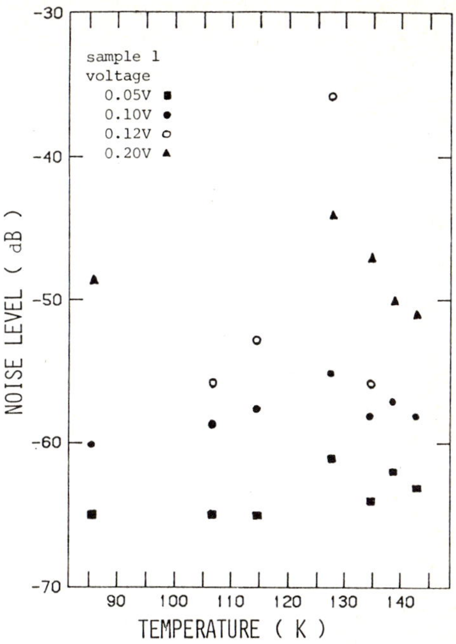

Figure 5. Plot of noise level (dB) vs. temperature for various sample voltages.

These changes are illustrated in Figures 4-6 which show that at T = 114 K, α changes from 0.65 to 1.8 to 0.9 as the sample voltage changes from 0.11 to 0.12 to 0.13 volts respectively.

d. T < 108 K, α = 0.9 to 1.2 but the noise does not scale with V_s^2.

Similar experimental results have been obtained by Richard, et al.[7] near the lower transition where in addition they found for $T \gtrsim T_L$ the noise had a white region extending from $\sim 10^3$ to 10^5 Hz. We have not completed experiments in this frequency range.

IV. DISCUSSION

An explanation for the frequency dependence of the spectral noise density near a CDW transition has been given by Papoular[8]. He proposes that the noise is due to propagation of phase kinks (solitons) in the CDW. Several properties of CDW compounds can be explained with a phenomenological equation for the phase, Φ, of the CDW:[9,10]

$$\gamma\dot{\Phi} - c_o^2 \frac{\partial^2 \Phi}{\partial x^2} + \omega_o^2 \sin \Phi = F$$

where the CDW is overdamped so that the inertial term can be neglected, γ is a damping constant, c_o^2 is an elastic restoring force, ω_o^2 is proportional to the restoring force of the lattice and impurities, and F is proportional to the externally applied electric field. This equation admits large amplitude solitary wave solutions (kinks and antikinks or solitons). When the temperature is near the phase transition temperature, T_c, the energy necessary to create kink-antikink pairs will be small (the transitions are other than first order). At lower temperatures kinks and antikinks will be created by the field at favorable regions near impurities or, more likely, at contact regions.[11] At low applied fields the kinks and antikinks will not propagate (and dV/dI is nearly flat).

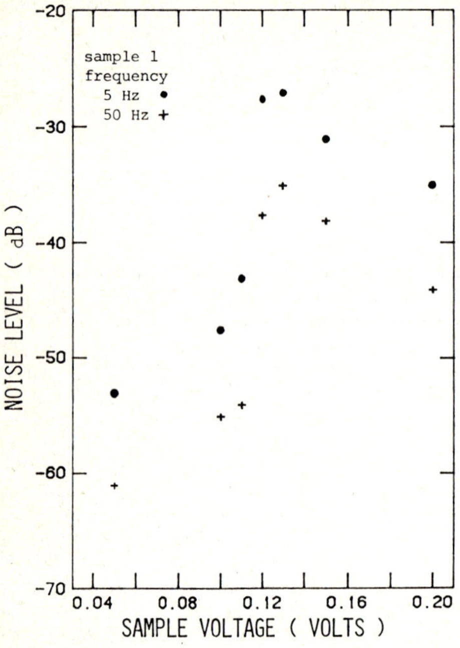

Figure 6. Plot of noise level (dB) vs. sample voltage at frequencies of 5 and 50 Hz. (T = 127K)

They will be able to diffuse, however, and this leads to a standard Brownian $1/f^2$ dependence. For high applied fields the kinks and antikinks can propagate. They will move until stopped by impurities, domain boundaries, or contacts. Here they repel each other and act like a dislocation pile-up. The kinks in such a pile-up may be destroyed by many different processes which may have a broad spectrum of relaxation times τ which leads through reasoning similar to that of McWhorter[12] or Dutta and Horn[13] to 1/f noise. Near T_c the kinks and antikinks will be plentiful and will usually annihilate one another before propagating to a discontinuity. This leads to generation-recombination noise which is white ($S_V \sim f^o$). Thus Papoular finds three regions:

a. low T and low E, $1/f^2$ noise
b. low T and high E, 1/f noise
c. high T, f^0 noise.

We conclude by saying that our data agree with Papoular's explanation for regimes (a) and (b) above, and that we have insufficient data to test regime (c).

REFERENCES

(1) Hodeau, J.L., Marezio, M. Roucau, C., Ayroles, R., Meerschaut, A., Rouxel, J. and Monceau, P., J. Phys. C: Solid State Phys. 11 (1978) 4117.

(2) Haen, P., Monceau, P., Tissier, B., Waysand, G., Meerschaut, A., Malinie, P. and Rouxel, J., Proc. Fourteenth Int. Conf. on Low Temperature Physics, Otaniemi, Finland 5 (1974) 445.

(3) Wilson, J.A., Phys. Rev. B 19 (1979) 6456.

(4) Monceau, P., Ong, M.P., Portis, A.M., Merschant, M. and Rouxel, J., Phys. Rev. Lett. 37 (1976) 602.

(5) Fleming, P.M. and Grimes, C.C., Phys. Rev. Lett. 42 (1979) 1923.

(6) Ong, M.P., Brill, J.W., Eckert, J.C., Savage, J.W., Kanna, S.K. and Somoano, R.B., Phys. Rev. Lett. 42 (1979) 811.

(7) Richard, J., Monceau, P., Papoular, M. and Renard, M., J. Phys. C: Solid State Phys. 15 (1982) 7157.

(8) Papoular, A.M., Phys. Rev. B 25 (1982) 7856.

(9) Buttiker, B.M. and Laudauer, R., Phys. Rev. Lett. 43 (1979) 1453.

(10) Gruner, C.G., Zawadowski, A., and Chaikin, P.M., Phys. Rev. Lett. 46 (1981) 511.

(11) Ong, N.P., Verma, G. and Maki, K., preprint.

(12) McWhorter, A.L., Semiconductor Surface Physics, edited by R.H. Kingston (University of Pennsylvania, Philadelphia, 1952).

(13) Dutta, P. and Horn, P.M., Rev. Mod. Phys. 53 (1981) 487.

NOISE IN PHYSICAL SYSTEMS AND 1/f NOISE
M. Savelli, G. Lecoy and J-P. Nougier (eds.)
© *Elsevier Science Publishers B.V., 1983*

1/f NOISE IN ZnO VARISTORS

Andrzej Kusy

Departament of Electrical Engineering
Technical University of Rzeszów
Rzeszów, Poland

The relative power spectral density of 1/f current fluctuations, S_I/I^2 in ZnO varistor in the ohmic region has been calculated, based on the mobility fluctuation explanation of the 1/f noise. The derived equation predicts a linear relationship between S_I/I^2 and the varistor resistance, R_o in the ohmic region. The predicted linear relationship has been experimentally confirmed changing S_I/I^2 and R_o by changing the temperature of the varistors in the range 270-420K. The calculated 1/f noise power spectral density is in agreement with the observed density.

1. INTRODUCTION

1/f noise in many semiconductor devices has been successfully explained based on the thesis that mobility fluctuations of the free charge carriers are the origin of this phenomenon [1] [2]. In particular this approach has been used by Kleinpenning [1] for the description of 1/f noise model in Schottky barrier diodes. He has found experimentally that 1/f noise originates when deviations from the ideal Schottky diode take place, like edge currents and diffusion effects in the depletion region. When such deviations are present then the I-U characteristic of the Schottky barrier diode is mobility dependent [3].

In this paper we present experimental results as well as a model of 1/f noise in zinc oxide varistors[1/] using Kleinpenning's model of 1/f noise in singular Schottky barrier diodes. Two types of commercially available varistors: Siemens and General Electric have been taken into account in the investigation [4]. In this paper we present the results only for Siemens SI 14K 170N varistors. The dimensions of the varistors are: the electrode distance L=0,1 cm and the electrode area A=1 cm^2. The size of ZnO grains in the varistor structure ranges from 10 µm to 25 µm. It is known from the literature that the conduction mechanism of ZnO varistors is governed by the number of back-to-back Schottky barriers [4]. In our model of the conduction mechanism in the ohmic and first nonlinear regions we consider the thermionic emission accross the barriers, neglecting the other ways of charge transport, like e.g. tunneling. Some of the necessary characteristics of the conduction mechanism have been found on the basis of our experiments and on the basis of literature. The 1/f noise model worked out has been found usefull for description of the 1/f noise experimental characteristics of the varistors in the ohmic region.

2. MODEL

The zinc oxide varistors constitute of a number of ZnO grains separated by a thin interface layer of insulating, Bi_2O_3 based oxides. It is also known that ZnO is a n-type semiconductor and that on the border between the grains and the interface layer the energy barrier is formed to restore the charge neutrality. The simplified band diagram model of two adjacent ZnO grains has been presented in Figure 1. In real varistor structure the thickness of the intergranular layer has been estimated as 30-100 Å, while in Figure 1 it has been assumed as equal to zero. This simplification is justified by the fact that we have

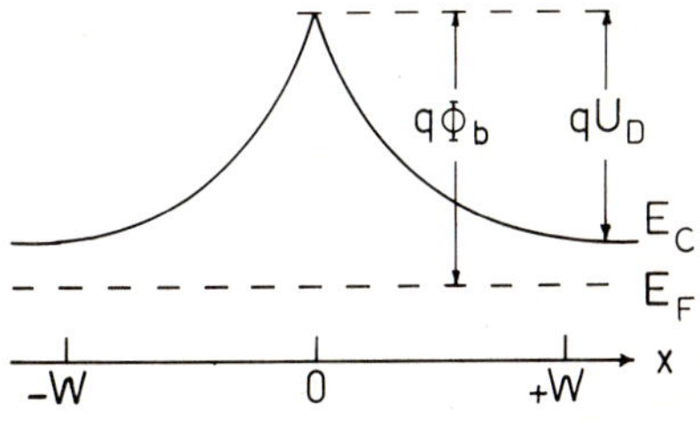

Figure 1 : Band diagram of two adjacent back-to-back Schottky barriers in the ZnO varistor

found it possible to describe the I-U and C-U characteristics of the varistors in the ohmic and in the first nonlinear regions using the thermionic emission as the only way of charge transport accross the barrier [4]. Thus we approximate the varistor as an array of back-to-back Schottky barrier units. The I-U characteristic in the ohmic region of one unit, taking into account diffusion in the depletion regions and edge effects, can be written as follows

$$I_g = \frac{A_g A^* T^2}{1 + \frac{v_R}{v_D}} \exp\left(-\frac{q\Phi_b}{kT}\right)\frac{qU}{gkT} , \qquad (1)$$

where v_R is the recombination velocity, v_D the so-called diffusion velocity, A_g the cross-section area of one barrier, A^* the Richardson constant, Φ_b the barrier height (Figure 1), g the number of grains connected in series between the two varistor electrodes and U the voltage applied to the waristor. The velocities v_R and v_D are given by

$$v_R = \left(\frac{kT}{2\pi m^*}\right)^{1/2} = \frac{A^* T^2}{q N_c}, \tag{2}$$

$$v_D \simeq \mu E_m, \tag{3}$$

where m^* is the effective mass of electrons in ZnO, N_c the effective number of states in the ZnO conduction band, μ the mobility of electrons in ZnO and E_m is the maximum value of the field in the depletion region. In our case we have $v_R \ll v_D$.

According to Hooge etal. [2][5] the 1/f noise in the electrical conductivity is caused by mobility fluctuations. The only quantity in equation (1) which depends on mobility is v_D. Inserting v_D fluctuations in equation (1) it can be shown that the spectral density of the current fluctuations given by

$$S_{I_g} = \left(\frac{v_R}{v_D}\right)^2 \left(\frac{I_g}{2W}\right)^2 \iint_{-W}^{W} \frac{S_\mu(x,y)}{\mu^2} dxdy, \tag{4}$$

where W is the width of depletion layer (Figure 1). Following Hooge et al. [2] one can further assume that

$$\frac{S_\mu(x,y)}{\mu^2} = \frac{\alpha}{f} \frac{\delta(x-y)}{A_g n(x)}, \tag{5}$$

where n(x) is the concentration of free electrons at the spot x inside the depletion layer of the Schottky barrier and α is the Hooge parameter. Inserting equation (5) into (4) one can show that 4

$$\frac{S_I}{I^2} = \frac{v_R^3 R_o C_o}{4fg A_{eff} U_D^2 N_D \mu^2} \tag{6}$$

where R_o and C_o are the ohmic region resistance and capacitance of the varistor respectively, A_{eff} is the total cross section of the barriers in a plane parallel to the electrodes, U_D the diffusion voltage (Figure 1) and N_D the concentration of donors in ZnO grains.

3. EXPERIMENTAL RESULTS

To check the theoretical predictions given by equation (6) we have performed 1/f noise and conduction mechanism experiments. They included the measurements of the spectral density of the 1/f current fluctuations as a function of frequency, temperature and current flowing through the sample, as well as the measurements of I-U characteristics (as a function of temperature) and C-U characteristics (at room temperature). In Figure 2 the I-U characteristics of Siemens varistor

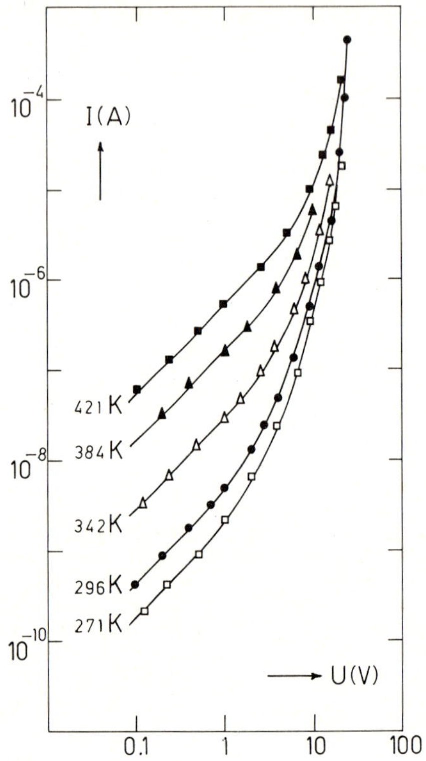

Figure 2 : Current-voltage relations of Siemens SI 4K 170N ZnO varistor (sample 1) at different temperatures

(sample 1) have been shown at 5 different temperatures. It can be seen from Figure 2 that the lower the temperature of the varistor during measurements the smaller the observable ohmic region in I-U characteristics. In figure 3 S_I vs I relations of the Siemens varistor (sample 2) have been shown at frequency f=2Hz. From Figure 3 one can see that in the ohmic region the relation $S_I \sim I^2$ is fulfilled and that at the constant frequency the relative 1/f noise power spectral density decreases with increasing of temperature. Basing also on S_I vs frequency measurements we can state that low frequency noise in the ZnO varistors in the ohmic region obey the relation

$$S_I/I^2 = C_{1/f} \cdot f, \tag{7}$$

where $C_{1/f}$ is the coefficient which does not depend on current and frequency. The quantitative illustration of the stated dependence of 1/f noise spectral density on temperature has been given in Figure 4, where S_I/I^2 at f=1Hz has been plotted as a function of ohmic region resistance

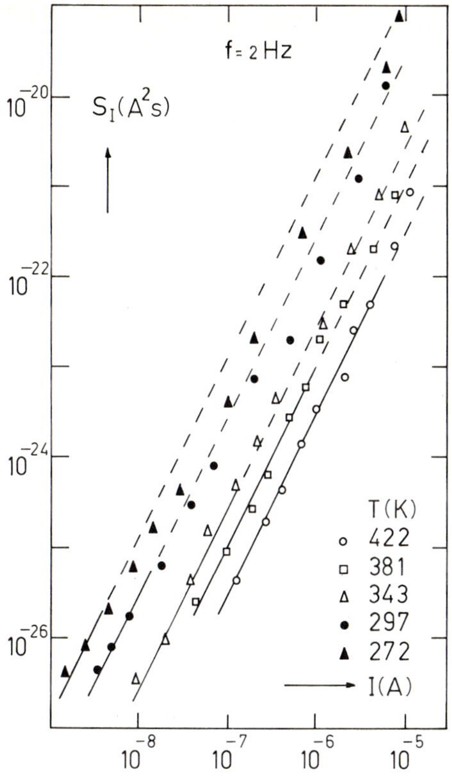

Figure 3 : The 1/f current noise S_I vs I characteristics at f=2Hz and at different temperatures of Siemens varistor (sample 2). The full lines with slope 2 represent the ohmic region

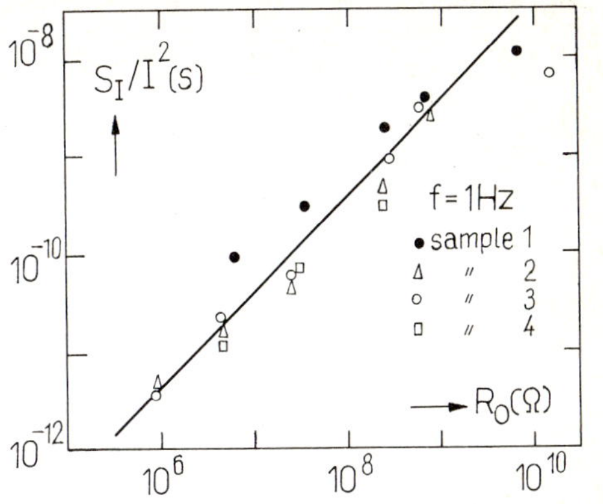

Figure 4 : Relative 1/f noise power spectral density S_I/I^2 at f=1Hz as a function of the ohmic region resistence R_o of 4 different Siemens varistors. R_o is varied by temperature

R_o of 4 Siemens varistors. As can be seen the relation S_I/I^2 (1Hz) vs R_o represents a straight line which is a confirmation of derived equation (6). A deviation from the straight line can be seen in Figure 4 for the values of the spectral density corresponding to the highest values of R_o and to the lowest temperatures. This deviation is probably due to the fact that these results were obtained in the nonlinear regions; in that temperature range 1/f noise was undetectable in the ohmic region.

It is necessary to mention here that the straight line in Figure 4 has been drawn for g=22, N_D=$6\cdot10^{18}$ cm^{-3}, U_D=0,7V which have been estimated on the basis of conduction mechanism investigation, for R_o, C_o and A_{eff}=0,3 cm^2 estimated directly from the measurements, μ=200 cm^2/Vs and m^*/m=0,27 found from the literature [3] and α = $=10^{-5}$. This value of α is about an order of magnitude lower than the values which are usually found in semiconductors being in the range 10^{-4}--10^{-3}. The observed discrepancy in α values may be ascribed to several factors. For instance: (i) the ZnO grains are not equal, (ii) the electron mobility depends on lattice and impurity scattering (the latter is substantial for so high concentration of donors) which results in reduction of α, (iii) the uncertainty concerning the estimation of g, A_{eff}, U_D and N_D, (iv) the structure of ZnO varistors is very complex.

4. CONCLUSIONS

It has been found in the investigation that the relative 1/f noise power spectral density of ZnO varistors measured in the ohmic region decreases when temperature of the varistor increases in the range 270-420 K. This observation meets the theoretical prediction of S_I/I^2 behaviour based on the mobility fluctuation approach to 1/f noise problem. In particular, using the 1/f noise model of Schottky barrier diodes by Kleinpenning 1 and modeling ZnO varistor by an array of back-to-back Schottky barrier units the equation (6) has been derived. This equation shows that in the ohmic region the relative 1/f noise power spectral density is a linear function of the resistance of ZnO varistor. This linear dependence has been confirmed by changing the varistor temperature. To obtain the agreement between theoretical straight line and the experimental data one would have to assume for the Hooge parameter $\alpha=10^{-5}$. Taking into account the typically observed values of α, in semiconductors usually in the range 10^{-4}-10^{-3}, we consider that the obtained in the paper value of α is quite acceptable. The observed one order of descrepancy can be explained if one takes into account e.g. the influence of impurity scattering caused by high donor concentration in ZnO and some other aspects of very complex structure of ZnO varistors.

1/ The researches were carried out in the Group of Electronic Materials, Departament of Electrical Engineering, Eindhoven University of Technology, Eindhoven, Netherlands.

ACKNOWLEDGEMENTS

The author is indebted to Prof. dr. F.N. Hooge and to Prof. dr. T.G.M. Kleinpenning for bringing the research project to his notice and also for helpful discussions.

REFERENCES

[1] Kleinpenning, T.G.M., Low-frequency noise in Schottky barrier diodes, Solid State Elektron., 22 (1979) 121-128.

[2] Hooge, F.N., Kleinpenning, T.G.M. and Vandamme, L.K.J., Experimental studies on 1/f noise, Rep. Prog. Phys., 44 (1981), 479-532.

[3] Sze, S.M., Physics of semiconductor devices, 2 nd ed. (Wiley-Interscience, New York, 1981).

[4] Kusy, A., Kleinpenning, T.G.M., On conduction mechanism and 1/f noise in ZnO varistors, J. Appl. Phys., to be published.

[5] Hooge, F.N., Discussion of recent experiments on 1/f noise, Physica, 60 (1972), 130-144.

NOISE IN DIODES AND TRANSISTORS

NOISE IN PHYSICAL SYSTEMS AND 1/f NOISE
M. Savelli, G. Lecoy and J-P. Nougier (eds.)
© Elsevier Science Publishers B.V., 1983

GENERATION-RECOMBINATION NOISE IN Si JFETS

T. S. Nashashibi, M. A. Carter and S. Taylor

Thorn-EMI Central Research Laboratories, Hayes, Middlesex, England

ABSTRACT

The equivalent input noise voltage of Si n-channel JFETs has been studied in the temperature range 100K to 350K. Generation-recombination noise is prominent at low frequency in all the devices measured. More than one type of imperfection is found in most devices. Activation energies and capture cross-sections are derived for the centres and suggestions are made as to the nature of the defects involved.

INTRODUCTION

Several authors [1,2,3] have attributed the low frequency excess noise in JFETs to charge generation-recombination (g-r) at traps in the space charge region of the gate junction. Lauritzen [4] has developed a theory where the charge fluctuations at Shockley-Read-Hall (SRH) centres in the depletion region produce fluctuations in the channel height and therefore noise in the channel current. K. Kandiah and his co-workers [3] have pointed out that only SRH centres within a small distance from the depletion region edge are significant in the noise generation. They present evidence from their work on four-terminal devices where the depletion region boundary can be made to move in the space between the two gate junctions. When the transition region encounters an SRH centre a maximum is produced in the noise vs. gate bias curve. Hiatt [1] observes g-r noise in his specially prepared samples and attributes the noise at high temperatures to depletion g-r noise, and at low temperatures to channel g-r noise.

It is sometimes difficult to carry out accurate analysis on g-r noise in JFETs. The effect is often obscured by a high thermal noise component in the device, or by SRH centres with a distribution of energy levels and capture cross-sections, or by a high 1/f noise component arising from surface effects. Lauritzen [4] showed some characteristic g-r noise spectra in JFETs where a high level of gold impurity has been added.

THEORETICAL BACKGROUND

Lauritzen [4] considered the case of a single level SRH centre with a concentration N_t, an energy level E_t below the conduction band and a characteristic lifetime τ. In the case of an abrupt junction FET, the equivalent input noise voltage per unit bandwidth at a frequency f is given by

$$\overline{e_n^2} = K.F_t(1 - F_t)N_t \frac{\tau}{1 + (2\pi f \tau)^2} \qquad (1)$$

where K is a parameter that is dependent on device geometry and bias conditions. In our case $K=10^{-21}\ V^2cm^3$. F_t is the probability of trap occupation and is given by

$$F_t = \frac{1}{1 + e^{(E_t - E_f)/kT}} \qquad (2)$$

where E_f is the quasi fermi level energy and T is the absolute temperature. τ is strongly dependent on temperature and, for a centre where the capture cross-section for electrons is larger than that for holes, it is given by

$$\tau = \frac{e^{(E_c - E_t)/kT}}{N_c v A_n} \qquad (3)$$

N_c is the effective density of states in the conduction band, v is the electron thermal velocity and A_n is the electron capture cross-section. τ is obtained according to equation (1) from the measured e_n spectrum at the frequency where e_n has fallen by 3dB below its plateau value. At that point $\tau = 1/2\pi f$. We can also assume, without a significant loss of accuracy, that the peak in the e_n - T curves at a frequency, f, occurs when $\tau = 1/2\pi f$. This assumption is valid provided that variations of F_t over the restricted temperature range of a peak are small compared with the variations in τ. The activation energy and the capture cross-section of the trap are obtained from a plot of ln τ vs 1/T according to equation (3).

The function $F_t(1 - F_t)$ is extremely sensitive to the energy difference $E_t - E_f$. It has a maximum of 0.25 when the quasi fermi energy coincides with the trap energy, but falls very rapidly for a very small energy difference. The effect becomes more exagerated at low temperatures. The only SRH centres in the fully depleted region that are likely to contribute to noise are, therefore, those with energies close to mid gap. Shallower traps become effective in the region between the fully depleted region and the neutral channel, where the quasi fermi level is higher up in the gap.

EXPERIMENTAL

The JFETs were fabricated on p^+-type silicon substrates with an n-type epitaxial channel and a diffused p^+ top gate. The top gate is joined to the substrate through the isolation diffusion. The samples were chosen from two processing batches, 17 and 34. A typical device had a pinch-off voltage (V_{po}) of 2V, an IDSS of 1mA and transconductance of 0.5 mS. Samples used in this investigation have low thermal and 1/f noise components and are well suited to g-r analysis.

The equivalent input noise voltages at five spot frequencies: 10Hz, 120Hz, 1kHz, 10kHz and 100kHz were measured using a noise analyser. The outputs at the different frequencies were fed to five digital integrators which averaged the signals for three minute periods. The e_n - T scans were made at a constant 6V drain to source voltage and a constant 0.3mA drain current. The transistors were always operated well in the saturation region where the drain current is independent of the drain voltage.

THE TEMPERATURE DEPENDENCE OF THE NOISE VOLTAGE: The e_n - T curves at 10Hz for three devices are shown in figure (1). Each of the devices measured exhibits a peak at 165K, (peak # 2), of approximately equal magnitude. At higher temperatures, the 10Hz noise for devices from batch 34 has a peak at 273K (peak # 1) and those from batch 17 at 255K (peak # 4). The magnitudes of these peaks vary from device to device. In device 17-21 the peak is just visible. At low temperatures only devices from batch 34 exhibit a peak at 115K (peak # 3). This maximum appears in devices where peak # 1 is present.

The temperature dependence of e_n at 10Hz, 120Hz and 1kHz for device 34-10 and device 17-42 are shown in figures (2) and (3) respectively. The curves are typical of generation-recombination noise in JFETs. Below 100K the noise at all frequencies begins to rise indicating the onset of carrier freeze out.

The e_n spectra at various temperatures near peak # 4 from device 17-42 are shown in figure (4). The curves highlight the g-r noise characteristic of a low frequency plateau which extends to higher frequencies as the temperature is increased. At high frequency the g-r noise component becomes small compared with the thermal noise and the curves become flat.

THE BIAS DEPENDENCE OF THE NOISE VOLTAGE: Figure (5) shows the dependence of e_n at 10Hz on the gate bias for device 34-5 at the three temperatures 273K, 165K and 115K that coincide with the three peaks # 1, # 2 and # 3 respectively. At all three temperatures the noise rises as the device is operated close to cut-off but at 115K there is a marked dependence of the magnitude of peak # 3 on gate bias. This dependence has been observed on all devices exhibiting this peak. The effect is not due to a shift in temperature through power dissipation in the device. The curves are essentially unaltered when the drain voltage, and therefore the power dissipation is changed by a factor of 3 while still keeping the transistor in the saturation region.

DISCUSSION

The experimental results shown in figures (1) to (5) indicate that the primary cause of excess low frequency noise in the JFETs studied is g-r noise. The peaks labelled # 1 to # 4 in the e_n - T curves are associated with four different SRH centres. The characteristic parameters of the centres, i.e. their energy levels and capture cross-sections, are obtained from plots of ln τ against 1/T as shown in figures (2) and (4) insets. The results are shown in table I.

TABLE I

Peak	Peak Temp. at 10Hz	Capture cross section	Activation energy
# 1	273 K	10^{-15} cm^2	0.53 eV
# 2	165 K	5×10^{-15} cm^2	0.34 eV
# 3	115 K	10^{-15} cm^2	0.18 eV
# 4	255 K	10^{-16} cm^2	0.43 eV

Near room temperature, the excess noise in devices from batch 34 appears to be caused by a different defect from those from batch 17. The corresponding peaks are # 1 and # 4. In devices such as 17-21 and others where there is no peak in the e_n - T curve near room temperature, the low frequency noise voltage is still higher than the thermal noise by about a factor of 2. It is not clear what the source of this noise component is, but it does not conform to the theory of g-r noise due to a single trapping level. The four peaks listed above will now be discussed separately.

PEAK # 1
The energy level for this defect is very close to the intrinsic fermi level. The product, $F_t(1 - F_t)$, is therefore near its maximum value in the depleted region and the centres are very efficient in producing depletion g-r noise. There are several impurities with energies around 0.5eV [5]; for example oxygen - 0.51eV, Fe - 0.51eV and Au - 0.53eV. All give rise to donor levels with energies measured below the conduction band. There is also a Cu acceptor

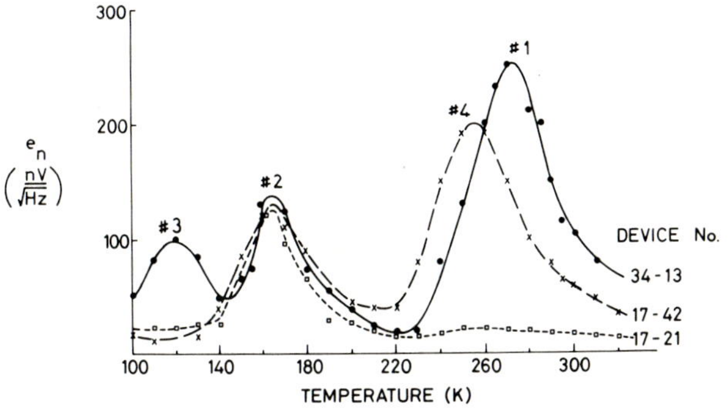

Fig.(1) The noise voltage at 10 Hz vs. temperature plots for three devices.

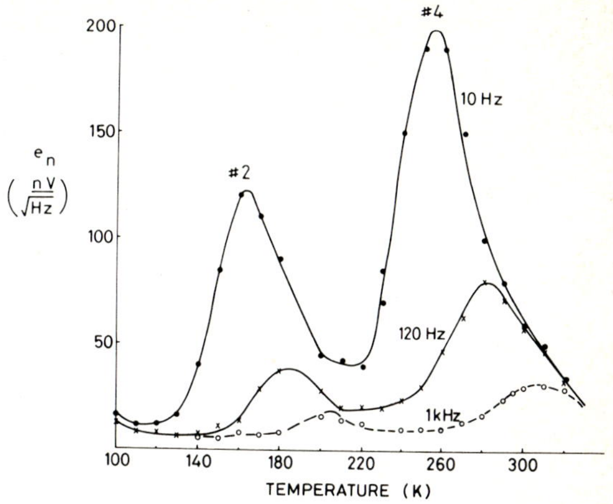

Fig.(3) The noise voltage vs. temperature plots for device 17-42 at three frequencies.

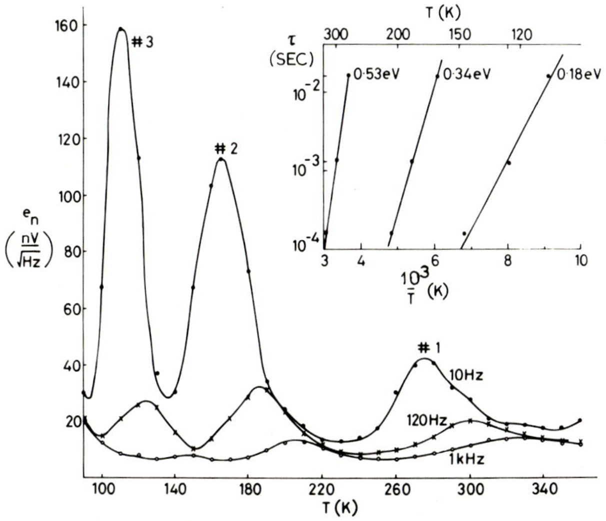

Fig.(2) The noise voltage vs. temperature plots for device 34-10 at three frequencies. Inset: Activation energy plots for the three peaks #1, #2, and #3 for device 34-10.

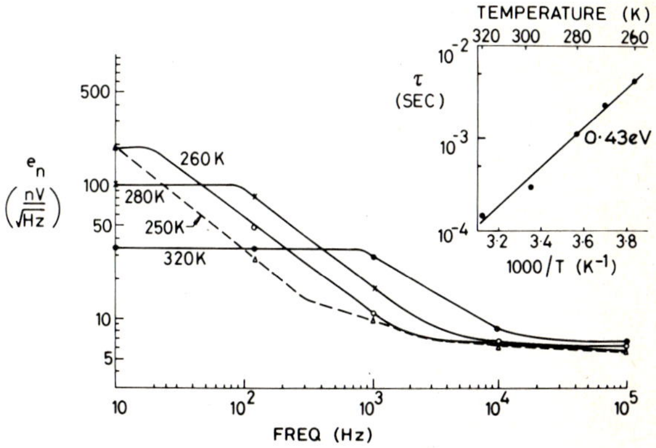

Fig.(4) The noise voltage spectra at different temperatures for device 17-42.
Inset: Activation energy plot for peak #4 for device 17-42.

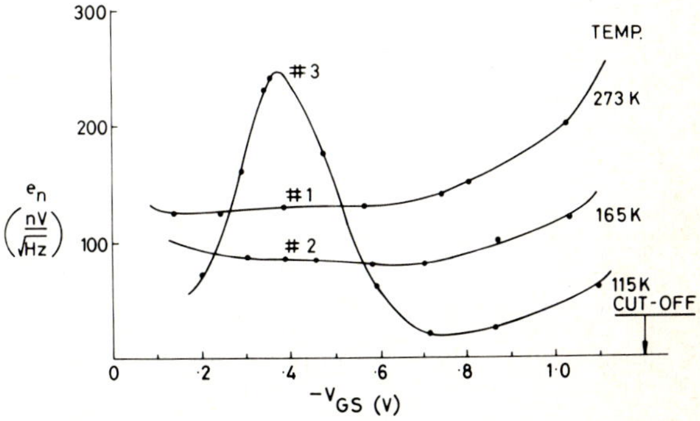

Fig.(5) The noise voltage at 10 Hz vs. gate bias for device 34-5 at three temperatures.

level 0.53eV above the valence band. Gold is a possible candidate due to its high diffusion rate in silicon and its wide use in industry. Published data for the electron capture cross-section for a gold impurity vary from $8.5\times10^{-17}cm^2$ to $2.2\times10^{-15}cm^2$ [6]. It is also possible that dislocations are responible for the excess noise. According to Glaenzer and Jordan [7], these have E_c-E_t = 0.52eV and A_n = 9×10^{-16} cm^2, very close to our estimated parameters for peak # 1. The estimated defect density in device 34-13 is $3\times10^{10}cm^{-3}$ and in device 34-10, it is 10^9cm^{-3}.

PEAK # 4

The energy level for this peak is about 0.1eV away from the intrinsic fermi-level and the product $F_t(1 - F_t)$ is 4×10^{-3}. In order for peaks # 1 and # 4 to be equal in magnitude, the concentration of SRH centres responsible for peak = 4 needs to be about 50 times larger than that responsible for peak # 1. The estimated impurity concentration is 10^{12} cm^{-3} in device 17-42 and is 3×10^9 cm^{-3} in device 17-21.The SRH centres responsible for peak # 4 could be Fe interstitials. These have an energy level at 0.43eV above the valence band and a capture cross-section of 3×10^{-17} cm^2 [8], in good agreement with our data for peak # 4.

PEAK # 2

The trap energy for this peak is 0.34eV and the product $F_t(1 - F_t)$ for the relevent SRH centres in the depletion region is 4×10^{-7}. Such centres are unlikely to make a significant contribution to noise generation at 10Hz unless the concentration is unusually high, $N_t>10^{15}$ cm^{-3}. The SRH centres are most effective when they are situated in a region where there is only partial depletion and the quasi fermi level is close to 0.34eV below the conduction band.

Peak # 2 occurs in all the devices measured and is always of approximately the same magnitude. If the centres responsible for this peak are due to an impurity introduced during the device fabrication, one would expect the impurity concentration and consequently the magnitude of the peak, to vary from device to device as is the case with peaks # 1 and # 4. It is therefore likely that the centres are due to defects that are introduced during material preparation. Such defects might be silicon vacancies or they could be oxygen impurities which are always present in large quantities in CZ single crystal silicon. Both produce acceptor levels with energies about 0.34eV below the conduction band [5].

PEAK # 3

There are two interesting features about this peak. Firstly, it is only present in devices where peak # 1 is present. Secondly, its magnitude is very sensitive to the gate bias.

It appears that the two peaks # 1 and # 3 are somehow related. It could be that both centres are due to the same impurity but with two energy levels. Fe for example has two donor levels 0.51eV and 0.14eV below the conduction band and oxygen has three levels: two donor levels 0.51eV and 0.16eV and an acceptor level 0.38eV below the conduction band.

The effect of gate bias requires explanation. One possibility is that there is a non-uniform distribution of centres in the epitaxial layer. In this case the only SRH centres effective in noise generation are those where the quasi-fermi level is at about 0.18eV below the conducting band, only a very narrow zone in the transition region could satisfy this condition. Kandiah and Whiting [3] observe several maxima in the e_n vs. gate bias curve at different temperatures which they attribute to individual centres in the transition region. We could well be observing a similar effect, only in our case there seems to be a region near the centre of the metallurgical channel where there is a large concentration of traps whilst there is a neglible concentration everywhere else. Hiatt[1] also observes a peak at about 110K which he attributes to channel generation-recombination noise. It is difficult, however, to explain the observed dependence on gate bias by this model.

CONCLUSION

G-r noise was found to be prominent in all the devices investigated. Four types of SRH centres have been detected with the following energy levels:0.53eV, 0.43eV, 0.34eV, 0.18eV. The 0.53eV level could be associated with dislocations or Au impurities, and the 0.43eV level has been tentatively identified as Fe interstitials.

ACKNOWLEDGEMENT

This work was carried out with the support of the Procurement Executive, Ministry of Defence, sponsored by DCVD.

REFERENCES

[1] Hiatt, C.F., Noise in Junction-gate Field Effect Transistors at Low Temperatures., Ph.D. Thesis, Dept. of Elect. Eng., The University of Florida, (1974).

[2] Haslett, J.W. and Kendall, E.J.M., IEEE Trans Electron Devices, Ed- 19 (Aug 1972), 943 - 950.

[3] Kandiah, K. and Whiting, F.B., Solid State Electronics, vol 21 (1978) 1079 - 1088.

[4] Lauritzen, P.O., Solid State Electronics, vol. 8 (1965). 41-58.

[5] Sze, M.S., Physics of Semiconductor Devices (Wiley, International Edition, 1969) p.30.

[6] Lang, D.V., Grimmeiss, H.G. Meiger, E. and Joros,M., Phys. Rev. ,B22 (1980) p 3971.

[7] Glaenzer,R.H. and Jordan, A.C.,Solid State Electronics, Vol. 12 (1969) 247-258.

[8] Wunstel,K. and Wagner,P., Appl. Phys.A (Germany) Vol. 27 (1982) 207-212.

NOISE IN PHYSICAL SYSTEMS AND 1/f NOISE
M. Savelli, G. Lecoy and J-P. Nougier (eds.)
© Elsevier Science Publishers B.V., 1983

JFET GATE-CURRENT NOISE

J.D.Stocker and B.K.Jones

Department of Physics, University of Lancaster,
Lancaster, LA1 4YB, UK.

The gate-current noise in n-channel silicon JFETs has been measured using the device as the preamplifier. The gate current variation with bias has also been measured. Current components due to generation current in the gate-channel depletion region and carrier impact ionization current from the channel show only shot noise. Extra current components, perhaps due to surface or bulk leakage channels show excess noise.

INTRODUCTION

Noise generated within a junction field effect transistor, (JFET), can be represented by an equivalent voltage noise source in series with the gate and an equivalent current noise source between the gate and source terminals. The voltage source represents the noise sources which produce fluctuations in the conductivity of the channel. The current source represents the fluctuations in the gate current.

EXPERIMENT

The gate current noise has been measured over the frequency range 0.2 Hz → 10 kHz using a capacitive feedback technique with a limit given by an equivalent shot noise source of 3.0×10^{-13} amps. The devices tested were biased in a common source configuration. The dc output of the capacitive circuit was stabilised using an opto-feedback system which sets the low frequency limit of the noise measurements. The voltage noise was measured by conventional techniques.

The devices used were experimental versions of BF 800 and BF 818 n-channel silicon devices, biased well above pinch off. The BF 818 has a ∿ 23 times larger top gate area and hence a large generation current. The gate current of both types of device exhibited a similar variation of gate current, measured along the load line. A typical example is given in fig.1. The gate current is made up of two components; the gate-channel diode reverse current, which could also include any surface leakage current, and the channel carrier impact ionization current. At small drain bias the positive temperature dependence of the impact ionization current is apparent. The impact ionization current arises

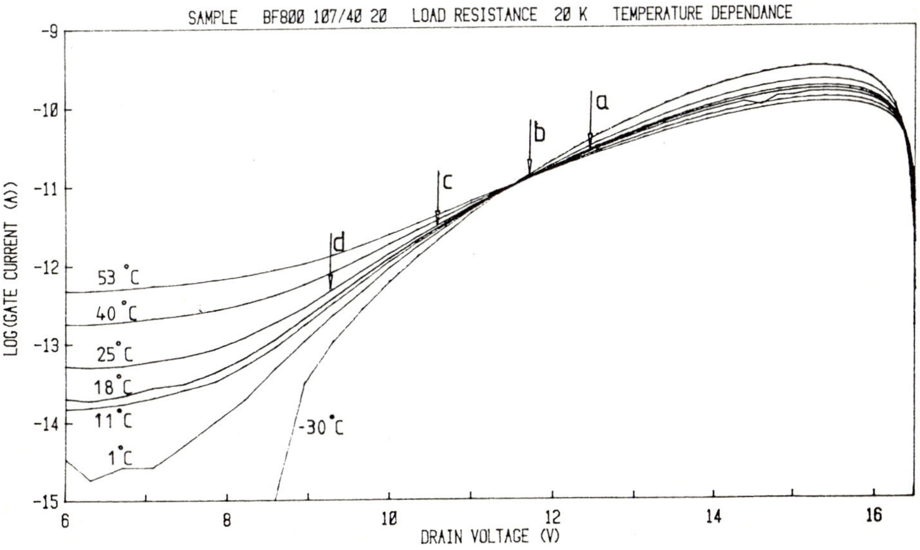

Figure 1 : The gate leakage current variation with drain voltage for a BF 800. The measurement is taken with a series load resistance of 20 k Ohm. The temperature is the parameter. The letters represent the bias points for the data in fig.2.

when the electric field along the channel near the pinch-off region accelerates the carriers in the channel sufficiently to generate electron-hole pairs. The minority carrier of this pair is then swept sideways into the gate-channel depletion region to give an extra gate current component. This current component is proportional to the drain current, (I_D), with a voltage dependent constant.

$$I_{G_I} = I_D \left(1 - \frac{1}{M_n}\right) = I_D \int_0^W \alpha_n \, dx$$

where α_n is the electron ionization rate, defined as the number of electron-hole pairs created by an electron per unit distance travelled.

The small negative tmeperature dependence is in agreement with previous observations[1]. As the drain voltage approaches the power supply voltage along the load line, the impact ionization current component rapidly decreases because the drain current reduces to zero.

The current noise was measured for various drain biases, see fig.2. This shows current noise, at 25°C, of the device (BF 800 107/40 20) used in fig.1. The arrows represent the estimated shot noise of the measured gate current at the bias points indicated. The shot noise estimates agree with the measured current noise spectrum. Excess current noise is not evident in either the generation current (from the depletion region bulk) or the impact ionization.

One of the BF 800's (107/40 21) and some of the BF 818's showed anomalous I-V characteristics. A typical example is shown in fig.3 for the BF 818 002 12. The gate current of this device increases with bias in a similar manner to that of devices which exhibit the effect of impact ionization. However, the current is several orders of magnitude larger and it continues to increase as the drain voltage approaches that of the supply. This excess gate current on further examination was shown to be dependent on the gate-drain voltage and not the drain current. It also has a positive temperature dependence, which implies that the process which generates this current is thermally activated. A typical activation energy is in the region of 0.3 eV. A mechanism that might be used to explain this effect could be surface or bulk conduction channels between the gate and the drain. The noise associated with this current contains an excess component, see fig.4. The excess current noise intensity scales as the current squared.

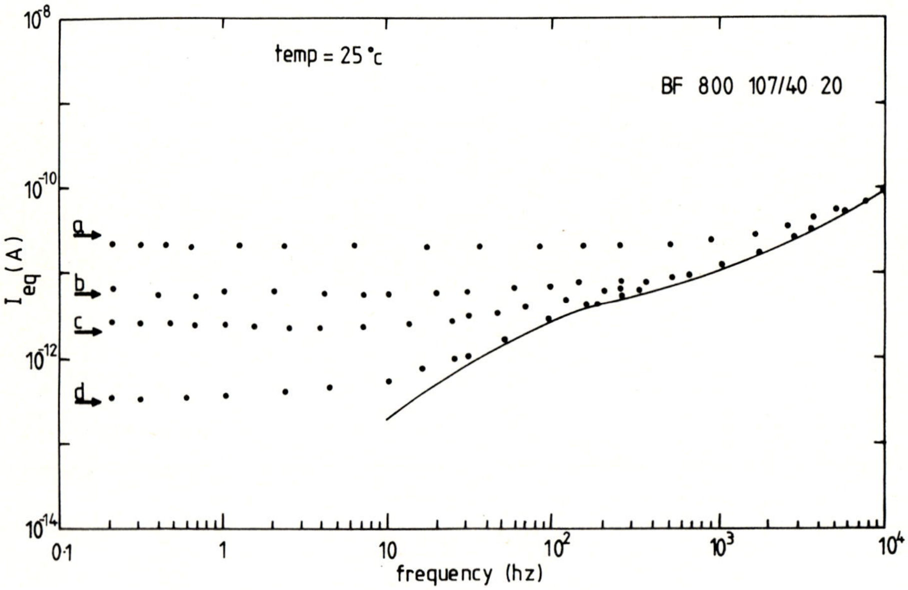

Figure 2 : The equivalent gate current noise spectrum of a BF 800 at 25°C in units of shot noise equivalent current. The arrows represent the expected shot noise for the bias positions indicated in fig.1. The right boundary represents the limit to the current noise measurement produced by the measured voltage noise of the device channel.

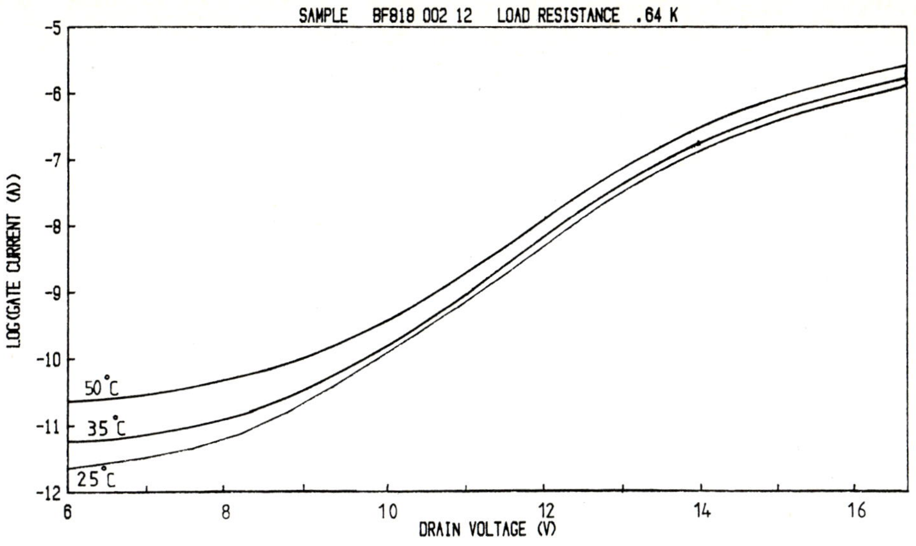

Figure 3 : The gate leakage current variation with drain voltage for a BF 818. The measurement is taken with a series load resistance of 0.64 k Ohm. The temperature is the parameter. This data should be compared with the data of fig.1. The variation indicates that the device has some leakage between gate and drain.

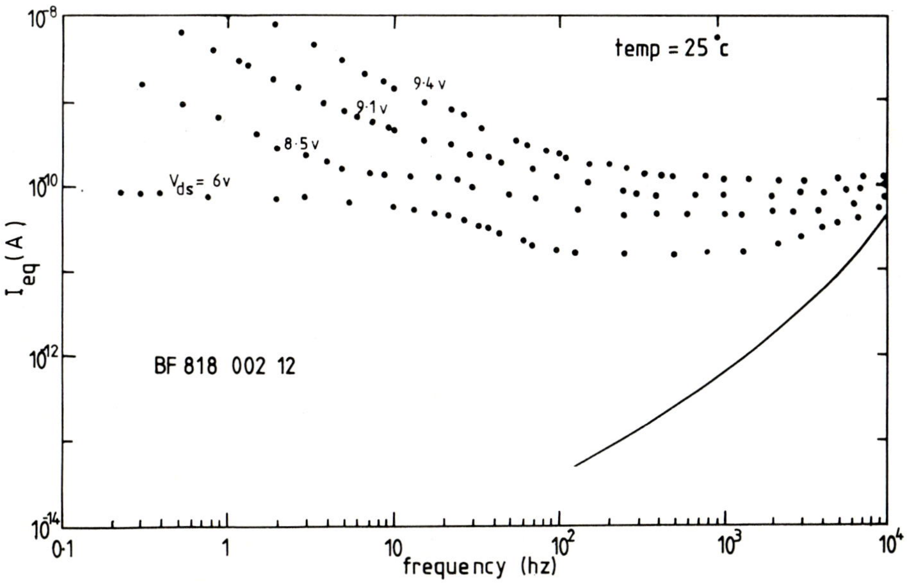

Figure 4 : The noise spectrum of the device in fig.3. The noise is expressed in units of equivalent shot noise current. The excess noise can be ascribed to the leakage current shown in fig.3.

CONCLUSION

The equivalent current noise due to generation current from the bulk depletion region and due to impact ionization current were observed not to exhibit an excess component above the shot noise level. The excess current of devices which suffered premature gate-drain junction breakdown contain an excess current noise component. Surface leakage is also suspected to be responsible for the excess current noise occasionally observed at low drain bias. The form of the dc characteristics shows a high correlation with the presence of excess noise and may therefore also be used to assess the quality of the device.

REFERENCES

[1] E.P.Fowler, Effect on operating conditions on reverse gate current of junction FETS, Electronics Letters, May 1968, 4, No.11, pp.216-217.

This work was done in collaboration with Thorn-EMI and supported in part by SERC.

NOISE IN PHYSICAL SYSTEMS AND 1/f NOISE
M. Savelli, G. Lecoy and J-P. Nougier (eds.)
© Elsevier Science Publishers B.V., 1983

MICROSCOPIC SIMULATION OF MULTIPLICATION NOISE

Lippens Didier, Nieruchalski Jean-Luc, Constant Eugène

Centre Hyperfréquences et Semiconducteurs
Laboratoire Associé au C.N.R.S. n° 287 - Bât. P4
Université des Sciences et Techniques de Lille I
59655 VILLENEUVE D'ASCQ CEDEX - FRANCE

We present a microscopic simulation noise in ultra short avalanche regions. The calculation include the threshold energies for ionization and the non stationary hot carrier transport. This novel model may be used for the study of avalanche devices in which the noise is strongly dependent on the real space band structure such as multilayer avalanche photodiode.

1. INTRODUCTION

In the majority of published papers devoted to the theoretical analysis of the impact ionization process in semi conductors (IMPATT - avalanche photodiode) the region over which the avalanche occursis very long compared to the mean free path for an impact ionization to occur.

As semi conductor devices become smaller it is expected that the relevant temporal and spatial time scales become sufficiently small that the classic approach of the avalanche noise is no longer valid [1] [2].

One one hand, traditional approaches assume that the response of the carriers to any applied force occurs simultaneously with the applied force even though the system may undergo subsequent relaxation to a non equilibrium state.

One the other hand, as has been previously demonstrated by OKUTO and CROWELL [3], since it is true that to originate an ionizing scattering a carrier has to have excess of a threshold energy E_{th}, it is obvious that a carrier has no possibility of making any ionization scattering right after it is generated. One consequence of the reduction in size of semi-conductor is that the distance (the dark space) which corresponds to the threshold energy becomesa sizeable portion of the avalanche region.

The aim of this paper is to propose a simulation of the impact ionization in precisely these ultra short avalanche regions. The difficulty of including the new features which are not included in the common previous theories was overcome by performing a microscopic numerical analysis of the ionization process which we describe in section 2. In section 3, we mention the results obtained for Silicon and Gallium Arsenide p-i-n diodes.

2. STATISTICAL DESCRIPTION

As a model, we use a reverse p-n junction of semi conductor of thickness δ in which a strong electric field causes impact ionization and multiplication of moving charges. Presently, not much is known about the transport of carriers at fields near breakdown.
Due to this lack of information, we have used simplifying assumptions for the transport and the ionization. In spite of these simplifications, which should not be considered as restrictions of the method itself, the simulation provides some insight into how the noise properties are influenced by the non stationary ionization process.

The method used to describe the high field transport keeps track of each particle (electron and hole) until it reach the device boundaries.

The motion is studied only in the real space in the following way :

First, the position inside the sample is determing by the equation of motion

$$x_{k_i}^{j} = x_{k_i}^{j-1} + \mu_k(\varepsilon_{k_i}^{j-1})\, E(x_{k_i}^{j-1})\, \Delta t \pm \sqrt{2D_k(\varepsilon_{k_i}^{j-1})\, \Delta t}$$

k = n for electrons k = p for holes. E is the field acting on the particle and is calculated by solving Poisson's equation. μ and D are respectively the mobility and the diffusion coefficient. The stochastic motion of carriers consist consequently of a single displacement due to the electric field and a random walk with a root mean square radial displacement equal to $\sqrt{2D\Delta t}$

Second, the energy is calculated by means of the balance equation :

$$\varepsilon_{k_i}^{j} = \varepsilon_{k_i}^{j-1} + qE(x_{k_i}^{j} - x_{k_i}^{j-1}) - \frac{\varepsilon_{k_i}^{j-1} - \varepsilon_0}{\tau_\varepsilon(\varepsilon_{k_i}^{j-1})}\Delta t$$

where ε_0 is the thermal energy and τ_ε is the energy relaxation time.

Since all the parameters of transport are taken function of the individual energy of carriers the relaxation effects such overshoot velocities right after ionization are taken into account. The functions $\mu_k(\varepsilon)$, $D_k(\varepsilon)$, $\tau_k(\varepsilon)$ are determined using the results obtained by Monte Carlo simulation in steady state condition [4].

The impact ionization is treated as an additional scattering mechanism. The impact ionization probability is calculated from the experimental ionization rate α applied individually for each carrier. Practically suitable random trials can be achieved to find out whether an ionizing collision has occured during the time step Δt.

In addition there is no ionization probability for any carrier until it gains the threshold energy after it is generated. This requirement garantees the non localized single carrier ionization probability introduced in [3].

This treatment give us a way to illustrate the influence of the energy threshold on the ionization rate. Fig. 1 depicts the results obtained. In full line, when impact ionization occurs, the energy gain during the drift since the last ionization, calculated by accumulating a differential distance v Δt is reinitialized to zero.

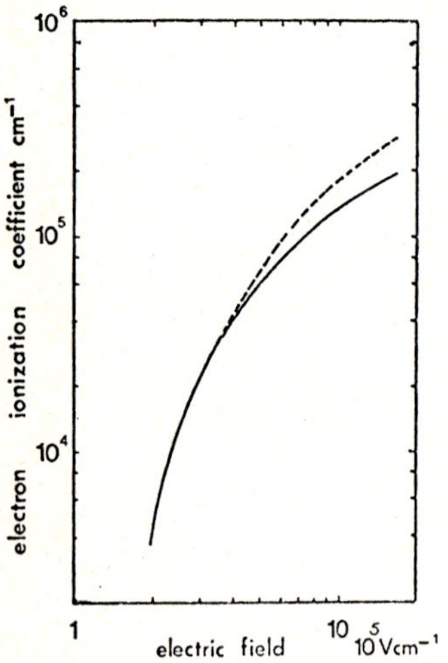

Figure 1 : Ionization coefficient as a function of the electric field for electron. The heavy line is our prediction. The dotted line is the classical approach (Si 300° K).

In dotted line are given the ionization rate when the threshold energy is set to zero. As can be seen the present analysis reproduces the bulk dark spaces and can be applied in a given structure.

3. RESULTS AND ANALYSIS

3.1. Multiplication calculation

It has already been pointed out [1] [3] that the calculation of DC multiplication by neglecting the threshold energy is inadequate to describe the multiplication in ultra short diodes in which the dark spaces constitute a non negligible portion of the length of the diode. Our method by keeping track of the past history of each carriers determines unambiguously the fraction of charge carriers which are able to ionize at each distance of the sample. The comparison between the usual method of performing the DC multiplication and our approach of including the threshold energy is given in Fig. 2. From these results it is apparent that when the threshold energies for ionization are taken into account the DC multiplications are smaller than those usually calculated. This is understandable because the dark spaces introduce inactive regions leading to a shortening for the integral of the ionization rates and consequently a diminution of the DC multiplication.

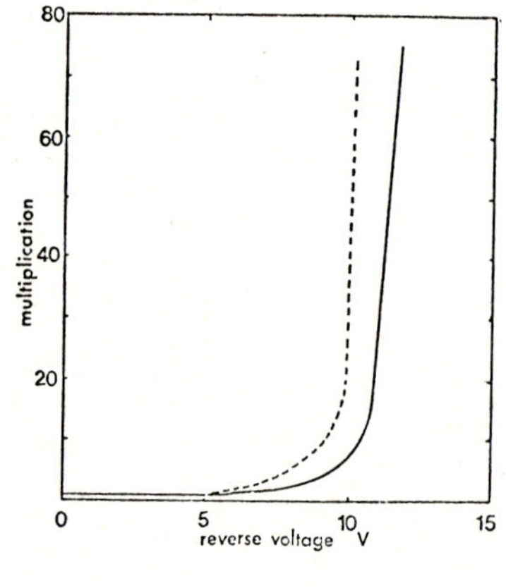

Figure 2 : Multiplication coefficient versus reverse voltage: conventional calculation --- calculated by including the bulk and boundary dark spaces ——. (Si 300° K)

The importance of the threshold energies as a function of width of the avalanche i region for Si p-i-n is illustrated in Fig. 3. Note a discrepancy with the previous work of OKUTO [3] when the region become smaller. However it remains unclear to what extent the functional dependence of the experimental rates on the electric field used to calculate the ionization probability implicity reflects the influence of the threshold energies for ionization.

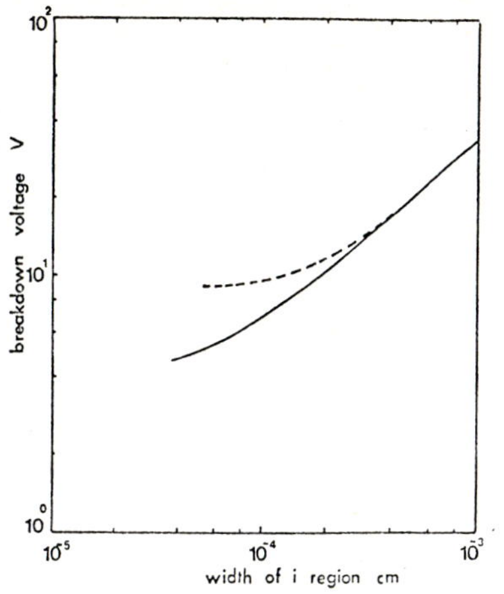

Figure 3 : Breakdown band bending as a function of width of i region
——— OKUTO and CROWELL prediction
------ our work
(Si 300° K)

3.2. Current fluctuations

Consider now the random nature of the current. In Fig. 4 the variations of the particle current versus time for a GaAs p-i-n like diode ($N_D = 10^{15}$ at/cm^3) is shown. The time serie is of 100 ps and the sampling time of 5 10^{-2} ps. The time autocorrelation function can be estimated in a straight forward fashion and is given in Fig. 5. This graph contains two interesting features.

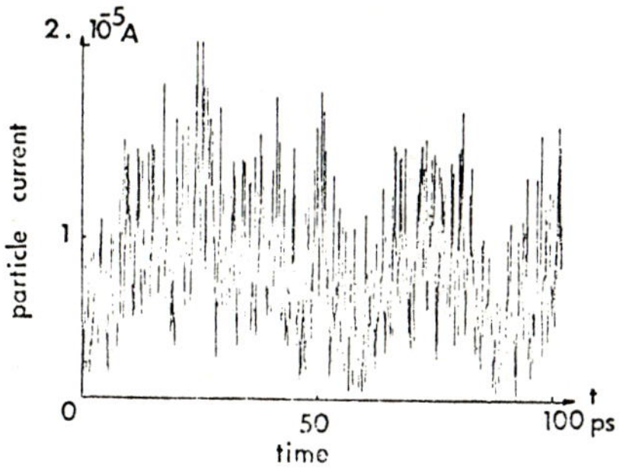

Figure 4 : Variation of the particle current versus time for a pin GaAs diode ($N_d = 10^{15}$ At/cm^3, L = 0,2 μm).

First the current fluctuations decrease very drastically in a time comparable to the scattering time. This is characteristic of velocity fluctuations and affect the high frequency tail of the spectral density.

Second, the fluctuations are found to decay more slowly in a time of the order of several picoseconds. The low frequency spectral density exhibit a low pass characteristic with a time constant proportional to the multiplication coefficient and the intrinsic build up time of the avalanche.

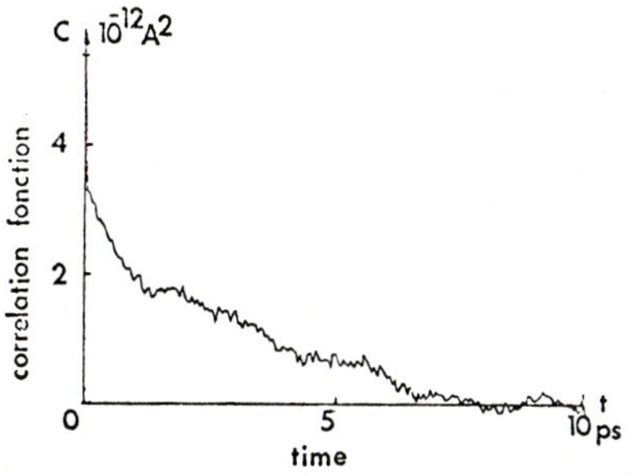

Figure 5 : The time auto-correlation function of the fluctuations in this current.

3.3. The noise versus the multiplication

Of particular interest is the variance of the current fluctuations versus the multiplication factor M. Fig. 6 depicts the results obtained for a 0.1 μm Silicon PIN diode for a pure electron injection when the dark spaces are included (full line) and neglected (dotted line). The results are normalized to the value of the variance obtained for a DC multiplication equal to one. We obtain for these two assumptions similar results inside the statistical error.

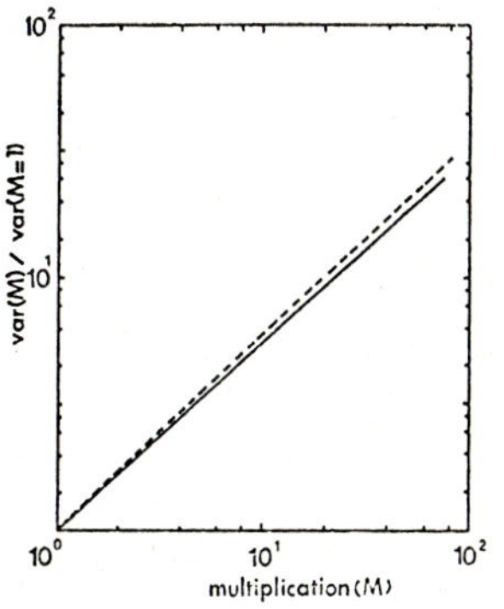

Figure 6 : The variance of current fluctuations versus the D.C. multiplication Factor
dark spaces included —— neglected ----

In common noise theory [5], the dependence of shot noise on multiplication factor is currently given as the excess noise ratio. According to Mc Intyre theory the excess noise factor for electrons is

$$F = M\,[1 - (1 - k)\,(\frac{M-1}{M})^2\,]$$

where k is the ionization rates ratio β/α. The values of F calculated from this expression are plotted in Fig. 7 for some k values and compared with our results for a 0.2 Si p-i-n diode. It can be seen from this figure that the results obtained are in reasonable agreement with the existing noise theories.

It seems, consequently, that the threshold energies for ionization affect mainly the average multiplication coefficient by limiting the ionizing scattering that charge carriers can undergo in one transit accross the diode and, if the DC multiplication is redefined taken into account this feature, the basic understanding of the noise is not questionable.

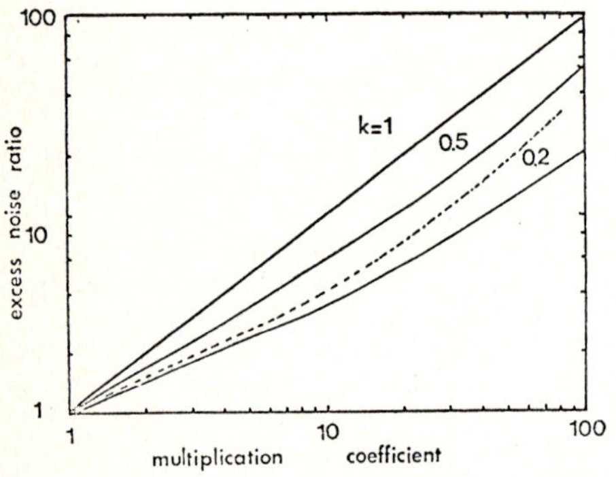

Figure 7 : Calculated excess noise ratio comparison with Mc Intyre theory 0.2 Si PIN diode.

3.4. Multilayer avalanche photodiode

To assess the possibility of application of our analysis to devices consider now a graded band gap multilayer avalanche photodiode. Recent experiments [8] have established that in this type of device, the band edge discontinuity at the layer interface assist the impact ionization.

Fig. 8 shows a typical band diagram of the graded gap multilayer material (assumed intrinsic) at zero applied field. Each stage is linearly graded in composition from a low (E_{g1}) to a high (E_{g2}) bandgap, with an abrupt step back to low bandgap material. The materials could be chosen for a conduction band discontinuity (ΔE_c) which provide a part or entirely the electron ionization energy. The carrier move through the layers of the detector under the combination of the bias field and the grading field. When an electron traverse a step, the conduction band discontinuity modify the impact ionization probability according with the chosen materials.

The gain for a 4 stages detector having each layer 400 Å thick as a function of the single carrier probability of ionization is given Fig. 8.

The operation voltage is 5V. Therefore the gain observed is provided by the conduction band discontinuity rather than via the applied field as in a conventional APD. The gain continuously varied from unity to the total gain 2^4 of the structure when each electron impact ionize after each conduction step.

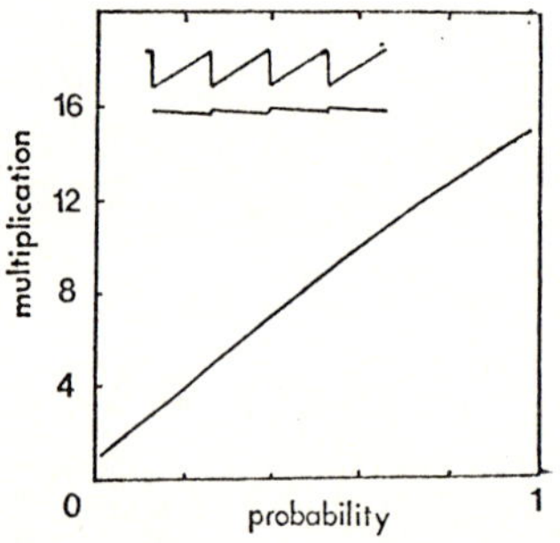

Figure 8 : Gain as a function of the single carrier ionization probability.

CONCLUSION

A microscopic simulation of the avalanche multiplication that include the threshold energy for ionization and the non stationary hot carrier transport has been described. The method has been used to study the multiplication noise in Si and GaAs ultrashort PIN diodes.
It has been concluded that the main effect resulting of the reduction in size of the avalanche region is the existence of a reverse voltage threshold for the onset of the impact ionization which affect the noise properties by limiting the number of electron hole creations that carriers can produce in one transit across the devices.
It allows us to study avalanche devices for which the impact ionization is strongly dependent on the real space band structure. Calculations on this subject are in progress.

REFERENCES

[1] LUKASZEK W., VAN DER ZIEL A., CHENETTE R. Solid State Electronics vol. 19 (1976) 57-71
[2] VAN VLIET K., RUCKER L. IEEE Trans. on Electron devices Vol 26 (1979)
[3] OKUTO Y., CROWELL CR. Physical Review B Vol. 10 (1974) Solid State Elect. Vol.18 (1975)
[4] SCHICHIJO H., HESS K. Physical Review B Vol. 23 (1981) n° 8.
[5] Mc INTYRE RJ IEEE Trans. Electron Dev. ED 13 n° 1 (1966)
[6] CAPASSO F. et al. Appl. Phys. Lett. Vol. 40 (1982).

1/f NOISE IN DIODES

NOISE IN PHYSICAL SYSTEMS AND 1/f NOISE
M. Savelli, G. Lecoy and J-P. Nougier (eds.)
© Elsevier Science Publishers B.V., 1983

NOISE BEHAVIOUR OF VARIOUS Au-InP SCHOTTKY DIODES UNDER HEAT-TREATMENT

J.M. Peransin and M. Abdelali

Laboratoire de Physique Appliquée,
U.S.T.L., 34060 Montpellier - France

Experimental investigations of noise behaviour and contact quality on Au/InP n type diodes are reported. Two different etching solutions are used for the surface cleaning and variations of contact characteristics with heat-treatment have been observed in connection to the device preparation. The interface is differently degraded by the presence of trap levels which are detected at low temperature. During the heating cycles, measurements of noise maxima between 77 K and 300 K give the positions of traps within the forbidden energy band. Changes in the G-R effects against biais current are attribuated to inhomogeneous state densities along the junction.

1. INTRODUCTION

Thermal aging experiments have been made on Schottky barrier contacts and the noise measurements are used as a tool for the investigation of interface transformations (1). InP is a well know III-V material inadequate to obtain ideal Schottky diodes but promising for various devices. The aim of our study is to link the role of the surface preparation with device noise and to find accurate correlation between noise and the parameters of contact. To prevent additional effects of various origins on the noise density occuring in several samples (2) we have used only a typical sample in successive operations to alter its contact properties. The process consists in heat-treatment by cycles of 10 minutes at 250°C.

2. EXPERIMENTAL PROCEDURE

The material used was bulk n type InP with a carrier concentration of 4 x 10^{16} cm^{-3}. Gold contacts are put on (100) wafers, cleaned using two etching procedure. All the samples were chemically etched for 90 seconds in Br_2-CH_3OH (3 : 100) solution. A part of them, called A type, is directly used after this first treatment. Then the second part is etched for 90 seconds more in HCl-H_2O (5 : 100) giving type B samples. To form Schottky barrier, gold coating of thickness of 2000 Å was thermaly evaporated, maintainning the substrate at 200°C. From the gold film, a pattern of circular dots with a diameter of 200 µm was produced by standard photolithographic technique. The ohmic contact was produced by Au-Ge alloying for 20 seconds at 300°C. Electrical connection on the rectifying contact was assured by means of a thermocompressed bounding. The heat-treatments were donne in a quartz tube furnace under 0.2 torr pressure of N_2 gaz.

3. RESULTS AND DISCUSSION

The influence of thermal treatment was first examined by the classical parameters such as : ϕ_B, J_S, n from I-V and C-V measurements at room temperature. Figure 1 shows the behaviour of ϕ_B for both A and B types. In the first series, the barrier height of samples A decreases continuously with the number of heating cycles. After similar treatment in the case of type B samples, gradual increase of ϕ_B is observed during a first step from about 0.44 to 0.45 ev. The curve indicates a maximum for 60 mn of annealing. Then ϕ_B decreases to an abrupt drop corresponding to a large diode degradation. The behaviour of ideality factor n is similar giving strict linear relationship between these parameters.

In type A samples at room temperature the spectra contain a dominant 1/f component rising steadily with the annealing time (figure 2). In the second series of experiments using type B samples, the evolution of the spectrum shows a

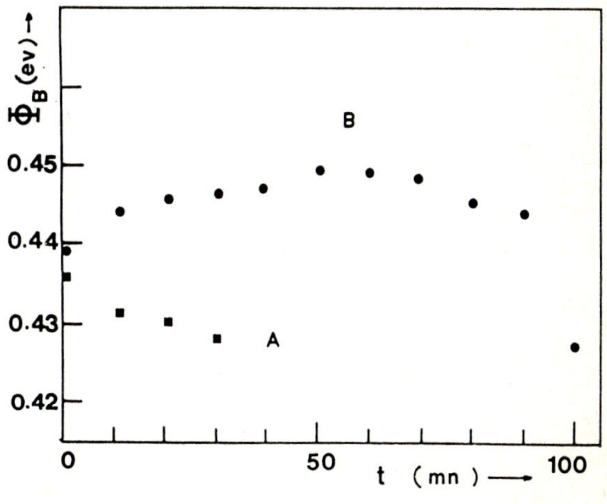

Figure 1 : Schottky barrier heights versus annealing time for type A (Hcl-etched) and type B (Br-etched) samples .

very different behaviour in figure 3. Particulary, the cutt-off frequency reaches a minimum value during the initial heating cycles (1 KHz for N = 3). The general noise level increases significantly with annealing time only after 60 mn, corresponding to the decrease of ϕ_B. A typical discrete level G-R spectrum is then observed with a well marked corner frequency at 120 Hz.(3).

In order to examine the relationship between the 1/f noise and biais current, the equivalent noise currents in a device B before and after heat-treatment are plotted in figure 4 and 5 as a function of forward current. The data indicate a linear relationship of the form : $Si = \alpha I_F^{\beta} f^{-\delta}$ where β appears as an empirical parameter depending weakly on frequency. The Ieq versus I_F characteristic (figure 4) gives β = 1.7 remaining constant over the whole range of polarisation. From figure 5, obtained after complete aging, the dependence of current on excess noise is separated in two parts, β = 2 and 0.8 respectively. This behaviour is clearly correlated to noise increase at N = 9, and consequently it is attribuated to a discret level G-R effect for the first value and to a discontinuous G-R process for the second part.

The results of the same measurements made on devices A are plotted in figures 6 and 7. At room temperature and after each heating cycle, β is found to be 1.9, 1.6, 1.5 and 1.2 respectively as ϕ_B decreases and n deviates more from unity. The main noise difference is caused by the first heating cycle. A strong correlation between Ieq and n factor has been already experimentally found, depending on the n factor origin (2). It should be observed here that there is correlation between β and n due to a specific interdiffusion effect.

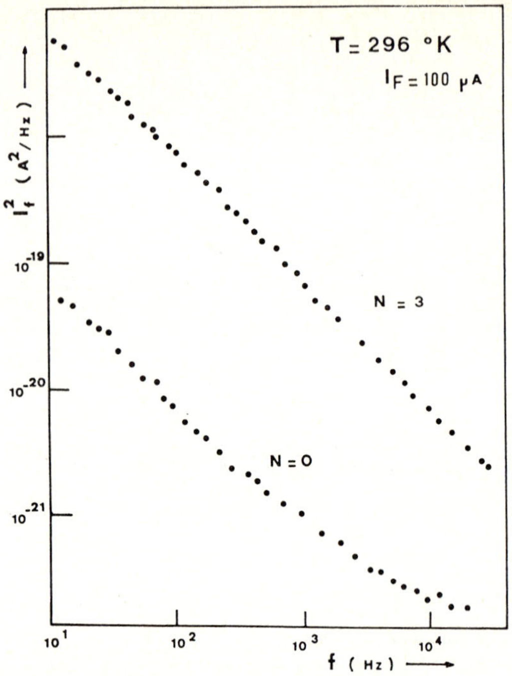

Figure 2 : Noise spectra of a device A
N is the number of heating cycles .

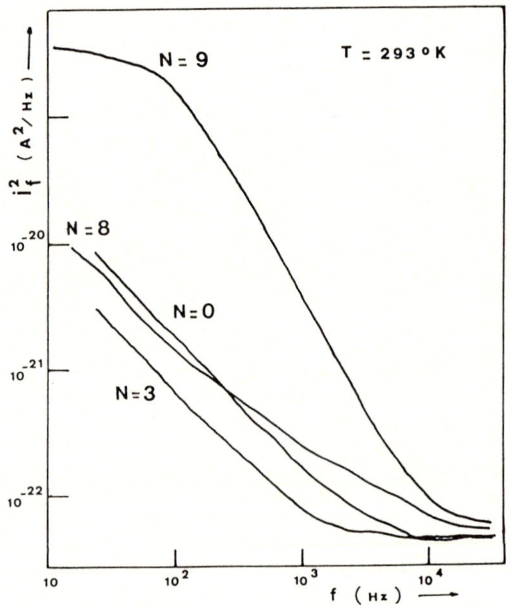

Figure 3 : Noise spectra of a device B with N as a parameter .

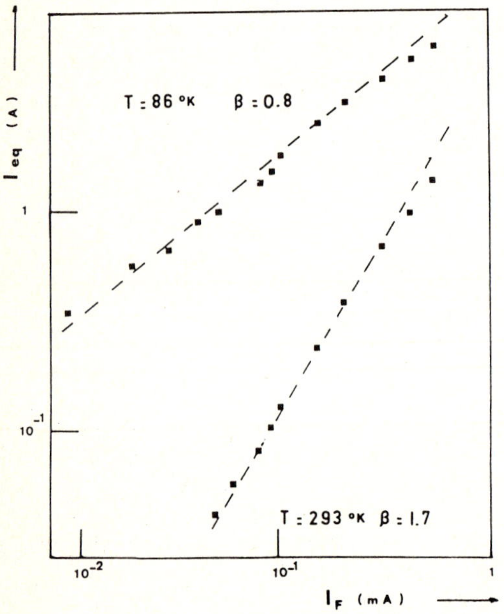

Figure 4 : Ieq versus I_F characteristics before heat-treatment (N=0) of a device B .

At lower temperature, the current noise density is again separated in two part as shown in figure 7. Ieq is found to be proportional to I_F^2 at low currents. But after a significant biais threshold voltage, β decreases and takes a negative value. This behaviour is strongly dependant on sample temperature. Thus the coefficient β lies between 1 and -0.2 when the temperature is between 77 and 300°K. To understand the effects induced by temperature, series of Si(T) dependences are obtained at various biais current.

Figure 8 shows such measurements performed at 10 μA exhibing strong variations of noise spectra with increasing time of heat-treatment. The fluctuations of the trap level population are maximal when its appropriate filling is reached (4). Thus the curves contain maxima corresponding to various G-R centers located near the contact in the depleted region. The main observed deviation of Si(T) dependence is between N = 1 and N = 2 operations. For N = 0 and N = 1 one first well marked level E_1 is observed at about 220°K. A second level E_2 appears at 150°K for N = 2. The similarity between N = 0 and N = 1 situations is due to a continuous aging effect after the thermocompression treatment. The relative decreasing of E_1 peak at N = 2 may be attribuated to a reduction of trap density induced by inhomogeneous diffusion.

Important changes are observed at higher operating current in the shape of the thermally induced noise intensity. The experimental results, S_V versus T, are plotted in Figure 9 for I_F = 60 μA. They are recorded in the same conditions as for 10 μA. A comparison of these characterics shows a new maximum marked as E_3 and a noticeable drop in E_1 component. To a first approximation, the active zone producing G-R noise follows the width variation of the space charge region and the above results indicate that additional trap levels are located near the interface. To describe the current dependance at E_1 maximum, it should be considered that the state density is not uniformly distributed.

Extensive measurements combining frequency and temperature were performed to evaluate the position of trap levels in the gap. It is assumed that the time constant of a discrete G-R process shifts with temperature following a simple exponential form $\tau_i = \tau_0 \exp(E_i/kT)$. $E_i = E_C - E_T$ is the activation energy of the trapping level i below the conduction band.

From this assumption we have determined the following values : E_1 = 0.68 ev, E_2 = 0.49 ev and E_3 = 0.44 ev. Other traps are included in the interface of devices but with too low concentration and thus these are not easily observed.

4. CONCLUSION

From the results presented here, it appears that the surface preparation and the metallic diffusion play a significant role in determining the aging behaviour. Each annealing operation gives

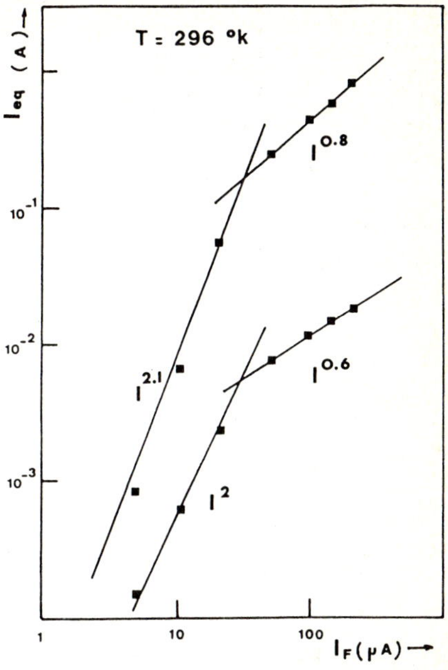

Figure 5 : Ieq versus I_F of a device B (at N=9) for 10^2 and 10^3 Hz.

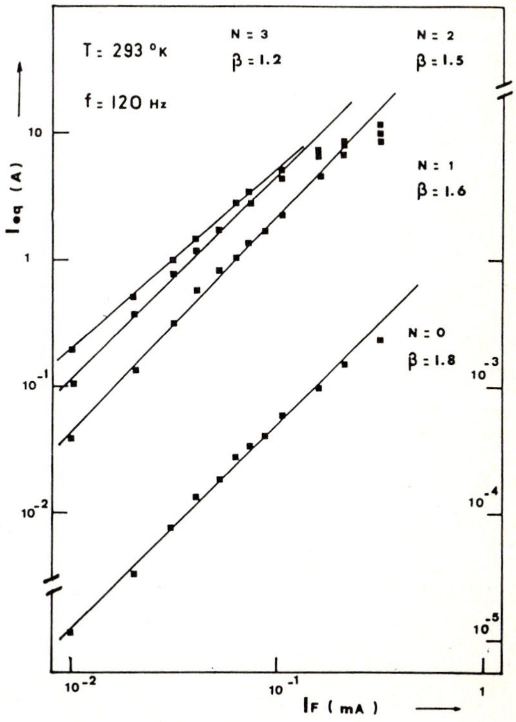

Figure 6 : Ieq versus I_F of a device A after each heating cycle (N = 0 to 3)

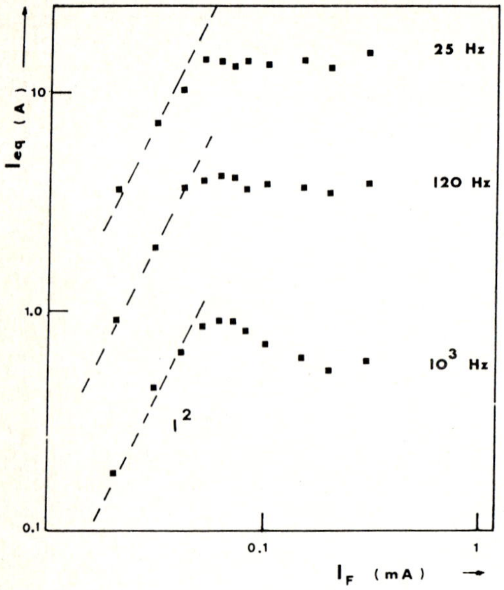

Figure 7 : Ieq vs I_F , device A at 86°K and N= 3

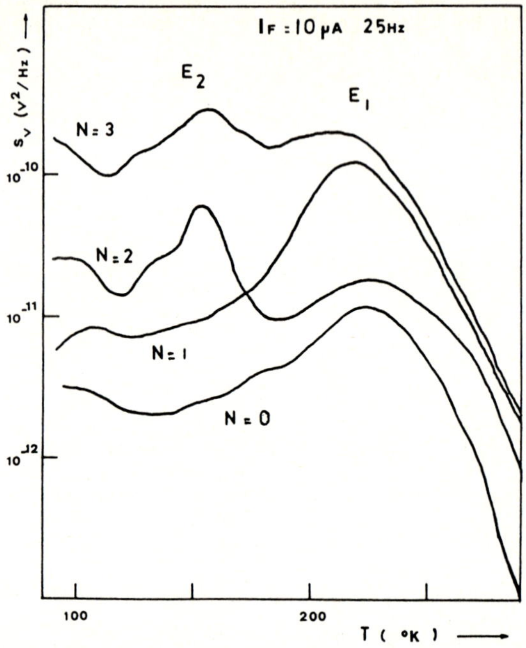

Figure 8 : Noise voltage as a function of temperature at low current after each heating cycle

a new interface complexion inducing important changes in the noise characteristics. However, in the first step of heat-treatment, the surface composition is more important than the deposited metal, presumably due to a layer of native oxyde, such as In PO_4. Consequently low noise devices corresponding to a low 1/f cutt-off frequency may offer better thermal stability than noisy devices. The comparative study shows that the decrease of noise density at high biais current is partly due a new impurities arrangement, this is justified by changes in the spectroscopic repartition against temperature as a function of polarisation.

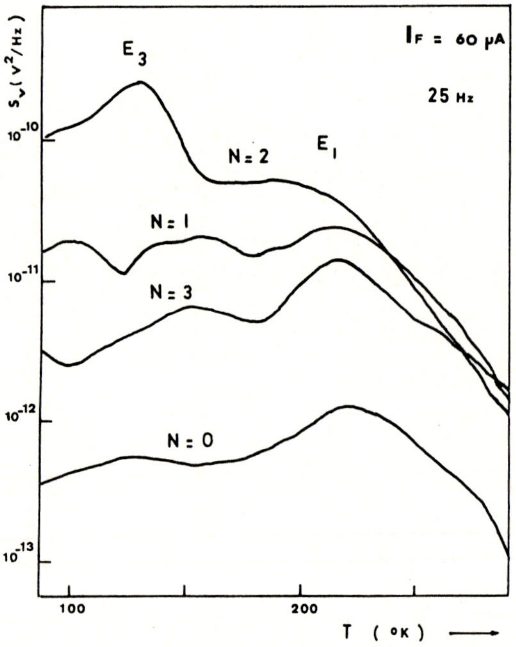

Figure 9 : Noise voltage as a function of temperature at high current .

REFERENCES

(1) Morgan D.V., Howes M.J. and Delvin W.J., A study of gold/n-InP contact, J. Phys. D : Appl. Phys. 11 (1978) 1341-1350.

(2) Kleinpenning T.G.M., Low-frequency noise in Schottky barrier diodes, Sol. St. Electronics 22 (1979) 121-128.

(3) Peransin J.M., Mesbah M., Groubert E., Effects of heat-treatment on the noise spectrum in Au-InP Schottky barriers, Electronics letters 17 (1981) 757-758.

(4) Sheng T. Hsu, Low-frequency Excess noise in Metal-Silicon Schottky Barrier diodes, IEEE Trans. Elect. dev. 17 (1970) 496-506.

NOISE IN PHYSICAL SYSTEMS AND 1/f NOISE
M. Savelli, G. Lecoy and J-P. Nougier (eds.)
© Elsevier Science Publishers B.V., 1983

NOISE PROPERTIES OF BULK-BARRIER-DIODES

Christian Hanke

Lehrstuhl für Technische Elektronik
Technische Universität München
Arcisstraße 21
D-8000 München 2, FRG

The Bulk-Barrier-Diode is a barrier controlled majority carrier device. A short description of its electrical properties is given. Two excess-noise models, based on mobility fluctuations and on charge fluctuations at traps, are given. The experimental results support the trap-noise model. At higher frequencies the diode shows shot-noise for forward and reverse bias.

1. ELECTRICAL PROPERTIES OF BULK-BARRIER-DIODES

Bulk-Barrier-Diodes or camel-diodes have recently gained growing interest for the application as photodiodes/1/, mixers/2/ and as temperature sensors. A short description for the currentflow in this device is given; for a more detailed description see /3,4/.
The Bulk-Barrier-Diode (BBD) is a device consisting of a three layer structure similar to a bipolar transistor.

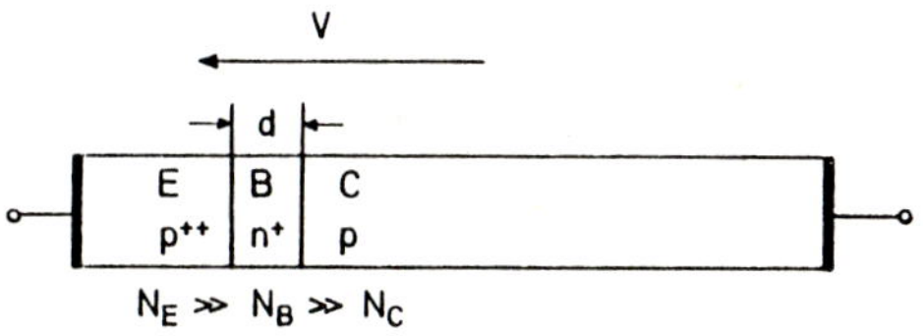

Fig.1. Structure of a BBD

Homogeneous doping and abrupt junctions are assumed for modeling the electrical behaviour of the diode. The main characteristic of a BBD is a very thin basewidth d. Therefore the base is punched through and depleted of free carrier-charge even at zero bias.

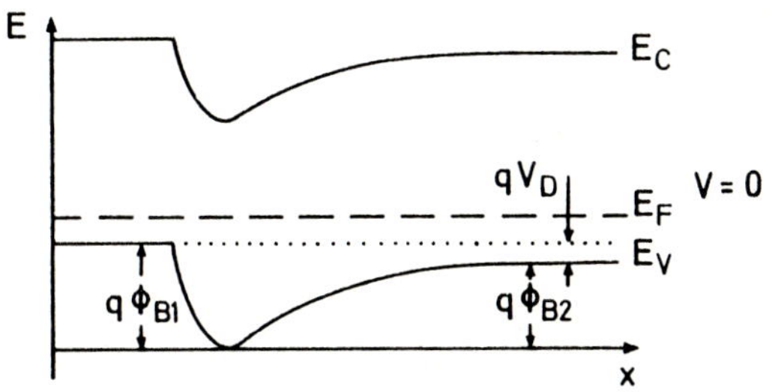

E_F:Fermi level V_D: Diffusion potential

Fig.2. Energy diagram for thermal equilibrium

The energy diagram shows a potential barrier for the majority carriers, in this case for the holes. The electrical behaviour can be described by two barrier heights Φ_{B1} and Φ_{B2}. In thermal equilibrium the hole currents emitted over the barrier from the collector and the emitter are equal. Because Φ_{B1} and Φ_{B2} are voltage dependent, the energy diagram changes for forward and reverse bias.

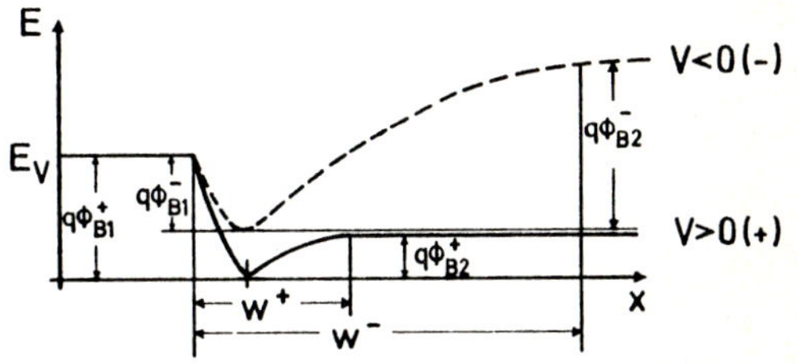

Fig.3. Energy diagram for forward (+) and reverse(-) bias

In particular Φ_{B2} decreases almost linearly with increasing forward bias whereas Φ_{B1} increases a little bit. More holes can then cross the barrier from the collector to the emitter than in the opposite direction. The current will increase nearly exponentially with the applied forward voltage. At reverse bias, Φ_{B2} increases rapidly whereas Φ_{B1} decreases slowly. There is a reverse hole current from the emitter to the collector. The bias dependence of the characteristic barriers is controlled by the doping densities in the three layers. Due to the asymmetrical doping of the emitter and the collector, an asymmetrical I/V-characteristic results.
If the holes cross the potential barrier, where the electrical field is zero, without collision, the thermionic emission theory applies. If not, the diffusion- or a mixed emission-diffusion-theory is valid. The I/V-characteristic is determined by a saturation current I_s and by the dependence of the barrier Φ_{B2} on the applied voltage. One gets:

$$I = I_s \exp\left(-\frac{\Phi_{B2}(V)}{V_T}\right)\left(1 - \exp\left(-\frac{V}{V_T}\right)\right) \quad (1)$$

In the case of a diffusion controlled current flow, the saturation current I_s is determined by (2)

$$I_s = \frac{D_p N_C A}{\sqrt{2\pi}\, L_D} , \qquad (2)$$

where D_p: Diffusion constant for holes, A: Diode area, N_C: Doping density of the collector, L_D: Debye-length in the base.

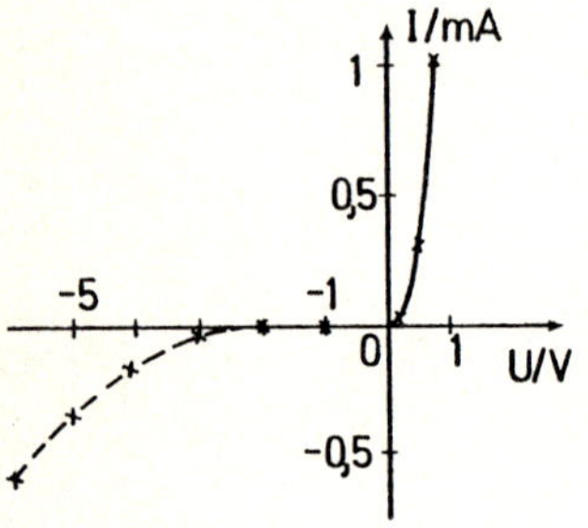

Fig.4. Measured I/V-characteristic of a BBD

From this characteristic and from theoretical evaluations the validity of the diffusion-theory results for our diodes.

2. NOISE-MODELS FOR THE BBD

The current flow in a BBD is similar to that in a Schottky diode. Therefore two excess-noise models for those diodes were applied to the BBD. The first model, based on mobility fluctuations, follows /5/, whereas the second model is based on charge fluctuations at traps according to /6/.

2.1. Mobility model

The mobility model uses the empirical formula of Hooge for mobility fluctuations.

$$S_\mu = \frac{\mu^2 \alpha}{f N} \qquad (3)$$

S_μ is the spectral density of the mobility fluctuation, α is Hooge's constant, f the frequency and N the total number of charge carriers. This formula is valid for homogeneous samples. In our device the carrier density is inhomogeneous. Assuming that a mobility fluctuation $\Delta\mu(x)$ at spot x leads to a current fluctuation ΔI, formula (4) results:

$$\Delta I = \Delta\mu(x) \frac{I}{\mu} . \qquad (4)$$

Using the cross-correlation spectral density for the one-dimensional case

$$S_\mu(x,x') = \frac{\alpha \mu^2}{f\, A p(x)} \delta(x-x') \qquad (5)$$

we get the current spectral density:

$$S_I = \frac{I^2 \alpha}{w^2 f A} \int_0^w \frac{1}{p(x)} dx , \qquad (6)$$

where w is the depletion layer width and p(x) the hole density in the depletion layer. Then

$$p_{eff} = \left(\frac{1}{w} \int_0^w \frac{1}{p(x)} dx \right)^{-1} , \qquad (7)$$

where p_{eff} is an effective hole density resulting from an averaging over the depletion layer. This density can be related to the diode current according to the drift-diffusion theory.

$$p_{eff} = \frac{I w}{q A D_p} . \qquad (8)$$

Inserting (8) into (6) we get:

$$S_I = \frac{q \alpha D_p I}{w^2 f} . \qquad (9)$$

According to this model the current spectral density is proportional to the diode current I and it is inversely proportional to the square of the depletion layer width w, which is bias-dependent. There is no dependence on the diode area.

2.2 Trap model

The trap model assumes that the fluctuation of charged traps or generation-recombination centers in the depletion layer modulate the electrical field if the diode is ac-shorted. Consequently the barrier height is modulated, yielding a fluctuating current.

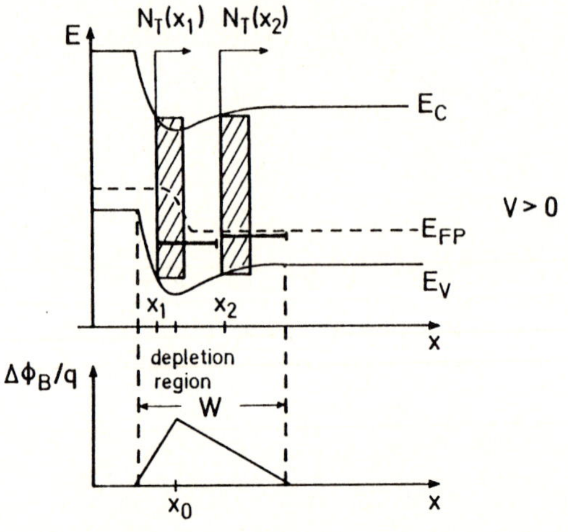

Fig.5. Trap noise model (see text)

Figure 6. shows the energy diagram for a BBD with forward bias. E_{FP} is the quasi-fermi level for holes. At two positions (x_1, x_2) there is a trap-density distribution N_T over the bandgap. Homogeneous distribution in energy and space is assumed including a discrete level. The occupancy of the traps is ruled by the quasi-fermi level. In the depletion region this level scans different energies. Therefore at a spot x only a part of the traps are active. Only those traps which are half occupied have a reasonable effect. That means that only those traps are active whose energetic position equals the quasi-fermi level. The characteristic time constant τ_t of the active traps depends on their energetic position according to the SRH-model. Because a localized charge has a different effect on the variation of the barrier height $\Delta\Phi_B$ depending on its position x in the depletion layer there is a weighing function $\Delta\Phi_B/q(x)$. Traps at the position of the potential maximum are most effective.

In this model a variation of the barrier height $\Delta\Phi_B$ leads to a variation of the diode-current ΔI:

$$\Delta I = \frac{\Delta\Phi_B}{V_T} I . \qquad (11)$$

After some calculations, the current spectral density, for a constant trap density N_T results:

$$S_I = \frac{I^2\, 4q^2\, N_T\, kT}{V_T^2\, \varepsilon^2\, A} \int_0^w \frac{\tau_t(x)}{1+(\omega\cdot\tau_t(x))^2}\, f(x)\, dx, \qquad (12)$$

where

$$f(x) = \begin{cases} (1-x_o/w)^2\, x^2 & 0 < x < x_o, \\ (1-x/w)^2\, x_o^2 & x_o < x < w . \end{cases} \qquad (12a)$$

In the above equation $f(x)$ is the weighting function already mentioned. The integral has been evaluated by numerical methods.

The noise spectral density is proportional to the square of the diode current I and it is inversely proportional to the diode area A. The evaluation of formula (12) yields spectra according to $1/f^{\delta}$, if a homogeneous trap density N_T is assumed. The values of δ are bias dependent. In the case of a single trap level the formula leads to a GR-spectrum which is also depending on the bias voltage.

3. EXPERIMENTAL RESULTS

Bulk-Barrier-Diodes with a mesa structure were fabricated. Ion implantation was used for the doping of the emitter and the base region. For the investigation of geometric effects, diodes with various mesa- and contact-diameters were fabricated using aluminum or gold as contact material.

The noise of the diodes was measured using a low-noise transimpedance amplifier and a Fast-Fourier-Transform spectrum analyzer with a measurement capability up to 50 kHz. The diodes were biased with constant current I ranging from 100 nA to 1 mA.

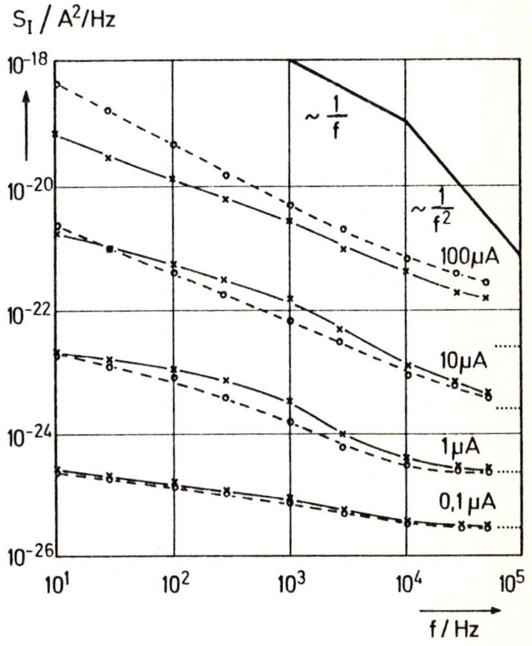

—— forward-, ---- reverse bias

···· shot noise 2qI

Fig.6. Current noise spectra of a BBD

On the right hand side of figure 6 the noise spectral density corresponding to shot-noise of the diode current is indicated. In the upper frequency range and for low currents the diodes exhibit shot-noise both in the forward and in the reverse direction. This behaviour is expected because the carriers cross a potential barrier.

In a wide range there is excess-noise. It is remarkable that the corner frequency, where the excess-noise reaches the shot-noise level, is not constant, but it depends on the diode current. In the reverse direction the noise spectral density can be well approximated by $1/f^{\delta}$ with δ-values from 0.6 to 0.8 depending on the reverse current. In the forward direction there is a deviation of the $1/f^{\delta}$-dependence at frequencies around 1kHz. This can be attributed to a discrete trap level. According to the trap model this influence vanishes for high and low current. From figure 5 it can be seen that a discrete level is effective only if the quasi-fermi level is at its energetic position.

Gold and aluminum were used for preparing the contacts. No difference in the I/V-characteristic and in the noise performance was measured. The noise is therefore not dominated by contact-noise, but the main effect lies in the active region of the device.

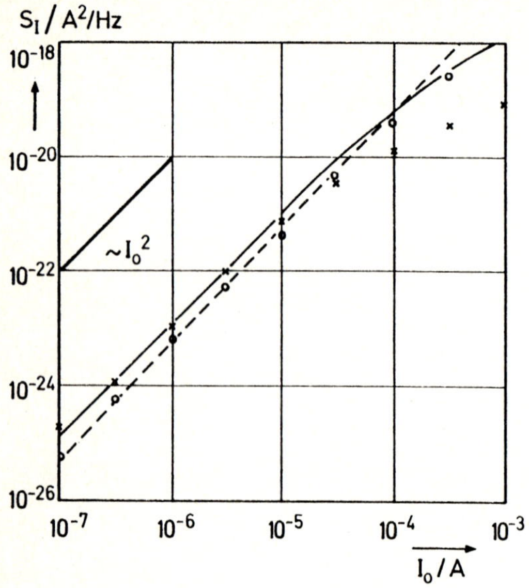

forward ××× } as measured; reverse ∘∘∘ } as measured; —— } fitted by the trap model; - - - - } fitted by the trap model

Fig.7 Current dependence of the noise spectral density (f = 100 Hz)

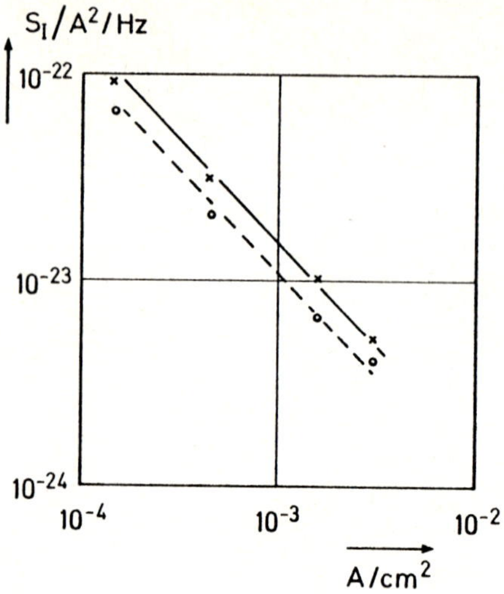

—— forward-, ---- reverse bias

Fig. 8. Area dependence of the noise spectral density (f=100 Hz, I= 1 µA)

The current dependence of the noise spectral density shows a characteristic to I^2 in a certain range. The curve according to the trap model is fitted using a trap density of $3 \cdot 10^{15}$ $eV^{-1}cm^{-3}$ and a capture cross-section of 10^{-15} cm^2. This high density may be explained by the technology of the diodes. Ion implantation was used for doping, which introduces damages in the crystal. If they are not completely annealed, there is a wide variety of traps in the active region. The trap model predicts a deviation from the I^2-characteristic at high forward current, which is also measured. This is due to the effect that the depletion layer gets smaller with increasing forward voltage. The mobility model predicts a dependence proportional to I, which is not measured.
Noise measurements were performed with diodes of different diameters.

The dependence proportional to 1/A according to trap model is evident. In contrast to that the mobility model has no area dependence.

4. CONCLUSION

The noise performance of Bulk-Barrier-Diodes was investigated theoretically and eperimentally. A mobility- and a trap model were applied to this type of diode. The experimental results show the shot-noise behaviour of the diode. In a wide range there is an excess-noise whose dependence on the diode current and the diode area supports the trap model. From this an assumed trap density homogeneously distributed in energy and space can be evaluated. Not annealed radiation damage, caused by ion implantation, is possibly responsible for that trap distribution. The effect of mobility fluctuations is not excluded, but it is not seen due to the high trap density.

Acknowledgement: The author is indebted to Prof. H. Mader for supplying the implanted wafers and to Mr. Krebs for preparing the diodes.

REFERENCES

/1/ Georgulas,N.,The camel diode as photodetector with high internal gain, IEEE Electron Device Letters, vol. EDL-3, pp.61-63,1982

/2/ Dixon,S., Malik,R.J., Subharmonic planar doped barrier mixer conversion loss characteristic, IEEE MTT-31, No.2, Feb. 1983

/3/ Mader,H., Electrical properties of Bulk-Barrier Diodes, IEEE ED-29, No.11 Nov. 1982

/4/ Shannon, J.M.,A majority-carrier camel diode, Appl. Phys. Lett., vol.35,p. 63, 1979

/5/ Kleinpenning T.G.M., Low-frequency noise in Schottky barrier diodes, Solid-State Electronics, vol.22, pp. 121-128, 1979

/6/ Hsu, S.T., Low-frequency excess noise in metal-silicon Schottky barrier diodes, IEEE, vol. ED-17 pp. 496-506, 1970

NOISE IN PHYSICAL SYSTEMS AND 1/f NOISE
M. Savelli, G. Lecoy and J-P. Nougier (eds.)

© *Elsevier Science Publishers B.V., 1983*

ON 1/f NOISE IN REVERSE-BIASED P-N JUNCTION DIODES

T.G.M. Kleinpenning

Eindhoven University of Technology,
Eindhoven, Netherlands

1/f noise calculations are presented for three types of p-n junction diodes: (i) a long diode where the current is determined by diffusion in the base, (ii) a short diode where the current is determined by diffusion in the base and by the contact recombination velocity, and (iii) a diode where the current is determined by generation-recombination (g-r) in the junction. Noise measurements are performed on reverse-biased diodes of type (i) and (iii); they can be interpreted in terms of mobility fluctuations. It is found that at fixed reverse bias the ratio of 1/f noise and shot noise decreases with increasing lifetime of minority carriers in the base or with increasing lifetime of carriers in the junction.

1. INTRODUCTION

During the last few years reverse-biased (HgCd) Te photodiodes have found widespread application as infrared detectors {1}. Reduction of the 1/f noise level is important in systems requiring high sensitivity. Therefore, an understanding of the 1/f noise in reverse-biased p-n diodes is necessary.
At present there is no generally accepted physical model for the 1/f conductivity fluctuations in metals and semiconductors. However, on the basis of experimental facts the 1/f noise in the conductance G of homogeneous materials can be described by the empirical relation {2}

$$S_G = \alpha G^2/fN \tag{1}$$

Here, α is an empirical constant of the order of 10^{-3}, f the frequency, and N the number of free-charge carriers. Experiments have shown that the 1/f noise is due to fluctuations in the free-carrier mobility. Starting from the empirical relation (1), the 1/f noise in forward-biased p-n diodes can be described very well {3}. The purpose of this paper is to investigate the 1/f noise in reverse-biased p-n diodes in view of relation (1).

2. CALCULATION OF 1/f NOISE IN P-N DIODES

2.1. Long p^+-n diode; diffusion-current dominated

Consider a p^+-n diode with cross-section A where the base extends from x = 0 to x = W. For x < 0 the material is p-type. The base width W is assumed to be larger than the hole diffusion length $L = \sqrt{D\tau}$ in the n-type base, so that most of the holes diffusing into the base do not reach x = W. Here τ is the hole lifetime in the base, and D the diffusivity of holes. If the diode is biased (reverse V < 0, forward V > 0), then the hole current I and the hole density p in the base are given by {4}

$$I(x) = -qAD\frac{dp}{dx} = \frac{qADp^o}{L}\left\{e^{qV/kT}-1\right\}e^{-x/L} \tag{2}$$

$$p(x) = \left\{p(o)-p^o\right\}e^{-x/L}+p^o \tag{3}$$

$$p(o) = p^o e^{qV/kT} \tag{4}$$

where p^o is the hole density in the base at V=0. Mobility fluctuations lead to diffusivity fluctuations {4,5} and so to fluctuations in I. According to {2,3,5} fluctuations in the hole mobility $\mu(x,t)$ at point x at time t lead to current fluctuations such as

$$\Delta I(x,t) = \tfrac{1}{2}\langle I(x)/\mu\rangle \Delta\mu(x,t) \tag{5}$$

with a spectral density

$$S_I(f) \cong \frac{1}{(2L\mu)^2}\int_0^L\int I(x)I(x')S_\mu(x,x',f)dxdx' = \frac{\alpha}{4AfL^2}\int_0^L \frac{I^2(x)dx}{p(x)} \tag{6}$$

The cross-correlation spectral density in μ is given by {2,3,6}

$$S_\mu(x,x',f) = \{\alpha\mu^2/fAp(x)\}\delta(x-x') \tag{7}$$

Substitution of eqs. (2,3,4) in eq. (6) yields

$$S_I(f) = \frac{\alpha q I_o}{4f\tau}\int_0^1 \frac{(e^{qV/kT}-1)^2 e^{-2z}dz}{(e^{qV/kT}-1)e^{-z}+1} \cong \frac{\alpha q I_o}{4f\tau}\left\{e^{qV/kT}-1-\frac{qV}{kT}\right\} \tag{8}$$

The saturation current I_o and the I-V relation are given by

$$I_o = qADp^o/L = qAn_i^2(D/\tau)^{\frac{1}{2}}/N_B \qquad (9)$$

$$I = I_o\{\exp(qV/kT)-1\} \qquad (10)$$

where N_B is the dope in the base, and $z=x/L$. The right hand side of eq. (8) gives an approximated expression for the 1/f current noise, it is obtained by taking the integration boundaries from 0 to ∞ instead of 0 to L in eq. (6). The shot noise is given by

$$S_I(f) = 2qI_o\{\exp(qV/kT)+1\} \qquad (11)$$

The corner frequency f_c where the 1/f noise equals the shot noise, can be obtained from eqs. (8) and (11). For reverse-biased diodes with $-qV/kT>>1$ we find

$$f_c = \frac{\alpha}{8\tau}\cdot\frac{-qV}{kT} \qquad (12)$$

2.2. Short p^+-n diode; diffusion current dominated

In short p^+-n diodes the base is so short, that recombination in the base can be neglected. Almost all holes then reach the contact at $x=W$, where they recombine. The hole current I and the hole density p in the base are given by {7}

$$I(x) = I = -qAD\frac{dp}{dx} = \frac{qAD}{W}\{p(o)-p(W)\} = qSA\,\{p(W)-p^o\} \qquad (13)$$

$$p(x) = p(o) - Ix/qAD\ ;\ p(o) = p^o\exp(qV/kT) \qquad (14)$$

Since the hole gradiënt dp/dx is determined chiefly by the contact recombination velocity S, we state {3} $I(x)\sim\mu(x)$. The current noise is found to be (cf. eqs. (5,6))

$$S_I(f) = \frac{1}{\mu^2W^2}\int_o^W\!\!\int I^2S_\mu(x,x',f)\,dx\,dx' = \frac{\alpha I^2}{fAW^2}\int_o^W\frac{dx}{p(x)} \qquad (15)$$

Using eqs. (13,14) we obtain

$$S_I(f) = \frac{\alpha I^2\ln\{p(W)/p(o)\}}{fAW\{p(W)-p(o)\}} = \frac{\alpha qID}{fW^2}\ln\left\{\frac{p(o)}{p(W)}\right\} \qquad (16)$$

$$\frac{p(o)}{p(W)} = \frac{D/W+S}{D/W+S\exp(-qV/kT)} \qquad (17)$$

$$I = qAp^o\left\{\frac{DS}{D+WS}\right\}\{\exp(qV/kT)-1\} \qquad (18)$$

For reverse-biased short diodes with $-qV/kT>>1$ and $S>D/W$ the corner frequenct f_c is found to be

$$f_c = \frac{\alpha D}{2W^2}\cdot\frac{-qV}{kT} = \frac{\alpha}{4\tau_D}\cdot\frac{-qV}{kT} \qquad (19)$$

where τ_D is the diffusion time for holes through the base region.

2.3. p^+-n diode; g-r current dominated

To calculate the 1/f noise, we follow the same procedure as in {3}. The main part of the g-r current in the junction is in the area where $p\cong n\cong n_i\exp(qV/2kT)$ and with an effective width W*. It is assumed that the regions outside this area do not contribute appreciably to the current and to the 1/f noise. Then the short-circuit current fluctuations are approximated by

$$\Delta I(t) = \frac{1}{W^*}\int_o^{W^*}\left\{\frac{I_p(x)}{\mu_p}\Delta\mu_p(x,t) + \frac{I_n(x)}{\mu_n}\Delta\mu_n(x,t)\right\}dx \qquad (20)$$

For the 1/f current noise we find {3}

$$S_I(f) = \frac{(2/3)\alpha I^2}{fAW^*n_i\exp(qV/2kT)} =$$

$$\frac{2\alpha qI_o}{3f\tau_j}\left\{e^{\frac{qV}{2kT}}-1\right\}^2 e^{-\frac{qV}{2kT}} \qquad (21)$$

where n_i is the intrinsic concentration and τ_j the lifetime of carriers in the junction. The saturation current I_o and the I-V characteristic are given by

$$I_o = \frac{qW^*An_i}{\tau_j}\ ;\ I = I_o\{\exp(qV/2kT)-1\} \qquad (22)$$

For reverse-biased diodes $(-qV/2kT>>1)$ the shot noise can be approximated by $2qI_o$, so the corner frequency is

$$f_c = \frac{\alpha}{3\tau_j}\exp\left(-\frac{qV}{kT}\right) \qquad (23)$$

3. EXPERIMENTAL RESULTS

Investigations have been made on two types of silicon p^+-n diodes at 426 K. The measurements were performed at rather high temperatures due to the fact that at room temperature the diode impedance is too high for noise measurements. The diffusion-current dominated diode (A) is a p^+-n diode with $A=0.05$ cm^2, $W=300$ μm, $N_B=10^{15}$cm^{-3} and $\tau=2.5$ μs. The I-V characteristic obeys eq. (10) with $I_o=10\mu A$. The calculated value of I_o from eq. (9) is $I_o=9\mu A$ at T=426 K.
The g-r current dominated diode (B) is the same p^+-n diode as treated in {3} with $A=2.5\times10^{-3}$cm^2. The width W* is calculated to be 0.5 μm {3}. The I-V characteristic can be approximated by eq. (22) taking into account that {4} $I_o\sim W^*\sim(V_o-V)^{\frac{1}{2}}$. The solid line in Figure 1 is the best fit with $V_o=75$mV. At 426 K the lifetime $\tau_j=qW^*An_i/I_o\cong 6\times10^{-8}$s.
The noise spectra, measured in the frequency range of 1-10^4Hz, show 1/f noise at low frequencies (see Figure 2). In Figures 3 and 4 we have plotted the experimental and theoretical relations between the 1/f current noise density and

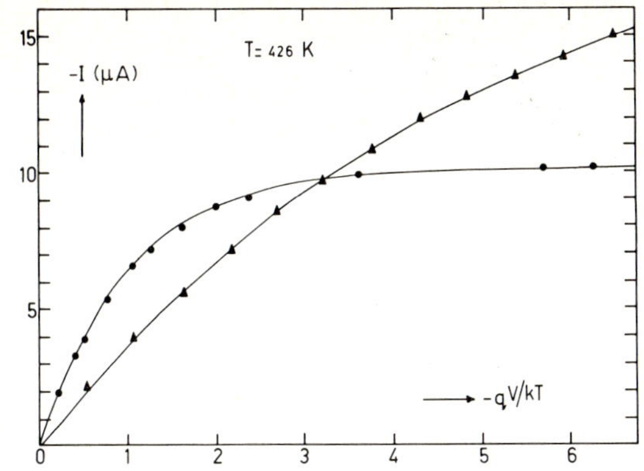

Figure 1: Reverse I-V characteristics at T=426K of diffusion current dominated diode A (●) and g-r dominated diode B (▲).

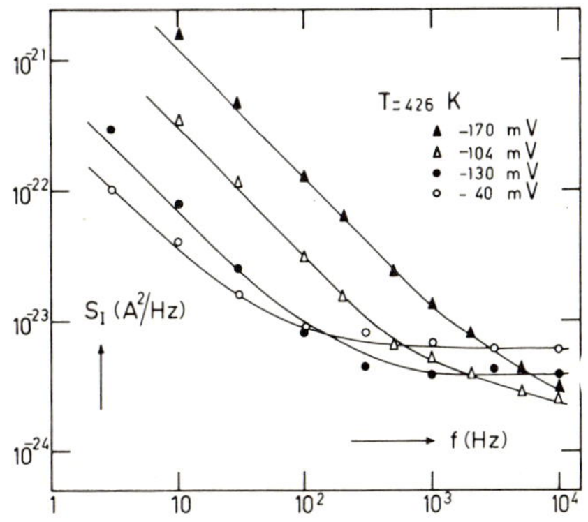

Figure 2: Noise spectra of the reverse-biased diodes A (○,●) and B (△,▲) at T=426K.

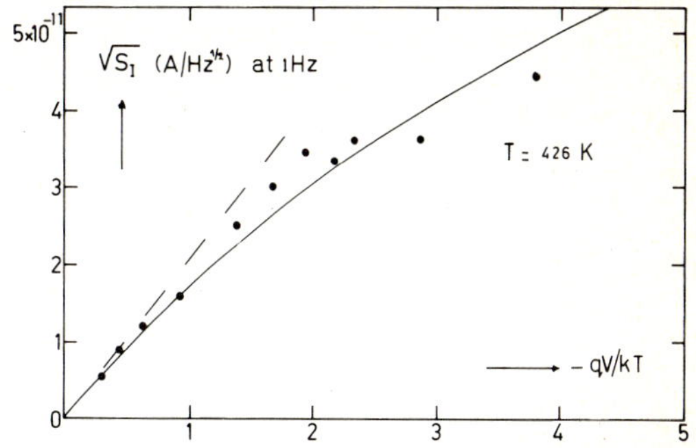

Figure 3: S_I-V characteristic of reverse-biased diode A at 426K. Solid line: theory eq. (8); broken line: extrapolation from ohmic region.

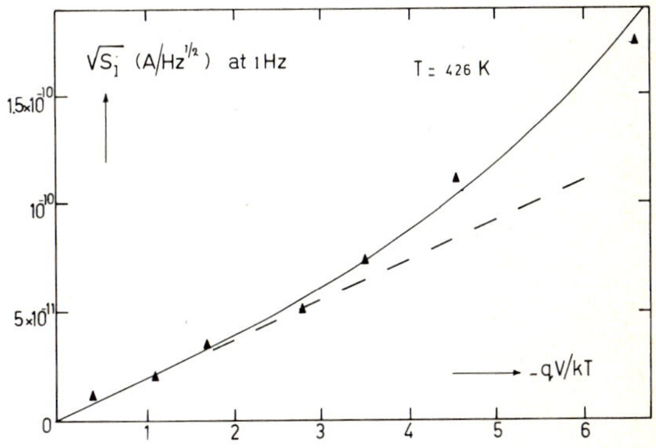

Figure 4: S_I-V characteristic of reverse-biased diode B at 426K. Solid line: theory eq. (21); broken line: extrapolation from ohmic region.

the reverse bias for both types of diodes. The solid lines are the best fits according to the theoretical predictions. For diode A we apply eq. (8) with $\alpha q I_0/4\tau = 10^{-21} A^2$, and for diode B eq. (21) with $2\alpha q I_0/3\tau_j = 1.4 \times 10^{-21} A^2$. The 1/f noise in diodes A and B can be fitted with $\alpha = 6 \times 10^{-3}$ and $\alpha = 10^{-4}$, respectively. From theoretical and experimental results it can be concluded that at high reverse bias (-qV>>kT) the 1/f noise of the diffusion-current dominated diode increases less than proportional to the square of the applied voltage. The 1/f noise of the g-r current dominated diode increases more than proportionally to the square of the bias voltage.
Recently, Tobin et al. {8} have investigated the 1/f noise in reverse-biased (HgCd)Te photodiodes. However, they did not discuss their results in terms of mobility fluctuations.
In figure 5b of ref. {8} the dark current and the 1/f current noise as a function of reverse bias are shown in normalised units for an almost diffusion-current dominated (HgCd)Te photodiode at T=193 K. In Figure 5 we have replotted their results; the solid lines are the best fits according to eqs. (8,10).

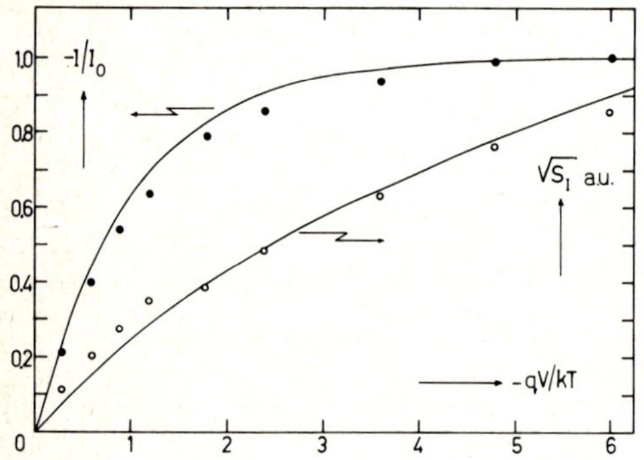

Figure 5: Normalised I-V and S_I-V characteristics of a reverse-biased (HgCd)Te photodiode at T=193 K. Solid lines: theory eqs. (8,10). Experimental data are from Tobin et al. {8}.

4. CONCLUSIONS

From the 1/f noise results presented here we can draw some conclusions:
- The 1/f noise in reverse-biased p-n diodes can be interpreted in terms of mobility fluctuations.
- The 1/f noise level at fixed reverse bias depends on saturation current and lifetime of charge carriers.
- The ratio of the 1/f noise and shot noise in diodes at fixed reverse bias depends only on the lifetime or diffusion time. It decreases with increasing lifetime or diffusion time.

REFERENCES

{1} Willardson, R.K. and Beer, A.C., Semiconductors and Semimetals Vol. 18 (Academic Press, New York, 1981).
{2} Hooge, F.N., Kleinpenning, T.G.M., and Vandamme, L.K.J., Experimental studies on 1/f noise, Rep. Prog. Phys. 44 (1981) 479-532.
{3} Kleinpenning, T.G.M. 1/f noise in p-n diodes, Physica 98B (1980) 289-299.
{4} Sze, S.M.,Physics of Semiconductor devices (Wiley Interscience, New York, 1969).
{5} Kleinpenning, T.G.M., On 1/f noise and detectivity in reverse-biased p-n junction photodiodes, Physica B to be published.
{6} Hooge, F.N., On expressions for 1/f noise in mobility, Physica 114B (1982) 391-392.
{7} Van der Ziel, A., Solid State Physical Electronics (Prentice Hall, Englewood Cliffs, N.J., 1968).
{8} Tobin, S.P.,Iwasa, S. and Tredwell, T.J., 1/f noise in (HgCd)Te photodiodes, IEEE-ED 27 (1980) 43-48.

NOISE IN PHYSICAL SYSTEMS AND 1/f NOISE
M. Savelli, G. Lecoy and J-P. Nougier (eds.)

© *Elsevier Science Publishers B.V., 1983*

1/f NOISE USED AS A RELIABILITY TEST FOR DIODE LASERS

L.K.J. Vandamme and L.J. van Ruyven

Eindhoven University of Technology, Eindhoven, Netherlands;
SSL Elcoma, Philips Research Laboratories Eindhoven, Netherlands.

The model, presented here, assumes a parallel connection of a bulk and surface diode. The experimentally observed relation between 1/f noise and degradation is qualitatively explained by the model.

1. INTRODUCTION

The forced life test is an important though time consuming way of testing as applied in semiconductor laser production. Each laser diode is tested at an elevated temperature (60°C) for about 20h and sometimes for a full week (168h). Criteria for rejecting diodes are based on the following properties before and after the life test: (i) laser efficiency (W/A), (ii) I-V characteristic, (iii) drive current at an optical output of 5mW per facet, (iv) degradation rate (≅dlogI threshold/dt).
It will be clear that this way of testing is expensive, in particular when large quantities of laser diodes have to be tested. Therefore, we have investigated whether a correlation exists between the 1/f noise at low and medium currents and the laser properties before and after testing. It has already been shown that 1/f noise is a good reliability test for solar cells {1}.

2. THE NOISE VOLTAGE OF A LASER DIODE

The voltage noise across a diode consists of a 1/f and a "white" contribution. In a first approximation, the corner frequency f_c, at which both contributions are equal, is independent of the forward current and diode area {2}. Both voltage noise contributions are inversely proportional to the current and the diode area. However, edge effects and series resistance cause an increase in f_c with increasing current. For our samples f_c was larger than 1kHz.
The 1/f noise of a diode consists of a contribution from the junction and one from the contacts. The contact contribution dominates the junction contribution only at very large currents. Below current densities of $10^4 A/cm^2$, the 1/f noise can give information about the lifetime of the carriers in the junction or i-region of the diode.
A double heterojunction laser consists of a small bandgap which is the active layer of 0.2μm in thickness sandwiched between two wide bandgap layers of 1.5μm in thickness each. One of the wide bandgap layers is n-type, the other p-type and the thin active layer has no intentional doping. Analysis of the I-V curves shows that in most cases the ideality factor is 2.2. More details of the investigated AlGaAs injection lasers can be found in {3}.

Kleinpenning's {2} results for diodes with an ideality factor n≅2 can be expressed as follows:

$$S_I \cong \frac{\alpha q I^2}{(I+I_o)\tau f}. \tag{1}$$

Here S_I is the spectral density of the current fluctuations, f the frequency, I the forward current, and α a dimensionless parameter for GaAs of the order of 10^{-3} {4}, while q is the elementary charge, I_o the saturation current, and τ the lifetime of the carriers in the optical active region. Equation (1) holds for PIN diodes and diodes for which the I_o is determined by recombination in the space charge region.
When a constant current I is passed through the diode, the open-circuit noise voltage S_V becomes $S_I.Z^2$. For $f<f_c$ the impedance Z is frequency independent, and is given by $dV/dI=nkT/q(I+I_o)$. This leads to an expression for the open-circuit noise at low frequencies

$$S_V \cong \frac{\alpha I^2 (nkT)^2}{q(I+I_o)^3 \tau f} \tag{2}$$

From eqs. (1) and (2) follows that for $I<I_o$, S_I and S_V are proportional to I^2. For $I>I_o$ we find $S_I \propto I$ and $S_V \propto I^{-1}$. At extremely high currents the noise generated in the recombination region is dominated by the noise contribution of the series resistance. S_V and S_I are then proportional to I^2 just as in the ohmic region when $I<I_o$.

3. EXPERIMENTAL RESULTS

(i) General: Most measurements have been made below the threshold current which is approximately 80mA corresponding to a current density of about $10^4 A/cm^2$. Fig. 1 represents the observed 1/f noise spectrum of a laser diode at 1.1 mA and 15mA. The curve at 15mA is shifted to lower spectral densities in accordance with eq. (2). Fig. 2 represents S_V at 300Hz as a function of I for a laser diode with a behaviour as predicted by eq. (2) for the full region of 10^{-5} - 10^{-2}A. The dotted line in fig. 2 represents the $S_V \propto I^2$ of a rejected diode. This is presumably due to a partially short-circuit junction.
In the experiments the 1/f noise was always observed at a constant current of 1mA and normalised at 1Hz. The normalised spectrum $fS_V(f)$ with $f<f_c$ equals the numerical value of S_V(1Hz).

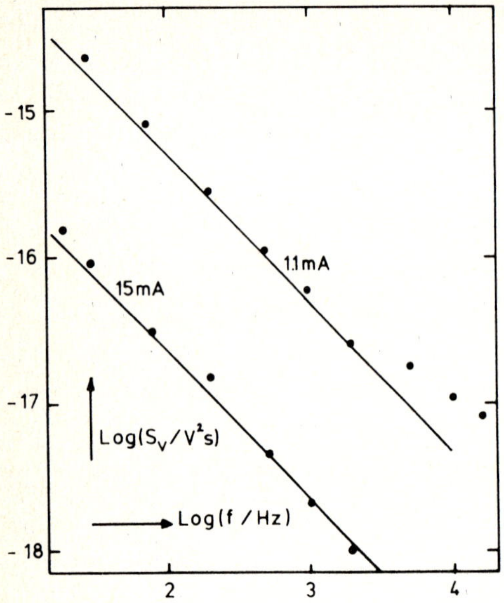

Fig. 1: 1/f Noise of a laser diode. S_V at I_1 = 1.1mA and 15mA.

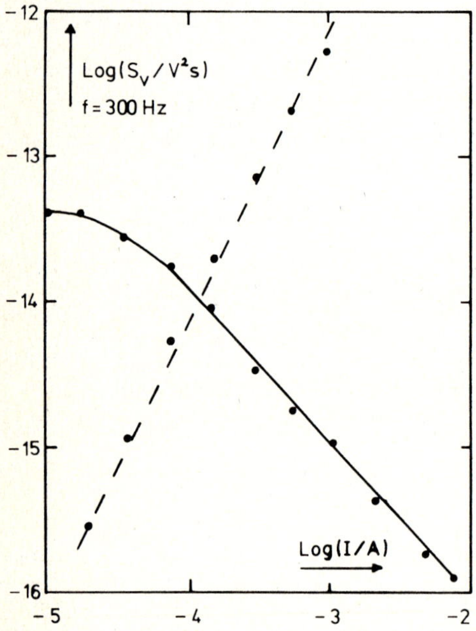

Fig. 2: S_V at 300Hz as a function of the current I_1. Solid line: S_V for an accepted diode following eq. (2). Dotted line: represents $S_V(I)$ at 300Hz of a rejected diode.

The background noise of the preamplifier is negligible. In general devices with a 1/f noise in excess of $5x10^{-14}V^2/Hz$ at 1mA and 1Hz are rejected for reasons of lifetest in standard measurements. On the other hand, not all rejected devices show an excess of 1/f noise.
From a set of 60 diodes we found the lower values of S_V(1Hz) at 1mA to be of the order of $5x10^{-15}V^2/Hz$. Using an α-value of 10^{-4} and n=2.2, we find $\tau \cong 10^{-8}$s, which is a reasonable value {5}.

(ii) Aging experiments
Of 12 diodes the 1/f noise was measured before and after the lifetest (168h). From another batch of 18 diodes the 1/f noise was measured after life tests of 5h, 20h and 168h. The results are presented in fig. 3.

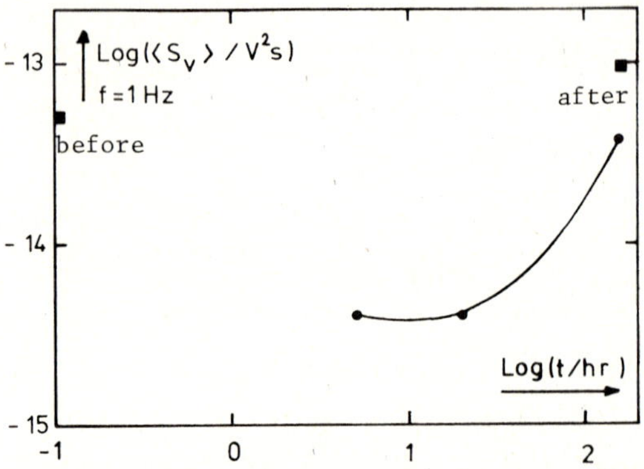

Fig. 3: Aging experiments.
■: noise, averaged over 12 diodes, before and after a lifetest of 168h.
●: noise, averaged over 18 diodes, after a lifetest of 5h, 20h and 168h.

After 5h the 1/f noise increases. An increase in noise of a factor 2 before and after a lifetest of 168h is often observed.
During the same lapse of time the threshold current shows a fairly rapid increase followed by a steady constant slope increasing at the rate of 5% per 1000 hours.

(iii) Static discharge experiments
A substantial fraction of the devices that showed a 1/f noise in excess of $5x10^{-14}V^2/Hz$ was rejected in the standard test of optical efficiency. This may seem surprising because the 1/f noise is observed at 1mA corresponding to a negligibly light output.
To check whether mirror degradation influences the 1/f noise level, some laser diodes were subjected to controlled static discharges through a capacitor of 1 µF charged to a preset voltage.
It is known that mirror damage gives rise to an increased threshold current and/or to lower optical efficiency. In addition, severe damage can be seen under the microscope. The results are summarised in table I.

Table 1 Dependence of noise level, threshold current increase and visual facet damage on capacitor voltage in controlled static discharge experiments.

capacitor voltage(V)	$fS_V(V^2)$ at 1mA	100mA	threshold increase	visible facet damage
<10	$5x10^{-14}$	-	yes	no
10<V<15	$5x10^{-10}$	B	yes	no
>15	B	B	yes	yes

The B in table I indicates burst noise. The table shows that visual facet damage goes hand in hand with burst noise at low and high currents. Burst noise across the diode always accompanies strong fluctuations in the light output. Its spectrum is $1/f^2$ in the frequency range of 1Hz-1kHz.

(iv) fluctuations in the light output

In an attempt to select diodes on only one criterion we observed the fluctuations of the light output S_p for the lasers biased at a fixed current of 100mA (above I_{th}). The light power P and the fluctuations were observed by a detector diode. In most cases the spectrum was 1/f for frequencies below 100Hz. The relative fluctuation in the light power $S_p(1Hz)/p^2$ increases with I_1. All devices with an $S_p(1Hz)/p^2 > 10^{-9}$s at I_1=100mA had also been rejected on standard life test criteria.

4. MODEL

In p-i-n type structures like semiconductor lasers, the 1/f noise is inversely proportional to the recombination time τ (eqs. (1), (2)). Therefore, if parts of the active layer, in which most recombination takes place, have a substantially shorter lifetime, a net contribution to the total 1/f noise is observed. The active region in a semiconductor laser has cleaved ends covered with Al_2O_3 for protection. The minority carrier lifetime close to the facets is controlled by the amount of surface states and therefore substantially shorter than the bulk lifetime. The model supposes a diode with a junction area A of which a small region at both ends has an area pA with a poor lifetime r.τ in which τ represents the bulk lifetime of the unperturbed area (1-p)A having a lifetime τ. The saturation current I_0 is proportional to the ratio of area to lifetime. The local current density in the trap-rich region at the mirrors is 1/r times as large as in the remaining part. This results in a large power dissipation to maintain the same optical output. Higher temperatures reduce the diode lifetime. In such a parallel connection of a bulk diode with a surface diode the total current is given by

$$I=I_h(1-p)+I_h p/r \cong I_h(1+p/r). \tag{3}$$

I_h is the current that would flow if the total area had a lifetime τ. (p=0). Assuming uncorrelated fluctuations, the current noise for $I>I_0$ is given by

$$S_I=S_{Ih}(p/r^2+1-p) \cong S_{Ih}(1+p/r^2) \tag{4}$$

where S_{Ih} is the expected noise for p=0.
If the damage at the facet reaches a level on which p/r^2 is comparable with 1, the observed 1/f noise will increase. In case of facet damage due to static discharges, r will abruptly assume a very low value resulting in a sudden increase of 4 decades, which gives an estimate for r of $r=10^{-3}$ with $p \cong 1\%$.
Internal degradation which occur at a much slower rate and is frequently caused by strain, climbing dislocations, local heating, crude bonding, etc. gives a steady increase in 1/f noise, resulting in an increase of a factor 2 during 168 h.

5. CONCLUSION

The observed 1/f noise (at 1mA) before life test shows a strong correlation with the outcome of the standard reliability tests.
A few percent of the samples rejected on optical grounds show a low 1/f noise.
All lasers with a 1/f noise observed at 1mA and 1Hz in excess of $5x10^{-14}V^2/Hz$ are rejected by standard criteria.
The relative noise in the light output $S_p(1Hz)/p^2$ at I_1=100mA seems a good criterion for accepting or rejecting lasers and will be further studied.
The model presented here, assumes a parallel connection of a bulk diode and a surface diode. It explains qualitatively the relation between 1/f noise and degradation.

REFERENCES

{1} Vandamme, L.K.J., Alabedra, R. and Zommiti, M., Solid State Electron. (1983) in press.
{2} Kleinpenning, T.G.M., Physica 98B+C (1980) 289-299.
{3} Finck, J.C.J., van der Laak, H.J.M. and Schrama, J.T., Philips Tech. Rev. 39 (1980) 37-47.
{4} Vandamme, L.K.J., Is the 1/f noise parameter α a constant?, in this conf. volume (1983).
{5} Sze, Physics of Semiconductor Devices, (Wiley-Interscience, New York, (1969)) p.58.

NOISE IN PHYSICAL SYSTEMS AND 1/f NOISE
M. Savelli, G. Lecoy and J-P. Nougier (eds.)
© *Elsevier Science Publishers B.V., 1983*

LOW FREQUENCY NOISE IN INGAAS/INP-PHOTODIODES

Joachim Kimmerle, Wolfgang Kuebart, Edgar Kuehn, Olaf Hildebrand,
Kurt Loesch*, Guenther Seitz

Physikalisches Institut, Teil 4, Universitaet Stuttgart, D 7000 Stuttgart-80 , FRG

* SEL Research Centre, D 7000 Stuttgart-40 , FRG

With low frequency noise measurements in different types of InGaAs/InP-single- and double-heterostructures we demonstrate that the heterointerface may introduce 1/f-noise sources. We find a very complex behaviour of noise spectrum and intensity of the dark and photocurrent in dependence of reverse bias and wavelength of the incident light. The results are discussed in terms of charge accumulation at the hetero-barrier and possible defect states at the heterointerface.

1. INTRODUCTION

InGaAs(P)/InP-photodetectors for long wavelength optical fibre communication [1] involve a heterojunction necessarily, since the ternary (quaternary) material has to be grown on InP substrates. With low frequency noise measurements (f≤ 1MHz) we demonstrate that this heterojunction may influence the noise behaviour of photodiodes, even if the heterojunction is located outside the depletion layer.
We have used two types of InGaAs/InP-diodes with the depletion layer (at zero bias) either located in the InGaAs ("pin-structure",cf.fig.1a) or in the InP ("SAM-structure"[2],cf.fig.1b). Then, with increasing reverse bias the depletion layer approaches the heterojunction either from the lower gap ternary InGaAs or from the wider gap binary InP layer (fig.1a and 1b). The according spatial variation of the conduction band minima and the valence band maxima at reverse bias is shown at the bottom of fig.1(Anderson model[3]). For these two types of diodes, we compare dark and photocurrent characteristics with the corresponding noise behaviour.

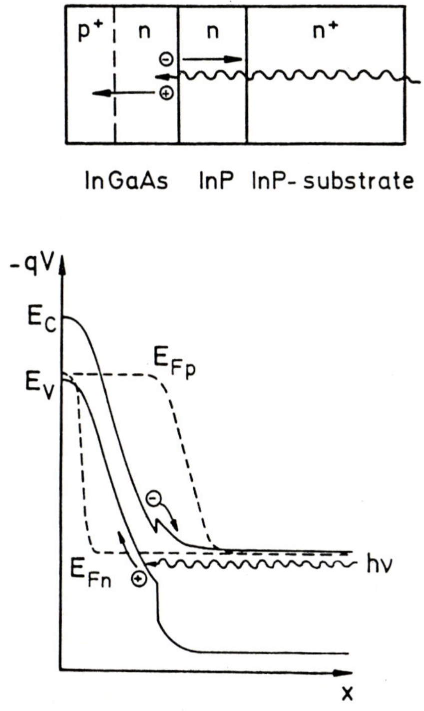

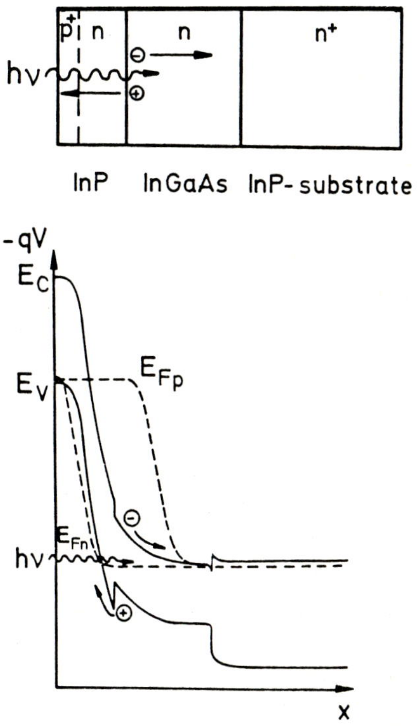

Fig.1) Schematic cross section and banddiagram at reverse bias V and the generation of electron-hole-pairs near the pn-junction with light hν. E_C, E_V, E_{Fn}, E_{Fp} denote conduction-, valence-band and imref-levels for electrons and holes resp..
The drawing assumes no interface states at the heterobarrier and no recombination in the space charge region.

a) Single heterostructure pin diode (SH) with the pn-junction in the InGaAs.

b) Double-heterostructure SAM diode (DH-SAM) with the pn-junction in the InP

2. DIODES FABRICATION

The pin-structures were made by LPE-growth of n-InGaAs onto (100)-oriented n-InP-substrates with subsequent Zn-diffusion into the InGaAs (cf.fig. 1a). For the SAM-structures, an additional n-InP layer was grown onto the InGaAs LPE-layer by VPE (double-heterostructure DH). The pn-junction was prepared by Cd-diffusion into the InP VPE-layer (cf.fig.1b,DH-SAM). Another type of SAM-structure was made by LPE-growth of n-InP on Zn-doped p-type InP-substrates with subsequent LPE-growth of n-InGaAs, resulting in a single-hetero-SAM-diode (SH-SAM) with a more or less graded gap [4].

3. EXPERIMENTAL SETUP

We measured the noise of the reverse current at room temperature with a high load impedance (100kΩ to 1MΩ) at the input of a low-noise JFET-preamplifier (LF357). The noise spectrum was analyzed with a Tektronix 3L5 up to 1MHz at a bandwidth of 240Hz. Thermal noise of metal film resistors was used for calibration. For comparison of noise with and without incident light we switched the light every few minutes. The current was measured simultaniously.
For completion of the data the capacitance-voltage characteristics of the heterophotodiodes was measured between 1kHz and 10MHz.

4. EXPERIMENTAL RESULTS

In fig.2 to 6 we show the spectral noise power density (SND) of the current at a frequency of 3kHz versus reverse bias and wavelength of the incident light. The SND is plotted in terms of the equivalent shot noise current I_{eq},

$$I_{eq} = S(I)/2q \quad , \qquad (1)$$

in order to allow a direct comparison with the current I.

4.1. DARK CURRENT NOISE

4.1.1. pin-structures

For a typical InGaAs/InP-pin-structure-diode (cf.fig.1a), the dark current and its SND are plotted in fig.2 :
For low reverse bias, where the depletion layer has not yet reached the heterojunction (U=0... -3.5V, homodiode operation), the SND of the dark current is **smaller** than the shot noise value $I_{eq}=I_d$ (compare [5]). Under these conditions, the corner frequency between 1/f- and white noise is less than 3kHz, and white noise dominates.
With increasing bias, as soon as the depletion region boundary crosses the heterojunction interface (arrow at V_c in fig.2), the dark current slightly increases, accompanied by a drastic increase in the 1/f-noise. The SND of this 1/f-noise is proportional to the square of the excess dark current I_{exc} (I_{exc} shall be defined as the deviation from the ideal dark current characteristic) :

$$S(I_d;1/f) \quad \simeq \quad I_{exc}^{\beta}/f^{\gamma} \quad , \qquad (2)$$

with $\beta \cong 2$ and $\gamma \cong 0.6$.
A similar behaviour is observed for all diodes with excess current; in those with nearly ideal dark current, however, no 1/f-noise could be detected at 3kHz.

4.1.2. SAM-structures

In SAM-structure diodes (cf.fig.1b), there cannot be observed such a distinct kink of the SND at V_c. The 1/f-noise dominates at any bias (cf.fig.3).
It should be mentioned that in all diodes investigated, the 1/f-noise shows bursts, which last several minutes with an amplitude of up to 20% of the corresponding noise level.

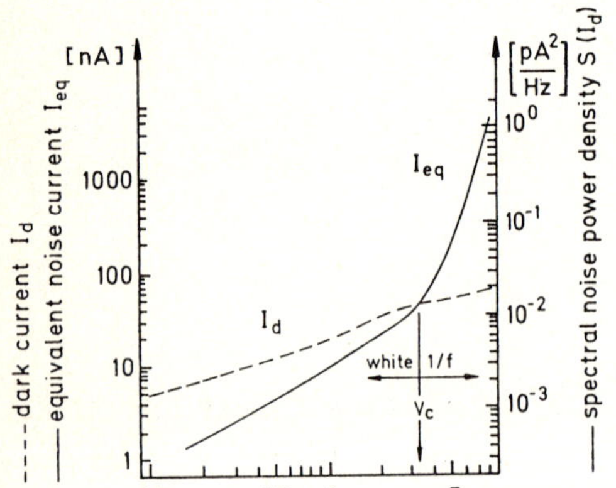

Fig.2) Dark current I_d and equivalent noise current at 3kHz vs. reverse bias of the the pin-diode.
At V_c the space charge region reaches the heterojunction. White noise dominates below V_c, 1/f-noise above V_c.
(The ordinate on the right-hand side directly shows the SND $S(I_d) = 2qI_{eq}(I_d)$.)

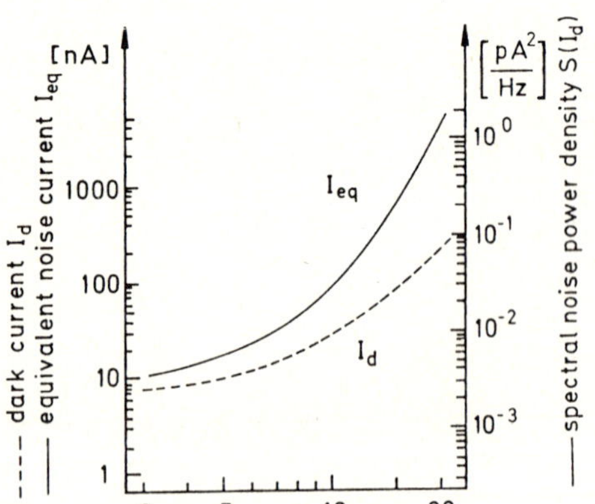

Fig.3) Dark current and noise for the DH-SAM-diode (in analogy to fig.2). 1/f-noise dominates over the entire voltage range.

4.2.PHOTOCURRENT NOISE

Fig.4 and 6 show the photo-SND, i.e. the difference between the total SND with and without illumination, and the corresponding photocurrent.

4.2.1.Pin-structures

For the InGaAs/InP-pin-structure-diodes (cf.fig. 1a) the photo-SND is different for various diodes of the same fabrication procedure. Fig.4 shows the photocurrent (which is typical for all diodes of this type), and the corresponding photo-SND of the same pin-diode, whose dark current behaviour is described above (cf.fig.2).

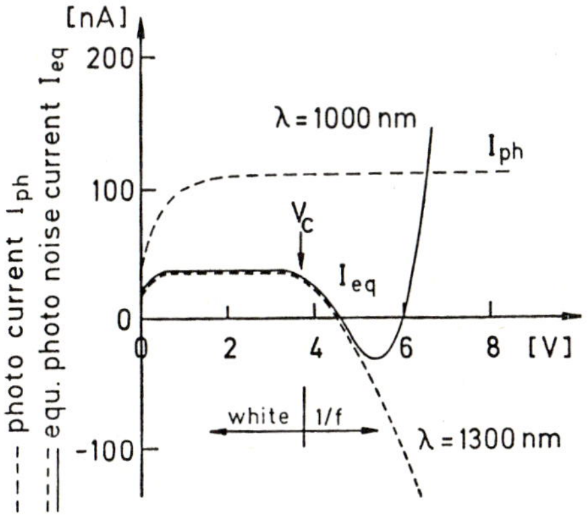

Fig.4) Photocurrent I_{ph} and equivalent noise current I_{eq} of the pin-diode at 3kHz vs. reverse bias. Above V_c the 1/f-noise is quenched (for λ=1300nm, and for λ=1000nm below 5.5V), or enhanced (for λ=1000nm above 5.5V).

The photo-SND characteristic can be divided into two regions below and above V_c :
For reverse bias below V_c, the photocurrent shows white noise, which is reduced compared to the expected shot noise value.
With increasing reverse bias, at voltages above V_c, the strong excess 1/f-darkcurrent-noise (cf. fig.2) is quenched, if the light is absorbed in the InGaAs layer only (i.e. λ=1300nm at any bias,and λ=1000nm below 5.5V).This effect leads to an apparent negative photo-SND, as shown in fig.4. The quenching becomes more effective with increasing light intensity.
When absorption occurs in the InP close to the heterojunction due to Franz-Keldysh-absorption (λ=1000nm above 5.5V in fig.4), the 1/f-noise is strongly enhanced.
These features can also be observed in the spectral response of the total SND, (cf.fig.5):
The 1/f-noise is strongly enhanced by irradiating light corresponding to the band gap energy of InP (λ≅940nm), and it is quenched by light absorption in the InGaAs (λ>940nm).
After exposure to air and light for several weeks, however, the 1/f-noise enhancement and quenching disappeared nearly completely,although the excess dark current and its 1/f-noise above V_c (cf.fig.2) was decreased by only 15%. This may indicate the existence of two different 1/f-noise sources.

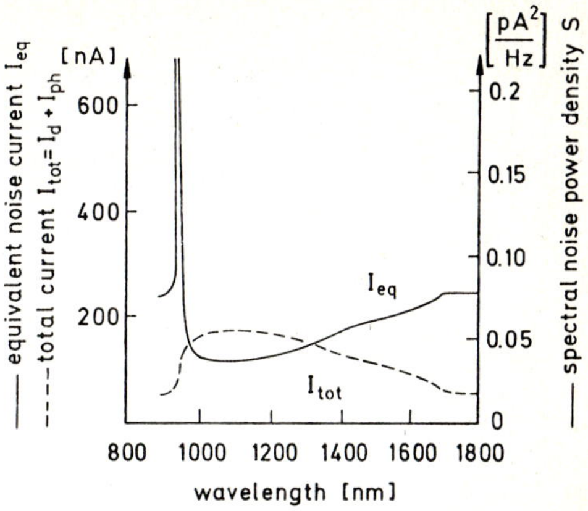

Fig.5) Spectral response of the total current of the pin-diode at 6V reverse bias and the equivalent noise current at 3kHz.

It should be noted that the spectral behaviour described above may be less pronounced for other diodes. It could not be observed in diodes which do not show 1/f-darkcurrent-noise at the frequency of investigation.

4.2.2.SAM-structures

For the double-hetero-SAM-structure-diode (cf. fig.1b), photocurrent and photo-SND are shown in fig.6.
At low reverse bias, the photocurrent is small, because the field at the heterojunction is too small for transporting the photo-generated holes from the InGaAs-hole-well (cf.fig.1b) across the energy barrier in the valence band. For this case, strong 1/f- noise is observed.
With increasing bias, the photocurrent saturates (arrow in fig.6): the field at the heterojunction is high enough for transporting the photo-excited holes across the barrier. When the photocurrent has saturated, it shows pure shot

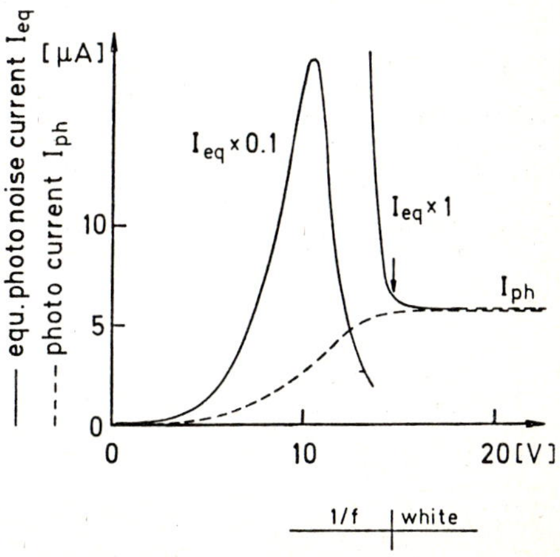

Fig.6) Photocurrent I_{ph}(λ=950...1700nm) and equ. noise current I_{eq} at 3kHz vs. reverse bias of the DH-SAM-diode (see text).

noise, and the photo-1/f-noise has completely disappeared (cf.fig.6).
With increasing light intensity, these effects occur at higher reverse bias.
Capacitance-voltage measurements show a photo-capacitance only before the photocurrent has reached its saturation value, which indicates a hole accumulation at the heterojunction.
In contrast, for the single-hetero-SAM-structure, the photocurrent shows pure shot noise, even before it has reached its saturation value, and no photo-induced capacitance is observed.

5. DISCUSSION

We have demonstrated that heterojunctions may be correlated with 1/f-noise sources in electronic devices. Our measurements show that already small excess currents, which may be introduced during device processing, can lead to a large 1/f-noise :
Even for comparable SAM- or pin-structure diodes very different 1/f-noise behaviour is observed, even if the diodes are made from the same wafer.

5.1. Noise in pin-structures

Especially in pin-heterodiodes, an excess dark current and 1/f-noise may occur, if the heterojunction is located within the depletion layer (cf.fig.2). This implies that their sources are located close to the heterointerface.
From the spectral response shown in fig.5 we we assume that the hole current generated in the InP is 1/f-modulated by the heterointerface, possibly caused by fluctuating charged defect states. Concerning the quenching of the noise above V_c by light absorption in the InGaAs (cf. fig.4,5), possible explanations are the saturation of the 1/f-noise-sources by the photocarriers, or the noise suppression due to space charge effects similar to observations in vacuum diodes [5].
From the different aging behaviour of the 1/f-noise dark- and photo-characteristic we assume at least two different types of noise sources. Possibly the fast aging one is located at the mesa-surface.
Finally, the shot noise quenching below V_c (cf. fig.4) might also be caused by space charge effets [5].

5.2. Noise in SAM-structures

Apparently, the 1/f-modulation of the photocurrent in the DH-SAM-structure is only effective, as long as the photogenerated holes are partially accumulated in the InGaAs-hole well near the heterojunction (cf.fig.1b). The explanation may be a 1/f-modulation of the heterobarrier involving hole traps similar to observations in Schottky diodes [5-7], or a self-modulation of the effective barrier height by the accumulated holes.
Up to now, from photocapacitance measurements we cannot distinguish these effects.

6. CONCLUSIONS

Our investigations show that heterojunctions may introduce 1/f-noise in semiconductor devices. The intensity of this 1/f-noise depends on the position of the p-n depletion layer relative to the hetero-interface and on the fabrication procedure. A distinct correlation is found between excess dark current and 1/f-noise. Evidence is given that 1/f-noise effects in SAM-heterodiodes can be avoided by graded heterojunctions.
In addition, low frequency noise measurements have shown to be a powerful tool, in combination with capacitance-voltage measurements, for the investigation and understanding of heterodiodes.

7. ACKNOWLEDGEMENTS

We would like to thank N.Kloetzer and M.J.Vilela for their helpful assistance and discussions, H.Eisele and H.Haspeklo for the growth of epitaxial layers , AEG-Telefunken (Ulm,FRG) and the SEL Research Centre (Stuttgart,FRG) for delivering some diodes, and M.H.Pilkuhn for helpful discussions.

8. REFERENCES

[1] Stillman, G.E., Cook, L.W., Bulman, G.E., Tabatabaie, N., Chin, R. and Dapkus, P.D., Long-wavelength(1.3-to 1.6-μm) detectors for fiber-optical communications, IEEE Trans. Electron.Dev. ED-29 (1982) 1355-1371.
[2] Ando, H., Kanbe, H., Ito, M. and Kaneda, T., Tunneling current in InGaAs and optimum design for InGaAs/InP avalanche photodiode, Japan.J.Appl.Phys. 19 (1980) L277-L280.
[3] Casey, H.C. and Panish, M.B., Heterostructure lasers, Part A, (Academic Press, New York, 1978).
[4] Forrest, St.R., Smith, R.G. and Kim, O.K., Performance of $In_{.53}Ga_{.47}As/InP$ avalanche photodiodes, IEEE J.Quantum Electronics QE-18 (1982) 2040-2048.
[5] van der Ziel, A., Flicker noise in elecelectronic devices, Adv.Electr.El.Phys. 49 (1979) 225-297.
[6] Hsu, S.T., Low frequency excess noise in metal-silicon Schottky barrier diodes, IEEE Trans.Electron.Dev. ED-17 (1970) 496-506.
[7] Hsu, S.T., Flicker noise in metal semiconductor Schottky barrier diodes due to multistep tunneling processes, IEEE Trans.Electron.Dev. ED-18 (1971) 882-893.

8 The photo-SND and equivalent photo noise current, respectively are defined as the difference between SND with and without illumination : $S(I_{ph}) = S(I_{tot}) - S(I_d)$
(according to the photocurrent $I_{ph}=I_{tot}-I_d$).

NOISE IN PHYSICAL SYSTEMS AND 1/f NOISE
M. Savelli, G. Lecoy and J-P. Nougier (eds.)

© *Elsevier Science Publishers B.V., 1983*

ON CURRENT NOISE LIMITED DETECTIVITY D* IN PHOTOVOLTAIC AND IN PHOTOCONDUCTIVE DETECTORS

T.G.M. Kleinpenning
Eindhoven University of Technology
Eindhoven, Netherlands

The influence of 1/f noise and white noise on the detectivity D* in photodiodes and in photoconductors is investigated. If the background temperature equals the temperature of the detector, the detectivity is found to be limited by 1/f noise and white noise. From the well known characteristics of the photon noise limited detectivity $D^*(\lambda_c,f)$ at peak wavelength λ_c and frequency f for various background temperatures, and from the 1/f noise and white noise limited $D^*(\lambda_c,f)$ characteristics for various detector temperatures, conclusions can be drawn about optimum performance of detectors.
For example, the photon noise limit is reached at detector temperatures $T_D \lesssim 250/\sqrt{\lambda_c}$ (K), with λ_c in μm.

1. INTRODUCTION

The detectivity of photovoltaic (PV) and photoconductive (PC) detectors is limited by three types of noise: (i) current noise in the detector, (ii) noise due to background photons (photon noise), and (iii) noise in the electronic system following the detector. The detectivity is defined as {1,2}

$$D^*(\lambda,f)=(A\Delta f)^{\frac{1}{2}}/NEP \quad (\mathrm{cmHz}^{\frac{1}{2}}/\mathrm{W}) \qquad (1)$$

where A is the area of the detector, NEP the noise equivalent power defined as the r.m.s. signal power of wavelength λ required to produce an r.m.s. signal voltage (current) equal to the r.m.s. noise voltage (current) in a bandwidth Δf, and f the frequency of modulation.
If the current noise and electronic system noise can be neglected, the detectivity is determined by the fluctuation in arrival rate of background photons. This photon noise appears as a random signal at the detector output. The photon-noise limited $D^*(\lambda_c,f)$ for a PV and PC detector at peak wavelength λ_c are shown in Figures 2 and 3 for two background temperatures {2}. The field of view is assumed to be hemispherical and the quantum efficiency to be η=1. At peak wavelength λ_c the detectivity is maximum. The wavelength λ_c is related to the bandgap E_g in PV detectors and to the relevant energy gap ΔE in PC detectors by

$$\lambda_c=hc/E_g \;;\; \lambda_c=hc/\Delta E \qquad (2)$$

In this paper we present calculations for current noise limited D* in PV detectors and in extrinsic PC detectors.

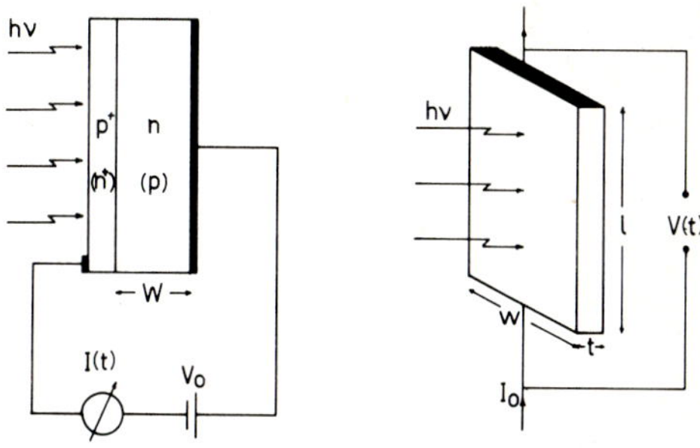

Figure 1: Geometrical models of a photovoltaic (PV) detector and of a photoconductive (PC) detector.

2. CALCULATIONS OF CURRENT NOISE LIMITED D*

2.1. Photovoltaic (PV) detector

The I-V characteristic for a monochromatically illuminated diode where the saturation current I_o is determined by diffusion in the base is given by

$$I=I_o\{\exp(qV/kT)-1\}-I_s \qquad (3)$$

where I_s is the photocurrent

$$I_s=\eta qU/h\nu=\eta q\lambda U/hc \qquad (4)$$

Here U is the incident radiant power, and hν the photon energy. Variations in U lead to variations in I or V. At very low light levels the current noise of a biased photodiode (reverse V<0, forward V>0) is {3,4}

$$S_I(f)=2qI_o\{\exp(qV/kT)+1\}\{1+f_c/f\}. \qquad (5)$$

Here f_c is the frequency where the 1/f noise equals the shot noise. Under weak illumination we have in the relevant frequency region

$$I_{s,rms}=(\eta q/h\nu)U_{rms} \;;\; I_{N,rms}=\sqrt{S_I(f)\Delta f} \qquad (6)$$

From eq. (6) the noise equivalent power NEP is found to be

$$NEP=U_{rms}=(h\nu/\eta q)\sqrt{S_I(f)\Delta f} \tag{7}$$

According to eqs. (1,7) we obtain the detectivity

$$D^*(\lambda,f)=(\eta q\lambda/hc)\sqrt{A/S_I(f)} \tag{8}$$

So D* is proportional to λ up to the peak wavelength λ_c. For $\lambda>\lambda_c$ we have $\eta=0$ and thus $D^*=0$. In order to evaluate eq. (8) we have to distinguish among several types of p-n photodiodes.
(i) Diodes with a base width larger than the minority carrier diffusion length ($W>L=\sqrt{D\tau}$). Here we have {1-3}

$$I_o = \frac{Aqn_i^2}{N_B}\left(\frac{D}{\tau}\right)^{\frac{1}{2}} = \frac{AqD^{\frac{1}{2}}N_cN_v\exp(-E_g/kT_D)}{N_B\tau^{\frac{1}{2}}} \tag{9}$$

The corner frequency f_c depends on the bias {3,4}

$$f_c(V=0)=0 \ ; \ f_c\left(\frac{-qV}{kT_D}>>1\right) = \frac{\alpha}{8\tau}\cdot\frac{-qV}{kT_D} \tag{10}$$

(ii) Diodes with $W<L=\sqrt{D\tau}$ and with a back electrical contact with a high recombination velocity S. Now we have {2,3}

$$I_o=Aqn_i^2D/N_BW \tag{11}$$

and {3,4}

$$f_c(V=0)=0 \ ; \ f_c\left(\frac{-qV}{kT_D}>>1\right) = \frac{\alpha D}{2W^2}\cdot\frac{-qV}{kT_D} \tag{12}$$

(iii) Diodes with W<L and $S\cong0$, then {2}

$$I_o=Aqn_i^2W/N_B\tau \tag{13}$$

Following the same procedure as in section 3.2 in ref. {5}, we find here for the 1/f current noise

$$S_I(f) = \frac{\alpha qI_o}{3f\tau}\left\{\exp\left(\frac{qV}{kT_D}\right)-1\right\}^2\exp\left(\frac{-qV}{kT_D}\right) \tag{14}$$

So we obtain

$$f_c(V=0) \ ; \ f_c\left(\frac{-qV}{kT_D}>>1\right)=\frac{\alpha}{6\tau}\exp\frac{qV}{kT_D} \tag{15}$$

The nomenclature is as follows, N_B is the base dope, N_c and N_v the densities of states in the conduction and valence band, τ and D the minority carrier lifetime and diffusivity in the base, T_D the detector temperature, and α the Hooge 1/f noise parameter ($\sim10^{-3}$).
With help of the eqs. (2,5,8,9,11,13) the detectivity at peak wavelength λ_c can be written as

$$D^*(\lambda_c,f)=\frac{\frac{\eta\lambda_c}{hc}\left(\frac{N_B}{2N_cN_v\chi}\right)^{\frac{1}{2}}\exp\frac{hc}{2kT_D\lambda_c}}{\{\exp(qV/kT_D)+1\}^{\frac{1}{2}}\{1+f_c/f\}^{\frac{1}{2}}} \tag{16}$$

where χ is $\sqrt{D/\tau}$, D/W or W/τ, depending on the type of diode.

The shot-noise limited $D^*(\lambda_c,f)$ at V=0 is shown in Figure 2 for various T_D. The curves have been calculated for typical parameters $\eta=1$, $m_e^*=m_h^*=m_o$, $N_B=10^{15}cm^{-3}$, and $\chi=1500$ cm/s. In the same figure, the detectivities of some commercially available detectors are also shown.

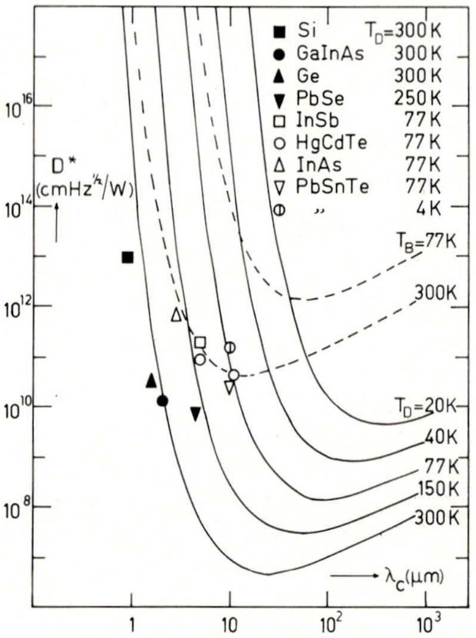

Figure 2: Noise limited $D^*(\lambda_c)$ of a PV detector. Broken lines: photon noise limited D* for two background temperatures T_B=77K and 300K and for free optical view 2πsr. Solid lines: shot noise limited D* for unbiased detectors at various detector temperatures T_D.

2.2. Extrinsic photoconductive (PC) detector

Extrinsic PC detectors require a material with an impurity level corresponding to the wavelength to be detected. Let us consider an n-type detector with area A=wl (see Figure 1). The I-V relation is

$$V=IR=Il^2/q\mu N \tag{17}$$

where μ is the electron mobility, and N the total number of free electrons. At very low light levels the voltage noise of an a.c. open-circuit detector with bias V is determined by Nyquist noise, 1/f noise and G-R noise

$$S_V(f)=4kTR + \frac{\alpha V^2}{fN} + \frac{(4\tau/N)V^2}{1+(2\pi f\tau)^2} \tag{18}$$

where τ is the electron lifetime. Here it is assumed $\langle\Delta N^2\rangle=N$. Under weak illumination we have in the relevant frequency region for the light-induced fluctuations

$$\Delta N_{rms} = (\eta\tau/h\nu) U_{rms} \quad ; \quad \Delta V_{rms} = (\eta\tau V/h\nu N) U_{rms} \tag{19}$$

The NEP is found to be

$$NEP = U_{rms} = (h\nu N/\eta\tau V)\sqrt{S_V(f)\Delta f} \tag{20}$$

The detectivity with eq. (1) is

$$D^*(\lambda, f) = (\eta\tau V/h\nu N)\sqrt{A/S_V(f)} \tag{21}$$

From eqs. (18) and (21) follows that the maximum value for D* is reached when G-R noise and 1/f noise exceed the Nyquist noise. Then we find at peak wavelength λ_c and at frequencies $f<(2\pi\tau)^{-1}$

$$D^*_{max}(\lambda_c, f) = \frac{\eta\lambda_c}{2hc}\sqrt{\frac{\tau}{nt(1+\alpha/4\tau f)}} \tag{22}$$

Here n=N/At is the electron density. The highest value for D* is obtained in the frequency range $f_c = \alpha/4\tau < f < 1/2\pi\tau$

$$D^*(\lambda_c) = \frac{\eta\lambda_c}{2hc}\left(\frac{\tau}{nt}\right)^{\frac{1}{2}} = \frac{\eta\lambda_c}{2hc}(\tau q \mu R_\square)^{\frac{1}{2}} \tag{23}$$

where $R_\square$ is the dark sheet resistance. For extrinsic detectors we have {1,2}

$\Delta E = E_I$ and $n = (N_c N_I)^{\frac{1}{2}} \exp(-hc/2kT_D\lambda_c)$

and for intrinsic detectors of n-type material

$\Delta E = E_g$ and $n > n_i = (N_c N_v)^{\frac{1}{2}} \exp(-hc/2kT_D\lambda_c)$.

Here N_I is the density of impurities. In Figure 3 $D^*(\lambda_c)$ is shown for various detector temperatures. The curves are calculated for $\eta=1$, $\tau=1\mu s$, $\Delta E=E_g$, $n=n_i$, $m^*_e=m^*_h=m_o$, and $t=1\mu m$.

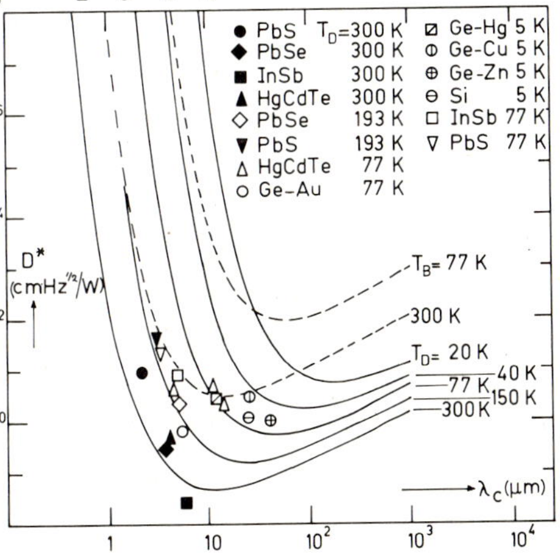

Figure 3: Noise limited $D^*(\lambda_c)$ of a PC detector. Broken lines: photon noise limited D* for T_B=77K and 300K. Free optical view 2π sr. Solid lines: current noise limited D* for detectors at various T_D.

3. DISCUSSION AND CONCLUSIONS

3.1. PV detector

Several remarks can be made here.
First, at low frequencies the current noise is determined by 1/f noise. In this case the detectivity is proportional to $\sqrt{f}$ up to $f=f_c$ where f_c is given by eq. (10), (12) or (15). In order to remove 1/f noise, we have to use the detector in the unbiased mode. See also Figure 4.
Secondly, if the detector is connected with an amplifier with input resistance R_A and temperature T_A, then the detectivity has an upper limit {3}

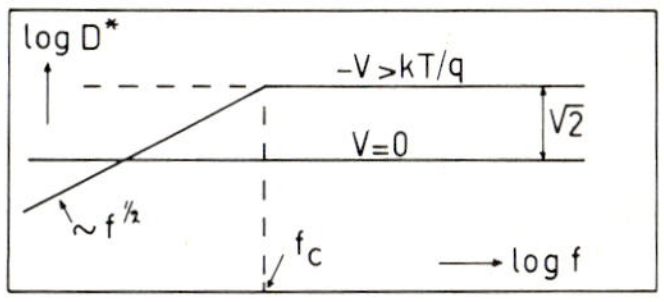

Figure 4: Frequency dependence of D* for a PV detector.

$$D^*(\lambda_c, f) \leq (\eta q \lambda_c/hc)(AR_A/4kT_A)^{\frac{1}{2}}$$

Thirdly, it can be concluded that for optimum D* the detector temperature has to decrease with increasing wavelength. Approximately one finds that the photon noise limit is reached if the detector temperature T_D has been chosen below

$$T_D \leq 250/\sqrt{\lambda_c} \quad (K)$$

with λ_c in μm.
Fourthly, the relation for D* in eq. (16) has been derived with the assumptions that (i) the donors (acceptors) in the base are completely ionized and (ii) in the base holds $pn=n_i^2$ i.e. Fermi-statistics may be replaced by Boltzmann-statistics. The latter condition can only be fulfilled if $E_G \gtrsim 6kT_D$.
More details on this topic can be found in ref. {3}.

3.2. PC detector

Here several remarks can also be made.
First, a similar expression as eq. (23) holds for intrinsic PC detectors of n-type materials provided that $\mu_n >> \mu_p$ and sweep-out effects may be neglected {2}. In such detectors we have $\lambda_c = hc/E_g$.
Secondly, for optimum performance the detector temperature T_D has to decrease with increasing λ_c. The photon noise limit is reached at $T_D \leq 250/\sqrt{\lambda_c}$ (K), with λ_c in μm.
Thirdly, the corner frequency $f_c=\alpha/4\tau$ (see Figure 5) is often found to be in the range of 10^2-10^3Hz which is to be expected, since $\alpha \cong 10^{-3}$ and mostly $\tau \cong 10^{-6}$s. (for example see the SBRC brochure on infrared detectors).
Fourthly, calculated detectivities with the help of eq. (23) agree very well with observed values for HgCdTe, Ge and other detectors in the SBRC brochure.

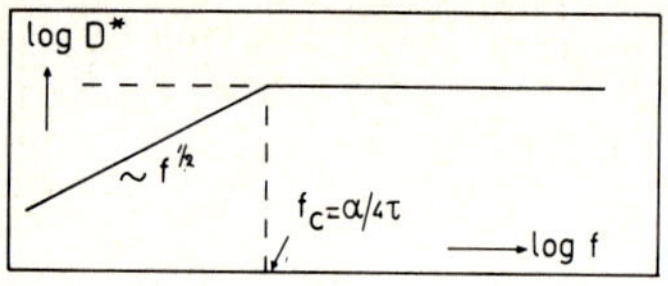

Figure 5: Frequency dependence of D* for a PC detector

REFERENCES

1 Sze, S.M., Physics of semiconductor devices (Wiley Interscience, New York, 1969).
2 Keyes, R.J. (ed.), Optical and infrared detectors, Topics in applied physics vol. 19 (Springer Verlag, Berlin, 1977).
3 Kleinpenning, T.G.M., On 1/f noise and detectivity in reverse-biased p-n junction photodiodes, Physica B to be published.
4 Kleinpenning, T.G.M., On 1/f noise in reverse-biased p-n junction diodes, 7th Int. Conf. on Noise in Physical Systems, Montpellier, 1983.
5 Kleinpenning, T.G.M., 1/f Noise in p-n diodes, Physica 98B (1980) 289-299.

NOISE IN PHYSICAL SYSTEMS AND 1/f NOISE
M. Savelli, G. Lecoy and J-P. Nougier (eds.)
© Elsevier Science Publishers B.V., 1983

1/F NOISE IN $p^+ n n^+$ DEVICES. INFLUENCE OF A MAGNETIC FIELD*

A. CHOVET, S. CRISTOLOVEANU, A. MOHAGHEGH, A. DANDACHE

Lab. de Physique des Composants à Semiconducteurs (ERA CNRS N° 659)
Institut National Polytechnique de Grenoble, ENSERG, 23, rue des Martyrs
38031 - Grenoble Cedex - France

For the first time, experimental results on 1/f noise in double injection diodes (from Si or Ge technologies) are presented. It is shown that this noise originates from bulk conductance fluctuations in the (long) base of the diode. In the semiconductor regime ($I \propto V^2$), this leads to a spectral density $S_V \propto V$, although the classical relation $S_V \propto I^2$ may prevail at the beginning of the injection (lower current *densities*). The variation of 1/f noise magnitude when a magnetic field is applied (magnetodiode effect) fully confirms these findings.

1. INTRODUCTION

The 1/f noise of injecting structures have not been so widely studied than their thermal or generation recombination noises [1]. In p-n junctions, the flicker noise was first explained (using the McWHORTER theory) as produced by fluctuations of the *surface* recombination ; but it appears (see [2]) that no new experimental results have been given in this direction for more than ten years, while the technology has highly improved the device quality. Recently the 1/f noise in p-n junctions has been considered by KLEINPENNING [3], who observed that this noise could be well described by *bulk* noise sources according to HOOGE's empirical relation , and explained by fluctuations in the mobility of free carriers [4]. Such a "mobility noise" has been previously shown to account for the 1/f noise in solid state single injection diodes (like $p^+ p p^+$ structures) [5].

No results have been published on double injection diodes (e.g. $p^+ n n^+$ structures with a "long" n-type base : $L >> L_D$, where L_D is the carrier diffusion length). In this paper, we present experiments on such diodes, made with silicon on sapphire as well as Germanium technologies, which allow us to get informations on the origin (bulk or surface) of their flicker noise. These results will be confirmed by the influence of a magnetic field on the 1/f noise level.

2. BASIC DISCUSSION ; RESULTS ON LOW-FREQUENCY NOISE OF SILICON MICRODIODES

Investigations were made on silicon on sapphire (SOS) $p^+ n n^+$ diodes, processed (by LETI-CEN Grenoble) on <100> substrates, 0.65 µm thick, using standard technology. The low base doping ($n_0 = 5.10^{15}$ to $10^{17} cm^{-3}$) was obtained by phosphorous ion implantation. Hole (p^+) and electron (n^+) injecting contacts were also realized by boron and phosphorous ion implantation (doping $\sim 10^{20} cm^{-3}$). Different dimensions of the base have been studied : the length was L = 10, 20, 30, 50 µm and the width was Z = 50, 100, 200 µm. Since the carrier lifetime is low enough in SOS films (τ < 10 ns), the carrier diffusion length L_D is of the order of µm, so that $L >> L_D$. The I(V) characteristics of such diodes (see fig. 1) can exhibit all the different successive conduction regimes of a "double-injecting" structure [6] : exponential (due to the p^+-n junction), ohmic ($I \propto V$), semiconductor ($I \propto V^2$), etc...

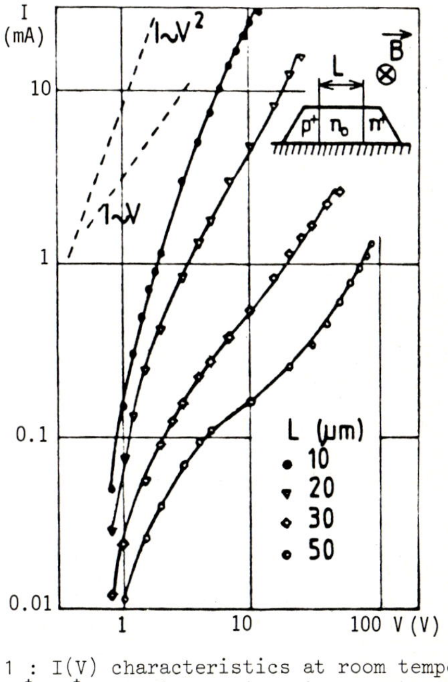

Fig. 1 : I(V) characteristics at room temperature of $p^+ n n^+$ SOS diodes with different length L. Base doping $n_0 = 7.10^{15} cm^{-3}$, Z = 50 µm.

Notice that the quadratic regime, which is of main interest for us, is easily reached in the short structures. The corresponding current

---
*Work partly supported by DGRST and GIS Mini-micro Informatique

density i_x can be derived from a simplified one-dimensional model [6] and is given, with standard notations, by :

$$i_x = \sigma(x)E_x(x) \simeq q(\mu_n+\mu_p)n(x)E_x(x) = $$
$$= (9/8)q\mu_n\mu_p n_0 \tau V^2/L^3 \qquad (1)$$

with :

$$p(x) \simeq n(x) \simeq \frac{3}{4}\, n_0\, \frac{\mu_n\mu_p}{\mu_n+\mu_p}\, \frac{V\tau}{L^2}\left(\frac{L}{x}\right)^{1/2} >> n_0 \qquad (2).$$

The ohmic regime corresponds to the case where the applied voltage is essentially developped over the base ; moreover the electric field is there uniform ($E_x(x) = V/L$ = const.) and the injected carrier density negligible with regard to n_0. In this regime (curve (1) in fig. 2), the measured 1/f noise is well described by HOOGE's empirical relation for homogeneous materials :
$S_R/R^2 = S_V/V^2 = \alpha/(Nf)$ (3),
with $\alpha \simeq 2.10^{-3}$ and N the total number of charge carriers ; here $N = n_0.A.L$, A being the cross-section of the device (A = Z x thickness), so that $S_V = (\alpha L/fA^3).(I^2/\sigma_0^2 n_0)$ (4).

In *the semiconductor regime*, the carrier distribution is no longer homogeneous (see eq.(2)), and relation (3) has to be applied to volume elements. After integration, we get, for any bulk conductance fluctuations (supposed, as usual, spatially uncorrelated) :

$$S_V(f) = \frac{2}{5}\frac{\alpha}{f}\frac{L}{A^3} \cdot \frac{I^2}{\sigma^2(L)n(L)} \qquad (5)$$

Notice that for mobility fluctuations alone, we should have had (using eq.(20) in [3]) the same result, multiplied by the ratio $(\mu_n^2+\mu_p^2)/(\mu_n+\mu_p)^2$; therefore it will be difficult to specify the origin of conductance fluctuations, if any. Besides, $\sigma(L) \propto n(L) \propto n_0 V/L^2$ and, using eq.(1),(2) and (5), conductance fluctuations will lead to :

$$S_V \propto LV/An_0 \qquad (6)$$

Experimental results

Noise measurements were performed at room temperature. SOS microdiodes operating in the semiconductor double injection regime exhibit 1/f noise below 10 kHz and generation-recombination or diffusion noises beyond.

It seems that for largest structures (Z = 200 µm), voltage fluctuations have a power spectral density S_V nearly proportional to I^2 (fig. 2 : curves (2) to (4)) which would disagree, with the conductance fluctuation hypothesis. In fact, the current density i_x is there relatively low so that the device operates just at the beginning of the semiconductor regime ($i_x \propto V^2$). Then the main noise source still corresponds to the flicker noise generated in the base by its *ohmic* resistance $R_0 \simeq L/q\mu_n n_0 A$ (owing to good technology and to the noise magnitude, the contribution of the electrical contacts to such a noise is negligible). Besides, the noise measured at lower currents is in quite fair agreement with eq.(3) or (4).

At higher current *density* i_x (narrow diodes or higher currents I : fig. 2, curves (3) to (7)), the noise varies with geometrical or electrical parameters according to eq.(5) and (6) : for example, for given Z and n_0 and at the same current I (i.e. V^2/L^3 = const.) S_V is proportional to $V^{5/3}$ (curves (6) and (7)), while for given L, n_0, I, we get $S_V \propto L^{5/2}/Z^{3/2} \propto V^3/L^2$ (curves (3) and (5)).

Notice that the data obtained for higher voltages applied to SOS diodes cannot be taken into account because hot carrier effects are no more negligible [7].

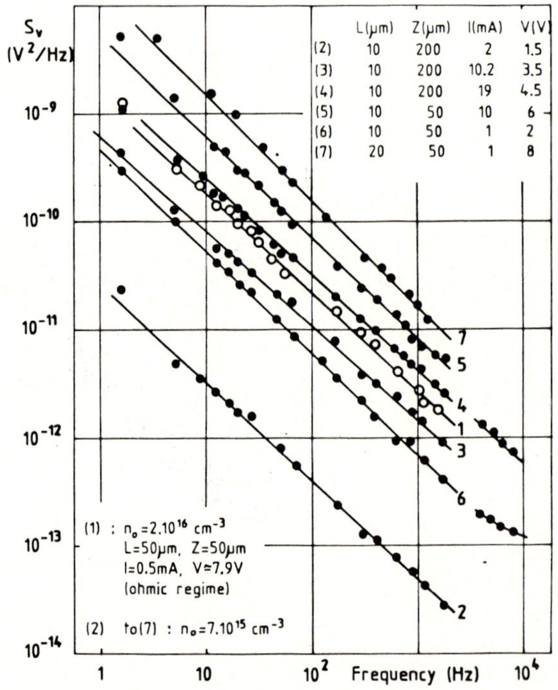

Fig. 2 : Noise spectral density $S_V(f)$ for several SOS $p^+ n\, n^+$ diodes.

3. 1/F NOISE IN DOUBLE INJECTION DIODES : A BULK EFFECT

Our experimental results on $p^+ n\, n^+$ SOS microdiodes are found to be consistent with the hypothesis of the bulk origin of the flicker noise. Other measurements have been made on germanium double injection diodes : $p^+ n\, n^+$ structures were realized from low-doped germanium ingot ($\sigma_0 < 1/30\ \Omega^{-1}cm^{-1}$), using In and Sn diffusions to make p^+ and n^+ injecting contacts. The carrier lifetime τ is about 10 µs, and $L_D \sim 0.3$ mm. Although the geometry is not as well defined as in SOS diodes, the I(V) characteristics of the semiconductor regime are very correctly described by eq.(1) ; moreover high injection levels are quite easily reached for L < 5 mm (care must be sometimes taken to prevent Joule heating). We then observe that the voltage noise S_V is proportional to the mean voltage V (see also fig. 3), in agreement with our analysis.

As regards the controverted hypothesis of a surface origin of the noise, we think that it has to be rejected for the following reasons :

i) Concerning the noise which could be generated at the p^+n junction by a surface effect, its magnitude S_V at a given current I would be roughly proportional to $1/Z^3$ (or 1/Z), since $S_I \propto 1/Z$; we don't observe such a behaviour. This is not really surprising since the static characteristics of a double injection structure are mainly governed by the bulk base properties (see eq.(1) and (2)) and not by the injecting diodes.

ii) Indeed surface recombination properties acts on the characteristics of a double-injection device, through the carrier lifetime τ, which is actually the carrier *effective* lifetime (taking into account bulk and surface recombinations) [6]. But in this case, it is shown that τ should not depend upon the geometrical parameters, except the film thickness.

iii) Since silicon-sapphire interface is well known for its drawbacks (as well as the interface of germanium with its oxide), it could be regarded as potential noise sources. But in such a case, the measured noise would be proportional to (Z.L), which is not experimentally observed.

4. OTHER EXPERIMENTAL PROOF : THE INFLUENCE OF A MAGNETIC FIELD

The forward current-voltage characteristic of double injection devices with convenient length and thickness can be very sensitive to a magnetic field B applied along Z direction (magnetodiode effect [8]). Moreover, if the surfaces where the carriers are deflected by the Lorentz force have very different recombination rates (as it is the case for SOS structures), we get devices which are also sensitive to the magnetic field direction [8][9]. In this respect, the noise appears as a practical limitation of their sensitivity.

Fig. 3 and 4 show that the flicker noise level variations with B qualitatively agree with the modifications (increase or decrease) of the device resistance. Moreover, the noise dependence on B is directly related to the sensitivity of the magnetodiode (i.e. to the *relative* variation of the device resistance when B is applied) : for example, when the voltage developped accross the SOS $p^+ n\ n^+$ device is modified by ± 3 % for B = ∓ 0.5 T, the voltage noise level is varied by approximately ± 30 % (fig. 3). But with a Ge double injection diode which exhibits a 100 times higher sensitivity, the noise power density S_V is multiplied by 30 when a magnetic field B = - 0.5 T is applied (fig. 4).

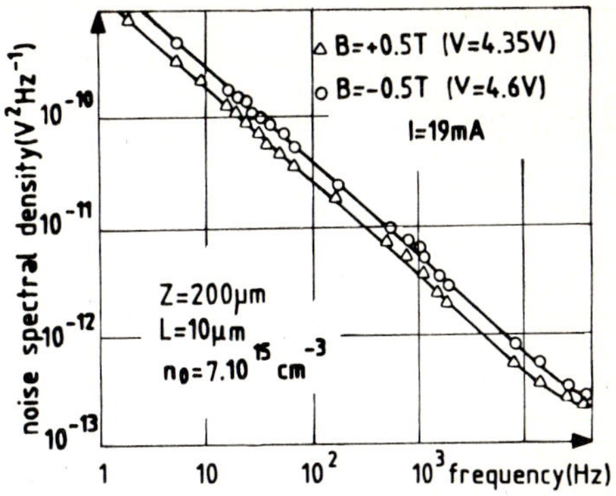

Fig. 3 : Magnetic field dependence of 1/f noise in a SOS $p^+ n\ n^+$ diode.

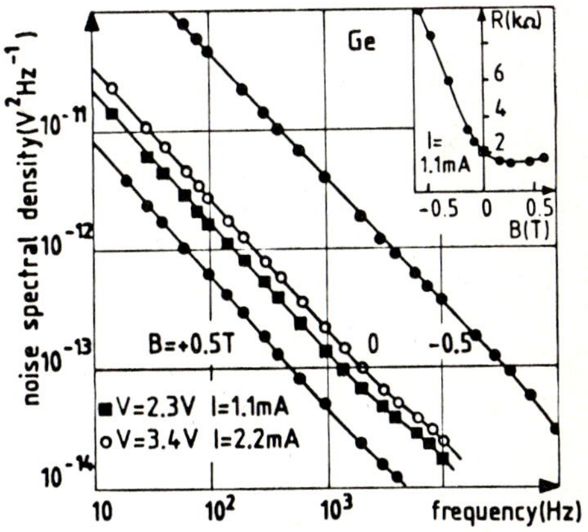

Fig. 4 : Low frequency voltage noise $S_V(f)$ of a germanium $p^+ \nu\ n^+$ diode (0.3 × 0.8 × 2 mm^3) at room temperature. Black circles correspond to the influence of a magnetic field B (at constant current I = 1.1 mA) on $S_V(f)$ and on the device static resistance R (insert).

These results also agree with temperature studies. At 220 K, the double injection regime of Ge structures is conserved ; when the resistance is multiplied by 2 with regard to room temperature, the noise is raised by about 10.

5. CONCLUSION

Noise measurements in $p^+ n\ n^+$ devices give evidence that the 1/f noise of double injection diodes has to be related to a bulk effect (conductance fluctuations) inside the base of the diode. The noise magnitude and its variations with geometrical and electrical parameters very well agree with HOOGE's empirical relation or its consequences. In particular we have shown that the voltage noise spectral density S_V in the semiconductor double injection regime ($I \propto V^2$) is proportional to V. However at lower current density a transition region with $S_V \propto I^2$ can be observed. The influence of a magnetic field on the noise has also confirmed the bulk origin ; it must be noted that the noise is lowered when the magnetic field decreases the device resistance. In spite of several phenomenological similarities, this is quite different from the case of flicker noise observed during carrier concentration by a magnetic field in near-intrinsic semiconductors [10], where the 1/f noise results from surface recombination fluctuations.

REFERENCES

[1] Nicolet M.A., Bilger R.H. and Zijlstra R.J.J, Noise in single and double injection currents in solids, Phys. Stat. Sol. (b) 70 (1975), 9-45 and 415-438.

[2] Van der Ziel A., Flicker noise in electronic devices, Advances in Electronics and Electron Physics, 49 (1979), 225-297.

[3] Kleinpenning T.G.M., 1/f noise in p-n diodes, Physica 98B (1980), 289-299.

[4] Hooge F.N., Kleinpenning T.G.M. and Vandamme, L.K.J., Experimental studies of 1/f noise, Rep. Prog. Phys., 44 (1981), 479-532.

[5] Kleinpenning T.G.M., 1/f noise in solid state single injection diodes, Physica 94B (1978), 141-151.

[6] Baron R. and Mayer J.W., Double injection in semiconductors, chap 4 in *Semiconductors and semimetals, vol. 6, Injection Phenomena* Willardson R.K. and Beer A.C., Ed., Academic Press, New-York (1970).

[7] Dufour M. and Cristoloveanu S., Hot carrier effects in double injection phenomena, Applied Physics A 29 (1982), 87-92.

[8] Cristoloveanu S., Magnetic field and surface influences on double injection phenomena in semiconductors, Phys. Stat. Sol. (a) 64 (1981), 683-695 and 65 (1981), 281-292.

[9] Mohaghegh A., Cristoloveanu S. and De Pontcharra J., Double injection phenomena under magnetic field in SOS films : a new generation of magnetosensitive microdevices, IEEE Trans. Electr. Devices, ED28 (1981), 237-242.

[10] Dilmi T., Chovet A. and Viktorovitch P., Influence of a magnetic field on 1/f noise in ambipolar semiconductors : evidence of its surface origin, J. Appl. Phys. 50 (1979), 5348-5351.

NOISE IN PHYSICAL SYSTEMS AND 1/f NOISE
M. Savelli, G. Lecoy and J-P. Nougier (eds.)
© *Elsevier Science Publishers B.V., 1983*

LOW FREQUENCY NOISE IN AlGaAs-GaAs 2-D ELECTRON GAS DEVICES AND ITS CORRELATION TO DEEP LEVELS

L.Loreck, H.Dämbkes, K.Heime (1); K.Ploog (2) and G.Weimann (3)

Universität Duisburg, FB9, Halbleitertechnik/-technologie, Duisburg, FRG (1)
Max-Planck-Institut, Stuttgart, FRG (2)
Forschungsinstitut der Deutschen Bundespost, Darmstadt, FRG (3)

Low frequency noise of AlGaAs-GaAs-heterostructures grown by molecular beam epitaxy was investigated. The temperature of the samples was varied between 100 K and 400 K. In the frequency range from 1 Hz to 25 kHz noise spectra could be described as superposition of several generation-recombination noise components. From a comparison of these time constants with those obtained by DLTS measurements three deep electron traps with activation energies 0.40; 0.42 and 0.60eV were detected.

1. INTRODUCTION

One of the most important progresses in the development of high speed electron devices was the discovery of a quasi two-dimensional electron gas located at the interface of certain semiconductor heterostructures /1/. The system AlGaAs-GaAs grown by molecular beam epitaxy is one of them and often used for fabricating field-effect-transistors (FET) which are called TEGFET /2/ (two dimensional electron gas FET), HEMT /3/ (high electron mobility transistor) or modulation-doped FET /4/. Applications are in the area of microwave or high speed digital integrated circuits. Electrons are transferred from the n-doped AlGaAs into the undoped GaAs and form a quasi two-dimensional gas at the interface. Therefore ionized shallow donors and free electrons are spatially separated and the Coulomb scattering is reduced. This results in a very high mobility of the electrons, especially at lower temperatures.

To achieve the performance required for these heterojunction devices, the epitaxial AlGaAs and GaAs materials must be free of defects and chemical contaminants.The deep level characteristics of these layers are therefore very important ,since deep electron traps reduce carrier concentration, affect electron transport properties and increase noise. Deep levels can be charaterized by low frequency noise measurements. Generation-recombination (g-r) noise is caused by fluctuations in the number of free electrons. Discrete energy levels in the forbidden gap are able to trap electrons or act as recombination centers. Theory explains that these fluctuations cause different Lorentzian shaped g-r noise contributions in the spectrum of a-c open-circuit voltage noise. /5,6/ Each trap has a charateristic spectrum defined by its low frequency plateau value (amplitude) and the corner frequency, which gives the appertaining time constant τ. From the variation of τ with temperature the activation energy of the deep level is deduced. The trap concentration is related to the amplitude. But only those energy levels contribute to noise which are not more than a few kT away from the Fermi level. By changing the temperature of the sample the Fermi level is shifted across the forbidden gap.

2. EXPERIMENTAL PROCEDURE

Two types of heterostructures were investigated. Table 1 gives the most important data of these samples.

Table 1

	S-3	D-1
Top Layer GaAs	22 nm p^- 10^{15} cm^{-3}	20 nm p^- $\approx 10^{14} cm^{-3}$
$Al_xGa_{1-x}As$ (Si-doped)	60 nm $1*10^{18}$ cm^{-3} X=0.25	70 nm $6*10^{17}$ cm^{-3} X=0.3
Spacer (undoped Al_xGa_xAs)	/	6 nm
undoped GaAs layer	1 μm p^- $\approx 10^{15}$ cm^{-3}	1.1μm p^- $\approx 10^{14}$ cm^{-3}
n_s/cm^2 300K 77K	$1.2*10^{12}$ $7*10^{11}$	$5*10^{11}$ $6.5*10^{11}$
μ_{Hall} cm^2/Vs 300K 77K	3600 55000	7900 130000

n_s = sheet concentration of the two-dimensional electron gas

The main difference between the two types is the existence of a so-called "spacer" in sample D1. This spacer-layer of undoped AlGaAs enhances the mobility of the two-dimensional electron gas.

A test structure allowed the measurement of electron mobility by Hall measurements and the contact resistance from a transmission line structure. For the noise measurements a simple arrangement was chosen, consisting of two ohmic contacts with 60 μm separation and 100 μm contact width. Ohmic AuGe/Ni contacts were used.

Samples were mounted in a variable temperature cryogenic system, so that measurements at temperatures from about 20K to 400K with 1K accuracy could be done. All measurement equipment was situated in a shielded cabin to avoid interferences from outside. A desktop computer HP 9845 B was connected to the measurement apparatus inside the cabin via fiber optic links. The experimental arrangement is shown in fig.1.

Variable bias current was supplied by a BURSTER Digistant 6425 T current source with high inner resistance and low output capacitance. So the sample noise could not be short circuited by the d.c. bias supply. A low noise preamplifier ITHACO 1201 was connected in series with the HP 3582 A spectrum analyzer in order to achieve a higher input sensitivity. Smallest detectable noise voltages were about 10nV(RMS). The frequency range was 1Hz to 25 kHz. A computer program allowed to control frequency range and temperature automatically. Furthermore, regular interferences such as line frequency and their multiples were suppressed by the computer evaluation.

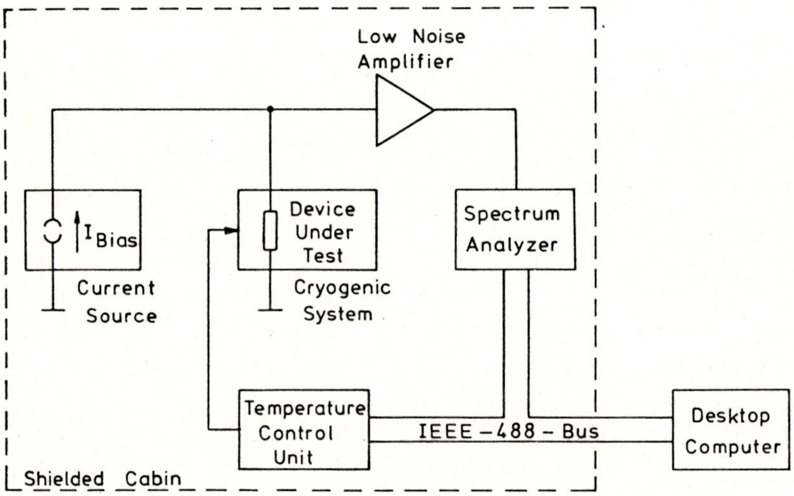

fig.1: block diagram of the experimental set-up

3. MEASUREMENTS

Ohmic behaviour of the samples was observed over the whole temperature range as long as the bias current was less then 5 mA. The spectrum analyzer bandwidth was switched to 726 mHz over the whole frequency range, so that the measurement frequency was larger than the bandwidth.

To be sure that no noise from the apparatus itself disturbed the measurements, a metal-film ohmic resistor of the same value as the sample was mounted. The measured white noise was far less than that of the heterostructure samples. Therefore no special calibration measurements were necessary.

Noise voltages were measured in dBV, where OdBV corresponds to 1V(RMS) single tone. Measurements were made at different temperatures with 10K steps and stored upon a magnetic tape for further evaluation.

4. RESULTS AND DISCUSSION

A characteric spectrum is shown in fig.2.

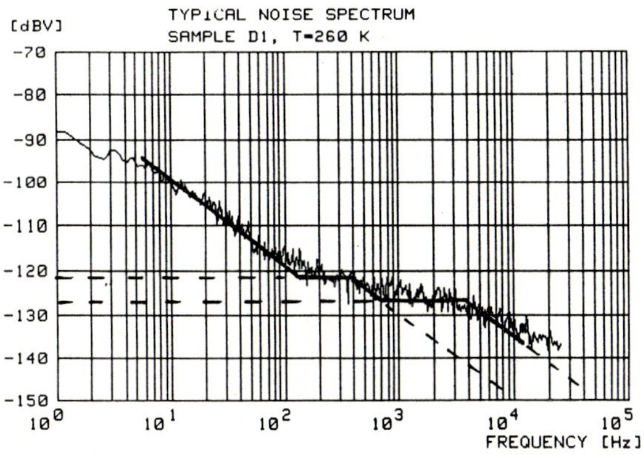

fig.2: typical noise spectrum

The curves are fitted by a superposition of several Lorentzian type g-r noise contributions.The amplitudes and characteristic corner frequencies of the different g-r-noise components varied with temperature.

It was tried to evaluate trap concentrations, capture cross sections and energy levels from the measured amplitudes and time constants by use of a model /5/ assuming that the noise is generated in the AlGaAs layer. The trial failed probably because the strong band bending in the AlGaAs-GaAs heterostructure and the electron transfer into the two-dimensional electron gas are not included in the model.

Since the time constants τ are related to the activation energies ΔE by

$$1/\tau = T^2 \exp(-\Delta E/kT)$$

the activation energies can be determined from an Arrhenius plot of $\tau \cdot T^2$ vs. 1/T. The result is shown in fig.3a,b.

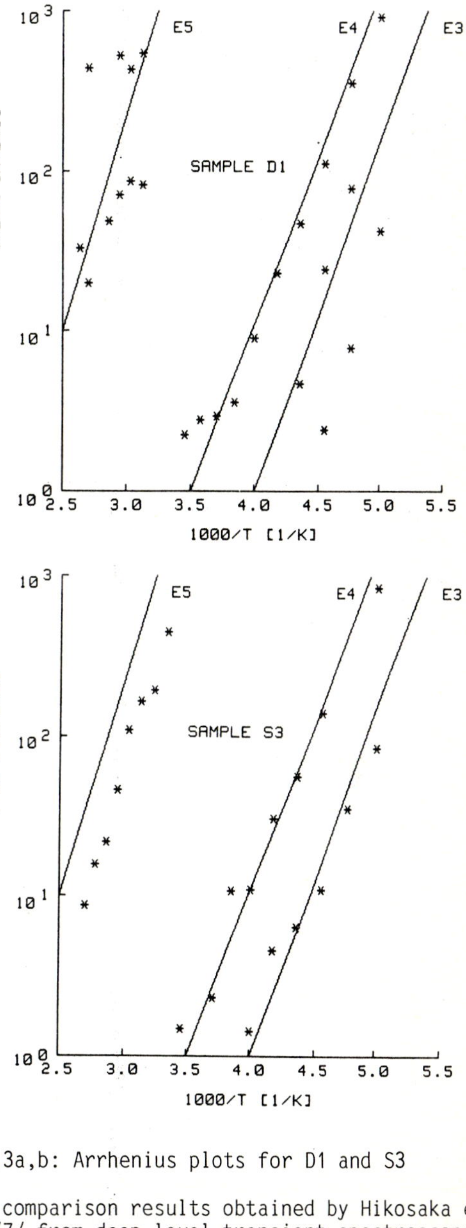

fig.3a,b: Arrhenius plots for D1 and S3

For comparison results obtained by Hikosaka et al./7/ from deep level transient spectroscopy (DLTS) measurements with similar AlGaAs-heterostructures are included in fig.3a,b as straight lines. The good agreement between the results from noise and DLTS measurements leads us to the conclusion that the levels in our samples are located in the AlGaAs layer.

Table 2

Level	Activation energy /eV/	
E 1	0.26	
E 2	0.28	
E 3	0.42	0.42
E 4	0.40	0.40
E 5	0.60	0.60
E 6	0.78	
	Hikosaka et al. (DLTS)	this work (noise)

5. Conclusion

It was shown that low-frequency noise measurements are a useful method for the detection of deep levels even in complicated heterostructures such as the AlGaAs-GaAs system. Electron trap levels were found at 0.42; 0.40 and 0.60eV and are identical to those measured by DLTS in similar samples. A modification of the existent g-r noise model /5,6/ is necessary which includes the probability of electron transfer into quantum wells.

Acknowledgement:

The authors would like to thank Prof.Dr.R.J.J.Zijlstra (Utrecht, NL) for providing the computer program and G.Howahl for the sample preparation.

References:

/1/ Dingle,R. et al., "Electron mobilities in modulation doped semiconductor heterojunction superlattices", Appl. Phys. Lett., vol.33, pp 665-667,Oct.1978

/2/ Delagebeaudeuf, D. et al., "Metal-(n) AlGaAs/GaAs two-dimensional electron gas FET", IEEE Trans. Electron Devices, vol ED-29, pp 955-960, June 1982

/3/ Joshin, K. et al., "Noise performance of microwave HEMT", priv. com., to be presented at IMS 83, Boston

/4/ Duh, K.H. et al., "1/f noise in modulation-doped field effect transistors", IEEE Electron Dev. Lett., vol E DL-4, pp 12-13, Jan. 1983

/5/ Bosman, G., Zijlstra, R.J.J., "Generation-recombination noise in p-type silicon", Solid St. Electron., vol 25, pp 273-280, no.4, 1982

/6/ van der Ziel, A.: "Noise in measurements" (John Wiley & Sons, Inc. New York, 1976)

/7/ Hikosaka, K. et al. "Deep electron traps in MBE-grown AlGaAs ternary alloy for heterojunction devices", Fujitsu Laboratories Ltd.

1/f NOISE IN
BIPOLAR TRANSISTORS

NOISE IN PHYSICAL SYSTEMS AND 1/f NOISE
M. Savelli, G. Lecoy and J-P. Nougier (eds.)
© Elsevier Science Publishers B.V., 1983

1/f NOISE IN BIPOLAR TRANSISTORS

C.T.Green and B.K.Jones

Department of Physics, University of Lancaster,
Lancaster, LA1 4YB, UK.

Measurements have been made of the 1/f, excess, noise and dc characteristics of low-noise bipolar transistors between 80 K and 375 K under a wide range of bias conditions. The excess noise can be represented by a single base current noise source $\overline{i_n^2} = \frac{k^1}{A} I_B'^2 \Delta f$ where k^1 is a temperature independent constant, A is the emitter junction area and I_B' is the non-ideal component of the base current.

INTRODUCTION

Detailed experiments have been performed on low-noise NPN bipolar transistors to determine the relationship between the noise properties and the dc characteristics. The dc emitter and collector junction properties and the transistor action were measured over a wide range of bias between 80 K and 375 K. The white and excess noise were measured between 10^{-2}Hz and 20 kHz for the same range of bias and temperature. The devices were biased in the common emitter configuration. All devices used were mainly type BC 413 but other devices have been studied to confirm the findings.

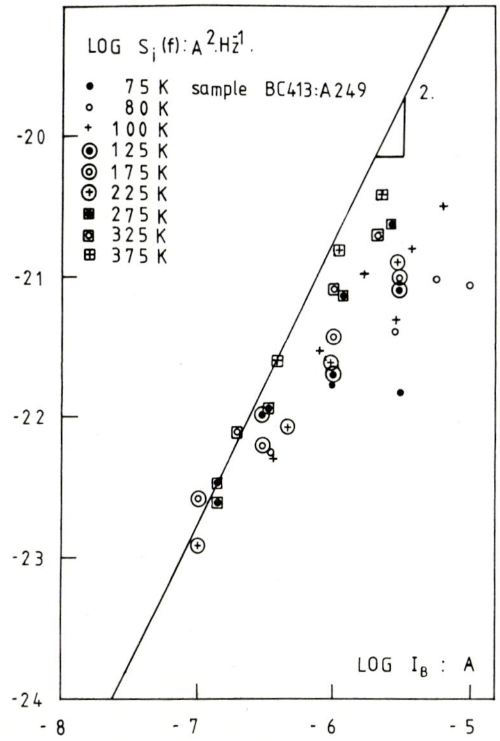

Figure 1(a) : The equivalent input excess current noise intensity as a function of total base current with temperature as parameter.

RESULTS

The low-frequency T noise equivalent circuit frequently used for the white noise analysis[1] was confirmed and used to determine the intrinsic base resistance, r_{bb}. The use of a wide range of source resistances showed that the excess noise could be represented as a base current generator, $\overline{i_n^2}$, with no observable voltage generator. The excess noise had a 1/f spectrum although the exponent was slightly temperature dependent.

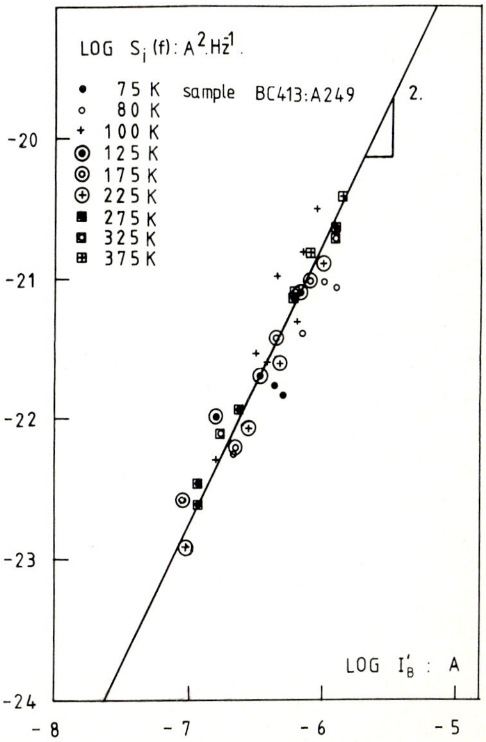

Figure 1(b) : The equivalent input excess current noise intensity as a function of non-ideal base current with temperature as parameter. The same data as in (a).

The magnitude of this excess current noise source was correlated with various dc parameters such as junction leakage current for a sample of 500 devices. Although trends were indicated no definitive results were suggested.

Since excess noise is generally a resistance fluctuation its magnitude was plotted against the base current, to determine any simple dependence. The considerable scatter resulting is shown in figure 1(a). However, a single curve was obtained by comparing the noise intensity with I_B' as shown in figure 1(b). Here I_B' is the non-ideal base current. Thus we can say that $\overline{i_n^2} = k\, I_B'^2\, \Delta f$ where k is temperature independent.

The non-ideal base current component was determined from the dc characteristics. The transistor base current can be considered as the sum of two components. The ideal component is that due to the diffusion of carriers over the base emitter junction with recombination in the base region and it follows the Schockley equation $I_B'' = I_{BO}''\,(\exp(eV/kT) - 1)$. The non-ideal component is due to carriers which are generated or recombine elsewhere and can be described by an equation $I_B' = I_B''\,(\exp(eV/mkT) - 1)$ where m is the ideality factor. There may be several such non-ideal current contributions due to different generation-recombination processes. It should be noted that, since $m \geqslant 1$, the non-ideal current dominates for small I_B and the slope of the log I_B - V_{BE} curve gives m at small I_B and the ideal factor 1 at large I_B with a transition region between. It was confirmed that the measured currents could be fitted to such a sum as shown by Ashburn[2].

This interpretation brings together the observations of Stoisiek and Wolf[3] and Higuchi and Ochi[4].

DISCUSSION

This simple relationship found for the excess noise suggests that a base-resistance fluctuation[3] is unlikely since the total base current would then be involved. Also measurements suggest that there is no dependence of the noise on the base resistance.

The excess noise is apparently caused by the same process which produces the excess current. In these specimens there is no indication of a surface component to the excess current so that recombination in the neutral emitter bulk or depletion region is suggested.

The constancy of the noise parameter k with temperature is shown in fig.2 and contrasts

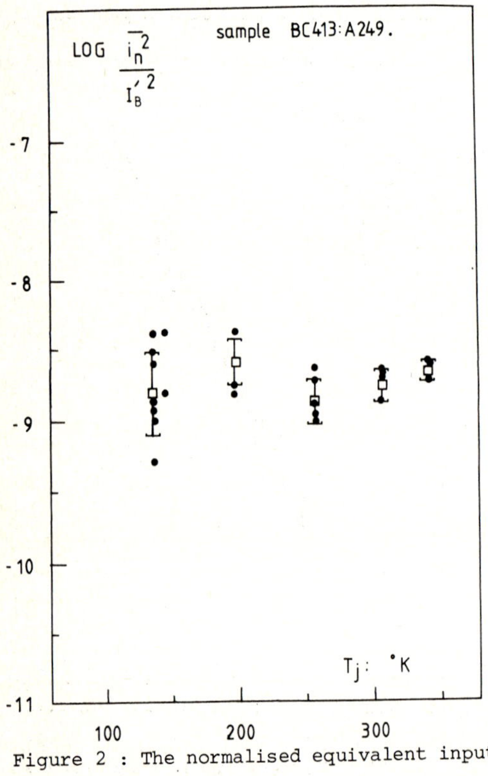

Figure 2 : The normalised equivalent input current noise, $\overline{i_n^2}/I_B'^2 = k$ as a function of junction temperature.

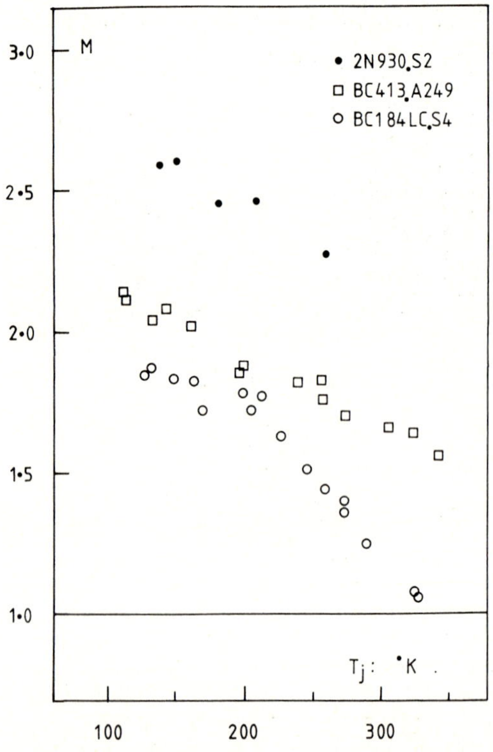

Figure 3 : The base-emitter junction ideality factor m as a function of device temperature for three different devices.

with the large variation of the ideality parameter, m, as shown in fig.3. The activation energy of the non-ideal current given by the temperature variation of I_{BO}' was found to be half to one-third of the band gap.

For a device with a uniform distribution of noise sources one might expect that $\overline{i_n^2} \sim 1/A$ where A is the emitter junction area[4]. This relationship is illustrated for four types of devices in figure 4. A reasonable 1/A fit is found through the device processing scatter. This is quite remarkable since the devices chosen did not have a common geometry, processing formula or even manufacturer.

CONCLUSION

The excess noise in the bipolar transistors studied can be expressed by a single base current generator $\overline{i_n^2} = \frac{k'}{A} I_B'^2 \Delta f$ where k' is temperature independent.

This work was supported in part by Ferranti Semiconductors Ltd and SERC.

REFERENCES

[1] Unwin, R.T., IEEE Proc., 127 (1980) 53-61.

[2] Ashburn, P., Solid State Electronics, 18 (1975) 569-577.

[3] Stoisiek, M. and Wolf, D., IEEE Trans., ED-27 (1980) 1753-7.

[4] Higuchi, H. and Ochi, S., Symposium on 1/f Noise, Tokyo (1977).

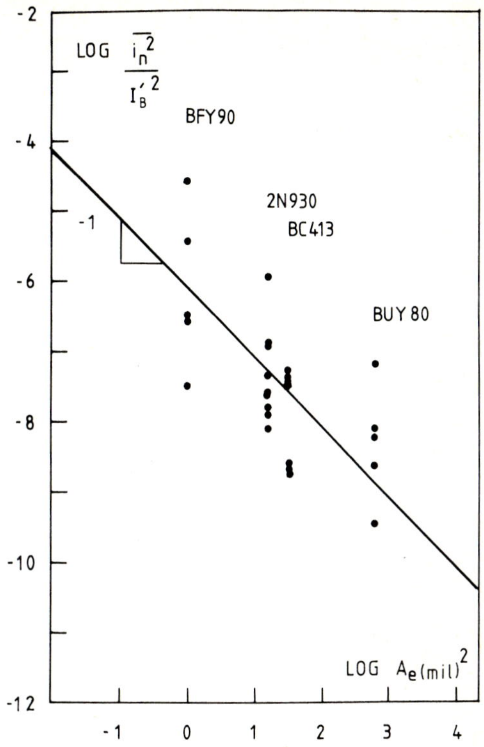

Figure 4 : The variation of the normalised equivalent input current noise $\overline{i_n^2}/I_B'^2 = k$, with emitter junction area. The data are for several samples of four different device types.

NOISE IN PHYSICAL SYSTEMS AND 1/f NOISE
M. Savelli, G. Lecoy and J-P. Nougier (eds.)
© Elsevier Science Publishers B.V., 1983

MOBILITY-FLUCTUATION 1/f NOISE IDENTIFIED IN SILICON P$^+$NP TRANSISTORS

J. Kilmer, A. van der Ziel and G. Bosman

Department of Electrical Engineering,
University of Florida, Gainesville, FL 32611, USA

The magnitude and location of mobility-fluctuation 1/f noise sources have been identified by means of biasing a PNP transistor in a common emitter configuration with first a high and then a low source resistance. Comparison of the two noise spectra at the same base currents shows the low source resistor bias isolates the collector noise sources, and the high source resistance isolates base noise sources. The magnitude of the observed collector 1/f noise gives an $\alpha \geq 2\times10^{-6}$ from Kleinpenning's mobility-fluctuation theory. The base 1/f noise gives an $\alpha \geq 10^{-7}$ due to an impurity mobility reduction factor of about 100.

I. INTRODUCTION

To date, three causes regarding the origin of 1/f noise in transistors prevail.

(1) Fluctuating occupancy of electrons in oxide surface traps (or dislocations) in the base or emitter space-charge region modulates the (surface) recombination velocity. 1/f noise due to fluctuating recombination velocity is represented as a recombination current I_R flowing from emitter to base.[1]

(2) Mobility fluctuations due to holes interacting with phonons cause 1/f noise in the hole current I_{Ep} diffusing from the emitter to the collector.

(3) Mobility fluctuations due to the electron current I_{En} injected from the base into the emitter may also cause 1/f noise.

1/f noise due to a fluctuating series base resistance r_b we do not consider since I_B is small in a high β transistor. The three possible causes are represented as current sources δI_R, δI_{En}, and δI_{Ep} in an equivalent circuit for a PNP transistor first drawn by Plumb and Chenette[2] and later modified by van der Ziel (see Fig. 1). Here we combine the two base current sources into an equivalent base 1/f noise source, i_{fb}, where $i_{fb} = -\delta I_R - \delta I_{En}$ and rename the collector current source, i_{fc}, where $i_{fc} = -\delta I_{Ep}$.

In older transistors the predominant 1/f noise source was the recombination current because those devices had large surface recombination velocities. The purpose of our present investigation is to determine whether 1/f noise due to mobility fluctuations, as presented first by Hooge[3] and recently by Kleinpenning,[4] is present in contemporary devices with small surface recombination velocities. Mobility

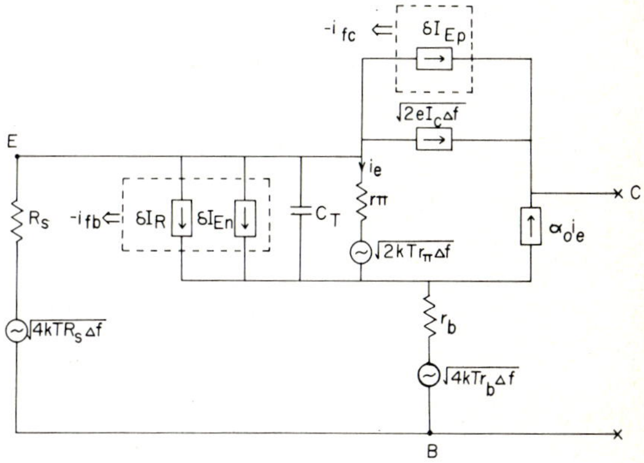

Figure 1. Equivalent common base circuit with noise sources.

fluctuations imply fluctuations in the diffusion constant D_p, since by the Einstein relation

$$q\delta D_p = kT\delta\mu_p. \tag{1}$$

Thus we may expect the mobility fluctuations to modulate the emitter-collector hole diffusion current and/or the base-emitter electron injection current.

Van der Ziel's appendixed derivation[5] of Kleinpenning's expression for the noise spectrum due to mobility fluctuations of emitter-collector hole diffusion in P$^+$NP transistors shows,

$$S_{I_{Ep}}(f) = 2qI_{Ep}\frac{\alpha_p}{4f\tau_{dp}}\ln\left[\frac{P(0)}{P(w_B)}\right] \tag{2}$$

where α_p is the Hooge parameter associated with hole current, $\tau_{dp} = w_B^2/2D_p$ is the diffusion time for holes through the base region, w_B the base width, and $P(0)$ and $P(w_B)$ are the hole concentrations for unit length at the emitter side and the collector side of the base, respectively. We see the magnitude of $S_{I_{Ep}}$ is inversely proportional to τ_{dp}, which means that $S_{I_{Ep}}$ is proportional to f_T since

$$f_T = \frac{1}{2\pi\tau_{dp}} . \tag{3}$$

Therefore, the hole mobility fluctuation 1/f noise source is larger in transistors with a large f_T (e.g. microwave transistors).

Also for electron injection from base to emitter we have, due to mobility fluctuations,[5,eq.(4)]

$$S_{I_{En}}(f) = 2qI_{En}\frac{\alpha_n}{4f\tau_{dn}}\ln\left[\frac{N(0)}{N(w_E)}\right] \tag{4}$$

where $\tau_{dn} = w_E^2/2D_n$, w_E the width of the emitter region, D_n the electron diffusion constant in the emitter region, whereas $N(0)$ and $N(w_E)$ are the electron concentrations for unit length at the base side of the emitter and at the emitter contact, respectively.

II. EXPERIMENTS TO DISCRIMINATE BETWEEN BASE AND COLLECTOR NOISE SOURCES

Since I_{En} is much less than I_{Ep} in a p^+np transistor, we expect $S_{I_{En}}$ to be smaller than $S_{I_{Ep}}$. For this reason we use the transistor in a common-emitter configuration so that an amplified $S_{I_{En}}$ can be measured at the output (collector) terminal. The latter signal and $S_{I_{Ep}}$ are comparable in magnitude.

Redrawing Figure 1 into a common-emitter configuration and squaring the noise sources so they represent spectral contributors (see Fig. 2), we see $(R_s + r_b)$ is now in parallel to the input (base) equivalent circuit of the transistor. Also in Fig. 2, the collector noise current sources have been referred to the input equivalent circuit as noise voltages sources by multiplying by $1/g_m = r_\pi/\beta$ (valid if $r_\pi \gg r_b$).

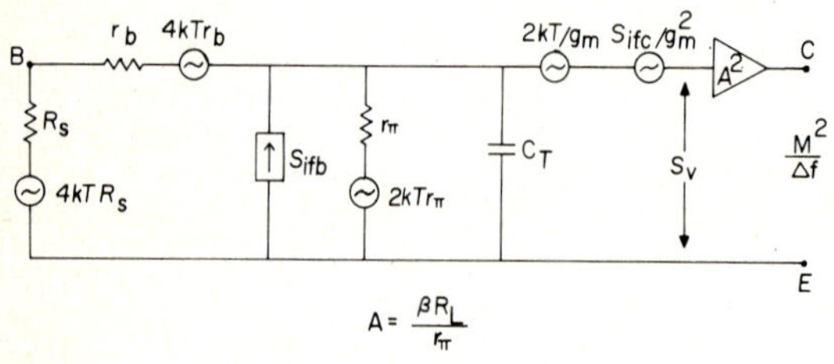

Figure 2. Equivalent common emitter circuit.

An HP3582A FFT spectrum analyzer measures the spectral density of the collector noise, $M^2/\Delta f$. Calculations from Figure 2 reveal

$$\frac{M^2}{\Delta f} = A^2\left\{\left(S_{R_s} + S_{r_b}\right)\left[\frac{r_\pi}{R_s + r_b + r_\pi}\right]^2 + 2kTr_\pi\left[\frac{r_b + R_s}{R_s + r_b + r_\pi}\right]^2 + S_{ifb}\frac{(r_b + R_s)^2 r_\pi^2}{(R_s + r_b + r_\pi)^2} + \frac{2kT}{g_m} + \frac{S_{ifc}}{g_m^2}\right\} . \tag{5}$$

If we use that $r_\pi \gg r_b$ and $\beta \gg 1$, then eq.(5) can be rewritten so that we obtain

$$\frac{M^2}{\Delta f} = A^2\left\{\left[\frac{r_\pi}{R_s + r_b + r_\pi}\right]^2 \left[2kT(2r_b + 1/g_m) + \frac{S_{ifc}}{g_m^2} + S_{ifb}r_b^2\right] + R_s\left[4kT + 2\frac{S_{ifc}r_\pi}{\beta^2} + 2S_{ifb}r_b\right] + R_s^2\left[\frac{2kT}{r_\pi} + \frac{S_{ifc}}{\beta^2} + S_{ifb}\right]\right\} . \tag{6}$$

We see that there are three regions to the magnitude of the measured noise versus R_s--an independent, a linear, and a quadratic regime.

Ideally, the mobility-fluctuation 1/f noise measurements should be made on microwave transistors biased with low currents for both high and low R_s. Unfortunately, microwave transistors usually do not have a high DC β. So the experiment was performed on low-noise PNP transistors (GE 82 185) with $\beta \simeq 350$ typically. A simple biasing scheme was used for the high R_s experiment,[6] and the noise was measured for three different I_B's. From eq.(6) and for the case of high R_s, we see that we measure with the spectrum analyzer,

$$\frac{M_{HI}^2}{\Delta f} = \beta^2 R_L^2\left[2eI_B + S_{ifb} + \frac{S_{ifc}}{\beta^2}\right] , \tag{7}$$

using $r_\pi = \frac{kT}{eI_b}$ where we have neglected the small r_b and r_π compared to a high R_s and the terms independent of and proportional with R_s. The measured high R_s noise spectra, $M_{HI}^2/\Delta f$, is now scaled down by $1/\beta^2 R_L^2$ so that the noise plotted in Fig. 3 (curves IV, V, II) represents the absolute magnitude of the physical noise sources (in amp^2 sec) referred back to the (base) input,

$$S_{HR_s} = \frac{M_{HI}^2}{\Delta f}\left[\frac{1}{\beta^2 R_L^2}\right] = 2eI_B + S_{ifb} + \frac{S_{ifc}}{\beta^2} . \tag{8}$$

The high-frequency roll-off, which each of the plots indicates, is attributed to the Miller

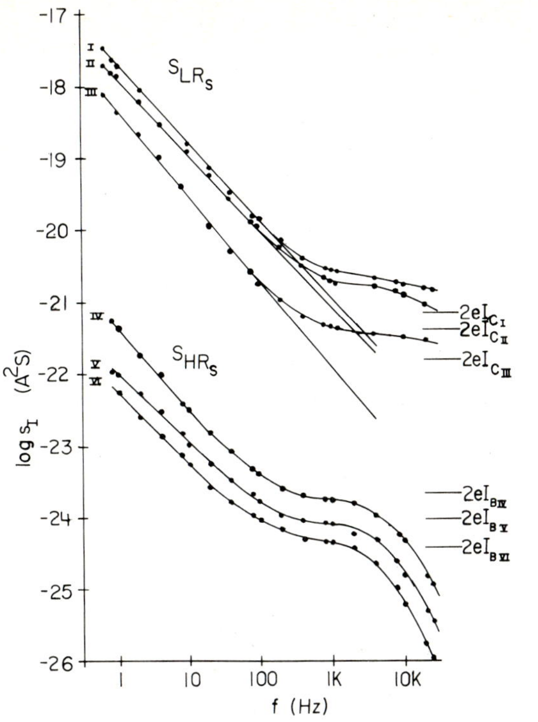

Figure 3. Measurement of high-source impedance spectra (S_{HR_s}) and low-source impedance spectra (S_{LR_s}). The base currents are 6 μA, 3 μA, and 1 μA.

effect of the capacitance C_T in the equivalent circuit (see Fig. 2) where

$$C_T = C_{beo} + C_{bco}(1 + |A_v|). \tag{9}$$

Since I_B is small, r_π is large, and the $f_m = 1/C_T r_\pi$ Miller cut-off frequency, is low ~2 KHz. Shot noise, low-pass filtered across the parallel combination of r_π and C_T, gives at sufficiently high frequencies,

$$S_{HR_s} = \frac{2eI_B}{1 + \omega^2 C_T^2 r_\pi^2}, \tag{10}$$

the observed $1/f^2$ roll-off.

Biased with a low R_s configuration,[6] we neglect the terms in eq.(6), which are proportional with R_s and R_s^2. Using $g_m = \beta/r_\pi$ and neglecting R_s and r_b with respect to r_π, we see that we can plot (again in amp^2 sec).

$$S_{LR_s} = \frac{M_{LO}^2}{\Delta f}\left[\frac{1}{R_L^2}\right] = 2eI_C + 4kTr_b g_m^2 + S_{ifc} + S_{ifb} r_b^2 g_m^2 . \tag{11}$$

This was done in Fig. 3 (Curves I, II, III) at the same three I_B's used in the high R_s experiment in order that the high and low R_s spectra can be quantitatively compared.

It should be noted that eq.(11) is only valid for $R_s \ll r_b$. In practice, however, R_s was of the same order of magnitude as r_b at low I_E ($r_s \simeq 5\Omega$). As a consequence, the thermal noise generated by R_s cannot be neglected and has to be incorporated in eq.(11). The expression for S_{LR_s} becomes

$$S_{LR_s} = 2eI_c + 4kT(r_b + R_s)g_m^2 + S_{ifc} + S_{ifb} r_b^2 g_m^2 . \tag{12}$$

We see that the low R_s measurement provides the means to measure r_b.[6] Using the magnitude of the white noise levels of S_{LR_s} in Fig. 3, the calculated values of r_b are indicated in Table 1.

III. EVALUATION OF α_p AND α_n

To calculate the magnitudes of S_{ifc} and S_{ifb}, we look only at the 1/f portion of our spectra (i.e. at f < 100 Hz) where we are above the shot-noise level and can write, at low f,

$$S_{LR_s} = S_{ifb}(r_b^2 g_m^2) + S_{ifc}, \tag{13}$$

and

$$S_{HR_s} = S_{ifb} + \frac{S_{ifc}}{\beta^2} . \tag{14}$$

Table 1. Data obtained

	Low R_S Data					High R_S Data			
Bias	Curve	I_E	β	r_b	$(\alpha_p)_{MIN}$	Curve	I_B	β	$(\alpha_n)_{MIN}$
$I_B \simeq 6\mu A$	I	2.25mA	340	4Ω	1.72×10^{-6}	IV	6.7μA	362	1.2×10^{-7}
$I_B \simeq 3\mu A$	II	1.3mA	420	9Ω	2.17×10^{-6}	V	3μA	363	6.6×10^{-8}
$I_B \simeq 1\mu A$	III	505μA	413	20Ω	1.86×10^{-6}	VI	1.2μA	307	9.2×10^{-8}

Having two equations involving the two unknown S_{ifc} and S_{ifb}, we solve for S_{ifc} and find

$$S_{ifc} = \frac{S_{LR_s}\left[\frac{1}{r_b g_m}\right]^2 - S_{HR_s}}{\left[\frac{1}{r_b g_m}\right]^2 - \frac{1}{\beta^2}} . \quad (15)$$

Now from inspection of Fig. 3, we see $S_{HR_s} \ll S_{LR_s}$ at 1 Hz, and since

$$\left[\frac{1}{r_b g_m}\right]^2 = \left[\frac{r_\pi}{\beta r_b}\right]^2 \sim 1,$$

we can neglect $1/\beta^2$ and (15) simplifies to

$$S_{LR_s} = S_{ifc}. \quad (16)$$

We see at low frequencies the low R_s bias configuration isolates S_{ifc}. Solving S_{ifb} we find

$$S_{ifb} = S_{HR_s} - \frac{S_{LR_s}}{\beta^2} . \quad (17)$$

From our data, S_{LR_s}/β^2 is a factor of ten less than S_{HR_s} at 1 Hz and we see the high R_s configuration essentially isolates S_{ifb}.

Now that the 1/f noise sources have been identified, we must apply the results of the mobility fluctuation theory eqs. (2) and (3) to calculate the magnitude of the Hooge parameter α. Since the low R_s experiment isolated collector noise, we have

$$S_{LR_s} = S_{ifc} \approx S_{I_{Ep}} . \quad (18)$$

Solving for α_p and using eq.(3), we have

$$\alpha_p = \frac{S_{LR_s} f}{\pi f_T q I_E \ell n\left[\frac{P(0)}{P(w_B)}\right]} . \quad (19)$$

To estimate the ratio $P(0)/P(w_B)$ according to Kleinpenning,[5, eq.(A7)] we have the inequality

$$P(0)/P(w) \leq w v_s/D_p \quad (20)$$

where v_s is the saturated drift velocity. Using this permits us to calculate a minimum value for α. For our silicon PNP with f_T = 200 MHz, w_B = 1.35 μm. Using $v_s \simeq 10^7$ cms^{-1}, we calculate $\ell n[P(0)/P(w_B)] \leq 4.75$. This value, used in eq.(19), gives the minimum values of α_p tabulated in the data table for S_{LR_s} evaluated at 1 Hz.

For the case of base 1/f noise, we have

$$S_{HR_s} \approx S_{ifb} \approx S_{I_{En}}, \quad (21)$$

and using the base to emitter expression (4) we have, for α_n,

$$\alpha_n = \frac{S_{HR_s} f\, 2\tau_{dn}}{q I_B \ell n\left[\frac{N(0)}{N(w_E)}\right]}, \quad (22)$$

since $I_{En} \simeq I_B$ in a P^+NP transistor if we neglect recombination. We saw for the case of holes the ℓn term in the denominator did not significantly affect the order of magnitude of α, and we expect a similar case for electrons. We take $\ell n\ [N(0)/N(w_E)] \leq 5$, since we expect the ratio of electrons in the emitter to be a few orders of magnitude greater than the ratio of holes in the base due to the high recombination of electrons in the heavily doped emitter. Using this and the approximation that $\tau_{dn} \simeq \tau_{dp}$ suggested by van der Ziel,[5] we calculated the minimum values of α_n which are tabulated in the table for S_{HR_s} evaluated at 1 Hz. Here we see the values of α_n are one or two orders of magnitude lower than α_p, which at first glance seems to imply that recombination current fluctuations, Cause 1, still account for base 1/f noise. However, we realize that we have a P^+NP device where the emitter is heavily doped and our observed α_n is diminished by an impurity mobility reduction factor. Kilmer et al.,[6] using a ratio of $\mu_{imp}/\mu_{latt} \simeq 1/10$, obtain a minimum value of $\simeq 2 \times 10^{-5}$ for (α_n) true.

Bosman et al.[7, Fig. 5] report α values ranging between 10^{-5} and 10^{-3} for electrons in n-type silicon. Hence we conclude that the 1/f noise in the base of transistors can also be attributed to a mobility-fluctuation mechanism, similar to the one causing the collector 1/f noise.

ACKNOWLEDGEMENTS

We thank Dr. C.M. Van Vliet for a critical reading of this manuscript and the support by the Air Force Office of Scientific Research, under grant #80-0050.

REFERENCES

[1] Fonger, W.H., Transistors I (R.C.A. Laboratories, New Jersey, 1956).

[2] Plumb, J.L. and Chenette, E.R., IEEE Trans. ED-10 (1963) 304.

[3] Hooge, F.N., Physica 60 (1972) 130; Physica 83b (1976) 14.

[4] Kleinpenning, T.G.M., Physica 98B (1980) 289.

[5] van der Ziel, A., Solid-St. Electron. 25 (1982) 141.

[6] Kilmer, J., van der Ziel, A. and Bosman, G., Solid-St. Electron. 26 (1983) 71.

[7] Bosman, G., Zijlstra, R.J.J. and van Rheenen, A., Physica 112B (1982) 193.

NOISE IN PHYSICAL SYSTEMS AND 1/f NOISE
M. Savelli, G. Lecoy and J-P. Nougier (eds.)
© Elsevier Science Publishers B.V., 1983

F^{-1} BULK CURRENT NOISE IN SHORT DIODES AND BIPOLAR TRANSISTORS

G. BLASQUEZ and D. SAUVAGE*

Labororatoire d'Automatique et d'Analyse des Systèmes du C.N.R.S.
7, avenue du Colonel Roche
31400 TOULOUSE - France

Experimental studies of f^{-1} noise in pnp integrated bipolar transistors showed that bulk current noise exists in the neighbourhood of the emitter base junction. Crowding effects were used to show that this noise is distributed and that its local value depends on the local value of the current. At low level biasing conditions, noise spectral density at the input is proportional to emitter current, and is independent of the emitter area.

The results of a theoretical model, attibuting the current noise in the bulk of short diodes and bipolar transistors to a trapping process of minority carriers, are given. In this model, the f^{-1} law comes from a distribution of lorentzian spectra. The physical origin of the distribution is the linear distribution of minority carriers injected within the base. The model satisfactorily explains the experimental observations.

1. INTRODUCTION

Over the last twenty years field effect controlled electronic devices have been used to demonstrate that surface regions generate f^{-1} current noise. During this period, various authors assumed that f^{-1} current noise came from current fluctuations within bulk regions of these devices. In most cases, up to now, the experimental proofs offered to support this thesis is debatable. Indeed, in one port devices, it is practically impossible to separate surface and bulk contributions. In two-port devices, this impossibility can be overcome in the case of surface effects because the latter can be easily modulated or amplified electrically. Conversely, means to make the bulk contributions preponderant are not known. It is therefore difficult to separate them from other noise sources and their existence cannot be easily proved. Consequently, the existence of f^{-1} bulk current noise is a postulate rather than a well established experimental fact. From the theoretical point of view, the validity of this postulate conditions the validity of all models based on the assumption of current fluctuations.

In what follows, it is demonstrated almost certainly that f^{-1} current noise exists within the bulk of pnp integrated transistors. Subsequently it is shown that trapping of minority carriers could be the cause of this noise.

2. EXISTENCE OF F^{-1} CURRENT NOISE IN THE BULK OF PNP INTEGRATED TRANSISTORS

Generally it is possible to represent any noise process in an electronic device by a distributed model containing elemental voltage and current noise generators. In bipolar transistors, following the procedure developed in [1] , the distributed model can be reduced, at low frequencies, to the lumped model given in Figure 1.

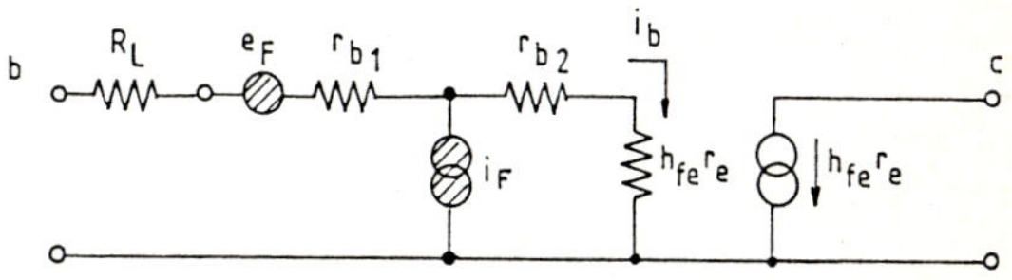

Figure 1 - Low frequency noise equivalent circuit in the case where surface effects and transverse injection at the emitter base junction are negligible.

In this model, resistor R_L simulates the inactive base region situated between the edge of the emitter base junction and the base contact. The intrinsic base resistance associated to the active region situated under the emitter is equal to $r_{b1} + r_{b2}$. It follows from this that the total base resistance is equal to $r_b = R_L + r_{b1} + r_{b2}$. To simplify it was assumed that injection in the lateral base and surface effects at the periphery of the EB junction were negligible. In the opposite case, they can be included in the previous model by following the procedure described in [2].

What are experimentally measurable, are the input equivalent generators e_N and i_N shown in Figure 2.

*Present address : CIT ALCATEL, Centre de Villarceaux, La Ville des Bois - 91000 (F)

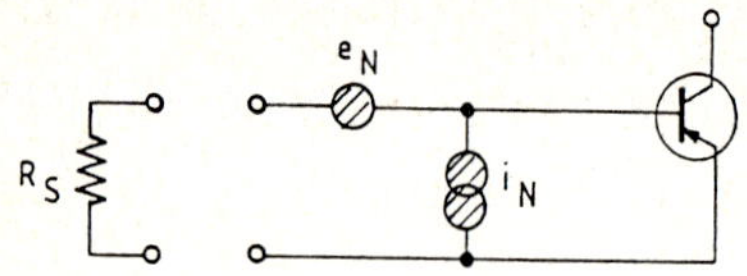

Figure 2 - Representation of noise by two equivalent noise generators at the input. Resistor R_S is placed at the input to measure noise figure.

Equivalence of Figure 1 and 2, yields :

$$e_N = e_F + (R_L + r_{b1})\, i_F \tag{1}$$

$$i_N = i_F \tag{2}$$

and therefore

$$\overline{e_N^2} = \overline{e_F^2} + (R_L + r_{b1})^2\, \overline{i_F^2} + 2(R_L + r_{b1})\, \overline{e_F\, i_F} \tag{3}$$

$$\overline{i_N^2} = \overline{i_F^2} \tag{4}$$

$$\overline{e_N i_N} = (R_L + r_{b1})\, \overline{i_F^2} + \overline{e_F i_F} \tag{5}$$

Practically, $\overline{e_N^2}$, $\overline{i_N^2}$ and $\overline{e_N i_N}$ are obtained from noise figure measurements. By definition, the noise figure F of the transistor connected to a resistor R_S producing thermal noise is equal to

$$F = 1 + \frac{\overline{e_N^2} + R_S^2\, \overline{i_N^2} + 2R_S\, \overline{e_N i_N}}{4\, k\, T_o\, R_S\, \Delta f} \tag{6}$$

where k is the Boltzmann constant, T_o = 290 K and Δf an elemental frequency bandwith.

Theoretically to obtain $\overline{e_N^2}$, $\overline{i_N^2}$ and $\overline{e_N i_N}$, three measurements of F for three distinct values of R_S have to be carrier out. Experimentally, to have an acceptable accuracy, about ten measurements have to be effected. Then, numerical regression procedure must be used to estimate $\overline{e_N^2}$, $\overline{i_N^2}$ and $\overline{e_N i_N}$.

Following such a procedure the accuracy obtained for $\overline{e_N^2}$ and $\overline{i_N^2}$ is approximately equal to 5% while it is of the order of 25% for $\overline{e_N i_N}$.

On the present case, measurements were carried out in integrated pnp transistors manufactured according to the usual process for analog integrated circuits* [3]. Figures 3 and 4 show an example of variations of equivalent noise generators as a function of frequency.

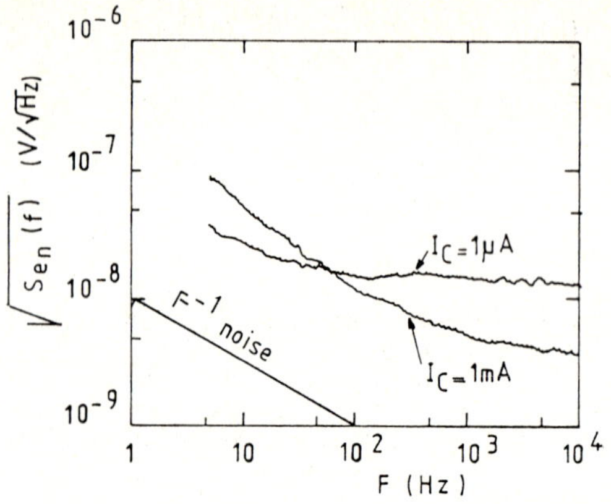

Figure 3 - Spectral Density of the input equivalent voltage noise generator. Experimental Conditions V_{CE} = 4 V, T = 20°C and R_S = 0 Ω

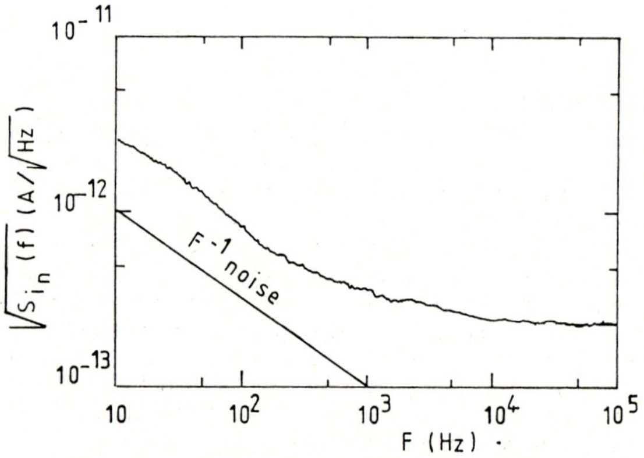

Figure 4 - Spectral density of the input equivalent current noise generator. Measurement conditions : V_{CE} = 4 V, T = 20°C, R_S = 1 MΩ, I_C = 10 μ A.

At low frequencies f^{-1} noise dominates. At higher frequencies, noise is white. A complementary study showed that the white noise behaved as expected by existing theories [4], [5] and [1]. Consequently, in what follows only f^{-1} noise is considered.

With the aim of following up the presentation of the experimental results and to facilitate their analysis, two resistances r_o and r_c were defined as :

$$r_o = \overline{e_N^2} \,/\, \overline{i_N^2} \tag{7}$$

$$r_c = \overline{e_N i_N} / \overline{i_N^2} \tag{8}$$

Substituting (3) to (5) into (7) and (8) yields:

*The transistors were manufactured by SESCOSEM, St Egrève (F). We would like to thank Mr. M. ROCHE for having both suggested the study and supplied the transistors.

$$r_o = \left[(R_L + r_{b1})^2 + \overline{e_F^2}/\overline{i_F^2} + 2(R_L + r_{b1})\overline{e_F i_F}/\overline{i_F^2} \right]^{1/2} \quad (9)$$

$$r_c = R_L + r_{b1} + \overline{e_F i_F}/\overline{i_F^2} \quad (10)$$

Figure 5 gives a typical example of results obtained for r_o and r_c as a function of the collector current.

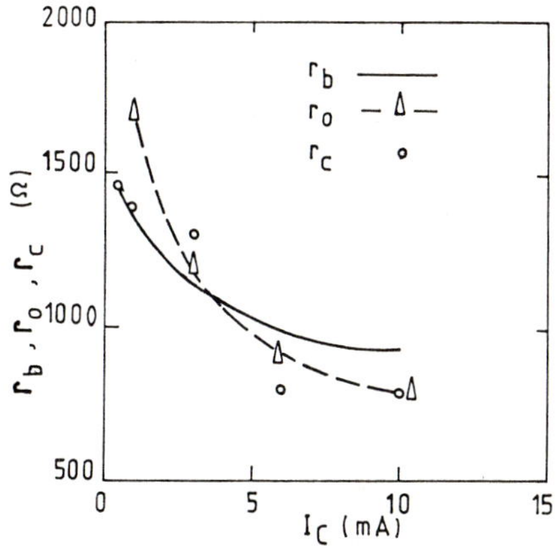

Figure 5 - Base resistance, optimal source resistance and correlation resistance as a function of collector current. Measurement conditions V_{CE} = 4V, T = 20°C, f = 10 Hz

As a comparison, the total base resistance r_b was calculated from the theoretical model given in [1] and drawn in Figure 5.

It is seen that r_b decreases when I_c increases. This behaviour is caused by crowding effects [1] In short, simplified theories of the transistor assume that the base potential, the emitter base voltage and the emitter current injected within the base are constant. This approximation is only valid as long as the product of the de base resistance by the base current does not exceed a few millivolts. For higher values, the base current flow is accompanied by a non-negligible voltage drop in the base. The emitter base voltage, and therefore the emitter current density injected within the base, are not uniform. They are higher at the emitter periphery than at its center. Subsequently, by increasing biasing conditions the passage from a uniform injection regime to a highly non-uniform regime can be effected. This phenomenon is accompanied, as can be seen in Figure 5, by a hyperbolic decrease in base resistance.

Moreover, Figure 5 shows that the values of r_c are very spread out. A numerical estimation indicates that the average error in r_c is about 30% while it is 10% for r_o. Bearing this in mind, it is seen that r_o is very close to r_c.

An examination of (9) and (10) indicates that if $r_o = r_c$, $e_F = 0$. According to [1] this implies that the distributed voltage noise sources, and, current noise sources at the collector base junction, do not exist. Owing to this, f^{-1} noise observed in pnp transistors is essentially current noise produced at the EB junction. In addition, r_o and r_c are very close to r_b. The f^{-1} current noise is thus generated in the intrinsic region of the transistors. In other words, f^{-1} current noise exists within bulk regions. Finally because crowding effects cause a decrease in r_o and r_c, the local value of this noise depends on the local density of the emitter base current injected in the base. Figure 6 and 7 shows current noise as a function of collector current and emitter area when crowding is negligible. It is seen that the noise is approximately proportional to the collector current and independent of the emitter area.

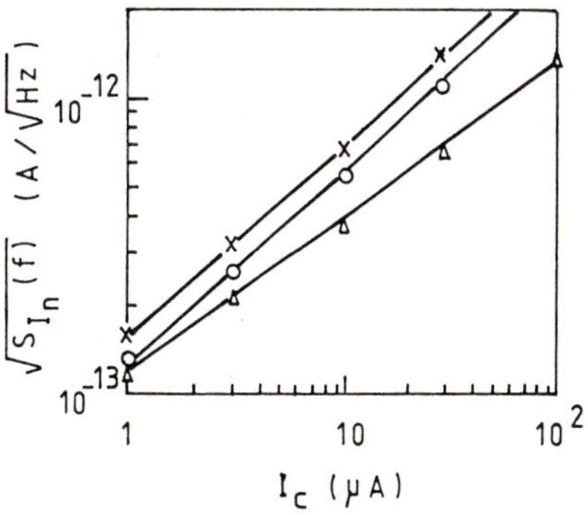

Figure 6 - F^{-1} input current noise vs collector current in integrated pnp transistors. Measurement conditions : V_{CE} = 4 V, T = 20°C, f = 10 Hz.

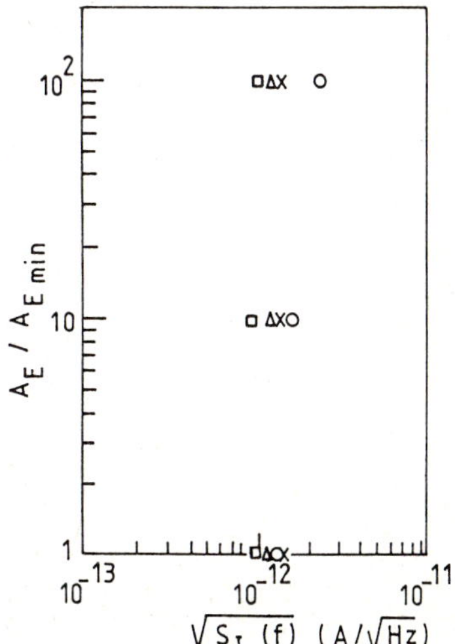

Figure 7 - F^{-1} input current noise vs. emitter area. Measurement conditions: V_{CE} = 4 V, T = 20°C, f = 10 Hz, I_c = 10 μA

3. CURRENT NOISE DUE TO MINORITY CARRIERS TRAPPING

Consider an abrupt short diode of the p+n type as shown in Figure 8

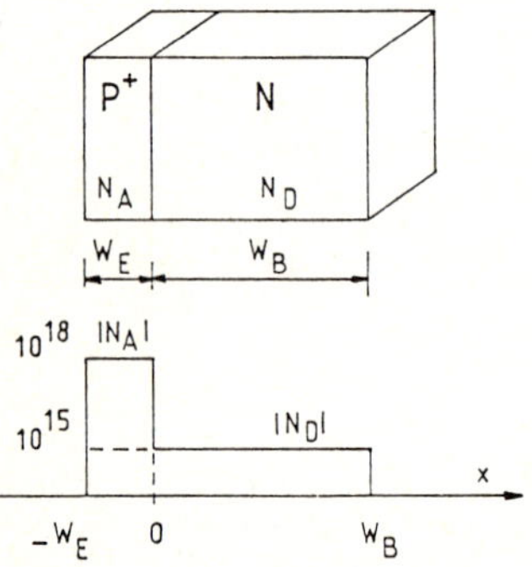

Figure 8 - Illustration of a short diode corresponding to the emitter base junction of integrated pnp transistors.

At low level biasing conditions the current flowing through the diode is a hole current and the hole distribution in the n region is practically linear. The traps situated within the n region modulate the local density $p(x)$ of the injected holes. Consequently, the current fluctuates and generates current noise. In the simplest case these traps are of the SHOCKLEY-READ-HALL type with a concentration N_t, a capture cross section C_p and an energy level E_t. Fundamentally trapping noise is of the Lorentzian type with a time constant τ_t equal to :

$$\tau_t^{-1} = C_p\ (p + p_1) \qquad (11)$$

where $p_1 = n_i \exp\ (E_i - E_t)/kT$,

n_i is the intrinsic concentration (10^{10} cm^{-3}) and E_i the intrinsic level (midgap). However since $p(x)$ is a decreasing function of x, τ_t is rather distributed than constant. By noting P the total number of holes within the base, the spectral density of P has to be written :

$$S_p(f) = \int_0^{W_B} 4N_t f_t (1-f_t) \frac{\tau_t}{1+\omega^2\tau_t^2} A_E dx \qquad (12)$$

where $f_t = \dfrac{p_1}{p+p_1}$, A_E is the junction area.

Tacking into account that :

$$p(x) \approx p(o) \left[1 - x/W_B\right] \qquad (13)$$

the calculation of $S_p(f)$ leads to :

$$S_p(f) = \frac{A_E\ W_B\ N_t\ p_1}{p(o)} \qquad \frac{1}{f} \qquad (14)$$

for $f_{min} < f < f_{max}$

where $f_{min} \approx \dfrac{C_p\ p_1}{2\pi}$ and $f_{max} \approx \dfrac{C_p\ p(o)}{2\pi}$

Expression (14) shows that the noise generated by this mechanism can be of the f^{-1} type. In order to make clear the frequency bandwidth in which this law can be observed, the case of the EB junction of pnp transistors is now considered. Typically $N_D \approx 10^{15}$ cm^{-3} and $W_B \approx 10$ µm. At low level injection,

$$p(o) = \frac{n_i^2}{N_D} \exp \frac{q\ V_{EB}}{kT}$$

At $V_{EB} \approx 480$ mV, $p(o) \approx 10^{13}$ cm^{-3}.

Assuming: $2\pi \times 10^{-12} < C_p < 2\pi \times 10^{-9}$ cm^3 s^{-1} and 10^{+7} cm^{-3} $< p_1 < 10^{10}$ cm^{-3}, the conditions associated to (14) give

$$1 < f_{max}/f_{min} < 10^9 , \quad 10^{-5} \text{ Hz} < f_{min} < 10 \text{ Hz}$$

These ranges of values are relatively spread out and consistent with the frequency bandwidth is which f^{-1} noise is usually measured.

If there are m independent trap levels, $S_p(f)$ must be written

$$S_p(f) = \frac{A_E\ W_B}{p(o)\ f} \sum_{i=1}^{m} N_{ti}\ p_{1i} \qquad (15)$$

The calculation of the spectral density of current fluctuations will be given elsewhere. The result is in full agreement with data shown in Figures 6 and 7

REFERENCES

1. G. Blasquez, J. Caminade and K. Van Vliet, An Accurate analysis of noise in rectangular bipolar transistors including current crowding, Solid State Electronics, 23, (1980) 423 - 431.

2. J. Martin, G. Blasquez et J. Caminade Une nouvelle représentation par schéma équivalent du bruit de fond des transistors bipolaires. In Acta of the International Conference "Le Bruit de Fond des composants actifs semiconducteurs", Toulouse (1971). Copy available at CNRS, Paris.

3. P.R. Gray, R.G. Meyer, Analysis and design of analog integrated circuits (J. Wiley, New-York, 1977).

4. A. Van der Ziel, Noise in junction transistors, Proc. IRE 46 (1958), 1019 - 1038.

5. K.M. Van Vliet, General transport theory of noise in pn junction like devices, Solid State Electronics, 15 (1972), 1033 - 1053.

NOISE IN PHYSICAL SYSTEMS AND 1/f NOISE
M. Savelli, G. Lecoy and J-P. Nougier (eds.)
© *Elsevier Science Publishers B.V., 1983*

LOW FREQUENCY NOISE DUE TO EMITTER-EDGE DISLOCATIONS IN NPN TRANSISTORS

M. Mihaila*, K. Amberiadis**, and A. van der Ziel

Department of Electrical Engineering
University of Minnesota
Minneapolis, Minnesota 55455

Noise measurements were conducted on npn bipolar transistors with different number of emitter-edge dislocations created via different phosphorus surface concentrations. It was found that the emitter-edge dislocations give rise to both 1/f and g-r noise. Moreover the resulting noise increases with the number of dislocations; the more dislocations, the more noise.

1. INTRODUCTION

Stojadinovic[1] and Mihaila[2,3] have shown that the low frequency noise of Si bipolar npn transistors is affected only by the emitter-edge dislocations and that it was of the 1/f type, increasing sharply with increasing density of emitter-edge dislocations.

In a further paper by Mihaila and Amberiadis[4] it was demonstrated that a threshold in both the noise figure and the dislocation density occurs for phosphorus surface concentration greater than 4.3×10^{20} cm^{-3} and that the dislocations affect the noise considerably when their density is in excess of 10^6 cm^{-2}. It was also shown that the ratios of the created bulk g-r centers were in very good agreement with the ratios of dislocations. This gave the evidence that dislocations can be a source of g-r noise.

To find out the exact shape of the noise spectrum we measured the noise equivalent current ($\overline{i_n^2}$) at the base of npn bipolar transistors with different numbers of dislocations; the results are reported in the next section.

2. EXPERIMENT - DISCUSSION

Usually, the correlation between noise-dislocations is established comparing the results of noise measurements of the encapsulated devices with the etch patterns of control wafers. Relating to this methodology some objections can be raised, one of which being the fact that the measured transistors could not be affected by the same type of defects as the control wafers. To remove this possible drawback a direct observation of the defects on the measured devices was made in the following way: five npn bipolar transistor batches with phosphorus surface concentration ranging from $(3.5-10) \times 10^{20}$ cm^{-3} were manufactured on n-type, 4-6 ohm-cm, (111) Czochralski silicon wafers. In order to reduce the effect of metallic contaminants, all the wafers had undergone a preoxidation gettering with Si_3N_4 on the back-side, in N_2 + 1% O_2, at 1100°C, for 3 hrs. Different phosphorus surface concentrations have been realized using different *in situ* oxidation times (prior to phosphorus deposition) and a single phosphorus bubble rate.

After the noise measurements a number of transistors from each group were decapsulated and fixed on a teflon support, thus being easy to handle. Then the Al-contacts were removed in orthophosphoric acid. In this way it was easier to remove the gold wires. After the Al-contact and the oxide removal, the transistor chips were Sirtl-etched and inspected by interference contrast microscopy and SEM. Emitter-edge dislocations have been generated by the phosphorus diffusion and they were by far the dominant defect observed.

Figure 1 shows the values of the equivalent noise current of three decapsulated transistors as a function of frequency. Comparing the etch patterns of the transistors with the corresponding noise levels, a direct correlation between noise-dislocations can be established. It is obvious that the transistor with many dislocations has an order of magnitude greater noise current than the other two transistors which have only a few dislocations. The noise has a $1/f^{\alpha}$ shape at low frequencies, with α slightly larger than unity, and a clear generation-recombination knee at higher frequencies. Both the 1/f noise and the g-r noise increase sharply with increasing number of dislocations.

The result is not surprising for it is known that dislocations are a source of noise[5,6,7]; the more dislocations, the more noise. The effect is more significant for (n-type) phosphorus emitters, because the solubility of phosphorus in silicon is very high. At high phosphorus concentration there is considerable lattice stress, which results in a large number of

---

* Permanent address: R&D Center for Semiconductors, str. Erou Iansu Nicolae 32B, Bucharest, 72996, Romania.

** Now at RCA Laboratories, Princeton, NJ 08540.

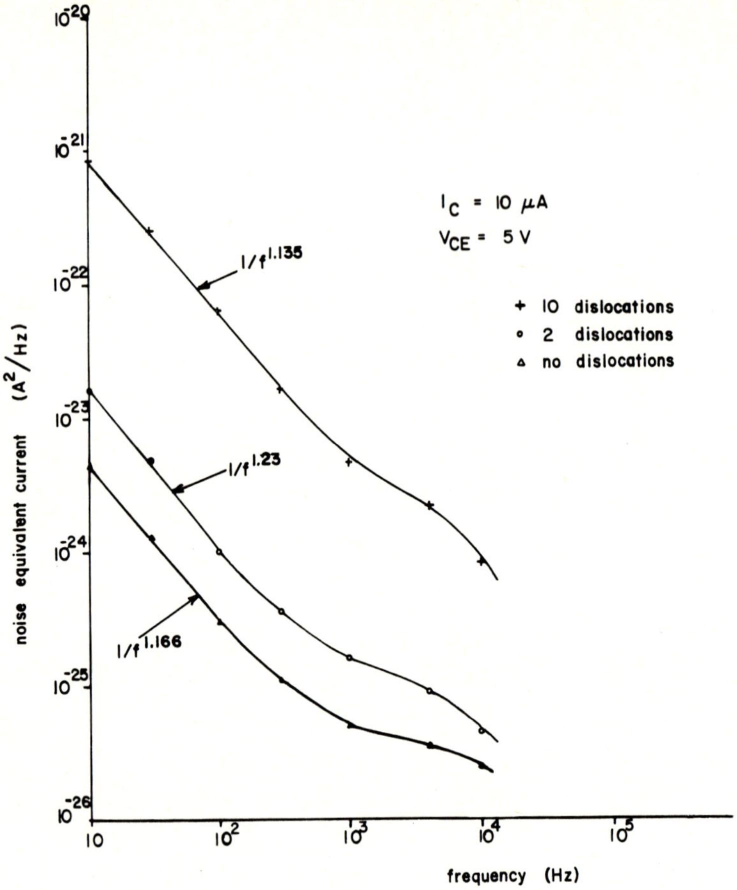

Figure 1 : Input base noise equivalent current ($\overline{i_n^2}$) of bipolar transistors with different number of dislocations as a function of frequency.

dislocations. The solubility of p-type impurities into p-type emitters is much smaller and hence the noise effect is not very significant.

It is also of interest to note that emitter-edge dislocations induce bulk g-r centers as well as surface states in the base region of npn transistors[4,7,8]. The g-r centers give rise to g-r noise and since the surface states accept and emit electrons, the 1/f noise can be surface number fluctuation noise. We mention here that the surface states and bulk centers have fluctuating charge, which can produce enhanced scattering fluctuations, and therefore in principle can give rise to mobility fluctuation 1/f noise.

In conclusion the emitter-edge dislocations of bipolar transistors produce both g-r and 1/f noise. It remains to be seen whether the 1/f noise is of the number or of the mobility fluctuation type. Detailed studies are currently conducted in this area.

REFERENCES

[1] N. Stojadinovic, Electron. Lett., 15, 340 (1979).
[2] M. Mihaila, EEA-Autom and Electron, 24, 49 (1980).
[3] M. Mihaila, Rev. Roum. Phys., 23, 975 (1981).
[4] M. Mihaila and K. Amberiadis, Solid-St. Electron., 26, 109 (1982).
[5] S. R. Morrison, Phys. Rev., 99, 1904 (1955); Phys. Rev., 104, 619 (1956).
[6] D. Green and A. G. Jordan, Int. J. Electron., 27, 159 (1969).
[7] M. Nishida, IEEE Trans., E-D 20, 221 (1973).
[8] M. Nishida, Jap. J. Appl. Phys., 11, 673 (1972).

1/f NOISE IN
FIELD EFFECT TRANSISTORS

NOISE IN PHYSICAL SYSTEMS AND 1/f NOISE
M. Savelli, G. Lecoy and J-P. Nougier (eds.)

© Elsevier Science Publishers B.V., 1983

1/f NOISE IN MODULATION-DOPED FIELD EFFECT TRANSISTORS

K. H. Duh, A. van der Ziel, A. Peczalski

Electrical Engineering Department
University of Minnesota
Minneapolis, Minnesota 55455 U.S.A.

H. Morkoc

Electrical Engineering Department
Univerisity of Illinois
Urbana, Illinois 61801 U.S.A.

Low frequency low noise measurements are reported in modulation-doped GaAs field effect transistors. The noise spectrum is 1/f and is relatively large. At a given frequency the equivalent saturated diode current varies as the square of the applied voltage V_d, as expected for a fluctuating resistor, and saturates when the characteristic saturates. Measurements on ungated devices show that the noise at room temperature is generated in the depletion region under the gate and is caused by the fluctuating occupancy of traps in the region; by removing those traps a very significant lowering of the 1/f noise might become feasible.

We report here on low-frequency noise measurements in "normally on" modulation-doped FETs (MODFETs or HEMTs)[1-3]. The noise was found to be of the 1/f type and was relatively large. The devices had a 1 μm gate length and a 4 μm distance between source and drain.

Figure 1 shows the equivalent saturated diode current, I_{eq}, versus frequency for V_d = 0.1 V, V_g = -0.2 V. We evaluated Hooge's parameter α and found a value of about $3x10^{-4}$, comparable to relatively noise MOSFETs. It is highly doubtful, however, that this parameter has physical signficance in this case (see below).

Figure 2 shows I_{eq} versus the drain voltage, V_d, at a frequency of 400 Hz. The noise varies as V_d at low V_d and saturates when V_d saturates. This agrees with what is found for MOSFETs; it is expected for any fluctuating resistor.

As said before, the 1/f noise resembles that of a MOSFET. The top n^+ -AlGaAs layer under the gate is depleted of electrons and electrons are on the gate side of the GaAs layer. In analogy with the theory of 1/f noise in MOSFETs one would expect the electrons in the channel to interact with traps in the undoped AlGaAs interface by tunneling. The 1/f noise should therefore be of the number fluctuation type, as in MOSFETs. However, as we shall see below, this is not the full picture, since modulation effects seem to be involved, and the undoped AlGaAs is not the source of the noise.

To elucidate the 1/f noise process further we investigated some structures in which no gate was applied (ungated structures). Figure 3 shows the cross-section of such a structure; not shown is how the source and drain penetrate the surface layers and make contact with the conducting channel in the undoped GaAs layer.

The ungated, normally on, devices were of two forms:

a) The depth of the recess was approximately 300 Å (for a 9 second etch). The n^+ -AlGaAs layer ($n \simeq 10^{18}/cm^3$) under the recess was depleted of electrons, so that conduction only occured in the GaAs channel.

b) The depth of the recess was approximately 190 Å (for a 6 second etch). The part of the n^+ -AlGaAs channel under the recess that was nearest to the undoped AlGaAs layer was not depleted of electrons, so that conduction occured not only in the GaAs channel but also in the undepleted part of the n^+ -region.

Figure 4 shows the (I,V) characteristic of the two devices at room temperature. We see that in case (b) the resistance at low drain bias is about a factor 2 smaller than in case (a), indicating that in case (b) the two conducting paths contribute about equally to the current.

Figure 5 shows noise spectra at room temperature at V_d = 0.1V for cases (a) (9 second etch) and (b) (6 second etch). We note in case (a) that I_{eq} is of the 1/f type, whereas it is of the $1/f^{0.5}$ type in case (b); moreover, below 100 KHz the noise is case (b) is smaller than in case (a). If the interaction of the electrons in the GaAs channel were with traps in the undoped AlGaAs layer, case (b) would have a larger noise than case (a), since in case (b) both the GaAs channel and the AlGaAs n^+ layer should contribute to the noise. We therefore conclude that the GaAs channel itself does not contribute significantly to the low frequency noise and that the noise of the n^+ -AlGaAs layer is of the $1/f^{1/2}$ type.

According to van Vliet and Fassett[4] a low frequency $1/f^{1/2}$ spectrum can arise in a one-dimensional diffusion of particles; there should be a transition to a $1/f^{3/2}$ spectrum at high frequencies that need not to be observed in all cases. An ungated MODFET is not a one-dimensional device, but Peczalski et al[5] have suggested that one-dimensional diffusion could occur if the particles (in their case ions) move along dislocation lines. It is well-known that n^+ -AlGaAs is full of electron traps, so that the occurence of ions trapped in dislocation lines and moving with an activation energy E_{ai} is not far-fetched.

That electron taps are signifcant in MODFET operation follows from the fact that many MODFETs show hysteresis loops in the (I_d, V_d) characteristics that can be removed by shining light on the device, or by using a small load resistance.

If the GaAs channel and the undoped AlGaAs layer do not contribute directly to the low-frequency noise in case (a), where then does the large amount of 1/f noise come from? The only source seems to be the depleted region under the recess. Due to the large number of traps present in the AlGaAs $-n^+$ layer, one would expect a fluctuating occupancy of the traps, fluctuating in a 1/f fashion, in the depleted region. This, in turn, modulates the conducting channel in the GaAs layer in a 1/f fashion. The mechanism is also present in case (b) but there the undepleted n^+ -AlGaAs layer shields the channel in the GaAs layer. One might object that the fluctuaing occupancy of traps in the depleted n^+ -AlGaAs layer would also modulate the width of the undepleted n^+ -AlGaAs layer. However, because that layer is so heavily doped, the modulation effect might be small.

We do not know at present why the occupancy of traps fluctuates in a 1/f fashion. Normally fluctuating occupancy of traps produces a series of generation-recombination noise spectra; at a given temperature only traps near the Fermi level contribute significantly to the noise. A uniform distribution of traps in a resistor usually gives two g-r spectra, one involving transitions between conductance band and Fermi-level and another involving transition between valence band and Fermi level[6]. But it may be that a uniform distribution of traps in a depletion layer behaves differently. At any rate, the experiments somehow require that the depletion layer under the gate is involved and this points to the electron traps in that region

Since the GaAs channel itself does not seem to be noisy, a considerable lowering of the low-frequency noise should come about if the deposited n^+ -AlGaAs layer could be made free of traps. This would make the MODFET operate as a low-noise device at much lower frequencies.

That the picture is even more complicated follows from Fig. 6. Here I_{eq} is shown for ungated devices as a function of frequency for V_d = 0.1 V and T = 77°K. We now see that case (a) and case (b) have the same amount of noise. This can be understood if the n^+ -AlGaAs layer and the depletion layer under the gate do not contribute because of freeze-out. The noise now comes from the channel in the GaAs layer itself. We presently do not know the interpretation of the complicated spectrum.

References:

[1] D. Delagebeaudeuf and N. T. Linh, IEEE Trans. El. Dev., ED-29, 955, 1982.

[2] T. Mimura, S. Hujamizu, K. Joshen and K. Hikosaka, Jap. J. Appl. Phys., 20, L 317, 1982.

[3] T. J. Drummond, W. Kopp, R. E. Thorne, R. Fister and H. Morkoc, Appl. Phys. Lett., 40, 879, 1982.

[4] K. M. van Vliet and J. R. Fassett, Fluctuation Phenomena in Solids (Ed. R. Burgess), Acad. Press Inc., New York, 1965, Chapter 7.

[5] A. Peczalski, A. van der Ziel and R. Zuleeg, Solid State Electronics, 26, 1983, in press.

[6] K. Lee, K. Amberiadis, and A. van der Ziel, Solid State Electronics, 25, 999, 1982.

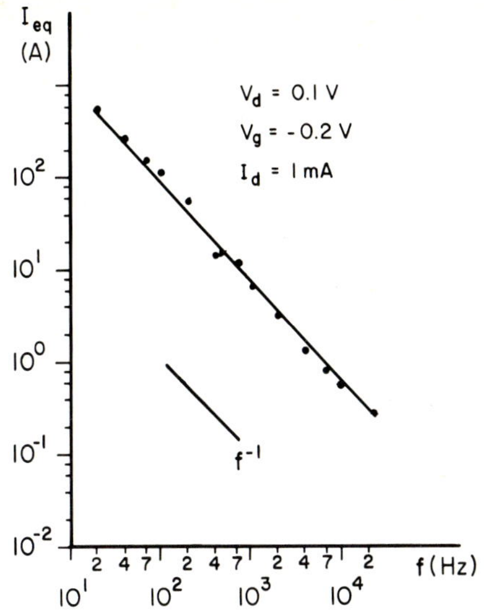

Fig. 1. I_{eq} versus frequency for GaAs MODFET $V_d = 0.1V$, $V_g = -0.2V$, $I_d = 1mA$

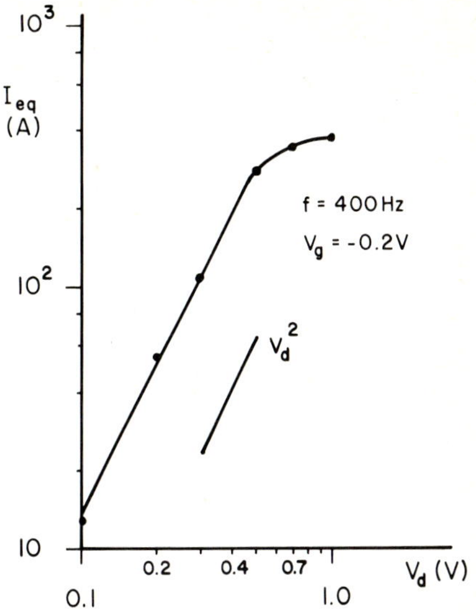

Fig. 2. Ieq versus drain voltage V_d for GaAs MODFET, f = 400Hz, $V_g = -0.2V$

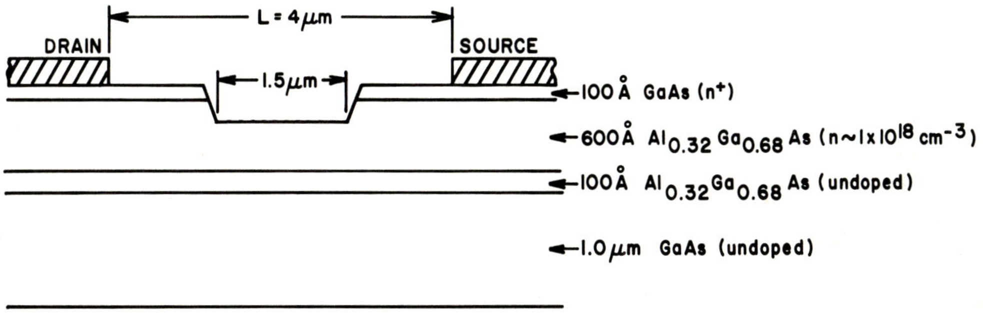

Fig. 3. Cross-section of ungated device

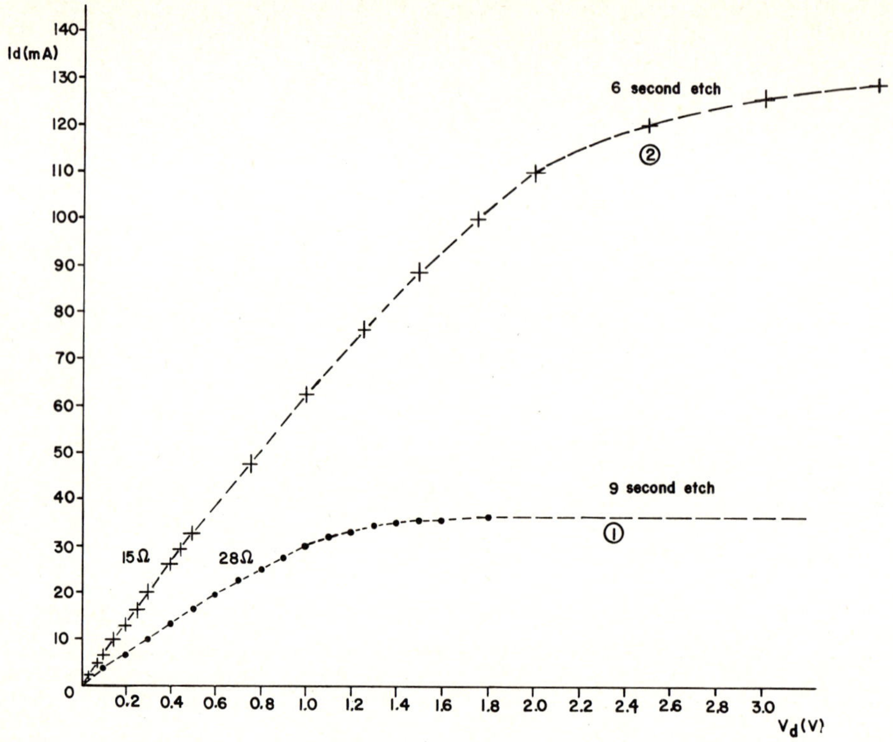

Fig. 4. I,V characteristic for ungated devices with 6 sec and 9 sec etch

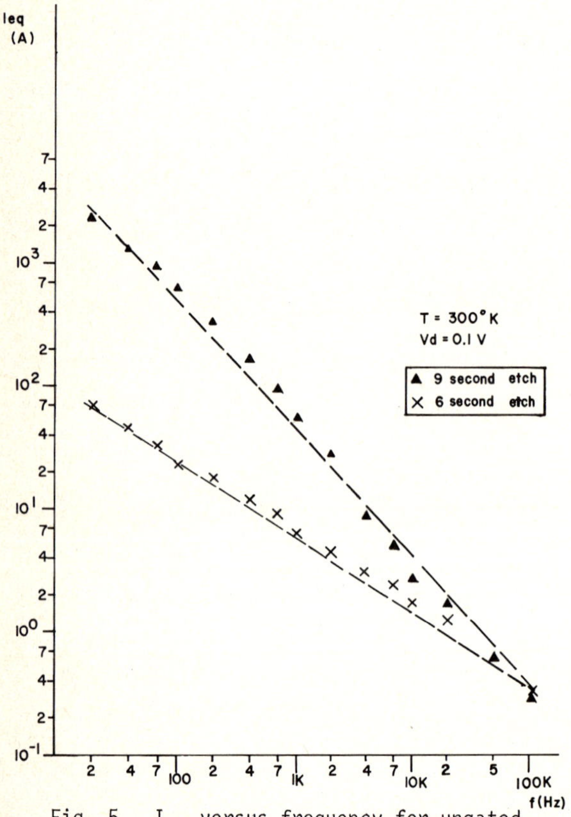

Fig. 5. I_{eq} versus frequency for ungated devices at 300°K and V_d = 0.1V

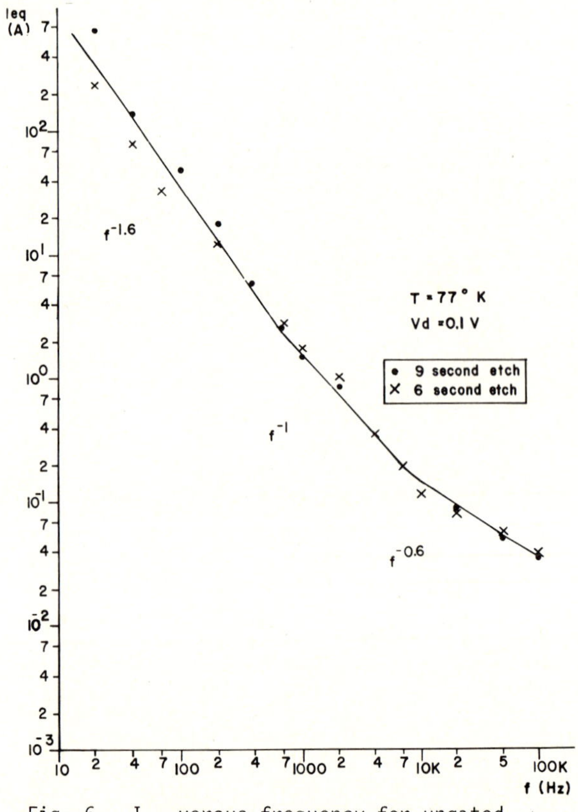

Fig. 6. I_{eq} versus frequency for ungated devices at 77°K and V_d = 0.1V

NOISE IN PHYSICAL SYSTEMS AND 1/f NOISE
M. Savelli, G. Lecoy and J-P. Nougier (eds.)
© Elsevier Science Publishers B.V., 1983

ENERGY LEVELS OF BULK DEFECTS RESPONSIBLE FOR L.F. NOISE IN Si JFETs

K. Kandiah

United Kingdom Atomic Energy Authority,
Harwell, Didcot, Oxon, U.K.

1. INTRODUCTION

It has been well established that low frequency noise in JFETs, in excess of the thermal noise of the channel, is mainly caused by charge fluctuations at defects in the bulk of the semiconductor. The physical model of the mechanism proposed by Sah(1) deals with the noise caused by midband defects at temperatures above about 250K. A model which covers the general case of a defect with an arbitrary energy level at any temperature has been described more recently (2).

We have made extensive measurements, on standard devices from a number of sources, of noise as a function of frequency in the range 10 Hz to 100 kHz and temperatures from 77K to 350K. Energy levels and capture cross-sections have been calculated from the noise measurements. The relevance of deep level transient spectroscopy (DLTS) as an alternative method for the identification of defects in JFETs is considered. Some defect types which commonly occur in many devices have been tentatively identified and the reasons for ambiguity when using noise measurements for this purpose are discussed.

2. CHARGE STATE TRANSITIONS AT POINT DEFECTS

The characteristic capture time t_1 and emission time t_2 of carriers at a point defect are given by

$$t_1 = 1/nVC \qquad (1)$$

$$t_2 = (1/NVC)\exp(E/kT) \qquad (2)$$

where V is the mean thermal velocity and n the density of free carriers, C is the capture cross-section of the defect, N is the density of states in the appropriate band, E is the depth of the defect level from the band edge, T is the temperature and k is Boltzmann's constant. In n-channel devices at temperatures below about 250K electron capture and emission at defects in the Debye region adjacent to the channel are the only processes leading to transitions of charge state and in p-channel devices capture and emission of holes are the only processes involved. At temperatures well above 250K transitions of charge state at defects can also occur by alternate emission of electrons and holes in the entire depletion region between the channel and the gate. We shall only discuss n-channel devices in the rest of the paper.

3. MAGNITUDE OF DRAIN CURRENT MODULATION CAUSED BY CHARGE TRAPPING AND EMISSION

Let us consider an acceptor type of defect i.e. one which is negative when in the charged state. The trapping or emission of an electron at a defect in or adjacent to the channel can be deemed to modulate the drain current in two ways. The loss of a conduction electron from the channel due to trapping will cause a reduction of the channel current for the duration of the charged state of the defect. Alternatively or additionally the field due to the fixed negative charge at the defect can be regarded as a barrier impeding the flow of free carriers around it resulting in a reduction in the drain current. The drain current waveform will resemble an asymmetric random telegraph signal (RTS) with dwell times of t_1 and t_2 in the high and low states respectively.

The amplitude ΔI_d of the drain current RTS has not been derived accurately but a crude analysis gives (3)

$$\Delta I_d = qu/l \qquad (3)$$

where q is the electronic charge, u is the free carrier drift velocity in the channel near the defect and l is the length of the conducting channel. It may be noted that ΔI_d is independent of gate width. Taking a channel length of 4 μm and considering a defect near the pinched down region where the carriers will have the saturation velocity of 10^7 cm/sec we obtain a figure of 4×10^{-9}A for ΔI_d. The maximum measured value of about 5×10^{-9}A is in remarkably close agreement with this prediction.

Using 4-terminal JFETs we have investigated the effect of changes in t_1 by varying the relative gate bias at constant temperature (therefore fixed t_2) and found that the value of ΔI_d does not change by more than a few % over the whole range of possible observations - t_1 ranging from about 100 μsec to over 1 sec. This result is interesting because it suggests that the drain current modulation is due to a field effect as it is known to be for defects in the depletion region.

4. NOISE SPECTRA

The spectrum of the drain current noise due to a single defect will be almost flat up to a corner frequency f where

$$2\pi f = 1/t_1 + 1/t_2 \qquad (4)$$

and fall as f^{-2} at higher frequencies. At a fixed temperature, when then is more than one defect of the same type in a number of positions in the Debye region we have to consider different values of t_1 for each defect depending on the value of n around it. The assumption of a single characteristic time constant for each type of defect for calculating the noise spectrum in JFETs is therefore not valid except for the case of defects occupying positions in the depletion range.

When there is a high density of more than one type of defect the spectrum will have multiple corner frequencies and varying rates of fall off as frequently reported in the literature.

5. EXPERIMENTAL DATA

In our measurements the drain current and voltage are kept constant while the temperature is varied and the noise is measured at a number of spot frequencies. The equipment maintains sample temperature to within 1K of any desired value between 77K and 350K and performs a direct calibration of noise at each measurement. The standard deviation for the measurement can be specified in advance and is limited solely by the available time. It is usually 3.5% at 10 Hz and better than 1% at frequencies greater than 300 Hz.

A plot of noise of a 3-terminal JFET as a function of temperature when the device contains only one type of defect is shown in Fig. 1. A few n-channel devices exhibit such simple characteristics but most devices contain many types of defects and Fig. 2 is more representative of the results obtained on devices with gate widths ranging from 0.3 mm to 10 mm. Fig. 3 shows an example of the noise characteristics of a JFET at low temperatures.

6. CHARACTERISTICS AT VERY LOW DEFECT DENSITIES

The temperature T_0 at the peaks and the magnitudes of the peaks in these plots depend on t_1, t_2 and f_0, the centre frequency of the filter, as well as on the distribution of the defects in relation to the Debye region when more than one defect is active in generating noise. We have found it convenient to categorise the defects according to the temperature at which the noise is a maximum in a 10 Hz filter. One or more of four types of defect occur in a large number of devices and the peaks due to these are seen in one or other of the above plots; the temperatures for the types 1 to 4 are approximately 85K, 99K, 170K and 256K respectively.

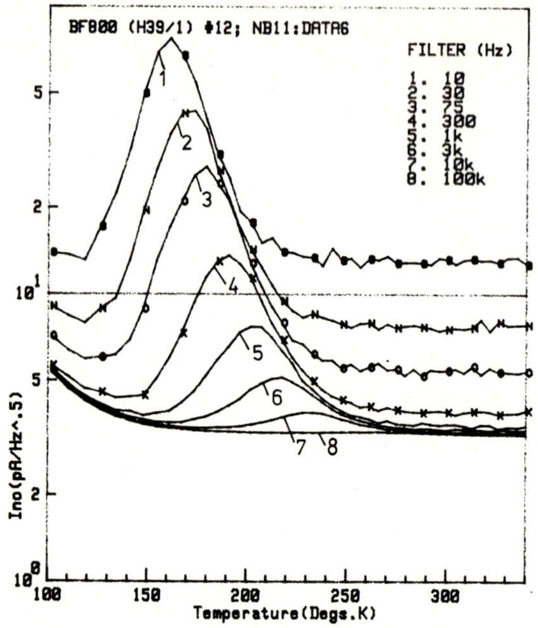

Figure 1: Current noise of JFET showing only one defect type

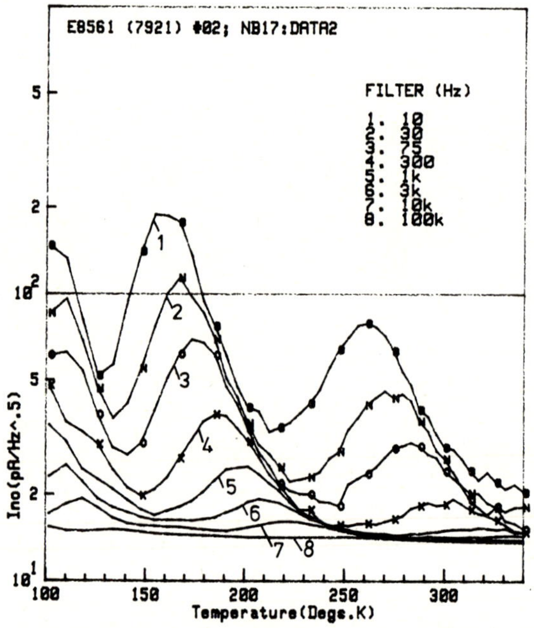

Figure 2: Current noise of JFET showing 4 types of defect

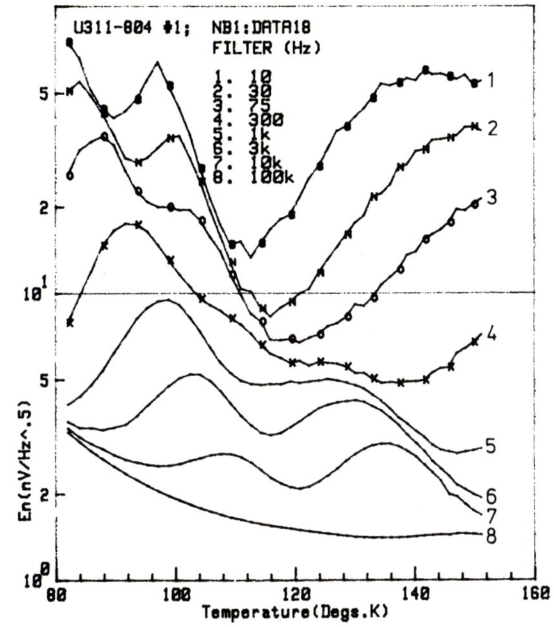

Figure 3: Typical behaviour at low temperature

Table 1

Frequency f_0 (Hz)	Temperature T_0 (Degs. K)
10	170.6
30	179.8
75	187
300	199
1K	212.5
3K	224.3
10K	239.3

In the simple case of a 4-terminal device in which only one defect is active and when the gate bias is also adjusted to maximise the noise at each filter frequency the peak occurs when $t_1=t_2=1/\pi f_\theta$ where the filter is assumed to have a narrow bandwidth. Table 1 gives the measured values of T_0 at each of seven frequencies for a single defect, of type 3, in a 4-terminal device.

Using equation (2) it is possible to calculate E and C from the values of T_0 at any 2 frequencies from data of the type shown in table 1 assuming that NV varies as T^2. In table 2 we present the calculated values of E from all the combinations of pairs of frequencies in table 1. We find that E=.32 eV and C=3×10^{-15} cm² for this defect.

Table 2
Frequencies in Hz and Energy Levels in eV

f_2 (Second Frequency)	30	75	300	1K	3K	10K
f_1 (First Frequency)						
10	.29	.31	.31	.32	.32	.32
30		.34	.34	.32	.33	.33
75			.34	.31	.32	.33
300				.29	.31	.32
1K					.35	.34
3K						.33

7. SOURCES OF ERROR AT VERY LOW DEFECT DENSITIES

An error of .5 deg in determining T_0 at f_1= 10 Hz would have resulted in an error of 10% at an energy level of .3 eV when f_2=30 Hz but the error decreases as f_2/f_1 increases. This sensitivity of the calculation to small errors in the temperatures is seen in the scatter of the results in table 2 at the lower frequencies because of the greater statistical error in the noise measurements. The corresponding values of the capture cross-sections vary smoothly between 1 and 5×10^{-15} if we ignore the unfavourable situations mentioned above.

The values of T_0 in table 1 were obtained with the relative gate bias optimised at each frequency. If the chosen bias was only .1V away from the optimum there could be an error in T_0 of a few degs. at the lower frequencies leading to an error in the derived energy level as high as 30%. This increased sensitivity to bias at the lower frequencies is a result of the Gaussian decay of the free carrier density n as a function of the distance from the conducting channel. An important consequence of this finding is that measurements of T_0 on 3-terminal JFETs will be expected to give different results depending on whether constant gate bias or constant drain current was maintained as the temperature was varied.

8. ERRORS AT HIGHER DEFECT DENSITIES

When the defect density is very high we need to consider a distribution of values of n from nearly zero to n_d where n_d is the doping level in the channel resulting in an equally wide range of values of t_1. Since the noise corner frequency will be determined by the smaller of the two times t_1 and t_2 it will be largely dependent on E and the temperature and will not be sensitive to bias. However the value of T_0 at each filter frequency will no longer coincide with that of the single defect case. The measurements indicate that E, at high defect densities, is usually lower than that for single defects of the same type.

For medium defect densities, when a few defects are active in the Debye region at or near the pinched down region of the channel, the spectrum at any temperature will be strongly dependent on drain current and bias. We have found large variations in the derived energy levels of 3-terminal devices of medium gate widths as a function of drain current due presumably to the variation in channel position as the temperature is varied.

9. DEFECT TYPES

Unambiguous identification of defect types is not expected from every determination of energy level from noise measurements. In the case of bias optimised measurements on single defects on a few 4-terminal devices with very low defect densities the derived energy levels are not consistent from one defect to another of the same type even in the same device. The suggestion by Lang et al (4) that some deep levels in semiconductors are the result of the association of an impurity with a dopant atom or other complexes may be relevant. These defects will show polarisation effects and be sensitive to the large differences in the electric field, carrier density and mean drift velocity between the various locations of the defect in a JFET.

The DLTS method has the advantage that the electric field is uniform over most of the measured volume, although somewhat higher than in our JFET noise measurements. However the sensitivity of this method is generally assumed to be poor for defect densities less than 10^{-4} of the dopant density. This is only marginally adequate for the estimated defect density in average devices. It has been reported (5) that current transient spectroscopy may provide adequate sensitivity for these levels of defect densities.

Energy levels of the 4 common types of defects are .19 eV, .22 eV, .33 eV and .53 eV in the majority of the measurements. In 3-terminal devices the third type of defect often gives a value of .29 eV and it has not been established whether this is a different type of defect or whether the different value of E is due to the problems outlined earlier.

It is exceptional for the .19 eV level to be absent (as in Fig. 1) and in general the density of the type 1 and type 2 defects is so high that interference between them leads to considerable scatter in the 2 values of E in any one device. We have not found any significant changes in behaviour at these low temperatures as a function of drain voltage and deduce that hot carrier effects are not present. It is seen that in examples like that in Fig. 1 there is an increase in the 100 kHz noise below about 120K which is attributable to carrier density fluctuations or generation-recombination at donors depending on the method of interpretation.

The common occurrence of three of the types of defects in devices from a variety of manufacturers would suggest that their nature may be related to basic materials or process characteristics. The type 1 defect could be the well known complex O-V A-centre which has a level of E_c-0.18 eV. Although the main sources of noise are defects in the Debye region next to the channel some noise will be generated by hole emission and capture at defects in the Debye region next to the gates at very high defect densities. The possibility of the type 1 defect being the E_v+0.19 eV level of Si interstitials should also be considered. The commonest defect in all n-channel JFETs is type 3 which is probably the Si interstitial with an energy level of E_c-0.34 eV. The type 4 defect energy level agrees well with the published figures for the acceptor type of Au impurity.

In a few cases we have seen other defect levels close to the 4th type of defect but not well enough resolved.

10. ACKNOWLEDGEMENTS

The author is indebted to colleagues M.O. Deighton, F.B. Whiting and C.E. Cox for valuable assistance and discussions during the preparation of the paper.

REFERENCES

(1) C.T. Sah, Proc. IEEE, 52, pp. 795-814, July 1964.

(2) K. Kandiah, M.O. Deighton and F.B. Whiting, in Proc. 6th Conf. on Noise in Physical Systems, NBS Publication 614, U.S. Department of Commerce, 1981.

(3) M.O. Deighton, Private communication.

(4) D.V. Lang- H.G. Grimmeiss- E. Meijer and M. Jaros, Phys. Rev. B, Vol. 22, p.3917, 1980.

(5) J.A. Borsuk and R.M. Swanson, IEEE Trans. Electr. Dev., Vol. ED27, p.2217 Dec. 1980.

NOISE IN PHYSICAL SYSTEMS AND 1/f NOISE
M. Savelli, G. Lecoy and J-P. Nougier (eds.)

© *Elsevier Science Publishers B.V., 1983*

THERMAL AND 1/f NOISE AT WEAK INVERSION AND LIMITING 1/f NOISE IN MOSFETs

K. H. Duh and A. van der Ziel

Electrical Engineering Department
University of Minnesota
Minneapolis, Minnesota 55455, U.S.A.

It is found experimentally that the equivalent saturated diode current I_{eq} for p-channel MOSFETs at weak inversion and at saturation is equal to the saturated drain current I_{sat}; this is explained theoretically. It is also found that the noise resistance R_{nf} of the 1/f noise at saturation is independent of gate bias; this too is explained theoretically. The independence of R_{nf} on bias holds also if the noise is due to dielectric losses in the oxide; however, this leads to a value of the loss tangent, tanδ, that is about a factor 10 larger than expected.

MOSFETs at low inversion have the advantages of very low 1/f noise, very high input impedance and high g_{max}/I_{sat} ratio, which will be useful as amplifiers at low frequencies in high impedance sensor applications.

Consider a MOSFET at weak inversion. The conductance $g(V_0)$ for unit length, according to van Overstraeten et al [1], is

$$g(V_0) = \frac{\mu w C_d}{\beta} \exp\left[\frac{\beta}{n^*}(V_g - V_g^*) - 1/2\beta\phi_F - \beta V_0\right] \quad (1)$$

where $\beta = q/kT$, ϕ_F the Fermi potential difference and

$$V_g^* = V_{FB} + \frac{3}{2}\phi_F + Q_b^*/C_{ox} \quad (1a)$$

$$n^* = (C_{ox} + C_d)/C_{ox} \quad (1b)$$

Here V_{FB} is the flatband voltage, Q_b^* the depletion change per unit area at the surface potential $\psi_s = 1.5\ \phi_F$, C_{ox} and C_d are the oxide and the depletion capacitances per unit area, respectivley. C_d is a weak function of applied voltage V_0 through the surface potential ψ_s and its value at $\psi_s = 1.5\phi_F$ is denoted by C_d; usually one puts $C_d = C_d$ throughout. The above expression is correct as long as the effect of the surface density N_{ss} can be neglected. Consequently

$$I_d = \frac{1}{L}\int_0^{V_d} g(V_0)dV_0 = I_{sat}[1-\exp(-\beta V_d)] \quad (2)$$

where

$$I_{sat} = \frac{\mu w C_d}{L\beta} \exp\left[\frac{\beta}{n^*}(V_g - V_g^*) - 1/2\beta\phi_F\right] \quad (2a)$$

is the saturated current, which is the I_d-value for $\beta V_d > 4$. Then the transconductance is

$$g_m = \frac{\partial I_d}{\partial V_g} = \frac{\beta I_d}{n^*} = \frac{qI_d}{kT}\left(\frac{C_{ox}}{C_{ox} + C_d}\right) \quad (3)$$

so that transconduction at saturation is

$$g_{max} = \frac{qI_{sat}}{kT}\left(\frac{C_{ox}}{C_{ox} + C_d}\right) \quad (3a)$$

Furthermore the drain noise spectrum is [2]

$$S_{I_d}(f) = \frac{4kT}{L^2 I_d}\int_0^{V_d} g^2(V_0)dV_0 = 2qI_{sat}[1+\exp(-\beta V_d)] \quad (4)$$

as is found after some manipulations. At saturation the high-frequency noise thus corresponds to shot noise of I_{sat}, even though the noise is, in fact, thermal noise.

Figure 1 shows I_{eq} versus frequency in the weakly inverted region in a p-type channel. It indicates that the flicker noise corner frequencies are much lower than in the strongly inverted region.

We can also express the high-frequency noise by an equivalent saturated diode current I_{eq}. Then

$$2qI_{eq} = 2qI_{sat}[1+\exp(-\beta V_d)],$$

or

$$I_{eq} = I_d \frac{1+\exp(-\beta V_d)}{1-\exp(-\beta V_d)} \quad (5)$$

so that $I_{eq} = I_{sat}$ at saturation. Figure 2 shows that this is indeed the case at 10 KHz at saturation for a device with a p-type channel.

Experimentally one finds that at weak inversion the flicker noise resistance R_{nf} of a p-channel MOSFET does not depend on V_g at saturation (Fig. 3). Since

$$4kTR_{nf} = \frac{[S_{I_d}(f)]_{sat}}{g_{max}^2} \quad (6)$$

and g_{max} is proportional to I_{sat}, if follows that $[S_{I_d}(f)]_{sat}$ varies as I_{sat}^2.

We can also derive this result from Park et al's calculation [3]. According to them

$$S_{I_d}(f) = \frac{q^2 w\mu^2}{fL^2} \int_0^L \left(\frac{dV}{dx}\right)^2 \frac{N_T(E_f)_{eff}}{\varepsilon} dx \quad (7)$$

where $N_T(E_f)_{eff}/\varepsilon$ is the effective trap density at the Fermi level, and ε is a tunneling parameter; $N_T(E_f)_{eff}/\varepsilon$ must be determined experimentally. At strong inversion one finds $N_T(E_f)_{eff}/\varepsilon = C(V_g-V_T-V_0)$ where C is constant. At low inversion equation (7) gives a meaningful result only if

$$\frac{N_T(E_f)_{eff}}{\varepsilon} = C\exp\{2[\frac{\beta}{n^*}(V_g-V_g^*)-1/2\beta\phi_F-\beta V_0]\}$$

$$= C\left(\frac{\beta}{\mu w C_d}\right)^2 g^2(V_0) \quad (8)$$

where C is a constant. Substitution into (9) yields

$$S_{I_d}(f) = \frac{q^2 w\mu^2}{fL} C \left(\frac{C_{ox}+C_d}{C_{ox}C_d}\right)^2 g_m^2 \quad (9)$$

$$= \frac{q^2}{flW} C\left(\frac{C_{ox}+C_d}{C_{ox}C_d}\right)^2 g_m^2 \quad (10)$$

which is independent of V_g and V_d. The experimental dependence of $I_{eq}(f)$ in the flicker noise regime is plotted versus I_d^2 for a given V_g in Fig. 4; it gives a straight line, as expected.

The independence of R_{nf} upon V_g and V_d would also result if the 1/f noise were caused by dielectric losses in the oxide [4]. Let the device have a capacitance C_{gs}, and let the oxide have an effective loss tangent, tan δ, then the equivalent noise resistance due to these losses is

$$R_{nf} = \frac{\tan\delta}{\omega C_{gs}} \quad (11)$$

From Fig. 3 we have at f = 100 Hz, R_n = 4MΩ, whereas C_{gs} = 0.47pF. This yields tan $\delta = 10^{-3}$, or

$$R_{nf} = \frac{3.7\times10^8}{f} \text{ ohms} \quad (11a)$$

The value of tan δ is about a factor 10 larger than expected, so that the interpretation does not seem to be adequate here. It is worthwhile to note, however, that the limiting 1/f noise given by (11) should remain even if one removes all the interface traps in the oxide. We conclude that present devices are within a factor 10 of the theoretical limit.

REFERENCES:

[1] R. J. van Overstraeten, C. J. Declerk and P. A. Muls, IEEE Trans. Electron Devices, ED-22, 282, 1975.

[2] G. Reimbold and P. Gentil, Sol. State Electron., 26, 1983.

[3] H. D. Park et al, J. Appl. Phys., 52, 296, 1981.

[4] A. van der Ziel, Solid State Electron., 18, 1031, 1975.

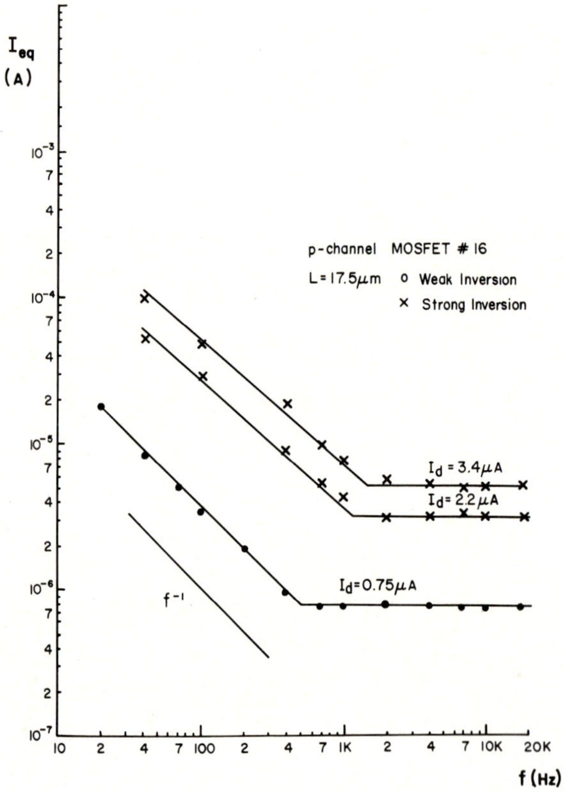

Fig. 1. Thermal noise and 1/f noise in p-channel MOSFET at saturation; I_{eq} versus frequency

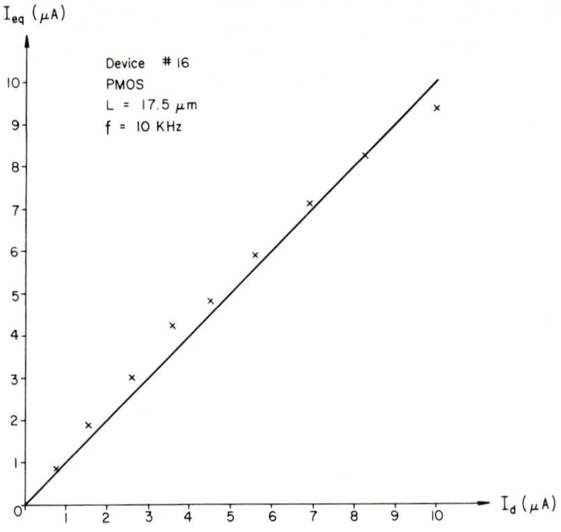

Fig. 2. Thermal noise at 10kHz (shot noise regime), showing $I_{eq} = I_d$ at saturation

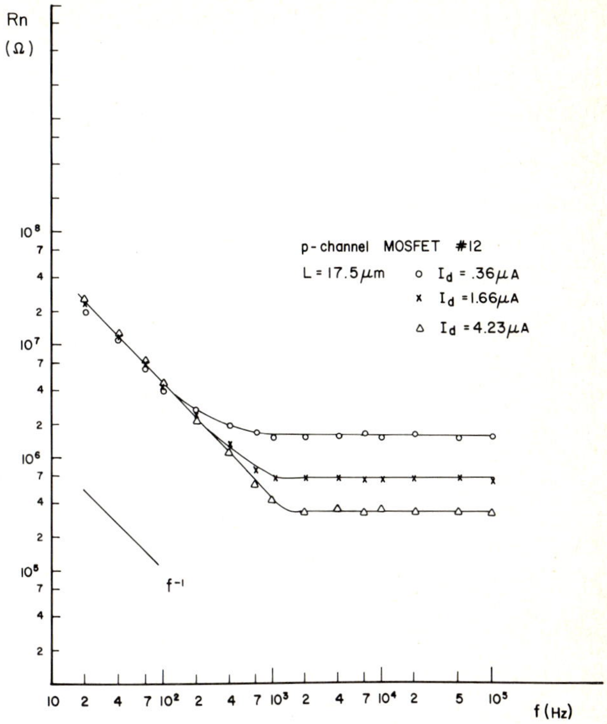

Fig. 3. Noise resistance R_n versus frequency in p-channel MOSFET at saturation

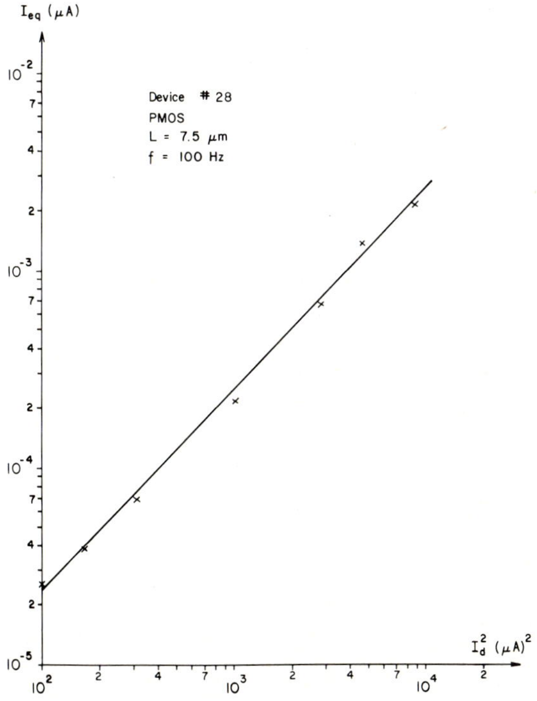

Fig. 4. I_{eq} versus Id^2 in flicker noise regime (100 Hz) showing I_{eq} varies as Id^2. Non-saturated device

NOISE IN PHYSICAL SYSTEMS AND 1/f NOISE
M. Savelli, G. Lecoy and J-P. Nougier (eds.)
© Elsevier Science Publishers B.V., 1983

1/F NOISE IN MOS TRANSISTORS BIASED FROM WEAK TO STRONG INVERSION

G. REIMBOLD, P. GENTIL and A. CHOVET

Lab. de Physique des Composants à Semiconducteurs (ERA CNRS N° 659)
Institut National Polytechnique de Grenoble, ENSERG, 23, rue des Martyrs
38031 - Grenoble Cedex - France

Under specific conditions, a plateau is observed in the variations of the relative spectral density of the drain current fluctuations S_{I_D}/I_D^2 versus the gate voltage V_G, in weak inversion, followed by a steep decrease in strong inversion. Based on the Mc Whorter trapping theory and using an exact description of charge fluctuations, a 1/f noise model valid in these regimes is derived. This model shows a very good agreement with the variations of all the investigated parameters : biases, temperature, geometries and technologies.

1. INTRODUCTION

In recent years, many works have been devoted to the 1/f noise of MOSFET's. However, most of them dealt with transistors operating in strong inversion and therefore with high drain currents. In this paper experimental results on low frequency 1/f noise in MOS transistors (*) are reported, for the whole range of the gate bias V_G from weak inversion regime to strong inversion regime.

Measured values of S_{I_D}/I_D^2 versus V_G at low drain voltage (I_D : drain current ; S_{I_D} : power spectral density of its fluctuations) exhibit a plateau in weak inversion followed by a steep decrease in strong inversion.

Using Mc Whorter's model and taking into account all the capacitive components of the small signal equivalent circuit, a new theoretical expression of S_{I_D}/I_D^2 at low drain bias is derived, which explains its dependence on V_G.
This expression is then extended to high drain bias in weak inversion.

To confirm the validity of this model, noise measurements are carried out on various n-channel Si gate MOS transistors. Different geometries and technologies are studied at room temperature and liquid nitrogen temperature.

The influence of the oxide capacitance C_{ox} and depletion capacitance C_D is then experimentally evidenced and the important role of the fast surface trap density N_{ss} on the noise characteristic S_{I_D}/I_D^2, is shown, especially at low temperature or for bad quality devices. Under these conditions, special peculiarities in S_{I_D}/I_D^2 curves are correlated to the static characteristic $I_D(V_G)$.

2. THEORY

As proposed in previous papers, the theoretical model assumes that 1/f noise arises from tunneling transitions between oxide traps located a few angstroms far from the interface between the MOST channel and the silicon dioxide.

Each local fluctuation of the number of occupied traps δn_t produces a fluctuation of the number of free carriers in the channel δn, giving elementary conductance and current density fluctuations δg and δI, the spectral densities of which are related by [1] :

$$S_{\delta I}(f) = \frac{I_D^2}{L^2Z^2}\frac{S_{\delta g}(f)}{g^2} = \frac{I_D^2}{L^2Z^2}\frac{S_{\delta n}(f)}{n^2} = \frac{I_D^2}{L^2Z^2}\frac{S_{\delta n_t}(f)}{n^2}\left(\frac{\delta n}{\delta n_t}\right)^2 \quad (1)$$

L and Z being respectively the channel length and width.
As shown by Jindal and Van der Ziel [2], the ratio $R = -\delta n/\delta n_t$ strongly varies with the bias, from weak inversion (typically $R \simeq 10^{-10}$) to strong inversion (R = 1). However, we can derive a more appropriate expression of R leading to an easy formulation and interpretation of the noise spectrum. When the trapped charge Q_t fluctuates, the charge conservation in the structure is such that :

$$\delta Q_G + \delta Q_{ss} + \delta Q_D + \delta Q_n + \delta Q_t = 0$$

where δQ_G, δQ_{ss}, δQ_D, δQ_n (respectively gate, fast surface states, depletion zone and channel charge fluctuations) can be related to the surface potential fluctuations by equations :

$$\delta Q_G = -C_{ox}\delta\phi_s \;;\; \delta Q_{ss} = -C_{ss}\delta\phi_s \;;\; \delta Q_D = -C_D\delta\phi_s,$$
$$\delta Q_n = -C_n\delta\phi_s$$

where C_{ss} and $C_n = -(q/kT)Q_n = -\beta Q_n$ are respectively surface state and charge channel capacitances per unit area.
It follows that :

$$R = -\frac{\delta Q_n}{\delta Q_t} = -\frac{\delta n}{\delta n_t} = \frac{-\beta Q_n}{C_{ox} + C_D + C_{ss} - \beta Q_n} \quad (2)$$

For weak inversion $|\beta Q_n| \ll C_{ox} + C_D + C_{ss}$ and R can vary over several orders of magnitude.
In strong inversion $|\beta Q_n| \gg C_D + C_{ox} + C_{ss}$ and $R \# 1$.
Using (1) and (2) and integrating all the fluctuations resulting from the trap distribution (supposed to be uniform over the whole channel) leads, through a classical calculation, to the

---
*Devices supplied by Thomson CSF (St-Egrève, France) which supported this work.

following general relation valid in weak and strong inversion at low drain bias :

$$\frac{S_{I_D}}{I_D^2} = \frac{K}{f} \frac{N_t}{(C_{ox}+C_D+C_{ss}-\beta Q_n)^2} \text{ with } K = \frac{q^4\lambda}{LZ} \quad (3)$$

λ being the transition tunneling constant for electrons and N_t ($eV^{-1}cm^{-3}$) the trap density at the Fermi level. For weak inversion $|\beta Q_n| \ll C_{ox} + C_D + C_{ss}$ and

$$\frac{S_{I_D}}{I_D^2} = \frac{K}{f} \frac{N_t}{(C_{ox}+C_D+C_{ss})^2} \quad (4)$$

Since C_D slowly varies with the biases, S_{I_D}/I_D^2 appears to be a constant as long as C_{ss} and N_t variations vs. biases remain small.
For strong inversion $|\beta Q_n| \gg C_{ox}+C_D+C_{ss}$; thus

$$\frac{S_{I_D}}{I_D^2} = \frac{K}{f} \frac{N_t}{\beta^2 Q_n^2} \quad (5)$$

and S_{I_D}/I_D^2 strongly depends on gate bias throught $1/Q_n^2$.

3. EXPERIMENTS

The noise measurement procedure has already been given in a previous paper [3]. The transistors are n-channel, enhancement type ; different technologies were available. Substrates have a doping density of 2.5 $10^{15}cm^{-3}$, the channel width is 100 µm and the channel length varies from 7.5 µm to 100 µm. The oxide thickness is 1 200 Å or 1 700 Å; some transistors have received a complementary boron implantation which modifies the threshold voltage and the depletion capacitance.

3.1. Noise in the linear region

Measured values of S_{I_D}/I_D^2 versus V_G for a transistor operating at low drain bias and T = 300 K are reported in Fig. (1). The curve presents a plateau in weak inversion, i.e. for our measurements at V_D = 20 mV drain currents from 1.8 nA to 125 nA, followed by a steep decrease in strong inversion. We reported on the same figure theoretical variations S_{I_D}/I_D^2 vs. V_G, assuming that N_t is a constant and modelling $Q_n(V_G)$ by taking into account all the different charge contributions in the solution of the Poisson equation. The two curves show a good agreement, particularly around the characteristic value $V_G(2\phi_F)$; the smoothly increasing difference between them in strong inversion can be attributed to an increasing value of N_t when the Fermi level approachs the conduction band, as is generally observed.

Noise dependence on C_{ox} and C_D

The plateau level has been precisely measured on many devices in order to study the noise dependence with C_{ox} and C_D. C_{ss} being negligible at T = 300 K (N_{ss} <2.$10^{10}eV^{-1}cm^{-2}$), eq. (4) reduces to :

$$S_{I_D}/I_D^2 \propto N_t(C_{ox} + C_D)^{-2}$$

Table (1) shows the comparison between theory and experiments for the noise plateau level of transistors with different C_{ox} and C_D, these transistors being available on a same chip. Two different technologies have been studied.

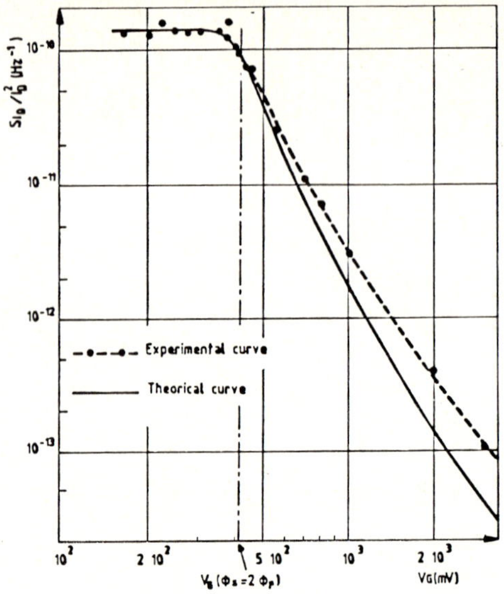

Fig.(1) : $S_{I_D}/I_D^2(V_G)$ in the whole range from weak to strong inversion regime, at low V_D ; T = 300 K ; f = 15 Hz. Transistor from "classical technology", not implanted ; Z/L = 200/15 ; wafer n° 2.

The different capacitances were measured by C(V) dynamic method at low frequency.

Device	C_{ox} (F/m²)	C_D (F/m²)	Measured plateau level S_I/I^2 (Hz^{-1}) at 15Hz	$\frac{(S_I/I^2)\text{not impl.}}{(S_I/I^2)\text{ impl.}}$	$\left[\frac{(C_{ox}+C_D)\text{impl.}}{(C_{ox}+C_D)\text{not impl.}}\right]^2$
"classical Technology", not implanted, Z/L=200/15 Wafer n°11	2.7 10^{-4}	1.93 10^{-4}	1.93 10^{-10}	2.2	2.15
"classical Technology", implanted Z/L=200/15	1.85 10^{-4}	4.98 10^{-4}	8.8 10^{-11}		
"LOCOS Technology", not implanted, Z/L=200/15	2.95 10^{-4}	2.75 10^{-4}	2.6 10^{-10}	2.95	2.36
"LOCOS Technology", implanted, Z/L=200/15	3.2 10^{-4}	5.53 10^{-4}	8.8 10^{-11}		

Table 1

These typical measurements show the fit of noise dependence on oxide and depletion capacitances, which has been observed for all the devices. They also allow to conclude that a small difference in the oxide growth and a complementary implantation of boron through the oxide didn't modify substantially the trap density.

3.2. Noise in the non linear region

When the conduction channel can't be any more considered as uniform, the integration from the source to the drain is given by :

$$S_{I_D} = \frac{q^4\lambda}{L^2 Z} I_D^2 \frac{N_t}{f} \int_0^L \frac{dy}{(C_{ox}+C_D+C_{ss}-\beta Q_n)^2} \quad (6)$$

However, in weak inversion, since $|\beta Q_n| << C_{ox}+C_D$ over the whole channel, assuming C_{ss} is a constant, eq.(6) leads again to eq.(4), which therefore remains valid at any drain bias.
Fig.(2) shows the experimental noise dependence $S_{ID}/I_D^2(V_D)$ in weak inversion for transistor implanted, or not implanted. As predicted by eq.(4) $S_{ID}/I_D^2(V_D)$ is constant in the whole range of drain bias from 10 mV to 10 V.

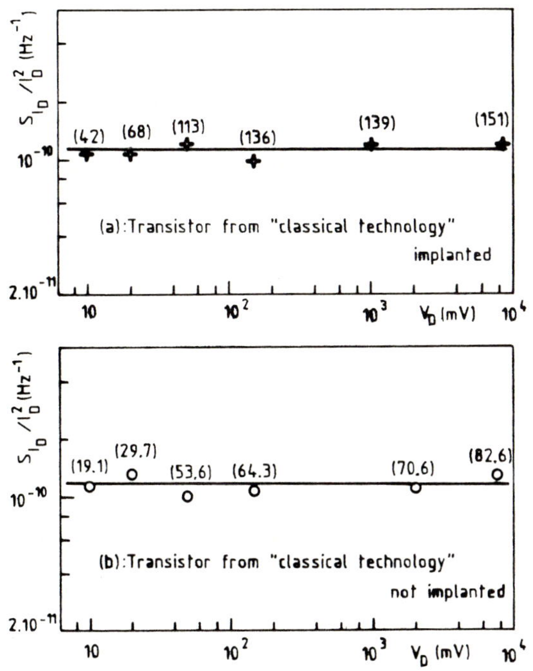

Fig.(2) $S_{ID}/I_D^2(V_D)$ from ohmic to saturation region in weak inversion. (Drain currents are indicated in nA inside brackets). T = 300 K, f = 15 Hz ; Z/L = 200/15.

3.3. Noise at low temperature

We have reported in Fig.(3) and (4) the experimental relative noise $S_{ID}/I_D^2(V_G)$ in weak and moderate inversion at low temperature (77 K) and low V_D for transistors of the two different technologies .Contrary to the results at 300 K, these curves don't present any plateau in weak inversion but a decreasing noise level, the slope of which depends on the device.

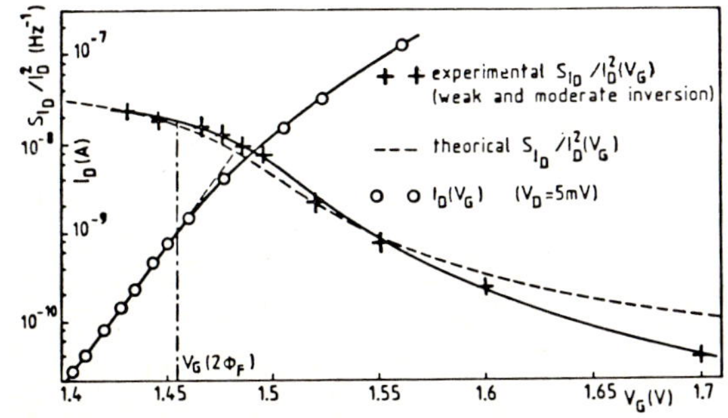

Fig(3) Transistor from "classical technology", not implanted. T = 77 K ; f = 0.2Hz ; Z/L = 200/15; low drain bias.

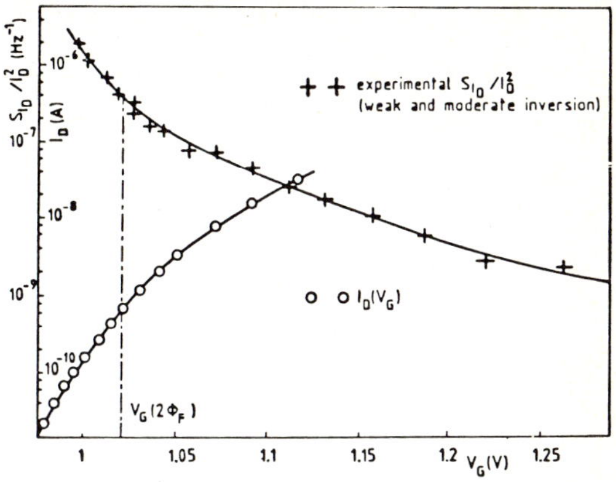

Fig.(4) Transistor from "LOCOS technology", not implanted. T = 77 K ; f = 0.2 Hz ; Z/L = 100/10 ; low drain bias.

For each transistor we have also reported the static $I_D(V_G)$ characteristic in the same range of gate bias. Transistor of Fig.(3), which has a Log(I_D) vs. V_G with a constant slope in weak inversion, presents a smoothly decreasing level of S_{ID}/I_D^2 in weak inversion. Transistor of Fig.(4) which exhibits a Log I_D vs. V_G characteristic without any constant slope presents a rapidly decreasing level of S_{ID}/I_D^2 in weak inversion.

Such typical results have been also observed for implanted transistors. They can be attributed to an increase of the fast interface state density N_{ss} when the Fermi level gets close to the conduction band (i.e. the region investigated at low temperature). This increasing N_{ss} decreases the Log I_D vs. V_G slope through the term $C_{ox}/(C_{ox}+C_D+C_{ss})$ and the noise level S_{ID}/I_D^2 through $1/(C_{ox}+C_D+C_{ss})^2$. Thus the transistor of Fig.(3) which shows a constant Log I_D vs. V_G slope, and also a low value of C_{ss}, must have a smoothly decreasing noise level. On the contrary, the transistor of Fig.(4) which shows greatly increasing values of N_{ss} with V_G exhibits a rapidly decreasing noise level.

A modelling of the noise dependence according to eq.(4), at low temperature has been done for the transistor of Fig.(3) by measuring C_{ss} in weak inversion by the $I_D(V_D)$ method [4]. A good agreement in the decreasing level of S_{ID}/I_D^2 is obtained in weak inversion.

It must be pointed out that such a decreasing $S_{ID}/I_D^2(V_G)$ may also be observed at 300 K for bad quality devices (high N_{ss}) ; this can explain why Aoki et al. [5] didn't get any plateau at 300 K on their transistors with $N_{ss} \sim 3$ to $4.5\ 10^{11} eV^{-1} cm^{-2}$.

Noise variation with V_D

When V_D increases from the ohmic region to the saturation region the channel electron density decreases from the source to the drain, and the pseudo Fermi level E_{F_n} varies along the channel. This behaviour has a same effect as a decrease of V_G at constant V_D and leads (as shown by Fig. (3) and (4)) to an increasing noise level. So we must observe a noise increasing with V_D up to saturation, and a constant level after saturation. Such a behaviour can be observed in Fig.(5). The noise increases rapidly up to saturation, i.e. $V_D \sim$ a few kT/q. For high V_D, we observe a very slowly increasing noise level.

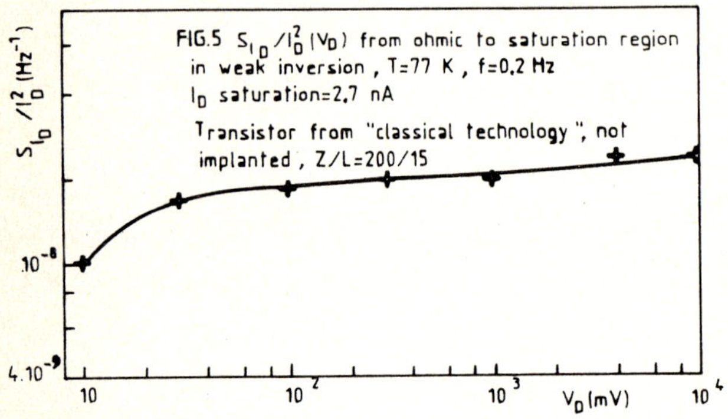

FIG.5 $S_{I_D}/I_D^2(V_D)$ from ohmic to saturation region in weak inversion, T=77 K, f=0.2 Hz
I_D saturation=2.7 nA
Transistor from "classical technology", not implanted, Z/L=200/15

4. CONCLUSION

A theoretical model has been presented which successfully accounts for the different behaviours of the noise function S_{I_D}/I_D^2 versus V_G, and especially the plateau observed in weak inversion at 300 K. The model is consistent with the experimental influences of the main parameters : gate and drain biases, oxide and depletion capacitance, quality of the device (through N_{SS}), and temperature. On the contrary, the mobility fluctuation model (which is usually described by the Hooge's empirical relation $S_{I_D}/I_D^2 = S_G/G^2 = S_{V_D}/V_D^2 = \alpha/Nf$ [6], [7]) seems unable to easily account for the experimental plateau because the number of carriers strongly varies with the gate voltage in the whole range of gate bias.

ACKNOWLEDGEMENTS

The authors thank Mr Munier, Mr Arques, Mr Reboul and Mr Thenoz from the DTE division of Thomson-CSF, St-Egrève, for helpful discussions and for supplying the devices.

REFERENCES

[1] Christensson S., Lundström I. and Svensson G., Low frequency noise in MOS transistor, Solid State Electronics 11 (1968) 797-812.

[2] Jindal R.P. and Van der Ziel A., Carrier fluctuation noise in a MOSFET channel due to traps in the oxide, Solid State Electronics 21 (1978) 901-903.

[3] Reimbold G. and Gentil P., White noise of MOS transistors operating in weak inversion, IEEE Trans. Elect. Dev. ED 29 (1982) 1722-1725.

[4] Van Overstraeten R.J., Declerck G.J. and Muls P.A., Theory of MOS transistor in weak inversion. New method to determine the number of surface states. IEEE Trans. Elect. Dev. ED 22 (1975) 282-288.

[5] Aoki M., Katto H. and Yamada E., Low frequency 1/f noise in MOSFET's at low current levels. J. Appl. Phys. 48 (1977) 5135-5140.

[6] Hooge F.N., 1/f noise, Physica, 83 B (1976) 14-23.

[7] Vandamme L.K.J., Model for 1/f noise in MOS transistors biased in the linear region. Solid State Electronics 23 (1980) 317-323.

NOISE IN PHYSICAL SYSTEMS AND 1/f NOISE
M. Savelli, G. Lecoy and J-P. Nougier (eds.)
© Elsevier Science Publishers B.V., 1983

1/f NOISE IN N-CHANNEL PNOS MEMORY TRANSISTORS

Herman.E. Maes° and Sabir Usmani

ESAT-Laboratory
Katholieke Universiteit Leuven
Kardinaal Mercierlaan 94
B-3030 Heverlee Belgium

(°) Senior Research Associate (Onderzoeksleider) NFWO

1/f noise behaviour in n-channel polysilicon-gate-nitride-oxide-silicon (PNOS) memory transistors is presented for the first time. It is found that the equivalent gate input noise voltage in these n-channel transistors can be very well explained by a recently proposed model developed for p-channel MNOS transistors, based on the carrier density fluctuation phenomenon. The mobility fluctuation model however can not account for all our experimental results. Using a large set of PNOS transistors the experimental 1/f noise dependence on gate voltage, geometry, process conditions, interface states density D_{it} and effect of Write/Erase cycling has been compared with model predictions. Finally, a comparison for 1/f noise in p-channel MNOS, n-channel PNOS and n-MOS is made.

INTRODUCTION

Metal-nitride-oxide-silicon (MNOS) transistors are electrically alterable non-volatile memory devices and have been used as such in a number of commercial products. Whereas most of these parts use a p-channel Al-gate technology(1) recently also n-channel polysilicon gate PNOS memories have been introduced(2). Recently the authors(3) presented an extensive study on the 1/f noise properties of thin-oxide p-channel MNOS transistors which are interesting tools for this study because their interface states density can be gradually increased by appropriate Write/Erase cycling and its effect on the 1/f noise can be readily studied. The validity of the two competing models to account for the 1/f noise in MOSFET's, i.e. the carrier density fluctuation model based on the Mc Whorter theory of tunneling(4) and the mobility fluctuation model based on Hooge's empirical relationship(5) can therefore be checked. It was shown that an adapted model based on the carrier density fluctuation phenomenon could account for all our experimental results.

The present paper presents for the first time experimental data on 1/f noise in n-channel silicon gate PNOS memory transistors. This study had to be done necessarily to the same extent as for p-channel devices since the technology for both types of devices is totally different. The dependence of the measured noise on gate and drain voltages, geometry, Write/Erase cycling, interface states density and memory state of the transistors was checked against the predictions of the two models. Finally the noise in p-channel MNOS, n-channel PNOS and nMOS silicon gate transistors obtained simultaneously in the PNOS process are compared.

THEORY

An expression for the equivalent gate input noise voltage in p-channel MNOS transistors in the saturation regime ($V_{GS}-V_{TH}=V_{DS}$), based on the Mc Whorter theory was derived in detail elsewhere(3) and is approximated here by

$$v_n^2 = \frac{2qkT}{WLC_{ox}^2} N_{CP} \cdot \beta^{*} \int_0^{x_t} \exp(-\frac{x}{\lambda})^n \frac{\tau(x)}{1+\omega^2\tau^2(x)} dx \quad (1)$$

where q is the electron charge, k the Boltzmann constant, T the absolute temperature, W the channel width and L the channel length and C_{ox} the oxide capacitance per unit area. The integration is carried out over a distance x_t within which the oxide traps are distributed following a $\exp(-\frac{x}{\lambda})^n$ law with λ and n constants. This distribution accounts for the slight observed deviations from perfect 1/f noise behaviour. N_{CP} (cm^{-2}) is the charge obtained in a charge pumping measurement (6,7) and is a measure for the average value of the interface states density around midgap. β^{*} finally is a parameter which mainly accounts for two effects, namely the proportionality between interface states and oxide trap distribution and the deviation between the trap densities at the quasi-fermilevel ($N_T(E_F)$) and at midgap ($N_T(E_i)$). This latter effect makes β^{*} also dependent on ($V_{GS}-V_{TH}$) where V_{GS} is the gate voltage and V_{TH} the threshold voltage.

EXPERIMENTAL RESULTS

a. Device technology

The PNOS transistors are fabricated using a double polysilicon gate process. The oxide thickness is 2 nm, the LPCVD silicon nitride thickness 45 nm. Source and drain are obtained by Phosphorus implantation and an hydrogen anneal at

800°C is performed to reduce the interface state density(1). The experimental conditions and procedure are identical to those described in Ref. 3. The interface states density is measured with the charge pumping technique (6,7). The W/E cycling conditions used were ± 28V, 50 usec.

b. Measurements

The 1/f noise in PNOS memory transistors was found to have the same basic characteristics as found for MOS and pMNOS transistors and can be summarized as :

(i) A 1/f noise behaviour is observed in both modes of operation, i.e. linear and saturation mode. Noise in saturation is larger than in the linear region (for the same V_{GS}-V_{TH}).

(ii) 1/f noise is proportional to 1/WL and comparison between experiment and eq(1) is excellent.

(iii) The shift of the threshold voltage in the written or erased states of these transistors has no influence on the 1/f noise.

(iv) Influence of degradation on 1/f noise

It is well known that W/E cycling causes an increase in the surface state density of MNOS transistors. Also the 1/f noise increases considerably with cycling as is shown on figure 1.

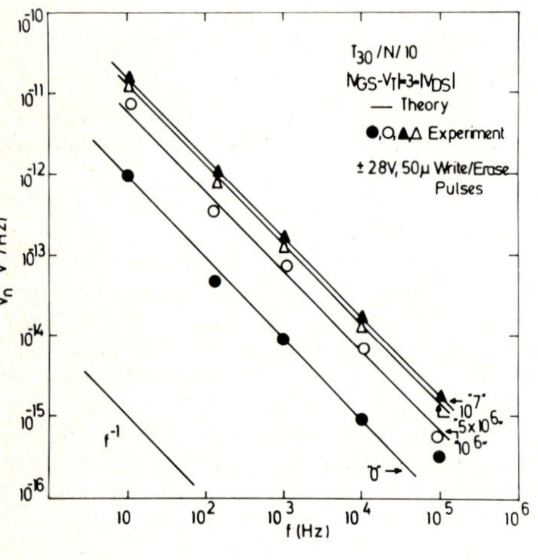

Figure 1 Increase of the 1/f noise in PNOS transistor after cycling

The 1/f noise behaviour is still observed after up to 10^7 cycles, in agreement with our results on p-channel devices but in contradiction with the results of Gentil(8). The experimental results can be explained by eq(1) when using a uniform trap distribution for the uncycled devices (n=0) and the same exponential profile for all the cycled devices (unique set of n, λ and β^*), see fig. 1. However it, can be concluded that after cycling the same physical mechanism remains responsible for the 1/f noise behaviour

(v) 1/f noise vs interface states density

Figure 2 shows the equivalent input noise mea-

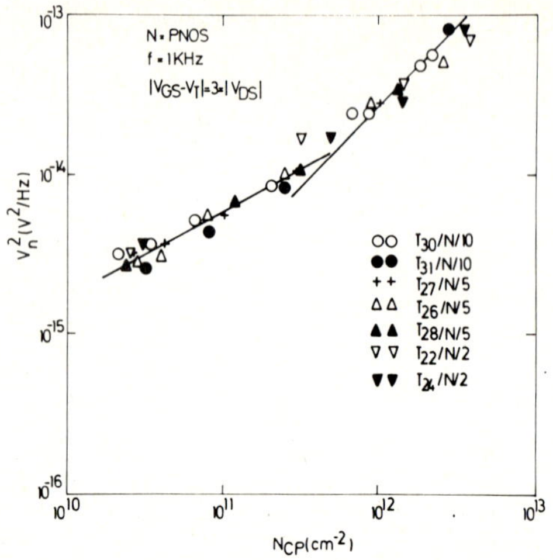

Figure 2 Input noise in saturation mode (V_{GS}-V_{TH}=V_{DS}=3V) versus charge pumping charge

sured in saturation (V_{GS}-V_{TH}=V_{DS}=3V) as a function of the charge pumping charge. It is seen that only for large N_{CP} a linear relationship between both quantities is obtained. For low N_{CP} values however we rather find v_n^2 $N_{cp}^{2/3}$. This is due to the large deviation which can be expected between $N_T(E_F)$ and $N_T(E_i)$ because of the measured U-shaped distribution of the interface traps. This deviation is a function of (V_{GS}-V_{TH}). For large interface states densities however this difference was found to become very small and therefore v_n^2 is also proportional to N_{CP}. In terms of β^*, this value is larger for low interface state densities but decreases and becomes constant for larger interface states densities. Its value however is dependent on (V_{GS}-V_{TH}). These considerations can also explain the dependence of the noise on the measured N_T as observed by Mikoshiba et al (9,10) and does not require a unification of both models as they proposed.

c. Comparison of the two models

Figure 3 shows the input noise in the linear mode and in the saturation mode (V_{GS}-V_{TH}=V_{DS}) as a function of V_{GS}-V_{TH} for an uncycled and a cycled PNOS transistor. The depicted behaviour can not be explained with the original mobility fluctuation model. In the linear regime this model rather predicts a linear increase with V_{GS}-V_{TH}. In the saturation regime a linear relationship is predicted. The large deviation from the linear relationship as seen in figure 3 can not accounted for by the simple mobility fluctuation model. The number fluctuation model can however account for this behaviour(3). A similar behaviour in the linear regime was also observed by Aoki et al(11).

d. Noise in pMNOS, nPNOS and nMOS

Figure 4 shows the 1/f noise spectrum of pMNOS, nPNOS and nMOS transistors. The three upper

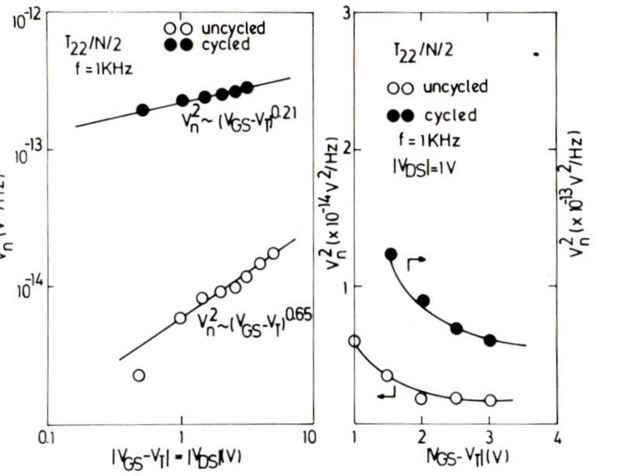

Figure 3 Input noise in saturation mode and linear mode as a function of $(V_{GS}-V_{TH})$ before and after cycling

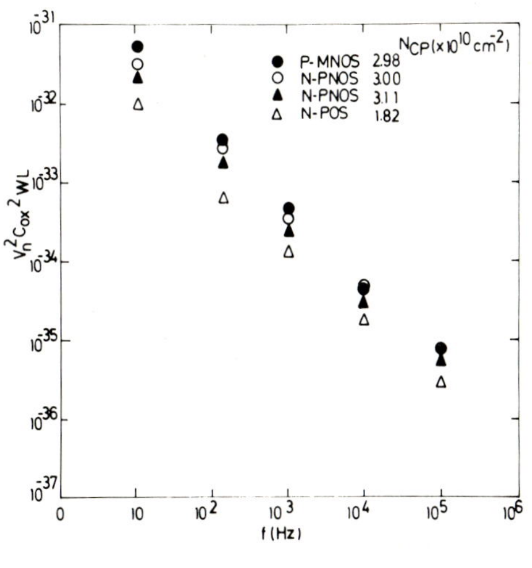

Figure 4 Comparison of 1/f noise in p-channel MNOS, n-channel PNOS and n-channel MOS

lines are for transistors with about the same N_{CP} value. The noise in p-channel and n-channel devices, although having totally different fabrication processes is almost the same. The noise in the n-channel MOS transistors obtained in the same process is lower than that of the memory devices but this can mostly be accounted for by the smaller N_{CP} value. Anyway the noise mechanism is the same in MOS and the double layer memory devices used in this study.

CONCLUSIONS

The 1/f noise behaviour in nPNOS memory transistors can be explained with the carrier density fluctuation model. The W/E cycling causes a degradation of the $Si\text{-}SiO_2$ interface which is predictably accompanied by an increase of the 1/f noise. The difference in the noise in pMNOS, nPNOS and nMOS is mainly determined by the interface states density.

REFERENCES

(1) H.E. Maes and G.L. Heyns, J. Appl. Phys. 51, 2706 (1980)
(2) T. Hagiwara, Y. Yatsuda, R. Kondo, S. Minami, T. Aoto and Y. Itoh, IEEE, J. Solid-St Circuits, SC-15, 346 (1980)
(3) H.E. Maes and S. Usmani, J. Appl. Phys. 54, 1937 (1983)
(4) A.L. McWhorter in Semiconductor Surface Physics ed. by R.H. Kingston (Un. of Pennsylvania, Philadelphia, 1957) p. 207
(5) F.N. Hooge, Phys. B 83, 14 (1976)
(6) G. Groeseneken, H.E. Maes, N. Beltran, R. De Keersmaecker, to be published
(7) H.E. Maes and S. Usmani, J. Appl. Phys. 53, 7106 (1982)
(8) P. Gentil and S. Chaussé, IEEE Trans. El. Devices ED-25, 1042 (1978)
(9) H. Mikoshiba, IEEE Trans. Electron Devices ED-29, 965 (1982)
(10) H. Mikoshiba, M. Sakamoto, Y. Hokari, Proc. IEDM, p. 662 (1982)
(11) M. Aoki, H. Katto and E. Yamada, J. Appl. Phys. 48, 5135 (1977).

NOISE IN PHYSICAL SYSTEMS AND 1/f NOISE
M. Savelli, G. Lecoy and J-P. Nougier (eds.)
© *Elsevier Science Publishers B.V., 1983*

ORIGINS OF 1/F NOISE IN MOS TRANSISTORS

G. BLASQUEZ AND A. BOUKABACHE

LABORATOIRE D'AUTOMATIQUE ET D'ANALYSE DES SYSTEMES
7, avenue du Colonel Roche
31400 TOULOUSE (France)

In the course of the last ten years,1/f noise has been modelled in terms of fluctuations in the number of carriers or in terms of fluctuations in mobility. To test the validity of these types of models, MOS transistors were irradiated whith low energy X-ray to modify oxide trap density. This experiment has shown that the increase in trap density is accompanied by an increase in noise and that radiation induced noise is perfectly described by the number fluctuations model. The experiment, in addition, also revealed a decrease in mobility. The simultaneous observations of the noise increases, and the decrease in mobility make the existence of a mobility fluctuation noise possible. The behaviour of this noise does not conform to the models elaborated to date.

1. INTRODUCTION

1/f noise in MOS transistors has frequently been attributed either to fluctuations in the number of carriers induced by oxide traps, or to fluctuations in the mobility of carriers colliding with the lattice[1],[2].

The results obtained in many cases cannot be used to draw definitive conclusions, and doubts remain as to the exact physical origins of 1/f noise in MOS transistors.

It is well known that low energy X-rays induce oxide traps and recombination centres at the Si - SiO_2 interface. Conversely, they have no effect on the silicon lattice. This suggests that low energy X-rays must cause selective 1/f noise increases by exclusively increasing the contributions resulting from the oxide traps. Thus, it can be hoped that the origins of the noise can be made clear by irradiating the transistors.

In what follows, the results of such a study are presented. Their interpretation provides new elements which support and clarify one of the models. They also show that reality is much more complex than assumed by the foregoing considerations.

2. F^{-1} NOISE MODELS IN MOS TRANSISTORS

In accordance with Christensson, Lundstrom and Svensson[1], 1/f noise in MOS transistors is due to fluctuations in the number of carriers in the channel caused by fluctuations in the occupation of the oxide traps situated near the Si - SiO_2 interface. In this model, the 1/f law arises from the distribution of oxide time constants. The transfer of carriers between the channel and the traps is carried out by tunnel effect in accordance with the Mc Whorter model. Now, consider an MOS transistor having a channel length L ; Z width, a C_{OX} oxide capacitance per unit area, an N_t density of traps more or less uniformly distributed in energy and space to a distance x_M from the interface. At strong inversion regime and for a low source drain voltage V_{DS}, the spectral density of the voltage noise at the drain is given, as a first approximation, by :

$$S_{V_D}(f) = \frac{1}{ZL}\left[\frac{qV_{DS}}{C_{ox}(V_G-V_T)}\right]^2 \frac{k\,T\,N_t\,x_M}{f\,\ln(\tau_{Max}/\tau_{Min})} \qquad (1)$$

where V_G is the gate voltage, V_T the turn-on voltage, τ_{Max}/τ_{Min} maximum and minimum values of trap time constant ; k, T and f have their usual meanings. In short, this noise is designated in what follows by "noise of the CLS type".

In accordance with Berz[3], the trapping of the carriers gives rise not only to a variation in the number of carriers, but also to the transverse electric field perpendicular to the current flow. Martinot, Rossel and Vassilieff[4]have shown empirically that carrier mobility in the channel is a function of the value of this component of the electric field and that this effect is described by the expression :

$$\mu = \frac{\mu_o}{1+\xi(V_G - V_T)} \qquad (2)$$

where μ_o is the mobility value when the perpendicular component of the electric field at the interface is low, and ξ is an empirical factor characteristic of a given transistor. In general, $\xi \ll 1$. It follows from this that variations in the traps occupancy modulate carrier mobility and give rise to a trap-induced mobility noise which is equal to :

$$S_{V_D}(f) = \frac{1}{ZL}\left[\frac{qV_{DS}}{C_{ox}(V_G-V_T)}\right]^2 \frac{k\,T\,N_t\,x_M}{f\,\ln(\tau_{Max}/\tau_{Min})}\left[\frac{1}{1+\xi(V_G-V_T)}\right]^2 \qquad (3)$$

In what follows, this noise is designated by the expression "noise of the B type".

In accordance with Vandamme[2], the 1/f noise arises from fluctuations in the intrinsic mobility of the carriers when they collide with the lattice. The physical mechanism of the fluctuations is unknown. Applying Hooge's, formula[5]to this noise gives :

$$S_{V_D}(f) = \frac{1}{ZL}\frac{V_{DS}^2}{C_{ox}(V_G - V_T)}\frac{q\alpha}{f} \qquad (4)$$

Where $\alpha = \alpha_L(\frac{\mu}{\mu_L})^2$, $\alpha_L = 2 \times 10^{-3}$, μ is the effective mobility, μ_L is the intrinsic mobility of the carriers in the absence of surface and impurities scatterings. This noise is called "type V noise" in what follows.

To summarize, it is seen that the noise observed in MOS transistors can be the result of these three types of noise. The present knowledge does not preclude the existence of other mechanisms.

3. EXPERIMENTAL CONDITIONS

To make clear the physical origin of the noise from its evolution after X-ray irradiation, the effect of the X-rays on the transistor in dc operating conditions and ac small signal operating conditions has first to be characterized. Given that g_m, the transconductance in the ohmic regime, is proportional to mobility, the evolution of the drain current I_D was first measured as a function of the gate voltage V_G. At the same time, and so as to show the creation of oxide traps, the high frequency capacitance of the transistor in the capacitor configuration was measured (source and drain open circuited). In addition, as there is a correlation between the variations in trap density and the total surface recombination velocity s_o^*[6], the total surface recombination current was measured (drain substrate and source substrate diodes forward biased and connected in parallel).

As far as the noise is concerned, given that the three models mentioned have different dependencies as a function of the effective gate voltage, the measurements were carried out as a function of this parameter to see if one of the contributions were predominant.

The noise measurement system was described in[7].

The general characteristics of the transistors[1] were : substrate of the p type ; doping impurity concentration $2 \times 10^{16} cm^{-3}$; channel width 700 μm; channel length 120 μm ; oxide thickness 2000 Å ; gate metallization chromium. The irradiations were made by means of an A-equivolt 300 C G R type generator which functions with a 150 KV voltage and 10 mA dc current. In all cases, the transistors were short-circuited under irradiation.

4. EXPERIMENTAL RESULTS

The main electric characteristics of the transistors prior to irradiation are given in the Table.

g_m	μ	V_T	s_o^*
5×10^{-6} A/V	930 cm^2/V.S	3 V	5 cm/S

1. The transistors were manufactured at the L.A.A.S. by F. Rossel, G. Lacoste and T. Do Conto according to a procedure established by G. Pierrel and P. Rossel. We would like to warmly thank them.

The salient features are that the total surface recombination velocity is relatively low, and, the electron mobility high. This tends to prove that the silicon surface is of good quality (low scattering by surface states and impurities). A noise spectrum example is given in Fig. 1. It is seen that there is 1/f noise for frequencies lo-

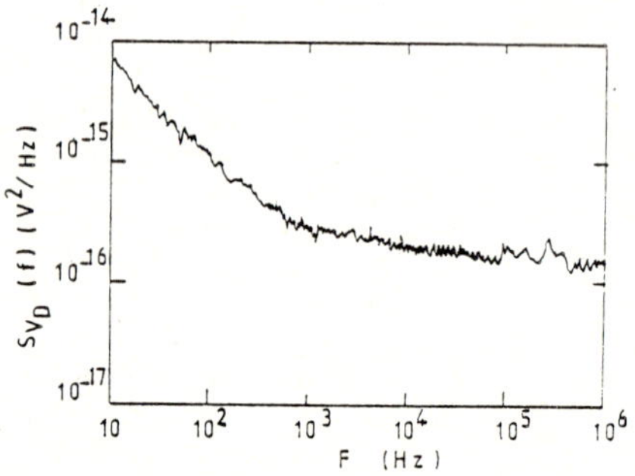

Figure 1 :Spectral density of the output noise. Experimental conditions : $V_D = 0.1$ V, $V_G - V_T = 1.1$V, T = 20°C, $I_D = 10\,\mu$A

wer than 100 Hz. The spectral density of this noise as a function of the effective gate voltage and of the drain voltage is described by the empirical relationship :

$$S_{V_D} \propto \left[\frac{V_{DS}}{V_G - V_T}\right]^2 \frac{1}{f} \qquad (5)$$

This agrees perfectly with the CLS model.

The effect of the X-rays are given in Figures 2 to 5. Subscript 0 indicates initial values before irradiation. The total surface recombination current I_{SM} is defined as the maximum difference between the values of the source-substrate and drain substrate diodes current when the surface is depleted of carriers and when the surface is accumulated[6].

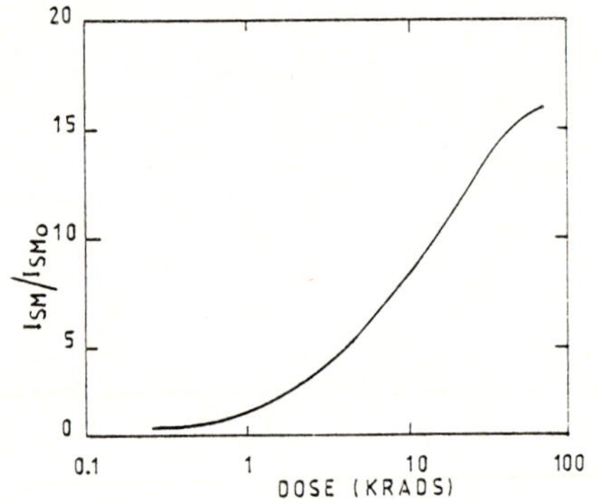

Figure 2 :Normalized increase of the total surface recombination current of the transistor in the gate controlled diode configuration. Experimental conditions : $V_D = 0.1$ V, T = 20°C, $I_{SMO} = 3.2 \times 10^{-11}$A.

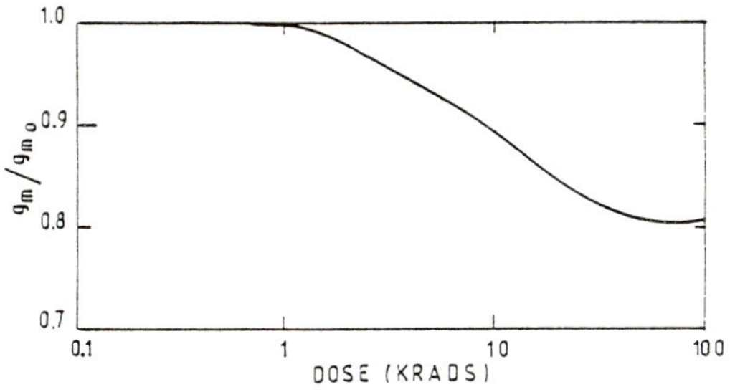

Figure 3 : Normalized decrease in transconductance as a function of the X-rays dose

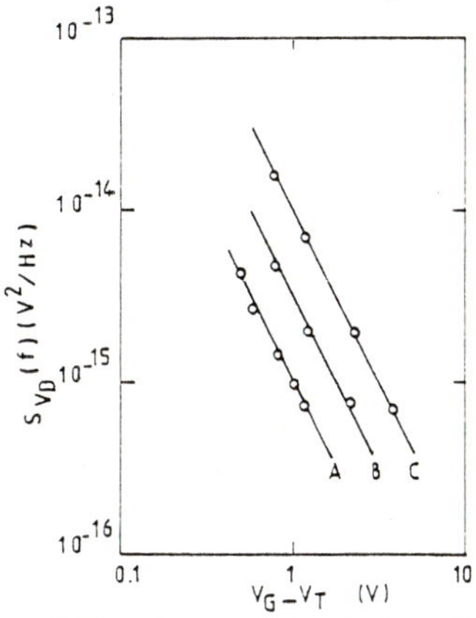

Figure 4 : 1/f noise spectral density as a function of the effective gate voltage before and after irradiation. Experimental conditions : V_{DS} = 0.05 V, f = 1 Hz, T = 20°C.
Doses : A = 0 krads, B = 6 krads and C = 34 krads

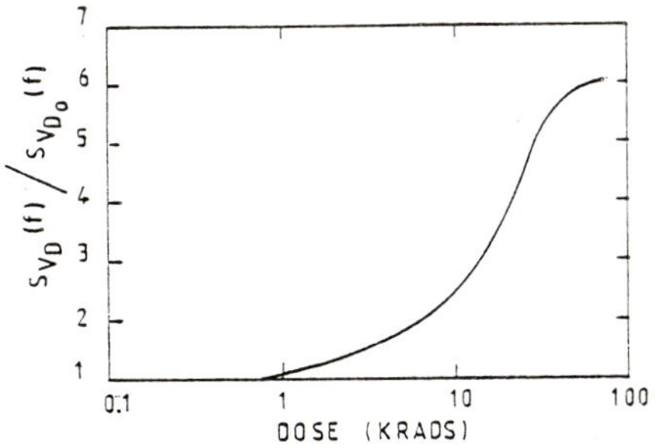

Figure 5 : Normalized increase in 1/f noise as a function of X-rays dose. Experimental conditions : V_{DS} = 0.05 V, f = 1 Hz, $V_G - V_T$ = 1.18 V

The analysis of Figures 2 to 5, show that, as a first approximation, the evolution of the electric characteristics consists of two quite distinct phases. Below 1 K rads[1], g_m (and therefore μ), $S_{VD}(f)$ are constant while I_{SM} slightly increases. For doses higher than 3 K rads, I_{SM} and $S_{VD}(f)$ increase rapidly and g_m decreases. In all cases (see Fig. 4), the noise follows a law of the form $S_{VD}(f) \propto (V_G - V_T)^{-2}$.

5. INTERPRETATION OF EXPERIMENTAL RESULTS

To facilitate interpretation, the value of the electric charge at the interface Si - SiO_2 equivalent to the total charge within the oxide was calculated from the shifts in the capacitance curves and was drawn as a function of the surface potential (see Figure 6).

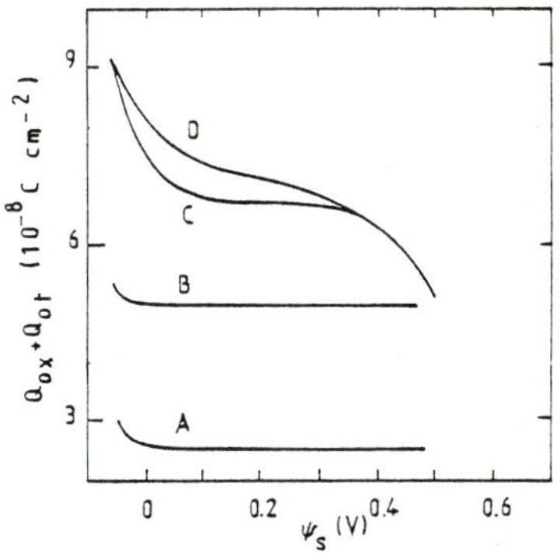

Figure 6 : Fixed oxide charge and oxide traps charge versus surface potential before and after irradiation. Doses : A = 0 krads, B = 0.5 krads, C = 8.5 krads and D = 24 krads.

The total oxide charge is independant of the surface potential and is also positive before and after irradiation for low doses. This signifies that low X-ray doses induce a fixed positive charge (which cannot capture or emit carriers) in the oxide. This charge does not influence the value of the transconductance and therefore does not affect mobility value.

Thus, the Columbian scattering effect of this charge on the carriers is slight or probably non-existent. The low increase in I_{SM} indicates, in addition, that recombination centres are created at the interface and at the surface of the silicon substrate. The fact that the noise is constant can have several meanings. In the first place, since the noise measurements verify relationship (1) it could be considered that this noise is of the CLS type. Thus, the recombination centres do not have the necessary characteristics to produce 1/f noise. Another possibility is that these centres produce noise. In this case, this contribution should be slight compared to the value of the noise before irradiation (because this contribution is not observable). Secondly, it is tempting to consider that since the noise and the mobility are both constant, the noise observed is of the V type. In this case, it would not obey Hooge's "fundamental" relationship since relationship (4) is not verified. For doses higher that 3 K rads, the oxide charge

1. The generator had not been calibrated for several years. It is possible that the absolute value of the indicated doses are erroneous. A calibration is being completed.

induced in the oxide (see Fig. 6) increases and is a monotonously decreasing function of the surface potential. This signifies that the radiation not only induces fixed oxide charges, but also traps which have a more or less uniform energy distribution. Given that the transconductance, and therefore the mobility, decreases, it is clear that this decrease is a consequence of carrier trapping. In accordance with expression (4), this decrease should be accompanied by a decrease in the noise since, a priori, μ_l is constant. It is concluded from this that if V type noise does exist, it is therefore not predominant. Comparing the evolution of I_{SM} and of $S_{VD}(f)$ reveals a great similarity in behaviour. Furthermore, since $S_{V_D}(f)$ follows (1), it is very probable that the noise induced by the X-rays is of the CLS type, and, that if B type noise does exist, it is negligible.

REFERENCES

[1] Christensson, S., Lundström, I. and Svensson, Low frequency noise in MOS transistors-I, theory, Solid State Electronics 11(1968), 797-812 .

[2] Vandamme L., Model for 1/f noise in MOS transistors biased in the linear region, Solid State Electronics, 23(1980), 317-323

[3] Berz F., Theory of low frequency noise in Si MOST's, Solid State Electronics, 13(1970), 631-647.

[4] Martinot, H., Rossel, P., Vassilieff, G., Mobility parameters and MOS transistors properties, Electronics Letters 24(1972), 599-600.

[5] Hooge, F., Kleinpenning, T., Vandamme L., Experimental studies on 1/f noise, Rep. Prog. Phys., 44(1981), 479-532 .

[6] Blasquez, G. and Roux-Nogatchewsky, M., Effets d'un rayonnement ionisant sur les mécanismes de bruit, Revue de Physique Appliquée, 11(1980), 1599-1605 .

[7] Blasquez, G., A high performance digital system for noise characterization in electronic devices, 9th IMEKO Congress Berlin 1982 .

x

x x

OSCILLATORS

NOISE IN PHYSICAL SYSTEMS AND 1/f NOISE
M. Savelli, G. Lecoy and J-P. Nougier (eds.)
© *Elsevier Science Publishers B.V., 1983*

PHASE AND FREQUENCY NOISES IN OSCILLATORS

(Invited Paper)

J.J. Gagnepain

Laboratoire de Physique et Métrologie des Oscillateurs du C.N.R.S.
associé à l'Université de Franche-Comté-Besançon
32, avenue de l'Observatoire - 25000 Besançon - France

The contribution of the different noise sources to the phase and frequency fluctuations of the output signal of an oscillator is described. Additive and parametric noises are distinguished. The limitations they introduce in the stability of various oscillators are evaluated. Comparisons are made between frequency and time domains. Finally the performances of the different types of frequency sources and standards are presented.

Introduction

An oscillator is supposed to deliver a signal with frequency, phase and amplitude as well defined as possible. But for many reasons these quantities can be perturbed and modified by stochastic or deterministic phenomena, which have their origin either in the fundamental properties of components of the oscillator, or in its environment.

Most of the undertaken efforts consist in trying to reduce the perturbations themselves when possible, or to minimize the sensitivities of the oscillator to the same perturbations. This is for instance the problems of the noise sources, ageing effects, etc... in the different components, or the problems of sensitivities to temperature fluctuations, mechanical vibrations, etc...

According to the location of the noise source inside the oscillator, or the way the perturbations will act, the consequences on the output signal can be different. Therefore an important task when designing an oscillator and when characterizing it consists on the one hand in evaluating the characteristics of the output signal, in terms of stability and spectral purity, from the characteristics of the noise sources and perturbations and on the other hand in identifying and localizing the different noise sources from the characteristics of the output signal.

Representation of the oscillator

An oscillator can be represented by an oscillating loop with output circuits in addition. As shown on Fig. 1 the oscillating loop is composed of a sustaining amplifier and a resonator (or an equivalent delay line or phase shifter). The main property of the resonator is to introduce in the loop a steep phase-frequency characteristic. The output circuit generally is used to amplify the signal and isolate the loop from external perturbations such as electrical load effects.

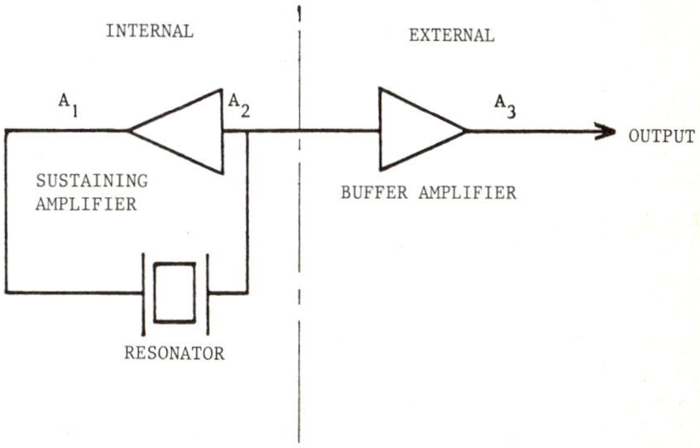

Fig. 1 : Schematic diagram of an oscillator

This kind of frequency source often is called an "active device". This is to distinguish with "passive" frequency sources, in which an oscillator, non necessary stable, is stabilized by phase-locking on an external resonator or spectral line (Fig. 2). Both active and passive configurations are used with quartz oscillators, atomic clocks and microwave frequency sources [1][2][3].

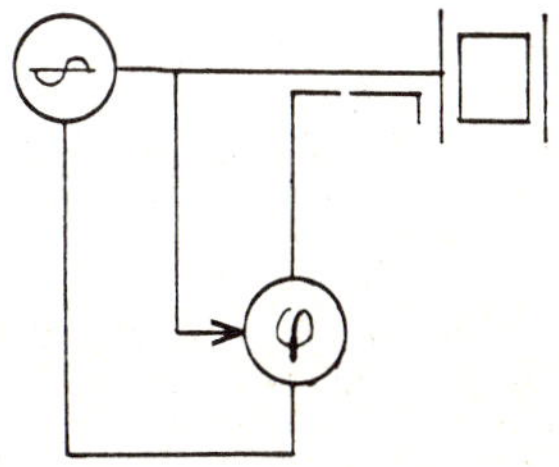

Fig. 2 : Passive system configuration

Modelization of the oscillator [4][5] can be made by considering simply the oscillator as a linear noise filter, or by using a more sophisticated representation of its nonlinear characteristics. The linear approximation gives results which generally agree with what is effectively observed. Therefore most of the properties of the output signal, in terms of stability and spectral purity, can be deduced from the phase-frequency response of the resonator and the linewidth of the output amplifier.

Classification of noise sources

According to the ways the noises will act on the signal, two main classes are distinguished :

- additive noise
- parametric noise.

Additive noise is generated by thermal noise sources. It is called internal additive noise when localized in the oscillating loop, and external additive noise when in the output circuits. The noise voltages induced by these noise sources are directly added to the signal, and therefore perturb the phase and the amplitude. Only the noise components with frequency close to the oscillation frequency will have a noticeable influence on the spectrum of the output signal.

Parametric noise is a low frequency noise which directly acts on the parameters of the oscillator. Noises of this type localized in the resonator and in the loop amplifier are preponderant. If there is parametric noise in the output amplifier, because it is a low frequency noise, it will have some influence only if it is converted to high frequency noise by nonlinear effects. Such a particular case will not be considered here.

Frequency, phase and amplitude fluctuations of the output signal induced by these different noise sources are characterized either in frequency domain or in time domain, respectively corresponding to spectral purity and stability. The mathematical tools, which are used, i.e. power density spectra, Allan variance, etc... are defined in the Appendix.

1) Internal additive noise

Additive noise sources in the oscillating loop induce phase and amplitude fluctuations by addition to the signal of the noise components with frequencies close to the oscillator center frequency.

In response to any phase perturbation in the oscillating loop there will be a frequency change in order to maintain the 2π phase condition of the loop. Such a readjustment will be mainly made by the element which presents the maximum phase-frequency rotation, i.e., the resonator.

This phase-frequency characteristic can be written

$$\phi \simeq -2Q \left(\frac{\nu - \nu_0}{\nu_0}\right) \qquad (1)$$

for $\nu \simeq \nu_0$, i.e. inside the resonator bandwidth. Q is the loaded quality factor and ν_0 the resonance frequency.

Wide band thermal noise sources are located on each side of the resonator, as indicated on Fig. 1 by points A_1 and A_2. The behaviors of these two sources will present some difference because the noise of the first source (point A_1), called input noise, will entirely be injected in the loop, whereas the noise of the second source (point A_2), called output noise, partly is injected in the loop and partly goes directly to the buffer amplifier.

The effect of the input noise was studied by many authors [6-11], and it was shown that it leads to the f^{-2} one-sided phase spectrum of the output signal.

$$S_\phi(f) = \frac{\nu_0^2 kT}{P\,Q^2\,f^2} \qquad (2)$$

where f_0 is the oscillator mean frequency, k the Boltzmann constant, T the effective temperature of the noise source, P the power in the loop and Q the resonator Q-factor.

The corresponding fractional frequency spectrum is

$$S_y(f) = \frac{kT}{PQ^2}$$

Using the relations given in the appendix the Allan variance over a time interval τ is

$$\sigma_y^2(\tau) = \frac{kT}{2PQ^2\tau} \qquad (3)$$

which shows for σ a $\tau^{-1/2}$ variation law.

The output noise contribution is given by a similar relation for the part of noise reinjected in the loop, and by a white spectrum for the part of noise going directly to the output [11].

$$S_\phi(f) = \frac{kT}{PQ^2}\left(\frac{\nu_0^2}{\nu^2} + 4Q^2\right) \qquad (4)$$

$$S_y(f) = \frac{kT}{PQ^2}\left(1 + 4Q^2\,\frac{f^2}{\nu_0^2}\right)$$

The two branches of these spectra intersect at $f = \nu_0/2Q$ which corresponds to the half bandwidth of the resonator.

The corresponding stability is

$$\sigma_y^2(\tau) = \frac{kT}{PQ^2}\left(\frac{1}{2\tau} + \frac{3Q^2 f_h}{\pi^2 \nu_0^2 \tau^2}\right)$$

σ follows $\tau^{-1/2}$ and τ^{-1} variation laws. To avoid divergence a frequency cut-off f_h must be introduced. Physically it corresponds to the limited bandwidth of the output elements and of the measurement system.

Quartz crystal oscillator

It must be pointed out that the $\tau^{-1/2}$ characteristic curve has never been observed. Therefore internal additive noise is not predominant in quartz oscillators and is covered by other noise sources, as it will be shown further.

Masers

The problem of masers differs from the previous one in the fact that must be taken into account the atomic line Q-factor (Q_ℓ) and the Q-factor of the loaded cavity (Q_c) where the interactions between the molecules and the electromagnetic field take place.

The contribution of the internal additive thermal noise to the phase and frequency spectra are [12]

$$S_\phi(f) = \frac{\nu_0^2\, kT}{P\, Q_\ell^2\, f^2} \cdot \frac{1}{1 + 4Q_c^2\, f^2/\nu_0^2} \qquad (5)$$

and $$S_y(f) = \frac{kT}{P\, Q_\ell^2} \cdot \frac{1}{1 + 4Q_c^2\, f^2/\nu_0^2}$$

P is the power delivered by the atomic beam to the loaded cavity.

The first term in the product of relations (5) is similar to relation (2). The second term introduces the filtering effect of the cavity.

Transformation in time domain gives

$$\sigma_y(\tau) = \frac{1}{Q_\ell}\sqrt{\frac{kT}{2P\tau}} \quad \text{for} \quad \tau > \frac{Q_c}{\pi\nu_0}$$

and

$$\sigma_y(\tau) = \sqrt{\frac{\pi^2\, kT\, \nu_0^2\, \tau}{6P\, Q_\ell^2\, Q_c^2}} \quad \text{for} \quad \tau < \frac{Q_c}{\pi\nu_0}$$

with respectively $\tau^{-1/2}$ and $\tau^{1/2}$ variation laws.

Microwave cavities

Electromagnetic cavities are also used to stabilize microwave oscillators. Among the different types of cavities, superconducting cavities in X-band are the devices with the highest Q-factors ever observed. For instance Q-factors of the order of 10^{10} to 10^{11} have been obtained with niobium cavities [13]. More recent is the development of dielectric loaded cavities, which, even if the Q-factor is much lower, have the advantages of size, price, resistance to environment, etc... [14].

The internal noise behavior of oscillators using such resonators is also well described by the model of Fig. 1.

From the previous relations it appears that the two main parameters involved in internal additive noise are the resonator Q-factor and the internal power. According to the type of resonator under use the values of these two parameters spread over very wide ranges.

Even if additive internal noise is not the predominant noise, the evaluation of the corresponding stability is of interest, because this thermal noise corresponds to the thermodynamic limitation, i.e. the ultimate stability, of the oscillator. This evaluation is made for various oscillators in Table I.

Type of oscillator	f_0	Q, Q_ℓ	P watts	$\sigma_y(\tau)$ for τ=1 s
Quartz	5 MHz	3.10^6	10^{-6}	$1,5.10^{-14}$
Masers H_2	1,4 GHz	2.10^9	10^{-12}	$2 \,.\, 10^{-14}$
Masers Rb	6,8 GHz	10^7	10^{-10}	$4 \,.\, 10^{-13}$
Masers NH_3	24 GHz	3.10^6	10^{-10}	$1,5.10^{-12}$
Super conducting cavity	9 GHz	10^{10}	10^{-3}	$1,5.10^{-19}$
Dielectric cavity	4 GHz	10^4	10^{-2}	$5 \,.\, 10^{-14}$

Table I : Evaluation of frequency instabilities due to thermal additive internal noise for various oscillators (T = 300°K)

2) External additive noise

This noise is associated with the output circuits, outside of the oscillating loop. The signal again is perturbed by addition of the noise components and thermal noise sources are involved. If the total noise power is small in comparison with the total signal power, half the noise power appears as phase modulation and half as amplitude modulation.

The broadband thermal noise voltage (with spectral density S_o) will be filtered by the limited bandwidth of the buffer amplifier, and can be represented at the output (point A_3 of Fig. 1) by the relation

$$S_N = \frac{S_o}{1 + 4\left(\frac{f - \nu_o}{f_h}\right)^2} \quad (8)$$

where f_h is the bandwidth of the circuit.

The phase spectrum of the output signal is

$$S_\phi(f) = \frac{2}{\pi f_h} \frac{P_N}{P_S} \frac{1}{1 + 4\,(f^2/f_h^2)} \quad (9)$$

P_S is the total signal power, and P_N the total noise power.

In time domain

$$\sigma_y(\tau) = \frac{\sqrt{2}}{2\pi\nu_o\tau}\sqrt{\frac{3\,P_N}{\pi\,P_S}} \quad \text{for } \tau > \frac{1}{\pi f_h}$$

and (10)

$$\sigma_y(\tau) = \frac{1}{2\pi\nu_o}\sqrt{\frac{\pi\, f_h\, P_N}{\tau\, P_S}} \quad \text{for } \tau < \frac{1}{\pi f_h}$$

If not using a very narrow band filter, the first of relations (10) will generally apply for usual values of τ. This shows that at constant signal to noise ratio the influence of external additive noise can be minimized only by increasing the operating frequency.

For 5 MHz quartz crystal oscillators with $P_N/P_S \simeq -110$ dB, the corresponding frequency instability at $\tau = 1$ sec is of the order of 10^{-13}.

In the case of hydrogen maser the frequency instability will also be of the order of 10^{-13} for $\tau = 1$ sec.

This shows that generally in active oscillators external additive noise dominates the internal one.

3) Parametric noise

The second type of noise sources is that which directly act on the parameters defining the frequency or the phase of the oscillating loop. Parametric noise is a slow varying noise.

Frequency noise

The sensitive element generally being the resonator, frequency parametric noise results from a modulation of the resonance frequency of this last one by internal or external perturbation.

If $S_{y_o}(f)$ is the power spectral density of the fractional frequency fluctuations of the resonator, the phase spectrum of the output signal is given by

$$S_\phi(f) = \frac{\nu_o^2\, S_{y_o}(f)}{f^2} \quad (11)$$

Two kinds of frequency noise are generally encountered : flicker noise and frequency random walk noise.

- **flicker noise** corresponds to

$$S_{y_o}(f) = A\, f^{-1} \quad (12)$$

and therefore to $S_\phi(f) \propto 1/f^3$.

In time domain

$$\sigma\,(\tau) \simeq (2\,A\,\ln 2)^{1/2} \quad (13)$$

This is independent of time, and often called "flicker floor".

Flicker noise is present in all the oscillators, even if sometimes flicker floor is not observed, because being hidden by other noise sources.

Flicker noise has been measured on quartz crystal resonators, independently of the influence of the electronics [15,16,17]. The results show a correlation between the levels of the 1/f frequency fluctuations and the unloaded Q-factor of the resonator, as shown on Fig. 3. Reduction of these data by linear regression gives the coefficient A of rel. (12)

$$A \simeq 1/Q^4 \quad (14)$$

Using this relation in (13) the flicker floor is

$$\sigma_y(\tau) \simeq \frac{1}{Q^2}\sqrt{2\ln 2} \quad (15)$$

At 5 MHz quartz crystal resonators can have Q-factors as good as 3×10^6. In an oscillator they will lead to a flicker floor of the order of 10^{-13}.

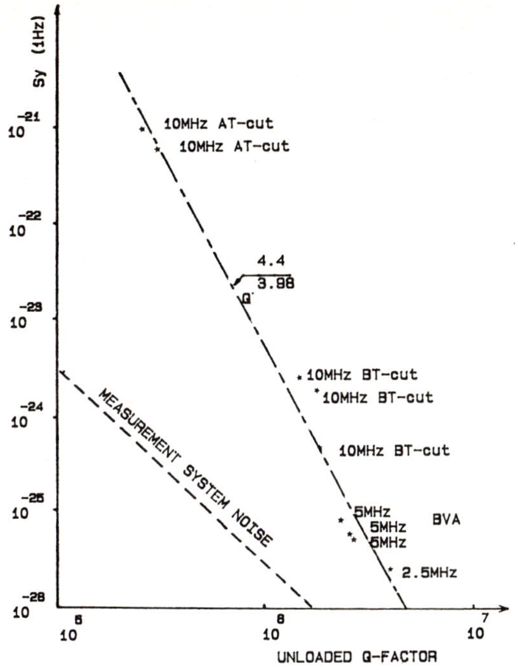

Fig. 3 : Relation between 1/f noise level and Q-factor in quartz crystal resonators

- **frequency random walk noise** corresponds to

$$S_{y_0}(f) = B\ f^{-2} \qquad (16)$$

and appears for the lowest Fourier frequency. It will give in time domain

$$\sigma(\tau) = \sqrt{2/3\ \pi^2\ B\ \tau} \qquad (17)$$

A correlation between frequency random walk and temperature fluctuations was shown in quartz resonators [18]-[19].

Phase noise

Parametric noise also induces phase fluctuations in the oscillating loop by changing parameters of the sustaining amplifier or of the impedances on both sides of the resonator.

As for thermal additive noise, these parametric phase fluctuations are transformed into frequency fluctuations in order to maintain the phase condition of the loop. Distinction can be made again between input and output phase noises. These low frequency fluctuations, essentially 1/f fluctuations, give therefore $1/f^3$ and 1/f phase spectra at the oscillator output. The two curves intersect at $f = f_0/2Q$, i.e. at half the bandwidth of the resonator. In time domain $\sigma_y(\tau)$ will respectively show a flicker floor and a $1/\tau$ variation law.

Therefore flicker floor has two origins : frequency noise of the resonator and phase noise of the circuit. Their respective contributions can be quite different from an oscillator to another one. For instance in quartz oscillator flicker floor was mainly due to phase noise. The progress in low noise electronics was such than today both contributions are generally equivalent, at least for ultrastable oscillators. At the opposite surface wave oscillators show the predominance of phase noise.

In table II are given the values of flicker floors for different oscillators.

type of oscillator	flicker floor σ_y
quartz	10^{-13}
H-maser	10^{-15}
Rb-maser	10^{-13}
S.C.O.	$2^{-6} \times 10^{-16}$
Cs	10^{-14} - 10^{-13}

Table II : Flicker floors of various oscillators

Frequency drift

Drift is a deterministic phenomenon, even if generally it is not determined, because its mechanisms are not well understood and therefore not predictible. However drift is due to ageing of the resonator and of the components of the oscillator. It appears as a slow monotonous variation of the frequency going all the time in the same direction.

Linear drift is represented by

$$y = \alpha t \qquad (18)$$

The corresponding Allan variance is easily obtained

$$\sigma_y(\tau) = \frac{\alpha\tau}{\sqrt{2}}$$

It can be seen from rel. (15) and (18) that frequency random walk and linear frequency drift give frequency instabilities $\sigma_y(\tau)$ which increase with τ following respectively $\tau^{1/2}$ and τ variation laws. These two slopes being quite close, it is in practice difficult to distinguish between them, if the measurements are not performed over large enough periods of time τ.

Comparisons between frequency and time domains

These different results are summarized on Fig. 4 and Fig. 5. For more clarity internal additive noise is represented apart, at a lower level.

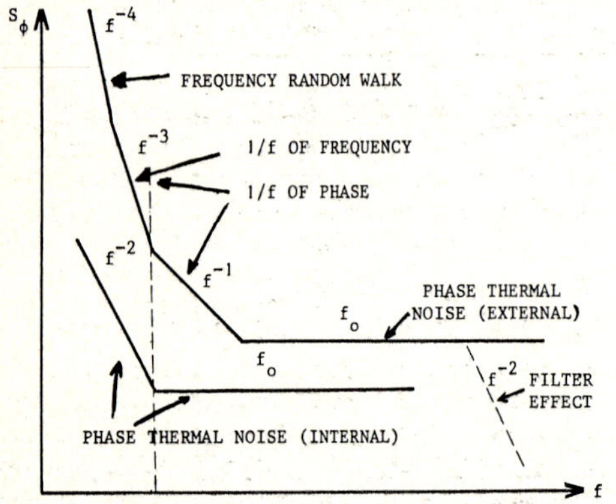

Fig. 4 : Frequency domain:
Contribution of the different noise sources to the phase power spectral density

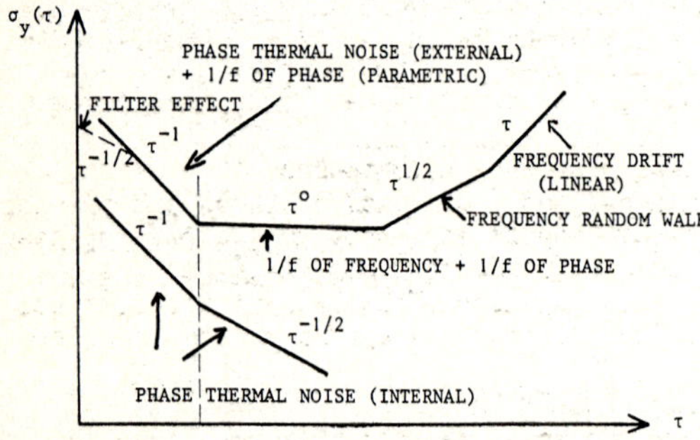

Fig. 5 : Time domain
Contribution of the different noise sources to the Allan variance

An oscillator being a nonlinear device, it presents amplitude-frequency effect, i.e. its frequency depends on the oscillation level. This means that amplitude noise (due for instance to additive noise) can be converted into frequency noise. But the effect is rather small and can generally be neglected.

It is to be remarked that different noise sources can give characteristic slopes which are identical in the spectrum or in the stability curves. This occurs for instance with 1/f of frequency and 1/f of phase, both giving a f^{-3} spectrum in frequency domain, and a flicker floor in time domain. Also the external additive noise and the 1/f of phase parametric noise give a τ^{-1} curve in time domain. This means that it is not possible to localize and identify all the noise sources by measuring the characteristics of the output signal.

Combined system with servo loop

All the previous discussion concerned "active" oscillators. However it is difficult not to say a few words on "passive" devices schematized on Fig. 2, because some of the most performing frequency standards are based on such a principle.

The interest is in combining two oscillators or an oscillator with a resonator, the one with good spectral purity and good short term stability, the other one with low drift and good long term stability. This can be done by phaselocking two quartz oscillators or a quartz oscillator to a passive quartz resonator (1) with a servo loop, or by phaselocking a quartz oscillator to an atomic line as in atomic clocks. By choosing carefully the attack time of the loop, it is possible to combine the best stabilities of each of the two devices following the relation

$$S_\phi(f) = S_\phi(f)_{osc} \frac{1}{|1+H(f)|^2} + S_\phi(f)_{ref} \left|\frac{H(f)}{1+H(f)}\right|^2$$

where H(f) is the total gain of the loop.

The combined stability of such a system is represented on Fig. 6 for $H(f) = jf_c/f$. In this case the reference could be an atomic line and the locked oscillator a quartz crystal oscillator.

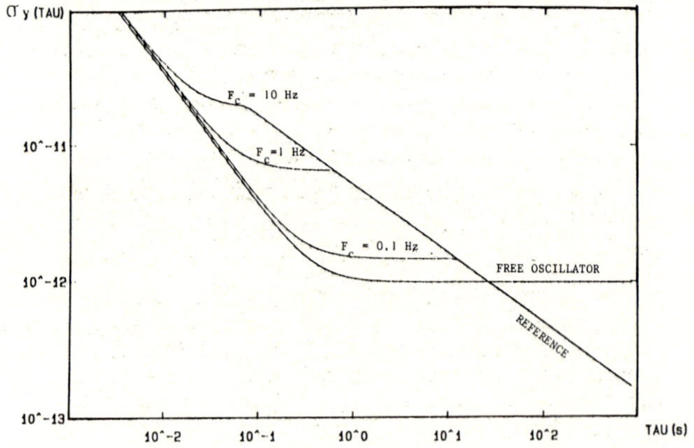

Fig. 6 : Combined stability of an oscillator phase-locked to a reference

$$S_\phi(f)_{osc} = \frac{7.10^{-26}}{f^3} + \frac{5.10^{-26}}{f} + 7.10^{-28}$$

$$S_\phi(f)_{ref} = \frac{5.10^{-23}}{f^2}$$

Conclusion

In the place of a conclusion are given on Fig. 7 typical stability curves of different oscillators : microwave oscillator stabilized by cavity, quartz crystal oscillator, passive Cs and Rb clocks, H-maser (linear drifts have been removed). This demonstrates quite clearly the contributions of the different noises, according to the oscillator under consideration, which can be passive or active.

The review of the different sources of frequency and phase fluctuations in an oscillator showed that the noise behavior of the output signal can be explained by using a simple schematic representation composed of an oscillating loop and an output amplifier, without complicate consideration on the details of the circuits. Only are involved mean frequencies, Q-factors (and bandwidths) and signal over noise ratios, which are sufficient to understand the differences observed between the different types of oscillators.

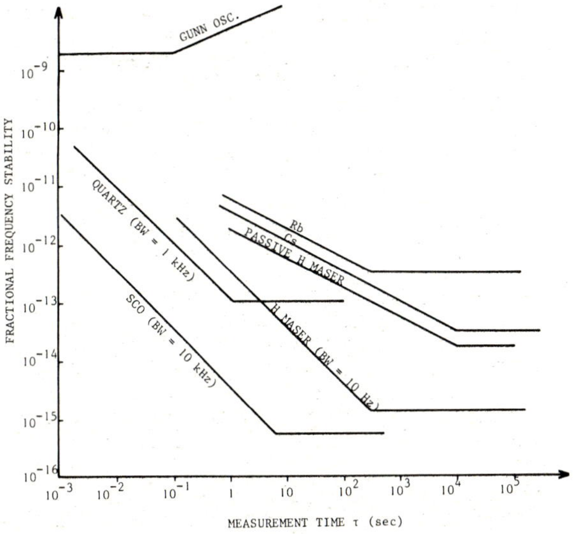

Fig. 7 : Frequency stability of various oscillators

Appendix

Characterization of frequency and phase fluctuations (see references 12 and 20)

The real signal observed at the output of an oscillator can be represented by the relation

$$V(t) = [V_0 + a(t)] \cos [2\pi\nu_0 t + \phi(t)]$$

where V_0 and ν_0 are the mean amplitude and frequency, and $a(t)$ and $\phi(t)$ are random functions of time corresponding to amplitude and phase noises.

The instantaneous frequency by definition is

$$\nu(t) = \nu_0 + \Delta\nu(t)$$

with $\Delta\nu(t) = \frac{1}{2\pi} \frac{d\phi(t)}{dt}$

For stable oscillators generally $|\Delta\nu(t)| \ll \nu_0$.

The instantaneous fractional frequency deviation is given by

$$y(t) = \frac{\Delta\nu(t)}{\nu_0}$$

a) frequency domain

If $y(t)$ is stationary (or made stationary) its autocorrelation function is independent of time

$$R_y(\tau) = \lim_{T\to\infty} \frac{1}{T} \int_t^{t+T} y(t')\, y(t'+\tau)\, dt'$$

By applying the Wiener-Khintchine theorem, the one-sided power density spectrum is obtained

$$S_y(f) = 4 \int_0^{\infty} R_y(\tau) \cos 2\pi f\tau \, d\tau$$

where f is the Fourier frequency.

Similarly the power density spectrum of the phase fluctuations can be introduced

$$S_\phi(f) = 4 \int_0^{\infty} R_\phi(\tau) \cos 2\pi f\tau \, d\tau$$

b) Time domain

The instantaneous value of the different quantities cannot experimentally be obtained, because a finite time is necessary for any measurement. Therefore only the mean value over a given time interval can be measured. Thus the time average of the fractional frequency deviation which corresponds to one frequency sample is

$$\bar{y}_k = \frac{1}{\tau} \int_{t_k}^{t_k+\tau} y(t)\, dt$$

The average $\langle\bar{y}_k\rangle$ is defined in term of ensemble average

$$\langle\bar{y}_k\rangle = \frac{1}{N} \sum_{k=1}^{N} \bar{y}_k$$

The true variance is given by

$$I^2(\tau) = \langle \bar{y}_k^2 \rangle - \langle \bar{y}_k \rangle^2$$

This mathematical definition corresponds to an infinite number of samples. In practice only a finite number is accessible. Therefore only an estimation of the variance is obtained. Several definitions can be used. The most common one is given for N samples by

$$\sigma_y^2\ (N,T,\tau) = \frac{1}{N-1} \sum_{k=1}^{N} (\bar{y}_k - \frac{1}{N} \sum_{j=1}^{N} \bar{y}_j)^2$$

where T is the period of repetition of the measures ($T = \tau$ + dead time).

The IEEE sub comittee for frequency stability proposed to use a two-samples variance, without dead time ($T = \tau$). This variance is often called "Allan variance", and it is noted $\sigma_y^2(\tau)$

$$\sigma_y^2(\tau) = <\sigma_y^2\ (2,\tau,\tau)> = < \sum_{k=1}^{2} (\bar{y}_k - \frac{1}{2} \sum_{j=1}^{2} \bar{y}_j)^2>$$

c) Relations between frequency and time domains

noise	$S_y(f)$	$\sigma_y^2(\tau)$	exponant of τ
frequency random walk	$h_{-2}\ f^{-2}$	$\frac{2}{3}\ \pi^2\ h_{-2}\ \tau$	+1
flicker frequency	$h_{-1}\ f^{-1}$	$2h_{-1}\ \ln 2$	0
white frequency	h_0	$\frac{h_0}{2\tau}$	-1
flicker phase	$h_1\ f$	$\frac{h_1}{4\pi^2\tau^2}\ (1{,}038 + 3\ln(2\pi f_h\ \tau))$	-2
white phase	$h_2\ f^2$	$\frac{3h_2 f_h}{4\pi^2\tau^2}$	-2

f_h : cut-off frequency (output amplifier, measurement system..)

Acknowledgements

The author would like to thank Prof. M. Olivier and Dr. J. Groslambert for their assistance in the preparation of this paper.

References

1) S.R. Stein et al., "A systems approach to high performance oscillators", Proc.* 32nd Annual Frequency Control Symposium, Philadelphia (1978).

2) H. Hellwig, "A review of precision oscillators", NBS Technical Note 662, National Bureau of Standards, Boulder, Col. 80402.

3) S.R. Stein, J.P. Turneaure, "The development of the superconducting cavity stabilized oscillator", Proc.* 27th Annual Frequency Control Symposium, Fort Monmouth (1973).

4) E. Hafner, "Noise in oscillators", Proc. IEEE, Special issue on frequency stability, vol. 54, n° 2 (1966).

5) R. Brendel, "Internal noise of a quartz crystal oscillator", Proc.* 29th Annual Frequency Control Symposium, Atlantic-City (1975).

6) A. Blaquière, "Spectre de puissance d'un oscillateur non linéaire perturbé par le bruit", Annales de Radioélectricité VIII, n° 32 (1953).

7) W.A. Edson, "Noise in oscillators", Proc. IRE, 48, n° 8 (1960).

* Copies of the proceedings are available from Electronic Industries Association, 2001 Eye Street N.W., Washington DC. 20006 USA.

8) P. Grivet, A. Blaquière, "Nonlinear effects of noise in electronic clocks", Proc. IEEE, 51, n° 11 (1963).

9) E. Hafner, "The effects of noise on crystal oscillators", IEEE NASA Symposium on short-term frequency stability, Goddard Space Flight Center, Greenbelt, Maryland (1964).

10) J. Rutman, "Sur les générateurs de fréquence de très hautes performances", ONERA publication n° 142, Chatillon 92 France (1972).

11) R. Brendel, "Bruit interne des oscillateurs à quartz", Doct. Ing. Thesis, n° 63, LPMO Besançon France (1975).

12) L. Cutler, C. Searle, "Some aspects of the theory and measurement of frequency fluctuations in frequency standards", Proc. IEEE, 54, n° 2 (1966).

13) S.R. Stein, "Application of superconductivity to precision oscillators", Proc.* 29th Annual Frequency Control Symposium, Atlantic-City (1975).

14) J.K. Plourde, Chung-Li Ren, "Application of dielectric resonators in microwave components", IEEE Trans. Microwave Theory and Techniques, MTT29, n° 8 (1981).

15) J.J. Gagnepain, J. Uebersfeld, "1/f noise in quartz crystal resonators", Proc. 1st Symposium on 1/f fluctuations, Tokyo (1977).

16) J.J. Gagnepain, J. Uebersfeld, G. Goujon, P. Handel, "Relation between 1/f noise and Q-factor in quartz resonators at room and low temperature, first theoretical interpretation", Proc.* 35th Annual Frequency Control Symposium, Philadelphia (1981).

17) M. Olivier, J.J. Gagnepain, "Chaotic states and anomalous noise in resonators", This Symposium.

18) J.J. Gagnepain, "Fundamental noise studies of quartz crystal resonators", Proc.* 30th Annual Frequency Control Symposium, Atlantic-City (1976).

19) Y. Noguchi, Y. Teramachi, T. Musha, "1/f frequency fluctuation of a quartz oscillator", Proc* 35th Annual Frequency Control Symposium, Philadelphia (1981).

20) J. Rutman, "Characterization of phase and frequency instabilities in precision frequency sources : fifteen years of progress", Proc. IEEE, 66, n° 9 (1978).

NOISE IN PHYSICAL SYSTEMS AND 1/f NOISE
M. Savelli, G. Lecoy and J-P. Nougier (eds.)
© *Elsevier Science Publishers B.V., 1983*

CHAOTIC STATES AND ANORMALOUS NOISE IN RESONATORS

M. Olivier and J.J. Gagnepain

Laboratoire de Physique et Métrologie des Oscillateurs du C.N.R.S.
associé à l'Université de Franche-Comté-Besançon
32, avenue de l'Observatoire - 25000 Besançon - France

Resonating systems, like many physical systems, exhibit characteristics which are not linear any more, when driven at high level. The nonlinearities induce resonance frequency variations versus driving power (amplitude-frequency effect), distorsions of phase and amplitude responses. Coming along with those effects are bi or multi-stability, harmonic and subharmonic generation, hysteresis. Finally unstable states, chaotic states, can take place.
In the present paper subharmonic generation and chaotic states are studied in the cases of resonators with high or low Q-factors. Modelization of such nonlinear systems is presented first. Then the results of measurements performed on real systems, quartz crystal resonators (high Q-factor) and phase lock loop (low Q-factor) are given.

I. INTRODUCTION

Resonant devices and systems do not remain linear when the driving power is increased, and nonlinear effects appear at levels which are peculiar to each kind of resonator. Among these nonlinearities harmonic generation, intermodulation, amplitude frequency effect are well known. The noise characteristics of the resonator also are influenced by the nonlinearities, and the behavior will apparently be different if the resonator is operated at low level (linear range), medium level (with distorsions of the amplitude and phase responses) or high level (chaotic states occuring). An important parameter also is the resonator Q-factor, which strongly influence the behavior according to the corresponding linewidth. Therefore high Q's and low Q's resonators are distinguished. These different aspects are examined in the present work.

II. HIGH Q-FACTOR RESONATOR

Quartz crystal resonators in the frequency range of 1 MHz to 10 MHz have very high Q-factors since of the order of 1.10^6 to 3.10^6. It has been shown that their resonance frequency presents fluctuations [1]-[5]. Most of these measurements were performed using one or two crystals in a π transmission network : in this case the resonance frequency fluctuations induce phase fluctuations of the driving signal, which are detected by means of a balanced phase bridge (Fig. 1).

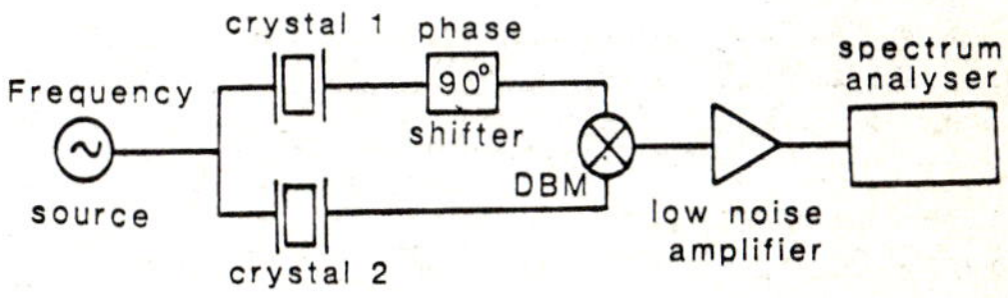

Fig. 1
Passive frequency noise measurement system

Two kinds of noises can be identified in the power density spectrum : flicker frequency noise and frequency random walk. If the last one can be attributed to temperature fluctuations and thermal stresses in the crystal, the origin of the first one is not well understood yet. However a correlation has been found between the level of flicker frequency noise and the resonator Q-factor [3-4] : this was confirmed by recent measurements [6] performed on 10 MHz, 5 MHz and 2.5 MHz resonators. As shown on Fig. 2 the flicker noise level (given at 1 Hz from the carrier) depends on the unloaded Q-factor value, following a $1/Q^4$ law. The 10 MHz resonators are particularly demonstrative. Due to the crystallographic orientation BT cut resonators have much higher Q's than AT-cut ones. A large difference in the noise levels is observed as exepcted from the $1/Q^4$ law. These measurements were obtained operating the resonators in their linear range (typically à 10 µW).

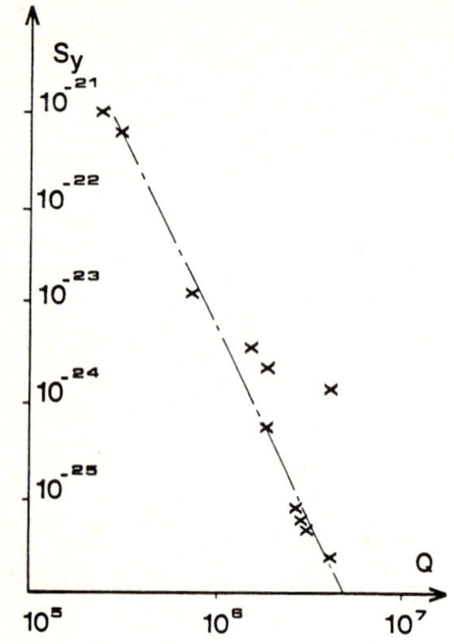

Fig. 2 : 1/F noise level (at 1 Hz from the carrier (as a function of the unloaded Q factor

When increasing the driving power the crystal experiences nonlinear effects due mainly to the higher order elastic constants. Distorsions appear in the amplitude and phase resonance curves, with bistability and hysteresis, as shown on Fig. 3. Therefore the spectrum of the phase fluctuations induced by the fluctuations of the resonator frequency (see Fig. 1) will depend on the position of the operating point on the resonance curve, i.e. the driving level and the driving frequency.

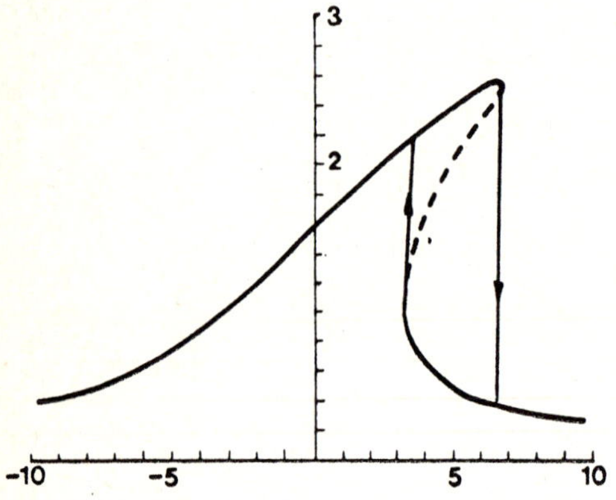

Fig. 3a : Amplitude of resonance curves

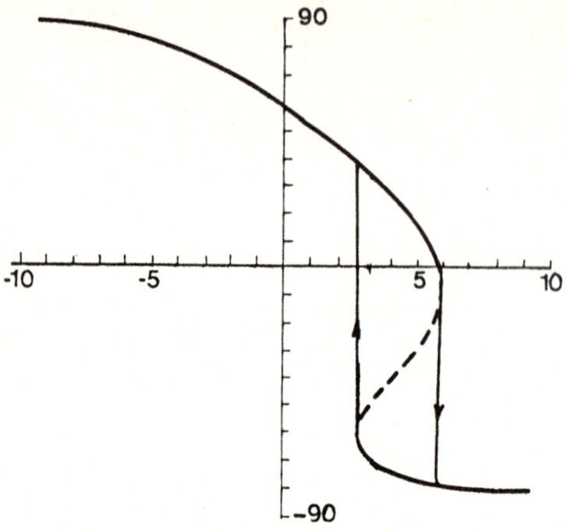

Fig. 3b : Phase of resonance curves

From such curves, it is obvious that when getting close to the jump frequencies the slope of the phase-frequency response becomes to be very high, and therefore very large phase fluctuations can be induced. The nonlinear behavior of the high Q quartz resonator, driven in transmission can be represented by the phenomenological relation :

$$\frac{d^2 i}{dt^2} + \frac{\omega_0}{Q}\frac{di}{dt} + \omega_0^2[1-\varepsilon(\Omega)\cos\Omega t]f(i) = F\cos\Omega t \quad (1)$$

where i is the courant through the crystal, Q the loaded Q-factor and F the amplitude of the driving force.

$\varepsilon(\Omega)\cos\Omega t$ introduces a modulation of the resonance angular frequency ω_0, which represents the frequency noise of the resonator. For the quartz resonator, $f(i)=i-Ki^3$ where K depends on the nonlinear third and fourth order elastic constants. This equation was studied by Hubertson and Cratchfield [7].

The application of a perturbation method gives the phase noise $\phi(\Omega)$ of the output signal : at high level, when driven near the jump frequency, the corresponding phase spectrum is given by

$$S_\phi(\Omega) = S_y(\Omega) \cdot \frac{\omega_0(\Omega^2 + \omega_0^2/4Q^2)}{\Omega^2(\Omega^2 + \omega_0^2/Q^2)} \quad (2)$$

$S_y(\Omega)$ is the frequency noise spectrum of the crystal.

This is to be compared to the corresponding spectrum obtained when the resonator is in its linear range

$$S_\phi(\Omega) = S_y(\Omega) \frac{\omega_0^2}{\Omega^2 + \omega^2/4Q^2} \quad (3)$$

Fig. 4 shows the evolution of this phase spectrum : all the curves asymptotically go to $S_{\phi} = \frac{S_y(\Omega)\omega_0^2}{\Omega^2}$: this asymptote corresponds to the maximum of noise generation : a large increase of noise is observed for the low frequency components at high level of excitation ; this shows, for instance, that the stability of an oscillator can be perturbed by such an anomalous noise in the case of frequency noise of the resonator.

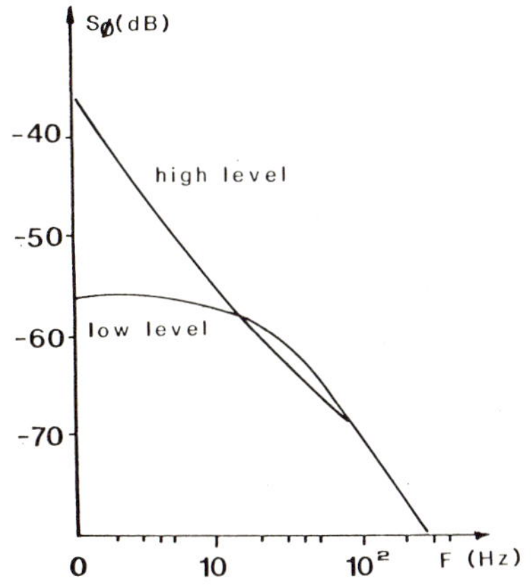

Fig. 4 : Phase noise spectrum generated in the nonlinear resonator at low and high power

It follows from its amplitude and phase characteristics that a quartz crystal rsonator will in principle reach a chaotic state for still higher driving level : we have driven a 5 MHz quartz resonator at very high level (2.5 mW) and a white phase phase noise was experimentaly observed (Fig. 5). This noise is a characteristic of the chaotic behavior of a nonlinear system, together with the generation of subharmonics : the latter phenomenon is not observed with quartz crystals. and this can be explained by the filtering action of the high Q-resonator : moreover, temperature effect and antiresonance make the study of the chaotic state rather difficult with the quartz crystal. Therefore, we have studied the occurence of the chaotic states mainly by simulation of low Q's resonant systems.

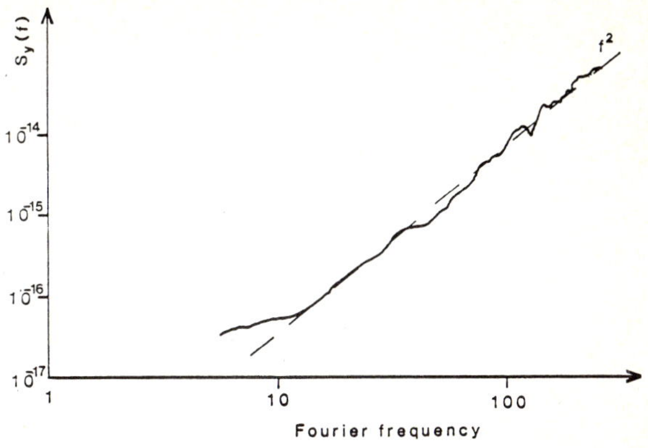

Fig. 5 : Frequency fluctuations generated during the chaotic state

III. LOW Q-FACTOR RESONATORS

The equation (1) which described the quartz resonator driven on transmission at high level can be associated to many other nonlinear physical systems : for instance the classical pendulum, Josephson junctions or phase lock loops [8][9][10]. In the last case the nonlinear term $f(\phi) = \sin\phi$ correspond to the voltage-phase characteristic of the double balanced mixer used for phase detection and locking, ϕ being the phase difference between the two locked oscillators. Such a system was studied by D'Humières et all [8].

The same system was used in the present work to study the noise generated by the chaotic state : schematically represented on Fig. 6a, two oscillators with frequency f_s equal to 5 MHz are phase locked, and a low frequency excitation signal is introduced in the loop which has a low attack time, corresponding to a low Q of the order of 2 to 10.

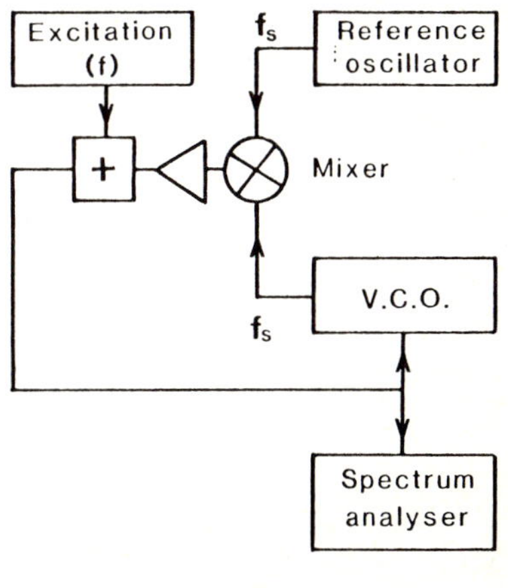

Phase look loop

Fig. 6a

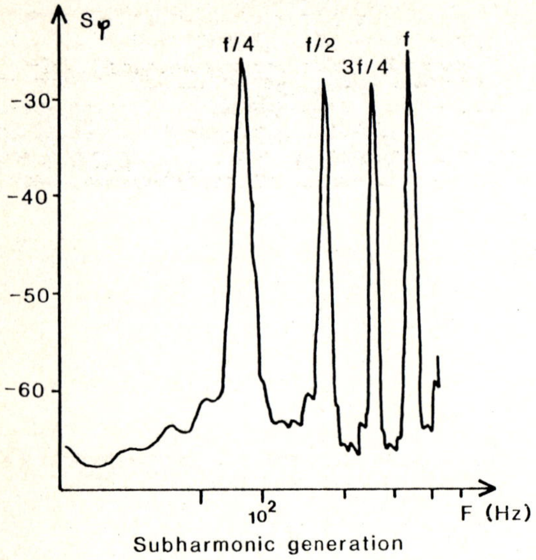

Fig. 6b

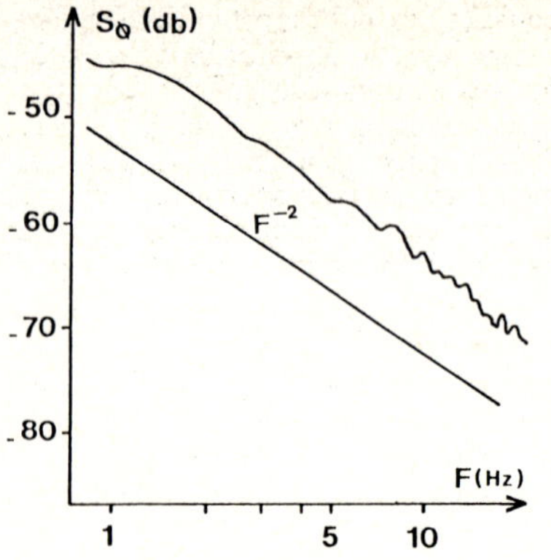

Fig. 7 : Phase spectral density of the nonlinear system with white frequency excitation noise

The characteristics of the chaotic behavior are observed on the error signal as the frequency or the level of the excitation signal vary. Subharmonics generation of the excitation signal appears first at medium level of excitation (Fig. 6b) : this phenomenon is explained by the well-known theory of cascading bifurcations [11]. The chaotic state, which show an extreme sensibility to the control parameters, appears at higher level of excitation and is characterized by a spectrum $S_\phi(\Omega)$ peaked up at the fundamental and subharmonics of the drive frequency.

We achieved a second phase lock loop to study the spectral characteristic of this noise : a synthesizer, operating at $f_S \pm f$, f being the frequency of the excitation signal or the frequency of a subharmonic, is locked to the frequency of the phase locked oscillator. The analysis of the filtered error signal shows the spectrum of the pedestal noise of each line. Such a system was also used for the study of the noise of the high Q quartz resonator.

The similarity between the equations describing the behavior of the phase lock loop and the nonlinear resonator shows us that, at medium excitation level, the phase noise of the output of the loop must have a F^{-2} spectral density : this is experimentally observed on Fig. 7, when the excitation oscillator is frequency modulated by a white noise source.

The chaotic behavior is observed by the onset of white phase noise which becomes predominant at high level of excitation (Fig. 8).

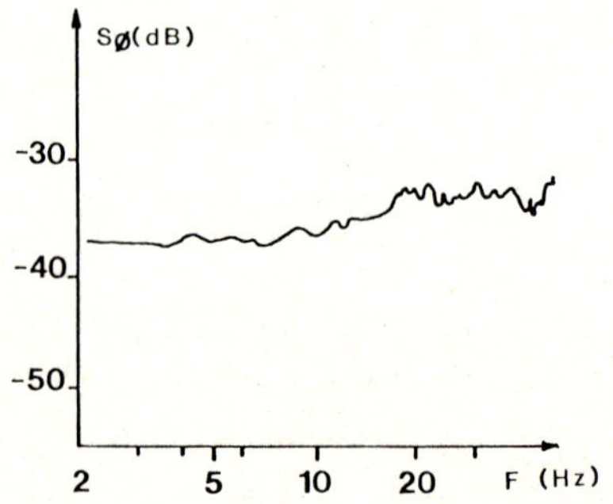

Fig. 8
Phase spectral density of chaotic states

IV. CONCLUSION

The noise behavior of nonlinear system is strongly dependent on the amplitude of the excitation, the corresponding nonlinearities and the Q-factor. For high Q quartz crystal resonator which exhibits an inherent 1/F noise spectrum, the nonlinearities can amplify this noise at medium power and gives larger phase fluctuations for the components of noise which are inside the bandwidth of the resonator.

For the low Q resonant system, like a phase lock loop, the chaotic state is caracterized by the onset of subharmonics of the excitation frequency and then by the generation of white phase noise. It was shown that for the high Q quartz resonator the latter phenomenon can also take place. These phenomenon have some important consequences, especially for the realization of oscillators of high spectral purity : higher drive level which seems to be a solution to improve the signal/noise ratio is not a good solution since the nonlinearity and the phase noise of the resonator increase with the excitation power.

Acknowledgements

The authors wish to thank J. Groslambert and G. Marianneau from LPMO for their suggestions and help in the noise measurements.

References

[1] F.L. Walls, A.F. Wainwright, Measurement of the short-term stability of quartz resonators and the implications for quartz oscillator design and applications, IEEE Trans. Instrum. Meas., IM-24, p. 15 (1975).

[2] J.J. Gagnepain, Fundamental noise studies of quartz crystal resonators, Proc. 30th Ann. Freq. Cont. Symp., p. 84 (1976).

[3] J.J. Gagnepain, J. Uebersfeld, 1/f noise in quartz crystal resonators, 1st Symp. on 1/f fluctuations, p. 173-181, Tokyo (1977).

[4] J.J. Gagnepain et al, Relation between 1/f noise and Q-factor in quartz resonators at room and low temperatures, first theoretical interpretation, 35th Ann. Freq. Cont. Symp., 1981.

[5] Y, Noguchi, Y. Teramachi, F. Musha, 1/f fluctuations of a quartz crystal oscillator, Proc. 35th Ann. Freq. Cont. Symp., p. 484 (1981).

[6] J.J. Gagnepain, F.L. Walls, to be published

[7] B.A. Hubertson, J.P. Cratchfield, Chaotic states of anharmonic systems in periodic fields, Phys. Rev. Letters, 43, n° 23, p. 1473 (1979).

[8] D. D'Humières, M. Bearley, B. Hubermann, A. Libchaber, Chaotic states and routes to chaos in the forced pendulum, Ginzton Lab. Report 3429, Stanford University (1982).

[9] N. Petersen, A. Davidson, Chaos and noise rise in Josephson junction, Appl. Phys. Lett., 39, n° 10, p. 830 (1981).

[10] B. Huberman, J.P. Cratchfield and N.H. Packard, Noise phenomena in Josephson junctions, App. Phys. Lett., vol. 37, n° 8, p. 750 (1980)

[11] I. Gumowski, C. Mira : Dynamique chaotique (Cepadues Editions - Toulouse, 1980).

NOISE IN PHYSICAL SYSTEMS AND 1/f NOISE
M. Savelli, G. Lecoy and J-P. Nougier (eds.)

© Elsevier Science Publishers B.V., 1983

NOISE PHENOMENA IN RINGLASERS

H. R. Bilger and M. Sayeh

Oklahoma State University
Stillwater, Oklahoma, USA

Fluctuations of the beat frequency Δf between optical oscillation frequencies in ringlasers can be represented by a power series of the power spectral density, $S_{\Delta f} = \sum h_i f^i$. The white noise term h_0 is theoretically well understood. Experiments in good ringlasers show this term to be dominant down to Fourier frequencies f of the order of millihertz. Measurements down to f = 1.2 μHz [(1/9) d^{-1}] demonstrate 1/f- and $1/f^2$ noise, the latter in agreement with other results obtained in the last two decades. The measured white noise, if expressed as a spectral density of the relative frequency fluctuation, is about four orders of magnitude below that of published hydrogen maser data.

1. INTRODUCTION

Since the published concepts (1) and the first implementation the ringlaser has been recognized as a device which allows interferometry of two or more laser modes with extremely high accuracy. This potential has been utilized in precision measurements of Fresnel drag (2), of Sagnac effect (3), and in other applications. The presence of at least two reasonably decoupled nondegenerate modes in one cavity, where arbitrary splitting can be introduced via rotation, drag, Faraday effect etc., is also ideal for the study of fundamental laser noise. A variety of publications adapted the theory of quantum fluctuations to ringlasers (4). The arduous task of reducing other, less fundamental, or "technical" fluctuations was mainly left to experimental researchers.

The theoretically attainable relative linewidth of an oscillation in the optical regime can be made quite small, for two major reasons: a) Cavities with large quality factors ($Q > 10^8$ is typical) can be constructed at optical wavelengths, b) the very high frequency (f_0 = 474 THz for the HeNe laser oscillating at a wavelength λ = 633 nm) makes the relative linewidth small, although quantum noise as such increases approximately $\propto f_0^{1/2}$ (see below).

2. THEORY

In the following, calculations of phase - and frequency fluctuations of optical oscillators are summarized. White frequency noise can be considered as a consequence of diffusion of phase, with

$$\overline{\delta\phi^2} = D\tau, \quad D = 4\pi^2 h f_0^3/2Q^2P. \qquad (1)$$

It is assumed here that $hf_0 \gg kT$, and that the cavity linewidth Δf_c is small compared to the spontaneous emission linewidth Δf_{atomic}. The oscillator frequency should then have white noise for Fourier frequencies $f \leq \Delta f_c = f_0/Q$. The Power spectral density (PSD) is

$$S_{\Delta f} = h f_0^3/Q^2P = h_0, \qquad (2)$$

where P is the total power loss, in Watt, of this particular cavity mode, including scattering and transmission at the mirrors, diffraction and absorption in the cavity. Equation (2) is the lower limit attainable with complete inversion. The Allan variance σ_A^2, in the case of white noise, is

$$\sigma_A^2(\tau) = (h_0/2)\tau^{-1}. \qquad (3)$$

Analogous to theoretical and experimental results obtained for maser noise, random-walk noise of laser frequency is also anticipated for various reasons, the most probable being power fluctuations giving rise to temperature effects on cavity dimensions. Fluctuations due to the interrogation (detection) process are less likely, because Si-junction detectors with near-unity quantum efficiency and very low shot noise are readily available in the optical regime. Random walk in frequency is given by a PSD and an Allan variance

$$S_{\Delta f} = h_{-2} f^{-2} \text{ and } \sigma_A^2 = (4\pi^2/6)\, h_{-2}\tau^{+1}. \qquad (4)$$

In this paper, 1/f noise is considered as a third ubiquitous noise phenomenon with

$$S_{\Delta f} = h_{-1} f^{-1} \text{ and } \sigma_A^2 = 2 \ln 2\, h_{-1}\tau^0. \qquad (5)$$

The latter formula shows the interesting property of the Allan variance of being constant in the presence of flicker noise, thus providing a "flicker floor".

The three noise phenomena presented here constitute a series

$$S_{\Delta f}(f) = h_0 + h_{-1} f^{-1} + h_{-2} f^{-2}, \qquad (6)$$
$$\sigma_A^2(\tau) = (h_0/2)/\tau + (2 \ln 2\, h_{-1}) + (4\pi^2/6)h_{-2}\tau.$$

In this series (which may include positive powers in f), the term h_0 is considered as the fundamental term which cannot be eliminated by technical perfection of the ringlaser, as opposed to the other terms where technical advances may produce substantial improvements.

With the definition $Q = 2\pi f_0 E/P$ (E = energy stored in cavity) and with $y = \Delta f/f_0$, equation (2) becomes

$$S_y = h\ P/4\pi^2\ f_0\ E^2. \tag{7}$$

Designing maser systems with roughly the same energy stored and power lost (the minimum power lost is the one necessary for the detection process), equation (7) predicts under optimum conditions an inverse relation of noise and laser frequency for oscillators with quantum-limited noise, at least in the frequency range where $h\ f_0 >> kT$.

To gain an estimate of the improvement of S_y(HeNe) over S_y(H), we replace f_0(H) by kT/h:

$$S_y(\text{HeNe})/S_y(\text{H}) \approx kT/h\ f_0(\text{HeNe}) = 0.013 \ , \tag{8}$$

i.e. two orders of magnitude.

3. EXPERIMENTAL CONSIDERATIONS

The ringlaser creates two or four countercirculating modes in the same cavity, which encloses a finite area. The noise is measured by recording the fluctuation of the beat frequency Δf between the modes. A variety of problems peculiar to ringlasers has to be overcome, among them: a) lock-in i.e. synchronization of two modes if the frequency difference becomes small b) unwanted drag effects, mainly due to Langmuir-flow in the discharge if dc-excitation of the plasma is used c) unwanted Sagnac effect due to inadvertent rotation of the cavity d) fluctuation of the bias introduced to split the mode frequencies.

4. INVESTIGATION OF THE BEAT FREQUENCY OF A RINGLASER

The data were achieved in a four-mode ringlaser with a four-mirror cavity made of Cervit (thermal expansion coefficient $\alpha \leq 10^{-7}\ K^{-1}$). The quality factor Q of the passive cavity was measured, by tuning the cavity through the resonance, as $Q = (2.9 \pm 0.2) \times 10^8$. The power lost was evaluated (see Dorschner, ref. 4) by measuring the output power through one mirror of known transmission (9.3×10^{-4} $\hat{=}$ 0.1%); with the Q given, P = 81 μW/mode. No pathlength control was used; instead, the cavity was thermally tuned in a thermostat with 100 μK short-term stability (over minutes) and a one-week standard temperature deviation of 420 μK. The three recorded runs are summarized in table I.

Table I: Records of $\Delta f(t)$

Run	A	B	C
Sampling interval t(s)	10	100	100
Total duration	11 1/2h	2 1/2d	9d
No. of points	4096	2048	7831
Percentage of outliers (%)	1.0	3.8	0.15
Std. deviation (Hz)	0.13	0.15	1.36

The following procedure was applied to the runs: a) Outliers which deviated more than 6% were replaced by the local average. This, somewhat arbitrary, measure affected only a small percentage of the points (see table I). b) The mean frequency is removed, but no trend-removal is applied. c) The data are Hamming-windowed and Fourier-transformed. d) The PSD is evaluated, after proper scaling of the absolute squares of the Fourier components. e) The spectral densities are averaged, up to 100 points, to achieve a random error down to 10% at the high-frequency end.

Figure 1 shows the data recorded of run B, after steps a) and b) are applied.

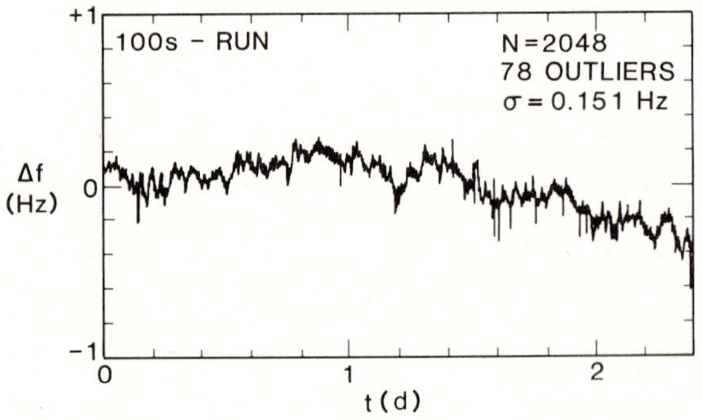

Figure 1: Beat frequency Δf (2048 points), sampled at 100 s - intervals. The rms frequency excursion over 57 h, after removal of 78 outliers, is σ = 0.15 Hz.

There are still clearly visible "outliers" not removed. The peak-to-peak frequency excursion is 0.74 Hz, but the rms deviation is only 0.15 Hz over 2 1/2 days. In figure 2 the PSD follows clearly a 1/f-line within a decade of the Nyquist frequency (5 mHz). Only 100 point averages are used in this figure. White noise is not reached in this run.

Run C lasted for 9 days, with an rms deviation of 1.36 Hz over this period. Its PSD barely reaches the white noise level (see below), but again shows clearly 1/f noise down to about

450 μHz, below which Fourier frequency $1/f^2$-noise is evident over 5 decades of PSD, down to the lowest Fourier frequency reached, 1.2 μHz, see figure 3. Run A, with a Nyquist frequency of 50 mHz, did show wite noise. It

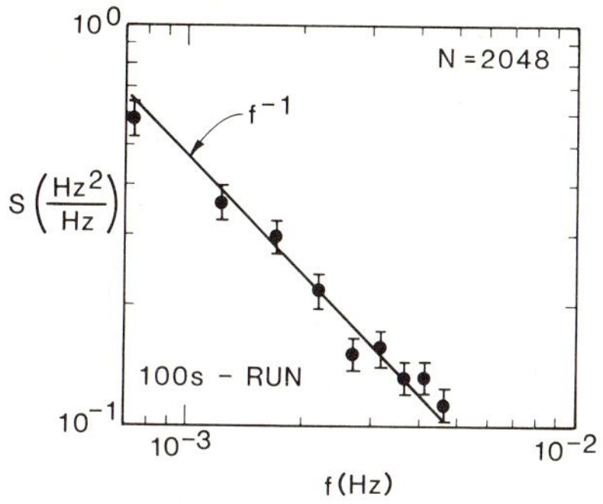

Figure 2: PSD of beat frequency given in figure 2. In the Fourier domain shown, it is well represented by 1/f noise.

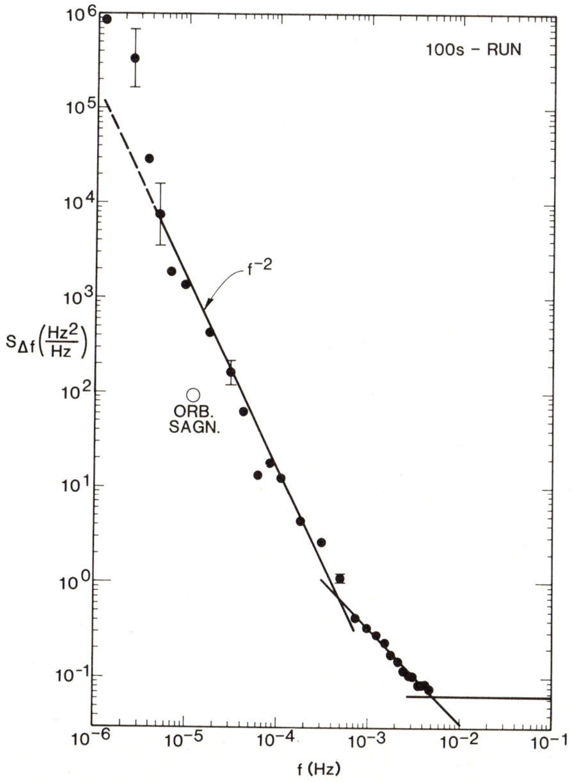

Figure 3: PSD of 8192 points of a beat frequency sampled at 100 s-intervals. The rms-deviation over the duration of 9 d is 1.36 Hz. The spectrum is well represented by $S_f = h_{-1} f^{-1} + h_{-2} f^{-2}$. The horizontal line at the lower right is from run A.

also demonstrated quantization noise, due to the fact that the .1 Hz resolution of the 10 s-run was somewhat larger than the white noise level. Correction for quantization noise resulted in a white noise level $h_{0,exp}$ = 0.063 $Hz^2/Hz \pm 10\%$.

5. SUMMARY OF RESULTS

The three runs, taken over a time span of about 3 months, can fairly well be described by the PSD:

$$S_{\Delta f,exp}(f) = 0.063 + 3.3\times10^{-4}/f + 1.5 \times 10^{-7}/f^2 \ (Hz^2/Hz)$$
$$= h_0 + h_{-1}f^{-1} + h_{-2}f^{-2} \ . \quad (9)$$

Of this series, the term h_0 can be compared to the theoretical white noise level, eq. 2. With the data of section 4, we obtain $h_{0,calc.}$ = 0.041 Hz^2/Hz.

Thus $h_{0,exp}/h_{0,calc}$ = 1.54, with an estimated error of about 20%. It is interesting to observe that this ratio is close to the one measured by Dorschner, ref. 4. There the correction factor $N_2/(N_2 - N_1)$ was advanced as a possible explanation (partial filling of the lower level N_1). Since no experiments were done which would allow to determine this ratio, we have no independent check at this time.

6. ALLAN VARIANCE

Figure 4 gives $\sigma_A^2(\tau)$ for two runs.

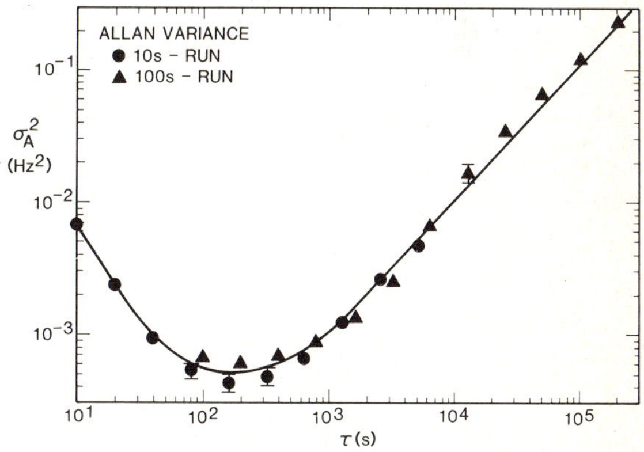

Figure 4: Allan variance of runs A and C. The minimum variance, at τ = 200 s, has a value of $\sigma_A^2 = 5 \times 10^{-4}$ Hz^2.

The broad minimum suggests again the presence of 1/f noise. The value of the minimum corresponds to $\sigma_A(200\ s)$ = 0.022 Hz, indicating that a relative frequency fluctuation of the order of $\Delta f/f_0 = \sigma_A/f_0 = 4.6 \times 10^{-17}$ can be reached.

7. DISCUSSION

To ascertain the explanation of the white noise level, additional runs with different power levels and different degrees of gain saturation should be made. It would also be interesting to test to which extent the 1/f- and $1/f^2$ components depend on the power levels. Investigations suggest that the lower the power level, the lower the low-frequency noise, which would be in keeping with observations on low-frequency noise in other devices.

The optimum (low) power level with respect to shot noise of the detector is not reached by far; in other words, the mirrors, including the output mirror can be made much tighter, in order to increase the quality factor, without introducing noticeable noise from the detection system, contrary to experiments in H-masers (6). The lower necessary power input will reduce temperature gradients. It appears also that a more direct attack on the "technical" fluctuations can be effected by using a different cavity material with higher thermal diffusivity than Cervit. With respect to technical and fundamental noise, the design of ringlasers appears yet just to be in the beginning.

Finally, figure 5 presents the data of this paper as $S_y(f)$.

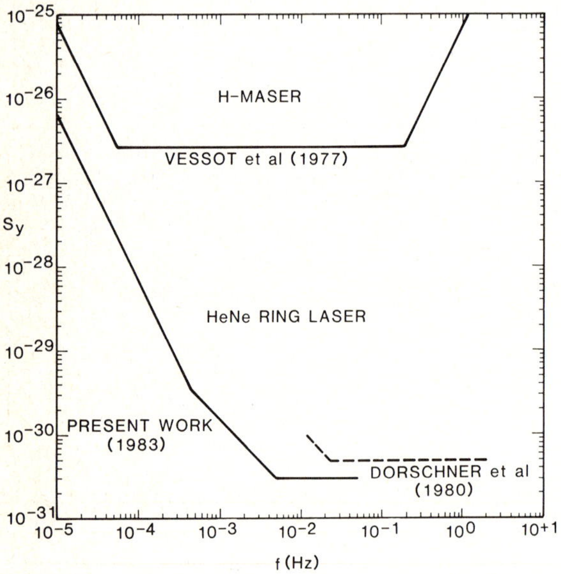

Figure 5: Comparison of recent measurements of $S_y = S_{(\Delta f/f_o)}$ in H-masers (f_o = 1.42 GHz) and in HeNe - ringlasers (f_o = 474 THz). The data of Dorschner et al. were obtained in a similar ringlaser but with different quality factor.

The differences in the white noise levels in the paper of Dorschner et al. is due to different quality factors and power levels. As mentioned above the ratio of calculated and experimental white noise is about the same in both papers.

A comparison to the H-maser (13) verifies that the white noise levels reached by the ringlaser at 474 THz is substantially below that of the H-maser.

The foregoing results let the ringlaser appear as an attractive tool for a variety of researchers, including studies of light propagation in accelerated systems (Belenov-Markin effect, e.g.), transversal light drag, drag in ionized gases, velocity noise, setting upper limits for preferred frames in gravitation theories. With future improvement and/or different versions (large ring interferometers, multiloop fiberoptic rings) the possibility of gravitation experiments opens up.

References

(1) Rosenthal, A. H., Regenerative circulatory multiple-beam interferometry for the study of light-propagation effects, J. Opt. Soc. Am. 52 (1962) 1143-1148.

(2) Bilger, H. R., and Stowell, W. K., Light drag in a ringlaser: An improved determination of the drag coefficient, Phys. Rev. A16 (1977) 313-319.

(3) Statz, H. et al., Four-frequency laser gyroscopes (Laser Handbook [ed. M. L. Stitch], North-Holland, Amsterdam 1983) to be published.

(4) Simpson, J. H., A fundamental noise limit to RLG performance, 1980 NAECON (IEEE, New York) 80-83; Klimontovich, Yu. L., Kovalev, A. S., and Landa, P. S., Natural fluctuations in lasers, Sov. Phys. Usph. 15 (1972) 95-113; Dorschner, T. A., et al., Laser gyro at quantum limit, IEEE J. Qu, El. QE-16 (1980) 1376-1379.

(5) Kuvatova, E. A., Experimental investigation of difference-frequency fluctuations in a ringlaser emitting at $\lambda = 1.15 \mu$, Sov. J. Quant. Electron. 6 (1976) 373-374; Dorschner, T. A., et al, ref. 4.

(6) Vanier, J., The active hydrogen maser: state of the art and forecast, Metrologia 18 (1982) 173-186.

NOISE IN PHYSICAL SYSTEMS AND 1/f NOISE
M. Savelli, G. Lecoy and J-P. Nougier (eds.)
© Elsevier Science Publishers B.V., 1983

ULTRA LOW NOISE GAAS MESFET MICROWAVE OSCILLATORS

J. GRAFFEUIL*, A. BERT**, M. CAMIADE**, A. AMANA*, J.F. SAUTEREAU*

* LABORATOIRE D'AUTOMATIQUE ET D'ANALYSE DES SYSTEMES DU C. N. R. S.
Université P. Sabatier - 7, avenue Colonel Roche 31400 TOULOUSE

** THOMSON-CSF-DCM Domaine de Corbeville 91405 ORSAY

It is shown that the phase noise of a GaAs FET oscillator is the result of a mixing process between the carrier and a noise voltage generator in series at the gate terminal. From the subsequent analysis rules for the selection of the most appropriate device and of the most suitable oscillator configuration are derived to decrease phase noise. As an illustration the designing of GaAs MESFET DRO exhibiting the best noise performances so far reported (- 105 dBc at 10^4Hz off the carrier) is given.

INTRODUCTION

Phase noise often limits the performance of today's microwave communication systems. For example, it degrades adjacent channel sensitivity. Therefore the high phase noise observed today in GaAs MESFET oscillators tends to limit their use. Indeed the single sideband phase noise of a dielectric resonator stabilized MESFET oscillator (FET D.R.O.) at 10 kHz offset from the carrier is hardly lower than - 90 dBc at 12 GHz. As a comparison, commercially available Gunn diodes oscillators exhibit a phase noise in excess of -100dBc in the same conditions but they have a very poor efficiency and a bad long term stability. On the other hand GaAs FET's are now common devices, which are universally used in most microwave systems, and FET's oscillators may have up to 40 % of efficiency. Therefore, the designing of ultra low noise FET D.R.O's has now become a major goal.

The high phase noise of GaAs FET DRO is the result of the carrier phase modulated by the low frequency (L.F) noise inherently present in the device. If the L.F noise at a frequency f is mixed with the signal carrier at a frequency F an F.M noise is produced at an offset frequency f from the carrier. This paper will investigate in turn the following points (i) L.F noise origins (ii) L.F noise up-conversion (iii) Designing of an ultra low noise FET DRO.

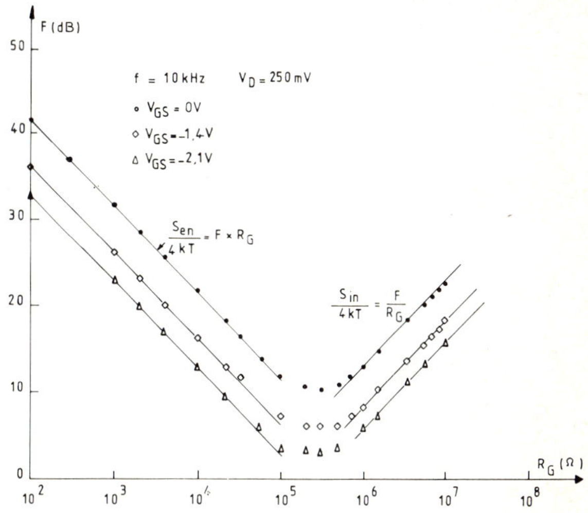

Fig. 1. Low frequency noise figure (f = 10 kHz) at low drain voltage (V_{DS} = 250 mV) of a common source GaAs MESFET versus input resistance R_G

I. LOW FREQUENCY NOISE ORIGINS AND PROPERTIES

a) L.F noise figure

Figure 1 represents the variations of the L.F noise figure (f = 10kHz) of a commercially available medium power device at low drain voltage (V_{DS} = 250 mV, common source) versus the input resistance R_G, the gate voltage V_{GS} being taken as a parameter.
From this figure, the following features are relevant :

(i) The L.F noise figure is a few times larger than the microwave noise figure at similar bias.

(ii) The parabolic shape of the curve provides evidence that noise is produced both by an input noise voltage generator e_n and an input noise current generator i_n. The minimum noise figure corresponds to an optimum R_{Go} approximately equal, at a frequency f, to $R_{Go} \simeq \sqrt{S_{en}(f)/S_{in}(f)}$ where $S_{en}(f)$ and $S_{in}(f)$ are the spectral intensity of e_n and i_n.

(iii) As can be seen in Fig. 1 the determination of the values of S_{en} and S_{in}, at the given frequency f, is straighforward.

b) Input noise voltage generator

The input noise voltage generator can be either measured directly or derived from the noise figure variations versus R_G.

As previously reported[1] e_n is very little dependant upon the drain voltage and the variations of its spectral intensity versus gate bias are given by :

$$S_{en}(f) = \frac{4Vp^2}{N_D\ Z\ a\ L}\left[X^4(1-X)^2 A(f) + X(1-X)^2 N_c(f) + \frac{(1-X)^3}{3} N_g(f)\right] \qquad (1)$$

where Vp is the pinch-off voltage, N_D,Z,a,L are the doping density, gate width, epilayer thickness and gate length. X is a dimensionless gate bias dependant parameter equal to :

$X = 1 - \sqrt{\frac{V_{bi} - V_{GS}}{Vp}}$ (V_{bi} : Schottky barrier built-in voltage)

$N_c(f)$ and $N_g(f)$ are frequency dependant noise coefficients linked to the basic physical phenomena responsible for the noise produced respectively in the conductive layer and in the depleted layer. A(f) is a more complex coefficient which depends not only on $N_c(f)$ and $N_g(f)$, but also on the physical characteristics of the lateral regions between gate and source or gate and drain. The first term into the square brackets corresponds to the noise produced in these lateral regions, the second term gives the noise produced in the bulk conductive region beneath the gate, the third term corresponds to the charge fluctuation noise in the depleted region beneath the gate. Under normal operating conditions, above 10^3 Hz,the noise is mostly flicker noise produced in the bulk conductive region [1] and relation (1) reduces to :

$$S_{en}(f) = \frac{4\ Vp^2}{N_D ZaL}\left[X(1-X)^2\ \frac{\alpha_c}{f}\right] \qquad (2)$$

with $10^{-5} < \alpha_c < 10^{-3}$

Therefore, according to relationship (2), an efficient way of ensuring a minimum input L.F noise is to select devices with a large Z (power devices), the largest L possible providing the expected frequency performances are met, and the smallest α_c.

Unfortunately α_c cannot yet be related to any physical parameters of the layer. Moreover, it has been found experimentaly that α_c could vary greatly from one device to another not elaborated in the same way. Therefore, the more efficient way of assessing a device for its α_c is to carry out L.F input noise generator measurements on a set of different devices previously selected for their geometrical features.

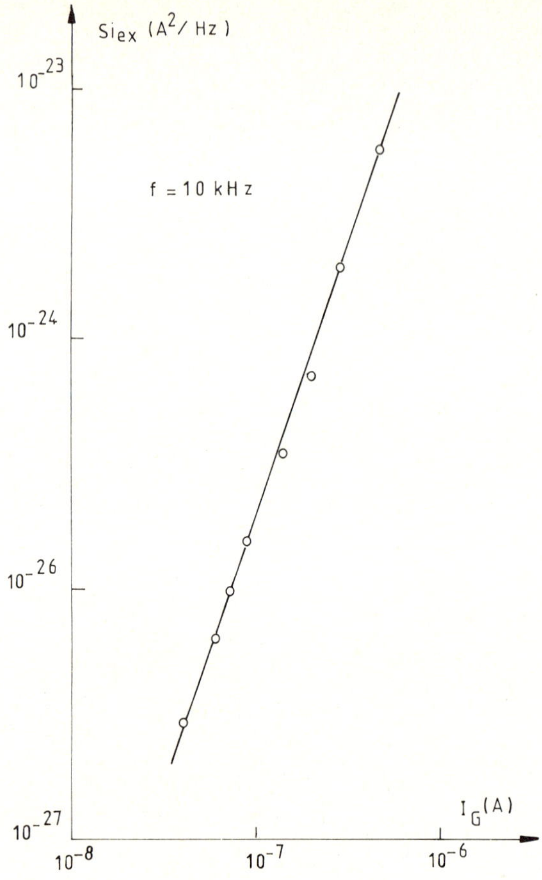

Fig. 2. Extra gate noise current spectral intensity versus gate leakage current I_g at a frequency equal to 10 kHz (V_{DS} = 250 mV)

c) Input current generator

The input noise current generator i_n is experimentally derived from the noise figure variations versus R_G (Fig. 1).

Its theoretical value, providing Y_{11} is the input admittance (including stray capacitances), for the grounded source GaAs MESFET (output short-circuited) is given by :

$$i_n = Y_{11}\ e_n + i_g$$

where i_g is the overall gate noise current produced not only by shot noise associated with the leakage current I_g across the gate, but the gate noise current could be also produced by excess noise sources such as the noise contributed by the tunneling of electrons into deep level defects[2].

In order to investigate further this excess noise the quantity $S_{iex} \simeq S_{in} - |Y_{11}|^2 S_{en}$, was plotted versus I_g at a given frequency f. It can be seen in Fig. 2, that this excess noise is proportional to I_g^2. Moreover, it appears to follow the $1/f^x$ law with x about 1 (Fig. 3). These features, together with the fact that leakage gate current is mostly surface current[3], support the hypothesis that this noise is surface 1/f noise.

Finally, it can be stated that in usual devices with small leakage current ($I_G \lesssim 1 \mu A$) this surface noise current i_g is small as compared to $Y_{11}e_n$. Hence, in most cases, the L.F noise behaviour of a GaAs MESFET is described correctly by a voltage noise generator e_n connected in series at the gate terminal. The present statement becomes invalid only if a large impedance Z_g ($Z_g > Z_{go}$) is connected at the gate terminal.

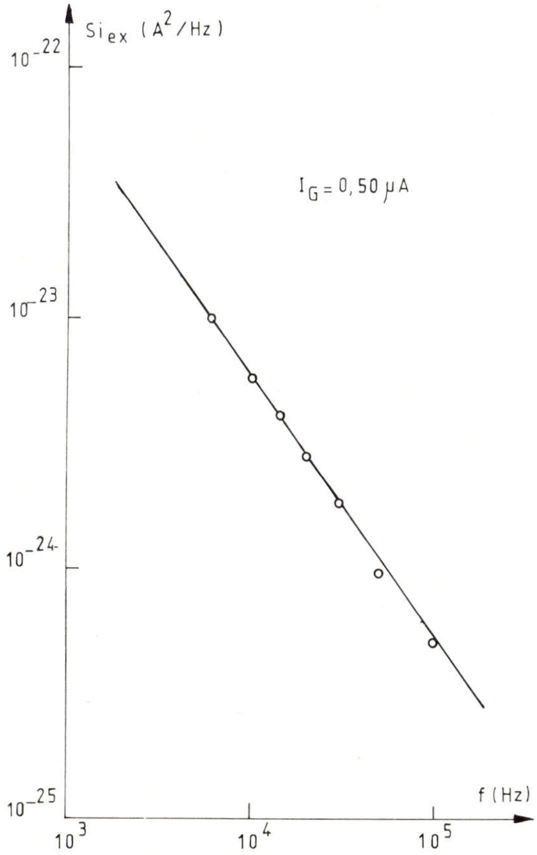

Fig. 3. Extra gate noise current spectral intensity versus frequency (I_G = 0.5 μA ; V_{DS} = 250 mV)

II. L.F NOISE UP-CONVERSION

As to investigate experimentally the conversion of this L.F voltage noise, an a.c low frequency generator of amplitude v(RMS) and frequency f($10 \leq f(Hz) \leq 10^6$) was connected in series at the gate terminal of several GaAs FET oscillators[4]. The subsequent FM modulation index m at the frequency f off the carrier was checked using standard techniques and in each case, the modulation sensitivity F_c = m.f/v (called the conversion coefficient) was derived in each case. It was found that coefficient is frequency independant in the measurement range[4] and its typical values are given in table 1.

Oscillator number	Oscillator configuration	F_c(Hz/v)
0	Free running	$14\ 10^7$
1	D.R.O (common source) Resonator on gate	$1.8\ 10^6$
3	D.R.O (resonator as a coupling element between gate and drain) Output on source	$0.6\ 10^6$
4	D.R.O (resonator as a coupling element between gate and source) Output on drain	$2.4\ 10^6$

Table 1 : Conversion coefficient for different types of oscillators

From Table 1, it appears that the use of a stabilization element, such as a dielectric resonator, greatly reduces the modulation sensitivity. Moreover, configuration 3 provides the smallest conversion coefficient and turned out to be the most suitable for low noise D.R.O.'s. Also, the conversion coefficient, measured on an oscillator made of two chips in parallel appeared to be twice as small as a single chip oscillator with the same output power. A possible interpretation invokes the smallest non-linearities occuring in the 2-chip oscillator as each chip is operated at a lower level. Moreover, if the L.F noise voltage generator e_n of a chip is measured "in situ" the product $S_{en}(f)$ by F_c corresponds closely to the F.M noise measured in a 1 Hz bandwith at an offset frequency f. These statements support the hypothesis that the F.M noise results from a mixing process between the carrier and the L.F noise voltage generator[5].

III. DESIGNING OF AN ULTRA LOW NOISE FET DRO

Devices to be used for the oscillator were previously assessed for their low L.F input noise voltage spectral intensity. Many power or medium power devices were investigated and some of the results are given in table 2.

Devices 1 and 2 exhibit the lowest e_n and are therefore the most suitable for the oscillator. Oscillators were subsequently elaborated on an alumina substrate with a resonator as a coupling element between gate and drain, the source being taken as the output.
F.M noise performances were finally investigated and the results are given in Figure 4 for devices 1, 2 and 5.

Device	Total gate width (μm)	Sen (V^2/Hz) f = 10^4Hz
1	800	3.1 10^{-15}
2	1500	9.8 10^{-15}
3	600	16. 10^{-15}
5	300	48. 10^{-15}
6	700	54. 10^{-15}
7	1400	33. 10^{-15}

Table 2 : L.F. noise voltage spectral intensity of seven different GaAs MESFET chips at 10^4 Hz (V_{GS} = 0 V, V_{DS} = 100 mV)

The phase noise obtained with device 5 is about 20 dB higher than the phase noise obtained with devices 1 and 2 assessed for their low noise.
It can also be seen that the phase noise of these two optimized oscillators is about - 105 dBc at 10^4 Hz off the carrier which is to our knowledge the best noise performances up to day reported for GaAs MESFET oscillators. Output power of the oscillator range between 10 and 30 mW.

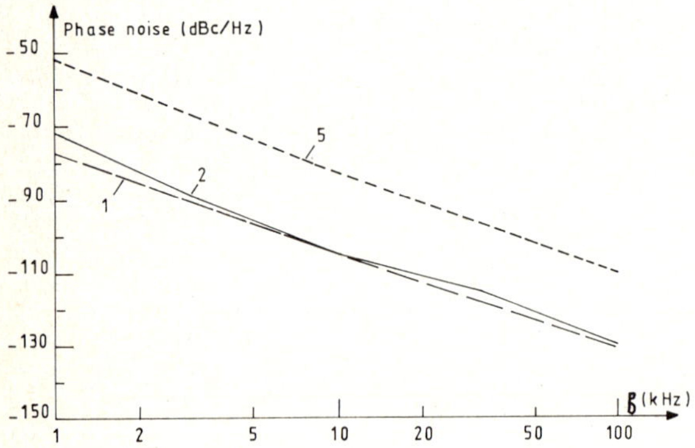

Fig. 4. F.M noise spectra of three GaAs MESFET oscillators. Oscillators 1 and 2 were built with devices assessed for their low LF noise

CONCLUSION

An oscillator was built with a power device previously assessed for its low frequency noise. A dielectric resonator was used as previously indicated and provided an external quality coefficient greater than 1000. The single sideband phase noise at 10 kHz offset was found to be - 105 dBc at 12 GHz. To our knowledge it is the best noise performance reported so far for GaAs MESFET oscillators.

As a result, FET DRO's can now compete successfully with bipolar or Gunn diodes oscillators and we believe that in the near future, they could be competitors with very low noise oscillators such as klystrons.

This work was supported by the "DIRECTION AFFAIRES INDUSTRIELLES & INTERNATIONALES" of the "DIRECTION GENERALE DES TELECOMMUNICATIONS" and by the "GRECO MICROONDES DU C.N.R.S.".

REFERENCES

1 GRAFFEUIL J, TANTRARONGROJ K and SAUTEREAU JF, "Low frequency noise physical analysis for the improvement of the spectral purity of GaAs FETs oscillators". Sol. St. Electron. 25.5 (1982), 367-374

2 DAS M.D. and GHOSH P.K., "Gate current dependance of low frequency noise in GaAs MESFET's". IEEE Electron. Device. Lett. 8 (1981), 210-213

3 OZEKI M, KODAMA K and SHIBATOMI A, "Surface analysis in GaAs MESFET by g_m frequency dispersion measurement". Gallium Arsenide and Related Compounds, Oiso (Japan) 1981 (Inst. Phys. Conf. n° 63, 323-328

4 SAUTEREAU J.F., GRAFFEUIL J, TANTRARONGROJ K, ABLART G, GOURDON J.C and VIGOUROUX B, "Large signal design and realisation of a low noise X-band GaAs FET oscillator". 11th European Microwave Conference Proceedings, Amsterdam (1981), 464-468

5 DEBNEY B.T and JOSHI J.S, "A theory of noise in GaAs FET microwave oscillators and its experimental verification", To be published

NOISE IN PHYSICAL SYSTEMS AND 1/f NOISE
M. Savelli, G. Lecoy and J-P. Nougier (eds.)
© Elsevier Science Publishers B.V., 1983

REDUCTION OF THE LOW FREQUENCY NOISE SIDEBANDS IN OSCILLATORS

H. B. Chen*, A. van der Ziel, and K. Amberiadis**

Department of Electrical Engineering
University of Minnesota
Minneapolis, Minnesota 55455

Low-frequency noise in an oscillator normally produces two l.f. noise sidebands around the center frequence f_o of the oscillator. It was shown experimentally that these noise sidebands can be eliminated by using an oscillator circuit with an odd-symmetrical characteristic. A simple theory based on the van der Pol approach to oscillators can explain the observed effects.

1. INTRODUCTION

In the course of an experimental program on noise in oscillators it was shown that oscillator circuits with odd-symmetrical characteristics eliminate the two l.f. noise sidebands normally produced in an oscillator. This property of low-frequency noise sideband cancellation is shared by any circuit that uses two or three-terminal devices that are so connected that they produce odd-symmetrical characteristics. We first give the theory and then discuss the experiments.

2. THEORY

We follow here van der Pol's famous treatment of the oscillator, in the form found in van der Ziel's 1954 book[1] and apply it to an arbitrary three-terminal device (bipolar transistor, junction field effect transistor, metal-oxide-semiconductor field-effect transistor, etc.).

The circuit is shown in Figure 1; v is the applied a.c. voltage, L and C provide the tuning, C incorporates the input capacitance of the active device and the mutual conductance M provides the feedback. It is assumed that all the l.f. device noise comes from the output current; in that case the only noise sources of interest are the thermal noise generator i_{na} associated with the resistance R of the tuned circuit and the l.f. output noise source i_n of the device. The time varying output current is represented by

$$i = \alpha v - \beta v^2 - \gamma v^3 \tag{1}$$

The equation for the oscillator circuit is then [see van der Ziel, eq. (10.52)][1]

$$\frac{d^2v}{dt^2} + \frac{d}{dt}\left\{\frac{1}{C}\left[\left(\frac{1}{R} - \frac{M}{L}\alpha\right) v + \frac{M}{L}\beta v^2 + \frac{M}{L}\gamma v^3\right]\right\} + \frac{v}{LC}$$

$$= \frac{1}{C}\frac{d}{dt}(i_{na}) + \frac{M}{LC}\frac{d}{dt}(i_n) \tag{2}$$

In the above equation $\omega_o = (LC)^{-1/2}$ is the center frequency of the oscillator's response. Van der Pol omitted the term with v^2 since it did not add any new feature in the amplitude limiting action of the oscillator; we take it into account since it plays a quite important role in the purity of the spectrum of the oscillator (it increases the linewidth).

Van der Pol replaces the random noise sources by their Fourier components in the frequency interval df; that is since

$$\overline{i_{na}^2} = \frac{4kT\,df}{R}$$

we have

$$i_{na} = \left(\frac{8kT\,df}{R}\right)^{1/2} \cos\omega t \tag{2a}$$

which drives the oscillator. If $S_I(f)$ is the l.f. noise spectrum of the active device we write

$$\overline{i_n^2} = S_I(f)df$$

or

$$i_n = [2S_I(f)df]^{1/2} \cos\omega_p t \tag{2b}$$

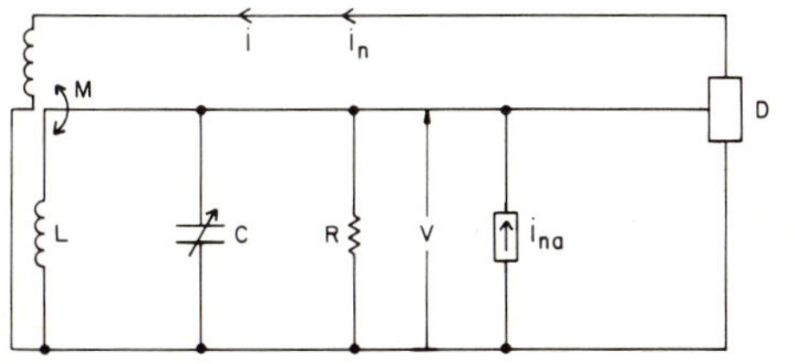

Figure 1 : Oscillator circuit with magnetic coupling (mutual inductance M) providing feedback; bias voltages are omitted and D represents the active device.

* On leave of absence from the Department of Information and Control Engineering, Jiaoton University, People's Republic of China.

** Now at RCA Laboratories, Princeton, NJ 08540.

We shall see that this l.f. noise source can modulate the oscillator.

The l.f. noise source i_n in (2) produces a low-frequency noise voltage v_n, and this, in turn, because of the term βv^2 in (1) and (2), produces noise sidebands $\omega_o \pm \omega_p$ around the oscillator frequency and so produces line broadening. In the same way an impressed low-frequency signal of frequency ω_p produces, because of the term βv^2 in (1) and (2), sidebands $\omega_o \pm \omega_p$ around the oscillator frequency ω_o.

Only if $\beta = 0$, that is, if the device is operating at an inflection point ($d^2i/dv^2 = 0$), so that it provides an odd-symmetrical characteristic, will the sidebands around ω_o be missing. This property holds for any device that either has an odd-symmetrical characteristic by its own nature or in which the odd-symmetrical characteristic is provided by the circuit.

In a balanced circuit (see Fig. 2) in which the two applied input voltages are v_1 and v_2 with $v_1 = -v_2 = 1/2\ v$, where v is the total signal we have

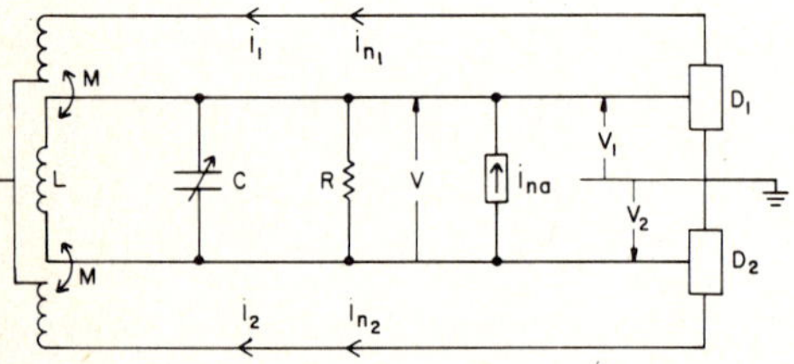

Figure 2 : Balanced oscillator circuit; bias voltages omitted. D_1 and D_2 represent the active devices.

$$i = i_1 - i_2 = (\alpha v_1 - \beta v_1^2 - \gamma v_1^3) - (\alpha v_2 - \beta v_2^2 - \gamma v_2^3)$$

$$= \alpha v - 2\gamma\left(\frac{v}{2}\right)^3 \quad (3)$$

so that the second order term with β in eqs. (1) and (2) is missing. Any balanced circuit thus automatically produces an odd symmetrical characteristic that eliminates the low frequency sidebands.

In Figs. 1 and 2 an arbitrary three-terminal device is shown. But it should be obvious that the results remain the same for any negative-conductance two terminal device such an Impatt or Gunn diode connected directly across the tuned circuit without feedback. In these cases balanced circuits might be particularly useful.

3. EXPERIMENT

In the experiment we demonstrated two things. First that the low frequency noise in an oscillator produces two sidebands centered around the oscillator frequency ω_o and second that these noise sidebands can be reduced by achieving odd-symmetrical characteristics. This can be shown by using the oscillator of Fig. 3. The variable potentiometer in the circuit is employed for changing the nonlinearities of the oscillator. When properly tuned the i-v characteristics of the circuit can be made odd-symmetrical. After the proper potentiometer adjustments the injected noise (side-lobes at 210 Hz from the center frequency) and the circuit noise are clearly reduced (Fig. 4). This results in improved oscillator output purity (linewidth reduction.

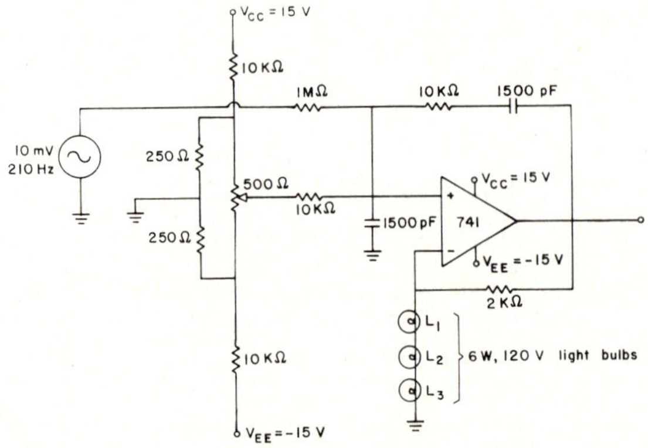

Figure 3 : Oscillator circuit using an op-amp; the 10 mV, 210 Hz signal is the impressed low frequency noise.

REFERENCES

[1] Balth. van der Pol, Proc. Inst. Radio Engrs., Vol. 22, 1051, 1934; we follow here A. van der Ziel, Noise, Prentice Hall, 1954, Section 10.5.

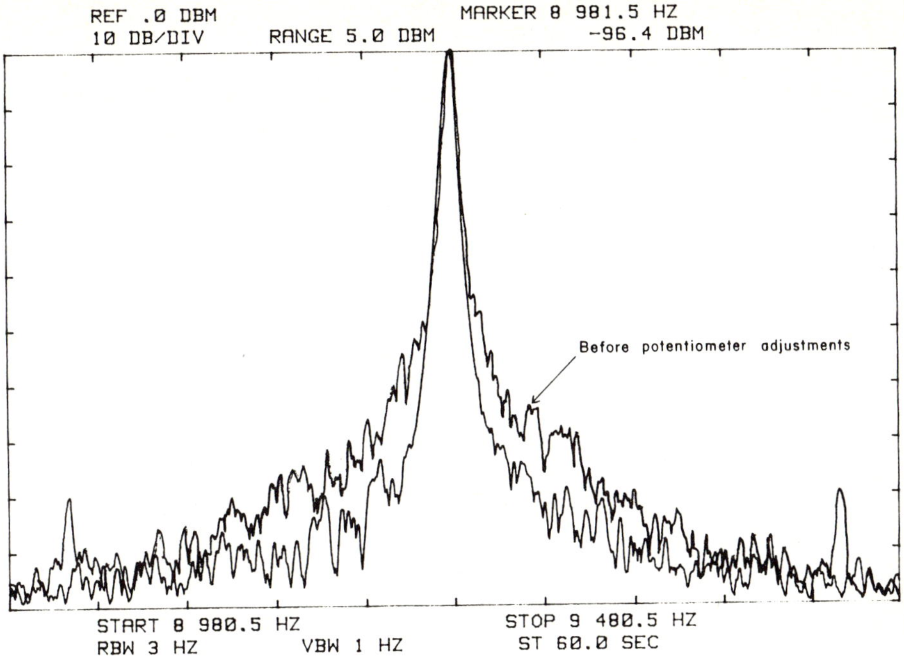

Figure 4 : Spectral intensity display of the output of the oscillator of Fig. 3 before and after the potentiometer adjustment. Injected noise and circuit noise are reduced considerably.

OTHER PHYSICAL SYSTEMS

NOISE IN PHYSICAL SYSTEMS AND 1/f NOISE
M. Savelli, G. Lecoy and J-P. Nougier (eds.)
© Elsevier Science Publishers B.V., 1983

A CURRENT NOISE INVESTIGATION ON STRESS RELAXATION MECHANISMS IN THIN METAL FILMS

Giorgio Bertotti and Fausto Fiorillo

Istituto Elettrotecnico Nazionale Galileo Ferraris
Gruppo Nazionale Struttura della Materia del C.N.R.
I-10125 Torino, Italy

Current noise gives information on the mechanical behavior of metal films on a substrate by providing direct evidence of lattice defect processes during stress relaxation. Experiments are reported in this communication showing that the deformation mechanisms involving dislocations are predominant in a wide interval of temperatures and strain rates. Diffusive relaxation, which consists in the transport of matter via a vacancy mechanism, plays a not negligible role at very low strain rates and high temperatures only. Evidence is also given of impurity-dislocation interaction, which gives rise, under specific experimental conditions, to anomalous noise intensity, according to the prediction of microserrations in the stress-strain curve.

1. INTRODUCTION

The metallurgical properties of metal films on substrates have received considerable attention in recent years, both from a technological and a fundamental point of view. For their study, fairly sophisticated experimental techniques have been developed.[1-3] As a general outcome, it is obtained that thin metal films present a very specific mechanical behavior, consequent to their uncommon grain size, texture, surface to volume ratio and, above all, to their interaction with the substrate. Actually, due to the thermal expansion mismatch, large stresses arise in the film upon a temperature variation. These stresses tend to relax through a variety of deformation mechanisms, involving the motion of dislocations and point defects. Based on a large body of experimental results in bulk materials, a classification of such mechanisms can be made.[4]

A common drawback of the above mentioned techniques lies in their inability to provide direct information on the kinetics of the defects operating in stress relaxation. Usually, one can only induce some information on the predominant relaxation mechanisms from the shape of the stress-strain curve, interpreted according to postulated constitutive equations which relate, for each mechanism, the strain rate to stress and temperature.[5]

In the present paper we show that, by measuring the current noise associated with the resistivity fluctuations exhibited by the metal film during thermal straining,[6] we gain insight into the microscopic aspects of stress relaxation and, in particular, into the features of the involved dislocation processes, in connection also with possible interactions with point defects. The case of an aluminum film deposited upon an oxidized silicon substrate, thermally cycled between -100°C and 400°C, is discussed. It is shown that the behavior of the noise power intensity during straining is directly related to the stress relaxation mechanisms involving the generation, motion and annihilation of dislocation lines. The role of diffusive relaxation, which is due to the motion of vacancies, and the peculiar effects related to the interaction of impurities with dislocations are also put in evidence.

2. EXPERIMENTS AND DISCUSSION

The investigation concerns aluminum films having a thickness of 1÷1.5 μm, a width of 10÷20 μm and a length of 2÷3 mm, vacuum deposited upon an oxidized silicon substrate about .3 mm thick. By virtue of the different thermal expansion coefficients α_m and α_s of metal and substrate, a temperature change gives rise to film straining. In particular, for a temperature rate of change $\dot{T}$ one gets a strain rate $\dot{\varepsilon}=(\alpha_m-\alpha_s)\cdot\dot{T}$. Since $\alpha_m \gg \alpha_s$, compressive and tensile straining are respectively accomplished upon heating and cooling. The experiments are performed by thermally cycling the sample at a constant rate (in the range $2.3\cdot10^{-7}\,s^{-1} \leq \dot{\varepsilon} \leq 1.4\cdot10^{-5}\,s^{-1}$) between two given temperatures. Corresponding to the widest investigated temperature interval (-100°C÷400°C), an overall strain $\varepsilon_{max}=1.2\cdot10^{-2}$ is obtained. As far as the deformation of the film is accomplished elastically, the thermally generated planar biaxial stress changes at a rate $\dot{\sigma}=\dot{\varepsilon}\cdot E(T)/(1-\nu)$, being E(T) the temperature dependent Young modulus and ν the Poisson ratio. When the elastic limit is attained, dislocation rearrangements take place, giving rise to resistance fluctuations in the sample. [6, 7] These can be sensed as voltage fluctuations by means of a sufficiently high sampling current. In papers [6-8] a description is given of the specifically devised apparatus for the noise detection and analysis, which allows the determination of excess noise spectral intensities as low as some $10^{-20}\ V^2/Hz$ above 10 Hz.

The essential features of the noise associated with the cyclic deformation of the metal film can be summarized as follows:
1) Noise intensity is strongly non-stationary along a given thermal loop, but it shows a reproducible behavior upon repeated cycling, similarly to the case of Barkhausen noise in cyclically magnetized ferromagnets. [9] Averages are thus made over the statistical ensemble, which is actually well approximated by a set of repeated identical thermal cycles. An example of cyclic dislocation noise behavior is presented in

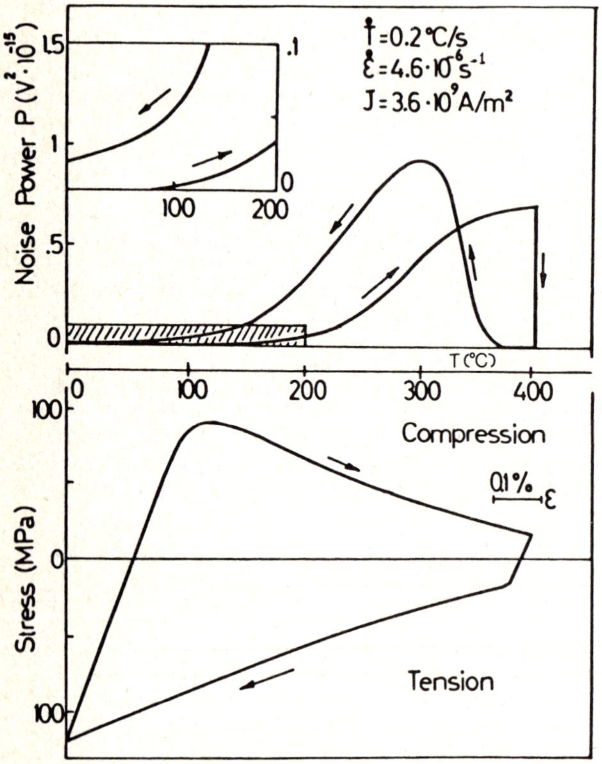

Figure 1 : a) Cyclic behavior of the noise power intensity upon thermal cycling between 0°C and 400°C;
b) Corresponding stress-strain loop.

Figure 1-a). One can notice the presence of two noise-free regions, indicating a corresponding elastic deformation of the material. An immediate consequence of such a result is the possibility of obtaining the actual stress-strain loop of the metal film upon thermal cycling, [6] as representatively shown in Figure 1-b). One is then able to see that the noise power P is, at each temperature, simply proportional to the plastic strain rate $\dot{\varepsilon}_p$:

$$P \propto |\dot{\varepsilon}_p| \; . \qquad (1)$$

2) The noise intensity is proportional to the square of the current density in the sample, as expected in the case of spontaneous resistance fluctuations.
3) Most of the noise consists of step-like bursts (rise time$<10^{-3}$ s) which are associated with avalanche-like rearrangements of the dislocation structure in the metal. This puts in evidence that stress relaxation through dislocations is an inherently discontinuous process, characterized by strong correlation effects between elementary movements of the dislocation lines.
4) Noise spectra are always of the $1/f^2$ type, in all the investigated range of frequencies (1 Hz-400 Hz).

Based on these results, we can try to work out a description of stress relaxation during thermal cycling in terms of typical deformation mechanisms, for which constitutive equations of the type $\dot{\varepsilon}_p = g(\sigma, T)$ exist. [5] To this end, we describe the time evolution of the overall dislocation resistance during plastic deformation as a random sequence of elementary events, each corresponding to the creation or annihilation of a dislocation segment. The noise bursts are then interpreted as the result of strong correlation effects between subsequent events. In strict analogy with the treatment of magnetization noise, [9] we arrive at the following expression for the dislocation noise power spectrum $\Phi(f)$:

$$\Phi(f) = \nu_0 \cdot (Jl\rho^{(d)}/\pi)^2 \cdot (\Delta n/f)^2 \; , \qquad (2)$$

where ν_0 represents the average number of clusters in unit time, J is the current density, l the sample length, $\rho^{(d)}$ the resistivity contribution of unitary dislocation density and Δn the average variation of the dislocation density in the sample associated with a cluster. If we assume that a dislocation rearrangement cooperatively involves a certain density of mobile dislocations n_m, characterized by a mean free path λ, the corresponding plastic strain variation can be written:

$$|\Delta\varepsilon_p| = b\lambda n_m\ , \tag{3}$$

b denoting the Burgers vector. Consequently, if a certain proportion $\Delta n=\theta n_m$ $(-1<\theta<1)$ of mobile dislocations is generated or annihilated during a rearrangement, we get:

$$\Delta n= |\Delta\varepsilon_p|\cdot\ \theta/\lambda b\ . \tag{4}$$

By substitution in Eq. (2) and subsequent integration vs. frequency (between 1 Hz and 400 Hz) we obtain:

$$P= (Jl\ \rho^{(d)}/\pi b)^2\cdot(\theta/\lambda)^2\cdot|\dot{\varepsilon}_p|^2/\nu_0 \tag{5}$$

according to the fact that $|\dot{\varepsilon}_p| = \nu_0\cdot|\Delta\varepsilon_p|$. On the basis of Eq. (5), applied to the results shown in Figure 1, we are now able to obtain information on the deformation mechanisms operating at different temperatures. From the stress-strain loop one finds that the plastic strain rate is fairly temperature independent. Moreover, pulse counting experiments show that this is approximately true for ν_0 as well. Thus, we can substantially attribute the observed temperature dependence of the noise power to the behavior of the quantity θ/λ . In the temperature region below 200°C (that is below $.5T_M$, with T_M the melting temperature) noise intensity is low ($\theta\to 0$, λ large) and the flow stress shows a nearly linear decrease with temperature. By increasing the temperature, the noise intensity rises dramatically ($\theta\to 1, \lambda\to 0$) and the flow stress behavior becomes non-linear. Then, the dominant stress relaxation mechanism at low temperatures must be one characterized by a fairly stable configuration of dislocations ($\theta\to 0$), which are able to run upon comparatively large distances in their glide plane (λ large). Dislocation glide is a mechanism of this type, for which a constitutive equation has been derived. [5] Accordingly, the predicted temperature dependence of the flow stress can be expressed as:

$$\sigma(T)=3\cdot\left[2(1+\nu)E(T)/l_0+k_B T\cdot\ln(\dot{\varepsilon}_p/\dot{\varepsilon}_0)/ba\right] \tag{6}$$

with l_0 the average spacing of obstacles in the glide plane, a the activation area and $\dot{\varepsilon}_0$ a known constant quantity. Equation (6) is in good agreement with the experimental flow stress behavior vs. temperature (see Figure 1) if we assume $l_0 \simeq 10^{-7}$ m and $a \simeq 2\cdot 10^{-17}$ m^2. Since $a\simeq bl_0$, it is inferred that dislocation glide is hindered by a system of very short range barriers. Such barriers are in effect dislocations themselves (forest dislocations), whose density in thin films is at least of the order of 10^9 cm^{-2}. [1]

The rapid increase of the noise intensity above $.5T_M$ indicates that more and more extensive dislocation rearrangements take place with increasing temperature. We have to deal now with a new predominant deformation mechanism, in which the formation and evolution of patterns in the dislocation structure takes place through a great deal of generation and annihilation processes ($\theta\to 1, \lambda\to 0$). Such a mechanism is basically thought to be regulated by the thermally activated climb of dislocations, in which temperature helps a trapped dislocation to jump to a more favourable glide plane, [10] where it can trigger an avalanche-like dislocation rearrangement. Unfortunately, due to surface and grain size limitations, typical of thin films, a quantitative interpretationof the noise results in terms of the classical constitutive equation of the climb-controlled creep [10] is not possible, because free development of the dislocation patterns, assumed in the theory, is strongly inhibited.

The description now given of stress relaxation in terms of dislocation processes applies to the range (actually very large) of strain rates where Eq. (1) holds. It can be seen however that at very low strain rates (less than about $5\cdot 10^{-7}$ s^{-1}) the noise intensity drops below the value expected on the ground of a linear dependence of P on $\dot{\varepsilon}_p$. This is clearly shown by the results presented in Figure 2, which suggest the presence of a nearly noiseless mechanism appreciably contributing to stress relaxation at such low rates. Arguably, this is the diffusion creep, which consists in the transport of matter through the motion of vacancies. As expected and experimen-

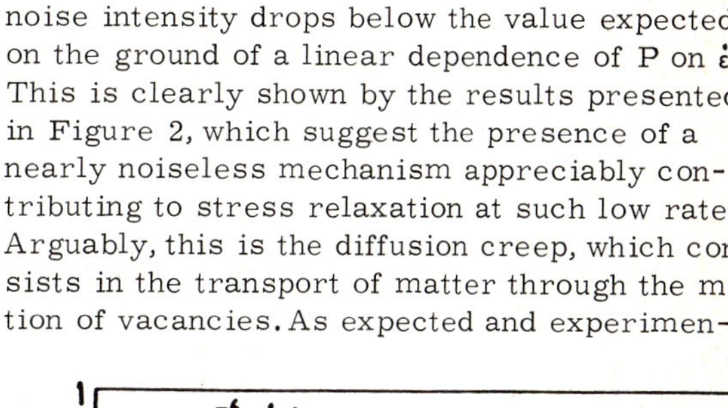

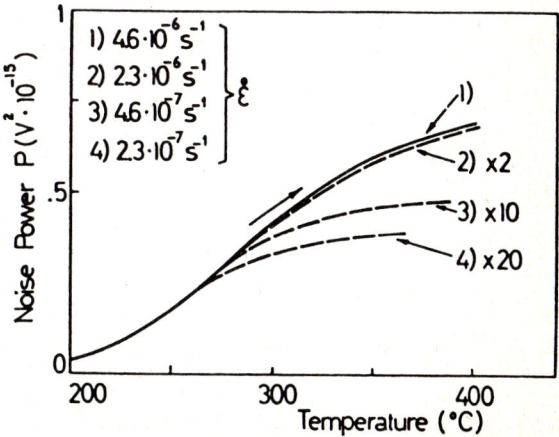

Figure 2 : Noise power behavior vs. temperature measured at different strain rates. Each curve is multiplied by the appropriate scaling factor.

tally verified, [8] the resistivity fluctuations associated with vacancy processes are much smaller than those due to the motion of dislocations and current noise is accordingly reduced. A further confirmation of the presence of a diffusive mechanism is obtained through high temperature annealing experiments, by which a complete stress relaxation can be achieved, as expected from the theory. [10]

Finally, we show that current noise represents a unique method of investigation of the dislocation-impurity interaction. In fact, it puts in evidence dynamic phenomena related to the release of the dislocations from surrounding locking atmospheres, formed by the precipitation of solute atoms (Portevin-LeChatelier effect). Conventional methods are capable of detecting dislocation unlocking only when macroscopic discontinuities (serrations) occur in the stress-strain curve, that is when the associated structural instabilities are extremely large. Experiments were performed where, during thermal cycling, straining was stopped at a given temperature, to be started again when a certain time interval t_a had elapsed. The modifications occurring in the noise intensity upon such a procedure are evident in Figure 3, where one can notice

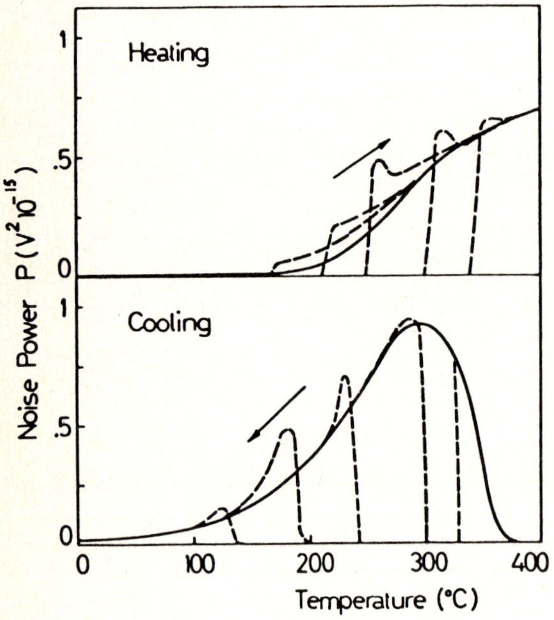

Figure 3 : Continuous line:noise power intensity upon continuous straining at a constant rate $\dot{\varepsilon} = 4.6 \cdot 10^{-6}$ s^{-1}. Broken line: straining is stopped at a given temperature and restarted at the same rate after a time interval t_{sat}.

the noise overshoot taking place upon restart. This fact can be naturally attributed to repeated dislocation breakaway from the solute atmospheres grown during the waiting time t_a and to the related anomalously large structural rearrangements. For each temperature, a saturation time t_{sat}, the minimum time required in order to have the maximum effect on the noise intensity, can be found. As expected, t_{sat} is a rapidly decreasing function of temperature. From the theory, [11] we get:

$$t_{sat} \simeq AT\exp(Q/k_BT) \qquad (7)$$

where the constant A incorporates all the temperature independent terms and Q is the activation energy for the solute diffusion in the host lattice. By introducing in Eq. (7) the values of t_{sat} found at two different temperatures (e. g. $t_{sat} \simeq 2 \cdot 10^3$ s at 200°C and $t_{sat} \simeq 30$ s at 300°C) we obtain $Q \simeq 1$ eV, which points to silicon as the interacting impurity. [12]

REFERENCES

[1] Hoffmann, R. W., in Hass, G. and Thun, R. E. (eds.), Physics of Thin Films (Academic Press, New York, 1966).

[2] Sinha, A. K. and Sheng, T. T., Thin Solid Films 48 (1978) 117.

[3] Murakami, M. M., Thin Solid Films 55 (1978) 101.

[4] Chaudari, P., IBM J. Res. Dev. 13 (1969) 197.

[5] Ashby, M. F., Acta Met. 20 (1972) 887.

[6] Bertotti, G., Celasco, M., Fiorillo, F. and Mazzetti, P., J. Appl. Phys. 50 (1979) 6948.

[7] Bertotti, G., Celasco, M. and Fiorillo, F., in Caglioti, G. and Ferro, A. (eds.), Mechanical and Thermal Behavior of Metallic Materials (North-Holland, Amsterdam, 1982).

[8] Celasco, M., Fiorillo, F. and Mazzetti, P., Phys. Rev. Lett. 36 (1976) 38.

[9] Bertotti, G., Fiorillo, F. and Sassi, M. P., J. Magn. Magn. Mat. 23 (1981) 136.

[10] Mukherjee, A. K., in Herman, H. (ed.), Treatise on Materials Science and Technology (Academic Press, New York, 1975).

[11] Friedel, J., Dislocations (Pergamon Press, London, 1964).

[12] McCaldin, J. O. and Sankur, H., Appl. Phys. Lett. 19 (1971) 524.

NOISE IN PHYSICAL SYSTEMS AND 1/f NOISE
M. Savelli, G. Lecoy and J-P. Nougier (eds.)
© Elsevier Science Publishers B.V., 1983

VACANCY NOISE IN METALS

Hermann Stoll

Max-Planck-Institut für Metallforschung
Institut für Physik
7000 Stuttgart 80, Fed. Rep. Germany

Measurement of current noise at high temperatures constitutes a novel technique for the quantitative study of vacancies in metals. Fluctuations in the number of vacancies in thermal equilibrium lead to electrical resistivity fluctuations. The measurement of the temperature dependence of the power spectrum of these resistivity fluctuations permits simultaneous determination of both the formation and the migration enthalpy of the vacancies in thermal equilibrium.

1. INTRODUCTION

Vacancies in crystals (= vacant lattice sites) constitute the simplest defects in a crystal structure. Quantitative knowledge of the properties of vacancies is of utmost importance for the understanding of a large number of processes in crystals, e.g. in the case of metals of diffusion, electro- and thermomigration, radiation damage, etc.

We shall report on the application of high temperature current noise measurements to the investigation of vacancies in metals, a technique first described by Celasco, Fiorillo, and Mazzetti [1].

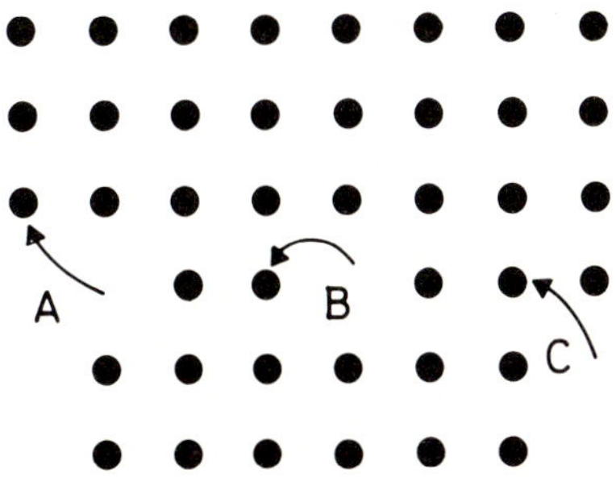

Fig. 1: A: production of a vacancy
B: migration of a vacancy
C: annihilation of a vacancy

Vacancies may be produced at inner and outer crystal surfaces (sample surfaces, grain boundaries, dislocations) (Fig. 1A). Such surfaces may also act as sinks for vacancies (Fig. 1C). A vacancy may migrate (diffuse) with a mean jump frequency $\nu(T)$ from one position on the lattice to another (Fig. 1B). The uncorrelated production, diffusion and annihilation of vacancies lead to a dynamic equilibrium of the vacancy concentration C_V (Fig. 2). C_V and its dependence on temperature T can be calculated in thermodynamic terms be minimizing the Gibbs' free energy of the crystal[2].

$$C_V(T) = \exp\,(S_V^F/k)\,\exp\,(-H_V^F/kT), \qquad (1)$$

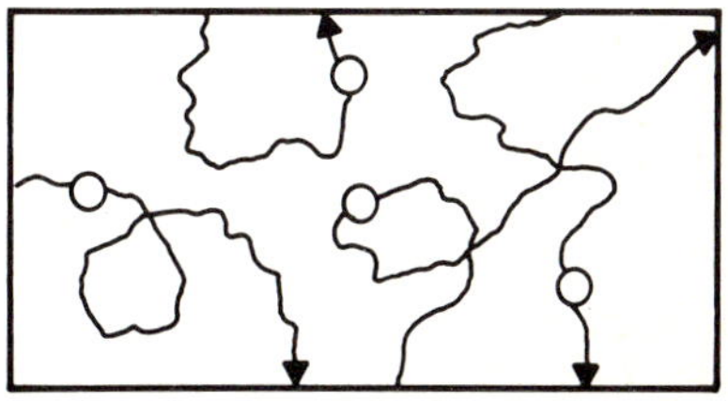

Fig. 2: Vacancies in a crystal in thermodynamic equilibrium

where H_V^F is the formation enthalpy, S_V^F the formation entropy and k is the Boltzmann's constant. The jump frequency of the vacancy also shows an exponential temperature dependence [3]

$$\nu(T) = \nu^o \exp\,(S_V^M/k)\,\exp\,(-H_V^M/kT). \qquad (2)$$

H_V^M is the migration enthalpy, S_V^M the migration entropy and ν^o is the attempt frequency. The diffusion coefficient $D_V(T)$ of the vacancies in cubic crystals may be expressed as follows [3]

$$D_V(T) = a^2\,\nu(T), \qquad (3)$$

where a is the lattice constant. Thus the temperature dependence of the vacancy concentration is determined by the formation enthalpy H_V^F, while the temperature dependence of the mobility is determined by the migration enthalpy H_V^M. The sum of these two activation enthalpies ($H_V^F + H_V^M$) can be found very accurately by measuring the self-diffusion coefficient in metals [3]. Other techniques (e.g. positron annihilation) allow the determination of the formation enthalpy H_V^F [4].

The measurement technique described here enables us to determine the difference ($H_V^F - H_V^M$) as well as H_V^M from a single type of thermal equilibrium measurement.

2. PRINCIPLE OF VACANCY NOISE MEASUREMENT

The principle of the technique is shown in Fig. 3. A vacancy which is formed spontaneously in a metal causes a resistivity increase ΔR during its lifetime τ since conduction electrons are scattered by the vacancy. When a current I_0 is passed through the sample we get a voltage pulse $\Delta U = \Delta R \cdot I_0$ of duration τ. As in thermal equilibrium many vacancies occur and disappear independently these uncorrelated pulses are superimposed on each other, producing a current noise.

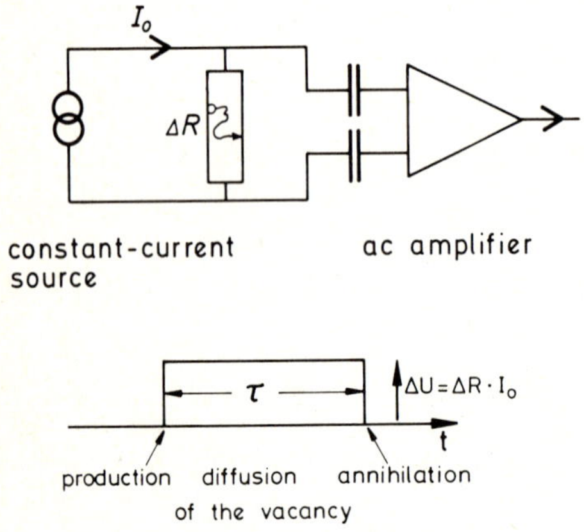

Fig. 3: Principle of the measurement

Fig. 4 shows a practical example of a circuit for measuring vacancy noise. The bridge of four metal strips decreases the dc component at the amplifier input. By correlation of the output of the two amplifiers the amplifier noise can be reduced.

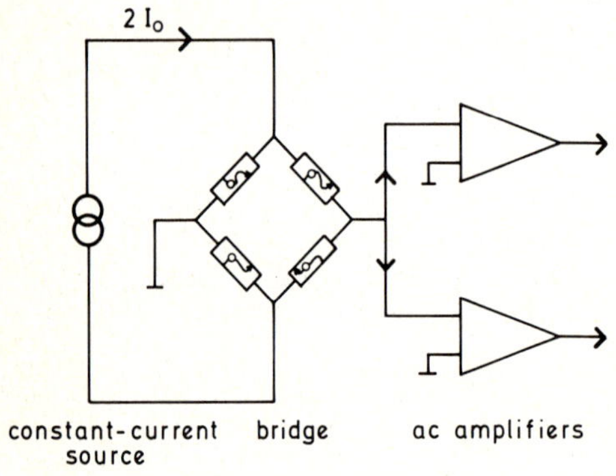

Fig. 4: Measurement circuit

3. CALCULATION OF THE POWER SPECTRUM OF VACANCY NOISE

The power spectrum $S_V(f)$ of the vacancy noise can be calculated by superimposing a series of statistically independent pulses with pulse amplitude ΔU and varying lifetime τ. The result is [5]

$$S_v(f) = 2\frac{\bar{N}}{\langle\tau\rangle}(\Delta U)^2 \frac{2}{(2\pi f)^2}\left[1-\int_0^\infty P(\tau)\cos(2\pi f\tau)d\tau\right] \quad (4)$$

$$(f > 0).$$

$\bar{N}$ is the average number of vacancies in the sample and $P(\tau)d\tau$ is the probability that a vacancy with a lifetime between τ and $\tau + d\tau$ will be formed. The lifetime distribution $P(\tau)$ can be determined from the diffusion equation of the vacancies in the sample. If we let $W(\vec{x},\tau)$ be the solution of the diffusion equation with boundary conditions given by the geometry of the sources and sinks of the vacancies we find

$$P(\tau) = -\frac{d}{d\tau}\int_V W(\vec{x},\tau)dV' \quad , \quad (5)$$

where V is the volume of the sample. By inserting $P(\tau)$ into (4) we find the power spectrum required. The schematic form of the power spectrum is shown in Fig. 5. The exact shape of the

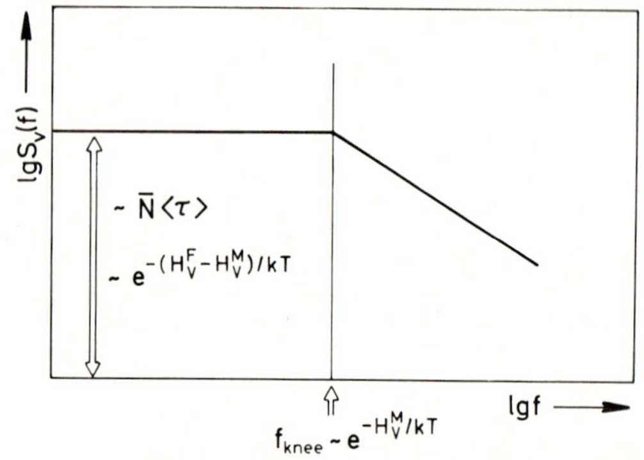

Fig. 5: Vacancy noise power spectrum and its dependence on temperature

spectrum depends on the geometry of the sources and sinks. According to [5] the drop of the spectrum above the knee frequency is proportional to f^{-2} if the sinks and sources are distributed homogeneously, and proportional to $f^{-3/2}$ if the vacancies diffuse from and to the surface of the sample.

At low frequencies (4) is found to be

$$S_V(f \to 0) \sim \bar{N}\langle\tau\rangle \sim \exp\left\{-(H_V^F - H_V^M)/kT\right\}. \quad (6)$$

The knee frequency has a temperature dependence

$$f_{knee} \sim \nu(T) \sim \exp\left\{-H_V^M/kT\right\}. \quad (7)$$

Thus by measuring the temperature dependence of the power spectrum, $(H_V^F - H_V^M)$ and H_V^M may be determined simultaneously. The power spectrum calculated for vacancy diffusion from and to the surface of a rectangular sample of width w and thickness d (where w and d are still large compared to the lattice parameter a) is shown in Fig. 6. For the case of a thin plate (w>>d) we find [5]

$$S_V(f) = \frac{8}{\pi}\bar{N}(\Delta U)^2(2\tau^*)^{-1/2}(2\pi f)^{-3/2}\frac{\sinh\beta-\sin\beta}{\cosh\beta+\cos\beta} \quad (8)$$

$$\text{with } \beta = \pi(\pi f\tau^*)^{1/2} \quad (9)$$

$$\text{and } \bar{N}(\Delta U)^2 = \frac{(\Delta\rho_V)^2}{C_A}\frac{\ell}{q}C_V\, j^2 . \quad (10)$$

$\Delta\rho_V$ is the resistivity change per unit concentration of vacancies, ℓ the length and q the cross-section of the specimen, j the electrical current density and C_A is the number of atoms per unit volume.

$$1/\tau^* = D_V\pi^2/d^2 \quad (11)$$

is the first eigenvalue of the diffusion equa-

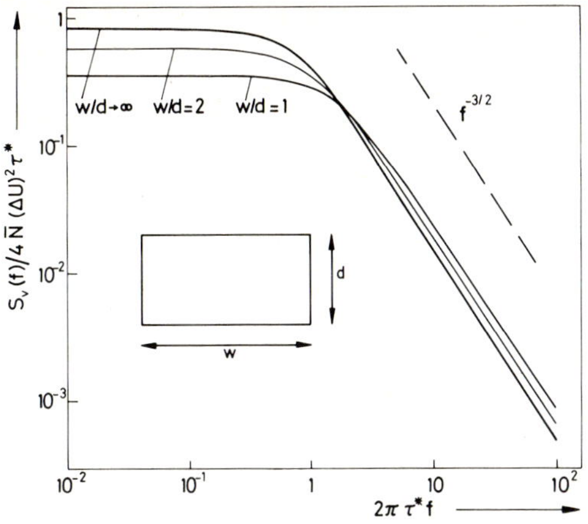

Fig. 6: Vacancy noise power spectrum for diffusion from and to the surface of a rectangular sample [6].

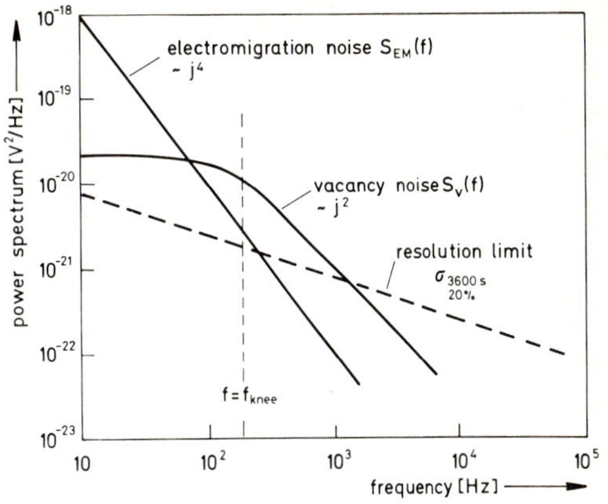

Fig. 7: Power spectrum of the vacancy noise and of the electromigration noise as calculated in [5].

tion. The spectrum (8) shows a plateau

$$S_v(f \to o) = 4\,\bar{N}\,(\Delta U)^2\,\tau^*\,\frac{\pi^2}{12} \qquad (12)$$

and a knee frequency

$$f_{knee} = (2\pi\,\tau^*)^{-1} \qquad (13)$$

above which there is a $f^{-3/2}$dependence of the spectrum (see Fig. 6).

Michel and Grabert [7] have derived the power spectrum of vacancy noise from a more general theory (hydrodynamics for metals). Where their findings overlap, they agree completely with our results derived from a series of single pulses caused by the production and annihilation of individual vacancies.

4. MEASURABILITY OF VACANCY NOISE

As in other current noise caused by fluctuations in resistivity, the power spectrum of vacancy noise is proportional to the square of the current density j in the sample. High current densities are required if a measurable noise is to be obtained, and these can only be realized in a thin film on a substrate. Below a certain thickness the surface of the film is the main source and sink of the vacancies. A further reduction of the thickness increases the knee frequency (13) but decreases the power spectrum (12), since $\tau^* \sim d^2$.

The power spectrum calculated for a typical gold thin film with thickness d = 2.2 µm, length ℓ = 1.28 mm and cross-section q = 22 $(\mu m)^2$ at a temperature of T = 1123K and a current density of $j = 5\cdot 10^9 A/m^2$ is shown in Fig. 7. The vacancy noise is found to be 6% of the Johnson noise of the sample. For this reason the Johnson noise must be determined in a second measurement without current passing through the sample, and subsequently subtracted. The standard deviation of this difference, with a measurement time of one hour for each run and a frequency resolution of 20%, is also shown in Fig. 7. The self-heating of the sample when the current is turned on causes an increase in Johnson noise which must be corrected. It should be possible to avoid the differential measurement with and without current passing through the sample, and the correction of the Joule heating, by using a measurement circuit as shown in [8].

5. THE EXTRANEOUS LOW FREQUENCY NOISE COMPONENT

Both in the measurements of current noise done by Celasco, Fiorillo, and Mazzetti on polycrystalline aluminium films [1], and in our measurements on polycrystalline gold films [9], an additional low-frequency noise component, approximately proportional to f^{-2}, was observed (see Fig. 7). In our gold thin films we found this noise component to be proportional to the fourth power of the current through the sample

$$S_{EM} \sim \frac{j^4}{f^2}\,. \qquad (14)$$

The current dependence (14) might be explained by a resistivity fluctuation Δr(t) which is itself proportional to the current in the sample. We suggest that this noise component can be explained by electromigration in the metal. According to [10] electromigration causes a flux of atoms J_a proportional to the current density j

$$J_a = \frac{Z^*e\; C_A\; D_V^{SD}\; \rho}{kT}\; j\,, \qquad (15)$$

where Z*e is the effective charge of an ion, ρ the electrical resistivity and D_V^{SD} the self-diffusion coefficient. In polycrystalline thin films electromigration is enhanced by grain boundary diffusion. Divergencies in the flux of atoms, caused by inhomogeneities in the film structure, produce local material accumulation and depletion and thus a resistivity fluctuation $\Delta r(t, j)$. If this explanation is confirmed in further experiments it would be possible to use current noise measurements to investigate not only production and random diffusion of vacancies but, in addition, also electromigration in metals.

REFERENCES

[1] Celasco, M., Fiorillo, F., and Mazzetti, M., Thermal-equilibrium properties of vacancies in metals through current-noise measurements, Phys. Rev. Lett. 36 (1976) 38-42.

[2] Seeger, A., The study of point defects in metals in thermal equilibrium, Cryst. Lattice Defects 4 (1973) 221-253.

[3] Seeger, A. and Mehrer, H., Analysis of self-diffusion and equilibrium measurements, in Seeger, A., Schumacher, D., Schilling, W., and Diehl, J. (eds.), Vacancies and Interstitials (North-Holland, Amsterdam 1970).

[4] Seeger, A., Investigation of point defects in equilibrium concentration with particular reference to positron annihilation techniques, J. Phys. F: Metal Physics 3 (1973) 248-294.

[5] Stoll, H., Vacancy noise measurement in metals: Calculation of the power spectrum, Appl. Phys. A 30 (1983) 117-122.

[6] Kienle, W., Personal communication.

[7] Michel, W. and Grabert, H., Calculation of vacancy noise in monocrystalline metals, these proceedings.

[8] Stoll, H., Analysis of resistivity fluctuations independent of thermal voltage noise, Appl. Phys. 22 (1980) 185-187.

[9] Stoll, H., Bestimmung der Eigenschaften atomarer Leerstellen im thermischen Gleichgewicht durch Messung der Fluktuation des elektrischen Widerstandes in Metallen, Dissertation, Universität Stuttgart (1981).

[10] d'Heurle, F.M. and Ho, P.S., Electromigration in thin films, in Poate, J.M., Tu, K.N., Mayer, J.W. (eds.), Thin films-Interdiffusion and reaction (John Wiley, New York, 1978).

NOISE IN PHYSICAL SYSTEMS AND 1/f NOISE
M. Savelli, G. Lecoy and J-P. Nougier (eds.)
© Elsevier Science Publishers B.V., 1983

VORTEX DENSITY FLUCTUATIONS DURING FLUX FLOW IN A TYPE II SUPERCONDUCTOR

H.Dirks, R.Dittrich and C.Heiden

Institut für Angewandte Physik der Justus-Liebig-Universität Giessen, D-63-Giessen FRG

The magnetic component of flux flow noise has been investigated on samples with inhomogeneous pinning. The local dependence of the flux density fluctuations is observed to correspond closely to that of the electric field fluctuations, revealing the same origin i.e. stochastic strain relaxation processes, of the vortex arrangement. Separating density and velocity fluctuations of vortices provides a valuable tool to distinguish between different noise mechanisms.

1. INTRODUCTION

Flux flow in a type II superconductor leads to an electric field $E=nv\phi_o$, where ϕ_o is the flux quantum and n the density of vortices moving with velocity v. Flux flow noise is therefore in general associated with fluctuations of n and/or v. A convenient experimental arrangement for flux flow measurements has been the use of thin films or foils in a perpendicular magnetic field B. The demagnetization coefficient is very close to unity, and the vortex density $n=B/\phi_o$ very uniform, provided that pinning and therefore the critical current density j_c is homogeneous throughout the sample. Applying a transport current with density j, vortex motion can be achieved, that can be described by the equation of motion $(j-j_c)\phi_o=\eta v$ where η is the viscosity coefficient [1].

There are various possibilities for the generation of flux flow noise [2]. Temperature fluctuations in the specimen for instance will lead in general to corresponding fluctuations of j_c which then will result in velocity fluctuations δv. Such noise, which usually exhibits a rather strong component at low frequencies ($f < 100$Hz) has been observed in the past [3,4]. Early attempts to detect a noise component in the density n however showed no results [5]. It is conceivable, that this type of noise, often called flicker noise, consists mainly of velocity fluctuations.

Another mechanism, that has been discussed for flux flow noise should involve predominantly fluctuations in n. It is the shot noise of flux bundles travelling with constant speed v across the sample. Although the high demagnetization coefficient of the aforementioned geometry does not favor the existence of individual vortex groups, but leads to the formation of a more or less homogeneous vortex arrangement (a flux line lattice in the ideal case), the motion of individual flux entities has been observed using a time of flight method [6]. It should be possible to observe a magnetic noise component in such cases.

Still another mechanism appears to be present in samples with inhomogeneous distribution of pinning sites. It consists of stochastic strain relaxations involving shear processes in the vortex lattice. Inhomogeneous pinning leads to an inhomogeneous vortex flow pattern which in turn will result in local compressions or dilatations of the vortex arrangement. In order to restore locally the flux density $n=B/\phi_o$ the strain can relax by shear processes. If they occur in a stochastic way, triggered for instance by local defects in the vortex lattice, a noise should be observable being composed of both, fluctuations in v and in n. Experimental evidence for such a mechanism was obtained by measurements on single crystalline niobium foils with low critical current in which suitable pinning structures were deposited in form of Nb_3Sn-films using a lithographic process [7]. Fluctuations of the local electric field were found to occur mainly in regions with plastic deformations of the vortex lattice (cf.Fig.1).

2. EXPERIMENTAL

In order to detect flux density fluctuations δB, a small pick up coil was made, whose position on the sample surface could be varied in a controlled way. To achieve a similar local resolution as for the electric noise component, the coil diameter was chosen of the same size as the distance between the voltage probes, i.e. ca. 0.3mm. Using a core of manganin wire, a few turns of 12μm insulated copper wire were wound on it resulting in a pickup coil with the desired dimensions. For ease of handling the finished coil was glued with araldite on a lucite plate. The winding area was parallel to the sample surface, the coil leads practically perpendicular to it. The amplified induced signal either was measured as such or fed to an electronic integrator whose output then was proportional to δ B(t) [8].

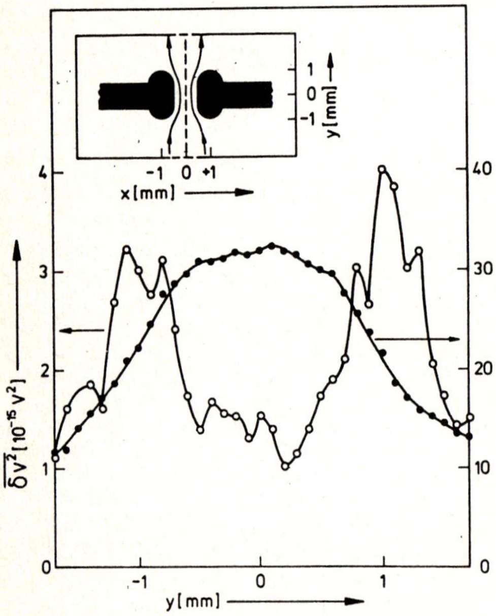

Fig. 1 Total noise power $\overline{\delta V^2}$ and dc-flux flow voltage V for different y-positions of midpoint between the voltage probes (separation 0.4 mm), which are aligned in x-direction. The gradient in the DC-velocity field corresponds to the slope of the V(y)-profile. Inset: Nb_3Sn structure as black area. Vortex flow indicated by two equipotential lines. Field: 0.175T, temperature: 4.2K, transport current 4A.

3. RESULTS

Measurements were performed on the same single crystalline niobium sample as in ref.[7], on which a dipole like thin film Nb_3Sn pinning structure was modelled. Fig.1 reproduces the results for the mean square noise voltage obtained from a pair of voltage probes,

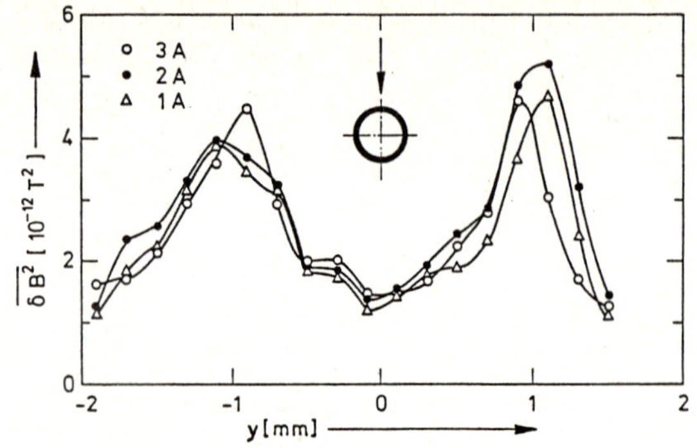

Fig. 2 Mean square flux density noise for positions of the coil center along the dashed line in fig.1. The diameter of the flat pick up coil is drawn to size. B=0.15T.

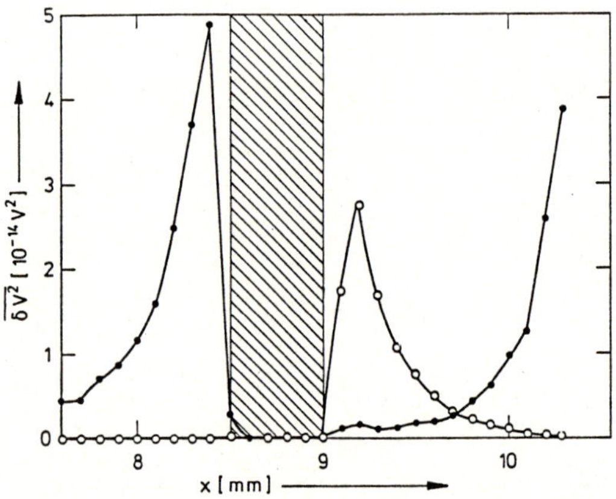

Fig. 3 Electric noise component near edge of a Nb_3Sn-strip for two directions of the transport current: ... I=+4A , ∘∘∘ I=-4A. Contacts were aligned parallel to the edge. B=0.15T.

whose position was varied along the dashed line.
A similar local dependence is observed for the magnetic noise component $\overline{\delta B^2}$ as shown in fig.2 for three different dc - transport situations. Both, electric and magnetic noise components therefore are likely to have the same origin, namely plastic deformations of the vortex lattice. This is corroborated by figs.3 and 4, which show electric and magnetic noise obtained from another sample on which a narrow Nb_3Sn-strip was deposited. Depending on the direction of vortex flow, either strong or weak noise is observed in the

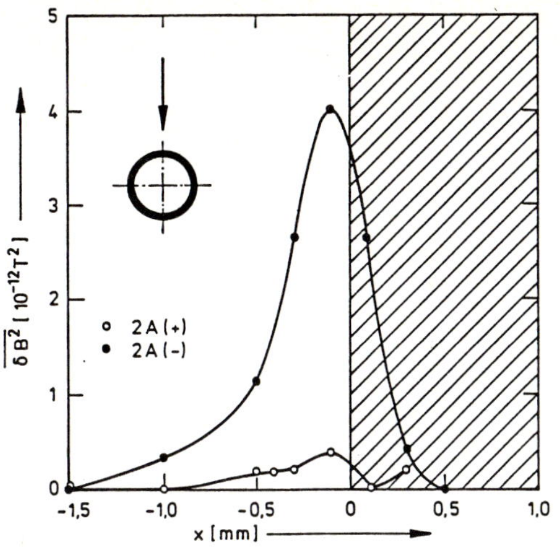

Fig. 4 Magnetic noise component near edge of the Nb_3Sn-strip for two directions of the transport current. B=0.15T.

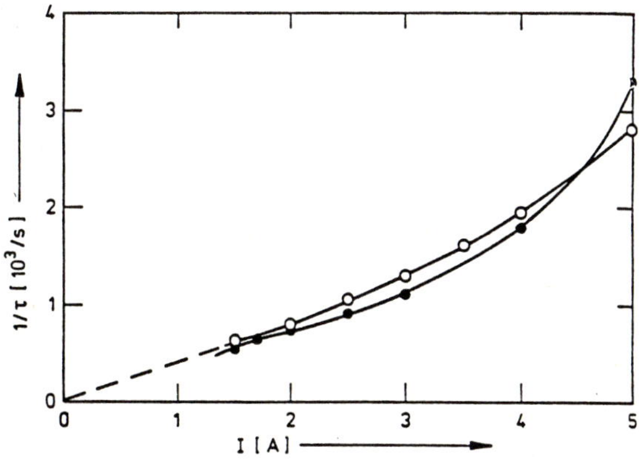

Fig. 5 Comparison between the time constants of electric (•••) and magnetic (∘∘∘∘) noise components. Data are from a constant position near the Nb_3Sn-strip. B=0.15T.

immediate neigbourhood of the strip edges. Although the reason for this asymmetry with regard to vortex flow is not quite clear at present, it is seen in both noise components. Fig.5 depicts the behavior of the time constant τ determined from both, the electric and magnetic noise of the Nb_3Sn strip. There is again a rather similar dependence on transport current for both signals.

4. DISCUSSION

Fluctuations in magnetic flux sensed by the pick up coil of course vary with the distance h of the winding area from the sample surface. To be able to deduce the value of $\overline{\delta B^2}$ in the sample, the dependence of $\overline{\delta B^2(h)}$ has been investigated. Fig. 6 shows preliminary results obtained from a rectangular small coil, whose winding area was placed vertically in order to have a better definition of the distance Δh between sample surface and the lower sensing fraction of the coil i.e. the wires parallel to the sample. The mean square induced noise voltage was measured at four different positions on the sample with the dipole like pinning structure and is plotted in a normalized way. Within the present accuracy, all data exhibit about the same Δh-dependence, the main decay of the signal taking place in a Δh - interval of ca. 0.4 mm. Although the involved measuring technique still has to be refined, for instance by using pick up coils made with the aid of lithography, one can conclude from the observed $\overline{V^2_{ind}(\Delta h)}$ - behavior, that flux density fluctuations in the sample can be determined. By measuring simultaneously at a given location $\overline{\delta E^2}$, $\overline{\delta B^2}$ and the product $\overline{\delta E \cdot \delta B}$, it then is possible to separate the electric and magnetic contributions, as was considered already by Jarvis and Park [9]. This leads to a valuable tool to distinguish between different noise mechanisms.

The profiles $\overline{\delta B^2(h)}$ can be used to obtain information on the average size of the area, in which a coherent flux change takes place. For this purpose, the flux density profile above the sample surface, originating from an inhomogeneity δB in the sample, can be modelled by an approach used earlier by Pearl [10] , who described the flux spreading from an isolated vortex by a magnetic monopole field. By superposition of fields from suitably arranged extra or defect vortices in the region of inhomogeneous flux density the measured profile $\delta B^2(h)$ can be reproduced.

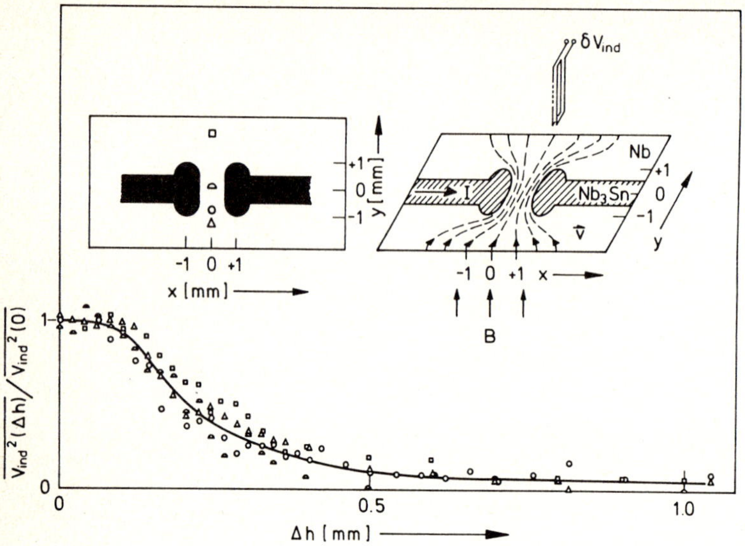

Fig. 6 Mean square noise voltage induced in a rectangular coil (13 turns of 20μm insulated Cu-wire wound on a saphire core, height 2mm, length 300μm, width 200μm) as function of distance Δh to the sample surface. B=0.15T, I=3.5A.

REFERENCES:

[1] Bardeen,J.and Stephen,M.J., Theory of the Motion of Vortices in Superconductors,Phys.Rev.140,(1965) A1197.

[2] Clem, J.R., Flux-Flow Noise in Superconductors,Phys.Rev. 75(1) (1981) 1.

[3] Van Gurp, G.J., Phys.Rev.166(1968) 436.

[4] Heiden, C., Kohake, D., Krings, W., and Ratke, L., J.Low Temp.Phys. 27(1977)1.

[5] Heiden, C., unpublished.

[6] Heiden, C., Defect Migration in a Moving Fluxon Array, Solid State Commun.18 (1976) 253. Beckstette, K. and Heiden, C., in: Noise in Physical Systems, ed. D.Wolf (Springer-Verlag,Berlin, (1978) p.234.

[7] Dirks, H.M., Beckstette, K.F., and Heiden, C., Flux-Flow Noise During Inhomogeneous Vortex-Motion. Dependence on DC-Velocity Field Beckstette, K.F., Dirks, H.M., and Heiden, C., Magnetic Field and Temperature Dependence, in: Proc.Sixth Int.Conf. on Noise in Physical Systems, eds. Meijer, P.A.E., Mountain, R.D., Soulen, Jr., R.J. (U.S.Department of Commerce/National Bureau of Standards,Washington, 1981)p.384.

[8] Dirks, H., Ph.D.Thesis, University of Gießen (1981).

[9] Jarvis, P. and Park, J.G., The Dependence of the Noise Voltage Associated with Flux Motion in Superconductors on the Arrangement of Potential Leads,J.Phys.F (1974) 1238.

[10] Pearl, J., Structure of Superconductive Vortices near a Metal-Air Interface, J.Appl.Phys. 37 (1966) 4139.

NOISE IN PHYSICAL SYSTEMS AND 1/f NOISE
M. Savelli, G. Lecoy and J-P. Nougier (eds.)
© *Elsevier Science Publishers B.V., 1983*

Contact Noise in Superionic Ceramics

James J. Brophy

University of Utah
Salt Lake City, Utah 84112

Voltage fluctuations at electrodes to polycrystalline sodium and silver beta alumina superionic conductors have been observed in the frequency range 10^0 to 10^5 Hz. Noise voltages in excess of bulk sample Nyquist noise exist in the absence of dc current and the noise power increases with current for both two- and four-terminal specimens. The noise spectra are strongly frequency dependent and spectra varying between f^{-3} and $f^{-0.5}$ are observed. Conductivity fluctuations in four-terminal measurements indicate a f^{-2} noise spectra.

1. INTRODUCTION

The objective of this study is to measure experimentally conductivity fluctuations in fast ionic conductors[1] and to interpret results in terms of transport mechanisms in the solid and at crystal-electrode interfaces. Electrode properties are expected to introduce experimental difficulties and it is well established that the electrochemical properties of electrical contacts to solid ionic conductors are conventionally avoided by employing high frequency ac techniques[2]. Conductivity fluctuations, on the other hand, are most easily examined under dc conditions and at low frequencies[3]. A series of experiments has been completed on the low frequency properties of contacts to sodium and silver beta alumina ceramics which give the noise characteristics of contacts alone and demonstrate that suitable electrodes for conductivity fluctuations can be devised.

2. SODIUM BETA ALUMINA

Sodium β" alumina[4] ceramic samples 1 cm x 0.5 cm x 0.2 cm shaped to provide either transverse or longitudinal potential electrodes use current contacts consisting of a sodium nitrate-sodium nitrite eutectic mixture or a sodium amalgam. Sodium β' polycrystalline "brick" specimens are less regularly shaped but can be examined in the transverse 4-probe electrode configuration. Samples are baked for several hours above 800 °C to eliminate absorbed moisture. Noise measurements using the eutectic contacts are carried out in the temperature range 250 °C to 300 °C where the current contacts are molten. Both blocking and conducting potential-probe electrode materials have been examined.

The eutectic current contacts exhibit a non-linear current-voltage characteristic, Figure 1, which suggests a large dc resistance at zero applied voltage compared to the sample resistance of 4 ohms. At temperatures above the melting point of the eutectic mixture, dc currents can be maintained for many hours with no indication of polarization effects. The dc voltage at the potential probes is a linear function of current through the current electrodes and the conductivity calculated from the observed resistance and sample dimensions (5 ohm-cm) agrees with published values for sodium β" alumina[4].

The non-linear I-V characteristic is ascribed to the interface between a platinum foil used to make contact with the external circuit and the eutectic. Other results indicate that an amalgam electrode is ohmic to both the Naβ" and Naβ' ceramic, Figure 1. The amalgam electrodes are more suitable than the eutectic, but the vapor pressure of mercury limits the experimental temperature range.

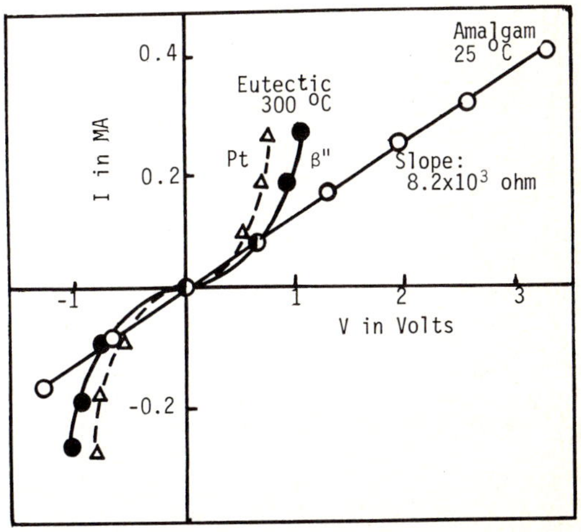

Figure 1. Current-voltage characteristics of contacts to Naβ" alumina. Dashed curve is for platinum-eutectic interface alone.

3. SILVER BETA ALUMINA

Similarly shaped silver β" alumina ceramic samples are provided with aqueous silver nitrate current electrodes since silver amalgam produces a high resistance ohmic contact. Potential probes of silver paint or silver amalgam exhibit similar noise spectra and both change significantly if the sample is etched in phosphoric acid before electroding.

4. CONTACT NOISE

Noise voltages are observed with a PAR 113 ac coupled preamplifier, a tunable electronic filter and an ac voltmeter. The system accurately measures the Nyquist noise of known resistors placed at the sample position. Initial measurements have been confined to the frequency range 10° to 10^4 Hz by thermal instability at low frequencies and amplifier noise at high frequencies. Current is supplied to the samples through a noiseless 10^5 ohm series resistor to reduce current fluctuations.

Electrical noise signals observed at the potential probes reflect the quality of the electrical contact, since thermal noise of the resistance associated with a blocking contact is always present. Noise voltages observed in the absence of current are many orders of magnitude in excess of Nyquist noise of the bulk sample and are dependent upon the electrode material at the potential probes, Figure 2. The sodium eutectic mixture exhibits the lowest, and the silver paste electrodes exhibit the highest noise levels to Naβ" alumina. The spectral noise density in these cases varies as f^{-3}. Additional noise accompanying the presence of dc current in the specimen shows the same frequency dependence, although the measurements are difficult because of instabilities in the current contacts.

The observed spectral shape might be accounted for by a simple model in which a distributed contact resistance is shunted by a distributed contact capacitance. This suggests that the contacts exhibit 1/f noise, which is not unexpected in view of the polycrystalline nature of these specimens. The conductivity fluctuations indicated by the current dependent noise indicates that a similar model may apply to the bulk phenomenon as well. Alternatively, noise may arise from reaction processes within the sample due to the basically nonequilibrium nature of these materials.

The sodium amalgam electrode noise decreases as $f^{-1.6}$, as also shown in Figure 2. A slope of -1.5 is usually associated with diffusion phenomena, so that this noise may be ascribed to such processes, either at the contacts or in the bulk. A remarkable difference is noted in the zero current blocking contact noise spectra for Agβ" ceramic, which shows a $f^{-0.5}$ spectrum, Figure 2. A surface etch treatment in phosphoric acid results in an increased noise level and a $f^{-1.6}$ spectra for both amalgam and silver paste electrodes. A $f^{-1.6}$ spectrum also observed for the ohmic aqueous contcts at zero current.

5. CURRENT NOISE

Stable two-terminal current noise measurements are observed using the sodium amalgam electrodes, Figure 2. The low frequency slopes are -1.6, which may be associated with ionic diffusion, while the slope of -2.1 at high frequencies suggests a Lorentzian characteristic with a time constant greater than $1/2\pi 300 = 5 \times 10^{-4}$ seconds. The current dependence is $I^{2.5}$, which is somewhat greater than the square law dependence expected for conductivity fluctuations.

Voltage fluctuations in excess of contact noise are also observed at transverse[5] potential contacts in four-terminal specimens of both Naβ" and Naβ' using amalgam electrodes. The spectral density varies approximately as f^{-2} while the current dependence is a power law for Naβ" (exponents range from 1.5 to 3.5) and increases exponentially in the case of Naβ'. This additional noise is associated with conductivity fluctuations in the bulk, including processes occurring at grain-grain contacts.

6. CONCLUSIONS

These experimental results show that the voltage fluctuations at electrical contacts to superionic conductors exhibit a rich variety of characteristics. Both conducting and blocking contacts to Naβ" ceramic have an f^{-3} spectral density, while ohmic contact current noise suggests both diffusion and Lorentzian processes. In favorable cases, suitable contacts can be devised to observe bulk conductivity fluctuations. The contact and bulk noise properties of Naβ' and Naβ" alumina appear to be quite similar and both are very different from Agβ" alumina. Fuller interpretation of the experimental results awaits additional data, and in particular extension of the measurements to very low frequencies in order to observe possible low frequency plateaus.[6].

7. ACKNOWLEDGMENTS

This research is supported in part by the Office of Naval Research. The author expresses his deep appreciation to Stephen W. Smith who carried out all of the

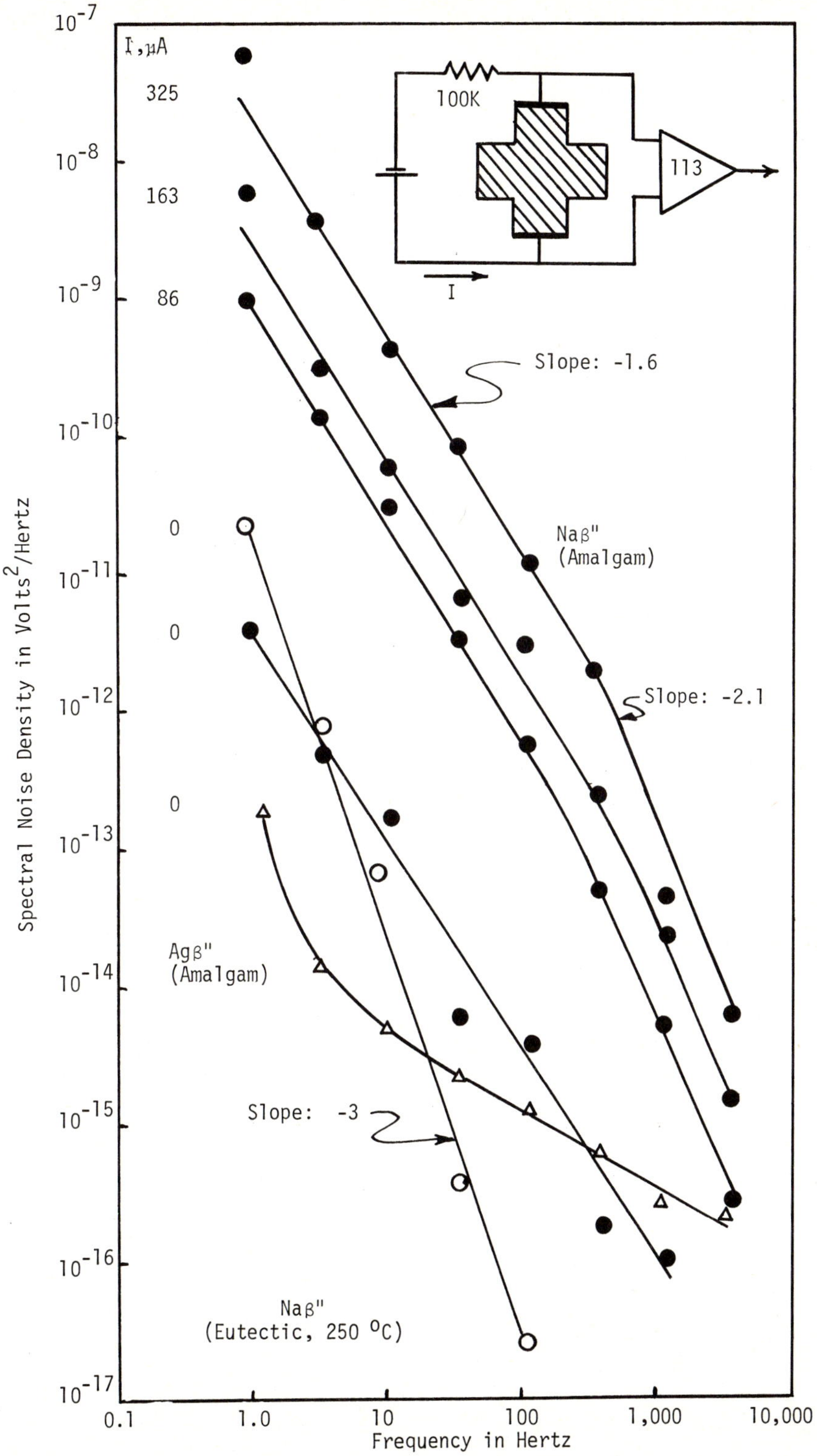

Figure 2 : Noise spectra of contacts to Naβ" and Agβ" ceramics.

experimental measurements and to A. V. Virkar, J. Janata, and G. R. Miller for many helpful suggestions.

8. REFERENCES

(1) Boyce, J. B. and Huberman, B. A., Physics Reports, 51, 189 (1979); Farrington, G. C., Sensors and Actuators, 1, 329 (1981).

(2) Radzilowski, R. H., Yao, Y. F. and Kummer, J. T., J. Appl. Phys., 40, 4716 (1969).

(3) Brophy, J. J., Phys. Rev., 106, 675 (1957).

(4) Kummer, J. T., Prog. Sol. State Chem., 7, 141 (1972).

(5) Vandamme, L. K. J. and Kamp, L. P. J., J. Appl. Phys., 50, 340 (1979).

(6) Feber, G. and Weissman, M., Proc. Nat. Acad. Sc. USA, 70, 870 (1973); Fleishmann, M., Labram, M., Gabrielli, C. and Starr, A., Surface Science, 101, 583 (1980).

NOISE IN PHYSICAL SYSTEMS AND 1/f NOISE
M. Savelli, G. Lecoy and J-P. Nougier (eds.)
© *Elsevier Science Publishers B.V., 1983*

SPACE-TIME CORRELATION PROPERTIES OF THE MAGNETIZATION NOISE AND MAGNETIC LOSSES

Giorgio Bertotti

Istituto Elettrotecnico Nazionale G. Ferraris,
Gruppo Nazionale Struttura della Materia del C.N.R.
I-10125 Torino, ITALY

The general relationship between eddy current losses and magnetization dynamics in ferromagnetic materials is investigated. A statistical model is worked out, in which the average loss per unit volume is expressed in terms of the power spectrum $S_{xy}(\vec{k},\omega)$ of the magnetization rate $\dot{I}(\vec{r},t)$ in the sample cross-section. The general properties of S_{xy} in the case where $\dot{I}$ is described as a sequence of elementary magnetization jumps are discussed. It turns out that the hysteresis, classical, excess contributions to the loss can all be given a proper statistical interpretation.

1. INTRODUCTION

A satisfactory characterization of the magnetization process in ferromagnets implies a detailed knowledge of the space-time properties of the magnetization rate $\dot{I}(\vec{r},t)$ in the whole material. The theoretical and experimental investigation of the behavior of $\dot{I}$ in different materials and in different conditions of domain structure, magnetizing frequency, etc., is a central problem of ferromagnetism, to which wide efforts have been devoted. To this end, statistical methods prove in general very helpful, since, due to the great complexity of the interactions controlling the dynamics of the magnetic domains, $\dot{I}$ shows an intrinsically stochastic character.

The largest amount of theoretical and experimental results certainly concerns the time properties of $\dot{I}$, usually characterized in terms of the power spectrum $S(\omega)$ of the magnetic flux rate $\dot{\Phi}(t)$, the space integral of $\dot{I}$ over the sample cross-section. Barkhausen noise investigations [1, 2] can provide valuable information on the properties of $S(\omega)$. They lead to describe the magnetization process as a random sequence of strongly correlated elementary magnetization changes, each corresponding to a sudden and localized jump of a domain wall segment in the material.[3] A natural step towards a better characterization of $\dot{I}$ is to take account of the longitudinal propagation of the flux Φ in the magnetization direction z. The statistical properties of the function $\dot{\Phi}(z,t)$ are properly characterized by the corresponding spectrum $S_z(k_z,\omega)$, and can be investigated through Barkhausen noise measurements on couples of windings placed along the sample length.[4, 5]

The main drawback of a characterization of the magnetization process in terms of the flux rate $\dot{\Phi}$ is that it cannot distinguish magnetization events occurring in different points of the sample cross-section. A case where it is essential to remove this limitation is when the problem of calculating the area of the hysteresis cycle, that is the magnetic loss, is considered. Actually, in conducting ferromagnets the loss basically originates from the decay through Joule effect of the eddy currents which develop around each elementary magnetization jump. The loss is not, however, simply proportional to the number of jumps occurring in unit time. Being quadratic in the eddy current density, it strongly depends on how the eddy current patterns produced by the single jumps superpose in space and time.[6] The main superposition effects take place in a same cross-section of the material and are therefore controlled by the space-time correlation properties of the function $\dot{I}(x,y;t)$ in a given cross-section. It is the aim of the present paper to give a theoretical analysis of this general connection between eddy current losses and magnetization dynamics. As a result, it will be shown that eddy current losses can actually be expressed in terms of the power spectrum $S_{xy}(k_x,k_y;\omega)$ of $\dot{I}(x,y;t)$.

It should be noticed that, contrary to the case of $S(\omega)$ and $S_z(k_z,\omega)$, it is quite difficult to obtain some experimental information on $S_{xy}(\vec{k},\omega)$. Only recently some new interesting results in this direction have been obtained by means of electrical and optical methods, in which either the local e.m.f.'s detected on the sample surface by means of point-like electrodes [7] or the Kerr effect induced fluctuations of intensity in a polarized laser beam reflected by the sample surface [8] are investigated.

2. THE MODEL

The general relationship existing between the average eddy current loss per unit volume P and the magnetization rate $\dot{I}(x, y; t)$ can be derived, under some reasonable simplifying assumptions, directly from Maxwell equations. Let us consider a rectangular sample of width ℓ (along x), thickness d (along y) and infinite length (along z), magnetized in the z direction. If we assume, as a first approximation, that any magnetization change is directed along the z axis and involves a cylindrical region of infinite length along the z direction, then the average loss P can simply be expressed as

$$P = -\lim_{\Delta T \to \infty} \int_{-\Delta T/2}^{\Delta T/2} \frac{dt}{\Delta T} \int_S \frac{d^2\vec{r}}{S} H(\vec{r}, t)\, \dot{B}(\vec{r}, t) \quad . \qquad (1)$$

S is the sample cross-section, while $H(\vec{r}, t)$ and $\dot{B}(\vec{r}, t)$ respectively represent the magnetic field generated by eddy currents and the induction rate, both directed along the z axis. Our purpose is to express explicitly eq. (1) in terms of the magnetization rate $\dot{I}(\vec{r}, t)$ associated with the irreversible jumps of the domain walls. To this end, let us note that, if we assume that the transient response of the material to a magnetization change is controlled by the reversible permeability μ , then

$$\dot{B}(\vec{r}, t) = \dot{I}(\vec{r}, t) + \mu \cdot \partial H(\vec{r}, t)/\partial t \quad . \qquad (2)$$

On the other hand, Maxwell equations imply, when displacement currents are neglected, that H satisfies the equation

$$\Delta H(\vec{r}, t) = \sigma \dot{B}(\vec{r}, t) \quad , \qquad (3)$$

where Δ represents the bidimensional Laplace operator and σ is the electrical conductivity of the material. From eqs (2) and (3), we get

$$\Delta H(\vec{r}, t) - \sigma\mu \partial H(\vec{r}, t)/\partial t = \sigma \dot{I}(\vec{r}, t) \quad , \qquad (4)$$

which shows that the space-time behavior of H is basically controlled by $\dot{I}$.

Equation (4) is conveniently solved by passing to space-time Fourier transforms. In doing so, it is necessary to take proper account of the boundary conditions of the problem, stating that both $\dot{I}$ and H are equal to 0 on the boundary of S. By using sine transforms, that is by assuming

$$\dot{I}(\vec{r}, t) = \frac{4}{S} \sum_{\vec{k}} \int_{-\infty}^{\infty} \frac{d\omega}{2\pi} I(\vec{k}, \omega) \sin(k_x x) \sin(k_y y) \exp(i\omega t) \qquad (5)$$

(and similarly for $H(\vec{r}, t)$), the boundary conditions simply imply that the summation over $\vec{k}$ in eq. (5) must be restricted to the values

$$k_x = \pi m/\ell \ , \ k_y = \pi n/d \ ; \ m, n = 1, 2, \ldots \qquad (6)$$

In the Fourier space $(\vec{k}, \omega)$, eq. (4) becomes

$$H(\vec{k}, \omega) = -\frac{\sigma \dot{I}(\vec{k}, \omega)}{k^2 + i\omega\sigma\mu} \quad , \qquad (7)$$

with $k^2 = k_x^2 + k_y^2$. From eqs (1), (2) and (7) and by applying the generalized Parseval's theorem, we obtain

$$P = \sigma \frac{4}{S} \sum_{\vec{k}} \int_{-\infty}^{\infty} \frac{d\omega}{2\pi} \frac{k^2 S_{xy}(\vec{k}, \omega)}{k^4 + (\omega\sigma\mu)^2} \quad , \qquad (8)$$

$S_{xy}(\vec{k}, \omega)$ representing the power spectrum of $\dot{I}$.

3. DISCUSSION

Equation (8) is the fundamental equation connecting eddy current losses with the microscopic dynamics of the magnetization process, characterized by the space-time properties of the magnetization rate $\dot{I}(\vec{r}, t)$. $\dot{I}(\vec{r}, t)$ can suitably be described as a random sequence of elementary magnetization changes $\dot{I}^{(s)}(\vec{r}, t; \vec{r}_i, t_i)$ taking place at different positions $\vec{r}_i$ and times t_i :

$$\dot{I}(\vec{r}, t) = \sum_i \dot{I}^{(s)}(\vec{r}, t; \vec{r}_i, t_i) \quad . \qquad (9)$$

Through eqs (8) and (9), the loss evaluation is thus reduced to the typical statistical problem of calculating the power spectrum $S_{xy}(\vec{k}, \omega)$ of $\dot{I}(\vec{r}, t)$ as a function of the shape and correlation properties of the elementary magnetization events of which $\dot{I}$ is built up. There is, however, an important point, concerning the definition of S_{xy} and implicit in our derivation of eq. (8), [9] which needs some comments. Actually, the concept of power spectrum usually refers to a process which extends up to infinity, both in space and time, in a stationary and ergodic way. On the other hand, the magnetization process we are considering is limited in space, being defined over the sample cross-section S only. We therefore ask : which is the process extending up to infinity associated with the power spectrum $S_{xy}(\vec{k}, \omega)$?

The answer is given by eq. (5), where, in order to take proper account of the boundary conditions of the problem, $\dot{I}$ has been expressed as a series of sine functions. Equation (5) directly defines an extension of the magnetization process in the whole x-y plane, which is antisymmetric with respect to the sample boundaries. This is equivalent to assuming that, when an elementary jump takes place, infinite other jumps of identical shape and alternate sign take place simultaneously at mirror positions with respect to the sample boundaries (see Figure 1). These "image" jumps force, simply by symmetry reasons, the component of the eddy current density normal to the lines ..., r, s, ... and ..., r', s', ... to be always equal to 0, thus correctly simulating the presence of the sample boundaries. Therefore, as far as the loss calculation is concerned, we can assume that the random process defined by eq. (9) extends over the whole x-y plane, and take account of the boundary effects related to the finite size of the sample through a non-local correlation of the type represented in Figure 1. Of course, in addition to this "boundary-induced" correlation, further correlation effects are expected to be present in the jump sequence, as a consequence of the dynamic interactions of the domain walls in the material. We shall call this type of correlation, which characterizes the physical properties of the domain structure dynamics, "magnetic" correlation.

Starting from these considerations, we are now ready to work out a general statistical interpretation of the behavior of eddy current losses. As well known, the typical behavior of the loss per cycle P/f_m vs. the magnetizing frequency f_m in a ferromagnetic material is usually interpreted as the sum of three contributions (see Figure 2) :

- hysteresis, assumed to be independent of f_m and equal to the limiting value of P/f_m when $f_m \to 0$. This contribution is associated with the occurrence of localized elementary jumps of the domain walls. [10]

- classical, which is evaluated assuming that the magnetization rate is perfectly homogeneous in the sample cross-section. It merely originates from the presence of the sample boundaries.

- excess or anomalous, determined by the physical properties of the magnetization process, such as the presence of a domain structure, dynamic interactions between the domain walls and so on.

In terms of our statistical approach, based on eqs (8) and (9), this general characterization of magnetic losses corresponds to some general properties of the power spectrum $S_{xy}(\vec{k},\omega)$. Actually, starting from eq. (9), the general mathematical structure of S_{xy} can be expressed as

$$S_{xy}(\vec{k},\omega) = \nu \overline{|i^{(s)}(\vec{k},\omega)|^2} + (\text{cross-products}), \quad (10)$$

where ν represents the average number of jumps in unit surface and time and $\overline{|i^{(s)}(\vec{k},\omega)|^2}$

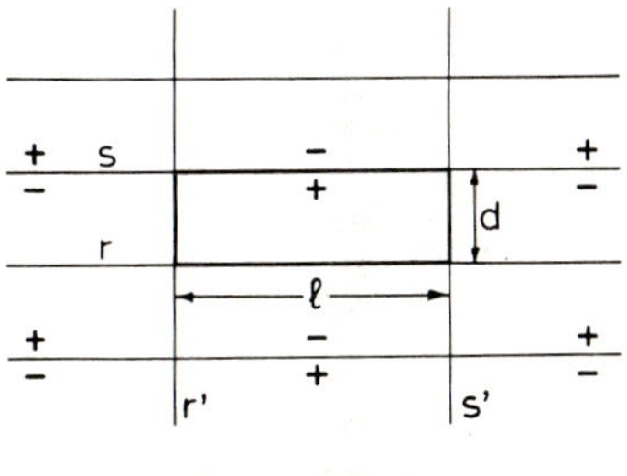

Figure 1 : Images of an elementary magnetization jump correctly simulating the presence of the sample boundaries.

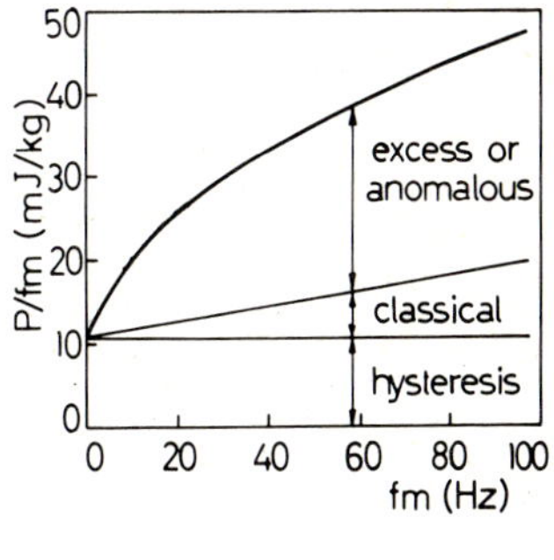

Figure 2 : Typical behavior of the loss per cycle P/f_m vs. the magnetizing frequency f_m in a ferromagnetic material.

is the average energy spectrum of a single jump. Since ν is proportional to f_m, when the first term of eq. (10) is inserted in eq. (8), a loss contribution per cycle independent of f_m is obtained. This is the hysteresis loss, associated, as expected, with the kinetics of the individual elementary jumps. The term schematically denoted by "(cross-products)" is instead determined by the ensemble average of the products $\dot{I}^{(s)*}(\vec{k},\omega;\vec{r}_i,t_i)\dot{I}^{(s)}(\vec{k},\omega;\vec{r}_j,t_j)$, which in turn depends on the behavior of the joint probability density $p(\vec{r}_i,t_i;\vec{r}_j,t_j)$ that the i-th and j-th jumps take place at $(\vec{r}_i,t_i)$ and $(\vec{r}_j,t_j)$ respectively. Let us remind, however, that two types of correlation, "boundary-induced" and "magnetic", are present in our process. If we assume that "magnetic" correlation effects vanish, which is equivalent to assuming that the magnetization process proceeds through statistically independent jumps, it can be verified that the loss contribution associated with the second term of eq. (10) simply becomes coincident with the classical loss. This result puts well in evidence the fact, previously mentioned, that the classical loss is in fact a "boundary-induced" contribution. When the physical interactions between the elementary magnetization jumps are also taken into account by introducing "magnetic" correlation effects, further contributions are found to add to the classical loss, corresponding to the so-called excess or anomalous loss (see Figure 2). We are therefore led to the following general scheme :

single elementary jumps
↓
hysteresis loss

statistically independent jumps ("boundary-induced" correlation only)
↓
hysteresis + classical loss

non-independent jumps ("boundary-induced" + "magnetic" correlation)
↓
hysteresis + classical + excess loss

Due to the lack of space, the previous discussion was, of necessity, somewhat qualitative, mainly providing a conceptual framework of interpretation, rather than a quantitative evaluation of magnetic losses in specific cases. A more detailed and complete analysis will be given in a forthcoming paper. [11] It is, in particular, of great interest to discuss the case where "magnetic" correlation effects in the jump sequence are described in terms of a Markov process. It turns out that, in the case of Markov correlated events, the loss can be expressed as a function of the transition density $M(\Delta\vec{r},\Delta t)$, representing the conditional probability density that two subsequent jumps take place at a distance $\Delta\vec{r}$ and with a time delay Δt. This is a result of great physical interest. Actually, if, by developing suitable statistical models, the main properties of $M(\Delta\vec{r},\Delta t)$ corresponding to different conditions of domain structure, magnetizing frequency, etc., can be predicted, it is possible, through our approach, to work out a direct and clear-cut connection between the loss behavior and the magnetization dynamics at a microscopic level.

REFERENCES

[1] H. Bittel, IEEE Trans. Magn. MAG-5 (1968) 359.

[2] G. Bertotti, F. Fiorillo and M.P. Sassi, J. Magn. Magn. Mat. 23 (1981) 136.

[3] P. Mazzetti and G. Montalenti, Proc. Int. Conf. on Magn. (Nottingham, 1964), p. 701.

[4] W. Grosse-Nobis and K. Jansen, in D. Wolf (ed.), Noise in Physical Systems (Springer, Berlin, 1978), p. 204.

[5] G. Bertotti, F. Fiorillo and M.P. Sassi, in P.H.E. Mejer, R.D. Mountain and R.J. Soulen (eds), Noise in Physical Systems (NBS, Washington, 1981), p. 324.

[6] P. Mazzetti, in D. Wolf (ed.), Noise in Physical Systems (Springer, Berlin, 1978), p. 192.

[7] G. Bertotti, F. Fiorillo and M.P. Sassi, IEEE Trans. Magn. MAG-17 (1981) 2852.

[8] M. Celasco, A. Masoero, P. Mazzetti and A. Stepanescu, IEEE Trans. Magn. MAG-18 (1982) 1475.

[9] P. Mazzetti, private communication.

[10] J.A. Baldwin, J. Appl. Phys. 39 (1968) 5982.

[11] G. Bertotti, to be published on J. Appl. Phys. .

NOISE IN PHYSICAL SYSTEMS AND 1/f NOISE
M. Savelli, G. Lecoy and J-P. Nougier (eds.)
© *Elsevier Science Publishers B.V., 1983*

NOISE ANALYSIS OF LASER LIGHT SCATTERED BY NEMATIC LIQUID CRYSTALS

Jitze P. van der Meulen, Rijke J.J. Zijlstra, Daan Frenkel and Maarten van Dort

Rijksuniversiteit Utrecht, Fysisch Laboratorium, afd. F.V.S.,
Princetonplein 5, 3584 CC Utrecht
The Netherlands

We present a study of the spectral and statistical properties of light scattered by director fluctuation modes in a nematic liquid crystal ("DIBAB"). The noise spectrum of the photomultiplier current is analysed on the assumption that the director fluctuations behave as a Gaussian-Lorentz process. Elastic and viscoelastic constants characterizing the "splay" and "twist" director fluctuations are obtained by measuring the light scattering spectrum as a function of angle. Measurements of the photocount probability distribution and the (half-open) time interval distribution support the assumption that director fluctuations behave as a Gaussian-Lorentz process.

1. INTRODUCTION

Nematic Liquid Crystals (NLC's) are fluids which are translationally disordered but orientationally ordered. On average the orientation of individual molecules is parallel (or antiparallel) to a preferred direction characterised by a vector $\vec{n}_0$, the nematic director. In any finite volume the instantaneous director $\vec{n}(\vec{r},t)$ fluctuates around its average value $\vec{n}_0$.

Light scattering is caused by fluctuations of the dielectric tensor $\underline{\underline{\varepsilon}}(\vec{r},t)$. In NLC's the off-diagonal elements of the dielectric tensor depend linearly on the fluctuations of the director. As a consequence NLC's exhibit strong depolarized light scattering.

The director fluctuations can be described by a set of linear ("hydrodynamic") relaxation equations. It can be demonstrated that these deformations are combinations of three special types of deformation, called the splay, twist and bend deformations, each having particular viscoelastic properties [3].

The Orsay Liquid Crystal Group has shown [4] that if the fluctuations are small, $\delta\vec{n}(\vec{r},t) = \vec{n}(\vec{r},t) - \vec{n}_0$ can be written as a sum of two uncoupled modes (α=1,2). These two uncoupled modes both decay exponentially and hence they predict a Lorentzian lineshape for the light scattering by either mode. The optical spectrum $S_E(\omega)$ of the electric field of laser light scattered by the long-wavelength orientational fluctuations is given by [4]:

$$S_E(\omega) = \beta \sum_{\alpha=1,2} \left[\frac{G_\alpha^2}{K_\alpha(\vec{q})} \; \frac{(1/\tau_\alpha^{-1})}{(\omega-\omega_0)^2 - (1/\tau_\alpha^{-1})^2} \right] , \quad (1)$$

where β is a function of the dielectric anisotropy, the scattering volume, the distance between this volume and the point of observation, the incoming electric field, the angular frequency of the incoming field, ω_0, and the temperature T. At constant temperature, β is a constant.

The optical geometry factor G_α is defined by:

$$G_\alpha = \hat{e}_\alpha . (\vec{f}\vec{i} + \vec{i}\vec{f}) . \vec{n}_0$$
$$= f_\alpha n_\gamma i_\gamma + i_\alpha n_\gamma f_\gamma \qquad (\alpha = 1,2; \gamma=1,2,3), \qquad (2)$$

where the vectors $\vec{i}$ and $\vec{f}$ stand for the chosen directions of the polarizer of the incoming light and the analyser of the scattered light respectively (*cf. e.g.* fig. 1).

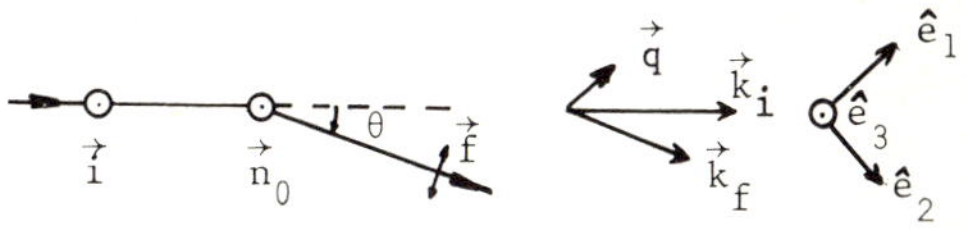

Fig. 1 - Schematic diagram of the optical configuration used in the experiments.

The subscripts α in eqs. (1) and (2) are associated with an orthonormal base $\hat{e}_i$ (i=1,2,3), defined such that $\hat{e}_3 = \vec{n}_0$ and $\vec{q}.\hat{e}_2 = 0$ (or $q_2 = 0$), where $\vec{q}$ is the scattering wave vector (see fig. 1), $\vec{q} = \vec{k}_i - \vec{k}_f$. The elasticity function $K_\alpha(\vec{q})$ is given by

$$K_\alpha(\vec{q}) = K_\alpha q_1^2 + K_3 q_3^2 \quad (\alpha=1,2), \qquad (3)$$

where K_α is a distortion elastic constant, which depends only on the physical properties of the NLC, and which is associated with the splay (K_1), twist (K_2) or bend (K_3) deformation. The relaxation time τ_α(α=1,2) is given by:

$$\tau_\alpha = \eta_\alpha(\vec{q})/K_\alpha(\vec{q}) \qquad (\alpha=1,2), \qquad (4)$$

where η_α is a rather complicated function of $\vec{q}$ and of the viscous properties of the NLC. η_α has the dimension of visocity [3].

In the special cases where $q_3 = 0$ or $q_1 = 0$, $\eta(\vec{q})$ is independent of $\vec{q}$ and related to the distortion viscosities, i.e.

$$\eta_1(\vec{q};q_3=0) = \eta_{splay},\ \eta_2(\vec{q};q_3=0) = \eta_{twist} \text{ and}$$

$$\eta_1(\vec{q};q_1=0) = \eta_2(\vec{q};q_1=0) = \eta_{bend}.$$

Therefore in these particular cases we have

$$\tau_\alpha = \frac{\eta_{bend}}{K_3} q^{-2} \text{ for } q_1 = 0\ (\alpha=1,2), \text{ and}$$
$$\tau_\alpha = \frac{\eta_\alpha}{K_\alpha} q^{-2} \text{ for } q_3 = 0 \quad (\alpha=1[\text{splay}],2[\text{twist}]). \tag{5}$$

The case where $q_3 = 0$ is shown in fig. 1.

Experimentally we find that $1/\tau_\alpha$ is very small compared to ω_0, *e.g.* $1/\tau_\alpha \simeq 10^{3\alpha}\ s^{-1}$, while $\omega_0 \simeq 3.10^{15}\ s^{-1}$. Therefore we use the technique of homodyne detection known as self-beat spectroscopy [5,6]. With this technique we can analyse the noise intensity spectrum of the scattered light, $S_{\Delta I}(\omega)$ with a photoelectric detector (PMT).

If we assume that the fluctuating process of the optical field is Gaussian, then $S_{\Delta I}(\omega)$ is equal to the auto-convolution integral over the optical spectrum $S_E(\omega)$ [6,7]. Thence,

$$S_{\Delta I}(\omega) = \beta^2 \left\{ 2 \frac{G_1^2\, G_2^2}{K_1(\vec{q})\, K_2(\vec{q})} \frac{(1/\tau_1 + 1/\tau_2)}{\omega^2+(1/\tau_1+1/\tau_2)^2} + \right.$$
$$\left. + \sum_{\alpha=1,2} \frac{G_\alpha^4}{K_\alpha^2(\vec{q})} \cdot \frac{2/\tau_\alpha}{\omega^2 + (2/\tau_\alpha)^2} \right\}. \tag{6}$$

Note that this spectrum consists of three Lorentzians with half-widths of comparable magnitude, which are functions of the two relaxation times τ_α $(\alpha=1,2)$. The relative weight of the Lorentzians is determined by the factors $G^2_\alpha/K_\alpha(\vec{q})$. The ratio $G^2_\alpha/K_\alpha(\vec{q})$ depends on the elastic and dielectric constants of the NLC, and on the optical configurations. In section 3 we show how the two relaxation times τ_α $(\alpha=1,2)$ can be extracted numerically from the experimental results with the help of eq. (6).

2. PHOTON STATISTICS

In the foregoing analysis of the laser light scattering experiments use was made of the assumption that the noise intensity spectrum $S_{\Delta I}(\omega)$ can be written as the auto-convolution of the optical spectrum $S_E(\omega)$. A sufficient condition for this relation to be satisfied is that the process responsible for the electric field fluctuations is a Gaussian-Lorentz process. In order to test whether such an assumption is warranted, we performed a number of experiments in which the photon statistics of the light scattered by DIBAB were analysed. Unlike measurements of the noise intensity spectrum, photon statistics measurements are sensitive to higher order cumulants of the scattered field and can therefore, in principle, be used to distinguish between Gaussian and non-Gaussian scattering processes.

Two types of measurements were performed: 1) a measurement of the probability distribution $P(n,T)$ of detecting n scattered photons in a time interval T and 2) the half-open time interval distribution, i.e. the probability density $P_f(t)$ of detecting a first photocount at time t if the observation was started at t=0. For a Gaussian-Lorentz process closed expressions for both $P(n,T)$ and $P_f(t)$ are known (see for instance ref. [9]). In order to test whether the scattering by DIBAB obeyed Gaussian statistics we first measured the relaxation time of the director fluctuation responsible for the light scattering. In this experiment we chose a configuration where only a single Lorentzian contributes to the light scattering. If director fluctuations obey Gaussian statistics, knowledge of this relaxation time is sufficient to predict both $P(n,T)$ and $P_f(t)$. The latter two quantities were measured experimentally using a Malvern K7026 correlator system (for $P(n,T)$) and a Tracor-Northern pulse height analysis system (TN1700) coupled to a time-amplitude converter (for $P_f(t)$). Great care was taken to ensure that the aperture of the detector was much, much smaller than one coherence area (typically by a factor of 100). The $P(n,t)$ and $P_f(t)$ measurements were corrected for small but significant dead-time effects, extra counts due to after-pulsing (only for $P(n,T)$) and a small uncorrelated background due to dark counts and stray light. All these corrections, except the one for stray light, were known *a priori*; the stray light level was left as an adjustable parameter (the only one) in the fit of the Gaussian predictions to the experimental results. This contribution always turned out to be less than 1%. As can be seen from fig. 2 we found no evidence for deviation from Gaussian behaviour in the temperature range studied. This result justifies the use of the auto-convolution approximation (eq. 6) in the interpretation of the homodyne light beating experiments.

3. ANALYSIS OF EXPERIMENTAL DATA AND RESULTING VISCOELASTIC RATIOS FOR DIBAB.

In this section we discuss the analysis of the measured noise intensity of laser light scattered by nematic DIBAB. DIBAB (*p*, *p'*-di-*n*-butylazoxybenzene) has the following structure:

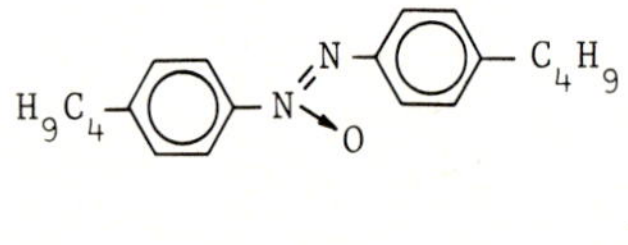

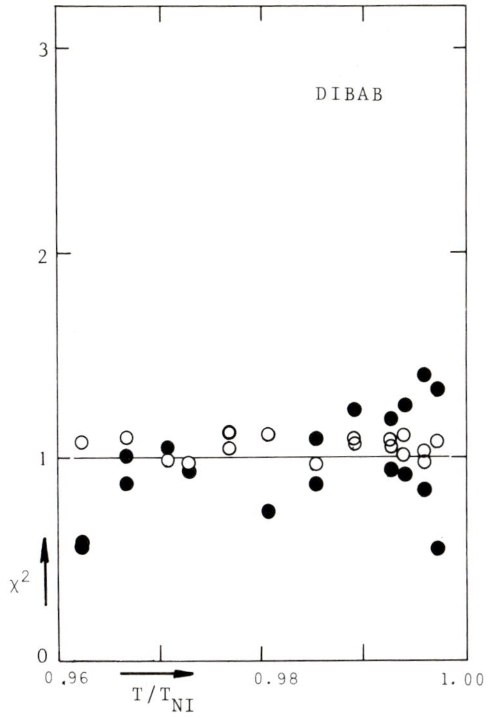

Fig. 2- Quality of the fit of experimental photocount distribution function P(n,T) (●) and half-open time-interval distribution $P_f(t)$ (○) to the corresponding theoretical predictions for a Gaussian-Lorentz process. Note that we obtain a satisfactory fit ($\chi^2 \simeq 1$) over the entire nematic range of DIBAB.

We studied the light scattering of DIBAB in its nematic range which extends from T_{KN} = 16 °C to T_{NI} = 31.9 °C at a pressure of 1 atm. We also re-analysed earlier measurements done on the same compound by van Eck *et al.* [3,8].

In order to extract τ_α (α=1,2) (eq. (6)), we tried two methods which we shall refer to as method 1 ("single mode" fitting) and method 2 (curve fitting). Method 1, the "conventional" procedure, is based on the idea that if

$$G_1 \gg G_2 \quad \text{or} \quad G_2 \gg G_1 \qquad (7)$$

only one Lorentzian contributes significantly to eqs. (1) and (6). The half-widths of these Lorentzians are related to the viscoelastic ratios through eq. (5). Although the "single mode" method has the advantage that it is very simple, condition (7) is rarely met in practice. We therefore developed a method that does not rely on condition (7) (method 2).

This method is based on the idea that the relative contributions of the Lorentzians (*cf.* eq. 6) can be changed by varying the scattering angle. Instead of using a limiting case the viscoelastic ratios are found by a computer fit to the light scattering data at various scattering angles. We measured the splay and twist viscoelastic ratios using the optical configuration shown in fig. 1. From eq. (6) it follows that, for the optical configuration of fig. 1, method 1 should work at scattering angles of $\sim 0^\circ$ and $\sim 25^\circ$ for the twist and splay ratios respectively. A typical spectrum $S_{\Delta I}(\omega)$ of the noise of the scattered light using the configuration of fig. 1 is given in fig. 3 at a scattering angle θ of 15° 37' and a reduced temperature of T/T_{NI}=0.9855. Note that, as expected, the datapoints do not fit a pure Lorentzian (eq. 6).
Instead of analysing the data on the noise spectral intensity we investigate a function $\tilde{s}(\omega)$, defined by

$$\tilde{s}(\omega) = \omega . S_{\Delta I}(\omega) \qquad (8)$$

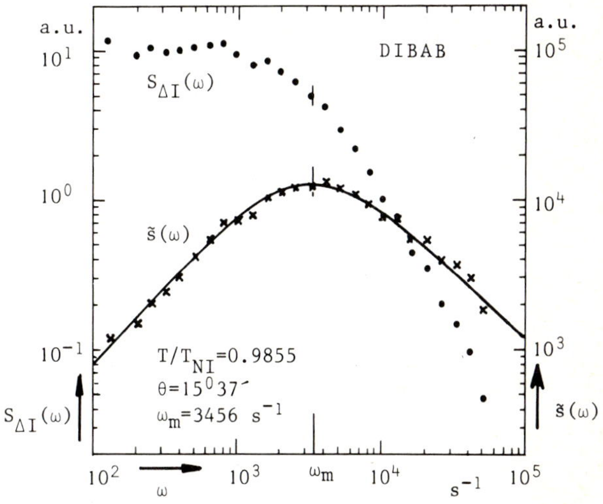

Fig. 3 - A typical result of $S_{\Delta I}(\omega)$ (dots) and $\tilde{s}(\omega) = \omega . S_{\Delta I}(\omega)$ (crosses and best fit) with its maximum at ω_m

We determine the angular frequency ω_m for which the best fit to $\tilde{s}(\omega)$ has a maximum. Then of course

$$\frac{d}{d\omega}\left[\omega . S_{\Delta I}(\omega)\right]_{\omega=\omega_m} = 0. \qquad (9)$$

Note that if $S_{\Delta I}(\omega)$ consists of one Lorentzian then ω_m equals the half-width of the Lorentzian.

If we substitute eq.(6) for $S_{\Delta I}(\omega)$ into eq. (6) we obtain an equation with the known variables ω_m and q_1^2 and the unknown η_{splay}/K_1, η_{twist}/K_2 and K_1/K_2. Note that K_1/K_2 can be obtained from a measurement of the angular distribution of the scattered light [3,8].

Measurements of the angular dependence of the spectral noise intensity $S_{\Delta I}(\omega)$ yield a set of ω_m and q-values with the corresponding eqs. of the form (9). Now we use a computer procedure to find the values for the viscoelastic and elastic ratios that fit best to the set of eqs. (9).
As starting values we used the results obtained by method 1. Fig. 4 shows a typical plot of ω_m versus q_1^2.
The solid line is a computer fit using eq. (9). The fitting procedures allow for a relative accuracy of 2% for the viscoelastic ratios. The experimental error for K_1/K_2 is about 30%, which is rather poor [10]. But K_1/K_2 can also be obtained with an experimental error of about 2% from measurements of the angular distributions of the scattered light, as stated before. Finally in fig. 5 the experimental results for the splay and twist viscoelastic ratios of DIBAB are plotted versus the reduced temperature T/T_{NI}.
The data given make a comparison of methods 1 and 2 possible.

Note that for η_{splay}/K_1 both methods yield comparable results. No such agreement is found in the case of η_{twist}/K_2. This poor agreement is due to the fact that in the present experiment the

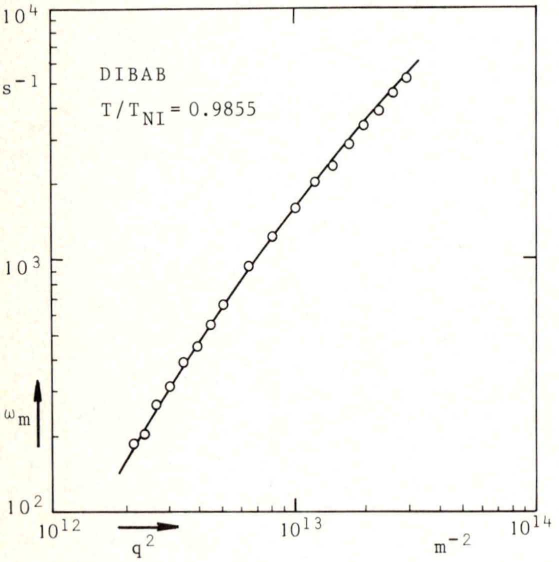

Fig. 4 - Typical plot of ω_m versus q_1^2. The curve through the datapoints is the best fit to eq. (9).

condition $G_2 >> G_1$ cannot be satisfied even at the smallest scattering angles studied. In fact, from our analysis it follows that the condition (7) is only satisfied for angles smaller than 2°. However, light scattering measurements at such small angles could not be carried out because of interference due to heterodyne mixing of the scattered light with the original laser light.

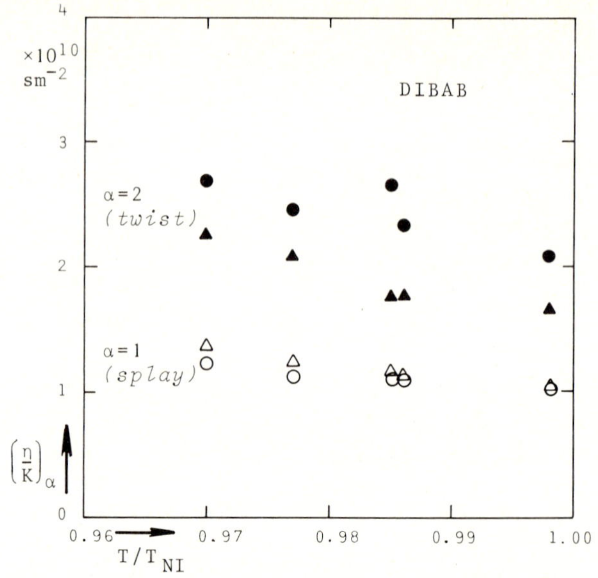

Fig. 5- Experimental data of the splay (○ : from method 2;△: from method 1) and twist (● : from method 2;▲ : from method 1) visoelastic ratio of DIBAB as function of the reduced temperature T/T_{NI}

We thank Sheila McNab for making linguistic improvements and Linda Dongen for typing the manuscript.
This work was performed as a part of the research programme of the <<Stichting voor Fundamenteel Onderzoek der Materie >> (FOM) with financial support from the << Nederlandse Organisatie voor Zuiver Wetenschappelijk Onderzoek >> (ZWO).

REFERENCES

[1] de Gennes, P.-G., *The Physics of Liquid Crystals* (Clarendon Press, Oxford) 1974.
[2] van der Meulen, J.P. and Zijlstra, R.J.J., *J. Physique* 43 (1982)411.
[3] van Eck, D.C. and Zijlstra, R.J.J. in *Noise in Physical Systems* (Wolf, D., ed.; Springer-Verlag, Berlin) 1978, pp. 270.
[4] Groupe d'Etude des Cristaux Liquides d'Orsay, *J. Chem. Phys.* 51(1969)816.
[5] van Eck, D.C. and Zijlstra, R.J.J., *J. Physique* 41(1981)351.
[6] Alkemade, C. Th. J., *Physica* 25(1959)1145.
[7] Berne, B.J. and Pecora, R., *Dynamic Light Scattering* (John Wiley and Sons, New York) 1976.
[8] van Eck, D.C. and Westera, W., *Mol. Cryst. Liq. Cryst.* 38 (1977) 319.
[9] Saleh, B., *Photoelectron Statistics* (Springer-Verlag, Berlin) 1978, pp. 185.
[10] van der Meulen, J.P., Zijlstra, R.J.J. and van Kooten, J.J., *Mol. Cryst.Liq. Cryst.* 82(1982)1.

NOISE IN PHYSICAL SYSTEMS AND 1/f NOISE
M. Savelli, G. Lecoy and J-P. Nougier (eds.)
© Elsevier Science Publishers B.V., 1983

NOISE IN METHYL - BUTYL MORPHOLINIUM $(TCNQ)_2$

H.F.F. Jos, R.J.J. Zijlstra, J. Ike

Rijksuniversiteit Utrecht, Fysisch Laboratorium, afd. F.V.S.
Princetonplein 5, 3584 CC Utrecht
The Netherlands

Noise data on quasi one-dimensional semiconducting Mn-BM $(TCNQ)_2$ crystals are described. As a tentative explanation for the noise a new mechanism for the semiconducting behaviour of TCNQ salts is proposed, based on the temperature dependence of the number of conducting channels. The observed noise is thought to be caused partly by fluctuations in the number of conducting channels.

1. INTRODUCTION

Compounds of Morpholinium derivatives and Tetracyanoquinodimethane (TCNQ) form crystalline organic, quasi one-dimensional electrical conductors. These organic conductors often show several phases i.e. an insulating, a semiconducting and a metallic phase (1). Their conductivity is anisotropic, i.e. the electrons (and holes) can move more freely along the stacks of TCNQ-molecules than in other directions. Interest in these crystalline organic conductors has been stimulated by the occurrence of typical one-dimensional effects and high conductivity (2).

To our knowledge information on current noise in these materials is scarce. The only noise publication known to us is on TTF-TCNQ (3), where the observed noise was 1/f noise. In order to obtain more insight into the conduction mechanisms it seemed worthwhile to make a detailed study of the noise properties of these materials at several temperatures. We have done experiments on the semiconducting crystal formed by compounds of one methyl butylmorpholinium molecule and two TCNQ-molecules (Mn-BM$(TCNQ)_2$. The morpholinium derivative acts as an electron donor, TCNQ as an acceptor.

Since the thermal expansion in TCNQ salts is usually large and affects the electrical conductivity (4), we also investigated the thermal expansion of Mn-BM $(TCNQ)_2$ in a temperature range around room temperature.

2. EXPERIMENTS AND RESULTS

2.1 Sample preparation

Typical crystal dimensions are $3 \times 0.2 \times 0.2$ mm^3. Ohmic contacts on the crystals were made with small copper filaments, glued to the ends of the crystal with Leitsilber 200. Because of the large thermal expansion the crystal was mounted at one point on a thread of normal, non-conducting glue. This arrangement kept the crystals strainless even when expanding thermally. The resistance of the crystals was found to increase in the course of time when kept in open air. This problem could be avoided by keeping the crystals in an atmosphere of nitrogen gas.

2.2 Noise measurements

The experimental set-up for the noise measurements was a conventional one for the determination of the equivalent Norton current noise generator. In the circuitry metal-film low noise resistors were used and the devices were fed from batteries. The low-noise pre-amplifier used has been described elsewhere (5). The spectral noise analysis was done with a Bruël & Kjaer Fast Fourier Transformer type 2033 (0 Hz - 20k Hz) and a Rohde & Schwarz selective microvoltmeter type USH 2 (10 kHz - 60 MHz). For the noise calibration we used Quan Tech type 420 (low frequency) and type 421 (high frequency) noise generators. The samples were cooled in a cold nitrogen gas stream. A Cryoson regulator type EA 2337 was used to regulate and stabilize the temperature within 0.1 K. The temperature was measured with a copper-constantan thermocouple.

2.3 Results

Mn-BM $(TCNQ)_2$ is semiconducting at least in the temperature range from 140 K to 310 K, which means that the conductivity decreases with decreasing temperature.
However, because of hysteresis in the conductivity -versus- temperature characteristic the conductivity at room temperature and its activation energy are not well defined (6). Typical values are $10(\Omega m)^{-1}$ -$10^2(\Omega m)^{-1}$ for the conductivity at room temperature and 0.1 eV - 0.3 eV for the activation energy. The current-voltage characteristics were ohmic up to electric fields as high as $2.\ 10^4$ Vm^{-1}, but for fields higher than 10Vm^{-1} Joule heating caused a lowering of the resistance of the semiconducting crystal. This could be avoided by applying voltage pulses of short duration and in a low duty cycle. In addition to the thermal noise we found an excess noise that varied quadratically with current at constant applied voltage, indicating fluctuations in the resistance. This excess noise consisted of 1/f noise and a contribution with a different frequency dependence. These results were obtained with a 4-probe technique as well as with a

2-probe technique. When the 4-probe technique was used the 1/f noise was slightly lowered, indicating that a part of the 1/f noise stems from the contacts. All results mentioned in this paper were obtained with the 2-probe technique, because in this configuration the relative noise S_I/I^2 could be determined unambiguously. The 4 probe technique is less suited to measure relative noise intensities in quasi one-dimensional conductors. We made an analysis of the extra noise in terms of Lorentz spectra of the form $S_I(o)/(1+\omega^2\tau^2)$. A typical computer analysis of a measurement is shown in fig. 1 where the noise has been decomposed into a 1/f component and three Lorentzians.

tion energies are indicated in the plots. The second Lorentzian could not be determined accurately, so no activation energy is given. The activation energies found for the resistance and the 1/f noise (the relative noise at a frequency of 1 Hz) in a single run with decreasing temperature are different for different samples, although the activation energy for the resistance is always about the same as for the 1/f noise. Typical values are 0.1 eV - 0.3 eV as mentioned before. The difference between different samples is ascribed to hysteresis effects.

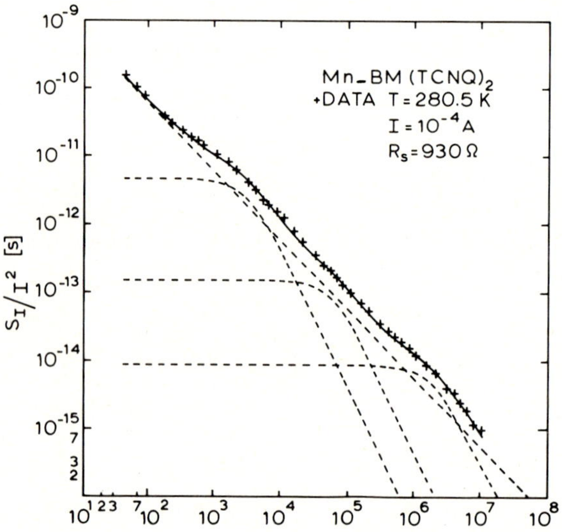

Figure 1: The relative noise in excess of thermal noise versus frequency
The dotted lines are the 1/f component and 3 Lorentzians
The solid line is the summation of these components

Because of the hysteresis we measured the excess noise at several temperatures in a single run with decreasing temperature. The analysis always showed three Lorentzians. The low-frequency noise levels of the three Lorentzians $S_I(o)/I^2$ are shown in fig. 2. In fig. 3 the temperature dependence of the relaxation times τ have been plotted. The solid lines represent exponential relations of the form $\exp.(-E_a/kT)$ where E_a is an activation energy. The corresponding activa-

2.4 Thermal expansion

We measured the thermal expansion of Mn-BM $(TCNQ)_2$ in the temperature range 273 to 290 K with the help of a Michelson interferometer. To that end a thin mirror was glued to one end of the crystal. This mirror served as one of the two mirrors in the interferometer. The thermal expansion coefficient for a direction parallel to the conduction axis was found to be:

$(1.40 \pm 0.07) \times 10^{-4}\ K^{-1}$. Note that this is a relatively large thermal expansion and that on the basis of this result one might expect a decreasing mobility with increasing temperature.

room temperature mobility of Silicon. The activation energies of $\frac{S_I(o)}{I^2}$ and of τ as obtained from plots then yield the activation

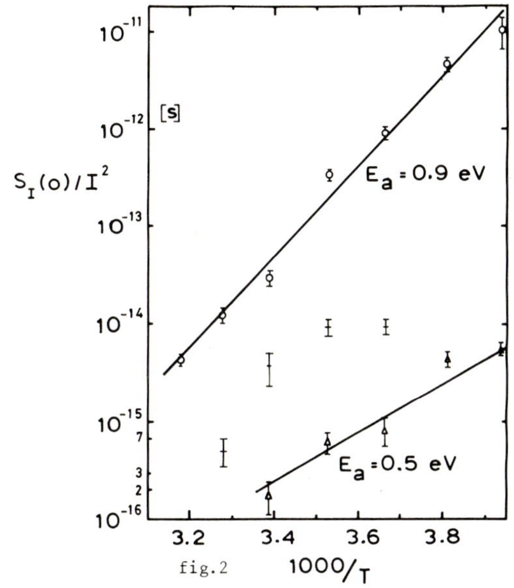

fig.2

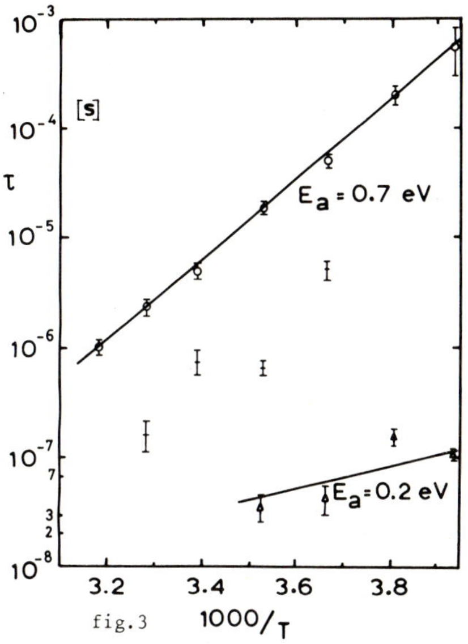

fig.3

Figure 2: The low frequency level of the relative noise of the 3 Lorentzians versus $10^3/T$. The activation energies belong to the solid lines.

Figure 3: The relaxation times of the 3 Lorentzians versus $10^3/T$. The activation energies belong to the solid lines.

3. DISCUSSION OF THE RESULTS

Since the observed current dependence of the noise was quadratic while samples were ohmic, the noise must be due to resistance fluctuations. This suggests an interpretation of the experimental results for the noise in excess of thermal noise, in terms of 1/f-noise and generation-recombination noise. In general the noise could be represented as was mentioned before, as the sum of a 1/f-noise component and three Lorentzians. As a first estimate we assume that the lowest frequency Lorentzian can be described by:

$$\frac{S_I(f)}{I^2} = \frac{4\overline{\Delta N^2}\tau/\bar{N}^2}{1+\omega^2\tau^2} \qquad (1)$$

where N is the number of charge carriers and τ the noise relaxation time. Then, assuming that $\overline{\Delta N^2} \simeq \bar{N}$ we found $\bar{N} \simeq 10^8$ for $T = 300$ K. Using this result and the corresponding value for the resistance we then obtain the following estimate for the carrier mobility: $\mu \simeq 10^3\ m^2/Vs$. This value is rather large, about 10^4 larger than the

energy for N: $E_N \simeq 0.2$ eV. This value is in good agreement with the activation energy for the resistance and for the relative 1/f-noise, if it is assumed that

$$\frac{S_I(f)}{I^2} = \frac{\alpha}{fN} \qquad (6)$$

The equation: $N(T) = N_o \exp(-E_N/kT)$, where N_o is the number of electrons on the sites formed by the dimers of TCNQ molecules, gives $N_o \simeq 10^{11}$. However, taking the crystal as well as the molecular dimensions into account, we found $N_o \simeq 10^{18}$. The same order of magnitude for N_o has been estimated on the basis of Hall-effect data on TTF-TCNQ (7). Although the interpretation of Hall-effect data on one-dimensional conductors is ambiguous, we still believe that the data give the correct order of magnitude for the density of charge carriers. Since the density of charge carriers in metallic TTF-TCNQ should be of the same order as the density of electrons on dimers in semiconducting Mn-BM $(TCNQ)_2$ we believe that our estimate of N_o obtained from these Hall-effect data is correct. The large discrepancy, found in the estimates of N_o, is a more important reason for discarding the interpretation for the noise data as given above than the unusually large value for μ.

Therefore we suppose that the observed noise is not related to the fluctuations in the number of individual carriers, but instead to fluctuations in the number of groups of carriers. $N_o \simeq 10^{11}$ would be the total number of groups of carriers. Now it so happens that this number corresponds to the estimated number of stacks in the crystals. A tentative explanation for the observed fluctuations is then that the number of conducting stacks fluctuate in a manner that can be described by a relaxation mechanism. We estimate the total number of electrons per stack (one electron per two TCNQ-molecules) to be 10^7. Hence the upper limit of the number of charge carriers contributing to the conduction is $N_t \simeq 10^{15}$, at T = 300 K. Then the calculated lower limit of the mobility is: $\mu \simeq 10^{-4}$ $m^2/V.s$, which seems to be a more acceptable value than the mobility found before.

This model, although simple, is consistent with the measurements, although the interpretation of the conduction mechanisms is quite different from the usual interpretation. In the usual interpretation the material is semi-conducting because the stacks are semiconducting, i.e. the electrons are localized on the dimers and have an activation energy for delocalizing. The number of conduction-electrons on a stack is therefore relatively low, but all stacks contribute to the conduction. In our model the stacks need not be semiconducting. At least part of the semiconducting behaviour is due to the fact that the number of stacks contributing to the conduction is exponentially dependent on the temperature. Blocking may occur because impurities or lattice imperfections obstruct a stack. In that case the complete stack does not contribute to the conduction. So the phenomena described above would be typical one-dimensional effects. Then the chance of obstructing a stack ($\frac{1}{\tau_1}$ per second) must be much larger than the chance of freeing stack ($\frac{1}{\tau_2}$ per second). We can show that this is consistent with our model. The rate of change in the free stacks N is:

$$\frac{dN}{dt} = \frac{N_o - N}{T_2} - \frac{N}{\tau_1}$$

In the stationary state we have

$$\frac{d\bar{N}}{dt} = 0 \Rightarrow N_o = \frac{\tau_2 + \tau_1}{\tau_1} \bar{N}$$

Hence, if $\frac{\bar{N}}{N_o} \simeq 10^{-3}$, then $\frac{\tau_1}{\tau_2 + \tau_1} \simeq \frac{\tau_1}{\tau_2} \simeq 10^{-3}$

and so $\frac{1}{\tau_1} \simeq 10^3 \frac{1}{\tau_2}$

The analysis above was based on one Lorentzian only, however, Mn-BM$(TCNQ)_2$ is known to have two inequivalent dimerized chains in parallel (8), which allows for the occurrence of at least two Lorentzians in the spectrum.

4. CONCLUSIONS

On the basis of the limited experimental data available, we come to the following tentative conclusion. Mn-BM (TCNQ) shows resistance fluctuations in the semiconducting range of temperatures. Part of the noise is 1/f-noise, the rest is caused by fluctuations in the number of conducting channels or at least in the number of groups of carriers contributing to conduction. The lower limit of the mobility of the carriers is of the order of 10^{-4} m^2/Vs. More research needs to be done on TCNQ salts, especially in the temperature region where phase transitions occur.

ACKNOWLEDGEMENTS

We thank Dr. J. de Boer of the Chemistry Department of the University of Groningen for supplying the crystals. This work was performed as a part of the research program of the Stichting voor Fundamenteel Onderzoek der Materie (FOM) with financial support from the Nederlandse Organisatie voor Zuiver Wetenschappelijk Onderzoek (ZWO).

LITERATURE:

(1) J.B. Torrance, The Difference between Metallic and Insulating Salts of TCNQ, Acc. of Chem.Res. 12 (1979) 79.

(2) P.M. Chaikin from: Synthesis and Properties of Low-Dimensional Materials, eds. J.S. Miller, A.J. Epstein, New York, Ac. of Science (1978) 128.

(3) A.N. Bloch from: Chemistry and Physics of 1D metals. NATO Advanced Study Institutes series. Ser. B. ed. H.J. Keller vol. 25 (1977) New York, Plenum Press.

(4) E.M. Conwell, Phys. Rev. B (USA), 19 (1979) 2409.

(5) F. Driedonks, J.J.M. van Gasteren, Physica 50 (1970) 606.

(6) F.N. Hooge, T.G.M. Kleinpenning and L.K.J. Vandamme Rep. Prog. Phys. 44(1981) 479.

(7) D. Jérome, H.J. Schulz, Adv. in Physics, 1982, 31 (1982) 361.

(8) S. Oostra, Chemistry Department, State University of Groningen, private communication.

NOISE IN PHYSICAL CHEMICAL AND BIOLOGICAL SYSTEMS

NOISE IN PHYSICAL SYSTEMS AND 1/f NOISE
M. Savelli, G. Lecoy and J-P. Nougier (eds.)
© Elsevier Science Publishers B.V., 1983

I/F NOISE AND FLUCTUATION ANALYSIS IN BIOLOGICAL MEMBRANES

(Invited Paper)

Denis Poussart

Department of Electrical Engineering
Université Laval
Québec, Canada

The study of spontaneous fluctuations in biological systems, especially with regards to electrical currents in the nerve membrane, has been a field of considerable activity in recent years. Following the observation of fluctuations in macroscopic variables such as the rate of firing of action potentials, new experimental techniques were developed for a more direct investigation of the underlying sources of randomness. 1/F processes where among the first to be recorded. They were subsequently shown to occur in a variety of simple ionic systems and are now believe to have little physiological significance. But their observation stimulated further studies and a refinement of techniques which lead to the identification of relaxation processes. Convincing evidence has shown that these are direct reflexions of the unique properties of the excitable membrane. The picture that has emerged does not consider membrane noise as an annoying reality, but rather one of the clear expression of the basically statistical nature of membrane mechanisms.
This short paper is intended as a general overview of membrane fluctuations for experimenters in other fields where noise is also an important probe. It proposes to convey some idea of the unique constraints of experimental work on biological membranes and to outline some important conclusions that have been reached.

1. INTRODUCTION

In a recent tribute to the major contributions of Kenneth Cole, Jakobsson and Guttman [1] presented a chart from which Figure 1 is freely adapted and which has been augmented to indicate where membrane fluctuations fit in studies of the excitable membrane. It shows the occurence of major advances such as the discovery of the action potential in nerve membrane, that of the concurrent changes in membrane impedance (which already indicated that changes in potential were mediated via changes in conductances), the invention of the voltage clamp method (which demonstrated that conductances were themselves potential dependent), key elements which culminated in the mathematical model of Hodgkin and Huxley. From thereon, two major directions could evolve: applying "holism" toward an explanation of the encoding of nervous information by sequences of action potentials, and applying "reductionism" toward an understanding of the fundamental support of such processes. The study of 1/F and other fluctuations in biological membrane has mostly followed a "top-down" path. Interest in fluctuation analysis, which was initially triggered by direct observation of 1/F processes, has gradually shifted to additional components which are more specifically linked to the funtionality of excitable membranes.

MICRO AND MACROSCOPIC VIEWS OF MEMBRANE

Studies of the electrical properties of nerve membrane hinge on models which operate at different levels of abstraction. These are basically associated to altogether different - and complementary - experimental approaches.

At a microscopic level, a current, general image to which fluctuation analysis has substantially contributed, is as follows (for instance, see [2]): the bulk constituent of membrane is a thin sheet of lipoid material, about 5-10 nm thick. Charge carriers on either side of this membrane are mostly hydrated sodium, potassium and chloride ions. The lipoid material is hydrophobic, and the flow of ionic charges is essentially restricted to distinct, localized "channels" which provide aqueous pathways of high (of the order of 1-10 pS) conductance. Channels consist of specialized macromolecular structures which undergo conformal changes in response to electric field. These changes can be functionnally represented by states and transition probabilities. The simple hypothesis of binary conduction, where the channel is modelled by a random switch, has been used often (see below). Channels appear to occur in distinct families which display ion and pharmacological selectivities, hence a nomenclature which refers to "potassium" or "sodium" channels. The channels, which have effective diameters comparable to those of hydrated ions ($<0,5$ nm), are very sparsely distributed over the lipoid substrate, with an interchannel distance of 50-300 nm. Fluctuation studies discussed below have precisely aimed at caracterizing the probabilistic "microevents" associated with these channels.

Conventional biological preparations provide areas of membrane which are very large in comparison with these dimensions and consist of a numerous collection (e.g. >50000) number of channels. Classical measurements are concerned with the sum (mean) ionic current flowing through this system, in particular with the dependence of its

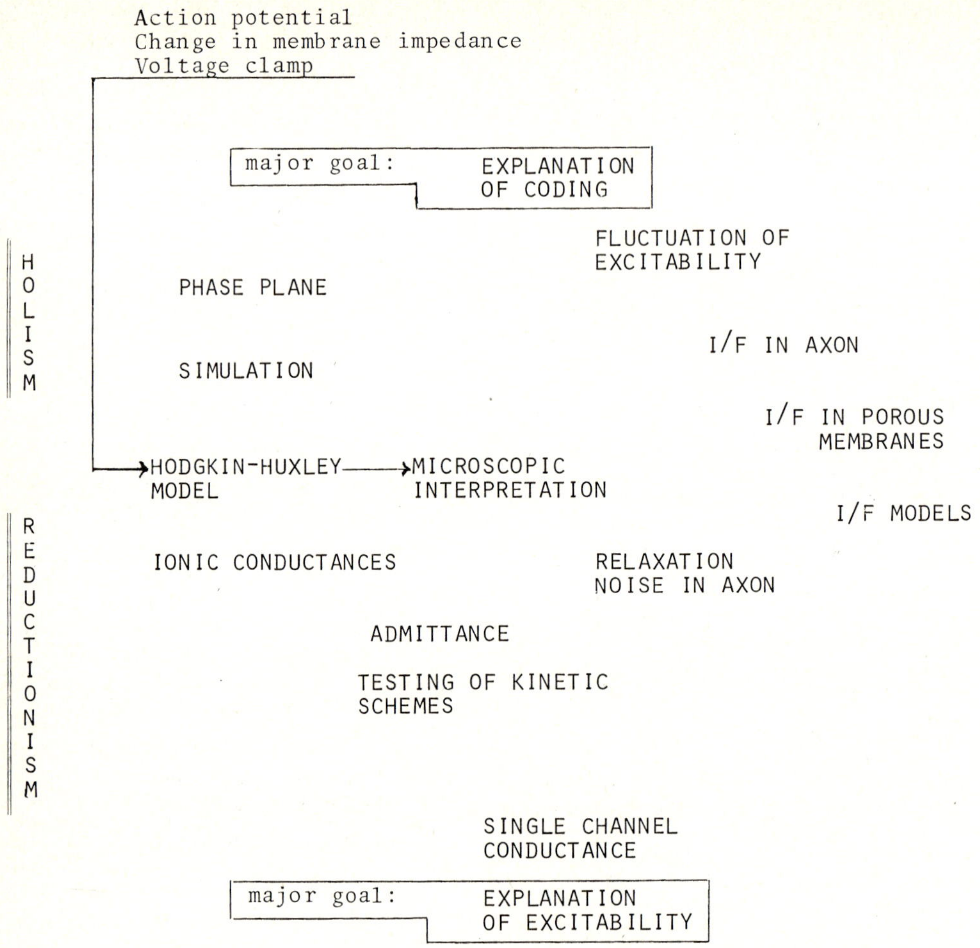

Figure 1: This chart adapted from [1] shows where 1/F and fluctuation analysis fit in studies of excitable membrane.

time course on membrane potential, and the effect of experimental parameters such as ionic environment, temperature, or presence of pharmacological agents.

The formalism of the Hodgkin - Huxley model has remarkably survived the test of time and continues to provide a basic conceptual framework. Mean membrane currents are represented as a sum of independent potassium, sodium, and non-specific components. These are driven by Nernst batteries (corresponding to the ionic gradients), through non linear, voltage dependent, time varying conductances. The thin lipid substrate accounts for a large, parallel capacitance of the order of 1 $\mu F/cm^2$.

In resting conditions, the potassium and non-specific conductances dominate, and drive the membrane toward the corresponding Nernst potentials, with a value of the order of -70 mV (the external bulk solution is taken as the reference point). If membrane potential is momentarily displaced toward a positive value (depolarized), the sodium conductance is activated. Because the normal value of the Nernst equilibrium potential for sodium is of the order of +45 mV, this process generates a current which tends to drive membrane potential toward further depolarized values. This evolves with a regenerative action which, if left to occur without constraint, produces the rising phase of the characteristic action potential (a purpose of the voltage clamp method is precisely to prevent this regenerative action to take place by maintaining membrane potential under electronic control). Within a brief time span, typically a millisecond, the sodium conductance self-inactivates, the potassium conductance fully activates, and membrane potential falls back toward its resting value, thus terminating the action potential.

A constant endeavour of fluctuations analysis has been to reconcile and link these two microscopic and macroscopic views. In particular, the efforts have dealt with the relationship between the global, lumped ionic conductances of the Hodgkin-Huxley formalism and their underlying support in the form of populations of channels with a probablilistic behavior.

1/F STUDIES

The electrical activity recorded in single cells, which encode and transmit information as sequences of actions potentials, form a point process which typically contains dominant stochastic components (for review, see [3]). A description of the relationship between stimulus and response, or of the instantaneous frequency of a spontaneously firing neuron often requires the inclusion of probabilistic elements (for a recent example of fluctuations of firing intervals of a single cell which happen to exhibit 1/F statistics, see [4]).

This observation provided the primary motivation for initial studies of membrane noise. Very small annuli of exposed nerve membrane, with areas of the order of 50 square microns and which are electrically insulated from adjacent tissues, occur naturally in myelinated fibers. These structures, which have unique characteristics of electrical access, have continued to be among major preparations for studies of excitable membrane. Verveen succeeded in recording the minute fluctuations superimposed on the mean, resting potential which occurs accross such a node of Ranvier [5]. These early studies were eventually extended to other preparations, including giant unmyelinated axons, and techniques were developed in order to carry observations in the form of current, rather than voltage fluctuations. Since membrane properties of interest are known to be voltage dependent, the voltage clamp method, which forces a prescribed field accross the membrane and records the resulting current, is a preferred approach. It yields the desired observable of charge transfer more directly, and automatically unfolds the presence of membrane impedance (including parallel capacitance).

The overall picture which has emerged to date from such observations on preparations in a "normal" environment is as follows (for review and discussion, see [6]):

- The flow of transmembrane current shows fluctuations with a gaussian amplitude distribution, the level of which is several order of magnitude above the thermal noise of an equivalent system assumed to be in equilibrium.

- Over the limited range of frequencies which can be explored (see below) the spectral density is largely of the 1/F type.

- Its magnitude reaches a minimum when the membrane potential is brought near the Nernst potential for potassium and increases approximately in proportion with the square of the mean potassium current, irrespective of the inward or outward direction of the current. There is a residual 1/F component even at the minimum.

- Its magnitude is lower in fresh preparations, and eventually increases, in parallel with the additional leakage current which develops with experimentation trauma.

- There is no obvious temperature dependance.

- The magnitude is not affected by inhibitors of the metabolic processes which are responsible for maintaining the normal ionic gradients across the membrane.

Noise studies on biological membranes were among the first to be conducted on ionic systems. Because naturally occurring membranes are complex and are typically imbedded in other tissues (e.g. the glial cells which surrounds non-myelinated nerve fibers), several studies have been carried out on "simpler" systems. Experiments on synthetic an porous membranes such as collodion, Mylar film with calibrated holes, or mica sheets [7, 8] have sought to mimic the restricted flow expected from the structure of biological membrane. These indeed yielded generally similar results, and consequently lead to ascribe 1/F noise as a process which, although observable in various excitable membranes, was most probably not directly involved in the specifics of the key mechanisms of excitability [9]. The absence of physiological significance has been reinforced by the finding [10] of an intracellular region of the squid axon, adjacent to but distinct from the excitable membrane, whose introduction in a glass microtip induces 1/F fluctuations in a previously "quiet" electrode.

A current interpretation of 1/F in excitable membrane is that it probably arises from restricted, passive flow of ions through open channels, a process which is not directly connected with the gating of the channels. In a normal ionic environment, the potassium ion is mainly involved, but a (variable) contribution from unspecific (leakage) patways is also present (hence the presence of a residual 1/F component even when the mean K current is zero). In cases where potassium conduction is suppressed, a 1/F component related to sodium flow may be detected.

A list of theories for 1/F processes which have been conjectured in the context of biological membranes include (for a critical review and original references, see [11]):

- Superposition of relaxation processes with ad hoc distribution of kinetics,

- Coupling between parameters (amplitude, duration, interval) of ion pulses passing through the membrane,

- Empirical relationship of Hooge, which states that the magnitude of flicker noise is proportional to the square of the mean current and inversely proportional to the number of charge carriers,

- $1/\sqrt{t}$ kinetics from one-dimensional and spherical three-dimensional diffusion,

- Coupling between a channel and surrounding membrane matrix,

- Interaction between adjacent channels,

- Turbulent convection in the depleted layer [12].

However, the experimental testing of models which require specific hypothesis is extremely difficult in face of the peculiar constraints which shroud even very carefully conducted experiments on biological membranes. It must also be fully appreciated that in the course of time, other fluctuations of direct relevance to excitability were discovered and have since captivated interest. Consequently no further concentrated effort appears to have been recently applied to studies of 1/F noise in the excitable membrane.

Some of the key experimental difficulties which are rather unique to biological systems include:

- A biological membrane is a very labile device. Changes in physical parameters which might be important for model testing, temperature for instance, are very restricted.

- Since it is necessary to accurately control numerous experimental variables, measurements are conducted on preparations which have been removed from their biological environment. The period over which stable operation can be expected is quite limited. This is especially significant in view of the prolonged lengths of time (of the order of minutes) which are required to obtain spectral estimates with acceptable statistical scatter.

- Complex protocols can be used to assess repeatibility, but slow drifts and decay are unavoidable. The stationnarity of a biological preparation must always be questionned since instabilities can so easily masquerade as 1/F noise.

- Current must be recorded from a precisely defined area of membrane which is maintained over a uniform potential. Contribution from fringe field areas must be restricted. This must be reconciled with the unfortunate fact that spatial guards magnify instrumentation noise and can introduce serious distorsions of frequency response [13, 14].

- The physical size of typical membrane preparation is very small. Electrical access to the area of interest is via microelectrodes and/or material with complex properties. Electrodes typically exhibit conflicting characteristics of noise and impedance and the corresponding access series voltage noise may be significant.

- The frequency range which can be reliably explored is limited. A typical lower limit is a few Hz, as a result of slow drifts and residual mechanical vibrations (if only from the flow of ionic solutions which continuously perfuse the preparation). Above a few hundred Hz, the admittance of the parallel capacitance of the membrane becomes progressively dominant. From the early development of low-noise voltage clamping it was shown that this shunt effectively magnifies the contribution of the artifactual voltage noise of the recording apparatus [15]. This typically limits the upper frequency of acceptable signal/noise ratio to 1-3 KHz.

- The range of membrane potential which can be maintained for periods required for suitable statistical estimates of the fluctuations is limited. The sustained passage of current can lead to change of ionic concentrations in the space which is immediately outside the membrane and continued depolarization induces modifications of the conductances. These effects can be alleviated by acquiring noise data during repetitive pulses interspersed between rest periods. But this has the otherwise unfortunate consequence of further consuming precious experimental time.

CHANNEL NOISE

As membrane noise data began to accumulate on a variety of preparations, and as experimental techniques improved, it became apparent that other spectral components could be detected. The 1/F process became to be considered as a background component, to be minimized by the choice of judicious experimental conditions. The existence of these additional components was predicted [16] as a statistical, microscopic interpretation of the Hodgkin-Huxley model for macroscopic currents. It was also suggested from other fluctuation studies on the neuromuscular junction and on artificial membranes.

The key ideas involved may be introduced from a simple model membrane which, for a single ionic species, is composed of an homogeneous population of N statistically independent channels. These channels are described by discrete states with corresponding transition probabilities (or rate constants) and conductances. The simplest variant, upon which other more complex models have been built, is that of a binary channel which flips randomly between a closed state S0 (assumed with nil conductance), and an open state S1 (assumed with conductance γ).

Two different types of microscopic information can be derived from an analysis of the current recorded from such a population placed across a potential difference V.

First, an estimate of the single channel conductance can be obtained from the variance.

Letting p be the unconditional probability of occurrence of state S1, then [17]:

elementary current: $i = \gamma V$

mean total current: $I = N p i$

total variance: $\sigma^2 = N p (1 - p) i^2$

Second, the power density spectrum of the deviations of the instantaneous current from its mean provides a direct reflexion of the kinetics of transitions. For instance, if in the simple model above the transitions between states S0 (closed) and S1 (open) occur with rate constants α and β, as depicted below:

$$\text{(closed state)} \quad (S0) \underset{\beta}{\overset{\alpha}{\rightleftarrows}} (S1) \quad \text{(open state)} \qquad \text{(a)}$$

then

$$p = \alpha / (\alpha + \beta) \qquad \text{(b)}$$

and the corresponding power spectral density consists of the Lorentzian

$$S(f) = \frac{4 I^2}{N} \frac{(1-p)}{p} \frac{1/wo}{(1 + w^2/wo^2)}$$

with $wo = (\alpha + \beta)$

In principle, fluctuation analysis can extract the pertinent parameters of such a model. However, biological membranes have more complex kinetics, and excitable membranes find the very support of their function in such complexity. The Hodgkin-Huxley model of axonal membrane considers the rate constants to be instantaneous function of membrane potential, and uses high-order products of first order components to represent the time courses of K and Na conductances. A basic, microscopic interpretation of this model for macroscopic currents takes the K channel as made up of four subunits which must all be activated for the channel to open (5 states). This leads to a weighted sum of four Lorentzian (with characteristic frequencies wo, 2wo, 3wo and 4wo) for the fluctuation spectrum. A corresponding, basic, interpretation of the Hodgkin-Huxley model for the Na sodium system involves a gate with activation (3 subunits) and inactivation (1 subunit) mechanisms which are assumed to be mutually independent, 8 distinct states, and a fluctuation spectrum which is a weighted sum of 7 Lorentzians.

Under suitable experimental conditions, it has been possible to identify spectral components in measured current noise of nerve membranes which are qualitatively compatible with the above vision (for a review, see [6]). This was first achieved with the potassium system, which can be activated in steady-state, in various degrees, by varying the clamping potential applied to the preparation. As mentionned earlier, in normal conditions, the sodium system is largely inactivated in steady-state. Therefore its fluctuations become directly observable only after the potassium system has been significantly disabled by pharmacological means, an operation which also suppresses the 1/F background. Indeed, recent work on the node of Ranvier has been able to report that "major contributions from this kind of noise (1/F) were absent... Fairly good fits of the data were obtained... with theoretical curves which contained no 1/F component" [18].

A dominant objective has been to estimate microscopic parameters (such as single channel conductance), evaluate the number N of channels, compare data from noise and macroscopic measurements (ionic and displacement currents and impedance) for kinetics parameters and assess models. Some of the difficulties in securing quantitative agreement are:

- A given model predicts precise fluctuations characteristics, but a given spectrum does not, obviously, necessarily implies a particular model. This is a fundamental difficulty with any inverse problem.

- Predicted spectra have generally smooth features: the overall shape of a sum of Lorentzians does not change critically with the number of terms. For instance, it is difficult to experimentally discriminate the sum of 4 Lorentzians predicted for potassium from a single Lorentzian. In the case of sodium, the composite spectrum practically reduces to a single low frequency Lorentzian, associated with the inactivation kinetics, and higher frequency terms, related to the activation kinetics. In the node of Ranvier, were data appears especially reliable [18], only a qualitative agreement with the double-humped prediction of the Hodgkin-Huxley model has been secured, and the resolution is not sufficient to discriminate between that model, where inactivation is independent from activation, and an alternate version which assumes some coupling.

- Experimental spectra are a mixed superposition of instrumentation and 1/F backgrounds, and ionic components. The background characteristics vary critically: 1/F is somewhat unpredictable, the contribution of instrumentation noise varies with the impedance of the preparation, which is itself very voltage and environment - dependent. Therefore the substraction of background data [19] and the mathematical templates used for curve fitting involve some ad hoc assumptions. In some cases, pharmacological procedures used for separating ionic contributions may themselves introduce additional noise [20].

In a situation where the background is relatively small, the variance - and consequently the channel conductance - can be directly estimated without resorting to spectral separation of components. This approach was successfully developed on the node of Ranvier by Sigworth [21] who measured the time course of the ensemble variance of the fluctuations of the sodium current which flows during a step change of membrane potential. Because a node contains a modest number of channels, the random deviations of the current from its mean time course are fairly large and can be directly resolved. As mentionned previously, this time course follows a transient, activation - inactivation sequence. This non-stationary situation, which is not amenable to conventional spectral analysis, is especially interesting since it is responsible for the excitability properties of the axonal membrane. The relationship between the variance and the mean of the recorded sodium current was found to be compatible with a simple open - closed model such as described previously and yielded a single channel conductance of 7.7 pS in a normal environment, and a number N in the range of 30000 to 60000. A similar analysis has recently been extended [22] over the fuller time course of the transient sodium current and again the data are compatible with an all - or - none switching of sodium channels. In most of these reports on the node of Ranvier the voltage dependence of the unit conductance seems to be slight, while on the squid axon, fitting based on macroscopic parameters obtained from admittance measurements indicate a significant voltage dependance [19].

CONCLUSION AND FUTURE DIRECTIONS

Fluctuations studies where initially motivated by the randomness which is commonly found in the encoding of nervous information, and first disclosed a large 1/F component. Eventually, interest focused on other components which could be directly linked with the physiological function of the membrane and used as probes of the corresponding mechanisms.

As a whole, the study of relaxation components of current noise in nerve axon has massively supported the vision of a membrane where ion movements occur through highly specialized structures which are sparsely distributed on the membrane matrix. Details vary with the preparations and the ionic environment, but estimated densities of 60 and 300 channels per square microns are representative numbers for the potassium and the sodium units [9]. Unit conductances are in the range of 2-15 pS.

As mentionned before, the node of Ranvier is perhaps the smallest structure which Nature provides with a favorable combination of electrical insulation from neighboring tissues and electrical access to the intracellular region. Ideally, one would like an experimental condition where the electrical activity of even a much smaller population, and eventually that of single channel, could be observed. Advanced patch techniques, where a glass microtip is gently pressed against the membrane in order to electrically insulate a very small area, have met with success on special preparations. There is much interest in performing such measurements on conventional preparations, such as squid axon, for which a whealth of related data has been accumulated. Already it has been possible, on this preparation, to define patches which contained very few channels and, in proper conditions, to observe the activity of single potassium channels [23]. This approach offers great potential and will receive considerable attention in the forthcoming years, as techniques further develop.

REFERENCES

[1] Jackobsson, E., Guttman, R., Continuous Stimulation and Threshold of Axons: the Other Legacy of Kenneth Cole., in The Biophysical Approach to Excitable Systems, W. J. Adelman and D. E. Goldman Ed., Plenum Press, New York, 1981.

[2] Holden A. V., Rubio, J. E., A Model for Flicker Noise in Nerve Membrane, Biol. Cybern. 24, 227-236, 1976.

[3] Feinberg, S. E., Stochastic Models for Single Neurone Firing: A Survey. Biometrics, 30, 399-427, 1975.

[4] Musha, T., Takeuchi, H., Inoue, T., 1/F Fluctuations in the Spontaneous Spike Discharge Intervals of a Giant Snail Neuron, IEEE Trans. Biomed. Eng., BME-30, 194-197, 1983.

[5] Verveen A. A., Derksen, H. E. Fluctuations in Membrane Potential of Axons and the Problem of Coding, Kybernetik, 2, 152-160, 1965.

[6] DeFelice, L. J., Introduction to Membrane Noise (Plenum Press, New York, 1981).

[7] DeFelice, L. J., Michalides, J. P., Electrical Noise from Synthetic Membranes, J. Membr. Biol., 9, 261-290, 1972.

[8] Dorset, D. L., Fishman, H. M., Excess Electrical Noise during Current Flow through Porous Membranes Separating Ionic Solutions, J. Membr. Biol., 21, 291-309, 1975.

[9] Conti, F., DeFelice, L. J., Wanke, E., Potassium and Sodium Ion Current Noise in the Membrane of the Squid Giant Axon, J. Physiol., 248, 45-82, 1975.

[10] Fishman, H. M., Material from the Internal Surface of Squid Axon Exhibits Excess Noise, Biophys. J., 35, 249-255, 1981.

[11] Neumcke, B., 1/f Noise in Membranes, Biophys. Struct. Mechan., 4, 179-199, 1978.

[12] Lifson, S., Gavish, B., Reich, S., Flicker Noise of Ion-Selective Membranes and Turbulent Convection in the Depleted Layer, Biophys. Struct. Mechan., 4, 53-65, 1978.

[13] Poussart, D. Moore, L., Fishman, H., Ion Movements and Kinetics in Squid Axon. I. Complex Admittance, Ann. N.Y. Acad. Sci., 303, 355-379, 1977.

[14] Fishman, H. M., Current and Voltage Clamp Techniques, in Techniques in the Life Sciences (Elsevier/North-Holland, Amsterdam, 1982).

[15] Poussart, D., Membrane Current Noise in Lobster Axon under Voltage Clamp, Biophys. J., 11, 211-234, 1971.

[16] Stevens, C. F. Inferences about Membrane Properties form Electrical Noise Measurements, Biophys. J., 12, 1028-1047, 1972.

[17] Neher, E., Stevens, C. F., Conductance fluctuations and ionic Pores in Membranes, Ann. Rev. Biophys. Bioeng., 6, 345-381, 1977.

[18] Conti, F., Neumcke, B., Nonner, W., Stampfli, R., Conductance Fluctuations of Sodium Channels in Myelinated Nerve Fibers, J. Physiol., 308, 217-239, 1980.

[19] Fishman, H. M., Leuchtag, H. R., Poussart, D., Nonlinear Single Channel Sodium Conductance in Squid Axon, to be presented at the Fourth Discussion Meeting of the Biophysical Society.

[20] Moore, L. E., Fishman, H. M., Poussart, D., Chemically Induced K Conduction Noise in Squid Axon, J. Membrane Biol., 47, 99-112, 1979.

[21] Sigworth, F. J., Sodium Channels in Nerve Apparently Have Two Conductance States, Nature, 270, 265-267, 1977.

[22] Neumcke, B., Stampfli, R., Sodium Currents and Sodium - Current Fluctuations in Rat Myelinated Fibres, J. Physiol., 329, 163-184, 1982.

[23] Conti, F., Neher, E., Single Channel Recording of K Currents in Squid Axons, Nature, 285, 140-143, 1980.

NOISE IN PHYSICAL SYSTEMS AND 1/f NOISE
M. Savelli, G. Lecoy and J-P. Nougier (eds.)
© Elsevier Science Publishers B.V., 1983

1/f NOISE IN JOSEPHSON JUNCTIONS-MEASUREMENTS AND PROPOSED MODEL

Roger H. Koch

IBM T. J. Watson Research Center
Yorktown Heights, New York 10598

Measurements of the spatial correlation length of the 1/f fluctuations in lead alloy Josephson junctions were made. The measured length is less than about 1 μm . Measurements of the magnitude and frequency dependence of the noise suggest a model based on thermal activation. A simple mechanism is proposed that includes the coulomb interactions between localized states in the oxide barrier.

INTRODUCTION

The origin of 1/f noise in Josephson junctions has been a topic of much recent research[1,2]. Previous work has suggested that the noise results from temperature fluctuations in the superconductors bordering the oxide barrier[1]. These models obtained the long-time correlation functions characteristic of 1/f noise by assuming the thermal fluctuations are spatially correlated over large enough distances to allow only slow changes in the local temperature. This paper reports, for the first time, measurements of the spatial correlation length of the fluctuations causing the 1/f noise in Josephson junctions. The measured length is less than about 1 μm , much smaller than the prediction of the thermal fluctuation model.

Detailed measurements of the magnitude and frequency dependences of the noise indicate a thermal activation process is the source of the noise. In the oxide barrier, electrons in localized states below the Fermi energy, can thermally activate up to the Fermi energy. The time constant for each the activation event is exponential in the energy[3], $\tau(E) = \tau_o \exp(E/k_B T)$, where $1/\tau_0$ is the attempt frequency. Inclusion of coulomb interactions between the localized states in the barrier can explain the origin of the time constants and distribution of activation energies. The physical quantity being measured is the fluctuating tunneling current. The localized states change the tunneling current by changing the barrier height[2,4]. Electrostatic interactions between the states and tunneling electrons change the exact barrier height as the states change their occupation.

EXPERIMENTAL PROCEDURE

The junctions measured were provided by the IBM Josephson group[5]. The base electrode was Pb-Au-In. RF oxidation formed the tunnel barrier through a 2.5 μm diameter window in SiO. The counter electrode was evaporated Pb-Bi. The barrier formed is a Schottky barrier composed of In_2O_5, a degenerate semiconductor[6]. The barrier height is about 1eV and the width through which tunneling occurs is about 20 Å . An external In-Au shunt was also evaporated.

The critical currents, I_o, varied from 20 to 50 microamps and the junction self-capacitance, C, was estimated to be about 0.5 picofarads. The shunt resistance, R, was 3 ohms, making the junctions non-hysteretic. To measure the 1/f noise the junctions were immersed in liquid helium and current biased. The quantity measured was the 1/f voltage fluctuations across the junction. In these experiments and previous ones, the noise measured has been found to result from 1/f fluctuations of the junction critical current. These fluctuations were coupled to the input coil of a rf SQUID using a superconducting transformer[7] in series with a resistor, R_s. The critical current noise spectrum, S_I, is related to the measured noise spectrum, S_M, at the primary of the transformer by

$$S_I = S_M\{I/I_o + VR_s/(I_oR^2)\}^{-2}, \qquad (1)$$

where V and I are the average voltage and bias current respectively. In the limit where $R_s << R$ and I_B is only just larger than I_o, Eq. 1 reduces to $S_I = S_M$. Since the Fermi levels across the junction are shifted by the applied voltage, which in these experiments is a maximum of 60 microvolts, modeling of the junction is facilitated by realizing there is effectively no applied voltage across the barrier.

A magnetic field in the plane of the junction was applied by a superconducting coil. The magnetic modulation of the junction critical current agreed well with the predictions for a round junction. The noise was measured by connecting the SQUID output to a spectrum analyzer, and in the later measurements, this was interfaced to an IBM personal computer. Magnetic shielding was provided using a superconducting lead shield and room-temperature mu metal shielding.

The spectrum was typically measured from 1 to 1000 Hz, and in most cases dominated by the junction noise. In all cases the noise magnitudes and slopes, as well as the junction critical currents, were remarkably stable when cycled to room temperature. Three junctions were measured carefully and several more briefly, and they all showed the same basic behavior.

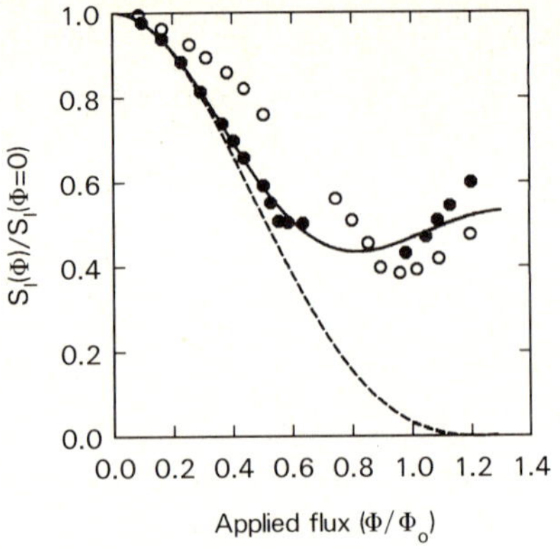

Fig. 1; Predicted noise vs. applied flux for spatially uncorrelated fluctuations (solid line) and spatially correlated fluctuations (dashed line). Measured noise for two junctions (dots and circles).

MEASUREMENT OF SPATIAL CORRELATION LENGTH

A magnetic field applied in the plane of a junction will modulate the magnitude of the $1/f$ noise resulting from critical current fluctuations, as well as modulating the average junction critical current. The amount of modulation of the noise depends on the applied flux and the spatial correlation length of the fluctuations. The expression for the normalized power spectrum as a function of applied flux when the fluctuations are totally correlated across the dimensions of a round junction is

$$\frac{S_I(\nu,\Phi)}{S_I(\nu,0)} = \left\{\frac{J_1(\Phi\pi/\Phi_0)}{\pi\Phi/(2\Phi_0)}\right\}^2 . \qquad (2)$$

In the case where the fluctuations are spatially uncorrelated the power spectrum is given by

$$\frac{S_I(\nu,\Phi)}{S_I(\nu,0)} = \frac{1}{2} + \frac{\Phi_0}{2\pi\Phi} J_1(2\pi\Phi/\Phi_0). \qquad (3)$$

These two expressions are plotted in Fig. 1. When the applied flux is 1.22 Φ_0, the average critical current, that is the local current density weighed by $\sin(\delta(x))$ over the entire junction, is zero. A uniform change in the critical current density over the junction, i.e. a spatially correlated fluctuation, will not change the total critical current through the junction. No noise would be measured. A spatially non-uniform change in the critical current density, i.e. a spatially uncorrelated fluctuation, would not average to zero and hence be observable.

Identification of the actual correlation distance was done by measuring the $1/f$ noise at 32 Hz as a function of applied flux. Data from two junctions is shown in Fig. 1. The results clearly fit the uncorrelated prediction better and imply the spatial correlation length is less than about 1 μm . These results are inconsistent with the thermal fluctuation models, which predict a much larger correlation length. The departures from the exact predictions are probably due to a non-constant density of noise sources across the plane of the junction.

MEASURED TEMPERATURE AND FREQUENCY DEPENDENCE

The magnitude of the measured critical current fluctuations vs. temperature is shown in Fig. 2. for three different frequencies. At 1 Hz the noise increases as the temperature is decreased from 4.2K to 3K. Fig. 3 plots the measured slope of the noise for the same junction as a function of temperature. The other junctions displayed similar behavior.

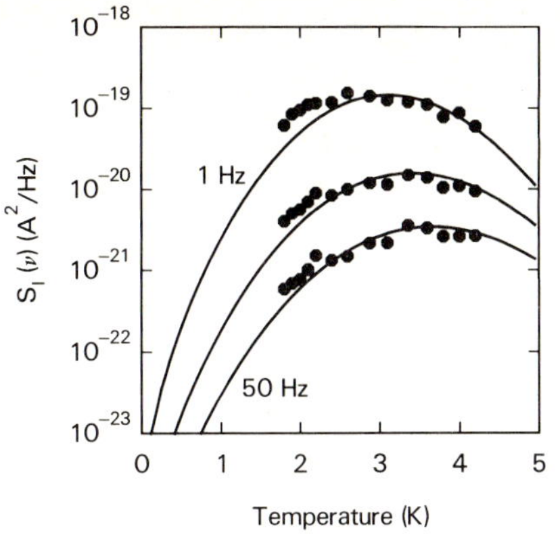

Fig. 2; Measured magnitude vs. temperature at three frequencies 1, 10, and 50 Hz (dots). Magnitude predicted using Eq. 4 and the density of states in Fig. 4 for the same three frequencies (lines).

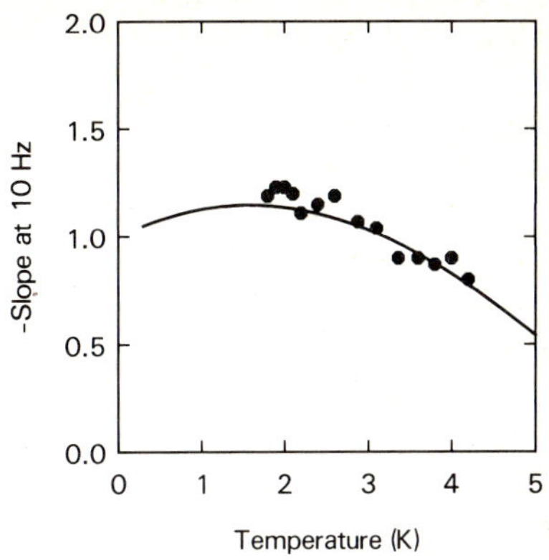

Fig. 3; Measured slope vs. temperature (dots). Slope predicted using Eq. 4 and the density of states in Fig. 4 (line).

PROPOSED MODEL

We propose a model in which the observed noise results from a thermal activation process with a range of activation energies, similar to that introduced by Dutta and Horn for metal films[3]. The time constant for activation from an energy, E, is given by $\tau(E) = \tau_o \exp(E/k_B T)$. The predictions are insensitive to the exact value of the attempt time, τ_0, here it is chosen to be a picosecond. This represents a typical tunneling and/or thermalization time. The total measured noise is composed of many Lorentzian spectra, each with a characteristic roll-off frequency of $1/\tau(E)$. When averaged over the distribution of energies, the total spectrum is[3]

$$S_I(\nu,T) = A^2 \int \frac{\tau(E)}{1 + (2\pi\nu\tau(E))^2} D(E)dE. \quad (4)$$

The constant A is the magnitude of the change in critical current resulting from one activation event. The density of states of the activation energies, D(E), can be determined from the noise measurements. A simple estimate, a gaussian, was obtained by trial and error. The measured data was found to be predicted well by the density of states shown in Fig. 4, using a value for A^2 of 5×10^{-18} Amps^2/Hz. These predictions are shown in Figs. 2 and 3.

The physical origin of this density of states and the factor A comes from localized states in the barrier. The states below the Fermi level in the oxide will be filled on average and those above empty. Electrons in states below can thermally activate up to the Fermi level and leave the oxide. The time constant for such a process will be $\tau(E) = \tau_o \exp(E/k_B T)$, where E is identified as the energy below the Fermi level. The change in electric field when a charged particle leaves the oxide will reduce the energy of many nearby traps from above to below the Fermi level. Equilibrium will quickly be restored when another electron tunnels into the oxide, occupies one of the empty states below the Fermi level, and electrostatically increases the remaining empty state energies to above the Fermi level. During this process the tunnel barrier will decrease drastically for a short time ,τ_0 , but more importantly the resulting final barrier height will be slightly different since the exact spatial positions of the charge in the oxide will be changed. In these junctions, an electron charge in the barrier changing its position by an average amount changes the tunneling current over a 40 Å radius region by about .07 of the region's average tunneling current. Over longer distances the interaction is effectively screened by the superconductors bordering the oxide barrier. The factor A^2 for a 20 microamp critical current junction and m ($\equiv$(junction area)/$(\pi(40\text{Å})^2) = 10^5$) independent fluctuating regions is estimated to be $A^2 = (.07I_o)^2/m = 10^{-17}$, in close agreement with the measured number.

The density of states of the localized states can be estimated using the techniques of many-body analysis developed in localization theory[8]. In an insulator, a gap of order $e^2/\bar{r}$, will appear in the single body density of states at Fermi level. This gap increases as the average distance between the states, $\bar{r}$, decreases. The gap results from the coulomb interactions between the charged states and represents a stability condition for a single particle in a many-body system. The resulting density of states has been estimated using a computer simulation that minimizes the ground state energy and satisfies stability constraints for all the localized electrons. A surface density of states of $1.5\times10^{13}/\text{cm}^2$, in a narrow energy distribution (2meV) randomly distributed in the oxide is initially chosen. After the charged particles are added, the resulting density of states below the Fermi level is shown in Fig. 4. The coulomb interactions have spread the initial narrow energy distribution into a wider distribution entirely consistent with the measured noise. While there is no independent measure of the state density for these barriers, the estimated number is quite consistent with the known properties of these junctions. Other surface densities or initial starting energies would give quantitatively different final densities of states, but with similar features to that shown in Fig. 4, and most producing 1/f like noise.

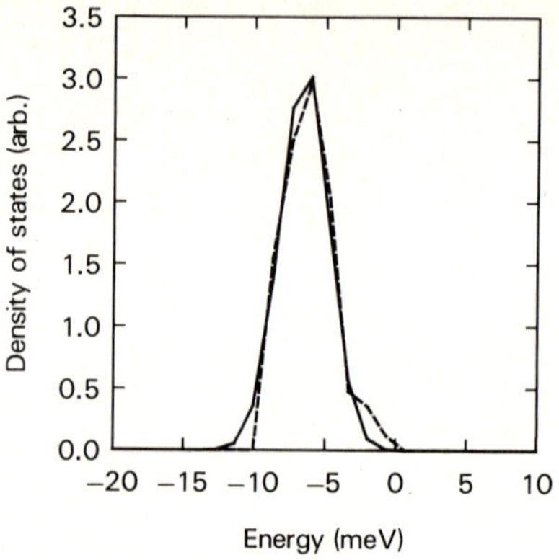

Fig. 4; Gaussian density of states predicted using Eq. 4 and measured noise data, <E>=-6.5 meV, σ_E=1.7 meV (solid line). Density of states predicted using proposed model with 1.5×10^{13} states/cm^2 (dashed line).

The defects most probably consist of excess oxygen atoms or metal atoms backscattered from the metallic electrodes during the sputter oxidation[6].

DISCUSSION

A new model has been presented that explains the origin of 1/f noise in Josephson junctions. The model predicts the measured frequency dependence and magnitude of the noise, its lack of measurable spatial correlations, and most other available data. It can be extended to almost any tunnel junction. The concept that coulomb interactions between localized states is important in understanding the origin of 1/f noise has been introduced. This kind of consideration could be significant in many other systems.

The author thanks Mark Ketchen, John Baker, Paul Horn, John Magerlein, and the IBM Josephson group for invaluable assistance and discussion.

[1] J. Clarke and G. Hawkins, Phys. Rev. **B14**, 2826 (1976); L. Krusin-Elbaum and R. F. Voss, Sixth International Conference on Noise in Physical Systems, NBS Special Publication 614, pp. 213, 1981.

[2] C. T. Rogers and R. A. Buhrman, IEEE Trans. Magnetics, to be published, (1982 Applied Superconductivity Conference).

[3] P. Dutta and P. M. Horn, Rev. Mod. Phys. **53**, 497 (1981).

[4] M. Celasco, A. Masoero, P. Mazzetti, and A. Stepanescu, Phys. Rev. **B17**, 2553 (1978).

[5] J. H. Greiner *et al.,* IBM J. Res. and Dev. **24**, No 2, (1980).

[6] J. M. Baker, C. J. Kircher, and J. W. Matthews, IBM J. Res. and Dev. **14**, 223 (1980); J. H. Magerlein, J. Appl. Phys, to be published.

[7] M. Ketchen and C. C. Tsuei, in the Proc. of the 2nd Int. Conf. on SQUID's, Berlin, 1980.

[8] A. L. Efros, J. Phys. C, **9**, 2021 (1976).

NOISE IN PHYSICAL SYSTEMS AND 1/f NOISE
M. Savelli, G. Lecoy and J-P. Nougier (eds.)
© Elsevier Science Publishers B.V., 1983

1/f NOISE FLUCTUATIONS IN α-PARTICLE RADIOACTIVE DECAY OF $_{95}AM^{241}$

J. Gong, C.M. Van Vliet, W.H. Ellis, and G. Bosman
Depts. of Electrical Engineering and Nuclear Engineering Sciences
University of Florida, Gainesville, FL 32611, USA

P.H. Handel
Department of Physics
University of Missouri - St. Louis, St. Louis, MO 63121, USA

Counting statistics of alpha particles from 95 Americium were determined over periods from 1 minute to 1,000 minutes. In particular, the two-sample variance or Allan variance was determined for many sample runs. According to a recent theorem, there is a unique relation between the particle flux spectral noise density and the Allan variance. It was found that for small counting periods the statistics were Poissonian, corresponding to shot noise of the particle flux. For long periods the counting statistics were found to be non-Poissonian, and a flicker floor of $\approx 10^{-7}$ was established. Good agreement with the quantum theory of 1/f noise was obtained. These experiments are the first quantitative confirmation of this theory.

1. INTRODUCTION

One of the general theories of 1/f noise is the quantum theory, based on infrared divergent coupling of the system to the electromagnetic field or other elementary excitations. It was mainly developed by Handel.[1] This theory is fundamental in the sense that it derives the 1/f spectrum from basic quantum physics at the level of a single charged particle subject to scattering with small energetic losses due to bremsstrahlung, although the final result depends essentially on the presence of many carriers; in this respect 1/f noise is similar to electron diffraction which is a one-particle effect, but which can be seen only if many particles are diffracted. In addition, the theory is universal in the sense that any infraquanta with infrared-divergent coupling to the current carriers will give a contribution to the observed 1/f noise proportional to their coupling constant. Such infraquanta are, for example, very low-frequency photons, various types of phonons, shallow electron-hole pairs on the Fermi surface of a metal, spin waves, correlated states, etc.

Recently, this theory has been reformulated with quantum-optical terminology and compound-Poisson statistics in a paper[2] written by Van Vliet et al., which led to the idea of verifying the theory on a "clean" system outside the domain of currents in solids: radioactive α-decay.

2. RESULTS OF THE THEORY

2.1 Quantum Approach to 1/f Noise[1]

It is known that upon scattering a beam of charged particles will emit bremsstrahlung. The spectrum W(f) of the emitted radiation per unit frequency interval is independent of frequency (W = constant) at low frequencies and decreases to zero at an upper frequency limit f_m which is approximated by E/h, where E is the kinetic energy of the electrons; h is Planck's constant. Consequently, the rate of phonon emission per unit frequency interval is $N(f) = W/hf$, i.e., proportional to 1/f. Therefore, we conclude that the fraction of charged particles scattered with energy loss ε is proportional to 1/ε, i.e., the relative squared matrix element for scattering with energy loss ε is $|b_T(\varepsilon)|^2 \propto 1/\varepsilon$. The subscript T indicates we consider time samples of duration T, where T^{-1} is smaller than the lowest frequency of interest.

If the incoming beam of electrons is described by a wave function $\exp[(i/\hbar)(\vec{p}\cdot\vec{r} - Et)]$, the scattered beam will contain a large nonbremsstrahlung part of amplitude a, and an incoherent mixture of waves of amplitude $ab_T(\varepsilon)$ with bremsstrahlung energy loss ε ranging from some resolution threshold ε_0 to an upper limit $\Lambda \leq E$, of the order of the kinetic energy E of the particles.

$$\psi_T = \exp\left[\left(\frac{i}{\hbar}\right)(\bar{p}\cdot\bar{r} - Et)\right] a\left[1 + \int_{\varepsilon_0}^{\Lambda} b_T(\varepsilon) e^{i\varepsilon t/\hbar}\, d\varepsilon\right]. \tag{2.1}$$

Here $b_T(\varepsilon) \equiv |b_T(\varepsilon)| e^{i\gamma_\varepsilon}$ has a random phase γ_ε which implies incoherence of all bremsstrahlung parts, and $|b_T(\varepsilon)|^2$ is proportional to 1/ε, as we saw above.

If the particle density fluctuation is defined by $\delta|\psi|^2 = |\psi|^2 - <|\psi|^2>$, its autocorrelation function will be

$$<\delta|\psi|_t^2 \, \delta|\psi|_{t+\tau}^2> = <|\psi|_t^2 \, |\psi|_{t+\tau}^2> - <|\psi|^2>^2$$

$$\approx 2|a|^4 \int_{f_o}^{\Lambda/h} h|b(\varepsilon)|^2 \cos 2\pi f\tau \, df \qquad (2.2)$$

which is proportional to $|b(\varepsilon)|^2$ and hence proportional to 1/f. Therefore, the spectral density of the particle density fluctuation (the Fourier transform of eq. (2.2) is proportional to 1/f.

The relative bremsstrahlung rate $|b(\varepsilon)|^2$ can be derived as follows. The constant spectral energy density can be written as $w(f) = 4e^2|\Delta\vec{v}|^2/3c^3\kappa$, where e is the charge of an α-particle, κ the dielectric constant of the medium, c is the velocity of light, and $\Delta\vec{v}$ is the velocity change in the scattering process. The relative scattering rate density with energy loss ε, $|b(\varepsilon)|^2$, is obtained by dividing W(f) by the energy of a photon $\varepsilon = hf$.

$$|b(\varepsilon)|^2 = \frac{4e^2(\Delta\vec{v})^2}{3c^3\kappa h\varepsilon} = \frac{\alpha A}{\kappa\varepsilon} \qquad (2.3)$$

where

$$A = \frac{8(\Delta\vec{\beta})^2}{3\pi}, \quad \Delta\vec{\beta} = \frac{\Delta\vec{v}}{c}, \qquad (2.4)$$

and $\alpha = q^2/\hbar c$ is the fine structure constant. q is equal to the magnitude of the electronic charge. In the M.K.S. system $\alpha = q^2/\hbar c\kappa_o$ where κ_o is the dielectric constant of vacuum. The spectral density of the relative fluctuations is from (2.2), (2.3), and the Wiener-Khintchine theorem,

$$S_{|\psi|^2}(f)/<|\psi|^2>^2$$

$$= 2[1+\alpha A \ln(\Lambda/\varepsilon_o)]^{-2}\alpha A/\kappa f \approx 2\alpha A/\kappa f. \qquad (2.5)$$

2.2 The Allan Variance Transform Theorem[3]

The main link between counting statistics and particle current noise is provided by MacDonald's theorem. Unfortunately, for 1/f noise MacDonald's theorem is not applicable, since the integral diverges. However, a useful concept in this case is the "two-sample variance" or "Allan variance."[4] Let $M_T^{(1)}$ be the total number counted in (t, t+T) and let $M_T^{(2)}$ be the total number counted in (t+T, t+2T). Then the Allan variance is defined by

$$\sigma_{M_T}^{A2} = \frac{1}{2} <\left(M_T^{(1)} - M_T^{(2)}\right)^2>. \qquad (2.6)$$

The variance σ^{A2} (which means $(\sigma^A)^2$) turns out to be finite for 1/f noise. The transform theorem reads[3]

$$\sigma_{M_T}^{A2}(T) = \frac{4}{\pi}\int_o^\infty \frac{S_m(\omega)}{\omega^2} \sin^4\left(\frac{\omega T}{2}\right) d\omega \qquad (2.7)$$

with inversion

$$S_m(\omega) = -\frac{1}{2\pi i}\int_{-i\infty+\beta}^{i\infty+\beta} \frac{dp}{\omega^{p-2}} \frac{\cos\frac{1}{2}p\pi}{1-2^{p-3}} \Gamma(p)$$

$$\int_o^\infty \frac{dT}{T^p} \sigma_{M_T}^{A2}; \qquad (2.8)$$

m is the instantaneous number of particles detected per second. (Slightly more complicated inversion forms using partial Mellin transforms are found in [3].) For Poissonian shot noise $S_m(\omega) = 2m_o$, where m_o is the average counting rate. Substituting $S_m(\omega)$ into eq. (2.7), one has $\sigma_{M_T}^{A2}(T) = m_o T$. For 1/f noise, with a spectrum of $S_m(\omega) = 2\pi C/|\omega|$, where C is a constant, the Allan variance $\sigma_{M_T}^{A2} = 2CT^2 \ln 2$. Now suppose that the noise is composed of shot noise and 1/f noise, i.e.,

$$\sigma_{M_T}^{A2}(T) = m_o T + 2CT^2 \ln 2. \qquad (2.9)$$

We recall that $<M_T> = m_o T$, so that a measurement of the relative Allan variance $R(T) = \sigma_{M_T}^{A2}(T)/<M_T>^2$ yields,

$$R(T) = \frac{1}{m_o T} + 2C' \ln 2, \qquad (2.10)$$

where $C' = C/m_o^2$ is the characteristic strength of the 1/f noise $S_m(f)/m_o^2$. For short-time intervals the term $1/m_o T$ is dominant, hence R(T) is proportional to 1/T. When T is long enough, the term $2C'\ln 2$ becomes dominant. For large T one cannot further reduce the relative accuracy by longer counting; $R(T_\infty) = 2C'\ln 2$ is therefore called the "flicker floor."

We conclude from the above that the presence of 1/f noise in counting statistics can be determined from a measurement of the Allan variance as a function of T.

3. EXPERIMENTAL METHOD AND RESULTS

The block diagram of the counting system being used to investigate 1/f fluctuation in the α-particle emission rate is shown in Fig. 1. The source is $_{95}Am^{241}$, which decays with a half-life of $T_{1/2} = 458$ years with the emission of 5.48 MeV α-particles into $_{93}Np^{237}$. The detector, a silicon surface-barrier detector, is reverse biased at 80 volts, and the dead times of the ND575 Analog to Digital Convertor (ADC) and ND66 Multi-Channel Analyzer are 60 n-seconds and 6 μ-seconds respectively. Therefore, no dead-time correction is necessary, as long as the counting rate is kept lower than 1,000 counts per second[5] (or the averaged time

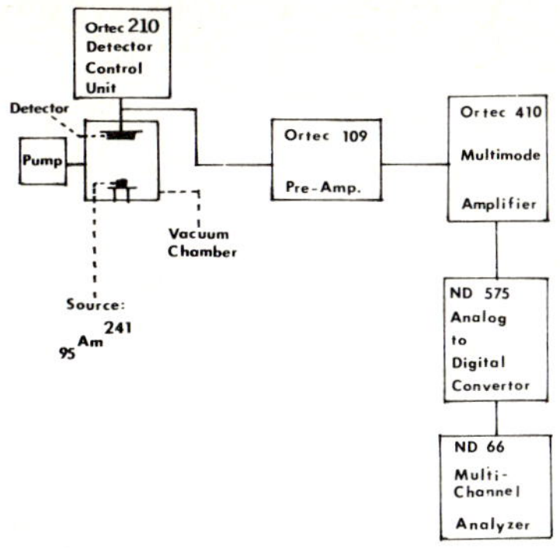

Figure 1. Block diagram of the counting system.

elapse between two counts is higher than 1,000 μ-seconds).

The counts M_T of adjacent time intervals can be read directly from the memory units of the ND66 Multi-Channel Analyzer; thus the Allan variance can be calculated by

$$\sigma^{A2}_{M_T} = \frac{1}{2}\left(\frac{1}{N-1}\right)\sum_{i=1}^{N-1}\left(M_T^{(i)} - M_T^{(i+1)}\right)^2; \qquad (3.1)$$

since

$$\langle M_T \rangle = \frac{1}{N}\sum_{i=1}^{N} M_T^{(i)}, \qquad (3.2)$$

the relative Allan variance, R(T), defined by eq. (2.10), can be found by using eqs. (3.1) and (3.2).

Because of the existence of "variance of variance" (or "variance noise"), the relative Allan variance itself is a fluctuating parameter. In order to obtain an accurate value of R(T), a sufficient number of measurements must be made, especially when T is short. Figure 2 shows R(T) versus the number of measurements, N, for T = 3 min. When N is small, R(T) is spread over a wide range. When N is increased, R(T) shows less spread and finally converges to a stable value.

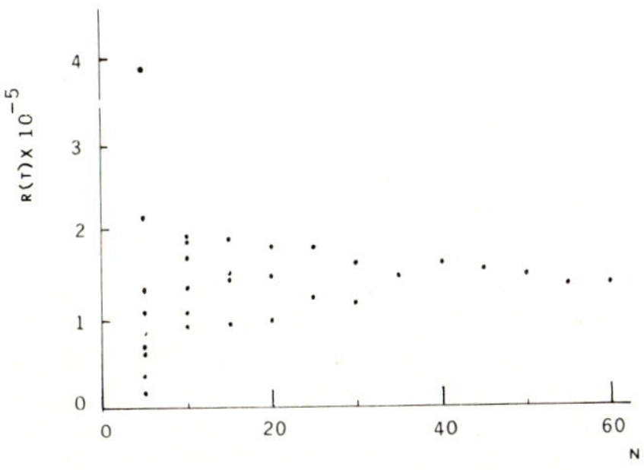

Figure 2. R(T) vs. N for T = 3 min.

Due to the presence of variance noise, the experimental results of R(T) contain a fluctuation term ΔR, i.e.,

$$R_{exp}(T) = \frac{1}{m_o T} + 2C'\ell n2 + \Delta R, \qquad (3.3)$$

where the average value $\langle \Delta R \rangle$ should be zero. The experiment was then repeated several times, and the average of $R_{exp}(T)$,

$$\langle R_{exp}(T) \rangle = \frac{1}{m_o T} + 2C'\ell n2 + \langle \Delta R \rangle, \qquad (3.4)$$

is obtained. Although the value of $\langle \Delta R \rangle$ in eq. (3.4) can hardly be exactly zero for a finite number of measurements, it should be reduced a great deal compared with the value of ΔR for single measurement. The value of $\langle R_{exp}(T) \rangle$ gives the best estimation to the true value of R(T).

The distance between radioactive source and detector was adjusted such that very close counting rates were obtained while repeating measurements. However, it is difficult to obtain identical counting rates. Therefore, the shot noise term in relative Allan variance, R(T), is slightly different between each series of measurement.

Before the average of $R_{exp}(T)$ is taken, the shot noise term, $1/m_oT$, should be normalized to the same counting rate. Here a rate of 18,000 counts per minute was chosen. The normalized relative Allan variance $R_n(T)$ is then

$$R_n(T) = R_{exp}(T) - \frac{1}{m_o T} + \frac{1}{18{,}000\ T}$$

$$= \frac{1}{18{,}000\ T} + 2C'\ell n2. \qquad (3.5)$$

The value $\langle R_n(T) \rangle$ is now used to estimate the true value of R(T).

Fig. 3 shows the experimental data for $\langle R_n(T) \rangle$. The theoretical curve (solid line) is based on $R(T) = (1/18{,}000 \times T) + 1 \times 10^{-7}$ versus 1/T, which suggests that the value of $2C'\ell n2$ is about 1×10^{-7}.

Fig. 4 shows the average value of the normalized Allan variance, $\langle \sigma^{A2}_{M_T n} \rangle$, which is obtained by

$$\langle \sigma^{A2}_{M_T n} \rangle = \langle R_n(T) \rangle \times (18{,}000 \times T)^2. \qquad (3.6)$$

The theoretical curve (solid line) is based on $\sigma^{A2}_{M_T} = 18{,}000\ T + 1 \times 10^{-7} \times (18{,}000 \times T)^2$. Very good agreement is obtained as shown in these figures.

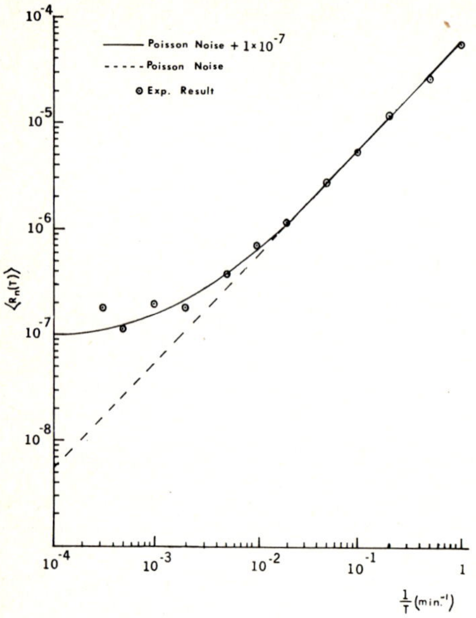

Fig. 3. $R_n(T)$ vs. $1/T$ for m_0 = 18,000/min.

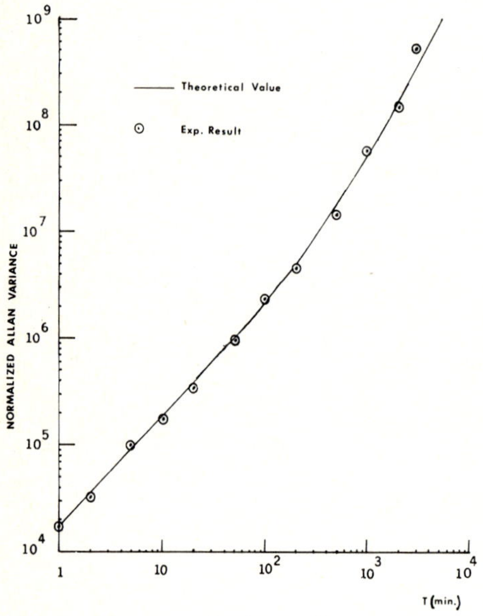

Fig. 4. $\langle\sigma^{A2}_{M_T}(T)\rangle$ vs. T for m_0 = 18,000/min.

4. DISCUSSION AND CONCLUSIONS

According to Handel's theory, the constant C' in the 1/f noise term in the relative Allan variance is, cf eq. (2.5),

$$C' = 2\alpha A\zeta/\kappa \quad (4.1)$$

where ζ is a coherence factor; α is the fine structure constant 1/137, and $A = 8(\Delta\vec{v})^2/3\pi c^2$ where $\Delta\vec{v}$ is the velocity change of the particles in the emission process, c is the velocity of light. For α-particles one finds

$$\frac{\alpha A}{\kappa} = 33.28\times10^{-7}\times\frac{E}{\kappa} \quad (4.2)$$

where E is energy in MeV and κ is the dielectric constant of the radioactive material. In these experiments E = 5.48 MeV, so that the value of $2C'\ell n2$ (the value of the flicker floor, F) is

$$F = 505.6\times10^{-7}\times\frac{\zeta}{\kappa}\,. \quad (4.3)$$

To the authors' knowledge, nobody has ever measured the dielectric constant of AmO_2, the α-particle source used in these experiments; thus the value of κ in eq. (4.3) is unknown. However, since the atomic structure of AmO_2 is similar to that for UO_2, the dielectric constant of UO_2 (21.0 ± 1[6]) can then be used as a reference.

If one considers the following factors that 1) the resulting αA in eq. (4.2) is the sum of contributions for all types of infraquanta participating in the energy transfer, and maybe only part of them were detected, i.e., the actual value for A in these experiments may be smaller than that given in eq. (4.2); 2) the coherence factor, ζ, is always less than unity; 3) the dielectric constant of AmO_2 may be larger than 21; then the measured value of $F = 1\times10^{-7}$ is in the right ballpark to verify Handel's theory. In particular, for ζ = 0.05, κ = 21, one obtains $F = 1.2\times10^{-7}$, in accord with the observed value of Figs. 5 and 6. We believe, therefore, that these experiments constitute the first experimental verification of Handel's quantum 1/f noise theory.

In conclusion, α-particle decay has statistics which for large counting intervals ($\simeq10^3$ min) are non-Poissonian. We established the presence of a flicker floor $R(T_\infty) = 1.0\times10^{-7}$. In the frequency domain, this indicates the presence of 1/f noise in the particle flux for frequencies of the order of 10^{-4} hz or less. The magnitude of the flicker floor can well be explained as electromagnetic quantum 1/f noise.

ACKNOWLEDGEMENT

This research was supported by the Air Force Office of Scientific Research, grant #80-0050.

REFERENCES

[1] Handel, P.H., Phys. Review 22A (1980) 745.

[2] Van Vliet, K.M., Handel, P.H. and van der Ziel, A., Physica 108A (1981) 511.

[3] Van Vliet, C.M. and Handel, P.H., Physica 113A (1982) 261.

[4] Allan, D.W., Proceedings IEEE 54 (1966) 221.

[5] Knoll, G.F., Radiation Detection and Measurement (Wiley and Sons, 1979) 95.

[6] Gesi, K. and Tateno, J., Jap. J. Appl. Phys. 8(11) (1969) 1358.

NOISE IN PHYSICAL SYSTEMS AND 1/f NOISE
M. Savelli, G. Lecoy and J-P. Nougier (eds.)
© *Elsevier Science Publishers B.V., 1983*

NOISE IN ELECTROCHEMICAL SYSTEMS

Claude Gabrielli, François Huet and Michel Keddam

G.R.4 du CNRS "Physique des Liquides et Electrochimie",
associé à l'Université P. & M. Curie, Tour 22, 4, place Jussieu
75230 Paris Cedex 05, France

INTRODUCTION

The electrochemical noise is the manifestation of the random fluctuations of the quantities (e.g. potential or current) which characterize the state of a metal-electrolyte interface [1-3]. The current noise is observed when the potential is maintained constant (potentiostatic regime) and the potential noise is observed when the current is maintained constant (galvanostatic regime). The specific character of the noise measurement techniques in electrochemistry have been described during the Sixth Conference on Noise in Physical Systems [4]. Hence in this paper we shall restrict ourselves to the theoretical framework in which the stochastic models of the electrochemical interface behavior can be derived and to some particular problems encountered in the interpretation of the electrochemical noise spectra.

A FRAMEWORK FOR THE STOCHASTIC MODELING OF THE INTERFACE BEHAVIOR.

The two main elementary processes involved at the metal-electrolyte interface are the heterogeneous electrochemical reactions at the electrode surface and the transport of the reactive species in the bulk of the solution. The laws governing these two kinds of quantity provide the framework for the stochastic models of the interface behavior around non-equilibrium steady states. This framework is on some points very close to the semi-conductor noise theory and likewise the involved processes are supposed to be governed by Langevin type equations.

Electrochemical reactions

In order to illustrate this approach a simple reaction mechanism is analyzed. A reactive species A adsorbs on the electrode surface M and gives a reaction intermediate X which gives M_{sol}. These electrochemical reactions involve an electron e and hence are potential activated.

$$M + A \underset{K_1'}{\overset{K_1}{\rightleftarrows}} X + e$$

$$X \xrightarrow{K_2} M_{sol} + e$$

where $K_i = K_i^o \exp \frac{\alpha_i e}{kT} V$

If θ is the surface coverage of the electrode surface by the intermediate X, the macroscopic equations which govern the system are respectively given by the X and the electrons balances.

$$\beta \frac{d\theta}{dt} = \phi_1 - \phi_1' - \phi_2 \qquad (1)$$

$$I_F = e\,[\phi_1 - \phi_1' + \phi_2] \qquad (2)$$

where β is the active site number, I_F the faradaic current which flows through the interface and the ϕ_i's are the fluxes of each species i.

$$\phi_1 = K_1(1-\theta) \quad ; \quad \phi_1' = K_1'\theta \quad ; \quad \phi_2 = K_2\theta$$

The Langevin equations are obtained from linearization of eq (1) and (2)

$$\frac{d\Delta\theta}{dt} = a\ \Delta\theta + b\ \Delta V + v_\theta$$

$$\Delta I_F = c\ \Delta\theta + d\ \Delta V + v_I$$

where v_θ and v_I are Langevin noise sources which act on the state variable $\Delta\theta$ and on the observable ΔI_F ; and

$$a = -\frac{\Sigma}{\beta} \ ; \ b = \frac{1}{\beta\Sigma}[K_1K_2(b_1-b_2) + K_1K_1'(b_1+b_1')]$$

$$c = -e\,(K_1+K_1'-K_2) \ ; \ d = \frac{eK_1}{\Sigma}\,[K_2(b_1+b_2) + K_1'(b_1+b_1')] = \frac{1}{R_t}$$

where $\Sigma = K_1 + K_1' + K_2$.

From these equations the faradaic admittance Y_F is calculated

$$Y_F(\omega) = \frac{bc}{j\omega - a} + d$$

In potentiostatic regime ($\Delta V = 0$) the power spectral density $S_{I_F}(\omega)$ of the current fluctuations is equal to

$$S_{I_F}(\omega) = S_{v_I} + \frac{c^2\, S_{v_\theta}}{\omega^2+a^2} - \frac{2ac\, S_{v_\theta v_I}}{\omega^2+a^2}$$

If the fluxes ϕ_i are supposed to fluctuate following a Poisson law such as [3]

$$\phi_i(t) = \langle\phi_i\rangle + \Delta\phi_i(t) + d_i(t)$$

where $\langle d_i(t)\rangle = 0$

$$\langle d_i(t)\, d_j(t-\tau)\rangle = \langle\phi_i\rangle\, \delta(\tau)\, \delta_{ij}$$

Hence $v_\theta(t) = \frac{1}{\beta}\,[d_1(t) - d_1'(t) - d_2(t)]$

$$v_I(t) = e\,[d_1(t) - d_1'(t) + d_2(t)]$$

i.e.

$$S_{v_\theta} = \frac{2K_1}{\beta^2\Sigma}[K_1' + K_2] \; ; \quad S_{v_I} = \frac{2e^2K_1}{\Sigma}(K_1' + K_2) \; ;$$

$$S_{v_\theta v_I} = \frac{2eK_1K_1'}{\beta\Sigma}$$

It has to be noticed that these Langevin sources have the fundamental properties that their correlations are equal to the 2nd order moment of the master equation [5] :

$$S_{v_\theta} = \lambda + \mu$$

where the birth rate $\lambda = \phi_1$ and the death rate $\mu = \dot{\phi}_1 + \phi_2$.

Transport noise

The reactive species of concentration $c(x,t)$ are supposed to be transported towards the electrode surface following the diffusion (where $\xi(x,t)$ are the bulk Langevin sources) equation

$$\frac{\partial\, \Delta c}{\partial t} = D\,\frac{\partial^2 \Delta c}{\partial x^2} + \xi\,(x,t)$$

with the boundary conditions :

- at $x = \delta_N$ (Nernst diffusion layer thickness)

$$\Delta c\,(x = \delta_N) = 0$$

- at $x = 0$

$$I_F = eD\,\frac{\partial c}{\partial x}\bigg|_{x=0}$$

hence

$$\Delta I_F = eK\,\Delta c\,(0,\omega) + \frac{\Delta V}{R_t} + \zeta$$

hence the stochastic b.c. at $x = 0$ has the general form (where ζ is the surface Langevin source)

$$\frac{\partial \Delta c}{\partial x}\bigg|_{x=0} = h\,\Delta c(0,\omega) + \nu$$

where $h = \frac{K}{D}$ and $\nu = \frac{1}{eD}\left(\frac{\Delta V}{R_t} + \zeta\right)$

This problem can be solved by standard techniques using the Green's formalism [6] :

$$\Delta c(x) = \int_0^{\delta_N} G(x,x')\,\xi\,(x')\,dx' + G(x,0)\nu$$

i.e. $$\Delta c(0) = \int_0^{\delta_N} G(0,x')\xi(x')dx' + G(0,0)\nu$$

The faradaic admittance is given by ($\xi = \zeta = 0$).

$$Y_F = \frac{1}{R_t}\left[\frac{K}{D}\,G(0,0) + 1\right]$$

and the current fluctuations in potentiostatic regime ($\Delta V = 0$)

$$S_{\Delta I_F} = (eK)^2\, S_{c_o} + S_\zeta + eK\,(S_{c_o\zeta} + S^*_{c_o\zeta})$$

as

$$S_{c_o} = \int_0^{\delta_N}\!\!\int_0^{\delta_N} G(0,x')G^*(0,x'')S_\xi(x',x'')dx'dx''$$
$$+ G(0,0)\,G^*(0,0)S_\nu$$

If the Λ- theorem is used [7]

$$\Lambda_x\,\Gamma(x,x') + \Gamma(x,x')\,\Lambda^T_{x'} = \frac{1}{2}\,S_\xi(x,x')$$

where $\Lambda_x = D\,\frac{\partial^2}{\partial x^2}$ and $\Gamma(x,x') = \langle\Delta c(x)\Delta c(x')\rangle$

$$S_{c_o} = 2\int_0^{\delta_N} [\Gamma(0,u)G^*(0,u) + \Gamma(0,u)\,G(0,u)]\,du$$
$$+ |G(0,0)|^2\, S_\nu$$

If $\Gamma(u,u') = \delta(u-u')$

$$S_{I_F} = 4e^2K^2\,\mathrm{Re}\,[G(0,0)] + \left[\frac{K^2}{D^2}\,|G(0,0)|^2 + \frac{K}{D}\,\mathrm{Re}\,[G(0,0)] + 1\right] S_\zeta$$

for a diffusion process the Green's function can be calculated by standard techniques, and

$$G(0,0) = \frac{\mathrm{sh}\,\gamma\,\delta_N}{\gamma\,\mathrm{ch}\,\gamma\,\delta_N + h\,\mathrm{sh}\,\gamma\,\delta_N}$$

where $\gamma^2 = j\omega/D$

EXPERIMENTAL

The measurement of the power spectral density (p.s.d.) of the potential fluctuations is performed through the cross-spectrum of the outputs of a two amplification channels arrangement which allows the studied spectrum cleared of the parasitic channel noises to be obtained. A FFT algorithm is used in the spectrum analyzer (Hewlett Packard 5451 C) in the 1-50 x 10^3 Hz range. All the polarization and amplification devices are battery-powered and a shielded box is used in order to minimize the parasitic pick-up noises. The impedance of the system is measured from the system response to a white noise superimposed on the polarization.

So far several electrochemical systems have been investigated either for analyzing fundamental processes or for obtaining some efficiency tests (batteries, corrosion ...). However only two problems will be emphasized in this paper : the nature of the noise at equilibrium and the possibility of a $1/f^{\alpha}$ noise generated at a metal electrolyte interface.

The p.s.d. of the potential fluctuations have been measured at the rest potential (I = 0) linked to two interfaces : i) An equimolar redox system $Fe(CN)_6^{3-}/Fe(CN)_6^{4-}$ in 1M KCl deoxygenated medium for the case when the overall reaction mechanism (Fe (II) $\rightleftarrows$ Fe (III) + e) evolves on a Pt electrode (fig. 1.a). ii) Corrosion of iron electrode (Johnson-Matthey) in 1M H_2SO_4 (fig.1.b). In the two cases the hypothetical thermal noise deduced from the Nyquist formula $S(\omega) = 4\,kT\,\mathrm{Re}[Z(\omega)]$ is plotted from the impedance measurement. The high frequency limit of the spectrum is in both cases equal to the thermal noise of the electrolyte resistance. For the redox system there is a fairly good agreement between the two curves where for iron corrosion a neat discrepancy appears. Hence a redox system at zero current can be considered as a thermodynamic equilibrium whereas corrosion of iron in acidic medium is not. This corresponds to the fact that a corrosion potential is a mixed potential where iron dissolves and hydrogen evolves simultaneously.

In Figure 2.a the p.s.d. of the potential fluctuations of the electrocrystallisation of silver in perchlorate bath have been plotted. The p.s.d. current fluctuations have been deduced by calculating

$$S_I(f) = S_V(f)/|Z|^2$$

and plotted in Figure 2.b for various polarization currents. The low frequency part of the spectrum of the current fluctuations varies like :

$$S_I(f) \sim A\,\frac{I^n}{f^{\alpha}}$$

where $1 \leqslant n \leqslant 2$ and $0.5 \leqslant \alpha \leqslant 2$. Whereas the spectra of the potential fluctuations slightly change with the current. For linear systems potential and current fluctuations behave similarly.

$$\text{e.g.}\quad \frac{\langle\Delta V^2\rangle}{V^2} = \frac{R^2\langle\Delta I^2\rangle}{V^2} = \frac{\langle\Delta I^2\rangle}{I^2}$$

whereas for nonlinear systems, R is a dynamical quantity which is equal to $(bI)^{-1}$ for exponential ($I = I_0 \exp bV$) system and thus for 1/f noise as [8]

$$\langle\Delta I^2\rangle = \frac{\alpha'}{Nf}\,I^2$$

$$\text{hence}\quad \langle\Delta V^2\rangle = \frac{\alpha'}{Nf}\cdot\frac{1}{b^2}$$

i.e. the potential fluctuation spectra are inde-

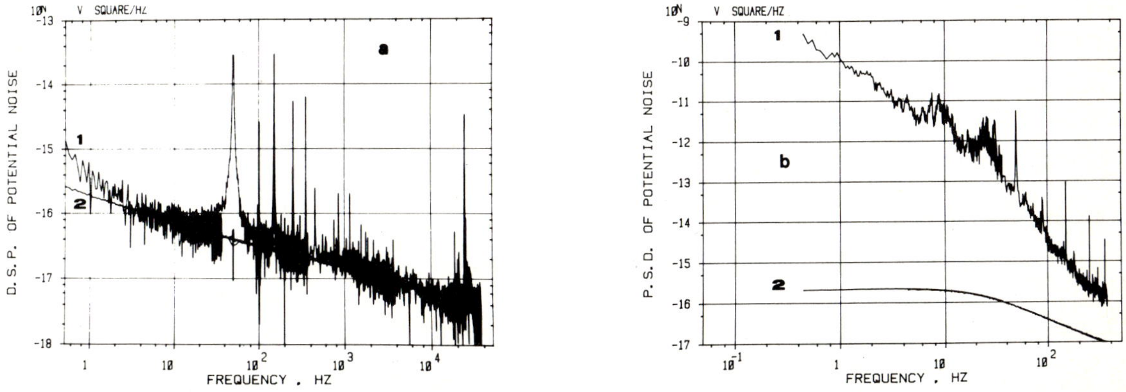

Figure 1 : Power spectral density of potential fluctuations (1) and hypothetic thermal noise (2) of a) redox system ; b) corrosion of iron.

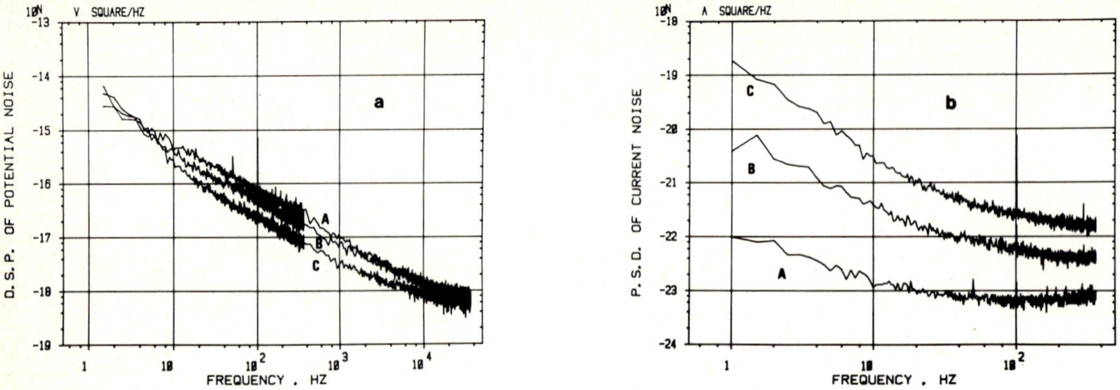

Figure 2 : Electrocrystallisation of silver (500 μm diameter) (A) 1 μA ; (B) 10 μA ; (C) 35 μA
a) p.s.d. of potential fluctuations. b) p.s.d. of current fluctuations.

pendent of the current flowing through the system. This situation seems to almost occur in the electrochemical systems under investigation as the processes involved are potential activated. This $1/f^{\alpha}$ like noise can be shown to be due to interfacial phenomena and not to the electrolyte.

In these conditions this noise and the high frequency electrolyte noise hinder the extraction of the spectrum of the elementary phenomena fluctuations which have been shown (from relaxation techniques) to take part in the overall electrochemical process.

Investigations are in progress in order to elucidate the origin of this low frequency noise which may be due to impurities or defects on the electrode surface and to model it. Then it will be possible to perform a quantitative comparison between the electrochemical stochastic models given in this paper and the experimental data.

REFERENCES :

[1] V.A. Tyagai, Elektrokhimija 10, 3 (1974).

[2] M. Fleischmann, J.W. Oldfield, J. Electroanal. Chem. 27, 207 (1970).

[3] G. Blanc, I. Epelboin, C. Gabrielli, M. Keddam, J. Electroanal. Chem. 75, 97 (1977).

[4] U. Bertocci, Sixth Intern. Conf. on Noise in Physical Systems, NBS Special Publication n° 614, 328 (1981).

[5] P. Hanggi, K.E. Shuler, I. Oppenheim, Physica, 107A, 143 (1981).

[6] K.M. Van Vliet, H. Mehta, Phys. Star. Sol. (b) 106, 11 (1981).

[7] K.M. Van Vliet, J. Math. Phys. 12, 1981, (1971).

[8] F.N. Hooge, Physica 83B, 14 (1976).

NOISE IN PHYSICAL SYSTEMS AND 1/f NOISE
M. Savelli, G. Lecoy and J-P. Nougier (eds.)
© *Elsevier Science Publishers B.V., 1983*

1/f NOISE IN AQUEOUS $CuSO_4$ SOLUTION

Toshimitsu Musha, Katsumi Sugita and Mitsutaka Kaneko

Tokyo Institute of Technology
Nagatsuta, Midoriku, Yokohama 227, JAPAN

1/f conductance fluctuations of aqueous $CuSO_4$ solutions were observed, and it is found that Hooge's α for the aqueous ionic solution is proportional to the ion concentration when it is larger than 0.001 mol/l; below this critical concentration α-value is independent of tho concentration. This observation is suggesting that the 1/f conductance fluctuation has a coherence length of 10 nm.

1. INTRODUCTION

The 1/f noise in aqueous ionic solutions was first observed by Hooge and Gaal (1) across a diaphragm which separated two compartments and had a small channel of diameter 10 μm and length 10 μm. They found that the alpha value in Hooge's empirical formula for 1/f noise in semiconductors (2) was proportional to the ion concentration which ranged from 0.05 to 10 mol/l. Afterwards, van den Berg and his colleagues repeated the same experiment with KCl solutions(3,4), but they observed Lorentzian voltage fluctuations only, which they attributed to ion number fluctuations and a volume flow generated by the applied voltage. The alpha value, if 1/f fluctuations existed, must be smaller than 0.0002 according to their experiment which covered frequencies down to 0.05 Hz.

We again repeated similar experiments with aqueous $CuSO_4$ solutions(5) to see if the alpha value goes further down as the ion concentration is extremely reduced. Contrary to negative results of van den Berg and his colleagues, we observed voltage fluctuations, which can be interpreted as conductance fluctuations, whose power spectral density was 1/f, and furthermore, alpha values we observed were all on a straight line on log-log plot of the alpha value vs. the ion concentration. This line was found to be connected to a line which Hooge and Gaal(1) already found. In ref(5) we reported a slight leveling off of this line for ion concentrations below 0.001 mol/l. In the present report we confirm this leveling-off with more observations, the result of which is strongly suggesting existence of the coherence length in 1/f conductance fluctuations of the ionic solution.

2. EXPERIMENTAL CONDITIONS

Structure of the glass chamber containing aqueous $CuSO_4$ solution is almost the same as was described in ref(5). The chamber is divided in two compartments by a thin membrane which has a small channel in it connecting the compartments. This has two pairs of electrodes; the outer pair supplies a dc current through the channel and the inner pair measures a voltage across the channel. All the electrodes are high purity Cu (99.99% to 99.999%)

We first used Nuclepore Membrane (Nuclepore Corp.) and pH electrodes, which consist of Ag-AgCl immersed in saturated KCl solutions, and observed very stable 1/f voltage fluctuations at small currents. Nuclepore Membrane has many channels of a given diameter which are randomly distributed, and hence there is a possibility that voltage fluctuations across channels may be correlated to some extent, which, therefore, complicates the analysis. On the other hand, saturated KCl solution may leak out into a very dilute $CuSO_4$ solution. For these reasons, they were not used for the further experiments. We came back to the original technique of making a channel with a tungsten wire of a sharp tip. The membrane thickness was 6- 10 μm. The voltage fluctuation was very sensitive to drift of a solution through the channel as was also reported in refs(3) and (4), and hence a free surface of the solution must absolutely be avoided. The glass chamber was put in a foamed plastic box to stabilize temperature, and the whole thing was placed on a shock-absorbing plate. The voltage electrodes were connected to a low-noise amplifier PAR 113. Resistance of doubly distilled water was larger than the maximum scale of the voltmeter (20 MΩ).

3. VOLTAGE (CONDUCTANCE) FLUCTUATIONS AND THE ESTIMATED ALPHA VALUES

With no dc voltages applied across the channel, one observes voltage thermal noise and amplifier 1/f noise at the lower frequency end. Thermal noise is cut off at a certain frequency which is determined by an electrostatic capacitance and resistance of the channel containing ionic solution. As a dc voltage across the channel was increased, voltage noise of 1/f spectrum appeared above the thermal noise background. When the voltage applied was too large the spectrum turned Lorentzian instead of 1/f. Was the current critical or the voltage? Over ion concentrations we tried which was from 0.0002 to 0.01 mol/l, the 1/f spectrum was no longer observed when the voltage exceeded 0.3 V across 10 μm, i.e., 300 V/cm. This critical voltage gives a critical ion drift velocity of 3 cm/s in the channel. Observation of 1/f fluctuations in a very dilute solution was difficult because of high resistance which required application of a relatively high voltage which killed 1/f noise. If the channel diameter was made large, the 1/f noise was buried in thermal noise as was discussed in ref(1).

Examples of the power spectral densities are plotted in Fig.1. The alpha value was evaluated for each of the observations and plotted in Fig.2 together with data which appeared in refs(1) and (5). It is sure that the alpha value is proportional to the ion concentration which is larger than 0.001 mol/l. Scatter of alpha values in ref(5) and in the present observation in this proportionality region is ± 3 dB. Judging from the scatter of the data it would be unreasonable to assume this proportionality down to an ion concentration as low as 0.0002 mol/l, and hence we should conclude that the alpha value leveled off below 0.001 mol/l at about 0.025.

4. HOW SHOULD THE CONCENTRATION DEPENDENCE OF THE ALPHA VALUE BE INTERPRETED?

The fractional variance of voltage or conductance fluctuations of the ionic solution is proportional to the degree of freedom of the ion motion. As long as the alpha value is proportional to the ion concentration, the fractional variance of voltage fluctuations stays independent of the number of ions in the channel. This was the case when the ion concentration was larger than 0.001 mol/l. Below this critical ion concentration, the degree of freedom of the ion motion was proportional to the number of ions in the channel. How should it be understood?

Suppose an individual ion has an influence sphere of diameter d, and the degree of freedom of the ionic motion is restricted when the influence spheres touch one another. It then follows that the diameter of this sphere is equal to the mean ion distance in a 0.001 mol/l solution, which is 10 nm. The Debye length of the Cu^{++} plasma is 8 nm which is very close to 10 nm. With increase of the ion concentration, however, the ratio of the Debye length to the mean ion distance decreases; thus mutual ion interactions are reduced. This is contrary to the observation. We cannot find any explanation for that the conductance fluctuations are attributable to ion number fluctuations in which our observation of the alpha value is taken.

As another possibility, suppose the mobility fluctuations are responsible for the observed voltage fluctuations, and the mobility fluctuations have a spatial cherence. Let the coherence length be equal to the mean ion distance in the critical ion concentration. If N ions are involved in a coherence volume, they will undergo the same mobility fluctuations, and hence the degree of freedom of the ionic motion will be degenerate N-fold. As a result,the degree of freedom of the ionic motion is equal to the number of coherence volumes in the channel and remains constant as long as N is larger than unity.

5. CONCLUSION

The 1/f conductance fluctuations were observed. From dependence of the alpha value on the ion centration it is quite probable that the conductance fluctuations are attributable to the mobility fluctuations and they have a spatial coherence or correlation length which was estimated to be 10 nm.

Inconsistency with other observations in which 1/f voltage fluctuations were not found is probably due too large voltage applied across the ion channel.

REFERENCES

(1) Hooge, F.N. and Gaal, J.L.M.,Fluctuations with a 1/f spectrum in the conductance of ionic solutions and in the voltage of concentration cells, Philips Research Reports 26 (1971) 77-90.

(2) Hooge, F.N., Physica 60 (1972) 130.

(3) van den Berg, R.J., de Vos, A. and de Goede J.,Electrical resistivity fluctuations in aqueous KCl solutions, Phys. Lett. 84A (1981) 433-434.

(4) van den Berg, R.J. and de Vos, A., Resistivity fluctuations in electro-osmotic flowing aqueous KCl solutions, Phys. Lett. 92A (1982) 203-206.

(5) Musha,T. and Sugita,K.,1/f conductance fluctuations in aqueous ionic solutions, J.Phys.Soc.Japan, 51 (1982) 3820-3825.

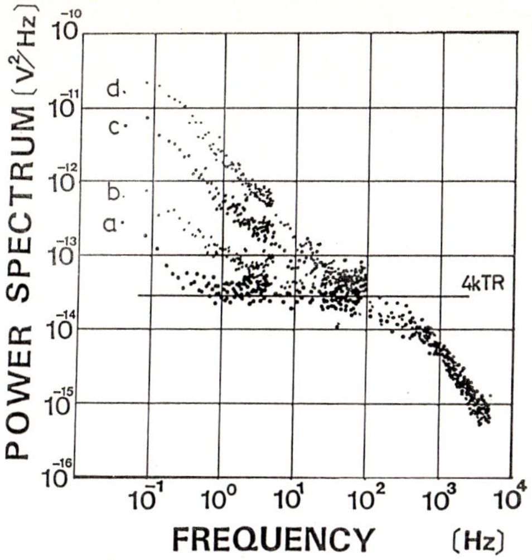

Fig.1A Power spectral densities of voltage fluctuations of 0.001 mol/l aqueous $CuSO_4$ solution.
A Neclepore membrane was used as part of a diaphragm separating two compartments of the ionic solution chamber; it had many cnannels of diameter 1 μm and length 10 μm. The electrodes for picking up voltage fluctuations were pH electrodes consisting of Ag-AgCl rods immersed in saturated KCl solution. Stable measurements can be performed with this type of electrode but KCl leaked out into the $CuSO_4$ solution. The resistance was 1.7 MΩ. Curves a, b, c, and d refer to dc currents 0, 0.03, 0.07, and 0.11 μA.

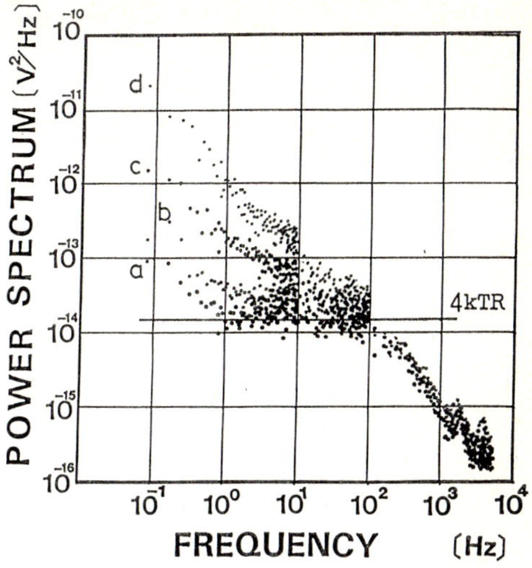

Fig.1B Power spectral densities of voltage fluctuations of 0.004 mol/l aqueous $CuSO_4$ solution.
The two compartments were separated by a membrane with an ion channel of diameter 10 μm and length 7 μm. Voltage fluctuations were picked up by Cu rods diameter 1 mm. Measurements were less stable as compared to with pH electrodes. The resistance was 870 kΩ. Curves a, b, c, and d refer to dc currents 0, 0.08, 0.23, and 0.39 μA.

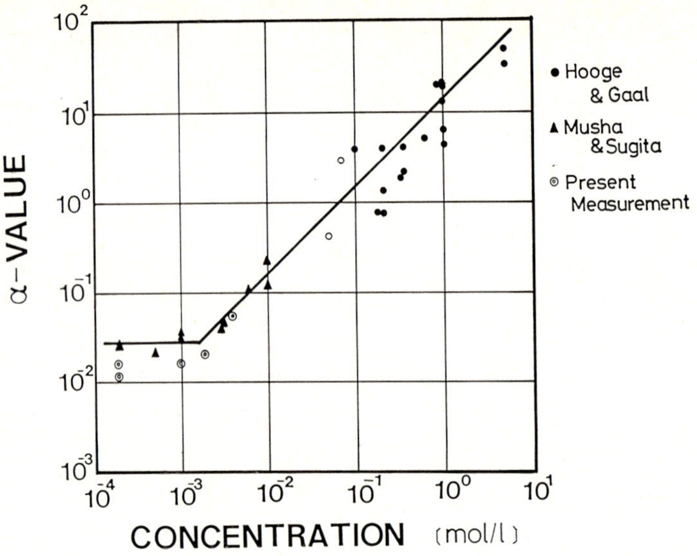

Fig.2 The α-values for 1/f conductance fluctuations of aqueous $CuSO_4$ solutions of various concentrations.
Solid and open circles have been taken from Hooge-Gaal's paper (1), triangles from Musha-Sugita's paper (5), and double circles refer to present observations. The α-value is proportional to the ion concentration for ion concentrations above 0.001 mol/l regardless of ion species, and it levels off at a constant value 0.025 for more dilute solutions.

NOISE IN PHYSICAL SYSTEMS AND 1/f NOISE
M. Savelli, G. Lecoy and J-P. Nougier (eds.)
© *Elsevier Science Publishers B.V., 1983*

ELECTRICAL RESISTIVITY FLUCTUATIONS IN SOLUTIONS OF POTASSIUM CHLORIDE

J. de Goede, N. Roos, A. de Vos and R.J. van den Berg

Laboratory of Physiology and Physiological Physics,
State University of Leiden, Wassenaarseweg 62
2333 AL Leiden, The Netherlands

Fluctuations in the electrical resistivity of solutions of a strong 1-1 electrolyte in a capillary connecting two reservoirs are considered and a simple physical model is proposed. This is based on a linearized convective diffusion equation for the concentration fluctuations of the dissolved electrolyte. It is shown that the model adequately describes the resistivity fluctuations measured in solutions of KCl in water or methanol.

1. INTRODUCTION

In the search for the physical mechanism of 1/f noise in electrolyte solutions we recently investigated the electrical resistivity fluctuations in aqueous KCl solutions in a capillary connecting two reservoirs [1-4]. However, instead of 1/f noise, our samples showed, beside Johnson noise, electrical noise which was characterized by spectral densities that were flat at low frequencies. Furthermore the shape of the spectral densities depended on fluid flow, induced either by the applied electric field (electro-osmosis) or an applied pressure drop across the capillary. This noise was apparently due only to concentration fluctuations since the variance $\langle \Delta V^2 \rangle$ of the voltage fluctuations under constant current was in good agreement with the formula

$$\langle \Delta V^2 \rangle = \frac{4\lambda^2 \langle V \rangle^2}{2n_s V_{eff}}, \qquad (1)$$

obtained from fluctuation theory. In eq. (1) λ is the coefficient relating resistivity fluctuations to concentration fluctuations, n_s the number density of the dissolved KCl molecules, $\langle V \rangle$ the average voltage drop across the capillary and V_{eff} the effective volume in which the measured noise is generated. For a capillary with length l and radius a this effective volume is approximately given by

$$V_{eff} = 20\pi a^3 (1+l/a)^2/(1+20l/a) \qquad (2)$$

In this contribution we propose a simple physical model to describe the electrical resistivity fluctuations in a solution of a strong 1-1 electrolyte. We show that for solutions of KCl in water or methanol theory and experiment are in good agreement.

2. FORMAL THEORY

Consider an isothermal, homogeneous solution of a strong 1-1 electrolyte. The system is in a non-equilibrium steady state with barycentric velocity field $\vec{v}(\vec{r})$ and uniform electric field $\vec{E}$. The density of the solution is constant so that

$$\vec{\nabla}\cdot\vec{v}(\vec{r}) = 0 \qquad (3)$$

From non-equilibrium thermodynamics [5] and assuming (local) electroneutrality, we find for deviations $\Delta n_s(\vec{r},t)$ from the average number density n_s, linearizing around the steady state the following convective diffusion equation:

$$\left[\frac{\partial}{\partial t} + \{\vec{v}(\vec{r}) + u\vec{E}\}\cdot\vec{\nabla} - D\nabla^2\right] \Delta n_s(\vec{r},t) = 0 \qquad (4)$$

where D is the ambipolar diffusion coefficient and u an effective mobility. The term with u will be neglected in what follows. For a volume Ω the correlation function $\Phi(t)$ of the number fluctuations

$$\Delta N(t) = \int_\Omega \Delta n_s(\vec{r},t)\,d\vec{r} \qquad (5)$$

is defined by

$$\Phi(t) = \frac{\langle \Delta N(t)\ \Delta N(o) \rangle}{\langle \Delta N(o)^2 \rangle} \qquad (6)$$

where the brackets denote an ensemble average. This correlation function can be written as [6]

$$\Phi(t) = \frac{1}{\Omega} \int_\Omega d\vec{r} \int_\Omega d\vec{r}\,'\ G(\vec{r},\vec{r}\,';t) \qquad (7)$$

in which $G(\vec{r},\vec{r}\,';t)$ represents the Green's function of eq. (4). Finally we introduce the spectral density

$$g(\omega) = \mathrm{Re} \int_0^\infty e^{-i\omega t}\Phi(t)\,dt \qquad (8)$$

where Re means: real part of.

3. OUTLINE OF THE MODEL

We now consider a circular capillary of length l and radius a. At first end effects will be neglected. A small pressure drop Δp across the capillary will give rise to a Poiseuille flow inside the capillary with a velocity profile [5]

$$v_p(r) = \frac{1}{4\pi\eta}(a^2-r^2)\frac{\Delta p}{l} \quad (9)$$

where r is the distance from the central axis and η the dynamic viscosity. In general an electric double layer in the fluid, at the interface with the capillary wall, will be present. A voltage <V> across the capillary will generate a transport of fluid (electro-osmosis) with a velocity profile [5]

$$v_e(r) = \frac{\varepsilon}{\eta}[\phi(r) - \zeta]\frac{\langle V\rangle}{l} \quad (10)$$

in which ε is the permittivity, ϕ the electric potential in the fluid at equilibrium and ζ the zeta potential. All other effects of the electric double layer will be neglected. The total velocity field is given by

$$v = v_p + v_e \quad (11)$$

To obtain an approximation to the spectral density of the number fluctuations in the capillary we proceed as follows. We first calculate the spectral density $g_1(\omega)$ of the number fluctuation in a one dimensional system with length l and and velocity v. This is given by [6]

$$g_1(\omega) = \mathrm{Re}\,\frac{1}{\pi}\int_{-\infty}^{+\infty} dz\frac{\sin^2 z}{z^2(i\omega+2ivz/l + 4Dz^2/l^2)} \quad (12)$$

For a closed form of the integral in eq. (12) see [6]. To account for a velocity profile in the capillary we subsequently average $g_1(\omega)$ over the cross section of the capillary. Physically, this amounts to neglecting all diffusion other than in the direction of the principal axis. Writing for the Poiseuille flow

$$v_p(y) = \bar{v}_p(1-y^2) \quad (13)$$

where

$$\bar{v}_p = \frac{a^2\Delta p}{8\eta l} \quad (14)$$

and neglecting ϕ in eq. (10) we obtain for the spectral density of the number fluctuations in the capillary

$$g(\omega) = \int_0^1 2y\, g_1\{\omega, v(y)\}dy. \quad (15)$$

For the spectral density $S_V(\omega)$ of the voltage fluctuations we now write

$$S_V(\omega) = \frac{4\lambda^2 \langle V\rangle^2}{2n_s V_{eff}} g(\omega). \quad (16)$$

Through introducing V_{eff} in eq. (15) the variance of the voltage fluctuations includes the end effects of a finite capillary.

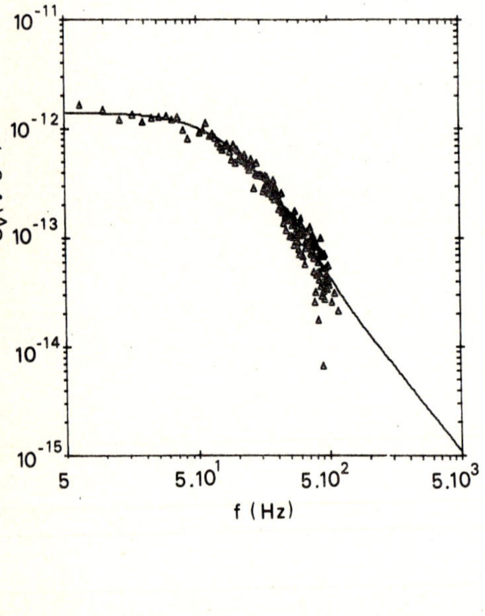

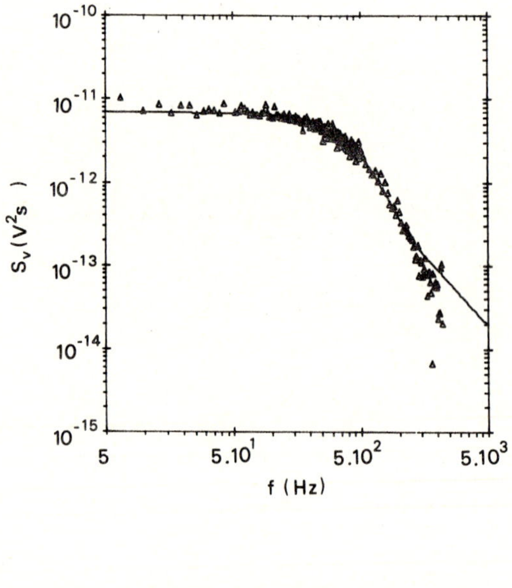

Figure 1 : Spectral densities of the voltage fluctuations across a capillary (G06) filled with 0.10 x 10^2 mol.m^{-3} KCl in water. Temperature, constant within 0.1°C, 20°C; D = 1.7 x 10^{-9} m^2/s; η = 1.00 x 10^{-3} Pa.s; dielectric constant 80.1; λ = 0.975; ζ = -25.6 mV. Solid lines: fit of eq. (15) to the observed data. The parameters with their standard deviations are as given in the table.
Left panel: V = 0.50 V; Δp = 0 N/m^2. Right panel: V = 2.0 V; Δp = 0 N/m^2.

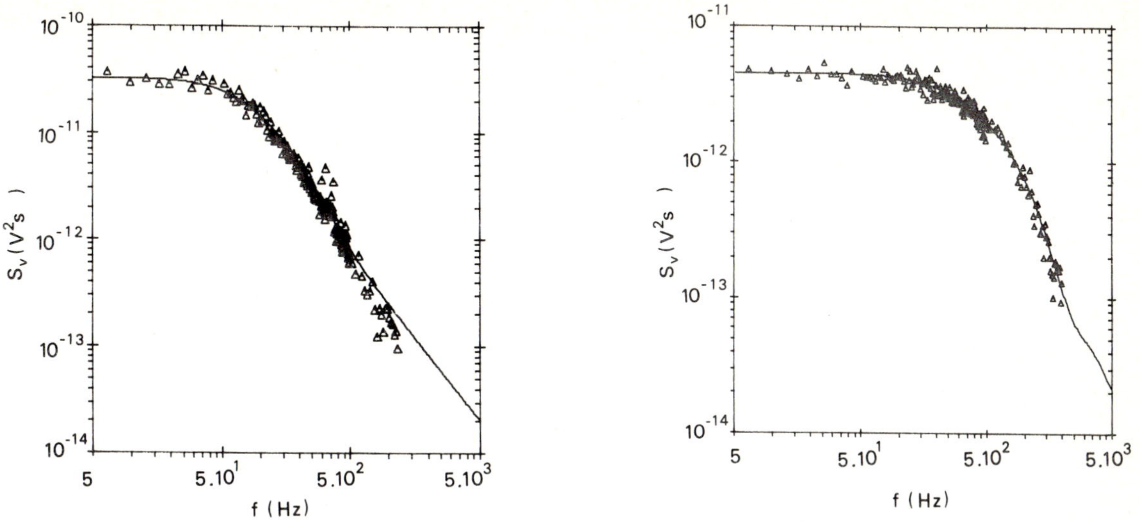

Figuur 2 : Spectral densities of the voltage fluctuations across capillary G06 filled with 0.10 x 10^2 mol.m^{-3} KCl in methanol. Temperature, constant within 0.1°C, 20°C; D = 1.2 x 10^{-9} m^{2}s; η = 0.60 x 10^{-3} Pa.s; dielectric constant 32.6. Solid lines: fit of eq. (15) to the observed data. The parameters with their standar deviations are as given in the table. Left panel: V = 2.0 V; Δp = 0 N/m^2. Right panel: V = 2.0 V; Δp = 1.2 N/m^2. The parameter D/l^2 has been kept fixed at the value obtained for the fit to the data shown in the left panel. The value of λ is 1.00.

4. SAMPLES AND EXPERIMENTAL METHODS

The sample cell consisted of a single cylindrical capillary connecting two reservoirs. The capillary was made by dielectric breakdown in a thin (Jena) glass membrane. Its radius, as measured with an electron microscope, was 1.70 µm and its length, as calculated from the measured resistance R and the formula

$$R = \rho\ [\frac{1}{2a} + \frac{l}{\pi a^2}] \qquad (17)$$

was found to be 5.4 µm. In eq. (17) ρ is the resistivity of the ionic solution. The cell was filled with a solution of KCl in water or methanol. The solutions were meticulously degassed and ultrafiltrated (0.02 µm PTFE). A pressure drop could be created by having a difference in fluid levels or by changing the pressure above the fluid levels with compressed N_2 gas. Electrical contacts were made with three pairs of Ag-AgCl electrodes. One pair was used to inject a constant current. The other pairs served to measure the mean voltage and the voltage fluctuations. Cross spectral densities were estimated from 0.05 to 5000 Hz. This procedure eliminates the instrumentation noise. From these spectral densities, the spectral density of the noise when the applied current was zero (which agreed with Nyquist's formula) was subtracted. Further details of the experimental methods can be found in [1-4].

5. RESULTS AND DISCUSSION

A comparison between the model spectra and the experimental ones is hampered by the fact that not all quantities needed for the calculations are accurately known. For instance, the adjustment to zero pressure drop is unreliable [2]. Since also the zeta potential was measured in a few experiments only we decided to estimate the model parameters

$$A = \frac{2\lambda^2<V>^2}{n_s\ V_{eff}}\ ,\ D/l^2,\ \overline{v}_p/l \text{ and } v_e/l$$

by curve fitting with a Marquardt type procedure for non-linear weighted least squares. In this way we also get information about the identifiability of the model parameters. In a first approach we estimated only three parameters at the same time keeping the fourth one fixed at an appropriate value.

Figure 1 depicts the spectral densities of the voltage fluctuations in an aqueous solution of KCl at two different applied voltages without an applied pressure drop. In Figure 2 the spectral densities of the voltage fluctuations are shown in a solution of KCl in methanol at the same applied voltage without and with an applied pressure drop. The fitted model and the experimental data are in good agreement, although there is some indication of a discrepancy at higher frequencies. The use of methanol instead of water as a solvent did not give rise to any important new aspects with respect to the resistivity fluctuations.

From the parameters collected in the table and the quantities quoted in the legends to the figures the capillary dimensions can be calculated. Using A and D/l^2 we find for the length values from 4.1 to 4.2 µm and for the radius values ranging from 1.7 tot 2.0 µm. With the parameters A and $\overline{v}_p$/l we obtain a length of

TABLE

solvent	<V> (V)	$\Delta p (N/m^2)$	$A \times 10^{+9} (V^2)$	$D/l^2 \times 10^{-2} (s^{-1})$	$\bar{v}_p/l \times 10^{-3} (s^{-1})$	$v_e/l \times 10^{-2} (s^{-1})$
water	0.50	0	13.1 ± 0.2	1.02 ± 0.08	-	4.1 ± 0.2
water	2.0	0	0.83 ± 0.02	0.98 ± 0.09	-	11.1 ± 0.2
methanol	2.0	0	18.6 ± 0.5	0.71 ± 0.05	-	3.7 ± 0.1
methanol	2.0	1.2	11.7 ± 0.1	-	1.20 ± 0.03	5.4 ± 0.3

7.5 μm and a radius of 1.7 μm. These estimated capillary dimensions are in reasable agreement with a length of 5.4 μm and radius of 1.70 μm as estimated by other methods (see section 4). Calculations also show that the introduction of V_{eff} in the model improves the estimates of the radius with respect to the value as measured by electron microscope.
The zeta potential can be obtained from D/l^2 and v_e/l. With water as a solvent we find for ζ the values -26.9 mV and 19.1 mV in reasonable agreement with the value -25.6 mV obtained from the streaming potential [4].
In the experiments shown in figure 1 there was only electro-osmotic flow. It is gratifying that a Poiseuille flow instead of an electro-osmotic one in the model gave a bad fit to these data.
These findings provide strong evidence that our physical model is basically correct. We therefore conclude that the electrical resistivity fluctuations measured in the KCl solutions have to be explained by concentration fluctuations as the sole noise source. Its kinetics is governed by volume flow and diffusion. It should be stressed that in all our experiments the volume flow was too large to observe the "universal $\omega^{-3/2}$ high frequency tail".
A noteworthy aspect of our experimental results is the absence of 1/f noise. It remains an important problem to establish under what experimental conditions 1/f noise can de detected in electrolyte solutions, if at all.

This research was supported by the Netherlands Organization for the Advancement of Pure Research (Z.W.O.).

REFERENCES

[1] Berg, R.J. van den, A. de Vos and J. de Goede, in P.H.E. Meyer, R.D. Mountain and R.J. Soulen, (eds.), Sixth Intern. Conf. on Noise in Physical Systems (National Bureau of Standards, S.P.614, Washington, 1981) 217-220.

[2] Berg, R.J. van den, A. de Vos and J. de Goede, Phys. Lett. 84A(1981) 433-434.

[3] Berg, R.J. van den, A. de Vos and J. de Goede, Phys. Lett. 87A(1981) 98-100.

[4] Berg, R.J. van den and A. de Vos, Phys.Lett. 92A(1982) 203-206.

[5] Groot, S.R. de and P. Mazur, Non-equilibrium Thermodynamics (North-Holland, Amsterdam, 1962).

[6] Lax, M. and P. Mengert, J. Phys. Chem. Solids 14 (1960) 248-267.

METROLOGY

NOISE IN PHYSICAL SYSTEMS AND 1/f NOISE
M. Savelli, G. Lecoy and J-P. Nougier (eds.)
© Elsevier Science Publishers B.V., 1983

INTRINSIC AND EXTRINSIC NOISE SOURCES IN AN RF BIASED R-SQUID

H. Seppä
Technical Research Centre of Finland, Helsinki, Finland

J. H. Colwell and R. J. Soulen, Jr.
Center for Absolute Physical Quantities
National Bureau of Standards, Washington, DC, USA

1. INTRODUCTION

An R-SQUID is a circuit consisting of a resistor R connected in series with a small inductor L_S and a Josephson junction. The resistor is in intimate thermal contact with a bath at temperature T. The Johnson noise developed in the resistor causes voltage fluctuations in R which are converted by the Josephson junction into fluctuations in frequency which occur in the audio frequency range (1-10^5 Hz). A small dc current I_o is also injected into this circuit: thus the audio frequency spectrum consists of a signal at a frequency $\nu_o = I_oR\ 2e/h = I_oR/\phi_o$ which is broadened by the thermal noise. The amplitude of this signal is too small to be used to measure the frequency, so the R-SQUID is driven by a radio frequency signal at frequency Ω, thereby amplifying the af signal by parametric upconversion. The amplified af signal is recovered for further processing by rf demodulation techniques. The rf current is injected into the R-SQUID by means of a tank circuit (a superconducting coil) also maintained at temperature T.

The experimental procedure used to measure the circuit noise consists of repeatedly measuring the audio frequency ν and calculating the variance δ^2. The long-term average of δ^2 is related to the Johnson noise and thus to T by the relationship (1)

$$\delta_T^{\ 2} = \sum_{i=1}^{N} \frac{(\nu_i-\bar{\nu})^2}{N} = \frac{2kRT}{\phi_o^{\ 2}\tau} \qquad (1)$$

where N is the total number of frequency counts acquired, k is Boltzmann's constant, and τ is the gate time of the frequency counter used for the measurements. It is this fundamental equation which is used to define a cryogenic temperature scale from 0.01 K to 0.5 K at the National Bureau of Standards. Recent improvements in the data acquisition system used to calculate $\delta_T^{\ 2}$ now permit this quantity to be measured with a statistical imprecision of 0.1 % in reasonable averaging times (5 hours or less). Systematic effects of comparable magnitude must thus be understood and accounted for. The effects of extraneous thermal emfs, temperature oscillations, noise due to I_o and the dc bias circuit, have been treated elsewhere (2) and have been shown to introduce systematic effects much smaller than 0.1 %. The influence of amplitude noise on the variance of the frequency of the af signal has also been treated (3); for the experimental conditions to be described in this paper a multiplicative correction of 0.8 % was applied to all of the data to account for this effect. The success of a recently developed model leads us to conclude that this multiplicative factor is known to 1 %, so that overall systematic errors introduced are far less than 0.1 %.

In this paper we identify yet another noise source - that generated by the Josephson junction - and demonstrate that the agreement between a model and experimental results is sufficiently good that we may account for this effect also to the 0.1 % level. It would thus appear that no impediment lies in the way of using the R-SQUID to define a temperature scale from 0.01 K to 0.5 K with statistical and systematic errors each less than 0.1 %.

2. THEORY

We summarize here the results of a more complete study /4/ for the influence of the nonlinearity of the Josephson junction on the variance of the frequency. Two effects contribute to that variance. First, the junction converts the noise voltage generated by R and the dc voltage drop into a harmonic series of thermally broadened spectral lines at ν_o. The linewidth (or variance) is the desired quantity from which the noise temperature may be determined. The nonlinearity of the junction converts noise at ν_o and its harmonics into additional low frequency noise. Furthermore, the junction also mixes rf noise in the tank circuit with the injected rf signal producing extra noise in the same audiofrequency range as the first effect, thus adding an unwanted af noise source which must be accounted for. The mixing is dependent on the dynamic impedance of the junction which in turn depends on the frequency and amplitude of the rf signal. We refer to this latter effect as mixed down noise.

Including all these processes described above and assuming that the SQUID works in the non-hysteretic mode, the normalized variance of the frequency fluctuation is given by /4/

$$\frac{\delta_\gamma^2}{\delta_T^2} = \frac{1}{1-\eta_e^2}\left[(1+\tfrac{1}{2}\eta_e^2)+(\tfrac{1}{2}+\eta_e^2)\,\alpha\frac{T_T}{T}\sum_{n=1}^{\infty} f_n^2\right] \quad (2)$$

where

$$\alpha = \tfrac{1}{2}k^2Q\varepsilon^2(R/\Omega L_S),\quad \eta_e^2 = \eta^2\sum_{n=1}^{\infty} g_n^2$$

$$\eta = I_c/I_o,\quad f_n = \partial g_n/\partial\hat{\varphi}_{rf} \text{ and } \varepsilon = 2\pi\frac{L_S I_c}{\Phi_o}$$

Here $\hat{\varphi}_{rf}$ and Ω denote the normalized amplitude and the angular frequency of the pumping signal. Q is the quality factor of the tank circuit and k coupling coefficient between the tank circuit and the SQUID loop. T_T denotes the equivalent temperature of the tank circuit which can be in practise thousands of times higher than the temperature of the SQUID circuit T. T_T is determined mainly by the first stage of the rf amplifiers. The coefficients of the Fourier series, g_n, depend on all the SQUID parameters and also on the frequency and the level of the rf signal. If the SQUID is operated very near resonance, g_n reduces to the rather simple form /4/

$$g_n = -2(-1)^n\frac{J_n(n\varepsilon)}{n\varepsilon}J_o(n\hat{\varphi}_{rf}) \quad (3)$$

from which we immediately find

$$f_n = 2(-1)^n\frac{J_n(n\varepsilon)}{\varepsilon}J_1(n\hat{\varphi}_{rf}) \quad (4)$$

Here J_n is the ordinary Bessel function of order n.

It should be pointed out here that the analytic expression for the normalized variance given in eg. 2 is accurate only if η_e^2 is much less than unity.

In the temperature measurement the dc and ac bias conditions are chosen so that $\eta \ll 1$ and $\hat{\varphi}_{rf}$ achieves the local maximum of the signal level $(J(\hat{\varphi}_{rf})=\max)$. Under these circumstances eq. 2 reduces to

$$\frac{\delta_\gamma^2}{\delta_T^2} = 1 + \tfrac{1}{2}\alpha\frac{T_T}{T}\sum_{n=1}^{\infty} f_n^2 \quad (5)$$

Thus the most important noise source arising from the nonlinearity of the Josephson junction is proportional to the tank circuit temperature and the sidebands of the rf voltage across the tank circuit. One may minimize this effect by decreasing the temperature of the tank circuit and increasing the pumping frequency. The approach of reducing this noise effect even further by reducing α (i.e., k^2Q and ε) unfortunately simultaneously decreases the af signal amplitude. This reduces the signal to noise of the af signal used to count the frequency, and thus increases the additive noise (another undesirable effect). Typical values of ε which strike a reasonable compromise between mixed down noise and additive noise are $\varepsilon = 0.5$.

3. EXPERIMENTAL PROCEDURES

The R-SQUID is mounted on the base of a He^3/He^4 dilution refrigerator capable of maintaining constant temperatures for very long times at temperatures as low as 0.01 K. A temperature control circuit with a resistance thermometer sensor is used to maintain the temperature constant for the several hours needed to average the noise data. The fluctuations in temperature recorded we found to be $\sim$10 μK at 10 mK increasing roughly linearly to 50 μK at 0.5 K (the highest temperature used for these experiments).

Since the spectral power density generated by the Johnson noise in the resistor ($\sim$10 μΩ) at the lowest T($\sim$10 mK) is very small [$\sim 6\times10^{-30}$ W/(Hz)$^{\frac{1}{2}}$] it is necessary to shield the R-SQUID from external noise sources. Accordingly, it is enclosed in a copper can maintained at T and which is lined with superconducting solder. This assembly is enclosed in a mu-metal shield at 4 K, which in turn is enclosed in a larger mu-metal can at room temperature. No coherent noise signals other than that generated by the R-SQUID have been found.

The variance data may be used to detect the presence of extraneous non-white spectral power densities. None were found to within the detectable limit of the experiments 10^{-33} W/(Hz)$^{\frac{1}{2}}$.

Generally, the gate time of the frequency counter was set at 10 ms, in which case the statistical imprecision of the measurement of the variance could be reduced to 0.3 % by accumulating data for approximately 30 minutes. Most data points were taken with this imprecision except when the effects to be described later were smaller. In this case, averaging times were increased to reduce the statistical imprecision to 0.1 %.

4. RESULTS

In this work we have studied experimentally the influences of the Josephson junction and the tank circuit on the variance of the frequency fluctuation. In the first experiment (see fig. 1) the temperature was controlled at 160 mK and the variance was measured for different values of the rf level. A cooled preamplifier (GAT 1/010) was used, thus minimizing the influence of the tank circuit noise. Under these conditions the theoretical model (eg. 2) for the normalized frequency variance reduces to the form

$$\frac{\delta_\gamma^{\ 2}}{\delta_T^{\ 2}} = \frac{1+\frac{1}{2}\eta_e^{\ 2}}{1-\eta_e^{\ 2}} \qquad (6)$$

The relation between $\hat{\varphi}_{rf}$ and the injected rf voltage $\hat{V}_{rf}$ is assumed to be linear. The coefficients of this linear equation are determined experimentally by recording the values of rf bias voltage for which the af signal disappears ($J_1(\hat{\varphi}_{rf}) = 0$). The theoretical curve given in eq. 6 is fitted to the experimental data assuming $\varepsilon = 0.6$ and varying η in order to find the best fit. The approximate value for ε was found by methods similar to that given by Erné et. al. /5/. A fitted value of $\eta^2 = 0.75$ is given in fig. 1 which shows rather good agreement with the experimental data. The fitting is not perfect and can be partly explained by the fact that the relation between $\hat{\varphi}_{rf}$ and $\hat{V}_{rf}$ is not exactly linear at low values of $\hat{\varphi}_{rf}$. We note that the theoretical curve does not depend strongly on ε and thus equally good fits to the experimental results can be found also with other values for ε varying by ± 0.1 from the value chosen here.

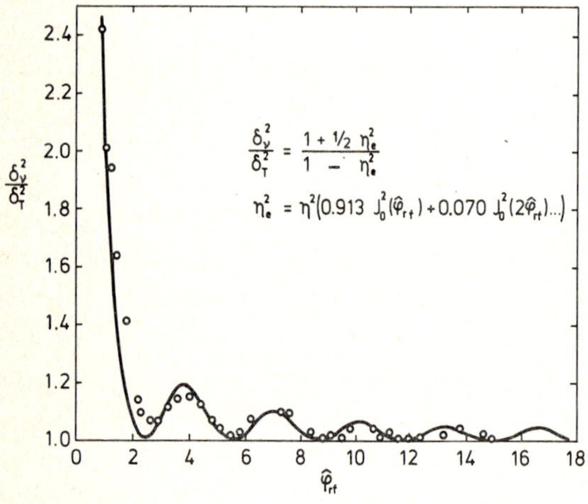

Figure 1: Normalized variance versus $\hat{\varphi}_{rf}$. The R-SQUID was maintained at a temperature of 160 mK as the variance was measured as a function of the amplitude of the radio frequency signal (proportional to $\hat{\varphi}_{rf}$). The statistical imprecision of the variance was approximately 0.2 %. A value of 0.6 was chosen for ε, and by fitting eq. 6 to the data, we obtained a value $\eta^2 = 0.75$.

In the other experiment, using the same Josephson junction, the temperature was set at 12.5 mK and an extra noise signal was fed into the tank circuit in order to increase the equivalent temperature of the tank circuit. The dc bias current was adjusted so that $\eta^2 = 0.015$. Under these circumstances the only important extra contribution to the frequency variance is the tank circuit noise. The theoretical curve given in eq. 5 is fitted to the experimental data by varying the term $\alpha T_T/T$. The result for $\alpha T_T/T = 7.5$ is shown in figure 2. We were obliged to make small correction to the data points near by the valleys of the $\delta_\gamma^2/\delta_T^2 - \hat{\varphi}_{rf}$-characteristic due to the poor signal to noise ratio. These corrections were made by using the procedure given in reference 3. This result given in fig. 2 shows clearly the existence of the down mixing effect and agrees rather well with the theoretical model.

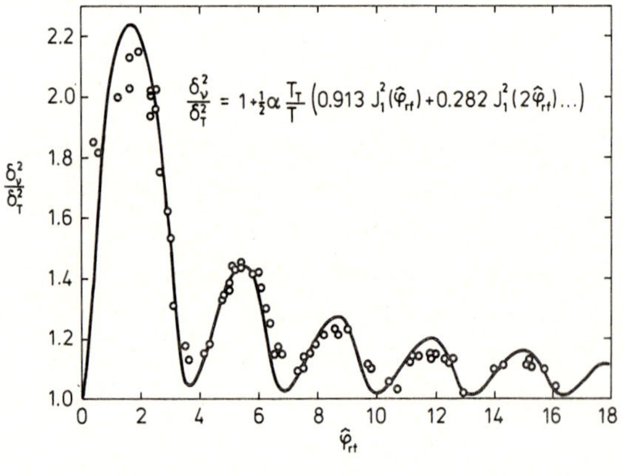

Figure 2: Normalized variance versus $\hat{\varphi}_{rf}$. The temperature of the R-SQUID was 12.5 mK and the statistical imprecision of the variance was approximately 0.2 %. An external rf noise source was used to inject noise into the tank circuit to enhance the mixed down effect. We assume $\varepsilon = 0.6$, and obtained a value for $\frac{\alpha T_T}{T} = 7.5$ by fitting eq. 5 to the data.

5. CONCLUSION

The function of the Josephson junction in an R-SQUID circuit is to convert the thermal noise generated by R into frequency fluctuations. The junction can, however, mix rf noise down into low frequencies. We have modelled this additional noise and shown that it has two limits. We have compared these predictions with experiments and found excellent agreement. Based on these results, we can choose operating conditions and operating parameters such that the contribution of this mixed down noise should contribute less than 0.1 % to the measured thermal noise. Thus noise thermometery with comparable inaccuracy should be possible, and experiments are being prepared to test this assertion.

REFERENCES

1. Kamper, R.A. and Zimmerman, J.E., Noise thermometer with Josephson effect. J. Appl. Phys. 42(1971), pp. 132-136.
2. Soulen, R.J. and Van Vechten, D., Noise thermometry at NBS using a Josephson junction, in Temperature, its measurement and control in science and industry. American Journal of Physics, Vol. 5, 1982, pp. 115-123.
3. Soulen, R.J., Van Vechten, D. and Seppä, H., Effect of additive noise and bandpass filter on the performance of a Josephson junction noise thermometer. Rev. Sci. Instrum. 53(1982), pp. 1355-1362.
4. Seppä, H., Noise properties in an rf-biased Josephson junction noise thermometer. To be published.
5. Erné, S.N., Hahlbohm, H.-O. and Lübbig, H., Theory of rf-biased superconducting quantum interference device for non-hysteretic regime. J. Appl. Phys. 47(1976), pp. 5440-5442.

NOISE IN PHYSICAL SYSTEMS AND 1/f NOISE
M. Savelli, G. Lecoy and J-P. Nougier (eds.)
© *Elsevier Science Publishers B.V., 1983*

OBSERVATIONS OF LOW-FREQUENCY CURRENT NOISE IN NIOBIUM FILMS: ELECTRODIFFUSION NOISE AND ABSENCE OF 1/f NOISE

John H. Scofield and W. W. Webb

School of Applied and Engineering Physics
Cornell University
Ithaca, New York USA

We report current noise measurements on bulk-like, sputter deposited, thin niobium films. The dominant excess noise in many specimens has a spectrum characteristic of one-dimensional diffusion, and is (apparently) associated with hydrogen that enters the specimen during chem-etching and ion-milling. The relative power spectral density $G_V(f)/\bar{V}^2$ of this noise was varied by three orders of magnitude by specimen treatments. No 1/f noise was observed even with the lowest diffusion noise level and was therefore at least one to two orders of magnitude below the level typically reported for other metal films. Ion-milling of films introduced additional excess noise above 1 Hz that is not yet understood.

INTRODUCTION

The room temperature low frequency excess noise of continuous thin films of various low-melting point metals (Au, Ag, Cu, Ni, Sn, Pb, In, Au_xAg_{1-x}) has been measured [1-8] and generally characterizes as resistance fluctuations with approximately 1/f spectra. The resistivities ρ of some of the films approach that of the bulk ρ_b at room temperature where phonon scattering dominates. The current noise levels are often of the order of magnitude given by the emperical formula $NfG_V(f)/\bar{V}^2 = \alpha \simeq 10^{-3}$ [1], where $G_V(f)$ is the power spectral density [9] of the fluctuations in voltage across a resistor with N atoms and a mean voltage drop $\bar{V}$. Amongst the more refractory metals the current noise in Cr [5] and Pt [7,8] at ambient temperature have been reported from specimens with resistivities 4 to 10 times those of the bulk metal. We have measured the current noise of "bulk-like" niobium films ($T_c \simeq 9.2$ K and $1.3 \lesssim \rho/\rho_b \lesssim 2$) at room temperature and have so far failed to observe any specimen 1/f noise with a detectability limit of $\alpha \simeq 3 \times 10^{-5}$. Instead, current noise observed in some preparations has a spectrum characteristic of one-dimensional diffusion, that is $G_V(f)$ = constant for $f \ll f_0$ and $G_V(f) \propto f^{-3/2}$ for $f \gg f_0$, with the corner frequency f_0 varying with the specimen length L as $f_0 \propto L^{-2}$ [9]. The origin of this noise has been traced to both chem-etching and ion-milling. By changing specimen preparation the noise level has been varied by three orders of magnitude. Evidence suggests that the conductance noise is due to a diffusing variable that is probably hydrogen.

EXPERIMENTAL METHOD

Current noise from 28 niobium specimens has been measured using the phase sensitive detection of the voltage that results when a fluctuating resistance modulates an alternating current. The technique avoids low-frequency preamplifier noise and allows the measurement of resistance fluctuations without the electrotransport that might occur with a direct current. While similar to AC methods for measuring 1/f noise [10] (the so-called "1/Δf noise") the use of balanced five-probe specimens allows contact noise to be distinguished from specimen noise with the sample in a Wheatstone bridge [2]. The bridge error signal measures the difference in resistance fluctuations of the two halves of the specimen, $\delta r_2(t) - \delta r_1(t)$ and is thus insensitive to correlated fluctuations such as those due to temperature drifts. Identical power spectral densities were obtained with this technique and DC methods in control experiments on gold films.

Specimen geometries (wires with large contact pads) were fabricated using photolithography followed by either chemical- or sputter-etching (ion-mill). Contacts were made either by soldering indium wire directly to the niobium pads (substrates N6 and N8) or by ultrasonically bonding 0.7 mil gold wire to 3000 Å thick gold films that overlapped the niobium contact pads (substrates N1, N2, N3, N4, N5, and N7). The specimen was then placed in an AC Wheatstone bridge otherwise composed of 5-watt, 2 ppm/°C stable wire-wound resistors. A PAR-124A lockin (with a 118 or 116 preamplifier) provides the bridge current and both amplifies and demodulates the bridge error signal. Operating the tuned amplifier in the bandpass mode with Q = 1, a carrier frequency f_c = 700 Hz, and accurately correcting for measured instrumental transfer characteristics, current noise has been measured for frequencies $f \lesssim 100$ Hz with less than 0.2 dB instrumental error. The technique has been used to measured the 1/f noise from Au, Cu, Ni, and Cr films in the frequency range 400 µHz $\lesssim f \lesssim$ 400 Hz [11].

THEORY OF DIFFUSION FLUCTUATIONS

Equilibrium fluctuations in diffusive systems have been extensively treated elsewhere [2,4,12-14]; we merely summarize the relevant results. The power spectrum $S_N(\omega)$ of the fluctuations in the number $N(t)$ of diffusing "particles" in a subvolume L of an infinite one-dimensional system is

$$S_N(\omega) = \frac{1}{2\pi^2}\frac{\langle\delta N^2\rangle}{\omega_0}\int_{-\infty}^{\infty} dx \frac{\sin^2 x}{x^4 + \theta^4/4} \quad (1)$$

where $\theta^2 = (\omega/\omega_0)$, $\omega_0 = 2D/L^2$, D is the diffusion coefficient, and $\langle\delta N^2\rangle$ is the variance of N in the subvolume L. Assuming that resistance and number fluctuations are linearly related (i.e. $\delta R = \gamma\delta N$) we obtain the usual spectrum

$$S_r(\omega) = \frac{\gamma^2}{2\pi}\frac{\langle\delta N^2\rangle}{\omega_0}\begin{cases}(\omega/\omega_0)^{-1/2}, & |\omega| \ll \omega_0 \\ (\omega/\omega_0)^{-3/2}, & |\omega| \gg \omega_0.\end{cases} \quad (2)$$

To include the cancelling effect of the bridge geometry in our system we calculate the spectrum of the difference in the number fluctuations of two adjacent subvolumes L, $\delta V(t) = I\gamma[\delta N_2(t) - \delta N_1(t)]$ and find that

$$S_N(\omega) = \frac{1}{2\pi^2}\frac{\langle\delta N^2\rangle}{\omega_0}\int_{-\infty}^{\infty} dx \frac{\sin^4 x}{x^4 + \theta^4/4}. \quad (3)$$

The only significant difference between Eq.(1) and Eq.(3) is that the latter becomes constant rather than diverging at low frequencies. Our simple model neglects the (small) effect of the niobium center contact on the diffusion problem. Taking $\langle\delta N^2\rangle = N_0 = c_0 N_a$ and including the effects of other circuit components, we expect the measured noise spectrum $G_V(f) = 4\pi S_V(2\pi f)$ to be given by

$$G_V(f) = \frac{c_0\beta^2 V_0^2}{3\pi f_0 N_a \rho^2 k}\begin{cases}1, & f \ll f_0 \\ 9/8(f/f_0)^{-3/2}, & f \gg f_0\end{cases} \quad (4)$$

where c_0 is the mean concentration of the diffusing variable, $\beta = \partial\rho/\partial c_0$, ρ is the film resistivity, the corner frequency $f_0 = \omega_0/(2\pi)$, N_a is the number of atoms in one half of the specimen, K is a numerical factor ($\simeq 1$) that accounts for other components in the bridge, and $(2)^{1/2}V_0\sin(2\pi f_c t)$ is the mean voltage measured between one of the voltage probes and the center contact.

EXPERIMENTAL RESULTS

The simplest noise spectra were obtained from specimens formed by chem-etching that had their gold contact films laid down prior to the niobium deposition. Noise spectra from four of these specimens of different lengths from 20 μm to 300 μm long on one substrate (N7) are plotted in Figure 1; these show the diffusion noise form of Eq.(4). Corner frequencies f_0 are consistent with an L^{-2} scaling and give a diffusion coefficient $D = (2 \pm 1) \times 10^{-6}$ cm^2s^{-1} as compared with 8×10^{-6} cm^2s^{-1} for α-phase hydrogen in bulk niobium and 1×10^{-6} cm^2s^{-1} for β-phase niobium hydride [15]. Thermal diffusion is several orders of magnitude too fast while vacancy diffusion is many orders of magnitude too slow to account for these corner frequencies.

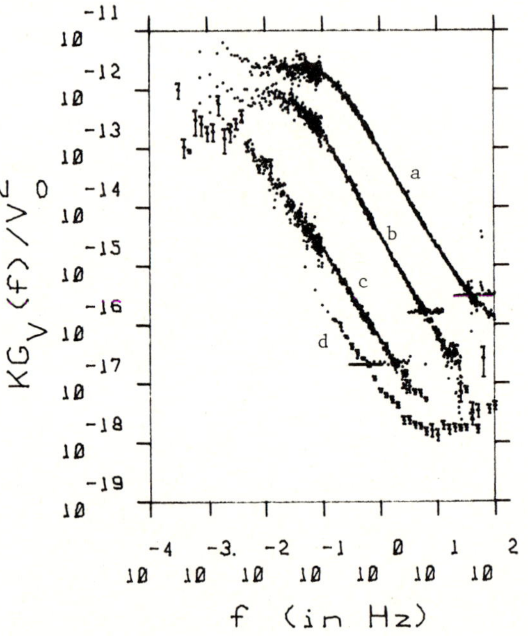

Figure 1: Noise spectra of 1200 Å thick specimens chem-etched from a $\rho \simeq 28$ μΩ.cm DC sputtered Nb film. Note that uncorrelated resistance fluctuations from the balanced noise specimens give $G_r(f)/r^2 = KG_V(f)/V_0^2$. (a) specimen N7-2 with L≃ 20 μm, w = (1.0±0.3) μm, and r ≃ 123 Ω. (b) specimen N7-1 with L≃ 40. μm, w = (2.3±0.3) μm, and r ≃ 56 Ω. (c) specimen N7-3 with L≃ 150. μm, w = (4.2±0.3) μm, and r ≃ 87 Ω. (d) specimen N7-4 with L≃ 300. μm, w = (9.3±0.3) μm, and r ≃ 75 Ω.

One would expect hydrogen to enter the specimen during chem-etching since metals are commonly loaded with hydrogen electrolytically in acids. Metal powders are known to form hydrides when placed in an acid [16]. However, diffusion noise was observed in sputter-etched specimens as well. Specimens on substrate N6 were fabricated using sputter-etching with contacts made by soldering indium wire directly to their niobium pads. The noise from one of these specimens is plotted as Fig.2(a). The structure above 1 Hz was usually observed in the noise of specimens that had been ion-milled and is not presently understood. The deviation from a pure diffusion spectrum below 1 Hz is probably due to the addition of 1/f noise from the soldered indium contacts [17]. The effect of ion-milling on the noise spectrum is illustrated by Fig.2(b) measured on the above specimen after its niobium contact pads had been bombarded with a 90 $mA{\cdot}s/cm^2$ dose of 500 eV Ar^+ beam. The specimen resistance increased by 15% while its current noise increased by three orders of magnitude. Below 1 Hz the spectrum has the form of the diffusion spectrum with the corner frequency giving $D = (5{\pm}2) \times 10^{-6}\ cm^2s^{-1}$.

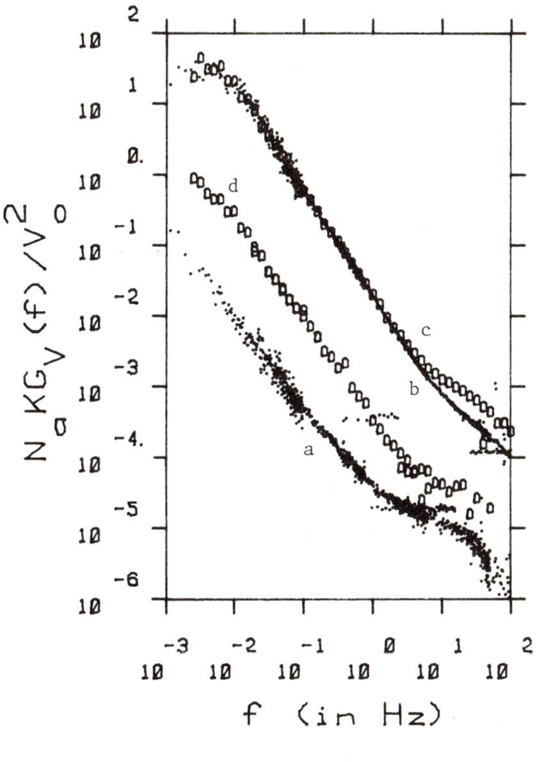

Figure 2: Normalized noise spectra of $L \simeq 146$ μm, $w = (6.5{\pm}0.3)$ μm, 2400 Å thick specimens sputter-etched from a $\rho \simeq 20.$ μΩ·cm RF sputtered Nb film. (a) as fabricated specimen N6-1, $r \simeq 24.2$ Ω. (b) specimen N6-1 after ion-milling the contacts, $r \simeq 28.5$ Ω. (c) specimen N6-5 after ion-milling the contacts, $r \simeq 29.1$ Ω. (d) specimen N6-5 after electrodiffusing, $r \simeq 24.4$ Ω.

Because hydrogen ions in niobium act as mobile charges we would expect to be able to move hydrogen from the specimen into one or more of its niobium contact pads with a large applied electric field. Indeed, this property has been used in developing resistometric techniques for studying hydrogen diffusion in metals [18]. A second specimen on substrate N6 (otherwise identical to the one above) had a large sixth contact connected to its center for this purpose. Its noise after ion-milling the contacts is plotted as Fig.2(c), and below 1 Hz, is identical to Fig.2(b). To drain the hydrogen its sixth contact was held at a -1 volt potential with respect to the other five contacts. After several days the specimen resistance stabilized 17% lower than it had started. The specimen was then severed from the sixth contact and its noise measured, shown as Fig.2(d). If the sixth "sink" contact is not severed the noise and resistance gradually recover toward their initial values.

Specimens on substrate N3 were chem-etched and their gold contact film deposited over the niobium contact pads after first ion-milling the niobium surface to sputter off the oxide layer. The noise of one specimen was measured both before and after heating it to 400 °C for 4 hours in a mid-10^{-8} Torr vacuum, a common method for removing hydrogen from bulk niobium. Both noise spectra looked similar to Fig.2(d), but the noise after heat treatment was a factor of 40 lower than before. The specimen resistance no longer changed with application of a DC field (as it had before).

DISCUSSION AND CONCLUSIONS

So far we have not been able to observe any 1/f noise that may be traced to the niobium specimens above the level of the diffusion noise. This places limits on specimen 1/f noise of $\alpha \lesssim 4 \times 10^{-5}$ for specimen N6-1 (Fig.1(d)), $\alpha \lesssim 2 \times 10^{-4}$ for specimen N7-4 (Fig.2(a)), and $\alpha \lesssim 3 \times 10^{-5}$ for specimen N3-3 (not shown), the last measured with continuing DC bias after being exposed to an electric field for several days. Under these conditions this specimen has a resistivity of 18 μΩ·cm so that α is at least a factor of 45 below that given by Hooge's formula [19] and a factor of 10 lower than the modified formula of Fleetwood and Giordano [7].

The noise spectra of Fig.1 imply diffusion noise associated with an entity diffusing with $D = (2{\pm}1) \times 10^{-6}\ cm^2s^{-1}$, comparable with the known values for hydrogen in niobium. The effect is enhanced by chem-etching and sputter-etching and reduced by vacuum annealing and electrodiffusion as expected for hydrogen. However, on calculating the hydrogen concentrations from the diffusion noise amplitudes and resistance changes, the expected solubility of H in α-Nb is substantially exceeded in some specimens. Nevertheless, consistency of the spectra with the theory of noise due to diffusion in solution implies that supersaturated solutions are retained. Hydrogen transport would be drastically altered in coexisting α- and β-phases so that Eq.(4) would not be applicable.

It is quite plausible, however, that there exists a supersaturated solution of α-phase hydrogen. This hypothesis requires furthur study.

Apparent hydrogen concentrations c_0 may be obtained by fitting Eq.(4) to the measured noise spectra using observed corner frequencies and $\beta = \beta_{\alpha-H} = 0.65$ μΩ·cm/at.% [20]. Measurements of resistivity changes also provide measures of c_0. The noise spectrum and resistivity for Fig.1(c) give $c_0 = (0.9\pm0.6)$ at.% and $c = (3\pm1)$ at.% and for Fig.1(d) give $c_0 = (1\pm1)$ at.% and $c_0 = (0\pm1)$ at.% respectively. The values are well within the solubility limits of 5 at.% at room temperature [16]. However, the noise spectra and resistivity changes of the more heavily charged specimens give hydrogen concentrations well above the 5 at.% solubility limit: $c_0 \simeq 40$ at.% and $c_0 \simeq 70$ at.% for Fig.1(a), and $c_0 \simeq (5\pm2)$ at.% and $c_0 \simeq 20$ at.% for Fig.1(b) respectively, if the solubility limit is ignored and the value of $\beta_{\alpha-H}$ is extrapolated. The data on ion-milled specimens also suggest high hydrogen concentrations; the noise spectrum of Fig.2(b) gives $c_0 = (40\pm20)$ at.% and the 15% resistance increase upon milling the contact surfaces gives $c_0 \simeq 7$ at.%. The noise level of specimen N3-3 gave $c_0 = (10\pm6)$ at.% while its 17% resistance decrease with a DC field indicates $c \simeq 5$ at.%. Note that concentrations deduced from diffusion noise amplitudes and resistance changes give better agreement for the lower hydrogen concentrations.

In ion-milling it is plausible that hydrogen in the background would enter the niobium film as its protective oxide layer is sputtered away. Free hydrogen might be made available from residual water vapor broken up as its oxygen reacts with the metal to form niobium-oxide. The reaction might take place spontaneously, or be aided by either the 500 eV Ar^+ ion or e^- neutralizing beams. The same process must occur during ion-milling to pattern the film but to a lesser degree. In this case hydrogen enters the specimen through the edges by diffusing under the photoresist from the unprotected film that is being sputtered away.

We conclude that subtractive patterning of the niobium films either by chem-etching or sputter-etching introduces hydrogen impurities and noise associated with its diffusion. Efforts are now underway to avoid hydrogen entirely by forming specimens using additive patterning (i.e. lift-off).

We thank B. D. Hunt for providing us with many of the niobium films, and both he and C. T. Rogers for assistance with sample preparation. Our thanks also go to R. A. Burhman, J. V. Mantese, M. Nelkin, and R. F. Voss for their useful discussions. This work was supported by the National Science Foundation and benefited from facilities of NRRFSS and the Materials Science Center at Cornell University.

REFERENCES

[1] F. N. Hooge and A. M. H. Hoppenbrouwers, Physica 45, 386 (1969).

[2] Richard F. Voss and John Clarke, Phys. Rev. B 13, 556 (1976).

[3] J. W. Eberhard and P. M. Horn, Phys. Rev B 18, 6681 (1978).

[4] J. H. Scofield, D. H. Darling, and W. W. Webb, Phys. Rev. B 24, 7450 (1981).

[5] R. D. Black and M. B. Weissman, Phys. Rev. B 24, 7454 (1981).

[6] D. M. Fleetwood and N. Giordano, Phys. Rev. B 25, 1427 (1982).

[7] D. M. Fleetwood and N. Giordano, Phys. Rev. B 27, 667 (1983).

[8] D. M. Fleetwood, J. T. Masden, and N. Giordano, Phys. Rev. Lett. 50, 450 (1983).

[9] $S_V(\omega)$ is the Fourier transform of the voltage autocorrelation function, defined for positive and negative angular frequencies $\omega=2\pi f$ while the measured power spectral density $G_V(f)$ is for positive natural frequencies only. The two are related by $G_V(f) = 4\pi\ S_V(2\pi f)$.

[10] see for instance, R. Rzemien, C. K. Iddings, and W. F. Love, Rev. Sci. Instr. 50, 488 (1979), and references therein.

[11] unpublished.

[12] Mark B. Ketchen and John Clarke, Phys. Rev. B 17, 114 (1978).

[13] K. M. van Vliet and J. R. Fasset, in _Fluctuation Phenomena in Solids_, edited by R. E. Burgess (Academic, New York, 1965) pp. 267-354

[14] Douglas Magde, Elliot Elson, and W. W. Webb, Phys. Rev. Lett. 29, 705 (1972).

[15] J. Volkl and G. Alefeld, in _Hydrogen in Metals I, Topics in Applied Physics_ v.28, edited by G. Alefeld and J. Volkl (Springer-Verlag, Berlin, Heidelberg, New York, 1978) pp. 321-348.

[16] T. Schober and H. Wenzle, in _Hydrogen in Metals II, Topics in Applied Physics_ v.28, edited by G. Alefeld and J. Volkl (Springer-Verlag, Berlin, Heidelberg, New York, 1978) pp. pp. 10-71.

[17] The noise of a another sputter-etched specimen with soldered indium contacts (N8-2) looked similar to Fig.2(a). Two noise measurements with different ballast resistors allowed us to decompose the spectrum into specimen diffusion noise and 1/f noise from the contacts.

[18] see for instance David J. Pines, Ph.D. thesis, Cornell University, 1982.

[19] F. N. Hooge and L. J. K. Vandamme, Physics Letter 66A, 315 (1978).

[20] D. G. Westlake, Trans. Metall. Soc. AIME 245, 287 (1969).

NOISE IN PHYSICAL SYSTEMS AND 1/f NOISE
M. Savelli, G. Lecoy and J-P. Nougier (eds.)
© *Elsevier Science Publishers B.V., 1983*

SYSTEMATIC ERRORS INTRODUCED BY QUANTISATION IN THE PRECISE MEASUREMENT OF NOISE AMPLITUDE BY DIGITAL CROSS-CORRELATION

C.P. Pickup
CSIRO Division of Applied Physics
Sydney, Australia 2070

For precise noise thermometry the effects of unwanted amplifier noise can be overcome by using two independent amplifiers with their outputs multiplied together or cross-correlated. This may be achieved by sampling and digitising the amplifier outputs followed by digital multiplication and averaging. It is shown that the presence of un-correlated noise enables relatively coarse quantisation, with accurately defined levels, to be used without causing serious systematic errors. Six bit conversion should be adequate for any noise thermometer.

1. INTRODUCTION

At present the most precise system for noise thermometry measures the ratio of noise signals from a single resistance at different temperatures using the correlation amplifier technique. The noise to be measured is applied in parallel to the inputs of two independent amplifiers whose outputs are multiplied together or cross-correlated. Unwanted noise signals originating in the amplifiers are un-correlated and, ideally, the averaged output of the correlator is proportional to the mean square of the input noise signal.

Realisation of a sufficiently accurate multiplier presents considerable practical difficulties. Analogue multiplier circuits have limitations with respect to both linearity and bandwidth; however, by introducing accurately calibrated attenuators into the amplifier channels when the larger noise source is being measured, the loading on the multiplier may be kept approximately constant. Using this arrangement a precision of 100 ppm has been obtained[1] however the ratio of correlated to un-correlated signals must change and it appears that better multiplier systems are required.

One possibility is to sample and digitise the outputs of the amplifier channels and perform the multiplication using high speed digital circuits. The statistical properties of noise signals suggest that it may be possible to retain sufficient precision whilst using only a relatively small number of accurately defined digitisation levels, thus significantly easing the design problems of the analogue-to-digital converter and multiplier circuits [2]

2. NUMERICAL ESTIMATION OF QUANTISATION ERRORS

Consider the analogue-to-digital converters to have $\pm K$ levels each of width Δ and to provide output equal to the mid-value of each interval for a sample falling within it except that all samples falling outside the largest levels return the same value.

At the digitiser the noise signal to be measured has RMS value x and it is assumed for convenience that the un-correlated amplifier noise has the same RMS value σ in each channel. Then, if K is not too large, it is possible to compute the expected value of the output as the sum of each possible product multiplied by its probability of occurrence given by the bi-variate normal distribution with variance $(x^2 + \sigma^2)$ and correlation coefficient $x^2/(x^2 + \sigma^2)$.

For a typical converter spanning ± 8 V with a total of 32 levels (5 bits, Δ = 0.5 V) the following results are obtained for a range of input amplitudes and amplifier noise.

TABLE 1
(Quantisation error of ideal 5-bit converter)

x	$\bar{x}^2$	σ = 0.125	σ = 0.2	σ = 0.3
0.6	0.36	0.36107423	0.36002288	0.36000001
0.8	0.64	0.64107423	0.64002288	0.64000001
1.0	1.00	1.00107423	1.00002288	1.00000001
1.2	1.44	1.44107423	1.44002288	1.44000001
1.4	1.96	1.96107412	1.96002274	1.95999980
1.6	2.56	2.56106809	2.56001596	2.55999125

At the two highest input levels the effect of peak clipping is evident, but otherwise it is seen that for each value of the un-correlated component σ the systematic error in the estimate of the mean square is constant, independent of the input x. This error decreases rapidly as the un-correlated component is increased, however normal designs minimise σ to avoid further increasing the already lengthy averaging time required for high precision.

In a practical precision noise thermometer using such a converter with low noise main amplifiers the value of σ could be as low as 0.2 V and in this case, to ensure negligible linearity errors, it would be preferable to use a 6 bit design having a total of 64 levels each of 0.25 V. With 6 bit conversion the multiplier can be very conveniently replaced by simple read only memories such as the 2732 EPROM which have 12 address lines.

3. EFFECTS OF IRREGULARITIES IN THE LEVELS

The computation outlined above may be performed for arbitrary irregularities in the boundary voltages separating the levels, and the errors arising from some of the imperfections other than quantisation effects which are likely to occur in a practical converter can be estimated.

For the relatively small number of levels being considered here it is quite possible to envisage a precise form of the "flash" converter in which a set of levels is established by a chain of nominally equal resistances and a separate comparator is provided for each level. This arrangement offers very high speed and also obviates the necessity for a sample and hold amplifier. Assuming ideal performance from the comparators the errors to be expected for several sets of uniformly distributed random errors in the divider chain were investigated. The results were rather difficult to characterise but the general conclusion was that the resistances forming the divider would all need to have precision comparable to that expected at the output of the correlator.

The more conventional system for accurate high-speed A/D conversion uses a sample and hold amplifier followed by a successive approximation converter. The model which was investigated comprised five precision current sources, nominally in binary ratios, switched by means of steering diodes together with a permanently applied negative off-set current, nominally equal to the MSB, to provide bi-polar measuring capability. The levels of the converter are thus composed of sums of the basic currents whose individual errors introduce a symmetric pattern of errors in the levels.

For example, consider a converter of the same range as in 2. above which has values for the current sources, expressed as equivalent input voltages, as follows. The second most significant source, in each converter, has an error of +10 ppm.

Converter-1	Converter-2
0.5	0.5
1.0	1.0
2.0	2.0
4.00004	4.00004 (+10 ppm)
8.0	8.0
-8.0	-8.0 (off-set)

As this is a 5 bit converter the un-correlated noise component is taken large enough (σ=0.3V) to make the quantisation errors negligible. The following results are obtained

TABLE 2
(+10 ppm error in second source)

x	x^{-2}	Output	Error (ppm)
0.6	0.36	0.36001632	+45
0.8	0.64	0.64002322	+36
1.0	1.00	1.00002994	+30
1.2	1.44	1.44003620	+25
1.4	1.96	1.96004116	+21
1.6	2.56	2.56003595	+14

For a similar +10 ppm error in the largest source with the others having their nominal values the errors are even larger, but of opposite sign.

TABLE 3
(+10 ppm error in MSB)

x	x^{-2}	Output	Error (ppm)
0.6	0.36	0.35996739	-90
0.8	0.64	0.63995395	-72
1.0	1.00	0.99994005	-60
1.2	1.44	1.43992672	-51
1.4	1.96	1.95991332	-44
1.6	2.56	2.55989167	-42

For a cryogenic noise thermometer the effects of these errors are not as serious as they might appear. If x = 1.6 V corresponds to the reference temperature 273.16 K then the absolute errors when measuring lower temperatures are not worse than 2.2 mK for either of the cases shown above.

Errors in the less significant current sources are much less important. Similarly errors in the negative off-set sources, which are also equivalent to DC off-sets in the output of the sample and hold amplifier, are not particularly serious. The general conclusion is that in order to ensure low overall linearity errors in the noise thermometer the more significant current sources of the converters should be in accurate binary ratios.

Other possible arrangements include a sign-magnitude structure for the converter which allows the elimination of both the most significant current source and the negative off-set source. Computations with this model did suggest that the resulting symmetry would improve the overall linearity somewhat when errors were present in the remaining current sources. This symmetry would be difficult to maintain in a practical system due to off-set errors in the absolute value amplifier which is necessary with this arrangement. These errors produce particularly serious non-linearity as shown by the following table for the case of 1 mV shift of all the negative levels relative to the positive.

TABLE 4
(Off-set between positive and negative levels)

x	$\bar{x}^2$	Output	Error (ppm)
0.6	0.36	0.36040577	+1127
0.8	0.64	0.64057754	+902
1.0	1.00	1.00074593	+746
1.2	1.44	1.44091180	+633
1.4	1.96	1.96107575	+549
1.6	2.56	2.56123016	+480

4. WEIGHTING OF EXTREME INTERVALS

Table 1 above shows peak clipping errors of several ppm at 1.6 V input which arise because samples exceeding the range of the converter are given a constant value. These larger input peaks are highly correlated and a significant improvement in the overall linear range can be obtained by simply increasing the value of the product corresponding to both inputs being in or above the largest level by approximately 10%. This is readily done if the products are all stored in ROM, and for the case considered above provides better than 1 ppm linearity to beyond 1.8 V RMS input.

5. GENERAL FORMULA FOR QUANTISATION ERRORS

The computations described above are rather expensive in terms of CPU time and it has been possible to obtain a formula which describes the quantisation error with great accuracy for the basic case of uniformly spaced levels in the absence of peak clipping.

Figure 1 represents a simple quantised measuring instrument which outputs the mean value for an interval for any input lying within it whose true value is defined by the parameter α as shown.

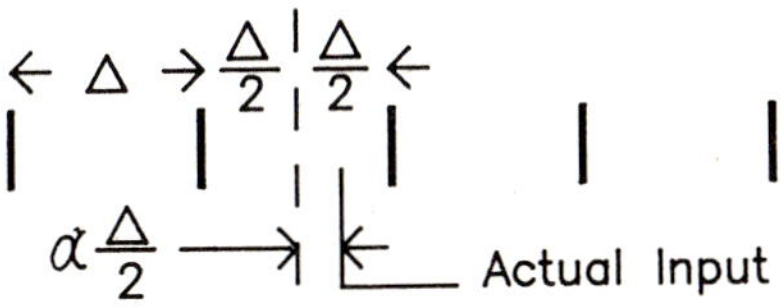

Figure 1 : Quantised measuring instrument

If a random disturbance with Gaussian distribution, zero mean, and RMS value σ is added to the input some output samples will come from adjacent levels and the averaged output will provide an estimate α' which approximates the true α. From Butterworth, et al[2]

$$\alpha' = \sum_{r=0}^{\infty} [\operatorname{erfc} \{\frac{(2r+1-\alpha)}{2\sqrt{2}} \frac{\Delta}{\sigma}\} - \operatorname{erfc} \{ \frac{(2r+1+\alpha)}{2\sqrt{2}} \frac{\Delta}{\sigma} \}] \qquad (1)$$

This relation was investigated numerically for a realistic range of σ/Δ and the error $\alpha'-\alpha$ was found to be quite accurately sinusoidal with peak amplitude decreasing rapidly as σ/Δ increases. By polynomial fitting to the logarithms of these amplitudes the following relation was obtained which represents (1) to better than 8 decimal digits for σ/Δ greater than 0.1.

$$\alpha' = \alpha - \frac{2}{\pi} \exp(-2\pi^2\sigma^2/\Delta^2) \sin \pi\alpha \qquad (2)$$

Thus the actual error of each converter is sinusoidal with peak value

$$\frac{\Delta}{\pi} \exp(-2\pi^2\sigma^2/\Delta^2) \qquad (3)$$

and the error in the final product of the outputs of the two converters is the mean square value

$$\frac{\Delta^2}{2\pi^2} \exp(-4\pi^2\sigma^2/\Delta^2) \qquad (4)$$

6. CONCLUSIONS

It appears that relatively coarse quantisation (6 bits) with accurately spaced levels should be satisfactory for the most precise noise thermometers employing digital cross-correlation. This should save time in the A/D conversion process allowing a greater sampling rate.

7. ACKNOWLEDGEMENT

I would like to thank Mark Berman of the CSIRO Division of Mathematics and Statistics for deriving the theory on which the table in 2. above is based.

REFERENCES:

[1] Klein H. H., Klempt G., and Storm L., Measurement of the thermodynamic temperature of He at various vapour pressures by a noise thermometer, Metrologia 15 (1979) 143-154.

[2] Digitalization of signals and the errors introduced has been studied in other contexts (see, for example, review article from IEEE). The content of this article is to indicate the specific influence for the specialized topic of noise thermometry.

[3] Butterworth J., MacLaughlin D.E., and Moss B.C., The use of random noise to improve resolution in analogue-to-digital conversion, J. Sci. Instrum. 44 (1967) 1029-1030.

CLOSING ADDRESS

NOISE IN PHYSICAL SYSTEMS AND 1/f NOISE
M. Savelli, G. Lecoy and J-P. Nougier (eds.)
© *Elsevier Science Publishers B.V., 1983*

Closing Address of the Joint Noise Conferences

Aldert van der Ziel

Electrical Engineering Department, University of Minnesota

Minneapolis, MN 55455, U.S.A.

I am sure that I speak for all the participants of the conferences when I express sincere thanks to Professors Savelli, Lecoy and Nougier for excellently organized and run conferences.

We learned with regret that the representatives of the USSR could not be with us. We hope that they will be able to attend future conferences, for we need interaction with them and they need interaction with us.

Much progress was made in various research areas. Clarke's paper on noise in Josephson junctions indicated how much understanding has been gained for these complicated devices. A new topic was the study of chaotic states; it is expected that it will be of great interest at future conferences. Dr. Voss gave his usual stimulating contribution. Dr. van Vliet gave an important paper on noise in non-equilibrium situations; this topic will also remain of interest at future conferences. Dr. Nougier gave an excellent paper on submicron devices, another topic of great current and future interest. Dr. Weissman's critical comments contributed much to the conferences, and his elegant experiments were received with interest.

Much progress was made in 1/f noise in MOSFETs. At low injection there is a plateau in the noise resistance that is independent of bias (Duh et al, Reimbold et al). Reimbold et al presented a theory of 1/f noise in MOSFETs, with only one adjustable parameter, that covers the whole bias range from very weak to very strong inversion. Maes's experiments on MOS silicon memory transistors open up the possibility of understanding and eliminating the deterioration with time of erasable proms. Blasquez's experiments clearly identified the 1/f noise in MOSFETs as number fluctuation noise.

The controversy of number fluctuation 1/f noise versus mobility fluctuation 1/f noise has subsided. In all fairness it should be admitted that initially a strong emphasis on mobility fluctuation 1/f noise was needed in order that that point of view received its rightful place. Earlier conferences suffered from the "true believer syndrome" in which there was only place for a single point of view. The dependence of the Hooge parameter α on the absolute temperature T and on the field F clearly indicate the presence of mobility fluctuation 1/f noise (vanDamme, this conference), whereas Zheng's sputtering experiment can decide which model is correct in individual cases (Physica B, 1983; this conference). Dr. Weissman made the interesting suggestion that number fluctuation noise can give rise to mobility fluctuation features; this unifies both points of view.

The true believer syndrome is still much alive with respect to Handel's theory of quantum 1/f noise; there are strong believers and equally strong non-believers. To the latter I suggest caution, since one never knows which theories may be needed in the future. To the former I suggest interaction with experimentalists, to devise approaches by which the theory can be checked. We have set up such an experimental program at the University of Florida in order to test the theory quantitatively, rather than giving a qualitative explanation of the experiments in terms of Handel's theory. This has led to interesting preliminary results.

1. Dr. Gong's measurement of the statistics of α-particle emission shows the presence of 1/f noise which can perhaps be quantitatively explained by Handel's theory (this conference).
2. Mr. Kilmer's measurement of the modified Hooge parameter in gold films at low temperature might perhaps be quantitatively explained by Handel's theory (this conference).
3. Mr. Hsieh (to be published) has measured partition 1/f noise in pentodes in which the cathode 1/f noise had been strongly reduced by feedback. He determined the dependence of the partition 1/f noise on the anode voltage and found a result in conflict with the elementary theory and with Handel's initial theory. But from the data, and guided by Handel's theory, we were able to guess an empirical formula that can describe the data. This formula must now be proved by modifying Handel's initial theory.

It was good to see so many old-timers from previous conferences; they provide the necessary continuity. It was equally gratifying to see so many new faces; the conferences need a constant supply of new researchers to keep the field alive and to provide an input of new ideas. It might be contemplated to relax the acceptance criteria for papers by some newcomers; there is no better way to generate enthusiasm than to be allowed to give a paper at an important conference such as ours.

There were no papers presented on SIS(superconductor-insulator-superconductor) microwave mixers. Most of us are probably unaware of this important cryogenic device that gives a low noise temperature combined with a power gain.

Moreover, those working in that area are not familiar with our conferences. It is to be hoped that at the next conference Prof. Richards of the University of California at Berkeley can give an invited paper on this important topic.

DETAILED PROGRAM

TUESDAY MORNING, ROOM A

Chairman : M. SAVELLI (Univ. Montpellier, France)

8.30 - 8.50 Opening addresses

8.50 - 9.35 C.M. Van Vliet (Invited Paper) (Univ. Florida, USA, and Univ. Montréal, Canada) Fluctuations around a non-equilibrium state.

9.35 - 10.20 A. Libchaber (Invited Paper) (Ecole Norm. Sup., Paris, France) Noise in chaotic fluid systems

THEORY

10.20 - 10.40 C.M. Van Vliet (Univ. Florida, USA, and Univ. Montréal, Canada), P.H. Handel (Univ. Missouri, USA) A new transform theorem linking spectral density and Allan variance

10.40 - 11.00 M. Sikulova, J. Sikula, P. Vasina, V. Nevecny (Tech. Univ. Brno, Czechoslovakia) Stochastic model of the burst noise

11.00 - 11.30 Coffee break

Chairman : A. VAN DER ZIEL (Univ. Minnesota, USA, and Univ. Florida, USA)

THEORY

11.30 - 11.50 J.P. Nougier, M. Gontrand, JC Vaissière (Univ. Montpellier, France) Microscopic spatial correlations

11.50 - 12.10 B. Boittiaux, E. Constant, A. Ghis (Univ. Lille, France) Simulation of diffusion noise in a device

12.10 - 12.30 C. Moglestue (GEC, Wembley, UK) Monte-Carlo particle modelling of noise in semiconductors

12.30 - 12.50 H. Grabert (Stuttgart Univ., FRG), P. Talkner (Basel Univ. Switzerland) Master equation for quantum Brownian movement

13.00 - 14.30 Lunch

TUESDAY MORNING, ROOM B

Chairman: G. BLASQUEZ (L.A.A.S., Toulouse, France)
I/f NOISE IN BIPOLAR TRANSISTORS

10.20 - 10.40 K.H. Duh, A. van der Ziel (Univ. Minnesota, USA) I/f noise in modulation - doped field effect transistors

10.40 - 11.00 M. Mihaila (I.C.C.E., Bucharest, Romania) Atomic impurities related low-frequency noise in bipolar transistors

11.00 - 11.30 Coffee break

Chairman : A. AMBROZY (Techn. Univ. Budapest, Hungary)

I/f NOISE IN BIPOLAR TRANSISTORS

11.30 - 11.50 C.T. Green, B.K. Jones (Univ. Lancaster, U.K.) I/f noise in bipolar transistors

11.50 - 12.10 J. Kilmer, A. van der Ziel, G. Bosman (Univ. Florida, USA) Mobility-fluctuation I/f noise identified in silicon P^+NP transistors

12.10 - 12.30 G. Blasquez, D. Sauvage (L.A.A.S., Toulouse, France) I/f current noise in the bulk of short diodes and bipolar transistors

12.30 - 12.50 M. Mihaila, K. Amberiadis, A. van der Ziel (Univ. Minnesota, USA) Low frequency noise due to emitter-edge dislocations in npn transistors

13.00 - 14.30 Lunch

TUESDAY AFTERNOON, ROOM A

Chairman: C.M. VAN VLIET (Univ. Florida, USA, and Univ. Montréal, Canada)

14.30 - 15.15 RF Voos, B. Mandelbrot (Invited Paper) (IBM, Yorktown Heights, USA) Why is nature fractal and when should noises be scaling ?

15.15 - 16.00 R. Lefever (Invited Paper) (Univ. Bruxelles, Belgium) Transition phonomena induced by multiplicative noise

16.00 - 16.30 Coffee break

Chairman : C. HEIDEN (Univ. Giessen, FRG)

THEORY

16.30 - 16.50 T. Munakata, D. Wolf (Frankfurt Univ. FRG) On the distribution of the level-crossing time-intervals of Random processes

16.50 - 17.10 A.M-S. Tremblay, F. Vidal (Sherbrooke Univ., Canada) Fluctuations in dissipative steady-states of thin metallic films

17.10 - 17.30 W. Michel, H. Grabert (Stuttgart Univ., FRG) Calculation of vacancy noise in monocrystalline metals

17.30 - 17.50 P. Talkner, D. Ryter (Inst. f. Physik, Basel, Switzerland) Lifetime of a metastable state at a weak noise

17.50 - 18.10 G. Graham, A. Schenzle (Essen Univ., F.R.G.) Stabilization by multiplicative noise

TUESDAY AFTERNOON, ROOM B

Chairman : B.K. JONES (Univ. Lancaster, UK)

I/f NOISE IN FET TRANSISTORS

16.30 - 16.50 K. Kandiah (AERE, Harwell, UK) Energy levels of bulk defects responsible for L.F. noise in Si JFETs

16.50 - 17.10 K.H. Duh, A. van der Ziel (Univ. Minnesota, USA) Thermal noise at weak inversion and limiting I/f noise in MOSFETs

17.10 - 17.30 G. Reimbold, P. Gentil, A. Chovet (ENSERG, Grenoble, France) I/f noise in MOS transistors biased from weak to strong inversion

17.30 - 17.50 H.E. Maes, S.H. Usmani (Kath. Univ. Leuven, Belgium) I/f noise in metal-nitride-oxide-silicon (MNOS) memory transistors

17.50 - 18.10 G. Blasquez, A. Boukabache (L.A.A.S., Toulouse, France) Origins of f^{-1} noise in MOS transistors

WEDNESDAY MORNING, ROOM A

Chairman : P.H. HANDEL (Univ. Missouri, USA)

8.30 - 9.15 J. Clarke (Invited Paper), R.H. Koch, J. Martinis, R.F. Miracky (Univ. Californie, USA) Chaotic noise in Josephson tunnel junctions.

9.15 - 10.00 M. Suzuki (Invited paper) (Univ. Tokyo, Japan) Long time tail relaxation noise

THEORY I/f NOISE

10.00 - 10.20 B. Pellegrini (Univ. Pisa, Italy) Flicker noise in non-electric systems

10.20 - 10.40 Y. Isawa (Tohoku Univ., Sendai, Japan) Theory of I/f noise in metals

10.40 - 11.00 H. Sato (Matsushita El. Co., Osaka, Japan) Unified model for I/f noise in semiconductors and metals

11.00 - 11.30 Coffee break

Chairman : D. WOLF (Univ. Frankfurt, FRG)

THEORY I/f NOISE

11.30 - 11.50 P.H. Handel (Univ. Missouri, and McDonnel Douglas Res. Lab., St Louis, USA), C.M. Van Vliet (Univ. Montreal, Canada, and Univ. Florida, USA), A. Van der Ziel (Univ. Minnesota and Univ. Florida, USA) Derivation of the Nyquist - I/f noise theorem

11.50 - 12.10 P.H. Handel (Univ. Missouri, and McDonnel Douglas Res. Lab., St Louis, USA) Any particle represented by a coherent state exhibits I/f noise

12.10 - 12.30 P.H. Handel (Univ. Missouri, and McDonnel Douglas Res. Lab., St Louis, USA), T. Musha (Tokyo Inst. Techn. Japan) Quantum I/f noise from piezoelectric coupling

12.30 - 12.50 S.M. Kogan, K.E. Nagaev (Inst. Radio Engg and Elect., Moscow, USSR) Low-frequency current noise and internal friction in solids

13.00 - 14.30 Lunch

WEDNESDAY MORNING, ROOM B

Chairman : J. CLARKE (Univ. California, USA)

QUANTUM NOISE

10.00 - 10.20 J.E. Mutton, R.J. Prance, T.D. Clark (Univ. Sussex, UK) A. Widom (Univ. Boston, USA) Influence of noise on the voltage versus current characteristics of AC-biased SQUID magnetometers

10.20 - 10.40 H. Seppa (Techn. Res. Cent. Helsinki, Finland), R.J. Soulen Jr (NBS, Washington, USA) Intrinsic and extrinsic noise sources in an R-SQUID

10.40 - 11.00 C.D. Tesche (IBM, Yorktown Heights, USA) Quantum noise limit constraints on the performance of DC SQUID linear amplifiers

11.00 - 11.20 V.V. Danilov, L.S. Kuzmin, K.K. Likharev, V.V. Migulin, A.B. Zorin (Moscow State Univ., USSR) Quantum, thermal and shot noise in Josephson junction parametric amplifiers and SQUIDs

11.20 - 11.30 Coffee break

Chairman : R.J. SOULEN (N.B.S. Washington, USA)

QUANTUM NOISE

11.30 - 11.50 H. Dirks, R. Dittrich, C. Heiden (Univ. Giessen, FRG) Vortex density fluctuation during flux flow in a type II superconductor

11.50 - 12.10 R.J. Prance, J.E. Mutton, H. Prance, T.D. Clark (Univ. Sussex, UK), A. Widom, G. Megaloudis (Univ. Boston, USA) Noise spectroscopy of a weak link constriction ring

12.10 - 12.30 A.N. Vystavkin, V.N. Gubankov, V.P. Koshelets, M.A. Tarasov (Inst. Radio Engg and Elect., Moscow, USSR) Noise increase due to fluctuations conversion in superconducting weak links

12.30 - 12.50 V.K. Kornev, V.K. Semenov (Moscow State Univ., USSR) Chaotic phenomena in superconducting quantum interferometers

12.50 - 13.10 Z.G. Ivanov, A.Y. Spassov, B.N. Todorov (Inst. Electron. Sofia, Bulgaria) Anomalous high noise in Josephson junction

13.10 - 13.30 A.N. Vystavkin, V.N. Gubankov, M.A. Tarasov (Inst. Rad. and Elec., Moscow, USSR) Shot noise and it's correlation with Josephson point contacts parameters and heating

13.30 - 14.30 Lunch

WEDNESDAY AFTERNOON, ROOM A

Chairman : L. DE FELICE (Emory Univ., Atlanta, USA)

14.30 - 15.15 D. Poussart (Invited Paper) (Univ. Laval, Québec, Canada) I/f noise in biologival membranes

METROLOGY

15.20 - 15.40 J.H. Scofield, W.W. Webb (Cornell Univ., USA) Observations of low frequency current noise in Niobium films - electrodiffusion noise and the absence of I/f noise

15.40 - 16.00 C.P. Pickup (CSIRO, Sydney, Australia) Systematic errors introduced by quantisation in the precise measurement of noise amplitude by digital cross-correlation

16.00 - 16.30 Coffee break

Chairman : H. SUTCLIFFE (Univ. Galway, Ireland)

NOISE IN DIODES AND TRANSISTORS

16.30 - 16.50 E. Calandra, G. Martines, M. Sannino (Univ. Palermo, Italy) Determination of microwave noise and gain parameters of C-band GaAs MESFET through noise figure measurements only

16.50 - 17.10 T.S. Nashashibi, M.A. Carter, S. Taylor (Thorn Emi, Hayes, UK) Generation - Recombination noise in SiJFETs

17.10 - 17.30 J.D. Stocker, B.K. Jones (Univ. Lancaster, UK) JFET gate-current noise

17.30 - 17.50 E.W. Wu, A. van der Ziel (Univ. Minnesota, USA) Noise and input admittance of HEMTs

17.50 - 18.10 D. Lippens, J.L. Nieruchalski, E. Constant (Univ. Lille, France) Microscopic energy conserving simulation of multiplication noise

WEDNESDAY AFTERNOON, ROOM B

Chairman : H. GRABERT (Univ. of Stuttgart, FRG)

THEORY I/f NOISE

15.20 - 15.40 H. Prance, R.J. Prance, T.D. Clark, J.E. Mutton (Univ. Sussex, UK) A. Widom, G. Megaloudis, G. Pancheri, Y. Srivastava (Univ. Boston, USA) Soft photon origin of I/f noise in electrical circuits

15.40 - 16.00 P.H. Handel, T.S. Sherif (Univ. Missouri, USA) Direct calculation of the Schrödinger field which generates quantum I/f noise

16.00 - 16.30 Coffee break

Chairman : H. THOMAS (Univ. Basel, Switzerland)

I/f NOISE IN PHYSICAL AND CHEMICAL SYSTEMS

16.30 - 16.50 R.H. Koch (IBM, Yorktown Heights, USA) Measurements of I/f noise in Josephson junctions

16.50 - 17.10 J. Gong, C.M. Van Vliet, W.H. Ellis, G. Bosman, P.H. Handel (Univ. Florida, USA) I/f noise fluctuations in α-particle radio active decay of 95 Americium 241

17.10 - 17.30 C. Gabrielli, F. Huet, M. Keddam (Univ. P. et M. Curie, Paris, France) Noise in electrochemical systems

17.30 - 17.50 T. Musha, K. Sugita, M. Kaneko (Tokyo Inst. Techn., Yokohama, Japan) I/f noise in aqueous $CuSO_4$ solution

17.50 - 18.10 J. de Goede, N. Roos, A. de Vos, R.J. van den Berg (Leiden Univ. The Netherlands) Electrical resistivity fluctuations in KCl solutions

THURSDAY MORNING, ROOM A

Chairman : T. MUSHA (Tokyo Inst. Techn., Japan)

8.30 - 9.15 J.J. Gagnepain (Invited Paper) (L.P.M.O., Besançon, France) Phase and frequency noises in oscillators

9.15 - 10.00 L.K.J. Vandamme (Invited Paper) (Eindhoven Univ. Techn., The Netherlands) Is the l/f noise parameter α a constant ?

I/f NOISE IN RESISTORS AND THIN FILMS

10.00 - 10.20 A. van Calster, L. van den Eede, S. de Molder, A. de Keyser (Ghent Univ., Belgium) l/f noise in cermet and metanet resistors

10.20 - 10.40 K. Zheng, K. Duh, A. van der Ziel (Univ. Minnesota, USA) Removal of l/f noise in HgCdTe by sputtering

10.40 - 11.00 M.B. Weissman, R.D. Black, P.J. Restle (Univ. Illinois, Urbana, USA) l/f noise in Silicon

11.00 - 11.30 Coffee break

Chairman : F.N. HOOGE (Eindhoven Univ. Techn., The Netherlands)

I/f NOISE IN RESISTORS AND THIN FILMS

11.30 - 11.50 D.M. Fleetwood, N. Giordano (Purdue Univ., West Lafayette, USA) Resistivity dependance of l/f noise in metal films

11.50 - 12.10 J. Kilmer, G. Bosman, C.M. Van Vliet, A. Van der Ziel (Univ. Florida, USA) l/f noise in metal films of submicron dimensions

12.10 - 12.30 R.S. Lear, T.M. Tritt, M.J. Skove, E.P. Stillwell (Clemson Univ., USA) l/f noise in the Trichalcogenides $NbSe_3$ and TaS_3

12.30 - 12.50 A. Kusy (Techn. Univ., Rzeszow, Poland) l/f noise in ZnO varistors

12.50 - 13.10 S. Hashiguchi (Yamanashi Univ., Kofu, Japan) Experimental estimation of the temperature fluctuation in a current-carrying metal microbridge

13.10 - 14.30 Lunch

THURSDAY MORNING, ROOM B

Chairman : G. SALMER (Univ. of Lille, France)

OSCILLATORS

11.30 - 11.50 M. Olivier, J.J. Gagnepain (LPMO, Besançon, France) Chaotic states and anormalous noise in resonators

11.50 - 12.10 H.R. Bilger (Oklahoma Univ. USA) Noise phenomena in Ringlasers

12.10 - 12.30 J. Graffeuil (LAAS, Toulouse, France) A. Bert, M. Camiade (Thomson CSF, Orsay, France) A. Amana, J.F. Sautereau (LAAS, Toulouse, France) Ultra low noise GaAs MESFET microwave oscillators

12.30 - 12.50 H.B. Chen, A. van der Ziel, K. Amberiadis (Univ. Minnesota, USA) Reduction of the low frequency noise sidebands in oscillators

13.00 - 14.30 Lunch

FRIDAY MORNING, ROOM A

Chairman : R.J.J. ZIJLSTRA (Univ. Utrecht, The Netherlands

8.30 - 9.15 A. Van der Ziel (Invited Paper) (Univ. Minnesota, USA, and Univ. Florida, USA) l/f, diffusion and thermal noise in GaAs devices

I/f NOISE IN DIODES

9.20 - 9.40 J.M. Peransin, M. Abdelali (Univ. Montpellier, France) Behaviour of various Au-InP Schottky diodes under heat-treatment

9.40 - 10.00 C. Hanke (Techn. Univ., München, FRG) Noise properties of bulk-barrier diodes

10.00 - 10.20 T.G.M. Kleinpenning (Univ. Tech., Eindhoven, The Netherlands) On l/f noise in reverse-biased pn-junction diodes

10.20 - 10.40 A. Peczalski, A. van der Ziel (Univ. Minnesota, USA) Diffusion and flicker noise in GaAs current limiters

10.40 - 11.00 B. Koktavy, Z. Chobola, V. Musilova, J. Sikula, P. Vasina (Techn. Univ., Brno, Czechoslovakia) l/f noise in Schottky diodes

11.00 - 11.30 Coffee break

Chairman : H.R. BILGER (Oklahoma Univ., USA)

I/f NOISE IN DIODES

11.30 - 11.50 L.K.J. Vandamme (Eindhoven Univ. Techn., The Netherlands) L.J. van Ruyven (SSL ELCOMA, Philips Res. Lab., Eindhoven, The Netherlands) l/f noise used as a reliability test for diode lasers

11.50 - 12.10 R. Alabedra, B. Orsal, M. Savelli, G. Lecoy (Univ. Montpellier, France) Noise in Ge avalanche photodiode at $\lambda = 1.3\ \mu m$

12.10 - 12.30 J. Kimmerle, W. Kuebart, E. Kuehn, O. Hildebrand (Univ. Stuttgart, FRG), K. Loesch (SEL Res. Cent., Stuttgard, FRG), G. Seitz (Univ. Stuttgart, FRG) Low frequency noise in InGaAs/InP-Photodiodes

12.30 - 12.50 T.G.M. Kleinpenning (Univ. Techn., Eindhoven, The Netherlands)
On current noise limited detectivity D* in photovoltaic (PV) and in photoconductive (PC) detectors

13.00 - 14.30 Lunch

FRIDAY MORNING, ROOM B

Chairman: P. MAZZETTI (Ist. Elett., Torino, Italy)

NOISE IN OTHER PHYSICAL SYSTEMS

9.20 - 9.40 G. Bertotti, F. Fiorillo
(Ist. Elett. Naz., C.N.R., Torino, Italy)
A current noise investigation on stress relaxation mechanisms in thin metal films

9.40 - 10.00 H. Stoll (M.P.I., Stuttgart, FRG)
Vacancy noise in metals

10.00 - 10.20 J.J. Brophy (Utah Univ., USA)
Contact noise in superionic ceramics

10.20 - 10.40 G. Bertotti
(Ist. Elett. Naz., C.N.R., Torino, Italy)
Space-time correlation properties of the magnetization noise and magnetic losses

10.40 - 11.00 J. Sikula, A. Cermakova, P. Vasina, M. Kliment
(Techn. Univ., Brno, Czechoslovakia)
Noise of dielectric materials

11.00 - 11.30 Coffee break

Chairman: J.P. NOUGIER (Univ. of Montpellier, France)

NOISE IN OTHER PHYSICAL SYSTEMS

11.30 - 11.50 J.P. van der Meulen, R.J.J. Zijlstra, D. Frenkel, M. van Dort
(Rijksuniv., Utrecht, The Netherlands)
Noise analysis of laser light scattered by nematic liquid crystals

11.50 - 12.10 H.F.F. Jos, R.J.J. Zilstra, J. Ike
(Rijksuniv., Utrecht, The Netherlands)
Noise in organic semiconductors

13.00 - 14.30 Lunch

FRIDAY AFTERNOON, ROOM A

Chairman: E. CONSTANT (Univ. Lille, France)

14.30 - 15.15 J.P. Nougier (Invited Paper)
(Univ. Montpellier, France)
Noise in submicron devices

HOT CARRIERS AND SUBMICRON DEVICES

15.20 - 15.40 V. Bareikis, A. Galdikas, I. Matulioniene, A.A. Samokhvalov, V.V. Osipov
(Inst. Semic. Phys., Vilnius, USSR)
Microwave electric and magnetic noise in magnetic semiconductors in high electric field

15.40 - 16.00 A. Chatterjee, P. Das (Rensselaer Polytech. Inst., Troy, USA)
Current noise in multivalley semiconductors

16.00 - 16.30 Coffee break

Chairman: G. LECOY (Univ. Montpellier, France)

HOT CARRIERS AND SUBMICRON DEVICES

16.30 - 16.50 D. Gasquet, H. Tijani, J.P. Nougier
(Univ. Montpellier, France)
Generation - recombination noise in n-Si at 77K

16.50 - 17.10 D. Gasquet, M. Fadel, J.P. Nougier
(Univ. Montpellier, France)
Hot carriers noise in n-type InP

17.10 - 17.30 R.R. Schmidt, G. Bosman, C.M. van Vliet, A. van der Ziel (Univ. Florida, USA)
L.F. Eastman, M. Hollis
(Cornell Univ., USA)
Noise in near - ballistic n nn and n pn gallium arsenide submicron diodes

17.30 - 17.50 E. Allamando, G. Salmer, E. Constant
(Univ. Lille, France)
B. Carnez
(Thomson-CSF, Orsay, France)
A new model of submicrometer dual gate MESFET

17.50 - 18.10 A. Van der Ziel, M. Savelli
Closing address

FRIDAY AFTERNOON, ROOM B

Chairman: L.K.J. VANDAMME (Eindhoven Univ. Techn., The Netherlands)

I/f NOISE IN DIODES

15.20 - 15.40 L. Loreck, H. Dambkes, K. Heime
(Duisburg Univ., FRG)
K. Ploog (M.P.I., Stuttgart, FRG)
G. Weimann (Forschungsinst. Bundespost, Darmstadt, FRG)
Low frequency noise in $Al_xGa_{1-x}As$ — GaAs two dimensional electron gas devices and its correlation to deep levels

15.40 - 16.00 A. Chovet, S. Cristoloveanu, A. Mohaghegh, A. Dandache
(ENSERG, Grenoble, France)
I/f noise in P^+NN^+ devices. Influence of a magnetic field

16.00 - 16.30 Coffee break

I/f NOISE IN RESISTORS

16.30 - 16.50 V.V. Potemkin, I.S. Bakshee
(Moscow State Univ., USSR)
I/f noise magnitude in fine-grained Aluminium films

16.50 - 17.10 V.V. Potemkin, M.E. Gertzenstein, I.S. Bakshee
(Moscow State Univ., USSR)
Zero-point lattice vibrations in I/f noise in metals

LIST OF THE PARTICIPANTS

R.	ALABEDRA	U.S.T.L.-C.E.M., Place E. Bataillon, 34060 MONTPELLIER (France)
E.	ALLAMANDO	Université de Lille,IEEA Bt P 4, 59655 VILLENEUVE d'Ascq Cedex (France)
A.	AMBROZY	Technical University, Dept of Electronics Technology, 1521 BUDAPEST (Hungary)
F.T.	ARECCHI	Istituto Nazionale d'Ottica, L.E. Fermi 6, 50125 FIRENZE (Italy)
J.P.	AUBERT	U.S.T.L. - C.E.M., Place E. Bataillon, 34060 MONTPELLIER (France)
L.	AUDAIRE	CENG/LETI/LIR, Avenue des Martyrs, 38041 GRENOBLE (France)
A.	BASSOMPIERRE	U.S.T.L., Place E. Bataillon, 34060 MONTPELLIER (France)
P.	BAUDET	L.E.P., 3 Av. Descartes, 94450 LIMEIL-BREVANNES(France)
D.	BELL	87 East End, Walkington, BEVERLEY HU17 8RX (U.K.)
G.	BERTOTTI	Istituto Elettrotecnico Nazionale G.Ferraris, 42 C. d'Azeglio, 10125 TORINO (Italy)
H.	BILGER	Oklahoma State University, School of El. and Comp. Engg, 310 ES, STILLWATER, Oklahoma 74078 (USA)
J.	BISSCHOP	Eindhoven University of Tech., 5600 MB EINDHOVEN (The Netherlands)
G.	BLASQUEZ	LAAS-CNRS, 7 Avenue Roche, 31400 TOULOUSE (France)
B.	BOITTIAUX	Université Lille 1, Centre Hyperfréquences et Semic.P3, 59655 VILLENEUVE d'ASCQ Cedex (France)
F.	BORDONI	C.N.R., Casella Postale 27, 00044 FRASCATI (Italy)
G.	BOSMAN	University of Florida, Dept of Electrical Engg, Larsen Hall, GAINESVILLE, Florida 32611 (USA)
M.	BOUGRINE	U.S.T.L. - C.E.M., Place E. Bataillon, 34060 MONTPELLIER (France)
A.	BOUKABACHE	L.A.A.S.-C.N.R.S., 7 Avenue Roche 31400 TOULOUSE (France)
J.J.	BROPHY	University of Utah, 304 Park Bldg, SALT LAKE CITY, UT 84112 (USA)
H.	BRUNING	Siemens Ag, 127 Henkestr., 8520 ERLANGEN (F.R.G.)
B.	CARBONNE	Université de Bordeaux 1, 33405 TALENCE Cedex (France)
M.	CARTER	Thorn Emi Central Res.Lab, Trevor Road, HAYES, Midd.UB3 1HH (U.K.)
J.M.	CHOPIN	S.N.Aerospatiale, 12 Rue Pasteur, 92153 SURESNES (France)
A.	CHOVET	ENSERG, Lab.PCS, 23 Rue des Martyrs, 38031 GRENOBLE (France)
T.D.	CLARK	Univ.Of Sussex, School of Math. & Phys.Sc.,Falmer, BRIGHTON, Sussex (U.K.)
J.	CLARKE	University of California, Dept of Physics, BERKELEY, Ca.94720 (USA)
J.	CLUZEL	CENG/LETI/LIR, Avenue des Martyrs, 38041 GRENOBLE (France)
E.	CONSTANT	Université de Lille 1, Centre Hyperfréquences et Semic., P3, 59655 VILLENEUVE d'ASCQ (France)
P.	DAS	Rensselaer Polytechnic Inst. Electrical Computer Engg Dt, TROY, NY 12181 (USA)
J.	DE GOEDE	Dept of Physiology and Physiological Physics, Wassenaarseweg 62, 2333AL LEIDEN (The Netherlands)
M.O.	DEIGHTON	AERE, Harwell, U.K. Atomic Energy Authority, Bldg 347.3,Inst. and A.P. Div., DIDCOT, Oxon OW11 ORA (U.K.)
A.	DE VOS	Dept of Physiology and Physiological Physics, Wassenaarsewef 62, 2333AL LEIDEN (The Netherlands)

R. DUPERDU — Ecole Supérieure d'Electricité, Plateau du Moulon, 91190 GIF / YVETTE (France)

C. ESCOFFRE — Motorola, 126 Bd Eisenhover, 31024 TOULOUSE (France)

M. FADEL — U.S.T.L. - C.E.M., Place Eugène Bataillon, 34060 MONTPELLIER (France)

F. FIORILLO — Istituto Elettrotecnico Galileo Ferraris, 42 C. d'Azeglio, 10125 TORINO (Italy)

D.M. FLEETWOOD — Purdue University, Physics Dept, WEST LAFAYETTE, In 47907 (USA)

V. FOGLIETTI — C.N.R. - I.E.S.S., 42 Via Cineto Romano, 00100 ROMA (Italy)

D. FRENKEL — Rijkuniversiteit Utrecht, Fysich Laboratorium, Princeton Pl. 5, 3508 TA UTRECHT (The Netherlands)

C. GABRIELLI — Université de Paris 6, Physique des Liquides et Electrochimie, 4 pl.Jussieu, T 22, 75230 PARIS Cedex 05, (France)

J.J. GAGNEPAIN — C.N.R.S., 32 Avenue de l'Observatoire, 25000 BESANCON (France)

J.F. GARNIER — CENG-LETI.MEM, 85 X, 38041 GRENOBLE (France)

D. GASQUET — U.S.T.L.-C.E.M., Place Eugène Bataillon, 34060 MONTPELLIER (France)

G. GINESTE — U.S.T.L.- C.E.M., Place Eugène Bataillon, 34060 MONTPELLIER (France)

M. GONTRAND — U.S.T.L.- C.E.M., Place Eugène Bataillon, 34060 MONTPELLIER (France)

L. GOUSKOV — U.S.T.L.- C.E.M., Place Eugène Bataillon, 34060 MONTPELLIER (France)

H. GRABERT — Universitat Stuttgart, Inst. Theor. Physik, 7000 STUTTGART 80 (F.R.G.)

J. GRAFFEUIL — L.A.A.S.-C.N.R.S., Avenue Colonel Roche, 31400 TOULOUSE (France)

R. GRAHAM — Universitaet Essen, Universitaets Str. 43 ESSEN (F.R.G.)

W. GROSSE - NOBIS — University of Munster, Inst. f. Angewandte Physik, Corrensstrasse 2/4, 4400 MUNSTER (F.R.G.)

F. GRUENEIS — Fraunhofer Gesellschaft - Inst.f.Hydroakustik, 41 Waldparkstrasse, 8012 OTTOBRUNN (F.R.G.)

H.D. HAHLBOHM — Physik Techn.Bundesanstalt, Abbe str.2-10, 1000 BERLIN 10 (F.R.G.)

P.H. HANDEL — University of Missouri & Mc Donnel Douglas, Physics Dept, ST LOUIS, Mo 63121, (U.S.A.)

C. HANKE — Lehrstuhl f. Techn.Elekt., Techn. Universitat, Arcisstrasse 21, 8000 MUNCHEN 2 (F.R.G.)

S. HARDING — Ferranti Electronics Ltd, Microelectronics Centre, Computer Rd, Hollinwood, Oldham, Lancs OL9 7JS (U.K.)

C. HEIDEN — University of Giessen, Heinrich Buff Ring 16, 63 GIESSEN (F.R.G.)

A. HERNANDEZ-MACHADO — Universidad Barcelona, Dpto Fisica Teorica, 647 Diagonal, BARCELONA 28 (Spain)

F.N. HOOGE — Eindhoven Univ.of Techn., Den Dolech 2, 5600 MB EINDHOVEN (The Netherlands)

Y. ISAWA — Tohoku University, 2-1-1 Katahira, SENDAI 980 (Japan)

B. JONES — University of Lancaster, Dept of Physics, LANCASTER LA1 4YB (U.K.)

R.I. JOS — Rijksuniversiteit Utrecht, 5 Princetonplein, 3508 TA UTRECHT (The Netherlands)

K. KANDIAH — U.K.A.E.A., AERE, Harwell, DIDCOT, Oxon OX11 0RA(U.K.)

J. KILMER — University of Florida, 216 Larsen Hall, GAINVESVILLE, FL 32611 (USA)

J. KIMMERLE — Inst.f.Physikalische Elek., Universitat Stuttgart, Boblingerstr., 7000 STUTTGART (F.R.G.)

W.J. KLEEN — Techn. Univ. Munchen, Arcisstrasse 21, 8000 MUNCHEN (F.R.G.)

T.G. KLEINPENNING — Eindhoven University of Technology, Den Dolech 2, 5600 MB EINDHOVEN (The Netherlands)

R. KOCH — IBM Research Center, P O Box 218, YORKTOWN HEIGHTS, NY 10598 (USA)

A. KUSY	Technical University of Rzeszow, ul. W. Pola 2, 35-959 RZESZOW (Poland)
G. LECOY	U.S.T.L. - C.E.M., Place Eugène Bataillon, 34060 MONTPELLIER (France)
R. LEFEVER	Université Libre de Bruxelles, Service Chimie Physique 11, CP 231, 1050 BRUXELLES (Belgium)
A. LIBCHABER	Ecole Normale Supérieure, 24 rue Lhomond, 75231 PARIS Cedex 05 (France)
D. LIPPENS	Université de Lille 1, Centre Hyperfréquence et Semic., P3, 59655 VILLENEUVE d'Ascq Cedex (France)
J.C.LOPEZ	U.S.T.L. - C.E.M. Place Eugène Bataillon, 34060 MONTPELLIER (France)
L. LORECK	Universitat Duisburg, Fachbereich 9,Halbleitertechnik, 60 Kommandantenstrasse, 4100 DUISBURG (F.R.G.)
G. LUZ	Standard Elektrik Lorenz Ag (ITT), Dept CS/FZW, 42 Hellmuth Hirth Strasse, 7000 STUTTGART 40 (F.R.G.)
H. MAES	Katholieke Universiteit, 94 Kardinaal Mercierlaan, 3030 HEVERLEE (Belgium)
J.C.MARTIN	Rectorat de Bordeaux, 5 rue Joseph de Carayon Latour, 33060 BORDEAUX Cedex (France)
P.E.MAY	University of Exeter, Engg Science Dept, North Park Road, EXETER EX4 4QF (U.K.)
P. MAZZETTI	I.E.N.G.F., 42 Corso Massimo d'Azeglio, 10125 TORINO (Italy)
H.H.MEHDI	University of Exeter, Engg Science Dept, North Park Road, EXETER EX4 4QF (U.K.)
W. MICHEL	Universitat Stuttgart, Inst. Theor. Physik, 7000 STUTTGART 80 (F.R.G.)
P. MIGNY	CENG-LETI/MEM, 85 X, 38041 GRENOBLE Cedex (France)
C. MOGLESTUE	G.E.C. Hirst Research Centre, East Lane, WEMBLEY, Middlesex HA9 7PP (U.K.)
R. MULLER	Lehrstuhl f.Techn.Elektronik, Technische Universitat, Arcisstrasse 21, 8000 MUNCHEN2 (F.R.G.)
T. MUNAKATA	Universitat Frankfurt a.M., Inst. f. Angewandte Physik, Robert Mayer Strasse 2-4, 6000 FRANKFURT (F.R.G.)
T. MUSHA	Tokyo Institute of Technology, Dept of Applied Electronics, 4259 Nagatsuta, Midoriku, YOKOHAMA 227 (Japan)
J.E.MUTTON	University of Sussex, Scholl of Mathematical and Physical Sciences, Falmer, BRIGHTON, Sussex BN1 9QH (U.K.)
T. NASHASHIBI	Thorn EMI Central Res. Labs, Trevor Road, HAYES, Middx (U.K.)
A. NEIDENOFF	Inst. of Radioastronomy-Bonn, 14 Lichtenvoorder Str., 4044 KAARST 2 (F.R.G.)
M. NELKIN	Cornell University, Dept of Applied Physics, Clark Hall, ITHACA, NY 14853 (USA)
J.P.NOUGIER	U.S.T.L.- C.E.M., Place Eugène Bataillon, 34060 MONTPELLIER (France)
M. OLIVIER	C.N.R.S., 32 Rue de l'Observatoire, 25000 BESANCON (France)
G. PALLOTINO	Ist. Plasma Spazio - CNR, Via Galilei, 4400 FRASCATI (Roma) (Italy)
J.F.PASCAL	Motorola Semiconducteurs, 126 Bd Eisenhower, 31024 TOULOUSE (France)
C. PEGRUM	University of Strathclyde, Dept of Applied Physics, GLASGOW G4 0NG (U.K.)
B. PELLEGRINI	Universita di Pisa, Istituto Elettronica e Telecomunicaz., 2 Diotisalvi, 56100 PISA (Italy)
J.M.PERANSIN	U.S.T.L. - C.E.M., Place Eugène Bataillon, 34060 MONTPELLIER (France)
C. PICKUP	CSIRO Division of Ap.Physics, P O Box 218, LINDFIELD NSW 2070 (Australia)
B. POURBAGHER	U.S.T.L. - C.E.M., Place Eugène Bataillon, 34060 MONTPELLIER (France)
D. POUSSART	Université Laval, Dept de Génie Electrique, QUEBEC G1W 2E3 (Canada)
R.J.PRANCE	Sussex University, 6 Friar Rd, BRIGHTON BN1 6NG (U.K.)
L. REGGIANI	Istituto di Fisica, Via Campi, MODENA (Italy)
D. RIGAUD	U.S.T.L. - C.E.M., Place Eugène Bataillon, 34060 MONTPELLIER (France)

D. ROUSSEAU — Centre d'Etude Nucléaire de Grenoblen LETI/MA, 85 X, 38041 GRENOBLE (France)

J.M.RUBI — Universidad Autonoma Barcelona, Depto Termologia Fac.Ciendad, Bellaterra, BARCELONA (Spain)

G. SALMER — Université de Lille 1, Centre Hyperfréquences et Semicond., 59655 VILLENEUVE d'ASCQ Cedex (France)

M. SAN MIGUEL — Universidad Barcelona, Dpto Fisica Teorica, 647 Diagonal, BARCELONA 38 (Spain)

H. SATO — Central Research Laboratory, Matsushita Electric Ind.Co Ltd, 3-15 Yakumo-Nakamachi, Moriguchi, OSAKA 570 (Japon)

M. SAVELLI — U.S.T.L. - C.E.M., Place Eugène Bataillon, 34060 MONTPELLIER (France)

J. SCOFIELD — Cornell University, Physics Dept, ITHACA, NY 14853 (USA)

E. SEEBALD — Fernmeldetechnisches Zentralamt, 3 Am Kavalleriesand, 6100 DARMSTADT (F.R.G.)

H. SEPPA — Techn. Research Centre of Finland, 5/I Otakaari, ESPOO 15 02150 (Finland)

D.T.SMITH — University of Oxford, Clarendon Laboratory, OXFORD OX1 3PU (U.K.)

D. SODINI — U.S.T.L. - C.E.M., Place Eugène Bataillon, 34060 MONTPELLIER (France)

I. SOSTAREC — Visa Tehnicka Skola, 2 A.Einstein, 24000 SUBOTICA (Yugoslavia)

R. SOULEN — National Bureau of Standards, Room B 128, Bldg 221, WASHINGTON DC 20234 (USA)

E.P.STILLWELL — Clemson University, Dept of Physics and Astronomy, CLEMSON, S.Carolina 20631 (USA)

H. STOLL — Max Planck Inst.f.Metallforsch., Heisenbergstr.1, 7000 STUTTGART 80 (F.R.G.)

H. SUTCLIFFE — University College, Dept of Electronic Engg, GALWAY, Ireland (Eire)

M. SUZUKI — University of Tokyo, Dept of Physics, Hongo, TOKYO 113 (Japan)

P. TALKNER — Universitat Basel, Inst.fur Physik, Klingebergstr.82, 4056 BASEL (Switzerland)

C. TESCHE — IBM Research Center, P O B.218, YORKTOWN HEIGHTS, NY 10598 (USA)

A. TOUBOUL — U.S.T.L. - C.E.M., Place E. Bataillon,34060 MONTPELLIER (France)

A.M.TREMBLAY — Université de Sherbrooke, Bd Université, Sherbrooke, QUEBEC J1K 2RI (Canada)

J. UEBERSFELD — Université P. et M. Curie, 4 Place Jussieu, 75005 PARIS (France)

R. UMATHUM — University of Cambridge, Darwin College, Silver Street, CAMBRIDGE CB3 9EU (U.K.)

J.C.VAISSIERE — U.S.T.L. - C.E.M., Place Eugène Bataillon, 34060 MONTPELLIER (France)

A. VAN CALSTER — Ghent State University, Lab. Of Electronics, Sint.Pietersnievwstraat 41, 9000 GENT (Belgium)

L.K.VANDAMME — Eindhoven University of Technology, 2 Den Dolech, 5600 MB EINDHOVEN (The Neth.)

R.J.VAN DEN BERG — Dept of Physiology and Physiological Physics, Wassenaarsew. 61, 2333 AL LEIDEN (The Neth.)

J. VAN DER MEULEN — Rijksuniversiteit, Fysich Lab., 5 Princetonplein, POB 80000, 3506 TA UTRECHT (The Netherl.)

A. VAN DER ZIEL — University of Minnesota, 123 Church Street S.E., E.E.Dept, MINNEAPOLIS, MN 55455 (USA)

L.J.VAN RUYVEN — Philips Research Lab., Parklaan 75, 5613 BB EINDHOVEN (The Neth.)

C.M.VAN VLIET — Université de Montréal (Canada) and University of Florida, Dept of Elec. Engg., GAINESVILLE, FL 32611 (USA)

R.F.VOSS — IBM Research Center, POB.218, YORKTOWN HEIGHTS, NY 10598 (USA)

J. WEHHOFER — Univ. of Frankfurt, Robert Mayer Str. 2-4, 6000 FRANKFURT, (F.R.G.)

M. WEISSMAN — University of Illinois, Physics Dept, 1110 W.Green Street, URBANA, IL 61801 (USA)

R.M.WESTERVELD — Harvard University, Div.of Ap. Sciences, CAMBRIDGE,MA 02138, (USA)

K. WIESINGER — Siemens Ag, UBB WIS TE BIP 21, 73 Balanstrasse, 8000 MUNCHEN 80 (F.R.G.)

D. WOLF — Universitat Frankfurt, Robert Mayer Str. 2-4, 6000 FRANKFURT (F.R.G.)

M. YOKOTA
Osaka City University,
Dept of Applied Physics,
3-138 Sugimoto 3 Chome
Sumiyoshi - ku,
OSAKA 558, (Japan)

R. ZIJLSTRA
Rijksuniversiteit Utrecht,
Fysich Laboratorium,
Princetonplein 5,
UTRECHT 3508 TA (The Neth.)

WORD INDEX

AUTHOR INDEX

UNIVERSITY OF RHODE ISLAND LIBRARY
3 1222 00272 7975

DATE DUE
1985
NOV 24
APR 22 1987
GAYLORD

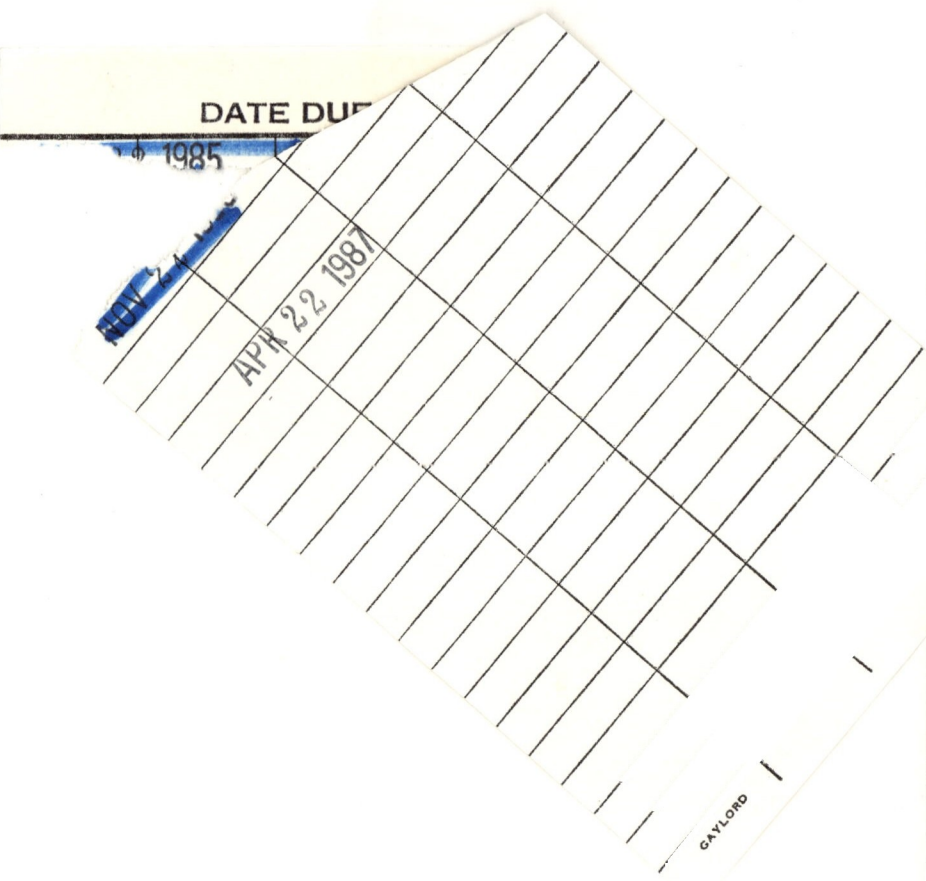